Simon Jungblut • Viola Liebich
Maya Bode-Dalby
Editors

YOUMARES 9 - The Oceans: Our Research, Our Future

Proceedings of the 2018 conference for YOUng MArine RESearcher in Oldenburg, Germany

Editors
Simon Jungblut
BreMarE - Bremen Marine Ecology, Marine Zoology, University of Bremen,
Bremen, Germany

Alfred Wegener Institute, Helmholtz Centre for Polar and Marine Research
Bremerhaven, Germany

Maya Bode-Dalby
BreMarE - Bremen Marine Ecology, Marine Zoology, University of Bremen
Bremen, Germany

Viola Liebich
Envio Maritime
Berlin, Germany

https://doi.org/10.1007/978-3-030-20389-4

To all young marine researchers

Foreword

YOUMARES 9, a conference from and for YOUng MArine RESearchers, is well-established and an format to present current research topics to early career scientists. This international conference represented a platform for early career scientists in Germany, Europe, and worldwide to build up a scientific network. At large congresses, young scientists often do not have the opportunity to present themselves. YOUMARES 9 was important, giving young researchers a place to discuss their research and engage in discussions on important research questions early in their scientific career.

YOUMARES 9 was organized by master's students and doctoral candidates as a bottom-up conference. The bottom-up concept of YOUMARES 9 was professionalized by a core organizational team and a local team provided by the host. The participants of the organizational team learned to organize conferences, communicate with different stakeholders, and moderate sessions or lead workshops. As a result, the team learned self-confidence and strengthened their key competencies besides their scientific work.

These kinds of conferences are indeed a very good way of supporting young researchers in their starting careers. Young researchers learn to present their work and discuss it with peers and network. To sum up, all participants learn the parts of "how to do research" that take place outside of the lab. During the conference, there is a spirit of curiosity, interest, and energy of young researchers and an open-minded atmosphere.

It was great to be the host of YOUMARES 9 under the theme "The oceans: our research, our future" from 11 to 14 September 2018 at the Carl von Ossietzky University of Oldenburg, ICBM. It was a pleasure to welcome over 300 participants to Oldenburg. Originally, YOUMARES 9 started with a zero budget, but with support from various sponsors from science and industry, it ended up being a prestigious conference.

As a future perspective, such conferences would be an essential link between industry, institutions, and universities to provide young scientists the best possibilities for future careers inside and outside the universities.

These proceedings, which include a peer-reviewed process, are an excellent summary of the research activities of young marine scientists and document the actual challenges in marine and social sciences. This book is the second that was published open access with Springer in the context of YOUMARES.

I congratulate the organizers of YOUMARES 9 for their enthusiasm, creativity, and engagement.

Dr. Ferdinand Esser

Institute for Chemistry and Biology of the Marine Environment (ICBM)
Carl von Ossietzky University of Oldenburg
Scientific Coordinator of Early Career Researchers (Doctoral Candidates and Postdocs) of the ICBM
Oldenburg, March 2019

Preface

This book is the final product of the YOUMARES 9 conference, held from 12 to 14 September 2018 in Oldenburg, Germany. From all areas of marine sciences, bachelor, master, and PhD students were asked to contribute. The oral and poster presentations of this conference represent the most recent research in marine sciences. All presentations were part of a topical session, which were also organized and moderated by early career scientists. Apart from handling the presentation abstracts, all session hosts were given the opportunity to write a review article on a topic of their choice in their area of research. These peer-reviewed articles and the corresponding abstracts are compiled in this book.

The 2018 edition of the YOUMARES series started with an icebreaker event at the State Museum for Nature and Man in the city center of Oldenburg. All participants were welcomed by Prof. Ursula Warnke (State Museum for Nature and Man), Prof. Oliver Zielinski (Institute for Chemistry and Biology of the Marine Environment, ICBM), and Prof. Dieter Hanelt (German Society for Marine Research, DGM). Some introductory games, food, and drinks indeed broke the ice, especially for the people who have not already been part of the YOUMARES family.

The scientific part of the conference was hosted by the Carl von Ossietzky University of Oldenburg and its Institute for Chemistry and Biology of the Marine Environment (ICBM). After some welcome words by Prof. Esther Ruigendijk (University of Oldenburg, Vice President for Early Career Researchers and International Affairs) and Prof. Oliver Zielinski (ICBM), we started a plenary discussion bridging marine sciences with ocean governance and conservation. The vivid discussion was moderated by James G. Hagan (Vrije Universiteit Brussel, VUB). The discussants on the podium were session hosts of the 2018 YOUMARES edition: Meenakshi Poti, Morgan L. McCarthy, Thomas Luypaert, and Liam Lachs (all VUB, experts in the field of environmental conservation), Pradeep A. Singh, and Mara Ort (University of Bremen, representing the field of ocean governance). They were joined by Prof. Zielinski (ICBM, University of Oldenburg) and Dr. Cornelia Nauen (Mundus Maris, Brussels). The opening morning was completed by a keynote talk of Prof. Frank Oliver Glöckner (Max Plank Institute for Marine Microbiology and Jacobs University Bremen) on the "Ocean Sampling Day, an Example for Science 2.0."

One afternoon was reserved for workshops and excursions. Participants could choose from workshops like "How to turn science into a story?," "Publishing in Natural Sciences," and "Knowledge transfer in marine science" as well as guided tours through the city center of Oldenburg or the Botanical Garden of the University Oldenburg and others.

The remaining time was filled with a diverse spectrum of talks and poster presentations of cutting-edge research results obtained by the conference participants. In total, 109 talks and 33 posters were presented in 1 of the 19 sessions. Including session hosts, helpers, presenters, and listeners, a total over 250 people contributed to YOUMARES 9.

We hope that this book is a source of knowledge and inspiration to the participants, session hosts, and helpers of YOUMARES 9, as well as to all young marine researchers and to everybody interested in marine research.

Bremen, Germany Simon Jungblut
Berlin, Germany Viola Liebich
Bremen, Germany Maya Bode-Dalby
March 2019

Acknowledgments

We would like to thank everyone who was helping with the preparation and realization of the conference in Oldenburg. Without the strong support of these volunteers, organizing such a big conference would be impossible. We would like to especially thank Jan Brüwer, Charles Cadier, Muhammad Resa Faisal, Lena Heel, Laura Hennings, Dorothee Hohensee, Patricia Kaiser, Elham Kamyab, Charlotte Kunze, Jonas Letschert, Veloisa Mascarenhas, Lea Oeljeschläger, Nora-Charlotte Pauli, Lena Rölfer, Lukas Ross, Yvonne Schadewell, Paula Senff, Joko Tri Wibowo, Nils Willenbrink, and Mirco Wölfelschneider.

We thank the State Museum Nature and Man and its director, Ursula Warnke, for providing the rooms and supporting the organization of a great icebreaker event and for offering free entrance to their exhibitions for all conference participants.

We are very grateful to the Carl von Ossietzky University of Oldenburg and its Institute for Chemistry and Biology of the Marine Environment (ICBM) for providing the space and rooms for the conference. Very special thanks go to Ferdinand Esser for all the organizational support during the preparation and during the actual conference.

The opening podium discussion received much attention. It was excellently moderated by James G. Hagan, and we would like to thank the discussants: Liam Lachs, Thomas Luypaert, Morgan L. McCarthy, Cornelia Nauen, Mara Ort, Meenakshi Poti, Pradeep A. Singh, and Oliver Zielinski.

Frank-Oliver Glöckner presented a stimulating keynote talk on the "Ocean Sampling Day, an Example for Science 2.0?" for which we thank him very much.

The workshops during the conference were organized by several people to whom we are all grateful: Alexandrine Cheronet, Lydia Gustavs, Daniel Hartmann, Marie Heidenreich, Thijs Janzen, Elham Kamyab, Veloisa Mascarenhas, Cornelia Nauen, Yvonne Schadewell, Tim Schröder, and Nils Willenbrink.

Several partners supported the conference financially, with materials or with special conditions for our conference participants. For any kind of support, we are very grateful to DFG-Schwerpunktprogramm Antarktisforschung, Reederei Laeisz GmbH, Briese Schiffahrts GmbH Forschungsschifffahrt, SubCtech, develogic GmbH subsea systems, Norddeutsche Stiftung für Umwelt und Entwicklung, Stadt Oldenburg, DFG-Sonderforschungsbereich Roseobacter, Institut für Marine Biologie, German Association for Marine Technology (GTM), Bornhöft Meerestechnik, Kraken Power GmbH, Die Flänzburch, and umBAUbar.

Springer Nature provided book vouchers to award the three best oral and the three best poster presentations and one voucher to raffle among all voters.

The Staats- und Universitätsbibliothek Bremen and the Alfred Wegener Institute, Helmholtz Centre for Polar and Marine Research, supported the open-access publication of this conference book.

Thanks go to Alexandrine Cheronet, Judith Terpos, and Springer for their support during the editing and publishing process of this book.

All chapters of this book have been peer-reviewed by internationally renowned scientists. The reviews contributed significantly to the quality of the chapters. We would like to thank all reviewers for their time and their excellent work: Martijn Bart, Kartik Baruah, Thorsten Blenckner, Hans Brumsack, Xochitl Cormon, Michael Fabinyi, Tilmann Harder, Enrique Isla,

Annemarie Kramer, Annegret Kuhn, Amy Lusher, Tess Moriarty, Elisabeth Morris-Webb, May-Linn Paulsen, Pamela Rossel Cartes, Chester Sands, Theresa Schwenke, Rapti Siriwardane-de Zoysa, Lydia The, David Thomas, Eva Turicchia, Benjamin Twining, Laura Uusitalo, Jan Verbeek, Ans Vercammen, Benjamin Weigel, and further anonymous reviewers.

We editors are most grateful to all participants, session hosts, and presenters of the conference and to the contributing authors of this book. You all did a great job in presenting and representing your (fields of) research. Without you, YOUMARES 9 would not have been worth to organize.

Contents

Contributors

Janis Ahrens Microbiogeochemistry Group, Institute for Chemistry and Biology of the Marine Environment (ICBM), Carl von Ossietzky University of Oldenburg, Oldenburg, Germany

Fanny Barz Thünen-Institute of Baltic Sea Fisheries, Rostock, Germany

Laura Basconi Ca Foscari University, Venice, Italy

Charles Cadier MER Consortium, UPV, Bilbao, Spain

Julius Degenhardt Paleomicrobiology Group, Institute for Chemistry and Biology of the Marine Environment (ICBM), Carl von Ossietzky University of Oldenburg, Oldenburg, Germany

Ulrike Dietrich UiT- The Arctic University of Norway, Tromsø, Norway

Leon Dlugosch Biology of Geological Processes Group, Institute for Chemistry and Biology of the Marine Environment, Carl von Ossietzky University of Oldenburg, Oldenburg, Germany

Hannah S. Earp Institute of Biology, Environmental and Rural Sciences, Aberystwyth University, Aberystwyth, Wales, UK

School of Ocean Sciences, Bangor University, Menai Bridge, Wales, UK

Mainah Folkers Institute for Biodiversity and Ecosystem Dynamics, University of Amsterdam, Amsterdam, The Netherlands

Gustavo Guerrero-Limón MER Consortium, UPV, Bilbao, Spain

University of Liege, Ulg, Belgium

James G. Hagan Faculty of Sciences and Bioengineering Sciences, Department of Biology, Vrije Universiteit Brussel (VUB), Brussels, Belgium

Faculty of Sciences, Department of Biology of Organisms, Université libre de Bruxelles (ULB), Brussels, Belgium

Brandon T. Hassett UiT- The Arctic University of Norway, Tromsø, Norway

Mara E. Heinrichs Paleomicrobiology Group, Institute for Chemistry and Biology of the Marine Environment, Carl von Ossietzky University of Oldenburg, Oldenburg, Germany

Elham Kamyab Institute of Chemistry and Biology of the Marine Environment, University of Oldenburg, Oldenburg, Germany

Matthias Y. Kellermann Institute of Chemistry and Biology of the Marine Environment, University of Oldenburg, Oldenburg, Germany

Špela Korez Alfred Wegener Institute, Helmholtz Centre for Polar and Marine Research, Bremerhaven, Germany

Andreas Kunzmann Leibniz Centre for Tropical Marine Research (ZMT) GmbH, Bremen, Germany

Faculty 02, University of Bremen, Bremen, Germany

Liam Lachs Marine Biology, Ecology and Biodiversity, Vrije Universiteit Brussel, Brussel, Belgium

Institute of Oceanography and Environment, Universiti Malaysia Terengganu, Kuala Terengganu, Terengganu, Malaysia

Department of Biology, University of Florence, Sesto Fiorentino, Italy

Philipp Laeseke Marine Botany, University of Bremen, Bremen, Germany

Jonas Letschert Leibniz Centre for Tropical Marine Research (ZMT), Bremen, Germany

Arianna Liconti School of Ocean Sciences, Bangor University, Menai Bridge, Wales, UK

School of Biological and Marine Sciences, Plymouth University, Plymouth, UK

Sara Doolittle Llanos Groningen Institute for Evolutionary Life-Sciences GELIFES, University of Groningen, Groningen, The Netherlands

Thomas Luypaert Faculty of Sciences and Bioengineering Sciences, Department of Biology, Vrije Universiteit Brussel (VUB), Brussels, Belgium

Faculty of Sciences, Department of Biology of Organisms, Université libre de Bruxelles (ULB), Brussels, Belgium

Faculty of Maths, Physics and Natural Sciences, Department of Biology, Università degli Studi di Firenze (UniFi), Sesto Fiorentino, Italy

Morgan L. McCarthy Faculty of Sciences and Bioengineering Sciences, Department of Biology, Vrije Universiteit Brussel (VUB), Brussels, Belgium

Faculty of Sciences, Department of Biology of Organisms, Université libre de Bruxelles (ULB), Brussels, Belgium

Faculty of Maths, Physics and Natural Sciences, Department of Biology, Università degli Studi di Firenze (UniFi), Sesto Fiorentino, Italy

School of Biological Sciences, The University of Queensland (UQ), St. Lucia, Queensland, Australia

Corinna Mori Microbiogeochemistry Group, Institute for Chemistry and Biology of the Marine Environment, Carl von Ossietzky University of Oldenburg, Oldenburg, Germany

Javier Oñate-Casado Department of Biology, University of Florence, Sesto Fiorentino, Italy

Sea Turtle Research Unit (SEATRU), Universiti Malaysia Terengganu, Kuala Terengganu, Terengganu, Malaysia

School of Biological Sciences, University of Queensland, St Lucia, QLD, Australia

Mara Ort artec Sustainability Research Center, University of Bremen, Bremen, Germany

INTERCOAST Research Training Group, Center for Marine Environmental Sciences (MARUM), Bremen, Germany

Lars-Erik Petersen Institute of Chemistry and Biology of the Marine Environment, University of Oldenburg, Oldenburg, Germany

Santiago E. A. Pineda-Metz Alfred-Wegener-Institut Helmholtz-Zentrum für Polar- und Meeresforschung, Bremerhaven, Germany

Universität Bremen (Fachbereich 2 Biologie/Chemie), Bremen, Germany

Meenakshi Poti Faculty of Sciences and Bioengineering Sciences, Department of Biology, Vrije Universiteit Brussel (VUB), Brussels, Belgium

Faculty of Sciences, Department of Biology of Organisms, Université libre de Bruxelles (ULB), Brussels, Belgium

Faculty of Maths, Physics and Natural Sciences, Department of Biology, Università degli Studi di Firenze (UniFi), Sesto Fiorentino, Italy

School of Marine and Environmental Sciences, University of Malaysia Terengganu (UMT), Terengganu, Kuala Terengganu, Malaysia

Sea Turtle Research Unit (SEATRU), Institute of Oceanography and Environment, Universiti Malaysia Terengganu (UMT), Kuala Terengganu, Malaysia

Natalie Prinz Faculty of Biology and Chemistry, University of Bremen, Bremen, Germany

Leibniz Centre for Tropical Marine Research (ZMT), Bremen, Germany

Anja Reckhardt Microbiogeochemistry Group, Institute for Chemistry and Biology of the Marine Environment (ICBM), Carl von Ossietzky University of Oldenburg, Oldenburg, Germany

Titus Rombouts Institute for Biodiversity and Ecosystem Dynamics, University of Amsterdam, Amsterdam, The Netherlands

Jessica Schiller Marine Botany, University of Bremen, Bremen, Germany

Peter J. Schupp Institute of Chemistry and Biology of the Marine Environment, University of Oldenburg, Oldenburg, Germany

Helmholtz Institute for Functional Marine Biodiversity at the University of Oldenburg (HIFMB), Oldenburg, Germany

Kai Schwalfenberg Marine Sensor Systems Group, Institute for Chemistry and Biology of the Marine Environment (ICBM), Carl von Ossietzky University of Oldenburg, Wilhelmshaven, Germany

Heike Schwermer Institute for Marine Ecosystem and Fisheries Science, Center for Earth System Research and Sustainability, University of Hamburg Germany, Hamburg, Germany

Stephan L. Seibert Hydrogeology and Landscape Hydrology Group, Institute for Biology and Environmental Sciences, Carl von Ossietzky University of Oldenburg, Oldenburg, Germany

Pradeep A. Singh Faculty of Law, University of Bremen, Bremen, Germany

INTERCOAST Research Training Group, Center for Marine Environmental Sciences (MARUM), Bremen, Germany

Tobias R. Vonnahme UiT- The Arctic University of Norway, Tromsø, Norway

Hannelore Waska Research Group for Marine Geochemistry (ICBM-MPI Bridging Group), Institute for Chemistry and Biology of the Marine Environment (ICBM), Carl von Ossietzky University of Oldenburg, Oldenburg, Germany

Yury Zablotski Thünen-Institute of Baltic Sea Fisheries, Rostock, Germany

About the Editors

Dr. Simon Jungblut Simon Jungblut is a marine ecologist and zoologist. He completed a Bachelor's Degree in Biology and Chemistry at the University of Bremen, Germany, and studied the international program, Erasmus Mundus Master of Science in Marine Biodiversity and Conservation, at the University of Bremen, Germany; the University of Oviedo, Spain; and Ghent University, Belgium. Afterward, he completed a PhD project entitled "Ecology and ecophysiology on invasive and native decapod crabs in the southern North Sea" at the University of Bremen in cooperation with the Alfred Wegener Institute, Helmholtz Centre for Polar and Marine Research, in Bremerhaven and was awarded the Doctoral title in Natural Sciences at the University of Bremen in December 2017.

Since 2015, Simon is actively contributing to the YOUMARES conference series. After hosting some conference sessions, he is the main organizer of the scientific program since 2017.

Dr. Viola Liebich Viola Liebich is a biologist from Berlin, who worked on invasive tunicates for her diploma thesis at the Alfred Wegener Institute Sylt. With a PhD scholarship by the International Max Planck Research School for Maritime Affairs, Hamburg, and after her thesis work at the Institute for Hydrobiology and Fisheries Science, Hamburg, and the Royal Netherlands Institute for Sea Research, Texel, Netherlands, she finished her thesis entitled "Invasive Plankton: Implications of and for ballast water management" in 2013.

For 3 years, until 2015, Viola Liebich worked for a project on sustainable brown shrimp fishery and stakeholder communication at the WWF Center for Marine Conservation, Hamburg, and started her voluntary YOUMARES work 1 year later. She is currently working as a self-employed consultant on marine and maritime management (envio maritime).

Dr. Maya Bode-Dalby Maya Bode-Dalby is a marine biologist, who accomplished her Bachelor of Science in Biology at the University of Göttingen, Germany, and her Master of Science in Marine Biology at the University of Bremen, Germany. Thereafter, she completed her PhD thesis entitled "Pelagic biodiversity and ecophysiology of copepods in the eastern Atlantic Ocean: Latitudinal and bathymetric aspects" at the University of Bremen in cooperation with the Alfred Wegener Institute, Helmholtz Centre for Polar and Marine Research, in Bremerhaven and the German Center for Marine Biodiversity Research (DZMB) at the Senckenberg am Meer in Wilhelmshaven. She received her Doctorate in Natural Sciences at the University of Bremen in March 2016.

Maya actively contributes to the YOUMARES conference series as organizer of the scientific program since 2016. Currently, she is working as a scientist at the Marine Zoology Department of the University of Bremen.

1 Science for the Future: The Use of Citizen Science in Marine Research and Conservation

Hannah S. Earp and Arianna Liconti

Abstract

Over the last decade, significant advances in citizen science have occurred, allowing projects to extend in scope from the ocean floor to the Milky Way and cover almost everything in between. These projects have provided cost-effective means to collect extensive data sets covering vast spatio-temporal scales that can be used in scientific research, to develop conservation policy and to promote environmental awareness. This review explores the current status of marine citizen science by examining 120 marine citizen science projects. Trends in geographic locations, focal taxa, participant demographics, tasks undertaken and data directionality (i.e. storage and publication) are highlighted, and the challenges and benefits of citizen science to marine research and conservation are reviewed. Marine citizen science projects act primarily at national levels (53.3%) and mainly focus on coastal ocean environments (49.2%) with chordates as the most popular focus taxa (40%). Some form of methodological training for participants is provided by 64.2% of projects, and the most popular tasks undertaken are field surveys (35.8%) and reporting of opportunistic sightings (34.2%). Data quality and participant motivation are among the most common challenges facing projects, but identified strengths include enhanced marine policy, increased scientific knowledge and environmental stewardship. In conclusion, marine citizen science lies at a crossroads of unresolved challenges, demonstrated successes and unrealized potential. However, should the challenges be addressed, the unique capacity of citizen science to broaden the scope of investigations may be the key to the future of marine research and conservation in times of global change and financial hardship.

Keywords

Volunteer · Public participation · Community-based monitoring · Environmental policy · Ecological surveying

H. S. Earp (✉)
Institute of Biology, Environmental and Rural Sciences, Aberystwyth University, Aberystwyth, Wales, UK

School of Ocean Sciences, Bangor University, Menai Bridge, Wales, UK

A. Liconti
School of Ocean Sciences, Bangor University, Menai Bridge, Wales, UK

School of Biological and Marine Sciences, Plymouth University, Plymouth, UK

1.1 Introduction

1.1.1 The History of Citizen Science

Citizen science, often described as amateur participation in scientific research and monitoring, has emerged as a powerful tool and popular activity in recent decades (Cohn 2008; Kullenberg and Kasperowski 2016; Burgess et al. 2017). However, this phenomenon is not new and extends back to before the professionalization of science, whereby most 'scientists' including Benjamin Franklin (1706–1790), Charles Darwin (1809–1888) and Margaret Gatty (1809–1873) made a living in different professions (Silvertown 2009). Yet, despite the evolution of science as a paid profession in the late nineteenth century, amateurs remained involved in many scientific disciplines such as archaeology, astronomy, meteorology and natural history (Silvertown 2009; Haklay 2015). On the verge of the twentieth century, the first 'citizen science project', the National Audubon Society Christmas Bird Count, was established (Cohn 2008; Bonney et al. 2009). It was, however, another 89 years before the first citation of 'citizen science' to describe the collection of rainwater

S. Jungblut et al. (eds.), *YOUMARES 9 - The Oceans: Our Research, Our Future*,
https://doi.org/10.1007/978-3-030-20389-4_1

samples by 225 volunteers as part of a National Audubon Society acid-rain awareness-raising campaign (Kerson 1989), and a further 15 years before its inclusion in the Oxford English Dictionary (OED) in 2014. Today, citizen science is widely defined as 'scientific work undertaken by members of the general public, often in collaboration with or under the direction of professional scientists and scientific institutions' (OED Online 2018a). However, as an evolving discipline, a transition from the primarily contributory paradigm whereby participants mainly collect data, to more collaborative and co-created approaches, where they are involved in additional elements of the scientific process has been observed (Bonney et al. 2009; Wiggins and Crowston 2011; Teleki 2012). Today, some citizen scientists work alone or through community-driven projects, as opposed to directly collaborating with scientists (Bonney et al. 2016a; Cigliano and Ballard 2018). Nevertheless, over the past 20 years, citizen science has boomed, with millions of participants from diverse backgrounds becoming involved in projects that have extended in scope from the seafloor to the Milky Way and covered almost everything in between (Foster-Smith and Evans 2003; Bonney et al. 2016b).

1.1.2 Marine Citizen Science

Although not as prevalent as their terrestrial counterparts (Roy et al. 2012; Cigliano et al. 2015; Theobald et al. 2015; Garcia-Soto et al. 2017), marine citizen science projects provide a cost-effective means of collecting and analysing extensive data sets across vast spatio-temporal scales, using conventional and new observation and simulation tools (Bonney et al. 2009; Silvertown 2009; Hochachka et al. 2012; Garcia-Soto et al. 2017). Wiggins and Crowston (2011) suggested that citizen science projects fall into five exhaustive groups: (1) action-orientated projects that encourage participation in local issues, for example, collecting and categorizing marine debris (e.g. Marine Conservation Society's Beachwatch available at www.mcsuk.org/beachwatch); (2) conservation projects that promote stewardship and management such as restoring coral reefs (e.g. Rescue a Reef available at sharkresearch.rsmas.miami.edu/donate/rescue-a-reef); (3) investigation projects that answer a scientific question including monitoring coral reefs (Marshall et al. 2012; Done et al. 2017), cetacean populations (Evans et al. 2008; Tonachella et al. 2012; Bruce et al. 2014; Embling et al. 2015) and invasive species (Delaney et al. 2008); (4) virtual projects that are exclusively ICT-meditated, for example, online photo analysis (e.g. Weddell Seal Count available at www.zooniverse.org/projects/slg0808/weddell-seal-count); and (5) education projects whereby outreach is the primary goal (e.g. the Capturing our Coast 'Beach Babies' survey available at www.capturingourcoast.co.uk/specific-information/beach-babies).

Thiel et al. (2014) examined 227 peer-reviewed studies involving volunteer-scientist collaborations and showed that developed nations including the United States of America (USA), Australia and the United Kingdom (UK) are hotspots for marine citizen science, with easily accessible areas including intertidal and subtidal regions among the most frequently surveyed environments. However, recent technological developments, often dubbed 'citizen cyberscience', have further elevated the accessibility of citizen science and may in turn alter these trends (Science Communication Unit – University of the West of England 2013). These developments have allowed volunteers from around the world to 'virtually' participate in marine research across international borders and in otherwise inaccessible environments (e.g. the deep sea) from the comfort of home. Examples include Seafloor Explorer (available at www.seafloorexplorer.org), where participants analysed over two million images of the seafloor (~250 m deep) in order to investigate the distribution of commercially important species such as scallops along the northeast United States continental shelf.

1.1.3 Citizen Science as a Tool in Research and Conservation

Despite the broad array of topics, the aims of citizen science projects remain similar: to gather data that answers scientific questions and/or drives policy (Cigliano et al. 2015; Bonney et al. 2016b; Garcia-Soto et al. 2017), to promote environmental awareness and literacy, and to empower citizens and communities (Danielsen et al. 2013; Garcia-Soto et al. 2017). Consequently, it has been suggested that citizen science processes and outcomes warrant acknowledgement as a distinct discipline (Jordan et al. 2015; Burgess et al. 2017; Garcia-Soto et al. 2017). Despite being incorporated into an increasing array of scientific literature, proposals and conference submissions (Cigliano and Ballard 2018), and evolving well-tested protocols and data validation techniques, citizen science has yet to be fully embraced by the scientific community, and questions remain surrounding best practices and data quality and/or verification (Cohn 2008; Silvertown 2009; Bonney et al. 2014; Burgess et al. 2017). This review builds on research by Thiel et al. (2014) that demonstrated trends across marine citizen science published in peer-reviewed journal articles, in order to highlight the diversity of current marine citizen science projects. This includes projects that have published their data in peer-reviewed journals, as well as those whose primary aims are to provide data that drives management or to educate and engage the public. 'Voluntourism' projects are excluded from our considerations as they primarily constitute 'voluntary work typically aiming to help others' (OED Online 2018b) as opposed to the 'scientific work' nature of citizen science. The selected marine citizen science projects were examined in order to highlight trends in terms of: geographic locations, focal taxa, participant demographics, tasks undertaken and data directionality (i.e. data publication and storage). Challenges and strengths arising from the review are then presented before suggestions for the future of citizen science in marine research and conservation are made.

1.2 Methodology

1.2.1 Project Selection

Marine citizen science projects were collated using: (1) Google searches using the keywords 'marine + citizen + science', (2) searches on the citizen science database SciStarter (available at www.scistarter.com) using the keyword 'marine', (3) the Wikipedia citizen science project list (available at www.wikipedia.org/wiki/List_of_citizen_science_projects), (4) social media searches on Facebook using the keywords 'marine + citizen + science', (5) projects mentioned in reviewed literature and (6) personal knowledge. Project websites were consulted, and a project was included in the review when it had a marine focus and involved citizen scientists. In cases where an organization coordinated multiple citizen science projects, each project was included individually (e.g. The Shark Trust coordinates; The Great Eggcase Hunt, Basking Shark Project and Angling Project: Off The Hook, available at www.sharktrust.org/en/citizen_science). In cases where a project organized multiple campaign style activities, the project alone was included (e.g. Capturing our Coast available at www.capturingourcoast.co.uk). A total of 120 projects, covering the majority of oceans, their associated flora and fauna, and several conservation issues met the selection criteria (see Appendix 1 for a list of reviewed projects). Data for each project was collected by combining information available from websites, newsletters, databases and email communications. Core data included lead organization, year of establishment, spatial coverage (i.e. international, regional, etc.), location, focus area/taxa, volunteer training requirement (i.e. written instructions, training programs), activity genre (i.e. fieldwork/online) and tasks undertaken (i.e. sightings, image/recording analysis, etc.). When available, information on the number of surveys undertaken by citizen scientists, data validation techniques (i.e. data quality checking), data directionality (i.e. storage location) and number of peer-reviewed scientific publications using the projects data set was also recorded.

1.3 Identified Trends Across Marine Citizen Science Projects

1.3.1 Geographic Location

1.3.1.1 Spatial Coverage

The reviewed projects occurred across multiple geographical scales, extending from local and regional levels (4.2%) to international and global coverage (42.5%). The majority of projects acted at national levels (53.3%) and spanned nine locations (Fig. 1.1), with the most being located in the USA (43.8%), followed by the UK (27.4%) and Australia (11%). A trend towards greater project abundances in developed

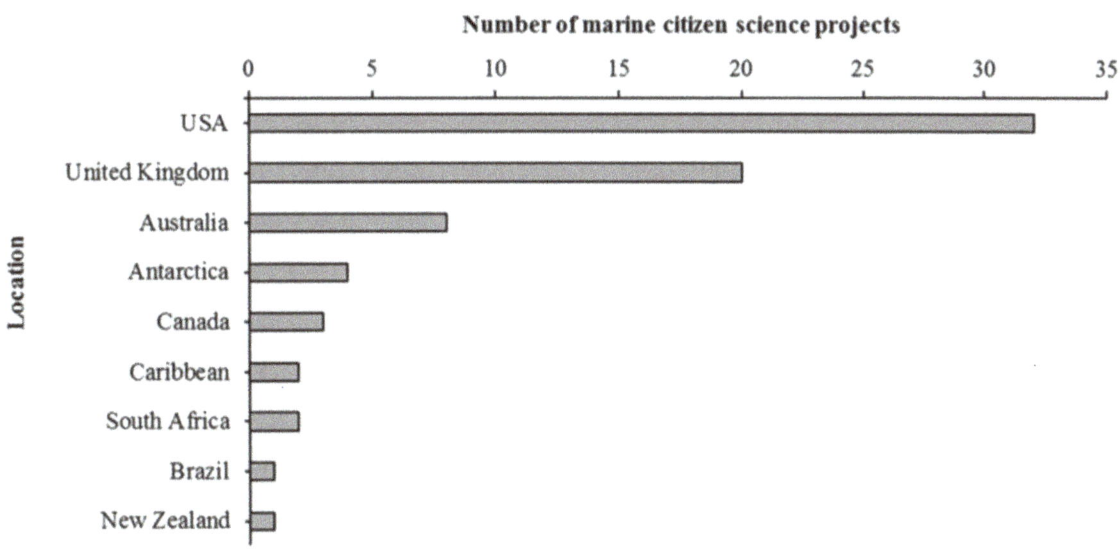

Fig. 1.1 Number of reviewed marine citizen science projects per location, excluding those operating on international (n = 10) or global (n = 37) scales

nations was observed, with only 6.8% of projects occurring in nations with developing economies (as defined by the United Nations Conference on Trade and Development (UNCTADstat 2018)), for example, Brazil and South Africa. A similar trend was reported by Thiel et al. (2014), although this may be attributed, in part, to the fact that projects incorporated in these reviews were selected based on their journal publications and websites, and consequently projects using other communication strategies to engage with citizen scientists (e.g. local community groups that may be more abundant in developing nations) are excluded.

1.3.1.2 Environmental Coverage

The most commonly investigated environment was the coastal ocean (depth < 200 m) (49.2%), closely followed by easily accessible coastline regions (34.2%) (Fig. 1.2). Although further divisions into zones such as the supralittoral, intertidal, subtidal, continental shelf and oceanic environments (similar to Thiel et al. 2014) were beyond the scope of this review, this information could provide a greater insight into hotspot environments for marine citizen science, as well as those with capacity for development. Interestingly, studies specifically focused on environments known for their roles in supporting ecosystem functions and services, including mangrove and kelp forests, seagrass meadows and wetlands, were limited (5% in total), demonstrating potential opportunities for expansion of citizen science in these environments. An exception was coral reefs that were the focus of investigation in 8.3% of projects, potentially due to their charismatic appeal, exotic location, alongside the relative ease of conducting research involving SCUBA diving in these environments, and the higher volume of visitors as potential citizen science participants (relative to colder oceanic environments).

The deep sea remained the least studied environment with only one project, Digital Fishers (available at www.oceannetworks.ca), focusing their investigations on the organisms inhabiting this remote and often inaccessible region. However, inaccessibility may not be the only reason for the lack of projects concerning this environment, as limited scientific knowledge and expensive technologies may also be factors. Despite large deep-sea video databases being available online (National Oceanic & Atmospheric Administration Ocean Explorer available at www.oceanexplorer.noaa.gov; Monterey Bay Aquarium Research Institute available at www.mbari.org; Japan Agency for Marine-Earth Science & Technology e-library of deep-sea images available at www.godac.jamstec.go.jp), the identification of deep-sea organisms remains complex and thus must be conducted by experts in this field. However, in order to enhance the identification process (i.e. make it quicker and easier), software is currently under development that can automatically identify deep-sea species, and in the case of Digital Fishers, citizen scientists are contributing to the development of this software by 'educating' it to count and identify different taxa (Ocean Networks Canada 2018).

The majority of reviewed projects (25.8%) focused on multiple taxa ('Diverse Taxa') (Table 1.1), through investigations on the intertidal or subtidal or on invasive species and planktonic communities. However, among the most popular individual taxa were the so-called charismatic megafauna, including marine mammals (15%), seabirds (8.3%) and

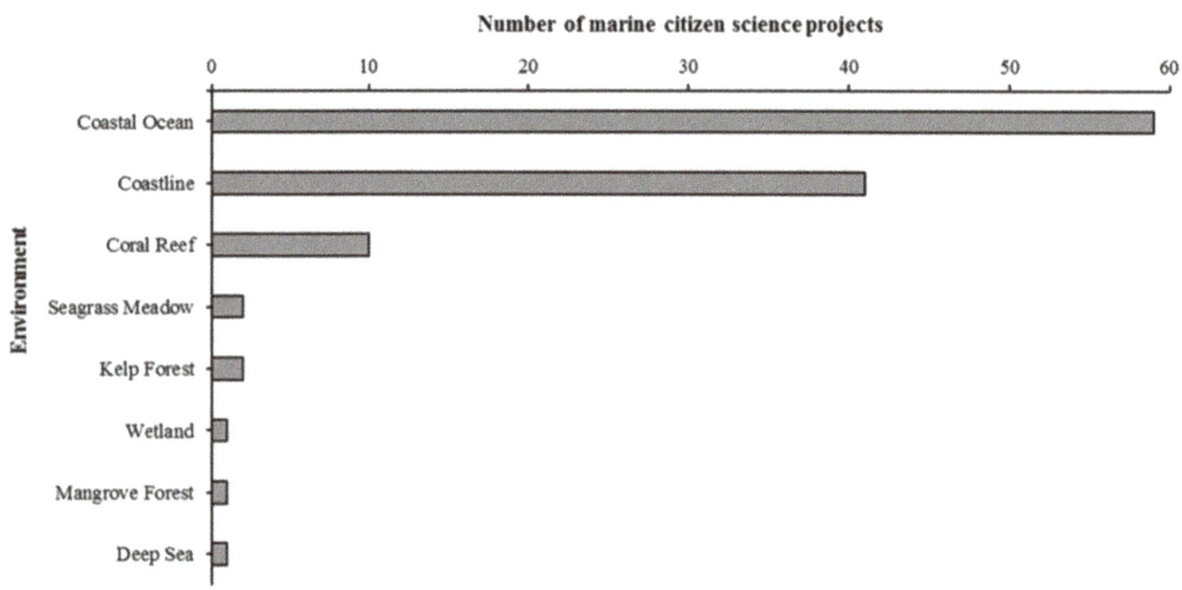

Fig. 1.2 Number of reviewed marine citizen science projects per environment, excluding those that focused on multiple environments (e.g. Redmap, available at www.redmap.org.au) (n = 3)

Table 1.1 Focus taxa of the reviewed marine citizen science projects, excluding those that focused on oceanography or pollution (n = 18). Diverse taxa includes projects focusing intertidal and subtidal flora and fauna, alongside those that focus on multiple invasive or planktonic taxa

Taxa		Number of projects
Chordata	Mammalia	18
	Aves	10
	Chondrichthyes	9
	Osteichthyes	8
	Actinopterygii	1
	Reptilia	1
	Diverse taxa	1
Cnidaria	Anthozoa	8
	Scyphozoa	1
Arthropoda	Crustacea	6
	Diverse taxa	1
Plantae	Angiosperma	4
Heterokontophyta	Phaeophyceae	3
Echinodermata	Echinoidea	1
Mollusca	Gastropoda	1
Diverse Taxa	General	29

sharks/rays (7.5%), which are often considered more newsworthy compared to projects focusing on seaweeds or plankton (Stafford et al. 2010). Surprisingly, sea turtles (also considered charismatic megafauna) were the focus of only one project (Seaturtle.org available at www.seaturtle.org), yet were highly popular among voluntourism projects (e.g. Sea Turtle Conservation available at www.volunteeringsolutions.com, www.frontier.ac.uk and www.gvi.co.uk; SEE Turtles available at www.seeturtles.org). In addition to the popularity of charismatic megafauna, charismatic sessile organisms, such as corals, are frequently investigated by marine citizen science projects, accounting for 6.6% of projects in this review. Despite the popularity of projects involving charismatic taxa, studies focusing on lesser charismatic organisms such as plankton (e.g. FjordPhyto available at www.fjordphyto.wordpress.com; Secchi Disk available at www.secchidisk.org) are growing in popularity, with estimates showing ~110,900 volunteers are engaged in the counting and identification of plankton in the Mediterranean Sea and California currents through Plankton Portal (www.planktonportal.org).

1.3.2 Participant Demographics

1.3.2.1 Participant Recruitment

At present, there is no quantification of the number of citizen scientists actively involved in scientific research. However, as it often entails limited/no cost, the number is likely to exceed that of voluntourists (estimated at 10 million people per annum by McGehee 2014). Citizen scientists involved in marine research descend from a diverse array of backgrounds and may have no formal training or qualifications in marine-related subjects (Thiel et al. 2014). Participant recruitment often occurs through collaborations with other established nature organizations including conservation groups and ocean water sport centres. These recreational users of the marine environment, especially SCUBA divers (Martin et al. 2016), often have enhanced interests in marine life and its preservation and are consequently attracted to opportunities whereby they can expand their knowledge base and participate in research (Campbell and Smith 2006; Cohn 2008). More recently, online tools (i.e. project websites and social media) have provided a low-effort method of recruiting both on- and off-site participants. This is partially due to the fact that those with an interest in nature conservation are usually connected with other like-minded people and/or groups online, and consequently a positive loop of information sharing is generated that benefits both citizen science outreach and recruitment.

The majority of reviewed projects are open to participants of any age, although several were noted to exhibit a preference for adult participants (i.e. aged 18 and over); however, this is often due to protocol complexity (see sect. 3.3 for a review). In cases where the protocol requires species identification, adult participants are often designated as final decisionmakers, although younger participants may assist under supervision (e.g. Capturing our Coast available at www.capturingourcoast.co.uk). In the case of projects that involve SCUBA diving, only participants that meet the minimum requirements (e.g. certification and/or experience level) are permitted to partake. However, some variation among minimum requirements is exhibited, for example, to certify as a Reef Check Ecodiver, participants must be comfortable with the use of a mask, snorkel and fins or be a certified SCUBA diver (Reef Check 2018), while the requirements to become a Seasearch Observer include being certified as a PADI Advanced Open Water Diver (or equivalent) and having > 20 dives, of which ≥ 10 should be in temperate waters (Seasearch 2018). Despite some background experience being required in these instances for safety, none of the reviewed projects required participants to have any educational background, as they become trained and therefore specialists in the task required (Hobson 2000). Furthermore, some projects allow participant development to a level whereby they can become project organizers, coordinators, or even lead authors in scientific publications and/or identification guides (see Bowen et al. 2011 for an example of an identification guide authored by citizen scientists). An example includes Seasearch (available at www.seasearch.org.uk) that coordinates general surveys that all participants may undertake, as well as a 'surveyor' level survey for participants that undertake advanced training, and 'specialist projects' created by marine biology experts and experienced volunteers. The latter may involve additional training but in some cases are open to experienced divers that have no previous Seasearch experience (Bunker et al. 2017; Kay and Dipper 2018). This demonstrates how well-designed and long-term projects can satisfy participants from varied backgrounds and allow for significant participant development.

1.3.2.2 Participant Training

Basic training of participants occurs across the majority of marine citizen science projects and extends from written instructions, to two–three-day training programs, especially in projects involving specific methodological techniques/protocols (Thiel et al. 2014). Within this review, 77 projects provided some form of participant training, of which 29.9% involved brief instructions, 53.2% involved basic training (i.e. an event where an expert introduced the protocol to be employed) and 16.9% included a ≥ one-day training course. Training of participants involved in projects that use simple protocols (i.e. count or presence/absence surveys) (see sect. 3.3 for a review) primarily occurs through basic written instructions on data sheets and at times video tutorials (Bravo et al. 2009; Ribic et al. 2011). However, in projects that require more complex protocols (i.e. quadrat or transect surveys) and species identification, participants often attend a compulsory ≥ one-day training course, and it was noted that many of these projects often also involve SCUBA diving. Participant capabilities are usually assessed throughout the training, although only six projects explicitly stated that they verified participant capabilities. In addition, complex survey techniques often require additional scientific equipment (e.g. quadrats, transects, diving slates, identification guides, etc.) that are costly, resulting in some projects (e.g. Reef Check California, Mediterranean Sea and Tropical available at www.reefcheck.org) requesting a fee to cover the cost of the training and tools. Although this may limit the project's accessibility, it also ensures training quality and often enhances the recruitment of highly motivated participants. Citizen scientists contributing financially to projects might consider it an investment, and they may in turn be more likely to continue participating. However, this theory has yet to be tested explicitly and represents the scope for future research. Despite the multiple benefits of training, 25.8% of projects required no training, and the majority of these are reliant on incidental sightings (i.e. stranded animals or marine debris) (McGovern et al. 2016). In the case of stranded animals, citizen scientists report the sighting, and professionals are then required for the subsequent removal, identification and autopsy (Avens et al. 2009).

For the most part, the projects considered in this review allow participants to conduct research without professional supervision. Consequently, full explanatory training is key to ensuring the collection of scientifically sound and high-quality data (see sect. 3.4 for a review), and the length of the training is somewhat correlated to the complexity of the protocol employed. Some projects further engage with participants through the organization of additional events and courses in order to maintain project engagement and allow for upskilling. An example of this is Capturing our Coast (available at www.capturingourcoast.co.uk) that organizes regular refresher events for trained participants to maintain their survey/identification skills and to enhance data quality, alongside engagement events such as 'Wine and Science' where participants are invited to talks by guest speakers that cover a range of marine science disciplines. Beyond training, many projects communicate with their participants through their websites, newsletters and social media in order to keep them up-to-date with the project progress and encourage further participation. In addition, 'group sourced identification forums' on websites and social media are growing in popularity and may assist in participant engagement and increase the accuracy of the citizen-collected data (Chamberlain 2018). Informal participant feedback has suggested that online engagement strategies are becoming increasingly important components of marine citizen science projects (E. Morris-Webb, personal communication). However, there is currently a lack of systematic reviews on the role of outreach tools in the retention of volunteers highlighting the potential for future research in this area.

1.3.3 Tasks Undertaken

In order for citizen science projects to investigate the diverse array of habitats and species mentioned previously, a heterogeneous range of methodologies are employed. Each project must use methods that are appropriate to the field of enquiry but that are within the capabilities of the participants recruited (Worthington et al. 2012). Among the most popular are field surveys (35.8%) and reporting of opportunistic sightings (34.2%) (Fig. 1.3), which aligns with the findings of Thiel et al. (2014). Field surveys primarily involve searches for both live (e.g. Reef Check Tropical available at www.reefcheck.org/tropical/overview) and deceased organisms (e.g. Beach COMBERS, available at www.mlml.calstate.edu/beachcombers), as well as ecological phenomena (e.g. Bleach Patrol available at www.ldeo.columbia.edu/bleach-patrol), during predefined time periods or within predefined areas such as transects and quadrats. Surveys generally require citizen scientists to report findings of abundance or presence/absence, although in some cases, parameters uniquely designed for that project are requested, for example, the reef coloration requested in the CoralWatch bleach-

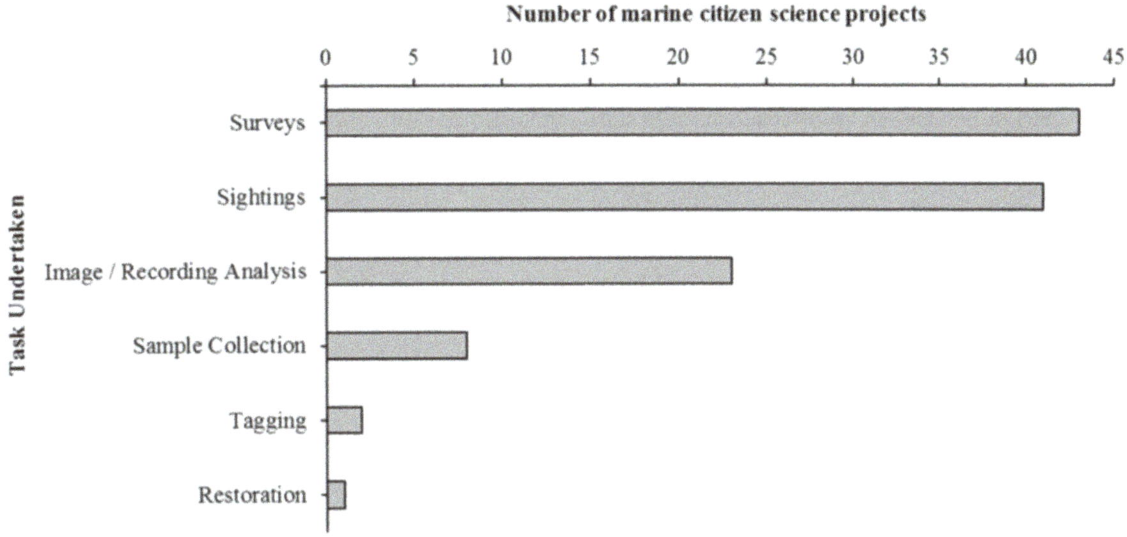

Fig. 1.3 Number of reviewed marine citizen science projects per primary tasks undertaken, excluding those that involved several tasks (n = 2)

ing protocol (available at www.coralwatch.org). Other surveys involve more novel methods, such as divers4oceanography (available at www.divers4oceanography.org) that asks SCUBA divers to report ocean temperatures recorded on their dive computers, and Smartfin (available at www.smartfin.org) that has designed a surfboard fin with sensors that allows surfers to collect real-time ocean parameters including temperature, location and wave characteristics (sensors that measure salinity, pH, dissolved oxygen and chlorophyll are under development). Surveying remains a key methodology of marine citizen science projects due to its cost-effectiveness, relative ease of implementation and ability to generate data across large spatio-temporal scales. Opportunistic sightings again allow data to be generated across vast scales and are at times a more time- and effort-efficient method compared to quadrat and transect surveys (Wiggins and Crowston 2011; Cox et al. 2012, 2015); although for the most part, they are employed by projects focusing on marine mammals, jellyfish and marine debris (including both field and online projects).

Technological developments have allowed an increasing number of projects to incorporate online citizen scientists to analyse vast data sets of images and recordings (19.1%), often through web portals such as Zooniverse. The popularity of this method lies in the fact that its only limitation is the often time-consuming preparation of the photos prior to being uploaded online. Finally, a combination of advanced technology and that fact that they are often focused on specific target organisms may explain why tagging (i.e. catch, tag and release of organisms) and restoration (i.e. environmental regeneration) were among the least used methodologies (1.6% and 0.8%, respectively) (Fig. 1.3).

1.3.4 Data Directionality

1.3.4.1 Data Quality

Citizen science strives to meet the same credibility standards as academic research and industry; however, it is often subject to limited resources and consequently faces trade-offs between data quantity and quality, protocol standardization and discrepancies in skills and expectations of participants and project facilitators/scientists (Robertson et al. 2010; Tulloch et al. 2013). To maintain data quality, some projects statistically compare results reported by citizen scientists to those of professional scientists as a means of data validation (Bell 2007; Worthington et al. 2012; Holt et al. 2013; Bird et al. 2014; Thiel et al. 2014; Earp et al. 2018b). Within this review, 19.2% of projects were found to validate their data in some way, which is much lower than the 55.1% reported by Thiel et al. (2014) in a similar investigation. However, an increasing body of research has shown that data collected by citizen scientists meets, or surpasses accepted quality standards, or detects important ecological trends (Cox et al. 2012; Forrester et al. 2015; Kosmala et al. 2016; Schläppy et al. 2017). In the study of Delaney et al. (2008), the accuracy of volunteers in identifying native and invasive crabs was assessed and found to be between 80 and 95% accurate for school children and even greater for those with a university education, suggesting that demographic variables such as age and educational background may be important drivers of data quality. As a result, choosing a research topic to suit the target participants is key to the success of a citizen science project. In other studies, increasing experience level (Jiguet 2009) and training of participants (Edgar and Stuart-Smith 2009) (see sect. 3.2.2 for a review) were shown to positively correlate with data quality. In other cases, citizen science data has been shown to demonstrate bias or inaccuracies (Courter et al. 2013; Forrester et al. 2015; van der Velde et al. 2017), but this can be minimized in data summaries by examining broader-scale trends (e.g. family level rather than species level) (Fore et al. 2001; Gouraguine et al. 2019) or excluding data from participants that differed substantially to data collected by scientists (Culver et al. 2010). Irrespectively, perceptions on data quality remain a key factor influencing the publication of citizen science data (Schläppy et al. 2017).

1.3.4.2 Data Publication

In recent years, an increasing number of peer-reviewed journal articles have focused on marine citizen science with many incorporating participant-collected data. This was the case for a minimum of 44 of the 120 reviewed projects that have contributed data to at least 1483 peer-reviewed journal articles. The majority of these publications (54%) were in relation to chordates (Fig. 1.4), of which 70% focused on marine mammals followed by seabirds (15.6%). Interestingly, only 5.2% of chordate publications focused on groups such as sharks and rays. Projects concentrating on diverse taxa were also highly likely to contribute to publications (29.1%), whereas <2% of publications focused on marine pollution.

Despite marine mammals and pollution being the focus of comparable project numbers, the publication frequency of marine mammal data is over 30 times greater than that of marine pollution. This discrepancy may be due to the fact that pollution is a relatively new trend in marine citizen science, whereas the majority of marine mammal projects are well established and commenced prior to 2008. Although the trend towards pollution-based studies has allowed for vast data sets to be generated in seemingly short time periods, the number of investigations (i.e. surveys) was shown to have less of an influence on publication frequency compared to project duration (Fig. 1.5a). Project durations vary from days to decades (Thiel et al. 2014), and of the projects that state their start date (n = 103), the greatest percentage (24.3%) are currently between 2 and 5 years in duration. Despite a limited correlation between project duration and publication frequency (Fig. 1.5b), short projects have a demonstrated capacity to be published, for example, in the 2-day 'bioblitz' undertaken by Cohen et al. (2011) in Sitka (Alaska), where citizen scientists collected data that confirmed a 1000 km northward extension of the colonial tunicate *Didemnum vexillum* (Sundlov et al. 2016).

1.3.4.3 Policy Development

It is important to note that peer-reviewed journal articles are not the only outlet for marine citizen science data, and in some cases, especially in terms of marine pollution, the data collected is more valuable for aspects such as informing policy or driving management (Newman et al. 2015; Burgess et al. 2017). Marine legislation is often underpinned by evidence from large data sets, and citizen science provides a cost-effective method for their generation (Crabbe 2012; Hyder et al. 2015). The importance of marine citizen science in delivering evidence to support decision-making in marine

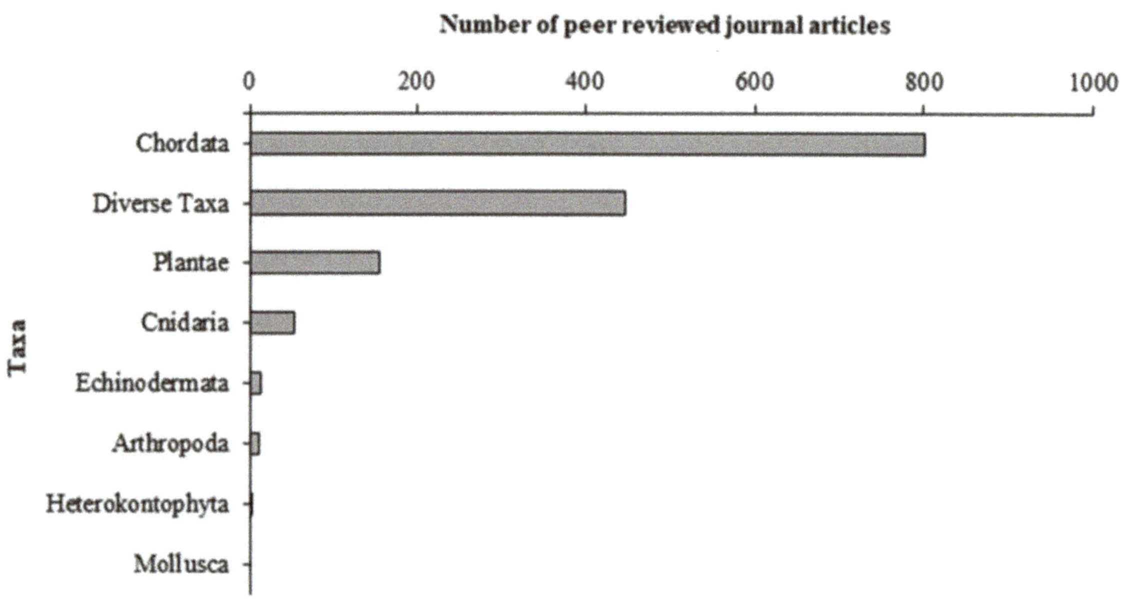

Fig. 1.4 Number of peer-reviewed journal articles (n = 1483) per focus taxa published by reviewed marine citizen science projects, excluding those that focus on oceanography or pollution (n = 18)

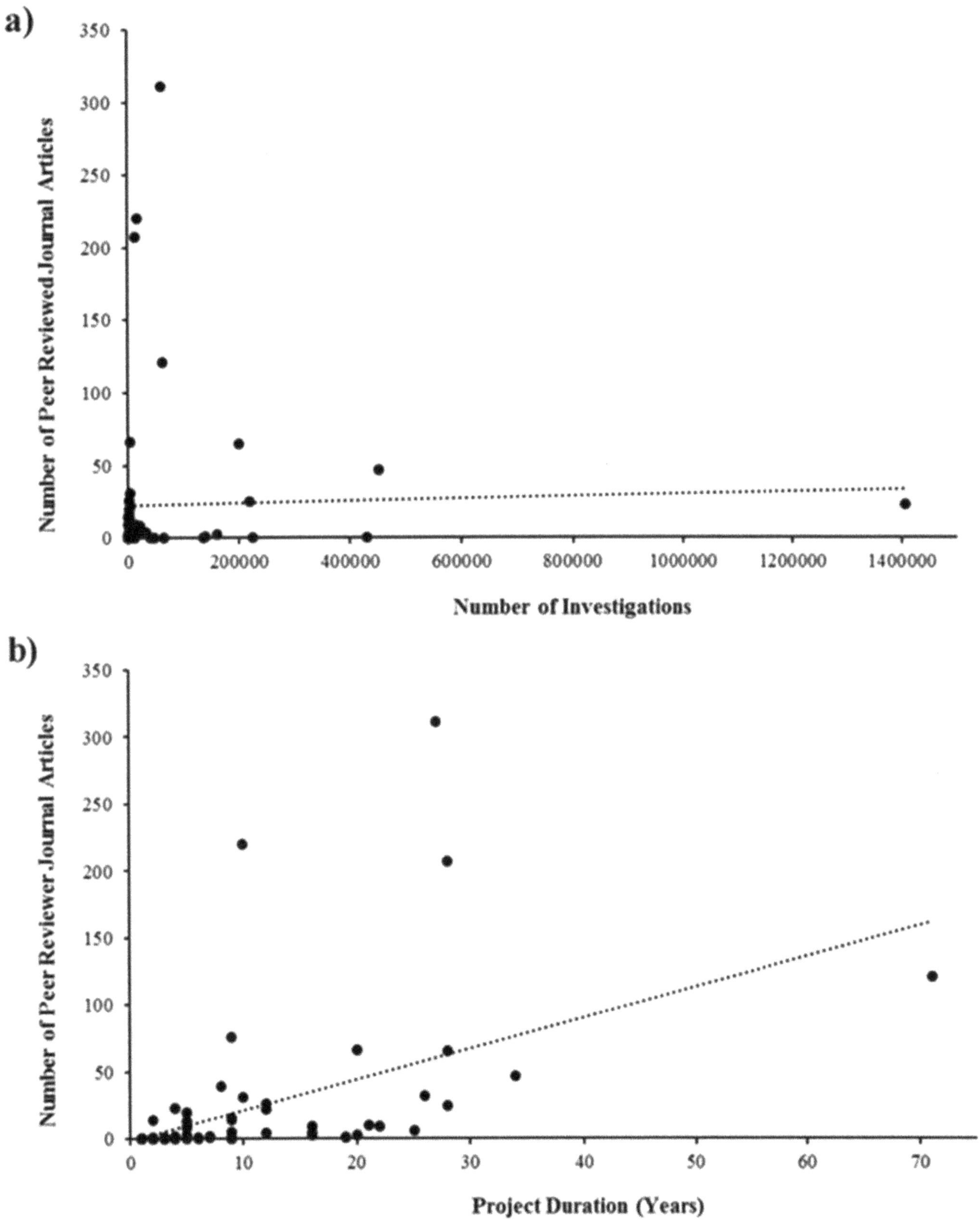

Fig. 1.5 Number of peer-reviewed journal articles published by reviewed marine citizen science projects (n = 1483) vs. number of project investigations (i.e. surveys) (**a**) and project duration (**b**)

Table 1.2 The primary policy area underpinned by reviewed marine citizen science projects and the total number of publications generated by projects in each of these areas

Policy	Area	Number of projects	Number of publications
Biodiversity	Species distribution	43	790
	MPA designation	7	166
	Invasive non-native species	6	0
	Stranding	4	211
	Threatened and rare species	4	42
	Other biological research	31	184
Physical environment	Oceanographical research	4	2
	Climate change	2	0
Pollution	Marine debris	10	15
	Water quality	4	15
Resource management	Fisheries	5	27

legislation was reviewed by Hyder et al. (2015), who classified four marine policy areas (biodiversity, physical environment, pollution and resource management) in which citizen science had played a valuable role. Within our review, the majority of projects (79.2%) were found to provide evidence underpinning biodiversity conservation policies (Table 1.2). Within this policy area, > 40% of projects investigated questions regarding species distribution, which was supported by findings from a study by Hyder et al. (2015). Other popular policy areas underpinned by the reviewed projects were: biological research (25.8%), marine debris surveying (8.3%), marine protected area (MPA) designation (5.8%) and invasive species tracking (5%) (Table 1.2).

1.3.4.4 Data Storage

Although the use of citizen science data varies, projects are encouraged to make their data publicly accessible, yet in this review, less than 10% of projects stored their data on a repository (e.g. NBN Atlas available at www.nbnatlas.org). In particular, all species survey data from reviewed citizen science projects in the UK is stored on NBN Atlas, which is accessible to the public and government for a range of purposes, although some data limitations exist regarding commercially sensitive/overexploited species. The majority of reviewed projects instead made their data available through the project website (50.8%), although 40% of projects kept their data private or failed to disclose its location.

1.3.4.4.1 Species Distribution

The long-term observational data sets generated by citizen science projects, which often extend beyond timescales of standard research programs (i.e. two-three years), are also of exceptional value in addressing questions regarding the spatio-temporal distribution of marine organisms (Ponti et al. 2011b; Mieszkowska et al. 2014). More recently these data sets have become important in assessing the response of organisms to climate change (Southward et al. 2005; Mieszkowska et al. 2014). Climate change can induce so-called shifts in marine species distributions, either towards more favourable seawater temperatures or because of trophic mismatches resulting from changes in prey phenology (Visser and Both 2005; Cheung et al. 2009). Mieszkowska et al. (2014) demonstrated some of the fastest geographic range shifts in marine species in 50 years using citizen science data collected as part of the Marine Biodiversity and Climate Change (MarClim) project. Lusitanian species, including *Phorcus* (previously *Osilinus*) *lineatus* and *Steromphala* (previously *Gibbula*) *umbilicalis* (topshells), *Chthamalus montagui* and *Perforatus perforatus* (barnacles), as well as the limpet *Patella depressa* and the macroalga *Bifurcaria bifurcata*, extended their range poleward, whilst Boreal species, such as the barnacle *Semibalanus balanoides* and the kelp *Alaria esculenta*, were shown to be important indicator species that fluctuate in abundance in response to periods of warming and cooling (Mieszkowska et al. 2014). Although only two projects in this review investigated physical components of climate change (i.e. changes in temperatures, pH and storm frequency), numerous projects (35.8%) addressed questions regarding species distributions with several of these providing information important for climate change studies (e.g. information on coral bleaching that results from climatic change).

1.3.4.4.2 Invasive Non-Native Species

Marine citizen science is not only a powerful tool in monitoring the distribution of native species but also the arrival and encroachment of invasive non-native species (Delaney et al. 2008; Parr and Sewell 2017), whose impacts on native ecosystems remain poorly understood (Ruiz et al. 1997). These species can go undetected for extended periods of time (i.e. years) (Geller et al. 1997; Lohrer 2001), meaning their subsequent eradication may be difficult, in part because of large population sizes (Bax et al. 2001). Consequently, monitoring for invasive species is of primary importance so that early eradication can be conducted (Delaney et al. 2008). In the USA, the Citizen Science Initiative: Marine Invasive Species Monitoring Organization (www.InvasiveTracers.com) used 190 groups of participants to survey 52 sites for a species of introduced Asian shore crab (*Hemigrapsus sanguineus*), previously known only to be present in Moores Harbour. They reported a notable expansion of the range of *H. sanguineus*, with one specimen being reported 60 km northeast of Moores Harbour (Delaney et al. 2008). Some invasive species are not only detrimental to native ecosystems, but also to local economies, for example, the carpet sea squirt (*Didemnum vexillum*), which was discovered in the UK for the first time by citizen scientists in a study conducted by Cohen et al. (2011). *D. vexillum* is detrimental to shellfish aquaculture with high abundances resulting in increased labour costs and reduced product value (Watson et al. 2009; Adams et al. 2011). Knowing the origin and arrival location of invasive species is very important to model their potential dispersion pathways, allow for early detection and in turn limit further colonization of new areas (Ricciardi et al. 2000). The spatio-temporal scale at which citizen science operates is therefore of exceptional value to invasive species monitoring and benefits both scientific research and industry. Although several of the reviewed projects (e.g. Capturing our Coast available at www.capturingourcoast.co.uk, Seasearch available at www.seasearch.org.uk and Reef Check Mediterranean Sea available at www.reefcheckmed.org) have trained volunteers to report sightings of non-native species, they were the primary focus of only 5% of reviewed projects (Table 1.2), therefore demonstrating scope for significant expansion of marine citizen science in this area.

1.3.4.4.3 Marine Debris

As part of an action to tackle a widespread and growing issue, citizen scientists are helping to investigate marine debris by contributing to vast global data sets that form the basis of both political decisions and conservation policies (Ryan et al. 2009; Eriksen et al. 2014; Hidalgo-Ruz and Thiel 2013, 2015; Nelms et al. 2017). Although this policy area was underpinned by only 8.3% of projects within this review (Table 1.2), it has grown considerably in the past decade, with more citizens sharing responsibility for the issue and contributing to projects aiming to provide solutions (Hidalgo-Ruz and Thiel 2015; Nelms et al. 2017). These projects often involve simple beach clean-ups that provide information on the distribution and abundance of marine debris items (Ribic 1998; Martin 2013). Because of the simplicity of the protocols, several citizen science projects underpinning different areas of policy (i.e. biodiversity policies such as species distribution and stranding) have organized events to tackle marine debris issues and contributed to litter recording databases. In the UK, citizen scientists reported and collected over 10,866 plastic bottles as part of the Marine Conservation Society Wild Bottle Sighting campaign (www.mcsuk.org/bottlesightings). The awareness raised and evidence collected through this, and other similar campaigns (OSPAR 2010; Van Franeker et al. 2011), were key to the decision of the UK government to develop a Deposit Return System for plastic bottles and aluminium cans as part of a plan to eliminate single-use plastic by 2042 (official press release available at www.gov.uk/government/news/deposit-return-scheme-in-fight-against-plastic).

1.3.4.4.4 Marine Protected Areas

One method to protect and promote biodiversity is the designation of marine protected areas (MPAs) and Marine Conservation Zones (MCZs); however to do this, patterns of species distributions across vast spatio-temporal scales (such as those covered by citizen science/scientists) are required (Dickinson et al. 2010; Cerrano et al. 2012; Crabbe 2012; Markantonatou et al. 2013; Branchini et al. 2015; Jarvis et al. 2015; Cerrano et al. 2017). Additionally, marine citizen science is also an effective tool for monitoring MPAs that is essential to support adequate management and to fulfil the requirements of the European Marine Strategy Framework Directive (Directive 2008/56/EC) (Ponti et al. 2011a; Cerrano et al. 2017; Turicchia et al. 2017). This review found that 5.8% of projects provided data that underpins MPA designation and/or monitoring (Table 1.2). Despite this low percentage, the majority of these projects were found to operate over extensive geographical ranges (e.g. Reef Check available at www.reefcheck.org; eOcean available at www.eoceans.co; SubseaObservers available at www.subseaobservers.com). Since its establishment in 1996, Reef Check data has contributed to the establishment and monitoring of several MPAs in regions with limited funding for conservation policies (Cerrano et al. 2012), and in the UK, the Seasearch data set that extends back to 1984 has contributed to the designation of 38 MCZs and several other MPAs including Lamlash Bay No-Take Zone (Seasearch 2018).

1.4 Challenges and Strengths of Marine Citizen Science

1.4.1 Challenges

The rapid expansion of marine citizen science, coupled with advancing possibilities and limited funding means, at present, limited guidelines for good practice are in place, and instead many facilitators are learning on the job (Silvertown 2009). Consequently, many projects face challenges, usually in the form of participant motivation and/or data issues.

1.4.1.1 Participant Motivation

At the organizational level, motivation is one of the most commonly referenced challenges facing citizen science (Conrad and Daoust 2008; Conrad and Hilchey 2011; Rotman et al. 2012), and it comes in two forms: (1) motivating outsiders to sign-up and begin participating and (2) motivating participants to continue or expand their participation

(Prestopnik and Crowston 2011; Rotman et al. 2012). To motivate participants to sign up, some projects focus specifically on either charismatic organisms (Bear 2016; Garcia-Soto et al. 2017) or accessible environments (Garcia-Soto et al. 2017), or incorporate an activity (e.g. SCUBA diving) into the protocol to engage participants who readily participate in this activity already (e.g. Seasearch available at www.seasearch.org.uk). However, these strategies generate issues including a data bias towards charismatic species and accessible nearshore environments that may be of limited ecological/scientific relevance, or if SCUBA diving is required, the project may become cost and/or experience prohibitive to certain participants. An often-unaddressed issue of citizen science is accessibility, especially for participants with impairments. For the most part, projects involve either a physical task in a somewhat hazardous (i.e. slippy) environment, a highly skilful and intense activity (e.g. SCUBA diving), or recording on small data sheets. Consequently, participants with impairments may be limited to online marine citizen science projects, unless a protocol can be adapted to suit their needs.

Motivating participants to continue or expand their participation is often a greater challenge, especially when the project involves reporting 'zero data', which may be of significant scientific importance, but it is often disengaging and might result in declines in participation (Bear 2016). One solution to maintain engagement that is also used to motivate sign-ups and initial participation is gamification (Prestopnik and Crowston 2011) and/or incentivization of the protocol (e.g. the 'Top Trumps' format of the Capturing our Coast 'Beach Babies' survey, available at www.capturingourcoast.co.uk/specific-information/beach-babies). For participants, the more fun, or the greater the benefit they receive from completing the work, the more likely they are to participate (Prestopnik and Crowston 2011). In addition, feedback to participants is of exceptional importance (Bonney et al. 2009; Silvertown 2009) and has been shown to increase and/or maintain participation by demonstrating the value of their contribution (Rotman et al. 2012). Research from Thiel et al. (2014) supports this view, stating that public recognition of participant effort is a significant motivator for participation but that factors including personal satisfaction (i.e. wellbeing impact of developing social connections and being outdoors) and development of a skill base (i.e. greater understanding of the scientific processes) are also important motivators.

1.4.1.2 Data Concerns

Another obstacle facing citizen science is perceptions regarding data quality (see sect. 3.4 for a review), despite several studies demonstrating that the data meets accepted quality standards (Cox et al. 2012; Forrester et al. 2015; Kosmala et al. 2016; Schläppy et al. 2017). These concerns often relate but are not limited to a lack of attention to project design and standardized data verification methods, limited participant training and sampling biases (Conrad and Hilchey 2011; Burgess et al. 2017). Riesch and Potter (2014) postulated that a lack of use of citizen science data in academia may stem from the belief of some scientists that the data would not be well received by their peers. In terms of policy development, the United States Congress excluded volunteer collected data from their National Biological Survey over concerns that it would be biased based on environmentalist agendas (Root and Alpert 1994). To address the issue of data perceptions, Burgess et al. (2017) suggest greater transparency and availability of methods and data attributes that will hopefully result from the generation of good practice guidelines and toolkits for citizen science (Silvertown 2009). However, despite the shortcomings, many benefits of marine citizen science have been documented. For this reason, the development of a standard 'impact assessment', to assess survey and data verification methods, as well as scientific and socio-psychological benefits would be highly beneficial to marine citizen science projects.

1.4.2 Strengths

The strengths of citizen science have been demonstrated to extend across scientific, social and economic boundaries, as well as underpinning several areas of marine policy (see sect. 1.3.4.3) (Delaney et al. 2008; Crabbe 2012; Mieszkowska et al. 2014; Hidalgo-Ruz and Thiel 2015; Hyder et al. 2015; Turicchia et al. 2015; Parr and Sewell 2017).

1.4.2.1 The Many Eyes Hypothesis

The 'many eyes hypothesis' has been used to describe the efficiency of marine citizen science in generating data across vast spatio-temporal scales and across multiple taxa (Hochachka et al. 1999; Dickinson et al. 2012; Thomas et al. 2017). This hypothesis, in the case of animal aggregations, shows a larger group size has extended individual feeding times and an increased likelihood of detecting predators compared to smaller groups (Lima 1995). When applied to citizen science, it demonstrates that a network of citizen scientists with clearly defined protocols and realistic survey aims is capable of surveying vast areas (Ponti et al. 2011b; Cerrano et al. 2012, 2017), which increases the chances of detection of a species/phenomenon, increases replication rates and decreases individual effort (Hochachka et al. 1999; Thomas et al. 2017). This 'many eyes' effect has allowed citizen science to benefit landscape ecology and macroecology research by covering extensive spatio-temporal scales (Parmesan and Yohe 2003; Southward et al. 2005; Dickinson et al. 2010, 2012; Mieszkowska et al. 2014; Schläppy et al. 2017), alongside providing an efficient means for detecting species with low abundances (e.g. rare or invasive species) (Delaney et al. 2008; Dickinson et al. 2010, 2012).

1.4.2.2 Marine Stewardship

Appropriately designed projects not only have the potential to broaden the scope of marine research and policy but also reconnect participants with nature that in turn increases their awareness of the current status of the marine realm and the threats it faces (Brightsmith et al. 2008; Dickinson et al. 2010; Koss and Kingsley 2010; Garcia-Soto et al. 2017; Cerrano et al. 2017; Schläppy et al. 2017; Turicchia et al. 2017). Through marine citizen science projects, participants may observe the impact of anthropogenic activities on marine environments, which may promote a sense of ownership and drive behavioural changes towards more sustainable actions (Branchini et al. 2015; Cerrano et al. 2017). This increased awareness may then be used to promote marine stewardship, and in many cases, participants often become advocates of marine conservation (Dickinson et al. 2010; Cerrano et al. 2017; Garcia-Soto et al. 2017). The enhanced ocean literacy, resulting from participating in marine citizen science projects, combined with exposure to science through other mediums (e.g. television documentaries and news articles) (Thiel et al. 2014) may also lead to greater support for scientific advances and policy, as opposition often results from a lack of understanding of the underlying science (Schläppy et al. 2017). This environmental stewardship also forms the basis of compliance with marine management policies such as MPAs and can indirectly enhance their efficiency (Evans et al. 2008; McKinley and Fletcher 2012).

1.5 Summary

At present, marine citizen science is at a crossroads of demonstrated sucesses, unresolved challenges and unrealized potential (Burgess et al. 2017). To resolve these challenges, and fulfill its potential, citizen science practitioners must be willing to acknowledge project shortcomings and work together to align objectives and methodologies that ensure the generation of high-quality data sets (Burgess et al. 2017). In addition, the accessibility and feasibility of the project to a diverse array of participants, as well as methods of disseminating project findings to participants, must be considered. Overall however, the projects investigated within this review demonstrated positive outcomes of collaborations between members of the public and scientists. Citizen scientists become specialized in the protocols used and in turn provide quality data that increase the spatio-temporal coverage of marine research (Thiel et al. 2014).

With our oceans and coasts in peril due to changing climatic conditions and increasing anthropogenic activities (Cigliano et al. 2015, Earp et al. 2018a), marine citizen science provides a unique platform to discover, innovate and address global challenges (i.e. species shifts and marine debris) for which data is significantly lacking (Bear 2016). As marine citizen science comes of age, although not panacea, if it successfully keeps pace with the changing contexts of marine ecological research, conservation needs and stakeholder interests, its capacity to increase ocean literacy may aid the development of culturally and politically feasible solutions for a more sustainable future (McKinley et al. 2017; Schläppy et al. 2017). With this in mind, it can be summarized that the current benefits of marine citizen science outweigh the challenges, and there is significant scope for the development and incorporation of 'science by the people' into marine research and conservation projects (Silvertown 2009).

Acknowledgements The authors are grateful for the constructive input of Dr. Siobhan Vye, Ms. Natalie Prinz and three reviewers, Dr. Ans Vercammen, Dr. Eva Turicchia and Ms. Elisabeth Morris-Webb that greatly improved the manuscript. We would also like to share our gratitude to the YOUMARES 9 committee for the opportunity to research a topic that is close to both our academic and personal experiences. Finally, our appreciation and recognition goes to all professional and citizen scientists involved in the reviewed projects for their outstanding work and commitment in making citizen science the science of the future.

Appendix

This article is related to the YOUMARES 9 conference session no. 1: "Could citizen scientists and voluntourists be the future for marine research and conservation?" The original Call for Abstracts and the abstracts of the presentations within this session can be found in the Appendix "Conference Sessions and Abstracts", Chapter "1 Could citizen scientists and voluntourists be the future for marine research and conservation?", of this book.

Supplementary Material

Table 1.A1 Name and website of the 120 marine citizen science projects reviewed in this manuscript

Marine Citizen Science Project	Website
Angling Project: Off The Hook	www.sharktrust.org/en/anglers_recording_project
B.C. Cetacean Sightings Network	www.wildwhales.org
Basking Shark Project	www.sharktrust.org/en/basking_shark_project
Beach Environmental Assessment, Communication & Health (BEACH)	www.ecology.wa.gov/Water-Shorelines/Water-quality/Saltwater/BEACH-program
Beach Watch	www.beachwatch.farallones.org
BeachObserver	www.beachobserver.com
Big Seaweed search	www.bigseaweedsearch.org
Birding Aboard	www.birdingaboard.org/index.html
Biscayne Bay Drift Card Study	www.carthe.org/baydrift
Bleach Patrol	www.ldeo.columbia.edu/bleachpatrol
Blue Water Task Force	www.surfrider.org/blue-water-task-force
Cape Radd Citizen Science Day	www.caperadd.com/courses/citizen-science-day
Capturing our Coast	www.capturingourcoast.co.uk
CARIB Tails	www.caribtails.org/home.html
Caribbean Lionfish Response Program	www.corevi.org
Chesapeake Bay Parasite Project	www.serc.si.edu/citizen-science/projects/chesapeake-bay-parasite-project
Clean Sea LIFE	cleansealife.it
Coastal Observation & Seabird Survey Team (COASST)	www.depts.washington.edu/coasst
Coastal Ocean Mammal & Bird Education & Research Surveys (Beach COMBERS)	www.mlml.calstate.edu/beachcombers
Community Seagrass Initiative	www.csi-seagrass.co.uk
Coral Reef Monitoring Data Portal	www.monitoring.coral.org
CoralWatch	www.coralwatch.org
Crab Watch	www.seachangeproject.eu/seachange-about-4/crab-watch
Delaware Bay Horseshoe Crab Spawning Survey	www.dnrec.alpha.delaware.gov/coastal-programs/education-outreach/horseshoe-crab-survey
Delaware Shorebird Project	www.dnrec.delaware.gov/fw/Shorebirds
Digital Fishers	www.oceannetworks.ca/learning/get-involved/citizen-science/digital-fishers
divers4oceanography	www.divers4oceanography.org
Earthdive	www.earthdive.com
eOceans	www.eoceans.co
Fish Watchers	www.fishbase.org/fishwatcher/menu.php
FjordPhyto	www.fjordphyto.wordpress.com
Floating Forests	www.zooniverse.org/projects/zooniverse/floating-forests
Follow & Learn About the Ocean & Wetland (FLOW)	www.amigosdebolsachica.org/flow.php
Global Microplastics Initiative	www.adventurescientists.org/microplastic
Gotham Whale	www.gothamwhale.org/citizen-science
Grunion Greeters	www.grunion.org
Happywhale	www.happywhale.com
Horseshoe crabs as homes	www.sites.google.com/site/epibiont
IHO Crowdsourced Bathymetry	www.ngdc.noaa.gov/iho/#csb
iNaturalist	www.inaturalist.org
Invader ID	www.zooniverse.org/projects/serc/invader-id
iSeahorse	www.iseahorse.org
JellyWatch	www.jellywatch.org
Kelp Watch	www.serc.si.edu/citizen-science/projects/kelp-watch
Long-term Monitoring Program & Experimental Training for Students (LiMPETS)	www.limpets.org

(continued)

Table 1.A1 (continued)

Marine Citizen Science Project	Website
Manatee Chat	www.zooniverse.org/projects/cetalingua/manatee-chat
MangroveWatch	www.mangrovewatch.org.au
Manta Matcher	www.mantamatcher.org/overview.jsp
Marine Debris Monitoring & Assessment Project	www.marinedebris.noaa.gov/research/monitoring-toolbox
Marine Debris Tracker	www.marinedebris.engr.uga.edu
Marine Metre Squared (Mm2)	www.mm2.net.nz
MCS Wild Bottle Sightings	www.mcsuk.org/bottlesightings
Mitten Crab Watch	www.mittencrabs.org.uk
Monitor Tupinambás	www.zooniverse.org/projects/larissakawabe/monitore-tupinambas
New England Basking Shark & Ocean Sunfish Project	www.nebshark.org
New York Horseshoe Crab Monitoring Network	www.nyhorseshoecrab.org
North Atlantic Right Whale Sightings Advisory System	www.nefsc.noaa.gov/psb/surveys/SAS.html
Ocean Sampling Day	www.microb3.eu/osd.html
Orcasound	www.orcasound.net
OSPAR Beach Litter	www.ospar.org/work-areas/eiha/marine-litter/beach-litter
Our Radioactive Ocean	www.ourradioactiveocean.org
Oyster Drills in Richardson Bay	www.serc.si.edu/citizen-science/projects/oyster-drill
Penguin Cam	www.penguinscience.com/education/count_the_penguins.php
Penguin Watch	www.penguinwatch.org
Plankton Portal	www.planktonportal.org
PlateWatch	www.platewatch.nisbase.org
Puget Sound Seabird Survey (PSSS)	www.seattleaudubon.org/sas/About/Science/CitizenScience/PugetSoundSeabirdSurvey.aspx
Redmap	www.redmap.org.au
Reef Check California	www.reefcheck.org/california/ca-overview
Reef Check Mediterranean Sea	www.reefcheckmed.org
Reef Check Tropical	www.reefcheck.org
Reef Environmental Education Foundation (REEF)	www.reef.org
Reef Life Survey	www.reeflifesurvey.com/reef-life-survey
Reef Watch	www.conservationsa.org.au/reef_watch
Rescue a Reef	www.sharkresearch.rsmas.miami.edu/donate/rescue-a-reef
Satellites Over Seals (SOS)	www.tomnod.com
Scuba Tourism For The Environment	www.steproject.org
Sea Star Wasting Disease	www.udiscover.it/applications/seastar
Seabird Ecological Assessment Network (SEANET)	www.seanetters.wordpress.com
Seabirdwatch	www.zooniverse.org/projects/penguintom79/seabirdwatch
Seagrass Spotter	www.seagrassspotter.org
Seagrass Watch	www.seagrasswatch.org
Sealife Survey	www.mba.ac.uk/recording/about
Seasearch	www.seasearch.org.uk
seaturtle.org	www.seaturtle.org
Seawatch Submit a Sighting	www.seawatchfoundation.org.uk
Send us your skeletons	www.fish.wa.gov.au/Fishing-and-Aquaculture/Recreational-Fishing/Send-Us-Your-Skeletons/Pages
Sevengill Shark Identification Project	www.sevengillsharksightings.org
SharkBase	www.shark-base.org
Sharkscount	www.sharksavers.org/en/our-programs/sharkscount
Smartfin	www.surfrider.org/programs/smartfin
Snapshots at Sea	www.zooniverse.org/projects/tedcheese/snapshots-at-sea
South Africa Elasmobranch Monitoring (ELMO)	www.elmoafrica.org
SubseaObservers	www.subseaobservers.com
Tag A Tiny	www.umb.edu/tunalab/tagatiny
Tangaroa Blue	www.tangaroablue.org

(continued)

Table 1.A1 (continued)

Marine Citizen Science Project	Website
TBF Tag & Release Program	www.billfish.org/research/tag-and-release
The Big Sea Survey	www.hlf.org.uk/our-projects/big-sea-survey
The Florida Keys BleachWatch Program	www.mote.org/research/program/coral-reef-science-monitoring/bleachwatch
The Great Eggcase Hunt Project	www.sharktrust.org/en/great_eggcase_hunt
The Great Nurdle Hunt	www.nurdlehunt.org.uk
The Plastic Tide	www.theplastictide.com
The Secchi Disk study	www.secchidisk.org
The Shore Thing	www.mba.ac.uk/shore_thing
The Wetland Bird Survey (WeBS)	www.bto.org/volunteer-surveys/webs
TLC Juvenile Lobster Monitoring Program	www.lobsters.org/volunt/volunteer.html
trackmyfish	trackmy.fish
Wakame Watch	wakamewatch.org.uk
WDC Shorewatch Programme	www.wdcs.org/national_regions/scotland/shorewatch
Weddell Seal Count	www.zooniverse.org/projects/slg0808/weddell-seal-count
Whale FM	whale.fm
Whale mAPP	www.whalemapp.org
Whale Track	whaletrack.hwdt.org
Whales as Individuals	www.zooniverse.org/projects/tedcheese/whales-as-individuals
Wildbook for Whale Sharks	www.whaleshark.org

All websites last accessed 24 June 2019 by the authors

References

Adams CM, Shumway SE, Whitlatch RB et al (2011) Biofouling in marine molluscan shellfish aquaculture: a survey assessing the business and economic implications of mitigation. J World Aquacult Soc 42:242–252. https://doi.org/10.1111/j.1749-7345.2011.00460.x

Avens L, Taylor JC, Goshe LR et al (2009) Use of skeletochronological analysis to estimate the age of leatherback sea turtles *Dermochelys coriacea* in the western North Atlantic. Endanger Species Res 8:165–177. https://doi.org/10.3354/esr00202

Bax N, Carlton JT, Mathews-Amos A et al (2001) The control of biological invasions in the world's oceans. Conserv Biol 15:1234–1246. https://doi.org/10.1111/j.1523-1739.2001.99487.x

Bear M (2016) Perspectives in marine citizen science. J Microbiol Biol Educ 17(1):56–59. https://doi.org/10.1128/jmbe.v17i1.1037

Bell JJ (2007) The use of volunteers for conducting sponge biodiversity assessments and monitoring using a morphological approach on Indo-Pacific coral reefs. Aquat Conserv 17:133–145. https://doi.org/10.1002/aqc.789

Bird TJ, Bates AE, Lefcheck JS et al (2014) Statistical solutions for error and bias in global citizen science datasets. Biol Conserv 173:144–154. https://doi.org/10.1016/j.biocon.2013.07.037

Bonney R, Cooper CB, Dickinson J et al (2009) Citizen science: a developing tool for expanding science knowledge and scientific literacy. Bioscience 59(11):977–984. https://doi.org/10.1525/bio.2009.59.11.9

Bonney R, Shirk JL, Phillips TB et al (2014) Next steps for citizen science. Science 343(6178):1436–1437. https://doi.org/10.1126/science.1251554

Bonney R, Phillips TB, Ballard HL et al (2016a) Can citizen science enhance public understanding of science? Public Underst Sci 25(1):2–16. https://doi.org/10.1177/0963662515607406

Bonney R, Cooper C, Ballard HL (2016b) The theory and practice of citizen science: launching a new journal. CSTP 1(1):1–4. https://doi.org/10.5334/cstp.65

Bowen S, Goodwin C, Kipling D et al (2011) Sea squirts and sponges of Britain and Ireland. Seasearch guides. Wild Nature Press, Plymouth

Branchini S, Pensa F, Neri P et al (2015) Using a citizen science program to monitor coral reef biodiversity through space and time. Biodivers Conserv 24:319–336. https://doi.org/10.1007/s10531-014-0810-7

Bravo M, Gallardo M, Luna-Jorquera G et al (2009) Anthropogenic debris on beaches in the SE Pacific (Chile): results from a national survey supported by volunteers. Mar Pollut Bull 58:1718–1726. https://doi.org/10.1016/j.marpolbul.2009.06.017

Brightsmith DJ, Stronza A, Holle K (2008) Ecotourism, conservation biology, and volunteer tourism: a mutually beneficial triumvirate. Biol Conserv 141:2832–2842. https://doi.org/10.1016/j.biocon.2008.08.020

Bruce E, Albright L, Sheehan S et al (2014) Distribution patterns of migrating humpback whales (*Megaptera novaeangliae*) in Jervis Bay, Australia: a spatial analysis using geographical citizen science data. Appl Geogr 54:83–95. https://doi.org/10.1016/j.apgeog.2014.06.014

Bunker FSPD, Brodie JA, Maggs CA et al (2017) Seaweeds of Britain and Ireland, Seasearch guides, 2nd edn. Wild Nature Press, Plymouth

Burgess HK, DeBey LB, Froehlich HE et al (2017) The science of citizen science: exploring barriers to use as a primary research tool. Biol Conserv 208:113–120. https://doi.org/10.1016/j.biocon.2016.05.014

Campbell LM, Smith C (2006) What makes them pay? Values of volunteer tourists working for sea turtle conservation. Environ Manag 38:84–98. https://doi.org/10.1007/s00267-005-0188-0

Cerrano C, Di Camillo CG, Milanese M et al (2012) Education through participation: the role of citizen science in marine habitat conservation. PIXEL New perspectives in science education, Florence

Cerrano C, Milanese M, Ponti M (2017) Diving for science – science for diving: volunteer scuba divers support science and conservation in the Mediterranean Sea. Aquat Conserv 27:303–323. https://doi.org/10.1002/aqc.2663

Chamberlain J (2018) Purple octopus. http://www.purpleoctopus.org/groupsourcing/. Accessed 9 Oct 2018

Cheung WW, Lam VW, Sarmiento JL et al (2009) Projecting global marine biodiversity impacts under climate change scenarios. Fish Fish 10(3):235–251. https://doi.org/10.1111/j.1467-2979.2008.00315.x

Cigliano JA, Ballard HL (2018) Citizen science for coastal and marine conservation. Routledge, Abingdon

Cigliano JA, Meyer R, Ballard HL et al (2015) Making marine and coastal citizen science matter. Ocean Coast Manage 115:77–87. https://doi.org/10.1016/j.ocecoaman.2015.06.012

Cohen CS, McCann L, Davis T et al (2011) Discovery and significance of the colonial tunicate *Didemnum vexillum* in Alaska. Aquat Invasions 6(3):263–271. https://doi.org/10.3391/ai.2011.6.3

Cohn JP (2008) Citizen science: can volunteers do real research? Bioscience 58(3):192–197. https://doi.org/10.1641/B580303

Conrad CT, Daoust T (2008) Community-based monitoring frameworks: increasing the effectiveness of environmental stewardship. Environ Manag 41(3):358–366. https://doi.org/10.1007/s00267-007-9042-x

Conrad CC, Hilchey KG (2011) A review of citizen science and community-based environmental monitoring: issues and opportunities. Environ Monit Assess 176(1–4):273–291. https://doi.org/10.1007/s10661-010-1582-5

Courter JR, Johnson RJ, Stuyck CM et al (2013) Weekend bias in citizen science data reporting: implications for phenology studies. Int J Biometeorol 57(5):715–720. https://doi.org/10.1007/s00484-012-0598-7

Cox TE, Philippoff J, Baumgartner E et al (2012) Expert variability provides perspective on the strengths and weaknesses of citizen-driven intertidal monitoring program. Ecol Appl 22(4):1201–1212. https://doi.org/10.1890/11-1614.1

Cox J, Oh EY, Simmons B et al (2015) Defining and measuring success in online citizen science: a case study of Zooniverse projects. Comput Sci Eng 17(4):28–41. https://doi.org/10.1109/MCSE.2015.65

Crabbe MJC (2012) From citizen science to policy development on the coral reefs of Jamaica. Int J Zool:102350. https://doi.org/10.1155/2012/102350

Culver CS, Schroeter SC, Page HM et al (2010) Essential fishery information for trap-based fisheries: development of a framework for collaborative data collection. Mar Coast Fish 2:98–114. https://doi.org/10.1577/C09-007.1

Danielsen F, Pirhofer-Walzl K, Adrian TP et al (2013) Linking public participation in scientific research to the indicators and needs of international environmental agreements. Conserv Lett 7(1):12–24. https://doi.org/10.1111/conl.12024

Delaney DG, Sperling CD, Adams CS et al (2008) Marine invasive species: validation of citizen science and implications for national monitoring networks. Biol Invasions 10(1):117–128. https://doi.org/10.1007/s10530-007-9114-0

Dickinson JL, Zuckerberg B, Bonter DN (2010) Citizen science as an ecological research tool: challenges and benefits. Annu Rev Ecol Evol Syst 41:149–172. https://doi.org/10.1146/annurev-ecolsys-102209-144636

Dickinson JL, Shirk J, Bonter D et al (2012) The current state of citizen science as a tool for ecological research and public engagement. Front Ecol Environ 10(6):291–297. https://doi.org/10.1890/110236

Directive 2008/56/EC of the European Parliament and of the Council (2008) Establishing a framework for community action in the field of marine environmental policy (Marine Strategy Framework Directive). Off J Eur Union 164:19–40

Done T, Roelfsema C, Harvey A et al (2017) Reliability and utility of citizen science reef monitoring data collected by Reef Check Australia, 2002–2015. Mar Pollut Bull 117(1–2):148–155. https://doi.org/10.1016/j.marpolbul.2017.01.054

Earp HS, Prinz N, Cziesielski MJ et al (2018a) For a world without boundaries: connectivity between marine tropical ecosystems in times of change. In: Jungblut S, Liebich V, Bode M (eds) YOUMARES 8 – oceans across boundaries: learning from each other. Springer, Cham, pp 125–144. https://doi.org/10.1007/978-3-319-93284-2_9

Earp HS, Vye SR, West V et al (2018b) Do you see what I see? Quantifying inter-observer variability in an intertidal disturbance experiment. Poster presented at YOUMARES 9, University of Oldenburg, Germany, 11–14 September 2018

Edgar GJ, Stuart-Smith RD (2009) Ecological effects of marine protected areas on rocky reef communitie – a continental-scale analysis. Mar Ecol Prog Ser 388:51–62. https://doi.org/10.3354/meps08149

Embling CB, Walters AEM, Dolman SJ (2015) How much effort is enough? The power of citizen science to monitor trends in coastal cetacean species. Glob Ecol Conserv 3:867–877. https://doi.org/10.1016/j.gecco.2015.04.003

Eriksen M, Lebreton LCM, Carson HS et al (2014) Plastic pollution in the world's oceans: more than 5 trillion plastic pieces weighing over 250,000 tons afloat at sea. PLoS One 9:e111913. https://doi.org/10.1371/journal.pone.0111913

Evans SM, Gebbels S, Stockill JM (2008) 'Our shared responsibility': participation in ecological projects as a means of empowering communities to contribute to coastal management processes. Mar Pollut Bull 57:3–7. https://doi.org/10.1016/j.marpolbul.2008.04.014

Fore LS, Paulsen K, O'Laughlin K (2001) Assessing the performance of volunteers in monitoring streams. Freshw Biol 46(1):109–123. https://doi.org/10.1111/j.1365-2427.2001.00640.x

Forrester G, Baily P, Conetta D et al (2015) Comparing monitoring data collected by volunteers and professionals shows that citizen scientists can detect long-term change on coral reefs. J Nat Conserv 24:1–9. https://doi.org/10.1016/j.jnc.2015.01.002

Foster-Smith J, Evans SM (2003) The value of marine ecological data collected by volunteers. Biol Conserv 113(2):199–213. https://doi.org/10.1016/S0006-3207(02)00373-7

Garcia-Soto C, van der Meeren GI, Busch JA et al (2017) Advancing citizen science for coastal and ocean research. In: French V, Kellett P, Delany J et al (eds) Position Paper 23 of the European Marine Board, Ostend, Belgium

Geller JB, Walton ED, Grosholz ED et al (1997) Cryptic invasions of the crab Carcinus detected by molecular phylogeography. Mol Ecol 6:901–906. https://doi.org/10.1046/j.1365-294X.1997.00256.x

Gouraguine A, Moranta J, Ruiz-Frau A et al (2019) Citizen science in data and resource-limited areas: A tool to detect long-term ecosystem changes. PloS One 14 (1):e0210007. https://doi.org/10.1371/journal.pone.0210007

Haklay M (2015) Citizen science and policy: a European perspective. The Woodrow Wilson Center, Washington, DC

Hidalgo-Ruz V, Thiel M (2013) Distribution and abundance of small plastic debris on beaches in the SE Pacific (Chile): a study supported by a citizen science project. Mar Environ Res 87:12–18. https://doi.org/10.1016/j.marenvres.2013.02.015

Hidalgo-Ruz V, Thiel M (2015) The contribution of citizen scientists to the monitoring of marine litter. In: Bergmann M, Gutow L, Klages M (eds) Marine anthropogenic litter. Springer, Cham, pp 429–447

Hobson B (2000) Recreation specialization revisited. J Leis Res 32:18–21

Hochachka WM, Wells JV, Rosenberg KV (1999) Irruptive migration of common redpolls. Condor 101:195–204

Hochachka WM, Fink D, Hutchinson RA et al (2012) Data-intensive science applied to broad-scale citizen science. Trends Ecol Evol 27(2):130–137. https://doi.org/10.1016/j.tree.2011.11.006

Holt BG, Rioja-Nieto R, MacNeil MA et al (2013) Comparing diversity data collected using a protocol designed for volunteers with results from a professional alternative. Methods Ecol Evol 4:383–392. https://doi.org/10.1111/2041-210X.12031

Hyder K, Townhill B, Anderson LG et al (2015) Can citizen science contribute to the evidence-base that underpins marine policy? Mar Policy 59:112–120. https://doi.org/10.1016/j.marpol.2015.04.022

Jarvis RM, Breen BB, Krägeloh CU et al (2015) Citizen science and the power of public participation in marine spatial planning. Mar Policy 57:21–26. https://doi.org/10.1016/j.marpol.2015.03.011

Jiguet F (2009) Method learning caused a first-time observer effect in a newly started breeding bird survey. Bird Study 56(2):253–258. https://doi.org/10.1080/00063650902791991

Jordan R, Crall A, Gray S et al (2015) Citizen science as a distinct field of inquiry. Bioscience 65(2):208–211. https://doi.org/10.1093/biosci/biu217

Kay P, Dipper F (2018) A field guide to marine fishes of wales and adjacent waters. Marine Conservation Society and Countryside Council for Wales, Marine Wildlife, Conwy

Kerson R (1989) Lab for the environment. MITS Technol Rev 92(1):11–12

Kosmala M, Wiggins A, Swanson A et al (2016) Assessing data quality in citizen science. Front Ecol Environ 14(10):551–560. https://doi.org/10.1002/fee.1436

Koss RS, Kingsley JY (2010) Volunteer health and emotional wellbeing in marine protected areas. Ocean Coast Manag 53:447–453. https://doi.org/10.1016/j.ocecoaman.2010.06.002

Kullenberg C, Kasperowski D (2016) What is citizen science? A scientometric meta-analysis. PLoS One 11(1):e0147152. https://doi.org/10.1371/journal.pone.0147152

Lima S (1995) Back to the basics of anti-predatory vigilance: the group-size effect. Anim Behav 49(1):11–20. https://doi.org/10.1016/0003-3472(95)80149-9

Lohrer AM (2001) The invasion by *Hemigrapsus sanguineus* in eastern North America: a review. Aquat Invaders 12(3):1–11

Markantonatou V, Meidinger M, Sano M et al (2013) Stakeholder participation and the use of web technology for MPA management. Adv Oceanogr Limnol 4:260–276. https://doi.org/10.1080/19475721.2013.851117

Marshall NJ, Kleine DA, Dean AJ (2012) CoralWatch: education, monitoring, and sustainability through citizen science. Front Ecol Environ 10(6):332–334. https://doi.org/10.1890/110266

Martin JM (2013) Marine debris removal: one year of effort by the Georgia Sea turtle-center-marine debris initiative. Mar Pollut Bull 74:165–169. https://doi.org/10.1016/j.marpolbul.2013.07.009

Martin VY, Christidis L, Pecl GT (2016) Public interest in marine citizen science: is there potential for growth? Bioscience 66(8):683–692. https://doi.org/10.1093/biosci/biw070

McGehee NG (2014) Volunteer tourism: evolution, issues and futures. J Sustain Tour 22(6):847–854. https://doi.org/10.1080/09669582.2014.907299

McGovern B, Culloch RM, O'Connell M et al (2016) Temporal and spatial trends in stranding records of cetaceans on the Irish coast, 2002–2014. J Mar Biol Assoc UK 98(5):977–989. https://doi.org/10.1017/S0025315416001594

McKinley E, Fletcher S (2012) Improving marine environmental health through marine citizenship: a call for debate. Mar Policy 36:839–843. https://doi.org/10.1016/j.marpol.2011.11.001

McKinley DC, Miller-Rushing AJ, Ballard HL et al (2017) Citizen science can improve conservation science, natural resource management, and environmental protection. Biol Conserv 208:15–28. https://doi.org/10.1016/j.biocon.2016.05.015

Mieszkowska N, Sugden H, Firth LB et al (2014) The role of sustained observations in tracking impacts of environmental change on marine biodiversity and ecosystems. Philos Trans R Soc A 372(2025):20130339. https://doi.org/10.1098/rsta.2013.0339

Nelms SE, Coombes C, Foster LC et al (2017) Marine anthropogenic litter on British beaches: a 10-year nationwide assessment using citizen science data. Sci Total Environ 579:1399–1409. https://doi.org/10.1016/j.scitotenv.2016.11.137

Newman S, Watkins E, Farmer A et al (2015) The economics of marine litter. In: Bergmann M, Gutow L, Klages M (eds) Marine anthropogenic litter. Springer, Cham, pp 367–394

Ocean Networks Canada (2018) Digital fishers. www.oceannetworks.ca/learning/get-involved/citizen-science/digital-fishers. Accessed 24 Sept 2018

OED Online (2018a) "citizen, n. and adj.". Oxford University Press. www.oed.com/view/Entry/33513?redirectedFrom=citizen+science. Accessed 22 Apr 2018

OED Online (2018b) "voluntourism, n.". Oxford University Press. www.oed.com/view/Entry/34245729#eid1151399030. Accessed 3 Oct 2018

OSPAR (2010) Guideline for monitoring marine litter on the beaches in the OSPAR maritime area. OSPAR Commission. www.ospar.org/ospar-data/10-02e_beachlitter%20guideline_english%20only.pdf. Accessed 30 May 2018

Parmesan C, Yohe G (2003) A globally coherent fingerprint of climate change impacts across natural systems. Nature 421:37–42. https://doi.org/10.1038/nature01286

Parr J, Sewell J (2017) Citizen sentinels: the role of citizen scientists in reporting and monitoring invasive non-native species. In: Cigliano JA, Ballard HL (eds) Citizen science for coastal and marine conservation. Routledge, London, pp 59–76

Ponti M, Rossi G, Bertolino M et al (2011a) Reef Check: involvement of SCUBA diver volunteers in the Coastal Environment Monitoring Protocol for the Mediterranean Sea Abstracts of the 9th International Temperate Reefs Symposium, Plymouth, p 164

Ponti M, Leoni G, Abbiati M (2011b) Geographical analysis of marine species distribution data provided by diver volunteers. Biol Mar Mediterr 18:282–283

Prestopnik NR, Crowston K (2011) Gaming for (citizen) science: exploring motivation and data quality in the context of crowdsourced science through the design and evaluation of a social-computational system. In: Proceedings of the 7th IEEE International Conference on e-Science Workshops, IEE Computer Society, Stockholm, pp 28–33

Reef Check (2018) EcoDiver courses and products. www.reefcheck.org/ecodiver/reef-check-ecodiver. Accessed 18 May 2018

Ribic CA (1998) Use of indicator items to monitor marine debris on a New Jersey beach from 1991 to 1996. Mar Pollut Bull 36:887–891

Ribic CA, Sheavly SB, Rugg DJ (2011) Trends in marine debris in the US Caribbean and the Gulf of Mexico 1996–2003. JICZM/RGCI 11:7–19

Ricciardi A, Mack RN, Steiner WM et al (2000) Toward a global information system for invasive species. Bioscience 50:239–244

Riesch H, Potter C (2014) Citizen science as seen by scientists: methodological, epistemological and ethical dimensions. Public Underst Sci 23(1):107–120. https://doi.org/10.1177/0963662513497324

Robertson MP, Cumming GS, Erasmus BFN (2010) Getting the most out of atlas data. Divers Distrib 16(3):363–375. https://doi.org/10.1111/j.1472-4642.2010.00639.x

Root T, Alpert P (1994) Volunteers and the NBS. Science 263(5151):1205. https://doi.org/10.1126/science.263.5151.1205

Rotman D, Preece J, Hammock J et al (2012) Dynamic changes in motivation in collaborative citizen-science projects. In: Proceedings of the ACM conference on computer supported cooperative work, CSCW, Seattle, pp 217–226

Roy HE, Pocock MJ, Preston CD et al (2012) Understanding citizen science and environmental monitoring: final report on behalf of UK-EOF. NERC Centre for Ecology & Hydrology and Natural History Museum, London, pp 1–170

Ruiz GM, Hines AH, Carlton JT et al (1997) Global invasions of marine and estuarine habitats by non-indigenous species: mechanisms, extent, and consequences. Integr Comp Biol 37:621–632

Ryan PG, Moore CJ, van Franeker JA et al (2009) Monitoring the abundance of plastic debris in the marine environment. Philos Trans R Soc B 364:1999–2012. https://doi.org/10.1098/rstb.2008.0207

Schläppy ML, Loder J, Salmond J et al (2017) Making waves: marine citizen science for impact. Front Mar Sci 4:146. https://doi.org/10.3389/fmars.2017.00146

Science Communication Unit, University of the West of England, Bristol (2013) Science for Environment Policy In-depth Report: Environmental Citizen Science. Report produced for the European

Commission DG Environment, December 2013. Available at: www.ec.europa.eu/science-environment-policy

Seasearch (2018) Seasearch diving. www.seasearch.org.uk/diving.html. Accessed 18 May 2018

Silvertown J (2009) A new dawn for citizen science. Trends Ecol Evol 24(9):467–471. https://doi.org/10.1016/j.tree.2009.03.017

Southward AJ, Langmead O, Hardman-Mountford NJ et al (2005) Long-term oceanographic and ecological research in the Western English Channel. Adv Mar Biol 47:1–105. https://doi.org/10.1016/S0065-2881(04)47001-1

Stafford R, Hart AG, Collins L et al (2010) Eu-social science: the role of internet social networks in the collection of bee biodiversity data. PLoS One 5(12):e14381. https://doi.org/10.1371/journal.pone.0014381

Sundlov T, Graham M, Davis T et al (2016) Sea squirt Invades BLM submerged lands in Southeast Alaska. U.S. Department of the Interior: Bureau of Land Management. www.blm.gov/blog/2016-10-14/sea-squirt-invades-blm-submerged-lands-southeast-alaska. Accessed 9 Apr 2018

Teleki KA (2012) Power of the people? Aquatic Conserv 22(1):1–6. https://doi.org/10.1002/aqc.2219

Theobald EJ, Ettinger AK, Burgess HK et al (2015) Global change and local solutions: tapping the unrealized potential of citizen science for biodiversity research. Biol Conserv 181:236–244. https://doi.org/10.1016/j.biocon.2014.10.021

Thiel M, Penna-Díaz MA, Luna-Jorquera G et al (2014) Citizen scientists and marine research: volunteer participants, their contributions, and projection for the future. Oceanogr Mar Biol 52:257–314. https://doi.org/10.1201/b17143-6

Thomas LM, Gunawardene N, Horton K et al (2017) Many eyes on the ground: citizen science is an effective early detection tool for biosecurity. Biol Invasions 19(9):2751–2765. https://doi.org/10.1007/s10530-017-1481-6

Tonachella N, Nastasi A, Kaufman G et al (2012) Predicting trends in humpback whale (*Megaptera novaeangliae*) abundance using citizen science. Pac Conserv Biol 18(4):297–309

Tulloch AI, Possingham HP, Joseph LN et al (2013) Realising the full potential of citizen science monitoring programs. Biol Conserv 165:128–138. https://doi.org/10.1016/j.biocon.2013.05.025

Turicchia E, Ponti M, Abbiati M et al (2015) Citizen science as a tool for the environmental quality assessment of Mediterranean Marine Protected Areas. In: Cerrano C, Henocque Y, Hogg K et al (eds) Proceedings of the MMMPA/CIESM international joint conference on Mediterranean marine protected areas: integrated management as a response to ecosystem threats. Ancona, Italy, p 29

Turicchia E, Abbiati M, Cerrano C et al (2017) May citizen science effectively support underwater monitoring programs? 3rd European conference on scientific diving book of abstracts. Funchal, Portugal, p 46

UNCTADstat (2018) United Nations conference on trade and development. Country classifications. unctadstat.unctad.org/EN/Classifications.html. Accessed 9 Oct 2018

van der Velde T, Milton DA, Lawson TJ et al (2017) Comparison of marine debris data collected by researchers and citizen scientists: is citizen science data worth the effort? Biol Conserv 208:127–138. https://doi.org/10.1016/j.biocon.2016.05.025

Van Franeker JA, Blaize C, Danielsen J et al (2011) Monitoring plastic ingestion by the northern fulmar *Fulmarus glacialis* in the North Sea. Environ Pollut 159:2609–2615. https://doi.org/10.1016/j.envpol.2011.06.008

Visser ME, Both C (2005) Shifts in phenology due to global climate change: the need for a yardstick. Proc R Soc Lond B Biol 272(1581):2561–2569. https://doi.org/10.1098/rspb.2005.3356

Watson DI, Shumway SE, Whitlatch RB (2009) Biofouling and the shellfish industry. In: Shumway SE, Rodrick GE (eds) Shellfish safety and quality. Woodhead Publishing Ltd., Cambridge

Wiggins A, Crowston K (2011) From conservation to crowdsourcing: a typology of citizen science. HICSS '11 Proceedings of the 44th Hawaii international conference on system sciences, pp 1–10

Worthington JP, Silvertown J, Cook L et al (2012) Evolution MegaLab: a case study in citizen science methods. Methods Ecol Evol 3(2):303–309. https://doi.org/10.1111/j.2041-210X.2011.00164.x

2

A Literature Review on Stakeholder Participation in Coastal and Marine Fisheries

Heike Schwermer, Fanny Barz, and Yury Zablotski

Abstract

Stakeholder participation is a fundamental component of many states' and local agencies' fisheries legislations worldwide. The European Common Fisheries Policy (CFP), as one example, increasingly adopted a holistic approach to managing marine living resources. An important component of such an ecosystem-based management approach is the consideration of knowledge, values, needs and social interactions of stakeholders in decision-making processes. However, despite that stakeholder participation is a widely used term, a great variety of definitions exist, which often cause misunderstanding. Stakeholder participation is often used as part of conducting research on stakeholders but not in the context of their participation in resource management. Here, we present the results of a comprehensive literature review on the topic *stakeholder participation* in coastal and marine fisheries. We identified 286 scientific publications in *Web of Science* of which 50 were relevant for our research questions. Publications were analysed regarding (i) definition of stakeholder participation, (ii) analysis of participating stakeholders, (iii) applied participatory methods and (iv) intention for participation. Stakeholder types addressed in the publications included, e.g. fishery (fishers and direct representatives, N = 48), politics (policymakers and managers, N = 31), science (N = 25) and environmental non-governmental organizations (eNGOs, N = 24). In total, 24 publications labelled their studies as stakeholder participation, while stakeholders were only used as a study object. We conclude that improving science and the practice of including stakeholders in the management of coastal and marine fisheries requires definitions of who is considered a stakeholder and the form of participation applied.

Keywords

Case survey method · Stakeholder types · Participatory methods · Multiple Correspondence Analysis

2.1 Introduction

Stakeholder participation is a fundamental component of many states´ and local agencies' fisheries legislations worldwide (NOAA 2015). As an example, the Common Fisheries Policy of the European Union increasingly adopted a holistic approach to managing marine living resources (Commission of the European Communities 2013). An important component of such an ecosystem-based management (EBM) approach is the consideration of knowledge, values, needs and social interactions of resource users and other interest groups in decision-making processes (Long et al. 2015). Aanesen et al. (2014) established that in the case of fisheries management, this implies having access to local ecological knowledge of fishers to complement scientific data which is often very limited. Furthermore, involving stakeholders is expected to increase the legitimacy of the management by creating understanding and support among the stakeholders for management measures such as new regulations (Aanesen et al. 2014). Moreover, stakeholders represent varying preferences about a resource and, therefore, ideally enable processes to reach sustainable management on different levels, such as ecological and social. But the terms 'stakeholder' and 'participation' have become 'buzz words' in environmental management (Voinov and Bousquet 2010). Deviating definitions and explanations of both terms occur, and it is often unclear what is actually meant by these concepts.

H. Schwermer (✉)
Institute for Marine Ecosystem and Fisheries Science, Center for Earth System Research and Sustainability, University of Hamburg Germany, Hamburg, Germany
e-mail: heike.schwermer@uni-hamburg.de

F. Barz · Y. Zablotski
Thünen-Institute of Baltic Sea Fisheries, Rostock, Germany

S. Jungblut et al. (eds.), *YOUMARES 9 - The Oceans: Our Research, Our Future*,
https://doi.org/10.1007/978-3-030-20389-4_2

We here reviewed worldwide case studies to investigate how stakeholder participation is applied in research projects concerning coastal and marine fisheries. The literature review creates an overview of current meanings and methods applied in this research field. The aim of our study is to highlight and to critically discuss the application of the term stakeholder participation and the significance of these findings for future research projects in general and particular in the field of coastal and marine fisheries. In our study, we developed and applied nine questions to review and analyse relevant publications. First, we investigated the publications regarding the use of the term *stakeholder*. Here, we focused on term definition, approach of analysing stakeholders as well as on the stakeholder types involved in the case study. Subsequently, we reviewed the publications in relation to the term *participation*, again first focusing on term definition, methods used related to the participation of stakeholders, description and intention for participation. Finally, we analysed all publications to evaluate whether the publications used participation as a tool for researching stakeholders (research tool) or for conducting true stakeholder participation (participation tool).

Our study revealed that only few publications in the research field of coastal and marine fisheries clearly defined the terms stakeholder and participation. Furthermore, the majority of publications labelled their studies as stakeholder participation, while stakeholders were only used as a study object. We conclude that improving the science and the practice of including stakeholders in the management of coastal and marine fisheries requires definitions of who is considered a stakeholder and the form of participation applied.

2.2 Material and Methods

We conducted a systematic literature review using the *case survey method* (Newig and Fritsch 2009), i.e. one article represented one analysis unit. Here, qualitative studies were transformed into semi-quantitative data, applying a coding scheme and expert judgements by multiple coders. The case survey method allowed us to synthesize case-based knowledge using at least two coders. We translated our research steps (RS) into a research protocol, adapted after Brandt et al. (2013), making RS repeatable and transparent. Our study included five working steps (WS): data gathering (WS 1), data screening (WS 2), data cleaning (WS 3), paper reviews (WS 4) and a statistical analysis of the collected data (WS 5) (Table 2.1).

In WS 1 we derived relevant publications from the *Web of Science* (WoS; www.isiknowledge.com), an extensive and multidisciplinary database covering a large number of scientific journals, books and proceedings in the field of natural science and technique, arts, humanities and social sciences (ETH Zürich 2018). We extracted articles published within the period from 2000 to 2018, considering the establishment of participation in (environmental) decision-making processes as a democratic right by the United Nations Economic Commission for Europe's 1998 Arhus Convention and an increased use (Reed 2008). To ensure an establishment in research publications, we started the review two years later. Publications were collected by using the basic search routine in the WoS (date of search: 16 May 2018) applying the following keyword strings: (i) stakeholder – participation – fishery, (ii) stakeholder – involvement – fishery and

Table 2.1 The five working steps (WS) of our literature review on stakeholder participation in the field of coastal and marine fisheries consisted of data gathering, data screening, data cleaning, paper review and statistical analysis. The review procedure and the results are presented for each WS

Working step (WS)	Review procedure	Result
1. Data gathering	Definition of Web of Science query (keywords: stakeholder, participation/engagement/involvement, fishery; 16 May 2018)	Bibliographical information of 286 potentially relevant publications
2. Data screening	Screening of publications guided by the question: Are all three keywords listed within the title, abstract or keywords of the publication?	A total of 81 publications were identified
3. Data cleaning	Cleaning of publications guided by the questions: i) Does the publication focus on coastal and marine fisheries? ii) Are the publications case studies?	A total of 50 relevant publications were identified
4. Paper review	Content analysis of relevant publications using a set of nine research questions concerning the term stakeholder participation	Different definitions and methods regarding the topic stakeholder participation in the field of coastal and marine fisheries were identified
5. Statistical analysis	Analysis of data using multiple correspondence analysis in R	Results are presented in this review publication

(iii) stakeholder – engagement – fishery. We additionally used the string 'fisheries' instead of 'fishery'.

In WS 2 we screened all publications derived in WS 1; we only further considered the publications that included all three keywords stakeholder, participation/involvement/engagement and fishery in (i) the title, (ii) the abstract or (iii) the keywords. We also included publications that either used the noun, the verb, i.e. to fish, to participate/involve/engage, or the adverb of the keyword, like 'fishing community'.

For the data cleaning (WS 3), we used an inductive approach to identify key issues of selected publications based on two characteristics:

1. Focus of the publication – fisheries, freshwater or estuarine ecosystems, recreational fisheries or marine protected areas; management (e.g. fishery, coastal management, EBM) or policy (e.g. Common Fisheries Policy (CFP), Marine Strategy Framework Directives (MSFD))
2. Study type of publication – a participation case study, a meta-analysis of participation studies or participation framework description

We here described policy as a set of rules or an established framework; management was defined by general environmental management approaches (e.g. ecosystem-based management (EBM), coastal management) or explicit management measures.

In WS 3 we excluded publications with focus on freshwater or estuarine ecosystems, recreational fisheries and marine protected areas. In addition, we discarded publications with focus on coastal management and EBM as well as publications looking at political frameworks (CFP, MSFD). All remaining publications focused on coastal and marine fisheries.

We further only analysed publications that presented a case study; in WS 3 we discarded studies that represented a meta-analysis or theoretical participation framework description. We here defined a case study as "[...], analyses of persons, events, decisions, periods, projects, policies, institutions, or other systems that are studied holistically by one or more methods. The case that is the subject of the inquiry will be an instance of a class of phenomena that provides an analytical frame — an object — within which the study is conducted and which the case illuminates and explicates" (Thomas 2011). For an evaluation of the regional distribution, we also extracted the continent where the case study has been conducted.

In WS 4 we analysed the content of the finally selected papers applying a mixed-method approach. We evaluated the publications based on 9s questions, investigating the terms *stakeholder* (questions 1–4) and *participation* (questions 5–8) first separately and subsequently in combination (question 9). The list of questions is shown in Table 2.2. We applied a quantitative approach to investigate naming and definition of both terms (questions 1–8, Table 2.2). Furthermore, we applied an inductive approach to generate categories for analysing derived data to elicit which type of stakeholders, participation tools and intention categories for participation were part of the research projects (questions 2, 6 and 8, Table 2.2) (Mayring 1988). Eight *stakeholder types* were distinguished in our analysis, i.e. science, politics, environmental non-governmental organizations (eNGOs), fisheries, fishery-related industry, recreational fisheries, public and others. Although we excluded publications that focus on recreational fisheries, this stakeholder type was part of the case studies focusing on coastal and marine fisheries and, therefore, was included as one stakeholder type within our analysis. The category 'others' included stakeholders that did not fit into any of the other categories but have been explicitly mentioned separately from them. We similarly analysed questions 6 and 8. Here, we distinguished between 11 *participatory methods*, i.e. workshop, interview, meeting, discussion, survey, questionnaire, modelling, coordination, mapping, presentation and conversation, and 10 *intention categories*, i.e. analysis, assessment, definition, description, development, establishment, evaluation, feedback, identification and improvement. Related to the description in the publications, we distinguished between active and passive participatory methods: active ones describing methods that directly involved stakeholders in decision-making processes; passive participatory methods had been described to support the participatory process but not to involve the stakeholders

Table 2.2 Nine questions used to review the identified case studies in coastal and marine fisheries management. The terms stakeholder (questions 1–4) and participation (questions 5–8) were investigated separately and in combination, i.e. stakeholder participation (question 9)

Term	Question
Stakeholder	1. How is the term stakeholder defined?
	2. Which types of stakeholder are part of the research project?
	3. Was a systematic approach used to analyse stakeholders?
	4. Which stakeholder analysis approach was used?
Participation	5. Was the term participation/engagement/involvement defined?
	6. Which participation/engagement/involvement methods were mentioned?
	7. How was the participation/engagement/involvement method described?
	8. What was the aim of using participation within this project?
Stakeholder participation	9. Is the described participation/engagement/involvement tool used for analysing stakeholders (research tool) or for involving stakeholders (participation tool)?

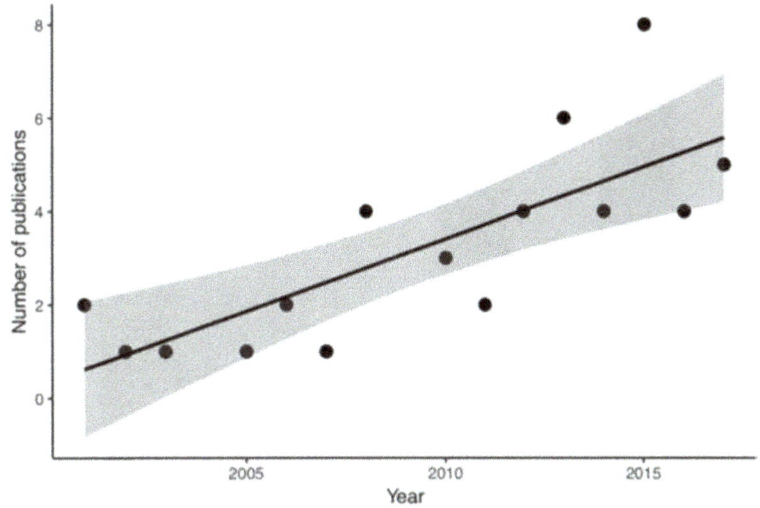

Fig. 2.1 Number of research publications published from 2000 to 2018 dealing with case studies in coastal and marine fisheries as found by Web of Science (keywords: stakeholder, participation/ engagement/involvement, fishery) as of May 2018. Black line represents the linear regression with 95% confidence intervals; the grey area indicates the confidence band ($R^2 = 0.6045$, $p = 0.000645$)

in research or management (decision-making processes). Participatory methods and intention categories were extracted according to the mention in the publications. Related to the participatory methods, we also determined whether preparatory work was done using an inductive approach.

Eventually, we investigated whether (i) the case studies conducted participation to gather knowledge from stakeholders but without engaging these stakeholders in a decision-making process (*research tool*) or (ii) stakeholders had a direct influence on data interpretation and decision-making processes (*participation tool*).

In the final working step (WS 5 – statistical analysis), we used Multiple Correspondence Analysis (MCA) to explore the relationships between stakeholder types. MCA is able to uncover correlations (i.e. similarities, grouping) in otherwise inconvenient survey data (Higgs 1991) and was designed to apply on multiple binary (or nominal) variables (e.g. our categories stakeholder 'science': absent = 0, present = 1; stakeholder 'public': absent = 0, present = 1), all of which had the same status (Abdi and Valentin 2007). MCA explores the patterns in data by measuring the geometric proximity between stakeholder types (e.g. science and public) using weighted least squares (Abdi and Valentin 2007) and graphically represents the proximity of the categories on a simple plane, i.e. correspondence map. Thus, MCA allows finding similarities between categories based on the chi-square distance between them and using the percentage of the explained variance to the new (reduced) dimensions. More details related to the method of MCA can be found in the original work Greenacre (1984). We used MCA to answer the question: Which stakeholder types often appear together in the reviewed publications?

2.3 Results

We identified in total 286 scientific publications, which we further analysed according to our review protocol (see Sect. 2).

Of 286 publications, in total 81 contained all keywords of which 56 publications had their emphasis on coastal and marine fisheries. 50 publications out of 56 were categorized as case studies and were further analysed in our study (detailed description in Table 2.A1 of the Supplementary Material).

The number of publications that focused on stakeholder participation significantly increased within the last 18 years (Fig. 2.1). In 2015, a maximum value of eight was reached. The majority of the case studies was conducted in Europe (N = 18), North America (N = 11) and Australia (N = 9).

2.3.1 Paper Review: Stakeholders

2.3.1.1 Term Definition

We identified four publications defining the term stakeholder (Brzezinski et al. 2010; Haapasaari et al. 2013; Tiller et al. 2015; Kinds et al. 2016) (Fig. 2.2a). Even though they defined the term more indirectly and in general, Brzezinski et al. (2010) stated stakeholders as members of a particular group that hold a personal stake. They referred to Olson (1965) to suggest that the increase of the personal stake of these members will lead to an increase of their participation in regulatory processes. Haapasaari et al. (2013) described stakeholders as a group of people having a stake and contributing towards a knowledge base for fisheries management. On the other hand, Kinds et al. (2016) focused on stakeholders as all people and organizations (here producer organizations), which are actively involved in the fishing sector. Tiller

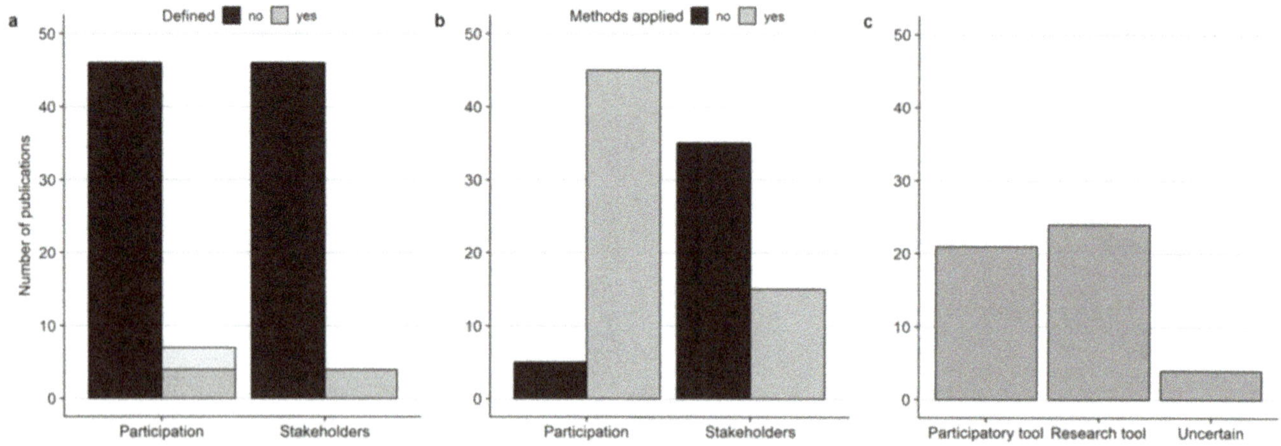

Fig. 2.2 Review of 50 research publications presenting case studies in coastal and marine fisheries (as of May 2018). (**a**) Term definition of participation and stakeholder; we distinguished between participation (grey) and participation-related terms (light grey), e.g. participatory management, participatory research and participatory action research; (**b**) method application for participation (e.g. interview, workshops and questionnaire) and stakeholders (e.g. snowball sampling); (**c**) the use of stakeholder participation, either as a participation or research tool

et al. (2015) took a deliberate look into the literature, referring to Freeman (2010). Freeman (2010) defined stakeholders as any group or individual who can affect, or is affected by, the achievement of the organization's objectives (Freeman 2010). Tiller et al. (2015) continued to criticize this definition as too broad; it allows the inclusion of nearly everyone as a stakeholder.

2.3.1.2 Stakeholder Analysis

Durham et al. (2014) stated that the selection of stakeholders strongly determines the outcome of the participation process. They, therefore, recommended to systematically select stakeholders based on the objective and impact of research. We, therefore, analysed the publications, looking for the description or reference of stakeholder analysis processes.

In our review corpus, 15 out of 50 publications applied methods to get an understanding of who their stakeholders are (Fig. 2.2b). We evaluated publications as using stakeholder analysis approaches if the case studies did not decide on stakeholder groups or stakeholder individuals (referring to Durham et al. 2014) but researched for them systematically. Three out of 15 publications defined stakeholder groups, three determined stakeholder groups as well as individual stakeholders and the remaining nine out of 15 case studies selected individual stakeholders out of a priori stakeholder groups.

Gray et al. (2012), Kinds et al. (2016), as well as Sampedro et al. (2017) evaluated stakeholder groups that were involved in past fisheries research and management; thereby, they have chosen the group of stakeholders they wanted to involve in their current research projects. Pristupa et al. (2016) applied three different approaches; on the one hand, they did not want to overlook a major stakeholder, and on the other hand, they aimed to identify the most knowledgeable individual within the appropriate stakeholder group: first they extracted information from reports and open-access information such as Marine Stewardship Council reports, interviews on specialized websites and scientific reports. Second, they identified stakeholders during a thematic conference, which was also used to establish contacts. Third, recommendations by fisheries experts were accumulated using the snowball approach (Pristupa et al. 2016). Different to the previous case studies, Miller et al. (2010) used two approaches to select relevant stakeholders. First, stakeholders were selected due to history, perspectives and relationships among those with a stake in a specific fishery (Miller et al. 2010). Second, relevant stakeholders should be knowledgeable and influential in their community as well as open minded for different views (Miller et al. 2010). Further, Mahon et al. (2003) analysed stakeholders based on public records before organizing discussion meetings where individual stakeholders were singled out.

Additionally, nine publications described methods that were applied to identify individual stakeholders, either within presumed stakeholder groups or randomly. Butler et al. (2015), Bitunjac et al. (2016) and Stratoudakis et al. (2015) based their choice of individual stakeholders on their long-time experience and their knowledge of the topic studied. Bitunjac et al. (2016) selected stakeholders of which the authors assumed to have a leading influence within their group and were, therefore, seen as representatives of their stakeholder group. Catedrilla et al. (2012), Kerr et al. (2006) and Murphy et al. (2015) had chosen fishers as individual stakeholders by sampling them from a registration list in their field of interest. Lorance et al. (2011) and Thiault et al. (2017) selected the stakeholders at random. Lorance et al. (2011) advertised workshops widely and, therefore, could not directly influence attendance; Thiault et al. (2017) did sampling among all households in their area of interest without focusing on a specific stakeholder group. Kittinger (2013) first conducted a snowball sampling followed by a 'purposive sampling approach' – a deliberately selective approach choosing knowledgeable individuals.

2.3.1.3 Stakeholder Types

Overall the stakeholder type 'fishery' had the highest frequency of appearance within all publications, followed by 'politics', 'science' and 'eNGO' (Table 2.3). In five publications, 'fishery' was considered as the only stakeholder (Clarke et al. 2002; Catedrilla et al. 2012; Eveson et al. 2015; Tiller et al. 2015; Thiault et al. 2017). Except for Catedrilla et al. (2012), these publications aimed at getting information about the spatial distribution of fishing grounds. Two case studies (Fletcher 2005; Dowling et al. 2008) did not name 'fishery' as a stakeholder but noted that fishers were involved in the conducted case study.

'Politics', 'science' and 'eNGO' were targeted in about half of the studies. Nonetheless, 12 case studies did not consider any of these three stakeholders at all (e.g. Mitchell and Baba 2006; Appledorn et al. 2008; Cox and Kronlund 2008). The stakeholder type 'others' mostly represented a business or the like (e.g. Carr and Heyman 2012; Butler et al. 2015). 'Public' stakeholders were mainly seen as community members (Kittinger 2013; Eriksson et al. 2016) or consumers (Mahon et al. 2003), who, therefore, did not have a primary economic or political interest in fisheries.

'Related industry' was described as processing and selling industry that was directly associated with fisheries and so depended on this stakeholder type (e.g. Cox and Kronlund 2008). 'Related industry' was considered 16 times in the reviewed case studies and differed widely in their topics in which context these stakeholder groups emerged, e.g. bycatch (Bojorquez-Tapia et al. 2016), stock assessment (Smith et al. 2001) or compliance (Garza-Gil et al. 2015). 'Recreational fishery' was represented in five publications, two in Australia (Fletcher 2005, Mitchell and Baba 2006) and three in North America (e.g. Miller et al. 2010; Gray et al. 2012; Murphy et al. 2015), all of them focused on management processes.

Table 2.3 Identified stakeholder types presented by case studies in coastal and marine fisheries (as of May 2018) and ranked by the frequency of their appearance (N). Description of stakeholder types corresponds to the one mentioned in the publication under review

Stakeholder type	Description of stakeholders	N
Fishery	Fishers and their direct representatives	48
Politics	Government officials, local and village officers	31
Science	Academic scientists	25
eNGO	Environmental non-governmental organizations	24
Others	E.g. local businesses, leaders of the tourism sector, leaders of other community-based associations	17
Related industry	Processing and selling businesses	16
Public	Community members, representatives from public organizations, consumers	7
Recreational fishery	Representatives of recreational fishery	5

2.3.1.4 Relationships Between Stakeholder Types

We applied a multiple correspondence analysis to evaluate the occurrence of certain stakeholder clusters. 48 publications included 'fishery' as a stakeholder, but this stakeholder type did not group with other stakeholders and, therefore, lessened the meaningfulness of other stakeholders. For this reason, we decided to exclude 'fishery' from the MCA, which resulted in a higher percentage of the variance explained by the dimensions. As a result, very similar variable clusters of categories appeared and were, therefore, easier to interpret. 'Others' were also excluded from the MCA; by definition this stakeholder type showed a great variety, and, therefore, interpretation of the data would be difficult.

Ideally, dimensions should be used to interpret the data whose eigenvalues exceed the mean of all eigenvalues (0.17). For this reason, we included three dimensions into our analysis, which together accounted for over 70% of the variance. Here, it is important that the dimensions obtained are hierarchical. Dimension 1 formed the strongest dimension (Dim1, Fig. 2.3, Table 2.A2 of the Supplementary Material), i.e. singled out 'science', 'eNGO' and 'politics', and explained 31.6% of the variance. Further, these three stakeholder types had the highest number of mentions after 'fishery'. Dimension 2 (Dim 2, Fig. 2.A2a of the Supplementary Material) focused on 'recreational fishery' and 'related industry', accounting for 21.8% of the variance. Although 'recreational fishery' was only considered in five case studies, this stakeholder type showed a strong contribution towards dimension 2. Also, explanatory power was increased by sharing contribution with 'related industry'. Less variance (17.7%) was explained by dimension 3, which was dominated by 'public' (Dim 3, Fig. 2.A2b, Table 2.A1 of the Supplementary Material).

Subsequently, MCA was applied separately to case studies from North America (N = 10) and Europe (N = 18) (Fig. 2.A3a-c of the Supplementary Material). The results of the MCA that was performed on North America case studies showed a similar picture as in Fig. 2.3, although these case studies did not dominate the review corpus. Even though 'science', 'eNGO' and 'politics' showed a strong contribution in different dimensions, these stakeholder types could

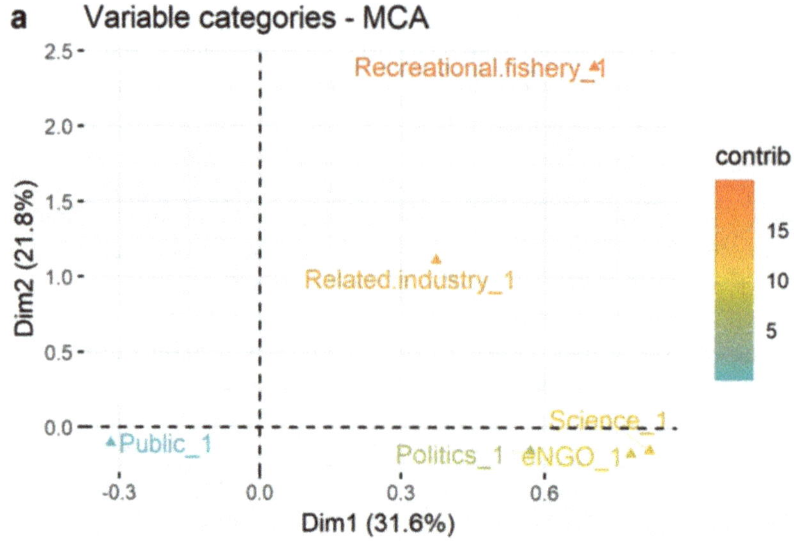

Fig. 2.3 Panel **a** of the visualization of correlation between dimension 1 (Dim1) and dimension 2 (Dim2), showing the variance of stakeholder types in 50 research publications of case studies in coastal and marine fisheries (as of May 2018) using multiple correspondence analysis (MCA). Figure shows which types of stakeholders are mostly correlated, i.e. regarding stakeholder participation in coastal and marine fisheries, 'eNGO', 'politics' and 'science' are often addressed together. Panels b and c of the MCA results for correlation of dimension 2 (Dim2) and 3 (Dim3) as well as dimension 1 (Dim1) and 3 (Dim3) are presented in Fig. A2 of the Supplementary Material

still be found as a group. In dimension 2 'related industry' and 'recreational fishery' were displayed in the negative area; the other stakeholder types moved from the negative area of dimension 2 into the positive area. 'Related industry' and 'recreational fishery' were also grouped together with a high contribution as seen in Fig. 2.A3a (Supplementary Material); 'public' was found apart.

We showed clearly that in European case studies, 'related industry' and 'public' as well as 'science' and 'eNGO' grouped together. 'Politics' was rather set apart and did not contribute much to dimension 1. 'Politics' solely dominated dimension 3; 'recreational fishery' did not appear in the case studies conducted in Europe.

2.3.2 Paper Review: Participation

2.3.2.1 Term Definition

In total, four publications defined the term participation (Brzezinski et al. 2010; Tiller et al. 2015; Pristupa et al. 2016; Sampedro et al. 2017); three publications described participation-related terms (Kittinger 2013; Hara et al. 2014; Trimble and Lazaro 2014) (Fig. 2.2a). After Sampedro et al. (2017), participation could take many different forms, e.g. from planning (Neis et al. 1999; Johannes and Neis 2007; Johnson and van Densen 2007) to co-management experiences (Berkes 2003; Wilson et al. 2003). Participation was described as a role that benefits the participating stakeholders (Brzezinski et al. 2010) and a strategy of involving the stakeholders in decision-making processes (Tiller et al. 2015). Further, dependent on the strategy of involvement, stakeholders could get further responsibilities in the results of the conducted participatory process (Tiller et al. 2015). Moreover, participation referred to the type and level of stakeholder or beneficiary involvement (Hickey and Kothari 2009; Pristupa et al. 2016). Pristupa et al. (2016) explained that countries had developed a whole range of formal mechanisms stipulating citizens and stakeholder participation, e.g. consultations, referendums and elections; the participation of the private sector was still challenging.

Within three case-study publications, participatory-related terms had been described, e.g. participatory management (PM, Hara et al. 2014), participatory research (Trimble and Lazaro 2014) and participatory action research (PAR, Kittinger 2013). PM or co-management was defined as an institutional and organizational arrangement for effective management between government and user groups (Hara et al. 2014). The function of PM was described as the sharing of power and the responsibility for the management decision-making, the encouragement of partnerships and provision of user incentives for sustainable use of resources (Wilson et al. 2003; Hara et al. 2014). Participatory research was defined as one way to create power sharing between researchers and communities for, e.g. developing resource management strategies (Arnold and Fernandez-Gimenez 2007; Trimble and Lazaro 2014). Related to the degree of participation or the relationships between researchers and the community, different modes of participatory research occurred (Trimble

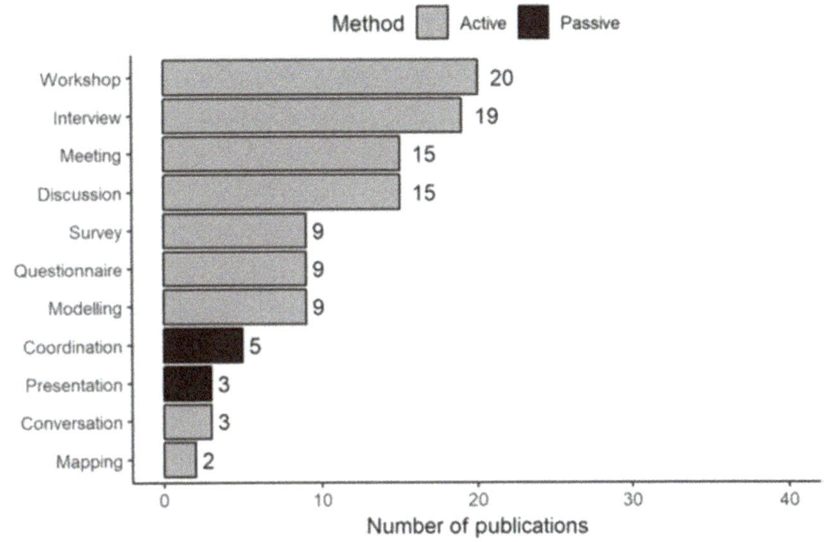

Fig. 2.4 Frequency of the appearance of active (workshop, interview, meeting, discussion, questionnaire, survey, modelling, conversation, mapping) and passive participation tools (coordination, presentation), related to stakeholder participation described in coastal and marine fisheries research publications (as of May 2018)

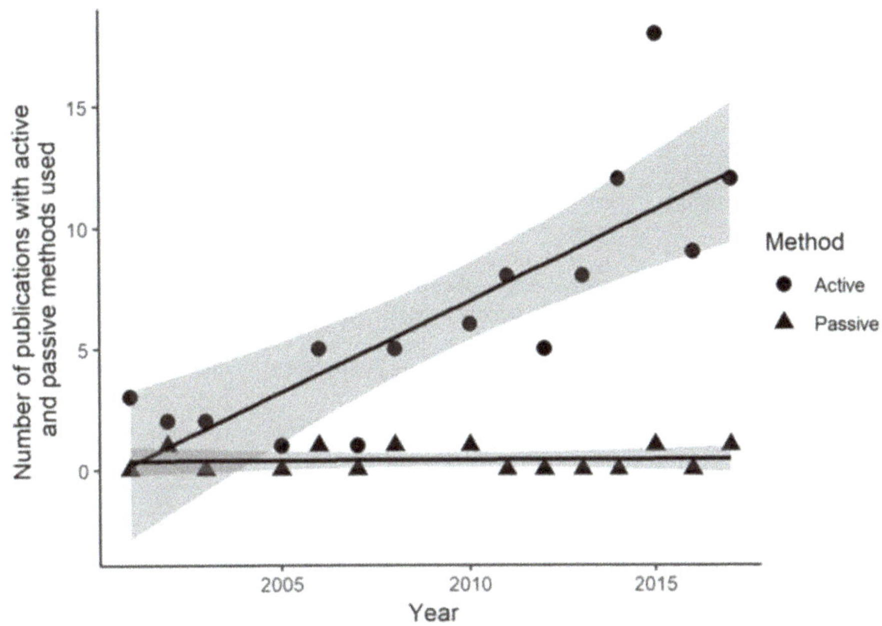

Fig. 2.5 Number of research publications with participation methods that occurred in coastal and marine fisheries research publications between 2000 and 2018; black lines represent the linear regression with 95% confidence intervals; grey area indicates the confidence band (Active: $R^2 = 0.670$, $p = 0.0001895$; passive: $R^2 = 0.00291$, $p = 0.8485$)

and Lazaro 2014), e.g. contractual, consultative, collaborative and collegiate (Biggs 1989), co-option, compliance, consultation, cooperation and co-learning (Kindon 2008). Kittinger (2013) used the term PAR, which is defined as a set of approaches related to the involvement of researchers and community members working collaboratively in the visioning, goal-getting, data gathering as well as assessment phases of research (Whyte et al. 1989; Kittinger 2013).

2.3.2.2 Participatory Tools

In contrast to the definition of participation, 45 publications focused on the description of participatory tools (Fig. 2.2b). We identified 11 participatory tools, which were divided into nine active and two passive participatory tools (Fig. 2.4). Active participation tools included workshops (N = 20), interviews (N = 19), meetings (N = 15), discussions (N = 15), questionnaires (N = 9), surveys (N = 9), modelling (N = 19), conversation (N = 3) and mapping (N = 2). Coordination (N = 5) and presentations (N = 3) represented passive participation tools (Fig. 2.4).

We detected no changes in the number of publications over time using passive methods. In contrast, we found a significant increase in the number of case studies applying active methods with a peak in 2015 (N = 19) (Fig. 2.5).

Table 2.4 presents utilized tools and related sub-tools. Here, the highest number of sub-tools was presented by interviews, including sub-tools (N = 14), e.g. semi-structured interview (e.g. Carr and Heyman 2012; Stöhr et al. 2014; Yates and Schoeman 2015; Rivera et al. 2017), unstructured interview (Hara et al. 2014) and key informant interview

Table 2.4 List of active as well as passive participation tools and including sub-tools, described within 50 research publications presenting case studies in coastal and marine fisheries (as of May 2018) between 2000 and 2018

	Participation tool	Participation sub-tool
Active	**Workshop**	Stakeholder workshop, 1-day workshop, 2-day workshop, participatory workshop, structured stakeholder workshop, value workshop, gaming workshop
	Meeting	Roundtable meeting, joint planning meeting, information meeting, face-to-face meetings, working group meeting, plenary meeting, group meeting, sub-group meeting, stakeholder group meeting, management group meeting
	Interview	Structured interview, semi-structured interview, unstructured interview, personal interview, key informant interview, in-depth interview, one-on-one interview, face-to-face interview, structured face-to-face interview, face-to-face semi-structured interview, open-end face-to-face interview, formal interview, informal interview, qualitative interview
	Conversation	Dialogue, informal conversation, focused conversation
	Discussion	Group discussion, focus group, forum discussion, open discussion, stakeholder advisory panel
	Questionnaire	Structured interview questionnaire, e-mail-based questionnaire, follow-up questionnaire
	Survey	Large-scale interview survey, face-to-face interview survey, online survey, in situ survey, attitudinal survey
	Modelling	Tool, participatory modelling, Bayesian belief network
	Mapping	Cognitive mapping, fuzzy cognitive mapping
Passive	**Coordination**	Voting, rating, evaluation
	Presentation	Video, poster, exhibition, tableaux

(Eriksson et al. 2016). Meetings and workshops showed the second and third highest number of sub-tools. Here, meetings were presented, with sub-tools (N = 10), e.g. roundtable meeting (Kerr et al. 2006), joint planning meeting (Kittinger 2013) and face-to-face meeting (Miller et al. 2010). However, workshops were shown, including sub-tools (N = 7), e.g. stakeholder workshop (Eriksson et al. 2016; Burdon et al. 2018) and participatory workshop (Bojorquez-Tapia et al. 2016). Passive participation methods included coordination, with sub-tools (N = 3), e.g. voting (Miller et al. 2010; Thiault et al. 2017; Zengin et al. 2018), rating (Goetz et al. 2015), evaluation (Cox and Kronlund 2008) and presentation, including sub-tools (N = 4), e.g. video (Clarke et al. 2002), poster (Kerr et al. 2006), exhibition (Kerr et al. 2006) and tableaux (Kerr et al. 2006). We also determined whether preparatory work was performed and described within the case studies under review. Among others, observations (Delaney et al. 2007; Granados-Dieseldorf et al. 2013; Trimble and Berkes 2013; Stöhr et al. 2014; Trimble and Lazaro 2014; Mabon and Kawabe 2015), fieldwork (Mabon and Kawabe 2015; Sampedro et al. 2017) and visits (Kerr et al. 2006) were carried out. Furthermore, newsletters (Kerr et al. 2006) and e-mails (Lorance et al. 2011) were sent out to call for participation within different stakeholder types. Moreover, telephone calls (Kerr et al. 2006) were made, and consultations took place, e.g. consultation with stakeholders (Cox and Kronlund 2008; Mapstone et al. 2008; Williams et al. 2011).

2.3.2.3 Intention for Participation

Within this review, we looked at the diversity of the intention for participation; we classified these intentions as types and sub-types (Table 2.5).

The intention types identification (N = 20), with sub-types, e.g. target species (Fletcher 2005), ways of communication (Zengin et al. 2018), stakeholder characteristics (Bojorquez-Tapia et al. 2016; Kinds et al. 2016; Burdon et al. 2018) and assessment (N = 12), with sub-types, e.g. management system (Lorance et al. 2011), knowledge (Carr and Heyman 2012) and data (Catedrilla et al. 2012) occurred most often (Fig. 2.6). Establishment (N = 5), development (N = 7), evaluation (N = 7) and improvement (N = 7) occurred moderately often (Fig. 2.6). Less widely used were analysis (N = 3), definition (N = 2), description (N = 2) and feedback (N = 2) (Fig. 2.6).

Establishment (N = 10), assessment (N = 7) and identification (N = 7) had the most sub-types within the case studies under review. Improvement (N = 5), development (N = 4) and evaluation (N = 4) showed a moderate diversity of sub-types, whereas feedback (N = 2), e.g. feedback from stakeholders on the meeting (Dowling et al. 2008), as well as description (N = 2), e.g. knowledge about socio-ecological systems (Gray et al. 2012) and management implications (Smith et al. 2001) presented the lowest diversity of sub-types.

2.3.3 Reflection on the Joint Term Stakeholder and Participation

In the final evaluation, we analysed the application of stakeholder participation as one term. We first evaluated whether

Table 2.5 Types and associated sub-types of intentions for participation determined within 50 research publications focusing on stakeholder participation in coastal and marine fisheries from 2000 to 2018

Type	Sub-type
Analysis	Stakeholders' perception
	Mental models
	Management system
Assessment	Management system (e.g. adaptive co-management, history of management implementation)
	Ideas of alternative livelihood
	Knowledge (e.g. fishers ecological knowledge), perception and attitude of stakeholders
	Method success
	Data (e.g. interviews, socioeconomic characteristics)
	Solution on regional level
	Effectiveness of collaboration between stakeholders
Definition	Criteria for evaluation
	Objectives
	Management implications
Description	Knowledge of socio-ecological system (SES)
	Management implications
Development	Consensus-building
	Comprehensive map
	Stakeholder-driven scenarios
	Criteria for participatory research
Establishment	Co-management mechanism
	Collective research agenda
	Vision for future fisheries management
	Comprehensive map of predicting fishing effort
	Guidance for scientists
	Scientific advice
	Theory of causal mechanisms
	Platform for information and decision-making
	Stakeholder-driven scenarios
	Clear and open views
Evaluation	Mental models
	Harvest policies
	Results from interview (cross-checking)
	Fishery and management system
Feedback	Forecast content
	Meeting
Identification	Stakeholders' characteristics (e.g. attitude, perception, wishes, concerns, knowledge (local ecological knowledge, fishers ecological knowledge)
	Information (e.g. socio-ecological)
	Target species
	Objectives (e.g. criteria, uncertainties, drivers, consequences, human dimensions, population needs, reference points)
	Weakness of fishery system
	Range of quantifiable objectives and strategies
	Ways of communication
Improvement	Stakeholder participation, relationships and requirements
	Management
	Socio-economic drivers
	Data
	Website

stakeholder participation was used for doing research on stakeholders or if the case studies were conducted with the participation of stakeholders. Overall, 24 publications utilized the term stakeholder participation for the research on stakeholders (research tool); 21 publications used stakeholder participation in their conducted case study (participation tool). Within five case studies, it was uncertain whether stakeholder participation was used or not (Fig. 2.2c).

In the case study conducted by Kinds et al. (2016), the term stakeholder participation was used to describe the development of a sustainability tool with the direct input from users, i.e. fishers. Here, the wishes and preferences of stakeholders were recognized and implemented to improve the output of the utilized tool but not to influence decision-making processes (*research tool*). Rivera et al. (2017) carried out semi-structured interviews to assess stakeholders' perceptions to identify management, biology and socioeconomic drivers related to the gooseneck barnacle fishery in Spain. This case study used the term stakeholder participation, but no influence on the management by stakeholders was mentioned (*research tool*). Tiller et al. (2015) applied an integrated approach of two methods, Systems Thinking and Bayesian Belief Networks, to elicit stakeholders' opinions through participatory engagement. Both methods were used to investigate, e.g. how stakeholders perceive the ecological system in the Trondheimsfjord, but with no further impact on decision-making processes (*research tool*). Through the method of Systems Thinking, shared mental models of the ecological system in the Trondheimsfjord were developed. Bayesian Belief Networks were further used for exploration of the priority issues as well as to represent causal relationships between defined variables. In contrast, Trimble and Berkes (2013) presented the concept of participatory research, i.e. involving fishers and policymakers as well as managers among other stakeholders in the case of a sea lion population and a fishery in Uruguay. Within this case study, stakeholders, e.g. fishers had an impact on decision-making processes related to the management of the sea lion population (*participation tool*). Williams et al. (2011) conducted a case study based on the participation of commercial fishers, defining various alternative management strategies related to the Torres Strait Finfish Fishery (TSFF) in Australia, i.e. sea-

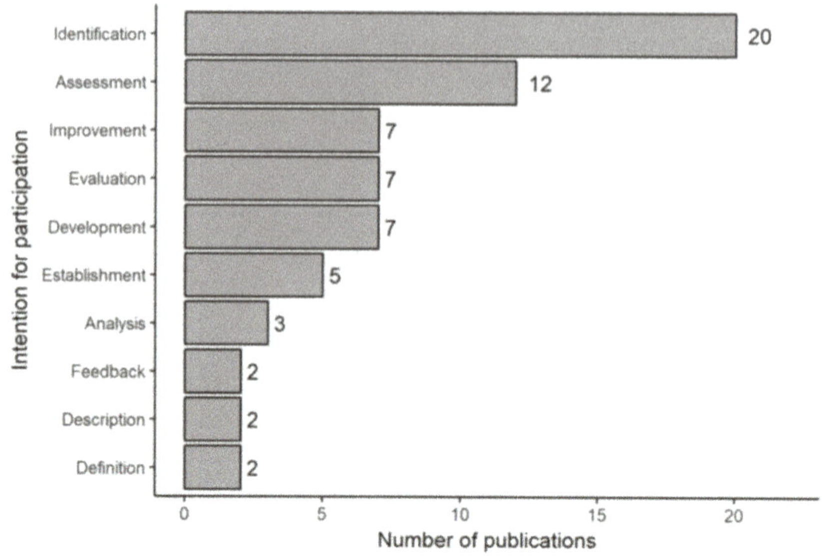

Fig. 2.6 Frequency at which types of intention for participation occurred in 50 research publication looking at coastal and marine fisheries (as of May 2018)

sonal closure, large minimum capture size, to provide a framework for impartial evaluation of management strategy performance (*participation tool*). In addition, Stöhr et al. (2014) described the concept of stakeholder participation by evaluating two case studies; only the Polish case had a coastal and marine focus. Within this case study, roundtables were applied to create a multi-stakeholder platform with the objective of informing and influencing decision-making processes (*participation tool*).

2.4 Discussion

In total, 50 case studies focusing on stakeholder participation in coastal and marine fisheries were identified and reviewed. Most of the publications did not define the term stakeholder or participation or described a systematical approach of selecting stakeholders. Moreover, stakeholder participation was mentioned in all 50 publications, but only half of the case studies involved stakeholders in the process of participation.

It should be noted that we could only show what has been described in the publications under review; here, we did not present a comprehensive overview of all relevant stakeholder types that would be possible in the respective contexts.

2.4.1 Stakeholder

2.4.1.1 Term Definition and Stakeholder Analysis

Four case studies defined the term stakeholder and, thus, showed a scientific examination of potentially concerned stakeholder types; 15 publications used a systematic description of how stakeholders were identified. The other publications used intuitive decisions to identify relevant stakeholders in their case study. This led to the fact that, related to our research focus on coastal and marine fisheries, the stakeholder type 'fishery' was mostly involved; 'public' stakeholders were only rarely involved.

All case studies included 'fishery' as a stakeholder type. Therefore, we proposed that 'fishery' is seen as the main stakeholder type in coastal and marine fisheries research. Mahon et al. (2003) supported this thesis literally by writing that within the conducted case study, the primary stakeholders are the fishers. In some publications, 'fishery' was even the only stakeholder type considered. Although at first sight this realization might seem logical, it can be discussed; fisheries are harvesting a common resource and, therefore, do not necessarily contribute towards the sustainable exploitation of coastal and marine fisheries resources, as most case studies consider stakeholder participation as a way of implementing more sustainable fisheries management (e.g. Wilson et al. 2003; Thiault et al. 2017). Although not all forms of fisheries were considered unsustainable, artisanal fisheries, for example, were often associated with having a small impact on fish stocks (Carvalho et al. 2011) but have been proven to cause impact beyond sustainable levels (Pomeroy

2012); they also deal with other sustainability issues such as bycatch of birds (Almeida et al. 2017).

Within our review, we used the category 'others' to classify stakeholders that did not fit into any other category. This fact shows very clearly that, on the one hand, there is great diversity of stakeholder types within the field of coastal and marine fisheries research; on the other hand, it describes existing discrepancies in the understanding of the term definition and the classification of corresponding stakeholders. We, therefore, suggest to clearly define the term stakeholder as well as to discuss their role in the specific context of the conducted case study. Although Tiller et al. (2015) criticized the stakeholder definition by Freeman (2010) as too broad, they did not give a clear term definition either in their own case study. We assume that there is a high risk of excluding relevant stakeholder types, when not applying a term definition for stakeholder as well as not using a stakeholder analysis tool to ensure that relevant stakeholders are approached. This could lead to the fact that, for example, no local ecological knowledge or fisheries ecological knowledge would be recorded for the corresponding case study, which is important inter alia for better understanding the marine ecology and making results more convincing for resource users (Davis et al. 2004) and, therefore, increase the legitimacy of resource management (Aanesen et al. 2014). It is not important to include all stakeholders available but to choose them carefully according to the objectives of the case study, which means applying a stakeholder analysis approach (Durham et al. 2014).

2.4.1.2 Stakeholder Clusters

We showed, with using MCA, that 'eNGO', 'politics' and 'science' are often addressed together within the strongest dimension. Therefore, we could conclude that these stakeholder types were considered important within many conducted case studies. This dimension described stakeholders that deal with a rather theoretical side in the field of fishery, i.e. in the form of regulations, research or campaigns. It can be argued that these stakeholders contributed towards research and management as well as towards different forms of sustainability; therefore, 'eNGO', 'politics' and 'science' have a more sustainability-oriented attitude. This finding is strongly supported by Aanesen et al. (2014); they concluded that, under the European Common Fisheries Policy, authorities, scientists and NGOs have a similar perspective on fisheries management. This is rather obvious for 'eNGO', as they are seen as representing the ecological sustainability. By contrast, 'politics' could be interpreted as representing the population, i.e. this stakeholder group acts in the interest of the sustainability of food, but is also driven by the economic sustainability. 'Science' could be seen as the representative and provider of research. We suggest that these stakeholder groups have a general interest in sustainable management and are not directly or financially dependent on the resource fish. Of course, it can be argued that certain jobs of eNGOs, scientists or politicians depend on the debate as well as on the public interest in fish and fishery. But this argument is to be classified as marginal in this context. One reason is that fish is one of the main protein sources for humans; even if the resource fish would shrink, it will always be of interest for certain stakeholder types.

In our sample of publications, 'politics', 'eNGO' and 'science' were mentioned most frequently after 'fishery'. For that reason, we can assume that these three stakeholder groups are deemed the second most important stakeholder groups. It can be discussed that 'politics', 'eNGO' and 'science' should have at least an equally strong stake in fisheries research compared to 'fishery'.

Another group displayed by MCA is formed by 'related industry' and 'recreational fishery'. Both stakeholder groups mostly occurred in the second strongest dimension, which can be interpreted as stakeholders who are handling the resource fish and, therefore, dealing with it in a practical way. Although they also have an interest in sustainable management, they, unlike 'science', 'politics' and 'eNGO', depend financially (especially 'fishery') or mentally (e.g. 'fishery' and 'recreational fishery') on the resource fish. Therefore, profit or benefit orientation can be seen as another factor describing dimension 2. This is supported by the fact that the two groups (dimension 1: sustainability vs. dimension 2: dependence) discussed are placed far away from each other in the MCA. Both stakeholder groups cannot be seen as independent from each other as their decisions are influencing each other's actions, e.g. if political regulations or campaigns led by 'eNGOs' resulted in decreasing harvest rates of fish, commercial and recreational fishers are negatively affected. We take a critical look at these stakeholder groups, as they are presented apart from each other in the conducted MCA and, therefore, are not engaged equally in the reviewed case studies. We recommend to engage these stakeholder types more equally. The cooperation between fishery-related stakeholders and scientists could lead to more informed stakeholders on both sides; therefore, a greater mutual understanding, trust as well as likelihood of long-lasting partnerships could be achieved (Hartley and Robertson 2006).

We showed that 'public' participation is relatively low in the field of coastal and marine fisheries research. This fact is reflected among other things in the low numbers of mention within the case studies. Here, 'public' as one stakeholder group contributed the least to the two strongest dimensions.

On the one hand, the low involvement could be interpreted as a lack of interest. On the other hand, we argue that public stakeholders were not directly addressed within the publications. In relation to the definition we used to classify 'public', it can be critically discussed that 'eNGOs' could also be seen as representatives of the civil society (e.g. Pristupa et al. 2016) and community leaders could include voted politicians (Rivera et al. 2017). But we decided to stick to the stakeholders as they were mentioned in the publications. The results showed that 'public' stakeholders are not part of any group; nevertheless, they dominated the weakest dimension and explained the high percentage of its variance.

Data from North American and European case studies resulted in different MCAs. This can be seen for example with 'recreational fishery'. Although this stakeholder type is part of the European Common Fisheries Policy, they are not considered as stakeholders in any of the case studies conducted in Europe. This is different for North American case studies; here 'recreational fishery' was seen as a stakeholder type. Even if this analysis gave only a small insight into the topic, regional differences related to stakeholder types could already be made clear here. These differences cannot be explained by different management systems, because both in Europe and in North America recreational fisheries are included in their regulations; the results further need to be investigated. Furthermore, we assumed different emphases of stakeholder types; therefore, when applying MCA to different regions, different interpretations of the dimensions have to be made. However, the small sample size for regional MCAs could reduce the significance of such interpretations.

Based on the application and analysis of the term stakeholder, we conclude that there were only a few case studies that critically assessed the concept of stakeholders. Nevertheless, our results provide an insight into how stakeholders were seen in the field of coastal and marine fisheries research, i.e. who is considered as important and which stakeholders are often consulted together.

2.4.2 Participation

2.4.2.1 Term Definition and Typologies

Out of 50 case studies focusing on the topic stakeholder participation in coastal and marine fisheries, only seven case studies defined the term participation or a participation-related term. However, there is a wide variety of definitions and typologies of stakeholder participation in the literature.

Green and Hunton-Clarke (2003) represented different typologies of participation regarding environmental decision-making. Five concepts of participation were listed and defined to increase the level of involvement. On the one hand, Arnstein's (1969) concept of stakeholder participation was described; this concept is based on eight levels: nonparticipation (manipulation and therapy), tokenism (informing, consultation and placation) and citizen power (partnership, delegated power and citizen control) (Luyet et al. 2012). On the other hand, the participation concept by Pretty and Shah (1994) was presented. Here, participation is classified by using six steps: passive participation, participation by information giving, participation by consultation, functional participation, interactive participation and self-mobilization. In Pristupa et al. (2016), participation was also described by the level of stakeholder involvement, but with regard to the concept of Arnstein (1969) and Pretty and Shah (1994), no further explanation was given of the different levels of participation in this case study.

In addition to Green and Hunton-Clarke (2003), Reed (2008) reviewed different typologies on stakeholder participation for environmental management. In this literature review, he defined the following typologies on which participation is based: (i) degrees of participation (e.g. Arnstein 1969), (ii) nature of participation (Rowe and Frewer 2000), (iii) theoretical basis (e.g. Thomas 1993) and (iv) participation based on objectives for which participation is used (e.g. Okali et al. 1994) (Reed 2008). The fourth typology was used in the case studies by Sampedro et al. (2017) and Tiller et al. (2015). Here, participation was described as the use for planning or co-management experiences (Sampedro et al. 2017) and as the strategy for involving stakeholders in decision-making processes (Tiller et al. 2015). Related to the case studies under review, we would add a fifth typology of participation, i.e. participation based on the opportunity to participate in relation to resources. Brzezinski et al. (2010) described and defined participation as a role benefiting participating stakeholders based on money and geographical proximity. The case study showed the connection between geographical closeness and the level of attendance, i.e. the closer stakeholders were to meetings, the higher was their level to attend at those meetings (Brzezinski et al. 2010).

As NOAA (2015) generalized, there is no 'one-size-fits-all' approach or definition of participation; the implementation and the process of participation is dependent on several aspects, e.g. issue at hand, stakeholders, geography, schedules, as well as on time frames. Furthermore, Green and Hunton-Clarke (2003) recommended selecting the type of participation suitable for the situation or the problem that needs to be solved. We argue, to create a successful resource management and increase the acceptance of management measures by resource users, a well thought out participation approach is essential with regard to the sustainable use of coastal and marine resources.

2.5 Conclusion

Stakeholder participation is and will continue to be of central importance when it comes to the management of ecosystems and its resources. Although our findings showed clear tendencies in stakeholder participation, they also opened several other questions.

The grouping of 'science', 'eNGO' and 'politics' was discussed critically, especially 'public', 'recreational fishery' and 'related industry' were presented far away in the MCA. We suggest that these groups should not be seen as opposed to each other but be included in a more integrated way in participatory research projects. Low involvement of 'public' stakeholders and their contribution towards decisions should be further discussed, because wild fish is widely seen and communicated as a common pool resource. As a consequence, 'public' stakeholders, i.e. representatives of the common, should also have a stake in the management of the resource also since ecosystem changes will affect all citizens.

We advise to include different stakeholder types and take advantages of their different experiences, although we recognize that pragmatic and methodological reasons such as the willingness to participate can constrain these efforts. While our analysis has not been profoundly focused on regional differences, it should be noted that there are regional differences between the relationships between and the contributions of stakeholders. Even though only done marginally in our analysis, dividing the data into different regions showed that relationships and contributions varied between stakeholder types. For further research, we advise to set a regional focus on stakeholder participation and discuss it under the light of different management regulations.

Although we presented stakeholder types carefully deducted from the texts, the perception of these types is always at risk to change throughout a paper review process. Soma and Vatn (2014), e.g. separated the role of stakeholders and citizens in participatory processes, not discussing citizens as stakeholders but also plead for the involvement of citizens in natural resource management; therefore, we categorized these stakeholders in the same manner.

Research projects and stakeholder participation processes apart from research projects are mostly restricted by resources, e.g. time, money (Angelstam et al. 2013), capacity (Mackinson et al. 2011), expertise, i.e. expertise of social researchers and the availability of researchers as well as of stakeholders in general. These limitations can be a reason for not including a systematic discussion of the term stakeholder or a scientific stakeholder analysis. Because only a few of the reviewed publications described a definition or an approach of analysing stakeholders, we conclude that there were also limitations of integration, i.e. the involvement of social scientists in the process of stakeholder participation. In addition, it is of great advantage to know which typology and degree of participation have been used and benefit from experienced advantages as well as disadvantages of applied methods (Luyet et al. 2012). This way, conflicts can be avoided and stakeholder participation can be implemented in a better way.

In times of interdisciplinary (Repko et al. 2011) as well as transdisciplinary research (Häberli et al. 2001), and the intention of further improving science in general, we call for an increasing involvement of social scientists regarding the processes of stakeholder participation in coastal and marine fisheries research; more funding opportunities are needed to support this kind of integrated research field.

Our review clearly showed that many different definitions of stakeholder participation exist, and so researchers need to be careful when they examine which one is applicable towards their research goal. Related to this great diversity of stakeholder participation definitions, we will not present *the* definition. Nevertheless, we advise to conduct a critical analysis of stakeholder types as well as on participation tools at the beginning of a new research project with the aim of involving stakeholders related to decision-making processes. Durham et al. (2014) and NOAA (2015) offer well-applicable and explained stakeholder participation guides, which can be applied at the process start of the project. A systematic and comprehensible consultation of the methods presented in these guides can lead to an improved transparency of the results and decreases the potential of overlooking stakeholder groups or participatory tools that fit the research goal.

Acknowledgements HS is funded by the PhD Scholarship Programme of the German Federal Environmental Foundation (Deutsche Bundesstiftung Umwelt, Osnabrück; no. 2017/480).

Appendix

This article is related to the YOUMARES 9 conference session no. 2: 'Towards a sustainable management of marine resources: integrating social and natural sciences.' The original Call for Abstracts and the abstracts of the presentations within this session can be found in the Appendix 'Conference Sessions and Abstracts', Chapter '2 Towards a sustainable management of marine resources: integrating social and natural sciences', of this book.

Supplementary Material

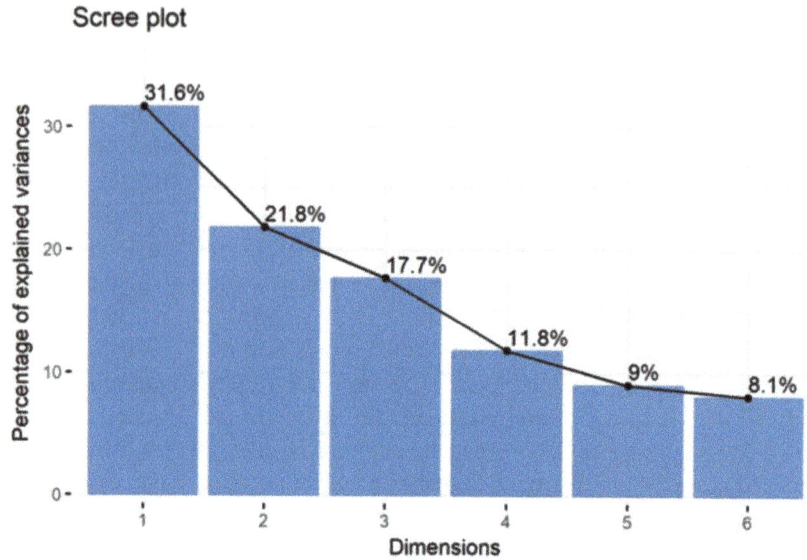

Figure 2.A1 Percentage of explained variance calculated by multiple correspondence analysis (MCA) showing the variance of stakeholder types within 50 research publications presenting 50 research publications of case studies in coastal and marine fisheries (as of May 2018)

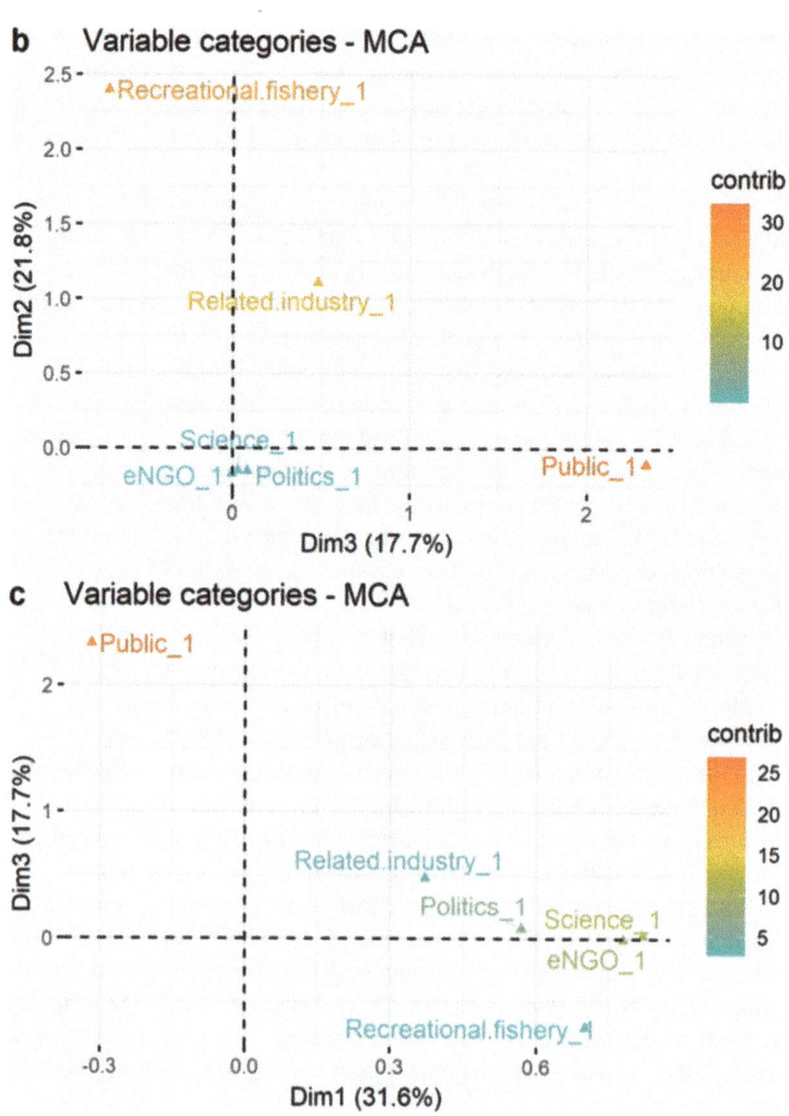

Fig. 2.A2 Panels **b** and **c** of the visualization of the correlation between dimension 2 (Dim2) and dimension 3 (Dim3) (**b**) as well as dimension 1 (Dim1) and dimension 3 (Dim3) (**c**) towards the variance of stakeholder types within in 50 research publications of case studies in coastal and marine fisheries (as of May 2018) using multiple correspondence analysis (MCA); (**b**) 21.8% of variance were explained; here 'recreational fishery' and 'related industry' were focused. (**c**) Less variance (17.7%) was explained; 'public' was the dominant stakeholder type

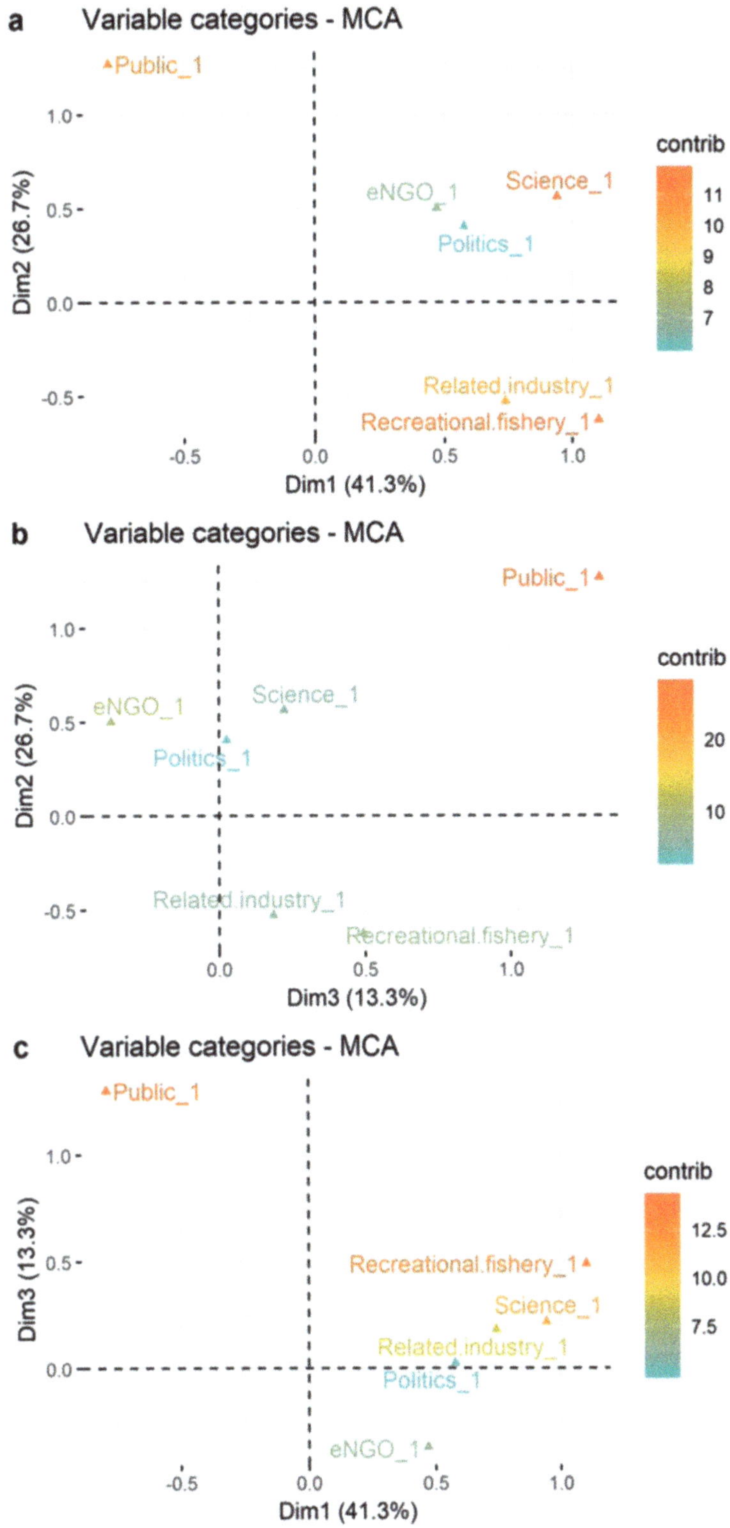

Figure 2.A3 Visualization of multiple correspondence analysis (MCA) results for case studies conducted in North America; here correlation between dimension 1 (Dim1), dimension 2 (Dim2) and dimension 3 (Dim3) is presented. (**a**) strong contribution was shown by 'science', 'eNGO' and 'politics'; 'related industry' and 'recreational fishery' were displayed in the negative area; (**b**) strong contribution was presented by 'science', 'eNGO' and 'politics'. (**c**) 'Public' was the dominant stakeholder type

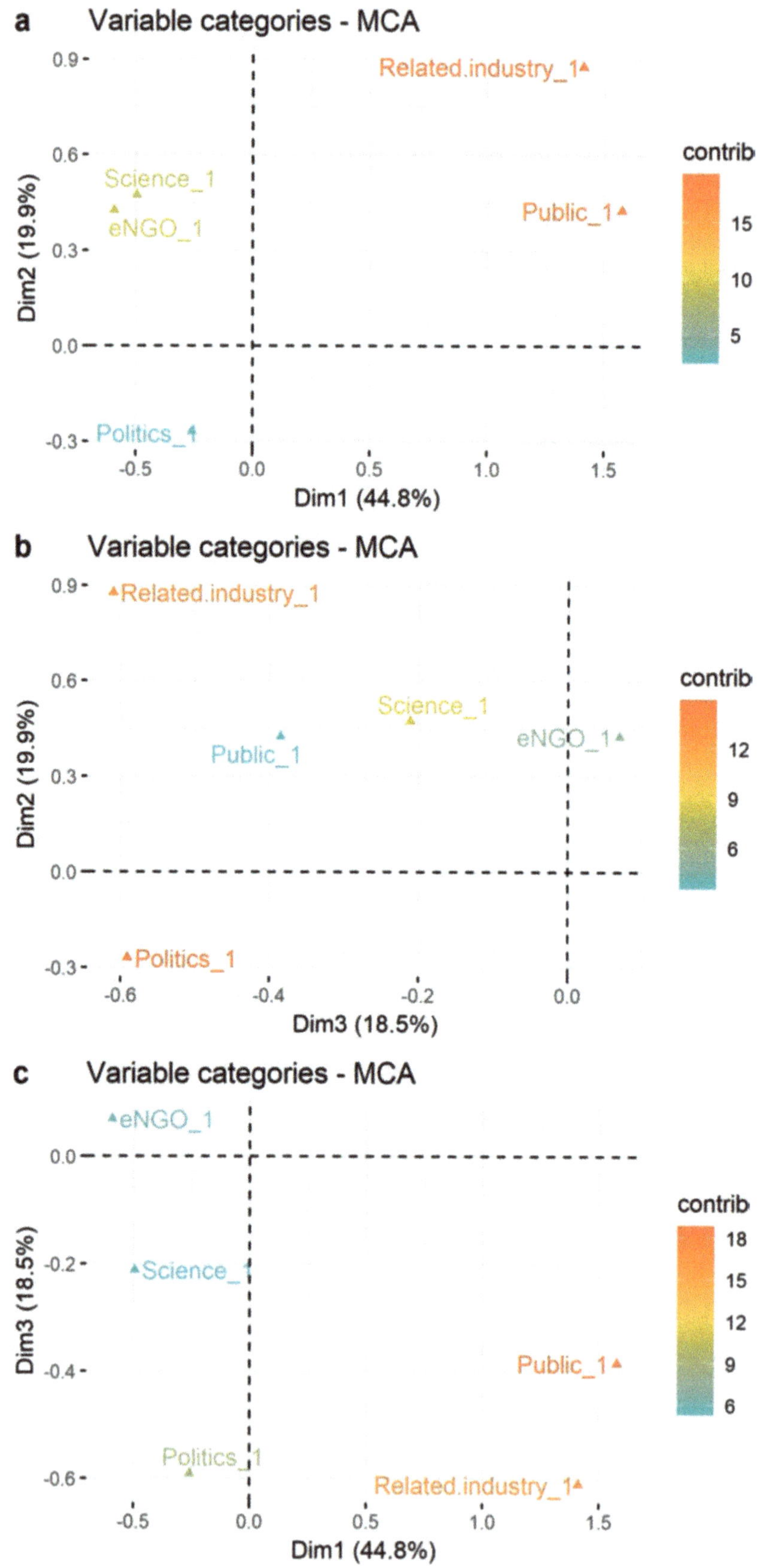

Figure 2.A4 Visualization of multiple correspondence analysis (MCA) results for research publications of European case studies in coastal and marine fisheries (as of May 2018); here correlation between dimension 1 (Dim1), dimension 2 (Dim2) and dimension 3 (Dim3) is presented. (**a**), (**b**) 'Related industry', 'public', 'science' and 'eNGO' were grouped together; 'politics' was rather set apart. (**c**) 'Politics' was the dominant stakeholder type

Table 2.A1 Contribution of variables, i.e. stakeholder types (measured in %) towards five dimensions using multiple correspondence analysis (MCA); stakeholder types, i.e. science, politics, eNGO, recreational fisheries, related industry and public occurred within 50 research publications presenting case studies in coastal and marine fisheries (as of May 2018)

	Dim 1	Dim 2	Dim 3	Dim 4	Dim 5
Science	35.37	1.69	0.07	0.38	0.28
Politics	27.89	2.88	1.02	36.32	19.43
eNGO	29.59	2.06	0.01	33.62	9.77
Recreational fishery	2.87	48.90	5.20	10.58	32.44
Related industry	3.41	44.34	10.16	11.79	30.04
Public	0.87	0.13	83.54	7.31	8.04

Table 2.A2 List of results related to the literature review focusing the topic stakeholder participation in the field of coastal and marine fisheries (type of stakeholder: S = science, PO = politics, E = eNGO, F = fisheries, RF = recreational fisheries, RI = related industry, PU = public, O = others; participatory method: MET = meeting, WOR = workshop, DIS = discussion, INT = interview, QUE = questionnaire, SUR = survey, CON = conversation, MOD = modelling, MAP = mapping, PRE = presentation, COO = coordination)

Author	Continent	Country	Definition of stakeholder?	Type of stakeholder	Stakeholder analysis approach?	Definition of participation?	Description of participatory method?	Which methods has been used?
Appeldoorn (2008)	North America	USA		F, O				
Bitunjac et al. (2016)	Europe	Adria		S, PO, E, F	x		x	DIS
Bojorquez-Tapia et al. (2017)	North America	Mexico		S, PO, E, F, RI			x	WOR, MOD
Brzezinski et al. (2010)	North America	USA	x	E, F		x		
Burdon et al. (2018)	Europe	Denmark, Germany		E, F, O			x	WOR, DIS, INT
Butler et al. (2015)	Europe	Scotland		PO, F, O	x		x	INT
Carr and Heyman (2012)	North America	USA		PO, E, F, O			x	INT, QUE
Catedrilla et al. (2012)	Asia	Philippines		F	x		x	DIS, INT
Clarke et al. (2002)	Asia	China, Hong Kong		F			x	MET, DIS, PRE
Cleland (2017)	Asia	Philippines		PO, E, F, RI			x	WOR
Coelho Dias da Silva et al. (2010)	South America	Brazil		F, O			x	MET, DIS
Cox and Kronlund (2008)	North America	Canada		F, RI			x	COO
Delaney et al. (2007)	Europe	NA		S, PO, E, F			x	INT
Dowling et al. (2008)	Australia	Australia		S, PO, O			x	MET, DIS
Eriksson et al. (2016)	Asia / Africa	Indonesia, Philippines, Solomon Islands, Tanzania		S, PO, E, F, PU, O			x	WOR, DIS, INT, SUR
Eveson et al. (2015)	Australia	Australia		F			x	DIS, SUR
Field et al. (2013)	Africa	South Africa		S, PO, E, F, O				

(continued)

Table 2.A2 (continued)

Author	Continent	Country	Definition of stakeholder?	Type of stakeholder	Stakeholder analysis approach?	Definition of participation?	Description of participatory method?	Which methods has been used?
Fletcher (2005)	Australia	Australia		S, PO, E, RF, RI, O			x	WOR
Garza-Gil et al. (2015)	Europe	Spain		S, PO, F, RI			x	QUE, SUR
Goetz et al. (2015)	Europe	Spain, Portugal		S, F, O			x	WOR, QUE, SUR, COO
Granados-Dieseldorf et al. (2013)	America	Belize		PO, F, O				
Gray et al. (2012)	North America	USA		S, PO, E, F, RF, RI, O	x		x	MAP
Haapasaari et al. (2013)	Europe	Central Baltic	x	S, PO, E, F			x	MOD
Hara et al. (2014)	Africa	South Africa		F, RI		x	x	MET, WOR, INT
Kaiser and Forsberg (2001)	Europe	Norway		F, RI, PU, O			x	WOR
Kerr et al. (2006)	Europe	Scotland, UK		S, PO, E, F, O			x	MET, INT, QUE, PRE
Kinds et al. (2016)	Europe	Belgium	x	PO, E, F	x		x	DIS, INT, MOD
Kittinger (2013)	North America	USA		S, E, F, PU	x	x	x	MET, INT
Lorance et al. (2011)	Europe	-		S, PO, E, F			x	WOR, DIS, INT, QUE, MAP
Mabon and Kawabe (2015)	Asia	Japan		S, PO, F, RI			x	MET, DIS, INT
Mahon et al. (2003)	North America	Barbados		PO, F, PU	x		x	WOR, CON
Mapstone et al. (2008)	Australia	Australia		PO, E, F, RI			x	MET, WOR, MOD
Miller et al. (2010)	North America	USA		S, PO, E, F, RF, RI	x		x	MET, WOR, DIS, MOD, COO
Mitchell and Baba (2006)	Australia	Australia		F, RF			x	QUE, SUR
Murphy et al. (2015)	North America	USA		F, RF, RI	x		x	SUR
Pristupa et al. (2016)	Europe	Russia		S, PO, F, RI, PU	x	x	x	INT
Punt et al. (2012)	Australia	Australia		S, PO, F, RI				

(continued)

Table 2.A2 (continued)

Author	Continent	Country	Definition of stakeholder?	Type of stakeholder	Stakeholder analysis approach?	Definition of participation?	Description of participatory method?	Which methods has been used?
Rivera et al. (2017)	Europe	Spain		PO, F, PU			x	DIS, INT, QUE
Sampedro et al. (2017)	Europe	Spain, Portugal		S, E, F, O	x	x	x	MET, WOR, INT, SUR, MOD
Smith et al. (2001)	Australia	Australia		S, PO, E, F, RI			x	MET, WOR
Stöhr et al. (2014)	Europe	Sweden, Poland		S, PO, E, F			x	MET, INT
Stratoudakis et al. (2015)	Europe	Portugal		S, PO, F, O	x		x	MET, WOR
Thiault et al. (2017)	Asia	French Polynesia		F	x		x	SUR
Tiller et al. (2015)	Europe	Norway	x	F		x	x	WOR, INT, QUE, MOD
Trimble and Berkes (2013)	South America	Uruguay		S, PO, E, F	x		x	MET, WOR, CON, INT
Trimble and Lazaro (2014)	South America	Uruguay		S, PO, E, F	x	x	x	MET, WOR, CON, INT, QUE
Watters et al. (2013)	NA	Scotia Sea, Drake Passage		F, O			x	MOD
Williams et al. (2011)	Australia	Australia		PO, F			x	WOR, DIS, MOD
Yáñez et al. (2014)	South America	Chile		S, PO, F			x	WOR, SUR
Zengin et al. (2018)	Europe	Riparian Countries		PO, F, RI, PU			x	DIS, PRE, COO

References

Aanesen M, Armstrong CW, Bloomfield HJ et al (2014) What does stakeholder involvement mean for fisheries management. Ecol Soc 19(4):35

Abdi H, Valentin D (2007) Multiple correspondence analysis. In: Salkind NJ (ed) Encyclopedia of measurement and statistics. Sage, Thousand Oaks

Almeida A, Ameryk A, Campos B et al (2017) Study on Mitigation Measures to Minimise Seabird Bycatch in Gillnet fisheries. European Commission. available at: https://publications.europa.eu/en/publication-detail/-/publication/f426200b-1138-11e8-9253-01aa75ed71a1/language-en

Angelstam P, Andersson K, Annerstedt M et al (2013) Solving problems in social-ecological systems: definition, practice and barriers of transdisciplinary research. Ambio 42(2):254–265. https://doi.org/10.1007/s13280-012-0372-4

Appeldoorn RS (2008) Transforming reef fisheries management: application of an ecosystem-based approach in the USA Caribbean. Environ Conserv 35(3):232–241

Arnold JS, Fernandez-Gimenez M (2007) Building social capital through participatory research: an analysis of collaboration on Tohono O'odham tribal rangelands in Arizona. Soc Nat Resour 20:481–495

Arnstein SR (1969) A ladder of citizen participation. J Am Plan Assoc 35(4):216–224

Berkes F (2003) Alternatives to conventional management: lessons from small-scale fisheries. Environments 31:5–19

Biggs S (1989) Resource-poor farmer participation in research: a synthesis of experiences from nine national agricultural research systems. OFCOR Comparative Study Paper 3, International Service for National Agricultural Research, The Hague

Bitunjac I, Jajac N, Katavic I (2016) Decision support to sustainable management of bottom trawl fleet. Sustainability 8:204

Bojorquez-Tapia LA, Pedroza D, Ponce-Diaz G et al (2016) A continual engagement framework to tackle wicked problems: curtailing loggerhead sea turtle fishing bycatch in Gulf of Ulloa, Mexico. Sustain Sci 12:535–548

Brandt P, Ernst A, Gralla F et al (2013) A review of transdisciplinary research in sustainability science. Ecol Econ 92:1–15

Brzezinski DT, Wilson J, Chen Y (2010) Voluntary participation in regional fisheries management council meetings. Ecol Soc 15(3):2

Burdon D, Boyes SJ, Elliott M et al (2018) Integrating natural and social sciences to manage sustainably vectors of change in the marine environment: Dogger Bank transnational case study. Estuar Coast Shelf Sci 201:234–247

Butler JRA, Young JC, McMyn IAG et al (2015) Evaluating adaptive co-management as conservation conflict resolution: learning from seals and salmon. J Environ Manag 160:212–225

Carr LM, Heyman WD (2012) "It's about seeing what's actually out there": quantifying fishers' ecological knowledge and biases in a small-scale commercial fishery as a path toward co-management. Ocean Coast Manage 69:118–132

Carvalho N, Edwards-Jones G, Isidro E (2011) Defining scale in fisheries: small versus large-scale fishing operations in the Azores. Fish Res 109:360–369

Catedrilla LC, Espectato LN, Serofia GD et al (2012) Fisheries law enforcement and compliance in District 1, Iloilo Province, Philippines. Ocean Coast Manage 60:31–37

Clarke S, Leung Wai-yin A, Mak YM et al (2002) Consultation with local fishers on the Hong Kong artificial reefs initiative. ICES J Mar Sci 59:171–177

Cleland D (2017) A playful shift: Field-based experimental games offer insight into capacity reduction in small-scale fisheries. Ocean Coast Manage 144:129–137

Coelho Dias da Silva AC, Comin de Castilhos J, Pinheiro dos Santos EA et al (2010) Efforts to reduce sea turtle bycatch in the shrimp fishery in Northeastern Brazil through a co-management process. Ocean Coast Manage 53(9):570–576

Commission of the European Communities (EC) (2013) Regulation (EU) No 1380/2013 of the European Parliament and of the Council. Brussels

Cox SP, Kronlund AR (2008) Practical stakeholder-driven harvest policies for groundfish fisheries in British Columbia, Canada. Fish Res 94:224–237

Davis A, Hanson JM, Watts H, MacPherson H (2004) Local ecological knowledge and marine fisheries research. The case of white hake (*Urophycis tenuis*) predation on juvenile American lobster (*Homarus americanus*). Can J Fish Aquat Sci 61(7):1191–1201

Delaney AE, McLay HA, van Densen WLT (2007) Influences of discourse on decision-making in EU fisheries management: the case of North Sea cod (*Gadus morhua*). ICES J Mar Sci 64:804–810

Dowling NA, Smith DC, Knuckey I et al (2008) Developing harvest strategies for low-value and data-poor fisheries: case studies from three Australian fisheries. Fish Res 94(3):380–390

Durham E, Baker H, Smith M et al (2014) The BiodivERsA stakeholder engagement handbook. BiodivERsA, Paris

Eriksson H, Adhuric DS, Adriantod L et al (2016) An ecosystem approach to small-scale fisheries through participatory diagnosis in four tropical countries. Global Environ Chang 36:56–66

Eveson JP, Hobday JA, Hartog JR et al (2015) Seasonal forecasting of tuna habitat in the Great Australian Bight. Fish Res 170:39–49

Field JG, Attwood CG, Jarre A et al (2013) Cooperation between scientists, NGOs and industry in support of sustainable fisheries: the South African hake *Merluccius* spp. trawl fishery experience. J Fish Biol 83:1019–1034

Fletcher W (2005) The application of qualitative risk assessment methodology to prioritize issues for fisheries management. ICES J Mar Sci 62:1576–1587

Freeman RE (2010) Strategic management: a stakeholder approach. Cambridge University Press, New York

Garza-Gil MD, Amigo-Dobano L, Surís-Regueiro JC et al (2015) Perceptions on incentives for compliance with regulation. The case of Spanish fishermen in the Atlantic. Fish Res 170:30–38

Goetz S, Read FL, Ferreira M et al (2015) Cetacean occurrence, habitat preferences and potential for cetacean–fishery interactions in Iberian Atlantic waters: results from cooperative research involving local stakeholders. Aquat Conserv Mar Freshw Ecosyst 25:138–154

Granados-Dieseldorff P, Heyman WD, Azuetad J (2013) History and co-management of the artisanal mutton snapper (*Lutjanus analis*) spawning aggregation fishery at Gladden Spit, Belize, 1950–2011. Fish Res 147:213–221

Gray S, Chan A, Clark D et al (2012) Modeling the integration of stakeholder knowledge in social-ecological decision-making: benefits and limitations to knowledge diversity. Ecol Model 229:88–96

Green AO, Hunton-Clarke L (2003) A typology of stakeholder participation for company environmental decision-making. Bus Strateg Environ 12:292–299

Greenacre MJ (1984) Theory and applications of correspondence analysis. Academic Press, London

Haapasaari P, Mäntyniemi S, Kuikka S (2013) Involving stakeholders in building integrated fisheries models using Bayesian methods. Environ Manag 51(6):1247–1261

Häberli R, Bill A, Grossenbacher-Mansuy W et al (2001) Synthesis. In: Thompson Klein J, Grossenbacher-Mansuy W, Häberli R et al (eds) Transdisciplinarity: joint problem solving among science, technology, and society. An effective way for managing complexity. Birkhäuser, Basel, pp 6–22

Hara MM, Rogerson J, de Goede J et al (2014) Fragmented participation in management of the fishery for small pelagic fish in South Africa – inclusion of small-rights holders is a complex matter. African J Mar Sci 36(2):185–196

Hartley T, Robertson R (2006) Stakeholder engagement, cooperative fisheries research, and democratic science: the case of the Northeast Consortium. Hum Ecol Rev 13:161–171

Hickey S, Kothari U (2009) Participation. International Encyclopedia of Human Geography, pp. 82–89

Higgs NT (1991) Practical and innovative uses of correspondence analysis. Statistician 40:183–194

Johannes RE, Neis B (2007) The value of anecdote. In: Haggan N, Neis B, Baird IG (eds) Fishers knowledge in fisheries science and management, Coastal Management Sourcebooks, 4. UNESCO, Paris, pp 41–58

Johnson TR, van Densen WLT (2007) Benefits and organization of cooperative research for fisheries management. ICES J Mar Sci 64:834–840

Kaiser M, Forsberg EM (2001) Assessing fisheries – using an ethical matrix in a participatory process. J Agric Environ Ethics 14:191–200

Kerr S, Johnson K, Side J et al (2006) Resolving conflicts in selecting a programme of fisheries science investigation. Fish Res 79:313–324

Kindon S (2008) Participatory action research. In: Hay I (ed) Qualitative research methods in human geography. Oxford University Press, Melbourne, pp 207–220

Kinds A, Sys K, Schotte L et al (2016) VALDUVIS: an innovative approach to assess the sustainability of fishing activities. Fish Res 182:158–171

Kittinger JN (2013) Participatory fishing community assessments to support coral reef fisheries co-management. Pacific Sci 67(3):1–43

Long RD, Charles A, Stephenson RL (2015) Key principles of marine ecosystem-based management. Mar Policy 57:53–60

Lorance P, Agnarsson S, Damalas D et al (2011) Using qualitative and quantitative stakeholder knowledge: examples from European deep-water fisheries. ICES J Mar Sci 68(8):1815–1824

Luyet V, Schlaepfer R, Parlange MB et al (2012) A framework to implement Stakeholder participation in environmental projects. J Environ Manag 111:213–219

Mabon L, Kawabe M (2015) Fisheries in Iwaki after the Fukushima Dai'ichi nuclear accident: lessons for coastal management under conditions of high uncertainty? Coast Manage 43(5):498–518

Mackinson S, Wilson DC, Galiay P et al (2011) Engaging stakeholders in fisheries and marine research. Mar Policy 35(1):18–24

Mahon R, Almerigi S, McConney P et al (2003) Participatory methodology used for sea urchin co-management in Barbados. Ocean Coast Manage 46:1–25

Mapstone BD, Little LR, Punt AE et al (2008) Management strategy evaluation for line fishing in the Great Barrier Reef: balancing conservation and multi-sector fishery objectives. Fish Res 94:315–329

Mayring P (1988) Die qualitative Wende. Arbeiten zur qualitativen Forschung. Augsburger Berichte zur Entwicklungspsychologie und Pädagogischen Psychologie, vol 32

Miller TJ, Blair JA, Ihde TF et al (2010) FishSmart: an innovative role for science in stakeholder-centered approaches to fisheries management. Fisheries 35(9):424–433

Mitchell R, Baba O (2006) Multi-sector resource allocation and integrated management of abalone stocks in Western Australia: review and discussion of management strategies. Fish Sci 72:278–288

Murphy RD, Scyphers SB, Grabowski JH (2015) Assessing fishers' support of striped bass management strategies. PLoS One 10:e0136412

Neis B, Schneider DC, Felt L et al (1999) Fisheries assessment: what can be learned from interviewing resource users. Can J Fish Aquat Sci 56:1949–1963

Newig J, Fritsch O (2009) The case survey method and applications in political Science: Paper Presented at the APSA 2009 Meeting, 3–6 September 2009, Toronto

NOAA (2015) Introduction to stakeholder participation. Office for coastal management. Available at: https://coast.noaa.gov/data/digitalcoast/pdf/stakeholder-participation.pdf. Accessed 18 July 2018

Okali C, Sumberg J, Farrington J (1994) Farmer participatory research. Intermediate Technology Publications, London

Olson M (1965) The logic of collective action. Harvard University Press, Cambridge

Pomeroy RS (2012) Managing overcapacity in small-scale fisheries in Southeast Asia. Mar Policy 36:520–527

Pretty J, Shah P (1994) Soil and water conservation in the twentieth century: a history of coercion and control. University of Reading Rural History Centre, 1

Pristupa AO, Lamers M, Amelung B (2016) Private informational governance in post-soviet waters: implications of the marine stewardship council certification in the Russian Barents Sea region. Fish Res 182:128–135

Punt AE, McGarvey R, Linnane A et al (2012) Evaluating empirical decision rules for southern rock lobster fisheries: a South Australian example. Fish Res 115–116:60–71

Reed MS (2008) Stakeholder participation for environmental management: a literature review. Biol Conserv 141:2417–2431

Repko AF, Newell WH, Szostak R (2011) Case studies in interdisciplinary research. Sage, Thousand Oaks

Rivera A, Gelcich S, García-Flórez L et al (2017) Trends, drivers, and lessons from a long-term data series of the Asturian (northern Spain) gooseneck barnacle territorial use rights system. Bull Mar Sci 93(1):35–51

Rowe G, Frewer L (2000) Public participation methods: a framework for evaluation in science. Sci Technol Hum Values 25:3–29

Sampedro P, Prellezo R, Garcia D et al (2017) To shape or to be shaped: engaging stakeholders in fishery management advice. ICES J Mar Sci 74(2):487–498

Smith DC, Smith ADM, Punt AE (2001) Approach and process for stock assessment in the South East fishery, Australia: a perspective. Mar Freshw Res 52:671–681

Soma K, Vatn A (2014) Representing the common goods – stakeholders vs. citizens. Land Use Policy 41:325–333

Stöhr C, Lundholm C, Crona B et al (2014) Stakeholder participation and sustainable fisheries: an integrative framework for assessing adaptive comanagement processes. Ecol Soc 19:14

Stratoudakis Y, Azevedo M, Farias I et al (2015) Benchmarking for data-limited fishery systems to support collaborative focus on solutions. Fish Res 171:122–129

Thiault L, Collin A, Chlous F et al (2017) Combining participatory and socioeconomic approaches to map fishing effort in small-scale fisheries. PLoS One 12(5):1–18

Thomas J (1993) Public involvement and governmental effectiveness: a decision-making model for public managers. Adm Soc 24:444–469

Thomas G (2011) A typology for the case study in social science following a review of definition, discourse, and structure. Qual Inq 17(6):511–521. https://doi.org/10.1177/1077800411409884

Tiller RG, Mork J, Liu Y et al (2015) To adapt or not adapt: assessing the adaptive capacity of artisanal fishers in the Trondheimsfjord (Norway) to jellyfish (*Periphylla periphylla*) bloom and purse seiners. Mar Coast Fish 7:260–273

Trimble M, Berkes F (2013) Participatory research towards co-management: lessons from artisanal fisheries in coastal Uruguay. J Environ Manag 128:768–778

Trimble M, Lazaro M (2014) Evaluation criteria for participatory research: insights from coastal Uruguay. Environ Manag 54:122–137

Voinov A, Bousquet F (2010) Modelling with stakeholders. Environ Model Softw 25:1268–1281

Watters GM, Hill SL, Hinke JT et al (2013) Decision-making for ecosystem-based management: evaluating options for a krill fishery with an ecosystem dynamics model. Ecol Appl 23(4):710–725

Whyte WF, Greenwood DJ, Lazes P (1989) Participatory action research: through practice to science in social research. Am Behav Sci 32:513–551

Williams AJ, Little LR, Begg GA (2011) Balancing indigenous and non-indigenous commercial objectives in a coral reef finfish fishery. ICES J Mar Sci 68(5):834–847

Wilson DG, Nielsen JR, Degnbol P (2003) The fisheries co-management experience: accomplishments, challenges and prospects. Kluwer Academic, Dordrecht

Yáñez E, Silva C, Barbieri MA et al (2014) Socio-ecological analysis of the artisanal fishing system on Easter Island. Lat Am J of Aquat Res 42(4):803–813

Yates KL, Schoeman DS (2015) Incorporating the spatial access priorities of fishers into strategic conservation planning and marine protected area design: reducing cost and increasing transparency. ICES J Mar Sci 72(2):587–594

Zengin M, Mihneva V, Duzgunes E (2018) Analysing the need of communication to improve Black Sea fisheries management policies in the Riparian Countries. Turk J Fish Aquat Sci 18:199–209

ETH Zürich (2018) Web of science core collection. Available at: www.library.ethz.ch/Ressourcen/Datenbanken/Web-of-Science-Core-Collection. Accessed 25 July 2018

3 Law and Policy Dimensions of Ocean Governance

Pradeep A. Singh and Mara Ort

Abstract

Human populations have relied on the oceans for centuries for food supply, transportation, security, oil and gas resources, and many other reasons. The growing prospects of the oceans, such as access to marine genetic resources and seabed minerals, to generate renewable energy and as a potentially enhanced carbon sink, are contributing to increased interests to control and exploit the seas. At the same time, human pressure on the oceans, both from land- and atmospheric-based sources and at sea, as well as from climate change, has led to unprecedented levels of stress on the oceans. The concept of ocean governance has developed as a response to this. This chapter explores ocean governance from the interdisciplinary perspectives of law and human geography.

We trace the development of ocean governance from first practices and legal concepts up to the emergence of contemporary ocean governance in recent decades and explore how it departs from traditional practices. Zonal and sectoral approaches, as well as their underlying norms, are discussed. We then take a more critical stance to shed light on the role neoliberalism plays in the forming of ocean governance and the effects this paradigm can have on governance outcomes. The cases of fisheries management and ocean grabbing illustrate some possible mechanisms and effects. In addition, the role of communities and indigenous people in ocean governance is discussed. Finally, the chapter addresses the shared or common concern surrounding the degradation of the marine environment, and the need for global and interdisciplinary cooperation in governing the oceans for mutual benefit.

Keywords

Marine environmental protection · Sustainable development · Common concern of humankind · Regime management and cooperation at sea · Governance tools · Local communities

3.1 Introduction

For centuries, the oceans have sustained life (both nonhuman and human), provided us with a host of nonliving resources, (renewable and nonrenewable), as well as fulfilled our transportation and recreational needs. Considering the vast importance of the oceans, it comes as no surprise that nations often demonstrate the desire to expand their sovereignty and gain control over the seas to exert control and dominance over it. In its centuries of development, the traditional law of the sea has witnessed the increased claim of state power and authority over marine spaces in a contested and competitive way. In recent decades, however, the global community started to realize the salient drawbacks from such a dominating approach – in particular, the depletion of living and nonliving marine resources as well as degradation of the marine environment. Thus, they declare in the United Nations Convention on the Law of the Sea 1982 (UNCLOS) that the "problems of ocean space are closely interrelated and need to be considered as a whole" (Preamble, UNCLOS 1982).

P. A. Singh (✉)
Faculty of Law, University of Bremen,
Bremen, Germany

INTERCOAST Research Training Group, Center for Marine Environmental Sciences (MARUM), Bremen, Germany
e-mail: pradeep@uni-bremen.de

M. Ort (✉)
artec Sustainability Research Center, University of Bremen,
Bremen, Germany

INTERCOAST Research Training Group, Center for Marine Environmental Sciences (MARUM), Bremen, Germany
e-mail: ort@uni-bremen.de

S. Jungblut et al. (eds.), *YOUMARES 9 - The Oceans: Our Research, Our Future*,
https://doi.org/10.1007/978-3-030-20389-4_3

The ocean governance discourse involves two presuppositions that renders it necessary to address ocean spaces and its uses in a holistic way. First, it acknowledges the existing and continuing division of ocean spaces (i.e., zonal/geographical aspect, in which maritime spaces are divided into zones that are subject to national jurisdiction and those beyond national jurisdiction) and the diverging regulations to control human activity in the oceans (i.e., sectoral/activity aspect, in which maritime activities are considered and regulated as a whole through regional or global regimes). Second, it strives to resolve governance problems arising from the zonal and sectoral aspects of the ocean spaces through a series of norms and tools implemented via increased cooperation and coodination across actors and institutions that operate within those spheres.

This chapter will discuss the concept of ocean governance – as the discourse is understood by lawyers and social scientists – and what that concept entails in terms of marine environmental protection. Specifically, this chapter traces the development of the concept in recent decades and explores how it departs from traditional practices where large parts of the seas were not subjected to national jurisdiction or any particular global or regional regime. In this context, it considers the zonal and sectoral aspects within the existing structure and its contribution toward achieving sustainable ocean governance. Next, this chapter will underscore a series of norms and tools that could bridge and harmonize the shortcomings arising from these realities. We then take a closer look at the role of neoliberalism as an influential hegemonic ideology in many parts of the world in the twentieth and twenty-first century that has left its traces in ocean governance, with examples from fisheries management and ocean grabbing. Furthermore, we touch upon the implications of ocean governance for local communities and indigenous people. Finally, this chapter ends by highlighting the importance of collaboration as an indispensable means to promote coherence in ocean governance.

3.2 Ocean Governance: A Conceptual Development

It is common knowledge that the oceans are fundamental to human life and our dependence on it continues to increase. This reliance necessitates international rules to govern the conduct of human activities in the oceans (Tanaka 2015a).

3.2.1 Historical Development

Before the twentieth century, the law of the sea was represented through a set of unwritten, customary rules that reflected the practice among states. The principle of freedom prevailed during this time, which meant that powerful maritime states were able to make the most use of the oceans and its resources (Bederman 2014). In the early days, the high seas were understood as being common property, open to all (Mansfield 2004; WBGU 2013). However, there existed two competing concepts about who had sovereignty over the oceans: *mare clausum* versus *mare liberum*, both dating back to the seventeenth century. *Mare clausum* was the Portuguese notion of giving the rights over the ocean to the coastal states. In contrast, the Dutch notion of *mare liberum* defended the understanding of the freedom of the ocean.[1] Both ideas were fueled by economic considerations: the Portuguese wanted to protect their role in trade, whereas *mare liberum* was developed by the Dutch jurist Hugo Grotius to establish the possibility of free trade for the Dutch East India Company (WBGU 2013). Over the centuries, many areas of the high oceans remained relatively open-access areas ("freedom of the seas"), whereas coastal waters were brought more and more under state control. The dawn of the twentieth century witnessed significant effort to codify the law of the sea (Harrison 2011). Alongside the emergence of new states, a greater balance between existing maritime powers and a shift in attention toward exploiting ocean resources for economic gains, the desire perceptibly swung toward establishing sovereignty or control over ocean resources while still maintaining the great freedom over navigation as previously enjoyed (since shipping remains an important interest for the maritime powers). Mainly from the 1950s to the 1980s, the primarily open-access system of the oceans became a system where coastal states gained sovereignty rights and controlled and exploited the marine resources (Steinberg 2001). This development was fueled by the wish to extend national claims over offshore resources, as the role of oceans for resource provision was rising (United Nations 2012; WBGU 2013). In 1954, the United States extended their jurisdiction over the whole continental shelf of their coast, in order to extract resources (United Nations 2012). Consequently, many other states followed this example and expanded their areas from traditionally 3 nautical miles to 200 nautical miles to control larger parts of the sea and the seabed to be able to claim fish stocks, mineral deposits, and oil. These expansions of territorial control were first only covered by customary law (United Nations 2012b) without much coherence; thus, the need for more binding international laws was apparent (WBGU 2013). After decades of negotiating, the UNCLOS (United Nations Convention on the Law of the Sea), otherwise known as the "constitution of the oceans" (Koh 1982), was signed in 1982 and codified these arrangements. Consequently, the 200 nm zone became known as exclusive economic zone (EEZ, where states have the rights over all living and nonliving resources), and states could expand their

[1] However, this is an understanding based specifically on a Western organization of space and society – actual history is much more complex than just these two categories (see Steinberg 2001).

sovereignty over parts of the continental shelf. The freedom of navigation, however, was mostly preserved.

3.2.2 The Concept of Ocean Governance

As the modern law of the sea is by itself not synonymous with ocean governance, it is useful to trace, within the design of the modern law of the sea, the development of "ocean governance" as a concept. In this context, Rothwell and Stephens (2016, pages 516–517) identify three significant phases in the development of the modern law of the sea that promotes ocean governance. During the first stage, starting from the early efforts to codify the law of the sea up until the adoption of four treaties in 1958, the focus was primarily on the desire of coastal states to exploit the resources of the ocean and not on crosscutting issues. Thus, although little heed was channelled toward advancing the protection of the marine environment, the seeds for the zonal and sectoral aspects of the law of the sea were established during this period. As the ocean spaces were divvied up into separate zones between nations, notably premised on the potential uses of these spaces, early steps were taken toward ascertaining who had jurisdiction over what. The second stage, from 1958 to 1982, demonstrates the increasing awareness of states of the need to regulate human activity based on sectors (i.e., shipping, fisheries, minerals, and so on) and the need to advance the protection of the marine environment from degradation and pollution arising from those activities. To this end, several instruments, including the 1972 Stockholm Declaration on the Human Environment, were adopted. These developments had a telling effect on the commencement of the third phase. In the year 1982, the UNCLOS was finally concluded after years of heated negotiations. This development withdrew the focus from sovereignty, jurisdictional rights, and freedoms for individual states and put emphasis on shared responsibilities for the protection and preservation of the marine environment (Freestone 2008). Thus, UNCLOS provided the foundation for integrated management and governance of the oceans, supplemented by diverse treaties and soft law instruments (Rothwell and Stephens 2016).

In other words, while simultaneously respecting traditional rights (i.e., freedom of navigation) and emerging rights (i.e., sovereignty and sovereign rights over resources in areas within jurisdiction), the third and current stage of the modern law of the sea reflects the interests of which the concept of ocean governance seeks to secure. It is crucial to further state that the concept of ocean governance is intertwined with parallel developments in international environmental law and human rights law (in so far as they related to the oceans). As such, contemporary notions such as sustainable development are interwoven into the wider concept of ocean governance. Concurrently, the proliferation of international organizations, intergovernmental bodies, and other actors since the mid-twentieth century also represents this movement toward asserting greater protection measures of the marine environment (Singh 2018).

Gradually, within the twentieth century, the concept of ocean governance evolved. However, even though "in use" for quite a while already, it is not clearly defined but remains as a rather fuzzy term (Future Ocean and International Ocean Institute 2015). In some fields (e.g., social and political sciences), ocean governance comprises all rules, laws, institutions, and political measures regarding the oceans (Mondré and Kuhn 2017). These can include formalized or informal processes, as well as top-down or bottom-up approaches. Additionally, research focuses on governance-related issues like security and political strategies of nation-states (see, e.g., Humrich 2017; Wirth 2017) or investigates how economic interests shape policies (see, e.g., Mansfield 2004; Ritchie 2014). From a law perspective, however, the concept is narrower and usually relates to regulating maritime activities with a particular focus on marine environmental protection (Singh and Jaeckel 2018). From the latter perspective, there is great interest in how ocean governance involves cooperation and coordination across zones and sectors to harmonize (insufficient or ineffective) laws and policies adopted within those respective realms to pursue sustainable environmental management goals that benefit present and future generations. Thus, the concept of ocean governance, from a law perspective, can usefully be seen as the collective attempt to govern the conduct human activity in the oceans in a sustainable and orderly manner with the overall objective of conserving and protecting the marine environment. Further, while topics of regime fragmentation (such as regulatory gaps and duplicity of efforts) continue to surface, there are instances where this may not result in negative consequences. Given that the discourse of fragmentation and its bearings on ocean governance is a matter that deserves specific attention on its own, it will not be the focus of discussion here.

As the UNCLOS plays a critical role in setting the scene for ocean governance, it would be necessary to examine to what extent it actually caters for this purpose. The UNCLOS provides the foundational basis for state action to protect the marine environment. In Part XII of UNCLOS, 46 provisions are specifically dedicated toward environmental protection, requiring states (or international organizations) to adopt and enforce measures to protect the marine environment from the harmful effects of human activities, including land- and atmospheric-based sources of pollution that impair or degrade the oceans. Harrison (2017) eloquently explains that through this, the UNCLOS provides the foundational basis for marine environmental protection by determining, among others, the jurisdictional mandate to take such measures, followed by the general principles, substantive rules, and procedural rules vis-à-vis human activities at sea and the protection of the marine environment.

3.2.3 Zonal and Sectoral Aspects to Ocean Governance

At this juncture, it is necessary to distinguish between the zonal and sectoral aspects in the modern law of the sea. As mentioned earlier, both these aspects are vital considerations in ocean governance. The zonal aspect simply refers to who possesses the mandate or the authority to prescribe regulations and enforce them (based on demarcated boundaries), whereas the sectoral aspect covers the type of activity sought to be regulated. As will be explained, jurisdiction in zones does not necessarily commensurate with sectors. This demonstrates that ocean governance is an intricate concept that requires addressing trans-boundary and cross-sectoral concerns.

In terms of zones, the manner in which the UNCLOS has carved out the oceans can generally be divided into two areas: areas within national jurisdiction and areas beyond national jurisdiction (Churchill and Lowe 1999). In the former category, UNCLOS recognizes several zones. In the territorial sea, which may extend up to 12 nautical miles from its baselines, coastal states enjoy sovereignty.[2] Save for restricted rights (such as the right of innocent passage for ships to navigate through) that must be accorded to other states,[3] the coastal state is free to exploit its living and nonliving resources, as well as to take measures to enforce its coastal security and environmental laws.[4] The contiguous zone follows the territorial sea, which may be declared for up to 24 nautical miles, in order to allow the coastal state to take measures specifically pertaining to custom, fiscal, and sanitary issues and enforce regulations in its territorial sea.[5] Apart from this, there exist another two other zones. Coastal states may declare an exclusive economic zone (EEZ), which may extend up to 200 nautical miles from its baselines, in which they have sovereign rights to exploit both living and nonliving resources for their own economic benefit.[6] However, in contrast to the territorial sea, the coastal state has limited noneconomic rights in the EEZ. As such, other states enjoy certain freedoms, such as rights to navigation and laying submarine cables, provided that these activities do not prejudice the rights of the coastal state to reap economic benefits from the resources therein.[7] Coastal states are also accorded sovereign rights to exploit and reap economic benefits from its continental shelf for up to 200 nautical miles and in some cases extending up to 350 nautical miles or beyond from its baselines.[8] Within the areas in which it exercises sovereignty, coastal states have absolute discretion to enact regulations to protect the marine environment. In areas where they enjoy sovereign rights, coastal states have wide-ranging jurisdiction to promulgate environmental laws to protect their economic interest.[9]

Additionally, many coastal states also play a crucial role in ocean governance through the exercise of sovereign control over their ports. The UNCLOS and various instruments created through regional and global regimes have strengthened the ability of port states to enhance marine environmental protection, particularly in compliance and enforcement of numerous international standards and regulations. Although the UNCLOS recognizes the rights of landlocked states and their interest to gain access to the oceans and its living resources (i.e., fisheries) through neighboring coastal states (which is largely a matter of the latter's discretion), landlocked states play a limited role in ocean governance and the protection of the marine environment. Their contribution is restricted chiefly through the exercise of jurisdiction over ships that fly their flags and participation in global or regional regimes.

In areas beyond national jurisdiction, the UNCLOS recognizes two distinct zones. The first zone is the high seas – here, in the water column that falls beyond the EEZ of any coastal state, all states enjoy numerous types of freedom, particularly with respect to fisheries (it will be recalled that all states enjoy certain freedoms, e.g., navigation, even within the EEZ of coastal states).[10] Pertinently, since it belongs to no one, no particular state has the jurisdiction to enact specific legislation, be it environmental or otherwise, in this zone. However, this does not mean that the entire high seas area is subject to lawlessness or anarchy. The flag state, i.e., the state to which the ship is registered, exercises jurisdiction over the ship flying its flag and the activity it carries out in the high seas.[11] Thus, ships are bound to the domestic legislation of their flag states, as well as other regulations to which that state decides to adopt at the international level.[12] The second zone created by the UNCLOS, the international seabed area, is simply known as the "Area".[13] It encompasses the seabed area that falls beyond the continental shelf area claimed by any coastal state and declares it as the "common heritage of mankind."[14] In the Area, the International Seabed Authority (ISA) is given the mandate, pursuant to UNCLOS,

[2]Articles 2–3, UNCLOS.

[3]Articles 17–19, UNCLOS.

[4]Article 21(1), UNCLOS.

[5]Article 33, UNCLOS.

[6]Articles 56–57, UNCLOS.

[7]Article 58, UNCLOS.

[8]Article 76, UNCLOS.

[9]Article 193, UNCLOS.

[10]Article 87, UNCLOS.

[11]Articles 91–92, 94, UNCLOS.

[12]Articles 116–117, 119, UNCLOS.

[13]Article 1(1)(1), UNCLOS.

[14]Article 136, UNCLOS.

to regulate the exploration and exploitation of nonliving resources, i.e., minerals, as well as to take necessary measures to protect the marine environment from the harmful effects of such activities.[15] As UNCLOS only prescribes a skeletal framework for the deep seabed mining regime, the ISA is specifically given the task to progressively develop detailed regulation.[16] Hitherto, three separate regulations for the exploration of mineral resources (specifically for polymetallic nodules, polymetallic sulfides, and cobalt-rich ferromanganese crusts) are in existence. Regulations for the exploitation of mineral resources are currently at an advanced draft stage (Brown 2018), and work to design a benefit-sharing mechanism is expected to commence in due course.

Moving from the zonal aspect to the sectoral aspect, the latter recognizes the fact that assigning jurisdiction to create environmental regulations premised on boundary lines is inadequate and needs to be complemented with laws that specifically apply to the activity in concern. As such, the modern law of the sea involves numerous sectoral regimes, e.g., fishing, shipping, mining, and recreational, subjecting each of them to separate sets of laws. The need for sectoral regulation becomes particularly obvious in trans-boundary scenarios. For example, shipping activities typically involve the crossing of multiple jurisdictions, including the high seas, and as such need to be regulated more delicately. Similarly, fishing and mining activities taking place in an adjacent area will inevitably affect, cumulatively, the marine environment of surrounding areas and impair the ability of others to meaningfully exercise and enjoy their rights (Markus and Singh 2016). As such, the sectoral aspect of the modern law of the sea has contributed to the proliferation of international organizations with carefully defined mandates to regulate particular activities (Harrison 2015). This includes, among others, the ISA to govern mineral mining activities in the Area, the International Maritime Organization (IMO) to regulate global shipping activities, the Fisheries and Agriculture Organization (FAO) to promote sustainable fisheries measures, and the International Whaling Commission (IWC) to control whaling activities. The functions and responsibilities of these international organizations are typically defined in international treaties signed by states, thereby giving them the mandate to create regulations or guidelines that would apply to their respective sectors or regimes (Harrison 2011). Regional regimes, particular in the case of fisheries, are also fast-growing, as can be seen in the number of Regional Fisheries Management Organizations (RFMOs) and Regional Seas Programs (RSPs) that are currently in existence (Singh 2018).

The above demonstrates that the jurisdictional mandate to regulate human activities at sea – specifically, to adopt necessary measures to protect the marine environment – is a critical consideration in the ocean governance discourse. This section also shows that, because a multitude of actors and institutions are involved in regulating human activities at sea – who often work independently and in isolation from each other – and compounded by the fact that the problems of the oceans are interconnected and interdependent, ensuring coherence in environmental protection is a complex and delicate subject. As such, there is a pressing need for a more effective response to the cumulative effects of human activities (and natural phenomena) to the marine environment (Scott 2015) – i.e., the *raison d'être* of the concept of ocean governance. To this end, reliance on a wide selection of norms and tools that are increasingly gaining recognition under international law, to assist in overall policy- and decision-making processes, goes a long way in contributing to the concept of ocean governance.

3.3 Norms and Tools Pertaining to Ocean Governance

The theme of ocean governance comprises of an array of norms and tools that serve to realize its objectives. At the outset, it would be useful to clarify what "norms" and "tools" mean in this context. By "norms," reference is generally made to legal principles, concepts, and doctrines that are recognized under international law (De Sadeleer 2002). Their normative values differ from one to another and depend, inter alia, on how they are perceived as binding (or merely as guiding) in practice by states and international organizations or as interpreted by international courts and tribunals (Tanaka 2015b). Under certain scenarios, non-binding norms may have a compelling effect (Winter 2018). However, as this chapter is not focused on ascertaining the exact status of these principles, doctrines, or maxims, the term "norms" is used generically here to refer to the handful of notions that apply to ocean governance. Concerning "tools," reference is made to the various measures and strategies designed with the aim to protect the marine environment from human activities at sea. In other words, "tools" here refer to the means that are adopted or deployed to attain the desired end. In the context of ocean governance, "tools" are both supplementary and complementary to "norms" and analogous in certain circumstances (Singh and Jaeckel 2018). In fact, the continuous and persistent practice of adopting certain measures or approaches, as a rule of thumb, may evince the normative value of a "tool." For instance, environmental impact assessments can be seen as a tool, in as much as it is an accepted norm or principle of international law.

[15] Articles 1(1)(2), 137, 145, 150, 153 and Annex III, UNCLOS.

[16] See Part XI and Annex III of the UNCLOS.

3.3.1 Norms Pertinent to Ocean Governance

Here, several norms that are relevant to the concept of ocean governance will be discussed. They comprise of the following: the "no-harm" or prevention of trans-boundary harm rule, environmental impact assessments, the precautionary approach, the ecosystem-based approach, sustainable development, and the polluter pays principle. One specific norm, the principle of cooperation, deserves special attention and will be considered later. This list is by no means exhaustive in the ocean governance discourse. Nevertheless, they have been selected for the present discussion because they either are grounded in the UNCLOS and other related agreements or have received significant acknowledgment and treatment by international and regional organizations. As such, they possess a certain degree of authoritative value and require specific consideration in decision-making processes as it has some bearing on the legitimacy of those outcomes.

The "no-harm" or prevention of trans-boundary harm refers to the obligation of states to ensure that activities within their jurisdiction or control do not cause significant harm to the environment of other states. In other words, states must ensure that the activities which they conduct, or which they permit, sanction or allow, are conducted in a manner that is environmentally sound, and that appropriate preventive measures are taken to ensure that the environmental harm arising from those activities are minimized – particularly if it is likely that such harm would be trans-boundary. It should be emphasized that this obligation is one of conduct and not of effect; it does not prohibit harmful effect in a strict or absolute sense, i.e., states are only liable for trans-boundary harm if they fail to exercise due diligence in controlling the activities (Birnie et al. 2009). The UNCLOS expressly affirms this obligation in the marine environment context.[17]

Environmental impact assessments (EIA) involve the practice of ascertaining the potential environmental effects of an activity prior to its conduct. This also extends to the requirement to monitor the environmental effects of the activity continuously throughout its life and upon cessation.[18] In this sense, while the practice of preparing an EIA is more appropriately seen as a tool to ensure adherence to the "no-harm" rule, its prevalence as a global practice in recent times has elevated its status to a customary rule of international law. Consequently, the absence of a proper EIA to ascertain the environmental harm of an activity, particularly where it is likely to cause trans-boundary harm, may be seen as a failure of the due diligence threshold and attract responsibility under international law (Birnie et al. 2009). Similarly, the UNCLOS also prescribes the requirement to conduct EIAs and continuously monitor the environmental effects of activities at sea.[19]

The precautionary approach, at its core, advocates for the exercise of caution in the face of uncertainty. More precisely, it excludes the reliance on lack of scientific certainty as a basis for inaction in adopting cost-effective measures to protect the environment. In other words, environmental threats should be effectively addressed even if the scientific certainty of the extent and effect of the threat is in question. A more contemporary interpretation goes further and requires that potentially harmful activities should be postponed until there is sufficient and reliable scientific verification that the potential environmental harm arising from its conduct can be convincingly averted or managed (Marr 2003; Jaeckel 2017).

Similar to the precautionary approach, the ecosystem-based approach also has its foundation rooted in science. Essentially, the ecosystem-based approach concentrates on "the protection of the ecosystem itself, including the structure, processes and functions of the community of biological organisms, and the interactions between them as well as between them and non-living components within a particular marine area" (Singh and Jaeckel 2018, page 624). Adopting this approach enables decision-makers to consider the cumulative impacts of human activities at sea on marine ecosystems in determining whether to permit a particular activity or not. The ecosystem-based approach closely resonates with the precautionary approach, and they both complement each other (Trouwborst 2009).

Sustainable development as a concept requires economic considerations to be weighed alongside social and environmental considerations in the context of development. It recognizes that economic progress involves certain detriment to the environment but is necessary for social development. At the same time, it is equally clear that under certain scenarios, the need to protect and preserve the environment may prevail over economic or social considerations (Tanaka 2015b). The sustainable development discourse has embraced the oceans, as can be seen in the Sustainable Development Agenda of 2030 and the Sustainable Development Goals of 2015 ("the SDGs") as adopted by the United Nations General Assembly in 2015.[20] SDG 14 (life below water) is specifically dedicated to the oceans and calls on states to "conserve and sustainably use the oceans, seas and marine resources for sustainable

[17] Article 194, UNCLOS.

[18] See the Pulp Mills on the River Uruguay (Uruguay v. Argentina) case, Judgment of 20 April 2010, International Court of Justice, available at: http://www.icj-cij.org/files/case-related/135/135-20100420-JUD-01-00-EN.pdf.

[19] Articles 204–206, UNCLOS.

[20] Resolution A/Res/70/1, Resolution adopted by the General Assembly on 25 September 2015, available at: http://www.un.org/ga/search/view_doc.asp?symbol=A/RES/70/1&Lang=E.

development."[21] It follows that, while the aspiration of states to develop progressively is valid and fundamental, marine development policies should not be carried out at the rapacious expense or in completed disregard of the marine environment. Also embedded in the proper implementation of this concept is the requirement to consider alternatives to prospective or ongoing projects, and even to exercise forbearance if the harm to the marine environment is demonstrably significant.

Finally, the polluter pays principle entails the requirement for the operator of the activity that causes harm to the environment to make the necessary reparations to address that damage. Effectively, this means that the operator of the activity is responsible to undertake restoration measures or to compensate for the harm caused to the environment. This should not be seen as consideration for a license to pollute or to deplete resources but rather as a basis to internalize those environmental costs (which would otherwise be treated as an externality) to ensure that profits are not made at the expense of the environment (Beder 2006).

3.3.2 Tools Pertinent to Ocean Governance

As mentioned earlier, some of the norms discussed above also take the form of tools, in so far as they are capable of serving as a means to advance the effective protection of the marine environment. In addition, several other measures or instruments may be adopted to give effect to the said norms. This includes environmental policy mechanisms and area-based management tools (ABMTs), including strategies such as maritime spatial planning (MSP) and marine protected areas (MPAs).

Environmental policy mechanisms involve the designing of internal policy mechanisms within states or regimes to ensure that marine environmental considerations are effectively taken into consideration in decision-making processes. This includes the setting up of scientific advisory bodies within the institutional setup as well as prescribing procedures to facilitate the flow of technical expertise into decision-making processes, as well as the use of incentives to promote technological advances or the adoption of higher environmental standards than that imposed by the regulator. Environmental strategies should be drawn up, providing for transparent decision-making processes, and include opportunities for stakeholder and public participation.

As the name suggests, ABMTs are area-specific and would suit as a perfect combination for the ecosystem-based approach. The MSP strategy, on the one hand, is premised on the fact that the stresses caused by human activities on the marine environment can be geographically mapped, thereby providing some useful indication on the vulnerabilities of that specific area to particular types of harm (Zacharias 2014). This enables decision-makers to permit activities in areas that are resilient enough to withstand (and recover from) the associated environmental harm and, conversely, to restrict activities in areas that are already subject to high levels of stress and would struggle to recover from the ensuing consequences of such activities (Markus et al. 2015). The creation of MPAs, on the other hand, is essentially a measure to protect certain areas (of environmental interests) from certain activities, such as shipping, fisheries, or mineral exploitation. This includes areas that have sensitive and fragile ecosystems, whether thriving or in despair, and breeding ground areas (Halpern et al. 2010). More importantly, ABMTs allow for cumulative effects (i.e. not only immediate impacts from the particular activity in concern, but the impacts arising from other related activities and natural causes) to be considered when determining the areas that require protection. However, due to regulatory gaps and the fragmentation nature of governance arising from the zonal and sectoral approach, states and competent organizations only have the mandate to designate MPAs in areas where they exercise jurisdiction or only concerning the activity it is tasked to regulate, respectively.

3.4 Critical Discussion of Ocean Governance

As reflected from the above, the oceans are not only diverse ecosystems; they are also economic and political spaces. In this fluid and boundary-less environment, diverse interests are articulated by nation-states, organizations, companies, local communities, and indigenous people. In this context, it seems obvious that ocean governance is a conflicting area (Vince 2014). To mediate these interests, various tools of ocean governance are applied. However, ocean governance does not develop in a neutral arena, as the previous section has shown. Rather, it is influenced by dominant rationalities. In the first part of this section, we focus on neoliberalism as a main normative foundation for many governance decisions in the last decades (Peet et al. 2011). The dominance of economic considerations over other issues, or the belief that the market mechanisms also help to solve environmental problems, however, leads to challenges and gaps in regard to communities' and indigenous peoples' needs and concerns. A closer look at these issues will form the second part of this section. Particularly, while we do not wish to undermine the importance of ocean governance as a concept, we would like to sensitize that there can also be problematic developments, which have to be investigated critically.

[21] Sustainable Development Goals (SDG 14), United Nations Department of Economic and Social Affairs (UN-DESA), Division for Sustainable Development Goals, accessible at: https://sustainabledevelopment.un.org/sdg14.

3.4.1 Neoliberalism and Property Rights as a Form of Ocean Governance

Even though ocean governance is often promoted by national agencies and international organizations under the premise of environmental protection, market forces also influence the setup of tools and norms. As observed earlier, in the last century, there was a move away from open access and freedom of the seas not only to national control of the oceans but also to the privatization of their resources (Mansfield 2004). Since the 1950s, a new political economy of the oceans emerged, with the question of the commons being a central one (Mansfield 2004). Mansfield identifies neoliberalism as a dominant form of ocean governance (Peet et al. (2011) observed this for trend in environmental governance in general). Using fisheries as an example, Mansfield outlines how a specific form of neoliberalism developed in ocean governance over the last 60 years. Often, there is a belief in neoliberalism as a helpful explanation as well as remedy in environmental governance: conservation is ineffective due to market failure, and market mechanisms will lead to more efficient environmental solutions in comparison to state-lead initiatives. Through the assignment of property rights and privatization, markets are created to govern the access to and use of ocean resources (Mansfield 2004). Neoclassical and neoliberal economists believe that economic efficiency will ultimately lead to social and environmental welfare as well (Mansfield 2004). In this tradition, conventional approaches like Gordon (1954) assume that market rationality is natural and that a lack of property rights leads to economic and environmental problems. In this argumentation, there are many parallels to the often repeated and much contested so-called tragedy of the commons by Hardin (1968), published 14 years later (for a contrary view see, e.g., Ostrom et al. 1999). Also, the German Advisory Council on Global Change adopts this notion and discusses how, without strong rules, a rationality of exploitation would prevail and free riders would enrich themselves at the cost of the community (WBGU 2013). Concerning fisheries specifically, Mansfield outlines Gordon's (1954) argumentation that "without property regimes that constrain individual behavior, people will overcapitalize and overuse resources because it is economically rational to do so" (Mansfield 2004, page 319). The problem of overfishing and overcapitalization is seen in the open-access regime and missing property rights. Commons are distinguished from open access in that they are managed through informal institutions, arrangements, etc. and therefore present a form of property rights. However, the underlying economic approach is not challenged, and critical voices on the economic backdrop and assumptions are not widely spread (Mansfield 2004). Critical research reviewed by Mansfield does not see the problem in the lack of property rights but in power relations established through colonialism, capitalism, and global markets (Mansfield 2004). Rather than applying economic models on ocean governance, it might be more helpful to investigate power relations that lead to the problems that have to be solved (i.e., overfishing, pollution, war). The following two case studies illustrate some of the problems that can arise from incorporating a neoliberal paradigm in ocean governance systems.

3.4.1.1 The Case of Fisheries and Quota Management Systems

Even though neoliberalism has the basic principle that the state should not interfere with the market, it still relies on the state to create and maintain property rights. These property rights are a form of enclosure (and therefore can enable primitive accumulation), which then allow privatization (Mansfield 2004). Regarding fisheries, market incentives are created as property is defined as the right to fish (Mansfield 2004). One mechanism is, for example, the individual transferable quota system (ITQ) to marketize the allocation of fish catch. Once publicly accessible, fishing grounds are now restricted to the use of a small group of companies and individual fisherpersons. Through privatization and commodification, a new market is created (Mansfield 2004). ITQs are "strong quasi-property rights" (Hersoug 2018, page 101). Hersoug assesses ITQ systems critically, using the example of Aotearoa New Zealand. In Aotearoa New Zealand, a Quota Management System (QMS) was introduced in 1984, during a neoliberal economic restructuring as a response to crowded and overfished inshore fisheries (Hersoug 2018). The market should regulate the situation. However, the system is still dependent on strong government regulations and is mainly managed in a top-down approach (Hersoug 2018). Government interventions are also important to cater for the interests of all stakeholders, as ITQs privilege just the groups that have a share. The research outlines how the introduction of QMS led to the exclusion of small-scale fishers, many of them being Maori (indigenous people of Aotearoa New Zealand) (Hersoug 2018). Even though Maori were included in the QMS process and got a share which was celebrated as a big success, the system failed in really enabling community participation, and no "trickle-down" effects, e.g., in terms of employment, could be observed (Hersoug 2018).

Hersoug is therefore quite critical of market-based solutions like the Quota Management System. He observes hardly any outcomes in terms of better stewardship for the environment: "ITQ solutions have not contributed to increased sustainability, neither in biological nor in social terms" (Hersoug 2018, page 109). Even though QMS and similar solutions do not solve all problems, they were recommended by most fisheries economists, as property rights were seen as an incentive for sustainable fisheries. Hersoug suggests that besides defining rights, also obligations should be formulated. Additionally, the marine environment should be framed as a public resource (Hersoug 2018).

3.4.1.2 The Case of Ocean Grabbing

When talking about property rights about ocean areas and resources, also the issue of ocean grabbing has to be mentioned. That ocean governance does not always only promote sustainable development but can also lead to (unintended) negative outcomes, e.g., for local communities, is discussed by several authors (Benjaminsen and Bryceson 2012; Bennett et al. 2015; Hill 2017). Bennett et al. (2015) refer to "ocean grabbing" as a process within which access and rights to marine resources and spaces are reallocated. This can happen through enclosure, appropriation, and disposession (Hill 2017). Ocean grabbing is described as "actions, policies, or initiatives that deprive small-scale fishers of resources, dispossess vulnerable populations of coastal lands, and/or undermine historical access to areas of the sea" (Bennett et al. 2015, page 61). Ocean grabbing can happen both intentionally and unintentionally and be conducted by private as well as state actors. Also, measures of ocean governance can lead to ocean grabbing. Zoning, the creation of MPAs or fisheries policies, can deprive local communities of resource use and access and lead to privatization or enclosure of marine resources and areas (Bennett et al. 2015) – at the cost of former custodians and in favor of more powerful actors (Hill 2017). In a case study in Malaysia, Hill (2017) found that through the establishment of a marine park, no-take zones were installed that divorce fisherpeople from their means of production. The protected areas now serve as "raw material" for capitalist production, to the benefit of state officials as well as commercial tourist operators. The local population was not included in the decision-making process prior to the establishment of the park. While the protection of our oceans still remains an essential goal, the social impacts of marine conservation have to be considered (Hill 2017).

3.4.2 Communities and Indigenous People and Ocean Governance

When dealing with laws, conventions, policy documents, and so on, it should not be forgotten that these decisions and rules have effects not only in an abstract way but can also affect the livelihoods of local communities and indigenous people in both positive and negative ways (Davies et al. 2018). This is especially problematic when governance tools are set up in a top-down manner and no real discussion is possible (Ritchie 2014).

Marine spatial planning and marine protected areas were discussed above as tools of ocean governance. In their research on marine spatial planning (MSP), Boucquey et al. (2016) analysed fisheries in the United States. They focus on the representation of communities in the setting up of the programs. Generally, Boucquey et al. (2016) found that management usually focuses on economic activities and does not take into account "the complex ways fishing communities are socially and emotionally integrated with marine species" (Boucquey et al. 2016, page 4). Rather, these interactions become simplified during the planning process. Communities are labelled and categorized, and this "data" is fed into the data system by "experts" such as planners or researchers. According to Boucquey et al., the narratives surrounding MSP and MPAs promote stabilization, categorization, and organization as a means to improve ocean management (Boucquey et al. 2016). This is underlined by the call for more scientific data and knowledge, which is understood to enable more efficient and effective governance (see, e.g., WBGU 2013). This will have the result that so-called expert knowledge shapes how oceanic actors are portrayed.

The movement toward MSP can also be understood as a proceeding neo-liberalization of nature, enforced through enclosure and exploitation (Boucquey et al. 2016). Ritchie (2014), for example, found in her research on MSP in the United Kingdom that there is a supremacy of economic development and the overarching goal to maximize economic growth. The mapping of oceans for particular users and uses during MSP processes sets up "the potential for ocean enclosures to privilege the most powerful actors" (Boucquey et al. 2016, page 2). These are seldom local communities or indigenous people, but rather large companies and nation-states (the ones with the biggest stake), a problem also perceived by Ritchie (2014).

Even though for the high seas outside of national jurisdictions, it might make sense to aim for overarching and general governance and even top-down solutions, local communities might get more out of small-scale, local, and bottom-up solutions, which are potentially also more sustainable. Stephenson et al. (2014, page 264) called for place-specific, local solutions rather than "top-down, one-size-fits-all policy solutions" that tend to "fail to achieve sustainable outcomes."

Participation and inclusion in decision-making is not only an issue concerning local communities in general but also indigenous peoples in specific. Indigenous peoples often have a deep cultural connection to the ocean. According to a report by the United Nations (United Nations Economic and Social Council 2016), they rely on oceans for food, health, economic activities, and cultural practices. Generally, many indigenous peoples do not distinguish so much between land and ocean; in their worldview all is one and connected (Nursey-Bray and Jacobson 2014; United Nations Economic and Social Council 2016). In addition, indigenous peoples often do not have access to traditional resources anymore; they were deprived of rights and access to resources during the colonial period and can also be affected by more recent governance processes (Stephenson et al. 2014). However, there is not much research on indigenous people involvement in MPA governance and management, as a literature review carried out by Ban and Frid (2018) found. They point to literature on ocean grabbing to show the contested nature of MPAs as places where conservation initiatives or the establishment of MSP or MPAs can deprive local communities of resources and/or undermine access to areas (Bennett et al. 2015). Traditional governance systems in postcolonial states were dominated by colonial forms of governance (Nursey-Bray and Jacobson 2014). In this regard, there is a "historical and ongoing collision between Western (and colonial) systems of law and governance and Indigenous modes of law and governance" (Nursey-Bray and Jacobson 2014, page 29).

In the UN study on the relationship between indigenous peoples and the Pacific Ocean, the authors state that generally indigenous people are not well included in decision-making for a like the UN organizations (United Nations Economic and Social Council 2016). Yet, the report finds that "[...] their [indigenous peoples of the Pacific] ability to meaningfully participate in decision-making on matters that will have a direct impact on oceans and their environments is limited" (United Nations Economic and Social Council 2016, page 3).

The situation in Aotearoa New Zealand provides another example. Customary fishery rights eroded during the colonial period (Bess and Rallapudi 2007). When a QMS was about to be established, many iwi (Maori tribes) objected to it due to concerns that even more rights would be alienated from them. Settlements (1992 Fisheries Deed of Settlement and Treaty of Waitangi (Fisheries Claims) Act 1992) found that Maori customary fishing rights had not extinguished (Bess and Rallapudi 2007). Specific legislation was passed to protect customary fishing rights (i.e., Taiapure areas, Mataitai reserves) (Bess and Rallapudi 2007). However, these are relatively small in size and number, and due to the lengthy and complicated establishment process, many Maori groups shy away from applying for them.

The examples in this section show that global trends, developments, and decisions have effects on local population, and these social and cultural impacts have to be taken into account in policy-making. In our opinion, especially the adoption of market mechanisms to ocean governance is problematic. Also, ocean governance is not independent from or free of hegemonic paradigms, norms, and political power structures. A thorough look is necessary to find out who is winning and who is losing from specific governance decisions and their complex effects. Taking that into account might lead to a more comprehensive analysis of ocean governance, as well as to more just decision-making processes and more sustainable outcomes.

3.5 Overcoming the Obstacle: Cooperation to Address a Common Concern and the Importance of Marine Scientific Research

As seen from the above, there are ample ways in which the oceans can be governed in a more orderly and sustainable manner. The adoption of ABMTs and other environmental strategies, for instance, can provide positive results over a short period (Hilborn and Ovando 2014). However, the primacy of the zonal and sectoral aspects to marine governance within the law of the sea sphere limits the potential of such measures, simply because states and international organizations are only empowered to act within their respective jurisdictions and mandates. To overcome this obstacle, two initiatives are relevant. First, there is a need for enhanced cooperation and coordination among states, between states and international/regional organizations, and among international/regional organizations. Industry and other private actors also need to feature in this setup (Singh 2018). Second, it is essential to view the protection of the marine environment as a shared concern or a "common concern of humankind," in which all subjects of international law have a legitimate interest in ensuring (Harrison 2017). Thus, any form of activity taking place in the ocean spaces of any jurisdiction is a matter of shared concern if it causes significant harm to the marine environment. Furthermore, this not only includes activities at sea but also extends to the pollution of the marine environment from land-based sources and from or through the atmosphere.[22]

There is some evidence to indicate that the first initiative is gaining traction. The proliferation of regional management regimes in different geographic areas with extraterritoriality jurisdiction illustrates that states are amenable to work together to resolve common concerns. Furthermore, the increased number in agreements to cooperate signed between international organizations signal that global and regional institutions are beginning to discover ways to harmonize the segregated regulation of activities within their respective sectors and to share their expertise. The involvement of industry within the setup of these organizations is also encouraging, as it ensures the wide acceptance of the results or outcomes of these internal processes. However, it should be noted that the proliferation of regional agreements and arrangements per se does not necessarily result in actual improvements, particularly where levels of compliance is low, enforcement is lax, or where non-parties and outliers to the arrangements remain active in the geographic area and undermine conservation efforts.

The second initiative is particularly troublesome due to the complications arising from sovereignty and sovereign rights. As states are entitled to explore and utilize the ocean

[22] Articles 207 and 212, UNCLOS.

resources within their jurisdiction pursuant to their domestic policies and interests, it is difficult for other states to intervene. Nevertheless, if the obligation to protect and preserve the marine environment is seen as an erga omnes obligation, this might encourage concerned states to commence dispute resolution proceedings against a delinquent state (Harrison 2017). As the dispute resolution process under the UNCLOS is mandatory, member states arguably have a recourse to compel other states to be more attentive toward marine environmental protection. It would be interesting to see how this plays out in the future, particularly in the light of the common but differentiated responsibility precept, which accepts that developing states are not legally required to adopt at par environmental measures as developed states.

3.6 Conclusion

This chapter describes ocean governance as a concept that embraces the sustainable and inclusive management of the oceans across the various maritime zones and sectoral divisions. It demonstrates the applicability of various norms and tools to advance different objectives in ocean governance and particularly promulgates the need to enhance cooperation across the multiple actors and institutions that participate in activities at sea. It should also have become clear that ocean governance is by no means a narrowly defined area, both within and across disciplines. This underlines the importance of interdisciplinary cooperation among different strands of research, e.g., law and human geography, as well as social sciences and natural sciences in general, to foster mutual understanding. In writing this chapter, we became even more conscious of the different approaches and perspectives adopted by our respective disciplines when dealing with the issue of ocean governance. Therefore, the goal of the chapter was not to circumscribe a common position but rather to emphasize topics in ocean governance that are important to us from our specific standpoints. We think that interdisciplinary approaches in research could help to improve analysis as well as the development of ocean governance, as perspectives from other disciplines might shed new light on many topics.

Obviously, different disciplines ask different questions. An integration of different forms of knowledge could therefore help to develop a more holistic perspective and address concerns that are not presently considered by any single discipline. For instance, although we did not touch upon it in this chapter, future research could benefit from a stronger inclusion of economic perspectives in the ocean governance discourse. Even if the application of economic mechanisms in fields like conservation is often disputed, it is necessary to critically consider the perspective of economists to have a truly comprehensive picture. Additionally, issues of participation and transparency in decision-making and agenda-setting (and the underlying power relations) at all levels also deserve attention from future research. After all, the success of realizing the Sustainable Development Goals, particularly SDG 14, largely hinges on successful local and regional level initiatives.

By viewing the protection of the marine environment as a common concern of humankind, in which we need representation of various disciplines, this chapter calls for greater coordination and coherence in a large shared area with competing claims and conflicting uses. Such an approach is critical in ensuring the rational and sustainable use of the oceans and its resources to meet current and future demands as well as interests.

Appendix

This article is related to the YOUMARES 9 conference session no. 4: "Law and Policy Dimensions of Ocean Governance." The original Call for Abstracts and the abstracts of the presentations within this session can be found in the Appendix "Conference Sessions and Abstracts", Chapter "3 Law and Policy Dimensions of Ocean Governance", of this book.

References

Ban NC, Frid A (2018) Indigenous peoples' rights and marine protected areas. Mar Policy 87:180–185

Beder S (2006) Environmental principles and policies: an interdisciplinary introduction. Earthscan, London

Bederman DJ (2014) The Sea. In: Fassbender B, Peters A, Peter S (eds) The Oxford handbook of the history of international law. Oxford University Press, Oxford, pp 359–382

Benjaminsen TA, Bryceson I (2012) Conservation, green/blue grabbing and accumulation by dispossession in Tanzania. J Peasant Stud 39(2):335–355

Bennett NJ, Govan H, Satterfield T (2015) Ocean grabbing. Mar Policy 57:61–68

Bess R, Rallapudi R (2007) Spatial conflicts in New Zealand fisheries: the rights of fishers and protection of the marine environment. Mar Policy 31(6):719–729

Birnie P, Boyle A, Redgwell C (2009) International law and the environment, 3rd edn. Oxford University Press, Oxford

Boucquey N, Fairbanks L, St. Martin K et al (2016) The ontological politics of marine spatial planning: assembling the ocean and shaping the capacities of 'Community' and 'Environment'. Geoforum 75:1–11

Brown C (2018) Mining at 2,500 fathoms under the sea: thoughts on an emerging regulatory framework. Ocean Sci 53(2):287–300

Churchill RR, Lowe AV (1999) The Law of the Sea, 3rd edn. Manchester University Press, Manchester

Davies K, Murchie AA, Kerr V et al (2018) The evolution of marine protected area planning in Aotearoa New Zealand: reflections on participation and process. Mar Policy 93:113–127

De Sadeleer N (2002) Environmental principles: from political slogans to legal rules. Oxford University Press, Oxford

Freestone D (2008) Principles applicable to modern oceans governance (editorial). Int J Mar Coastal Law 23(3):385–391
Future Ocean and International Ocean Institute (2015) Sustainable use of our oceans – making ideas work. Maribus, Hamburg
Gordon HS (1954) The economic theory of a common-property resource: the fishery. J Political Econ 62(2):124–142
Halpern BS, Lester SE, McLeod KL (2010) Placing marine protected areas onto the ecosystem-based management seascape. Proc Natl Acad Sci U S A 107(43):18312–18317
Hardin G (1968) The tragedy of the commons. Science 162(3859):1243–1248
Harrison J (2011) Making the law of the sea: A study in the development of international law. Cambridge University Press, Cambridge
Harrison J (2015) Actors and institutions for the protection of the marine environment. In: Rayfuse R (ed) Research handbook on international marine environmental law. Edward Elgar Publishing, Cheltenham, pp 57–80
Harrison J (2017) Saving the oceans through law: the international legal framework for the protection of the marine environment. Oxford University Press, Oxford
Hersoug B (2018) "After all these years" – New Zealand's quota management system at the crossroads. Mar Policy 92:101–110
Hilborn R, Ovando D (2014) Reflections on the success of traditional fisheries management. ICES J Mar Sci 71(5):1040–1046
Hill A (2017) Blue grabbing. Reviewing marine conservation in Redang Island Marine Park, Malaysia. Geoforum 79:97–100
Humrich C (2017) The Arctic Council at twenty: cooperation between governments in the global Arctic. In: Conde Pérez E, Iglesias Sanchez S (eds) Global challenges in the Arctic Region: sovereignty, environment, and geopolitical balance. Routledge, London, pp 149–169
Jaeckel A (2017) The International Seabed Authority and the precautionary principle: balancing deep seabed mineral mining and marine environmental protection. Brill, Leiden
Koh TTB (1982) A constitution for the Oceans, remarks by Ambassador Koh of Singapore, President of the Third United Nations Conference on the Law of the Sea, available at http://www.un.org/Depts/los/convention_agreements/texts/koh_english.pdf. Accessed 8 June 2018
Mansfield B (2004) Neoliberalism in the oceans: "rationalization," property rights, and the commons question. Geoforum 35(3):313–326
Markus T, Singh P (2016) Promoting consistency in the deep seabed: addressing regulatory dimensions in designing the international seabed authority's exploitation code. Rev Euro Comp Int Environ Law 25:347–362
Markus T, Huhn K, Bischof K (2015) The quest for seafloor integrity. Nat Geosci 8:164–165
Marr S (2003) The precautionary principle and the law of the sea: modern decision-making in international law. Brill, Leiden
Mondré A, Kuhn A (2017) Ocean Governance. In: Bundeszentrale für Politische Bildung (ed) Meere und Ozeane. Aus Politik und Zeitgeschehen 51–52:4–9
Nursey-Bray M, Jacobson C (2014) 'Which way'?: the contribution of Indigenous marine governance. Aust J Maritime Ocean Aff 6(1):27–40
Ostrom E, Burger J, Field CB et al (1999) Revisiting the commons: local lessons, global challenges. Science 284(5412):278–282
Peet R, Robbins P, Watts M (2011) Global nature. In: Peet R, Robbins P, Watts M (eds) Global political ecology. Routledge, London, pp 1–48
Ritchie H (2014) Understanding emerging discourses of Marine Spatial Planning in the UK. Land Use Policy 38:666–675
Rothwell D, Stephens T (2016) The international law of the Sea, 2nd edn. Hart Publishing, Oxford
Scott K (2015) Integrated oceans management: a new frontier in marine environmental protection. In: Rothwell D, Elferink AGO, Scott KN et al (eds) The Oxford handbook of the law of the Sea. Oxford University Press, Oxford, pp 463–490
Singh P (2018) Institutional framework for marine environmental governance. In: Solomon M, Markus T (eds) Handbook on marine environment protection. Springer, Heidelberg, pp 563–584
Singh P, Jaeckel A (2018) Future prospects of marine environmental governance. In: Solomon M, Markus T (eds) Handbook on marine environment protection. Springer, Heidelberg, pp 621–636
Steinberg PE (2001) The social construction of the ocean. Cambridge University Press, Cambridge
Stephenson J, Berkes F, Turner N et al (2014) Biocultural conservation of marine ecosystems: examples from New Zealand and Canada. Indian J Tradit Know 13(2):257–265
Tanaka Y (2015a) The international law of the sea, 2nd edn. Cambridge University Press, Cambridge
Tanaka Y (2015b) Principles of international marine environmental law. In: Rayfuse R (ed) Research handbook on international marine environmental law. Edward Elgar Publishing, Cheltenham, pp 31–56
Trouwborst A (2009) The precautionary principle and the ecosystem approach in international law: differences, similarities and linkages. Rev Euro Commun Int Environ Law 18(1):26–37
UNCLOS – United Nations Convention on the Law of the Sea (1982) 1833 U.N.T.S. 397
United Nations (2012b2012a) The United Nations Convention on the Law of the Sea (A historical perspective). Available at: http://www.un.org/Depts/los/convention_agreements/convention_historical_perspective.htm. Accessed 8 June 2018
United Nations Economic and Social Council (2016) Study on the relationship between indigenous peoples and the Pacific Ocean. E/C.19/2016/3
Vince J (2014) Introduction: oceans governance: where have we been and where are we going? Aust J Maritime Ocean Aff 6(1):3–4
WBGU (2013) World in transition: governing the marine heritage. Wissenschaftlicher Beirat Globale Umweltveränderungen/German Advisory Council on Global Change, Berlin
Winter G (2018) International principles of marine environmental protection. In: Solomon M, Markus T (eds) Handbook on marine environment protection. Springer, Heidelberg, pp 585–606
Wirth C (2017) Danger, development, and legitimacy in East Asian Maritime Politics: securing the seas securing the state. Routledge, London
Zacharias M (2014) An introduction to governance and international law of the oceans. Routledge, London

Status of Marine Biodiversity in the Anthropocene

4

Thomas Luypaert, James G. Hagan, Morgan L. McCarthy, and Meenakshi Poti

Abstract

Marine biodiversity plays an important role in providing the ecosystem functions and services which humans derive from the oceans. Understanding how this provisioning will change in the Anthropocene requires knowledge of marine biodiversity patterns. Here, we review the status of marine species diversity in space and time. Knowledge of marine species diversity is incomplete, with only 11% of species described. Nonetheless, marine biodiversity is clearly under threat, and habitat destruction and overexploitation represent the greatest stressors to threatened marine species. Claims that global marine extinction rates are within historical backgrounds and lower than on land may be inaccurate, as fewer marine species have been assessed for extinction risk. Moreover, extinctions and declines in species richness at any spatial scale may inadequately reflect marine diversity trends. Marine local-scale species richness is seemingly not decreasing through time. There are, however, directional changes in species composition at local scales. These changes are non-random, as resident species are replaced by invaders, which may reduce diversity in space and, thus, reduce regional species richness. However, this is infrequently quantified in the marine realm and the consequences for ecosystem processes are poorly known. While these changes in species richness are important, they do not fully reflect humanity's impact on the marine realm. Marine population declines are ubiquitous, yet the consequences for the functioning of marine ecosystems are understudied. We call for increased emphasis on trends in

T. Luypaert (✉)
Faculty of Sciences and Bioengineering Sciences, Department of Biology, Vrije Universiteit Brussel (VUB), Brussels, Belgium

Faculty of Sciences, Department of Biology of Organisms, Université libre de Bruxelles (ULB), Brussels, Belgium

Faculty of Maths, Physics and Natural Sciences, Department of Biology, Università degli Studi di Firenze (UniFi), Sesto Fiorentino, Italy

J. G. Hagan
Faculty of Sciences and Bioengineering Sciences, Department of Biology, Vrije Universiteit Brussel (VUB), Brussels, Belgium

Faculty of Sciences, Department of Biology of Organisms, Université libre de Bruxelles (ULB), Brussels, Belgium

M. L. McCarthy
Faculty of Sciences and Bioengineering Sciences, Department of Biology, Vrije Universiteit Brussel (VUB), Brussels, Belgium

Faculty of Sciences, Department of Biology of Organisms, Université libre de Bruxelles (ULB), Brussels, Belgium

Faculty of Maths, Physics and Natural Sciences, Department of Biology, Università degli Studi di Firenze (UniFi), Sesto Fiorentino, Italy

School of Biological Sciences, The University of Queensland (UQ), St. Lucia, Queensland, Australia

M. Poti
Faculty of Sciences and Bioengineering Sciences, Department of Biology, Vrije Universiteit Brussel (VUB), Brussels, Belgium

Faculty of Sciences, Department of Biology of Organisms, Université libre de Bruxelles (ULB), Brussels, Belgium

Faculty of Maths, Physics and Natural Sciences, Department of Biology, Università degli Studi di Firenze (UniFi), Sesto Fiorentino, Italy

School of Marine and Environmental Sciences, University of Malaysia Terengganu (UMT), Terengganu, Kuala Terengganu, Malaysia

Sea Turtle Research Unit (SEATRU), Institute of Oceanography and Environment, Universiti Malaysia Terengganu (UMT), Kuala Terengganu, Malaysia

S. Jungblut et al. (eds.), *YOUMARES 9 - The Oceans: Our Research, Our Future*,
https://doi.org/10.1007/978-3-030-20389-4_4

abundance, population sizes and biomass of marine species to fully characterize the pervasiveness of anthropogenic impacts on the marine realm.

Keywords

Extinction · Defaunation · Biotic homogenization · Conservation · Anthropogenic stressors · Ecosystem function · Ecosystem service · Marine threats

4.1 Introduction

Humans have impacted 87–90% of the global ocean surface (Halpern et al. 2015; Jones et al. 2018). Marine fish abundance has declined by 38% compared to levels in 1970 (Hutchings et al. 2010). The area of certain coastal marine habitats, like seagrass beds and mangroves, has been depleted by over two-thirds (Lotze et al. 2006). Anthropogenic activities have increased atmospheric carbon dioxide (CO_2) concentrations by over 40% relative to pre-industrial levels (Caldeira and Wickett 2003), reducing the global ocean pH by 0.1 unit in the past century (Orr et al. 2005). The scale of these human impacts has triggered the naming of a new geological epoch, the Anthropocene, where humans dominate biogeochemical cycles, net primary production, and alter patterns of biodiversity in space and time (Crutzen 2002; Haberl et al. 2007). These impacts have led to a loss of global biodiversity which is comparable to previous global-scale **mass extinction events** (see Box 4.1; Barnosky et al. 2011; Ceballos et al. 2015), suggesting that we are in a biodiversity crisis.

Addressing this human-induced biodiversity crisis is one of the most challenging tasks of our time (Steffen et al. 2015). The conservation of biodiversity is an internationally accepted goal, as exemplified by the United Nations Convention on Biological Diversity (CBD) Strategic Plan for Biodiversity 2011–2020, which aims to "take effective and urgent action to halt the loss of biodiversity in order to ensure that by 2020 ecosystems are resilient and continue to provide essential services" (CBD COP Decision X/2 2010). Furthermore, international policies with marine biodiversity targets are being adapted at national and regional levels (Lawler et al. 2006), as is reflected by the recent addition of Sustainable Development Goal 14: "Conserve and sustainably use the ocean, seas and marine resources" (United Nations General Assembly 2015). These goals focus on a multi-level concept of biodiversity: biological variation in all its manifestations from genes, populations, species, and functional traits to ecosystems (Gaston 2010).

The wide acceptance of these national and international policy goals reflects a growing understanding of the importance of biodiversity to humans (Costanza et al. 1997; Palumbi et al. 2009; Barbier et al. 2011). Marine ecosystems provide a variety of benefits to humanity. These **ecosystem services** (see Box 4.1) include the supply of over a billion people with their primary protein source, widespread waste processing, shoreline protection, recreational opportunities, and many others (MEA 2005; Worm et al. 2006; Palumbi et al. 2009). However, as marine ecosystems are degraded and biodiversity declines, the ability of ecosystems to deliver these ecosystem services is being lost (MEA 2005). Moreover, pressures on the marine realm may increase if the terrestrial environment continues to be degraded and humankind becomes increasingly reliant on marine ecosystem services (McCauley et al. 2015).

The provisioning of ecosystem services is strongly coupled to ecosystem functioning (Cardinale et al. 2012; Harrison et al. 2014). **Ecosystem functions** (see Box 4.1) refer broadly to processes that control fluxes of energy and material in the biosphere and include nutrient cycling, primary productivity, and several others. There is now unequivocal evidence that high biodiversity within biological communities enhances ecosystem functioning in a variety of marine ecosystems and taxonomic groups (reviewed in Palumbi et al. 2009; Cardinale et al. 2012; Gamfeldt et al. 2015). These studies mostly focus on local-scale diversity and on productivity as the ecosystem function. Nonetheless, similar patterns have been found at larger spatial scales (Worm et al. 2006) and for various other ecosystem functions (Lefcheck et al. 2015). Marine biodiversity is also linked to **ecosystem stability** through time (see Box 4.1; McCann 2000; Schindler et al. 2015). High fish diversity, for example, is associated with fisheries catch stability through time (Greene et al. 2010). Thus, diversity is not only linked to ecosystem functioning in marine communities but also improves the stability of these functions through time.

Several mechanisms have been proposed to explain the link between biodiversity, and ecosystem functioning and stability. For example, species occupying different niches is known as complementarity. "Complementarity effects" appear prevalent in marine ecosystems as niche partitioning is commonly documented (Ross 1986; Garrison and Link 2000). Diverse assemblages of herbivorous fishes on coral reefs, for instance, are more efficient at grazing macroalgae due to different feeding strategies (Burkepile and Hay 2008). Furthermore, diverse communities are also more likely to contain well-adapted species, or species which disproportionately affect ecosystem function (Palumbi et al. 2009). These "selection effects" can be important, as single species can have strong effects on ecosystem functions in marine systems (Paine 1969; Mills et al. 1993; Gamfeldt et al. 2015). The evidence for complementarity and selection effects suggests that the mechanisms driving the effect of diversity on ecosystem functioning are linked to **functional diversity** (see Box 4.1)—or the range of functions that organisms perform in an ecological community (Petchey and Gaston 2006). However, several species from the same functional group can be important for maintaining the stability of an ecosystem over time as species often respond differentially to temporal environmen-

Box 4.1 Glossary

Ecosystem-related terms

Ecosystem engineers: Species that regulate resource availability to other species by altering biotic or abiotic materials (Jones et al. 1994).

Ecosystem services: The benefits that humans derive from ecosystems (Cardinale et al. 2012).

Ecosystem function: Any ecological process that affects the fluxes of organic matter, nutrients and energy (Cardinale et al. 2012).

Ecosystem stability: The variability in an ecosystem property (e.g., biomass, species richness, primary productivity) through time (Schindler et al. 2015). Stable communities are those with low variability in ecosystem properties through time.

Keystone species: Species that affect communities and ecosystems more strongly than predicted from their abundance (Power et al. 1996).

Extinction and defaunation terms

Background extinction: The rate of natural species extinction through time prior to the influence of humans (Pimm et al. 1995).

Extinctions per million species years (E MSY^{-1}): The metric used to measure background extinction rates. This metric measures the number of extinctions per million species years. For example, if there are 20 million species and an extinction rate of 1 E MSY^{-1}, 20 species would be predicted to go extinct each year.

Mass extinction event: Substantial biodiversity losses that are global in extent, taxonomically broad, and rapid relative to the average duration of existence for the taxa involved (Jablonski 1986). The 'Big Five' are quantitatively predicted to have approximately 75% of species having gone extinct (Jablonski 1994; Barnosky et al. 2011).

Global extinction: When it is beyond reasonable doubt that the last individual of a taxon has died (IUCN 2017).

Local extinction: The loss of a species from a local community or in part of its geographical range.

Biotic homogenization: The process where native species (losers) are replaced by more widespread, human-adapted species (winners). These are often non-native species (McKinney and Lockwood 1999).

- **Loser species (Losers):** Species that are declining due to human activities in the Anthropocene. These are typically geographically restricted, native species with sensitive requirements, which cannot tolerate human activities.
- **Winner species (Winners):** Species that not only resist geographic range decline in the Anthropocene, but also expand their ranges. These are typically widespread generalists which thrive in human-altered environments.

Defaunation: The human induced loss of species and populations of animals, along with declines in abundance or biomass (Young et al. 2016).

Ecological extinction: Occurs when species are extant but their abundance is too low to perform their functional roles in the ecological community and ecosystem (McCauley et al. 2015).

Diversity terms

Local diversity: The number of species in an area at a local spatial scale.

Spatial beta diversity: The change in species composition across space, i.e., the difference in species composition between two local communities. It is frequently quantified as the change in species composition with distance (McGill et al. 2015).

Temporal beta diversity (turnover): The change in species composition through time, i.e., the difference in species composition in a local community at two points in time.

Functional diversity: The range of functions performed by organisms in an ecological community or an ecosystem (Petchey and Gaston 2006).

Table 4.1 The five most recent global estimates of the number of marine eukaryotic species based on a variety of methods (see Supplementary Material A for an overview of the different methods). The proportion of described species is calculated using 239,634 described species based on the number of species in the World Register of Marine Species database (WoRMS Editorial Board 2018, 15th April 2018; http://www.marinespecies.org/) and the mean or midpoint of the different estimates of total species numbers

References	Estimation method	Estimated species	Described species (%)
Mora et al. (2011)	Higher taxonomic extrapolation	2,210,000 ± 182,000[a]	11
Costello et al. (2010)	Expert opinion	1,000,000 – 1,400,000[b]	20
Appeltans et al. (2012)	Expert opinion	704,000 – 972,000[b]	29
	Past discovery rate extrapolation	540,000 ± 220,000[a]	44
Costello et al. (2012)	Expert opinion	295,000 – 321,000[b]	78

[a]means ± standard errors; [b]ranges

tal fluctuations (McCann 2000; Schindler et al. 2015). As such, having several species with similar functional roles can maintain ecosystem functioning through time.

Clearly, understanding how ecosystem function and service delivery will change through time requires knowledge of biodiversity and its temporal dynamics. As a contribution to this goal, we review the status of marine eukaryotic species (referred to as species hereafter) diversity in space and time in the Anthropocene. First, many biodiversity targets, such as those in the CBD, reflect known species. Thus, we briefly review the knowledge of global marine species diversity. Secondly, the current "biodiversity crisis" suggests there is a rapid loss of species diversity. As such, we examine trends in the loss of marine species diversity at multiple spatial and temporal scales, and its potential implications for ecosystem functioning and stability over time. Doing so, we summarize the main threats to marine biodiversity. Thirdly, we argue that focusing on losses of species diversity inadequately reflects the changes currently occurring in the marine realm. Therefore, we call for a greater emphasis on trends in abundance, population sizes, and biomass through time to better characterize the pervasiveness of anthropogenic impacts on the marine realm. Finally, we develop the greatest threats that are negatively affecting marine species in more detail, and discuss what measures are being employed to mitigate these risks. Our review focuses on species richness as a measure of biodiversity, as this is the most commonly used metric in conservation biology and ecology (Gaston 2010).

4.2 Global Marine Species Diversity

How many species inhabit the oceans and how many do we know about? The five most recent estimates of extant marine species using several indirect methods range from ~300,000 to 2.2 million, a full order of magnitude (Table 4.1; Supplementary Material A). Of these estimated species, approximately 240,000 have been described (WoRMS Editorial Board 2018 as of 15 April 2018). This suggests that between 11 and 78% of all marine species have been discovered and described, and reveals high levels of uncertainty in our knowledge of global marine biodiversity. This uncertainty is particularly prevalent in under sampled marine habitats such as the deep sea (Bouchet et al. 2002; Webb et al. 2010), and in taxonomic groups with few taxonomic experts (Costello et al. 2010; Griffiths 2010). Moreover, many marine species are small (<2 mm) and cryptic, and have only begun to be discovered with new molecular methods (de Vargas et al. 2015; Leray and Knowlton 2016). Thus, there is considerable uncertainty in estimates of how many marine species there are, along with potentially low levels of taxonomic knowledge about these species.

Incomplete knowledge of marine species diversity has serious implications for marine conservation. First, targeted conservation efforts that adequately represent local and regional biodiversity can only be effectively implemented with adequate biodiversity data (Balmford and Gaston 1999; Brito 2010). Without knowing how many species there are, there is no way to know whether we are effectively conserving marine diversity in different regions and marine groups. Indeed, decisions made using incomplete taxonomic knowledge have been shown to inadequately represent biodiversity if species are continually discovered and described (Bini et al. 2006; Grand et al. 2007). Second, a species must be described to be assessed by the International Union for Conservation of Nature (IUCN 2017; Box 4.2). Currently, the coverage of marine species on the IUCN Red List is severely incomplete (Fig. 4.1a). This is important as the IUCN is the global authority for assigning conservation statuses and assessing species extinction risks (IUCN 2017). If conservation efforts are to adequately represent marine biodiversity and understand the conservation status of marine species, it may be key to improve estimates of global marine diversity and marine taxonomic knowledge.

4.3 Trends in Marine Biodiversity Loss and its Consequences

The simplest, and perhaps most cited, type of biodiversity loss is **global extinction** (see Box 4.1). Global extinctions occur when the last individual of a species has died. While

Box 4.2 Spotlight on the IUCN Red List of Threatened Species

The IUCN Red List constitutes the most comprehensive database of the global conservation status of species (IUCN 2018). It contains a range of information related to species population size and trends, geographic distribution, habitat and ecology, threats, and conservation recommendations. The assignment of a conservation status is based on five categories: (i) population trends; (ii) geographic range size trends; (iii) population size; (iv) restricted geographic distribution; and (v) probabilistic analyses of extinction risk. The magnitude of these five categories places a species into a conservation category. The categories used in this review are detailed below. Species that are vulnerable, endangered, or critically endangered are considered threatened with extinction.

- **Data deficient (DD):** There is insufficient population and distribution data to assess the extinction risk of the taxon.
- **Least concern (LC):** Based on the available data, the taxon does not meet the criteria to be NT, VU, EN, or CR.
- **Near threatened (NT):** Based on the available data, the taxon does not meet the criteria to be VU, EN, or CR but is expected to qualify for one of the threatened categories in the future
- **Vulnerable (VU):** The available data suggest that the taxon faces a high risk of extinction in the wild.
- **Endangered (EN):** The available data suggest that the taxon faces a very high risk of extinction in the wild.
- **Critically endangered (CR):** The available data suggest that the taxon faces an extremely high risk of extinction in the wild.
- **Extinct in the wild (EW):** The taxon is known only to survive in cultivation, in captivity, or as a naturalized population well outside the past range.
- **Extinct (EX):** There is no reasonable doubt that the last individual of a taxon has died.

we may remain unaware of the current global extinction risk of many marine species, the fossil record can provide an insight into how long species typically survived before anthropogenic stressors became widespread. The oldest known horseshoe crab fossil dates back 445 million years (Rudkin et al. 2008). Known as a "living fossil," the horseshoe crab is part of a small group of organisms which have survived millions of years of Earth's history. The persistence of the horseshoe crab through time is, however, an exception. Over 90% of marine organisms are estimated to have gone extinct since the beginning of life (Harnik et al. 2012). This rate of natural species extinction through time is referred to as **background extinction** (see Box 4.1) and specifically refers to extinction rates prior to the influence of humans (Pimm et al. 1995). Widely accepted historic estimates range between 0.01 and 2 **extinctions per million species-years** (E MSY^{-1}; see Box 4.1, Table 4.2). The background extinction rate has been thoroughly investigated to contextualize the anthropogenic influence on accelerating species extinctions. It also provides a benchmark to differentiate intervals of exceptional species losses or "mass extinction events" from the prevailing conditions in Earth's history.

Current extinction rates of 100 E MSY^{-1} are at least 10–1000 times higher than the background rates, suggesting that we have entered the sixth mass extinction event in Earth's history (Pimm et al. 1995, 2014; Barnosky et al. 2011). Increases in atmospheric CO_2 and ocean acidification measured for the current proposed sixth mass extinction have been associated with three of the five previous mass extinctions (Kappel 2005; Harnik et al. 2012). It is, however, the first time that these changes are anthropogenically driven. The impacts of these anthropogenic stressors are observed directly in vertebrate extinctions. Over 468 more vertebrates have gone extinct since 1900 AD than would have been expected under the conservative background extinction rate of 2 E MSY^{-1} (Ceballos et al. 2015). This increased extinction rate argues in favor of anthropogenic causes for the current mass extinction.

The high extinction rates currently observed are largely due to the loss of terrestrial species (Barnosky et al. 2011; Ceballos et al. 2015). In contrast, estimated extinction rates of marine species have been closer to background extinction rates (Table 4.2; Harnik et al. 2012). Records from the IUCN indicate that only 19 global marine extinctions have been recorded in the last ca. 500 years (IUCN 2017). Conversely, 514 species from the terrestrial realm have gone extinct in the same timeframe (McCauley et al. 2015). The asymmetry in the number of extinctions between the marine and terrestrial environments has led to suggestions that **defaunation** (see Box 4.1), or human-induced loss of animals, has been less severe in the marine realm and may be just beginning (e.g., McCauley et al. 2015). Indeed, although marine resources have been harvested by humans for over 40,000 years, the intense exploitation of marine life is a relatively recent phenomenon compared to the terrestrial realm, only commencing in the last few hundred years (O'Connor et al. 2011; McCauley et al. 2015). Additionally, multiple biological factors have been proposed to explain the observed low extinction rates of marine species. Background extinction rates of marine species have decreased with time, which suggests that extinction susceptible clades have already gone extinct (Harnik et al. 2012). Moreover, marine species tend

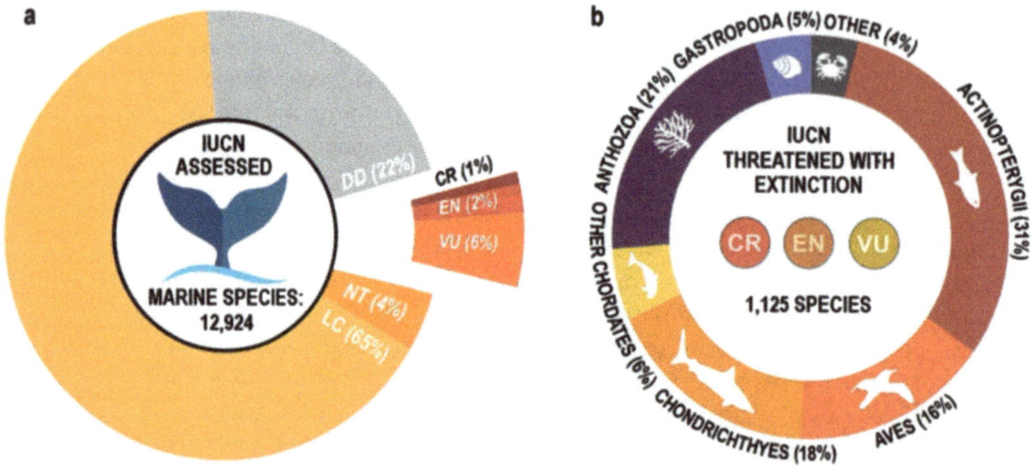

Fig. 4.1 (**a**) The conservation status of 12,924 IUCN-assessed marine species. Only 10,142 are in a category other than data deficient, an inadequate level of assessment (IUCN 2018). This represents 4.2% of currently described marine species (WoRMS Editorial Board 2018, 15 April 2018), and only between 0.5 and 3.3% of the estimated total marine species. Of the assessed species, 11% are either critically endangered (CE), endangered (EN), or vulnerable (VU), and are thus considered threatened with extinction. (**b**) Taxonomic distribution of marine species threatened with extinction, as defined by being classified CR, EN, or VU. Of these assessments, 64% are species from well-described groups (Webb and Mindel 2015). Furthermore, 64 of 88 recognized marine groups (groups as per Appeltans et al. 2012) had no IUCN assessed species (Webb and Mindel 2015). The "other chordates" category includes Mammalia, Myxini, Reptilia, and Sarcopterygii, and "other" category includes Polychaeta, Insecta, Malacostraca, Maxillopoda, Merostomata, Hydrozoa, and Holothuroidea. These were grouped due to low species availability. Data extracted from IUCN (2018)

Table 4.2 Various estimates of the background extinction rate using different methods and taxonomic groups. These rates are 10-1000 times lower than estimates of current extinction rates (Pimm et al. 1995, 2014; Barnosky et al. 2011). Extinction rates are measured in E MSY^{-1}, or the number of extinctions (E) per million species-years (MSY) (Pimm et al. 1995, 2014; Box 4.1)

References	E MSY^{-1} estimate	Taxonomic/animal group	Method
Pimm et al. (1995)	0.1–1	Marine invertebrates	Fossil record
Barnosky et al. (2011)	1.8	Vertebrates	Fossil record
Harnik et al. (2012)	0.01–0.27	Cetacea, marine Carnivora, Echinoidea, Chondrichthyes, Scleractinia, Gastropoda, Crustacea, Osteichthyes, Bivalvia, Bryozoa, Brachiopoda	Fossil and historical records
De Vos et al. (2015)	0.1	Chordata, Mollusca, Magnoliophyta, and Arthropoda	Molecular phylogenies

to have traits which are associated with a reduced extinction risk, such as larger geographic range sizes and lower rates of endemism (Gaston 1998; Sandel et al. 2011). Finally, on average, marine species can disperse further than terrestrial species and, thus, may respond better to environmental changes (Kinlan and Gaines 2003).

Nonetheless, the suggestion that extinction rates in the marine realm are lower than in the terrestrial realm is, however, not fully supported for several reasons. First, current marine and terrestrial extinctions may not be directly comparable. Species must be discovered and described taxonomically before they can be given a conservation status or shown to be extinct (Costello et al. 2013). As such, if the marine and terrestrial realms have variable rates of species discovery, taxonomic description, or conservation assessment, the detectability of their respective extinctions could differ (Pimm et al. 2014). The oceans are vast and largely inhospitable to humans. This makes marine systems particularly challenging to study. As a result, large parts of the marine realm are undersampled, there is a lack of taxonomic expertise for certain groups and, thus, marine extinction rates may have been underestimated (see Section 2). Although estimates of global species richness vary widely, there seems to be little difference in the proportion of marine and terrestrial species that have been described (ca. 30%, Appeltans et al. 2012; Pimm et al. 2014). However, of all taxonomically described species, the IUCN has assessed proportionally fewer marine than terrestrial species (3% vs. 4%), which might partially explain the discrepancy in extinction rates (Webb and Mindel 2015). This premise is supported by Webb and Mindel (2015), who found that there is no difference in extinction rate between marine and terrestrial species for marine groups that have been well-described and well-assessed. Thus, studies suggesting that marine extinction rates are low compared to terrestrial rates may be inaccurate.

Secondly, although global extinctions are important evolutionary events, and tools for highlighting conservation issues (Rosenzweig 1995; Butchart et al. 2010), they inadequately reflect the consequences of anthropogenic impacts on the ocean. There is little doubt that extinction rates are increasing at local and global scales (McKinney and Lockwood 1999; Butchart et al. 2010; Barnosky et al. 2011). However, local-scale time series (between 3 and 50 years) covering a variety of taxa and marine habitats around the world show no net loss in species richness through time (Dornelas et al. 2014; Elahi et al. 2015; Hillebrand et al. 2018). These time series analyses have recently received substantial criticism as synthetized by Cardinale et al. (2018). For instance, they are not spatially representative and do not include time series from areas which have experienced severe anthropogenic impacts such as habitat loss. Habitat loss and reduced habitat complexity due to anthropogenic disturbance may indeed reduce local-scale species richness and abundance in a variety of marine ecosystems (Airoldi et al. 2008; Claudet and Fraschetti 2010; Sala et al. 2012). Still, despite the limitations of the time series and the contradictory evidence from local-scale comparisons, the time series analyses illustrate an important point: changes in biodiversity at global scales may not always be evident in the properties of local-scale communities.

Focusing on extinctions and reductions in species richness can also hide changes in community composition (McGill et al. 2015; Hillebrand et al. 2018). There is increasing evidence that the destruction and modification of structurally complex habitats is leading to the rapid disappearance of the diverse communities they harbor at local, regional, and global scales (Lotze et al. 2006; Airoldi et al. 2008). For example, kelp forests and other complex macroalgal habitats have declined notably around the world, most likely due to overfishing and reduced water quality (Steneck et al. 2002). Similarly, 85% of oyster reefs, once an important structural and ecological component of estuaries throughout the world, have been lost (Beck et al. 2011). Globally, coastal habitats are among the most affected habitats (Halpern et al. 2015). This loss in habitat complexity can lead to geographic range contraction or **local extinction** (see Box 4.1) of associated resident species (**losers**; see Box 4.1), and the range expansion of a smaller number of cosmopolitan "invaders" with an affinity for human-altered environments (**winners**; see Box 4.1; McKinney and Lockwood 1999; Olden et al. 2004; Young et al. 2016). Thus, although local species richness may remain stable or even increase, there may be substantial changes in species composition through time, or **temporal species turnover** (see Box 4.1; Dornelas et al. 2014; Hillebrand et al. 2018). If resident species (losers) are going extinct locally and being replaced by these cosmopolitan invaders (winners), it is likely that adjacent communities in space will become more similar. This will result in declines in **spatial beta diversity**: the change in species composition across space (see Box 4.1). The consequence of this would be lower regional species richness, or large-scale **biotic homogenization** (see Box 4.1; McKinney and Lockwood 1999; Sax and Gaines 2003).

Biotic homogenization is not a new phenomenon, however, the process might have accelerated for several reasons (McKinney and Lockwood 1999; Olden et al. 2004). First, the breakdown of biogeographic barriers following the emergence of global trade in the nineteenth century (O'Rourke and Williamson 2002) has led to the widespread introduction of species out of their native range (Molnar et al. 2008; Hulme 2009; Carlton et al. 2017). As local habitat disturbance typically creates unoccupied niches, invasion by exotics is facilitated and, thus, local endemic species can be replaced with widespread species (Bando 2006; Altman and Whitlatch 2007; McGill et al. 2015). Although only a small fraction of non-native species successfully disperse and invade new habitats, the ecological and economic impacts are often significant (Molnar et al. 2008; Geburzi and McCarthy 2018). Secondly, similar types of habitat destruction or modification across space are leading to large-scale reductions in habitat diversity (McGill et al. 2015). For example, trawling activities have been reported on 75% of the global continental shelf area, which has considerably homogenized benthic habitats in space (Kaiser et al. 2002; Thrush et al. 2006). Moreover, vast dead zones emerge annually following the runoff of excessive nutrients and sediment from land, leading to eutrophication of coastal areas, increased algal blooms, and finally hypoxic aquatic conditions (Crain et al. 2009). This type of pollution can severely affect the growth, metabolism, and mortality of marine species (Gray 2002), and lead to large-scale homogenization of marine habitats and associated biological communities (Thrush et al. 2006). Finally, the rising temperatures associated with climate change, which is considered one of the most serious emerging threats to marine species and ecosystems (Harley et al. 2006; Rosenzweig et al. 2008; Pacifici et al. 2015), is causing species' range expansions and contractions, consequently altering spatial diversity patterns (Harley et al. 2006; Sorte et al. 2010). On average, marine organisms have expanded their distribution by approximately 70 km per decade in response to climate change, mostly in a poleward direction (Poloczanska et al. 2016). Polar species, which are unable to shift their range further poleward, are likely to be replaced by species expanding from temperate regions, leading to a reduction in both regional and global diversity.

Currently, there is insufficient evidence to quantify the extent of biotic homogenization in marine communities and whether there are trends through time (Airoldi et al. 2008;

McGill et al. 2015). There are, however, some examples of anthropogenic disturbances reducing spatial beta diversity in certain marine ecosystems. For instance, increased sedimentation has been shown to reduce spatial beta diversity between vertical and horizontal substrates in subtidal algal and invertebrate communities (Balata et al. 2007a, b). Furthermore, there is evidence that loss of habitat through bottom trawling does reduce spatial beta diversity, thus, reducing regional species richness (Thrush et al. 2006). Moreover, a recent analysis demonstrated declines in the spatial beta diversity of marine groundfish communities in the past 30 years, which is thought to be linked to recent ocean warming (Magurran et al. 2015). However, to our knowledge, this is the only explicit quantification of trends in spatial beta diversity through time in the marine realm. Moreover, the relative roles of species introductions, habitat loss and modification, and range shifts due to climate change are poorly known. Thus, understanding how biodiversity at broader scales is changing represents an important future challenge in marine species conservation.

The non-random distribution of winners and losers among taxonomic and functional groups is likely to worsen and intensify biotic homogenization. Certain ecological and life history traits influence the vulnerability of species to extinction (Roberts and Hawkins 1999; Dulvy et al. 2003; Reynolds et al. 2005; Purcell et al. 2014). For example, rare, large, highly specialized species with small geographic ranges are more likely to experience range contractions or local extinctions under human pressure (Dulvy et al. 2003). Conversely, smaller generalist species with a widespread geographic range and traits which promote transport and establishment in new environments tend to respond better to these pressures (McKinney and Lockwood 1999). Traits favoring either extinction or range expansion in the Anthropocene tend to be phylogenetically nested within certain groups of closely related species on the tree of life (McKinney 1997). Consequently, some taxonomic groups are more vulnerable to decline and extinction threats (see Fig. 4.1b, Lockwood et al. 2002). The unique morphological and behavioral adaptations within these groups are thus vulnerable to loss, especially if taxa are species-poor (McKinney and Lockwood 1999). Similarly, the winners of the Anthropocene tend to be clustered within certain taxa, further contributing to global biotic homogenization and losses in regional species diversity.

The implications of a loss in **local diversity** (see Box 4.1) for ecosystem functions and services are well-studied (Cardinale et al. 2012; Gamfeldt et al. 2015; Lefcheck et al. 2015), but the consequences of reductions in beta diversity at various spatial and temporal scales remain poorly understood. Studies on biotic homogenization usually describe the increased similarity in species composition between communities, driven by the replacement of many specialized species with few widespread generalist invaders (McKinney and Lockwood 1999; Olden et al. 2004). However, the presence of complementarity and selection effects on ecosystem functions suggests that the consequences of biotic homogenization on ecosystem functioning is best studied in terms of the diversity and composition of functional groups in the community (Olden et al. 2004; Palumbi et al. 2009). The spatial redistribution of taxonomic groups due to biotic homogenization may also alter the composition and variation in the functional groups of communities across marine habitats. The consequences of these changes across space for marine ecosystem functioning are currently not well-known.

The loss of specialists and replacement by generalist species or functional groups may negatively affect ecosystem functioning at multiple spatial scales. There is a trade-off between a species' ability to use a variety of resources and the efficiency by which each of these resources is used (Clavel et al. 2011). At local scales, specialists are more efficient at using few specific resources when the environment is stable (Futuyama and Moreno 1988; Colles et al. 2009). For example, specialist coral reef fish grow faster than generalists in a few habitats, but the growth rate of generalists is more consistent across a range of habitats (Caley and Munday 2003). Thus, the replacement of specialized species with generalists will lead to reduced ecosystem functioning on a local scale (Clavel et al. 2011; Cardinale et al. 2012). On a broader spatial scale, specialist species replace each other along environmental gradients, with each species optimally utilizing the resources in their specific environment (Rosenzweig 1995). For example, in marine systems, monocultures of well-adapted species had better ecosystem functioning than diverse communities on a local scale, even though diverse communities outperformed the average monoculture (Gamfeldt et al. 2015). This suggests that locally adapted specialist species are important for ecosystem functioning. As a result, the decrease in species turnover along environmental gradients following biotic homogenization may reduce the prevalence of locally adapted species, which in turn may cause a reduction in ecosystem functioning on a broader scale. In addition, functional homogenization between communities will likely reduce the range of species-specific responses to environmental change (Olden et al. 2004). Ecological communities will become increasingly synchronized when facing disturbance, reducing the potential for landscape or regional buffering of environmental change, and finally reducing the stability of ecosystem functions (Olden et al. 2004; Olden 2006). Thus, even though generalist species are more resilient to environmental change on a local scale, at broader spatial scales, the reduced number of specialists may negatively affect the stability of the system (Clavel et al. 2011). Improving our understanding of the functional consequences of biotic homogenization is, thus, key to understanding how current biodiversity trends will impact ecosystem functioning and the delivery of associated ecosystem services.

4.4 Looking Beyond Extinctions: Population Declines in the Marine Realm

Understanding extinctions and how biodiversity is changing through time and space is an important aspect of marine conservation. However, extinctions and declines in species richness do not fully reflect the extent of humanity's impact on the marine realm (McCauley et al. 2015). At the base of marine biological communities are populations of interacting species. In the marine realm, population declines are ubiquitous and often severe (Jackson et al. 2001). Therefore, to understand human impacts on marine biodiversity, and its consequences for ecosystem processes, it is crucial to understand the associated changes in species' population dynamics.

Between 1970 and 2012, the average size of 5,829 populations of 1,234 species of marine vertebrates has declined by 49% (WWF 2015). Overexploitation is the major cause of these declines, both through direct mortality of target species and multiple collateral effects on non-target species (Crain et al. 2009; WWF 2015). Harvested fish populations are routinely depleted by 50–70% (Hilborn et al. 2003), and losses of up to 90% are common (Myers and Worm 2005). These population declines are, however, not confined to vertebrates. The commercial exploitation of the white abalone in California and Mexico led to a reduction in population size to 0.1% of estimated pre-overexploitation levels (Hobday et al. 2000). Similarly, the increased demand for sea cucumbers as luxury food or traditional medicine in the last three decades has led to severe population declines, with 69% of sea cucumber fisheries now considered overexploited (Anderson et al. 2010). Thus, intensive harvesting of marine resources in recent times has caused widespread declines in several targeted marine species populations.

Overexploitation of targeted species can also affect populations of other marine species indirectly through bycatch, injury-induced mortality, or altered species interactions following population declines of target species (Crain et al. 2009). For example, global estimates suggest that as much as 39.5 million metric tons of fish may be caught as bycatch each year (Davies et al. 2009). However, fisheries bycatch is not restricted to other fish or invertebrates typically caught in industrial fisheries. Bycatch affects a variety of taxa such as seabirds, sea turtles, and marine mammals and has led to population declines of several well-known species including the Pacific leatherback turtle (*Dermochelys coriacea*), the Amsterdam albatross (*Diomedea amsterdamensis*), and the vaquita (*Phocoena sinus*) (Lewison et al. 2014). In addition, interactions with fishing vessels and other boats constitute an important cause of injury-induced mortality, especially for coastal air-breathing marine fauna such as marine mammals and reptiles (Shimada et al. 2017). In Moreton Bay (Queensland, Australia), for instance, the most commonly known causes of mortality for dugongs (*Dugong dugon*) were vessel strikes, trauma, and netting, and sea turtles are similarly impacted by boat strikes and discarded fishing gear (Lanyon 2019).

Notwithstanding the negative effects of bycatch and modern fishing methods, altered species interactions following the population depletion of targeted species, and the associated changes in food web structure, are also believed to have considerable collateral effects on non-targeted marine populations (Österblom et al. 2007; Estes et al. 2011). Reductions in the population size of a trophic level caused by overexploitation can induce correlated changes in the abundance of interacting species (Frank et al. 2005; Johannesen et al. 2012). For example, in temperate rocky reefs, reduced predation pressure due to overexploitation of herbivore predators can cause significant increases in the abundance and size of herbivorous invertebrates like sea urchins and chitons (Fig. 4.2, Ling et al. 2015). The associated increase in herbivory decreases macroalgal abundance (e.g., kelp) (Steneck et al. 2013). While these trophic cascades—or indirect effects on the population abundance of species at two/more trophic links from the primary one (Frank et al. 2005)—are frequently studied in relation to apex predator depletion and the associated loss of top-down control (Box 4.3), they are not restricted to high trophic levels. In the Barents Sea, changes in the abundance of the middle trophic level capelin (*Mallotus villosus*) can cause abundance changes in both high and low trophic levels (Johannesen et al. 2012). Thus, in marine systems, population sizes are generally coupled to populations of interacting species. As a result, any anthropogenically driven population reduction can indirectly affect population dynamics across trophic levels and over whole food webs.

In certain cases, changes in the abundance of different trophic groups can cause significant food web reorganizations (Baum and Worm 2009; Estes et al. 2011). Food web reorganizations may manifest as sudden shifts to a new ecosystem state, frequently termed regime shifts (Sguotti and Cormon 2018). Regime shifts are important because alternate ecosystem states can be maintained by internal feedback mechanisms which prevent a system from reverting back to a previous state (Scheffer et al. 2001). In the temperate rocky reef example, some areas that were previously dominated by macroalgae have shifted to a barren state dominated by sea urchin and crusting algae as a result of reduced predation pressure (Steneck et al. 2013). A combination of feedbacks including high juvenile sea urchin abundance, juvenile facilitation by adult sea urchins, and sea urchin-induced mortality of juvenile kelp maintain the system in this new state (Ling

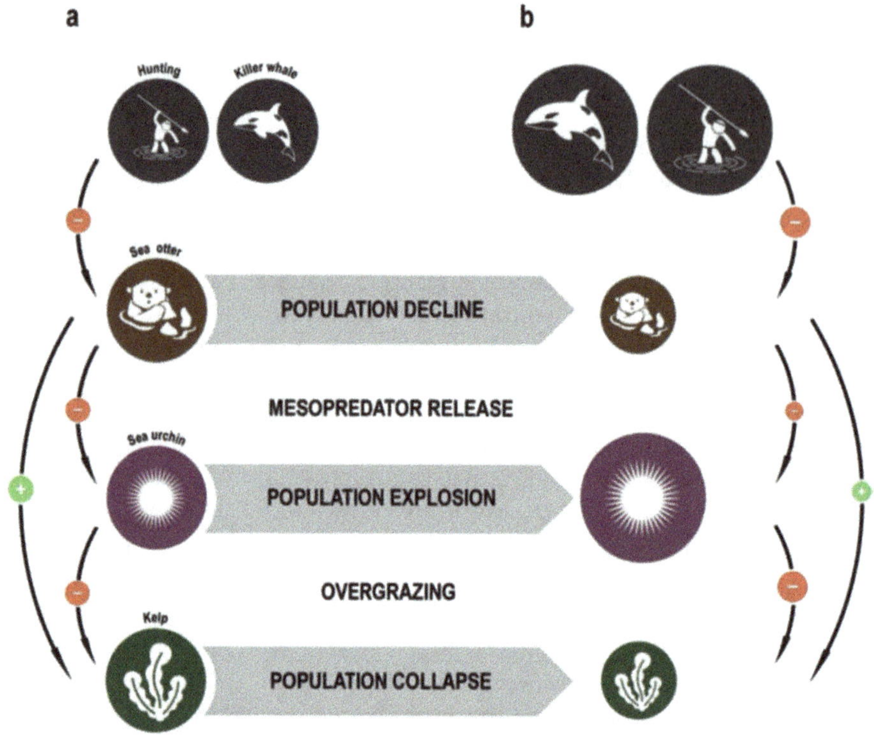

Fig. 4.2 Schematic representation of the changes in abundance between trophic groups in a temperate rocky reef ecosystem. (**a**) Interactions at equilibrium. (**b**) Trophic cascade following disturbance. In this case, the otter is the dominant predator and the macroalgae are kelp. Arrows with positive (green, +) signs indicate positive effects on abundance while those with negative (red, -) indicate negative effects on abundance. The size of the bubbles represents the change in population abundance and associated altered interaction strength following disturbance. Based on Estes et al. (1998)

et al. 2015). These feedbacks mean that even increasing predator abundance to historical highs may not shift the ecosystem back to a macroalgae-dominated state (Sguotti and Cormon 2018). Ecosystems can be resilient to such regime shifts if abundance declines in one species can be compensated by other species in a similar trophic level (Mumby et al. 2007). As a result, regime shifts may only occur long-after overexploitation has begun when all species of a trophic level have suffered declines in abundance and compensation is no longer possible (Jackson et al. 2001).

Overexploitation is, however, not the only anthropogenic stressor negatively affecting marine populations. Habitat modification and destruction, being the reduction in habitat quality or complete removal/conversion of ecosystems and their related functions, have driven drastic changes in marine habitats. These further result in declines in marine populations (Lotze et al. 2006; Knapp et al. 2017). Globally, coastal development has contributed to the widespread degradation or loss of coastal habitats. The annual loss of coastal habitat has been estimated to be between 1–9% for coral reefs (Gardner et al. 2003; Bellwood et al. 2004) and 1.8% for mangroves (Valiela et al. 2001), and seagrass beds have been disappearing at a rate of 7% annually since 1990 (Waycott et al. 2009).

Additionally, pollution can have important consequences for marine populations. For instance, the amount of debris in our oceans is rapidly increasing and is currently affecting an estimated 663 species through entanglement or ingestion, 15% of which are threatened with extinction (Derraik 2002; Secretariat of the Convention on Biological Diversity and the Scientific and Technical Advisory Panel 2012; Villarrubia-Gómez et al. 2018). Although evidence for population-level impacts is scarce, marine debris is believed to have contributed to the population decline of several threatened species such as the Northern fur seal (*Callorhinus ursinus*) and the Hawaiian monk seal (*Monachus schauinslandi*) (Franco-Trecu et al. 2017). The reported number of individuals affected by marine debris suggests this threat might be pervasive (Secretariat of the Convention on Biological Diversity and the Scientific and Technical Advisory Panel 2012; Wilcox et al. 2015). Furthermore, deaths from ingesting marine debris can happen in the open ocean with no evidence ever washing onto beaches. Thus, the frequency of mortality from debris may actually be higher than currently perceived.

Biological invasions also represent a serious disruption to the balance of ecosystems, which can have severe consequences for the population abundance of prey species or competitors. Albins and Hixon (2013) found that the introduction of the Indo-Pacific lionfish (*Pterois volitans*) on Atlantic and Caribbean coral reefs poses a serious threat to coral reef fishes, reducing prey fish recruitment and abundance compared to control sites by 79 and 90%, respectively. This loss of prey species increases competition for the same depleted resource base, negatively impacting native predators such as the coney (*Cephalopholis fulva*).

Finally, climate change driven by the anthropogenic emission of greenhouse gases represents an emerging threat to marine populations. Climate change has led to rising global atmospheric and ocean temperatures, in addition to increasing the ocean pH (ocean acidification), as the oceans take up greater amounts of CO_2 (IPCC 2013). The consequences for marine populations are wide and varied, altering species' morphology, behavior, and physiology (Harley et al. 2006; O'Connor et al. 2007; Rosenzweig et al. 2008). For example, warmer ocean temperatures have led reef-building corals to live in the upper limits of their thermal tolerance, and prolonged periods of thermal stress can result in mass coral bleaching and disease outbreaks (Scavia et al. 2002; Hoegh-Guldberg et al. 2007; Harvell et al. 2009). Furthermore, the recurring frequency of increasing thermal pressure reduces the capacity of coral reefs to recover between events, decreasing their resilience to future change (Baker et al. 2008). Additionally, calcareous organisms such as mollusks, crustaceans, some species of algae/phytoplankton, and reef-building corals are vulnerable to ocean acidification, as calcification rates decline notably in high pH environments, and calcium carbonate dissolves (Orr et al. 2005; Hoegh-Guldberg et al. 2007; Guinotte and Fabry 2008). Climate change is considered one of the most serious threats to marine species and ecosystems at present, as the potential for marine species to adapt to the changing environmental conditions, and the serious implications it may have on the functioning of marine ecosystems, still remain largely unknown (Pacifici et al. 2015).

These diverse anthropogenic stressors are expected to accelerate in the future and alter patterns of global marine biodiversity (Jones and Cheung 2014), with consequences for species survival, economics, and food security (Barange et al. 2014). Emerging evidence suggests that the various stressors may interact to increase the extinction risk of marine species (Brook et al. 2008; Knapp et al. 2017). This indicates that the synergistic effects of multiple stressors may exceed the additive combination of any single stressor (Crain et al. 2008; Harnik et al. 2012). For example, changes in ocean temperature and chemistry might increase the vulnerability of some species to overexploitation by altering demographic factors (Doney et al. 2012; Harnik et al. 2012). The amplifying nature and dynamics of these synergistic interactions are poorly known for most stressors and requires further investigation.

Despite these overarching negative trends in marine species populations, certain species, trophic levels, and body sizes are more susceptible to population declines than others. Both large and small-scale fisheries disproportionately target large species at high trophic levels (Myers and Worm 2003; Kappel 2005; Olden et al. 2007). Additionally, large bodied species tend to have smaller population sizes as well as slower growth and reproduction rates, making them intrinsically more vulnerable to overexploitation (Roberts and Hawkins 1999; García et al. 2008; Sallan and Galimberti 2015). Consequently, large and high trophic level marine species have declined more rapidly and severely, and have been found to be at greater risk of extinction (Olden et al. 2007). Over 90% of large pelagic fish have experienced range contractions (Worm and Tittensor 2011) and the biomass of commercially valuable large species such as tuna is estimated to have declined by 90% relative to pre-industrial levels (Myers and Worm 2003). In certain regions, shark populations have declined by over 90% (Baum et al. 2003;

Box 4.3 Consequences of Apex Consumer Loss on Biological Communities

From the beginning of human impact in the Pleistocene, the marine realm has seen a disproportionate loss of larger-bodied animals (Smith et al. 2003). There is mounting evidence that large fauna has a strong influence on the structure, functioning, and resilience of marine ecosystems for several reasons (Duffy 2002; Myers et al. 2007). Large animals are often apex consumers which exert top-down population control on prey communities through direct mortality or fear-induced behavioral alterations (Creel et al. 2007; Creel and Christianson 2008; Heithaus et al. 2008; Laundré et al. 2010). Additionally, large animals can be ecosystem engineers, increasing the structural or biogeochemical complexity of their ecosystem either behaviorally or morphologically (e.g., whale falls creating novel ecosystems, Mills et al. 1993; Jones et al. 1994; Coleman and Williams 2002). As such, the consequences of the defaunation of large fauna are not retained within the impacted group, but affect multiple trophic levels in the community.

The loss of top-down control following apex consumer decline is often followed by population increases of medium-sized vertebrate prey, known as mesopredator release (Baum and Worm 2009). The effects of this decline are, however, not just restricted to their immediate prey, but often propagate down the food web, resulting in inverse patterns of increase/decrease in population abundance of lower trophic levels – a process known as trophic cascading (Paine 1980; Baum and Worm 2009). For example, Myers et al. (2007) found that the near-complete eradication of shark populations on the eastern seaboard of the USA led to the population explosion of mesopredatory elasmobranchs such as the cownose ray, increasing in abundance with order of magnitude in four decades. Moreover, the effects of the removal of this functional group cascaded down the food chain. Cownose rays

(continued)

Box 4.3 (continued)

now inflict a near-complete mortality on their bay scallop prey populations during migration periods, which has led to the discontinuation of North Carolina's traditional scallop fisheries. Additionally, bay scallops are expected to be replaced by infaunal bivalves, which could cause the uprooting of seagrass and, consequently, the loss of this habitat's function as nurseries and feeding grounds. Thus, losses of apex consumers can destabilize ecological communities, and the consequences of these cascading effects may be multiple.

Although counterintuitive, the direct mortality inflicted by apex consumers on prey species may improve the long-term survival of these prey populations. Predators eliminate sick individuals from prey populations, thus increasing the overall health of the population (Severtsov and Shubkina 2015). Moreover, the prevalence of disease epidemics is strongly linked to population density (Lafferty 2004). As such, apex consumers may inhibit disease outbreaks by suppressing prey populations below the critical host-density threshold for effective disease transmission (Packer et al. 2003). For instance, local extinction of sea otters around the California Channel Islands reduced the predation pressure on the black abalone, leading to an abalone population outbreak (Lafferty and Kuris 1993). As a result, black abalone populations increased beyond the host-density threshold, which led to an outbreak of the previously unknown rickettsial disease. The final result was a collapse of the black abalone population to levels of probable extinction (critically endangered – IUCN 2017).

The loss of large marine fauna may also affect biogeochemical cycles between the ocean and the other major reservoirs (Crowder and Norse 2008). Predators limit herbivore abundance, thus buffering herbivore effects on autotrophic organisms (Hairston et al. 1960). Autotrophs hold a high proportion of the global non-fossilized organic carbon reserves in their tissues, and are able to convert inorganic carbon to organic carbon through their photosynthetic ability (Wilmers et al. 2012). Thus, the global apex consumer losses might have altered the carbon cycle considerably. Indeed, apex consumers have been shown to significantly alter carbon capture and exchange. In the absence of sea otters, for instance, trophic cascades (Fig. 4.2) reduced the net primary production and carbon pool stored in kelp forests 12-fold, resulting in reduced carbon sequestration (Wilmers et al. 2012). Similarly, with the reduced numbers of great whales, a large part of the primary production is consumed by smaller animals with much higher mass-specific metabolic rates. This greatly reduces the overall potential for carbon retention in living marine organisms compared to pre-industrial times (Pershing et al. 2010). In addition, the carbon sequestration resulting from the sinking of great whale carcasses to the deep sea declined from an estimated 1.9×10^5 tons C year^{-1} to 2.8×10^4 tons C year^{-1} – a decrease by an order of magnitude since the start of industrial whaling (Pershing et al. 2010).

Great whales also exhibit behavioral engineering functions, the loss of which can influence the global nutrient and carbon cycling (Roman et al. 2014). Through their diving/surfacing behavior and the creation of bubble nets, great whales often break density gradients (Dewar et al. 2006), allowing an influx of nutrients to the formerly stratified and nutrient-depleted photic zone. Additionally, great whales frequently release fecal plumes and urine near the surface which brings limiting nutrients such as N and Fe from the aphotic zone above the thermocline (Roman and McCarthy 2010). Finally, through their migration from nutrient-rich high-latitude feeding grounds to nutrient-poor low-latitude calving grounds, some great whales bring limiting nutrients to tropical waters (Roman et al. 2014). These processes increase the primary production in the nutrient-poor photic zone, and in turn, increase carbon sequestration through sinking algal blooms to the deep sea (Lavery et al. 2010).

Myers et al. 2007), and reef-associated predators have shown similar declines (Friedlander and DeMartini 2002). Furthermore, in a similar time period, global whale population abundance has declined by 66–99% compared to their pre-whaling estimates (Roman et al. 2014). The diminishing number of large apex consumers has led to a reduction in the mean trophic level and community body size of marine food webs, as species of progressively decreasing size and trophic level are targeted (Pauly et al. 1998; Jennings and Blanchard 2004). The effects of these population depletions are, however, not solely retained within their respective trophic levels or populations, but affect inter-species interactions and ecosystem stability and functioning.

Large apex consumers are often **keystone species** (see Box 4.1), performing crucial ecological roles within their communities (Paine 1969; Box 4.3). Other than the aforementioned potential top-down forcing they can exert on their communities, large animals can also be **ecosystem engineers** (see Box 4.1), acting as keystone modifiers of habitat features which are crucial to the survival of other species

(Mills et al. 1993; Jones et al. 1994). When these engineering functions are removed, habitats generally become less complex, which decreases the diversity they can sustain. The loss of these keystone species results in the reduction of structural and functional diversity and decreases ecosystem resilience to environmental change (Coleman and Williams 2002). Thus, the near-complete elimination of large apex consumers from their ecosystems represents a major perturbation with important and far-reaching consequences for the structure, functioning, and resilience of marine ecosystems (Duffy 2002; Myers et al. 2007).

Overall, the ubiquity of marine population declines is important, as analyses quantifying the consequences of anthropogenic stressors on biodiversity routinely focus on biodiversity trends without explicitly accounting for population abundance and biomass trends (Dirzo et al. 2014; McGill et al. 2015). For instance, much of the work on ecosystem functions and services focuses on changes in local diversity, while few studies explicitly consider population declines and the subsequent changes in relative species abundance, which may be equally important (Dirzo et al. 2014; Winfree et al. 2015). In some cases, population declines have been so high that species cannot functionally interact in communities across part or all of their range (Worm and Tittensor 2011; McCauley et al. 2015). These "**ecological extinctions**" (see Box 4.1) are more difficult to measure and may be more widespread than currently appreciated (McCauley et al. 2015). Furthermore, unlike complete extinctions, which usually occur slowly, population declines can be very rapid and, thus, can cause rapid ecosystem changes and even regime shifts (Säterberg et al. 2013). Incorporating population and biomass trends into biodiversity monitoring, and understanding how this affects marine ecosystem function and service delivery, will improve our understanding of anthropogenic impacts on the ocean (McGill et al. 2015).

4.5 The Distribution of Anthropogenic Stressors in the Marine Environment

It is clear that the growing human population has put increasing pressure on the world's oceans, leading to varying degrees of decline in marine populations (Kappel 2005; Crain et al. 2009). Anthropogenic stressors are increasing in global intensity and now impact nearly every part of the ocean (Jones et al. 2018). To mitigate these pressures, it is important to understand the main stressors on the marine environment and how they are distributed across the oceans and the marine tree of life.

To do so, we analysed the threats to marine species which are threatened with extinction globally using the IUCN Red List database (IUCN 2017; see Supplementary Material B for full methods). Species threatened with extinction are those that are listed by the IUCN as vulnerable (VU), endangered (EN), or critically endangered (CR) (Box 4.2). We grouped the threats listed by the IUCN into six categories (*sensu* Young et al. 2016): "habitat destruction and modification," "direct exploitation," "invasive species," "pollution," "climate change," and "other." Based on the frequency by which threatened marine species were affected by different threats, we found the most common threat to be habitat modification and destruction, followed closely by overexploitation (Fig. 4.3a). Pollution, climate change, and invasive species were less frequently observed as threats to threatened marine species (Fig. 4.3a).

The impact of these stressors is pervasive across the marine tree of life (Crain et al. 2009); however, as previously described for extinctions, the average number of threats affecting threatened species varies between taxonomic groups (Fig. 4.3b). Threatened anthozoan species (sea anemones and corals) experience the highest average number of threats (Fig. 4.3b). Coral reefs are impacted by diverse stressors such as overfishing, coastal development, and agricultural runoff (McLeod et al. 2013). Additionally, climate change and ocean acidification are emerging stressors that have caused widespread damage to reefs around the world (Hoegh-Guldberg et al. 2007). Nearly 75% of the world's coral reefs are affected by these stressors (Burke et al. 2011). Although threatened chondrichthyans (i.e., sharks, rays, and chimeras) are seen to have a relatively lower average number of threats, they are largely threatened by the effects of overexploitation through targeted fisheries and bycatch (Fig. 4.3b). The unique life history characteristics of these cartilaginous fishes (late maturation, slow growth, and low reproduction rates) make them particularly vulnerable to the impact of stressors (Worm et al. 2013).

Several marine areas suffer from high human impact, particularly those where human use of the ocean is the greatest (Halpern et al. 2015). However, the distribution and impact of different anthropogenic stressors vary geographically (Fig. 4.4). Habitat destruction and modification, and overexploitation remain the most important threats to threatened marine species across marine regions; however, their relative contribution to the overall threat level varies considerably between regions (Fig. 4.4). Previous analyses of the cumulative anthropogenic impact on the oceans have shown that the Central and North Atlantic, the Mediterranean, the East Indian, and the Central Pacific are heavily impacted marine regions (Halpern et al. 2015). Our analysis shows that threatened species in the Atlantic and Mediterranean regions are most frequently impacted by overexploitation. Conversely, in the East Indian and Central Pacific regions, habitat modification and destruction constituted the most prevalent threat. The Arctic region is also heavily impacted anthropogenically (Halpern et al. 2015). We found that threatened species in this region are heavily impacted by habitat modification

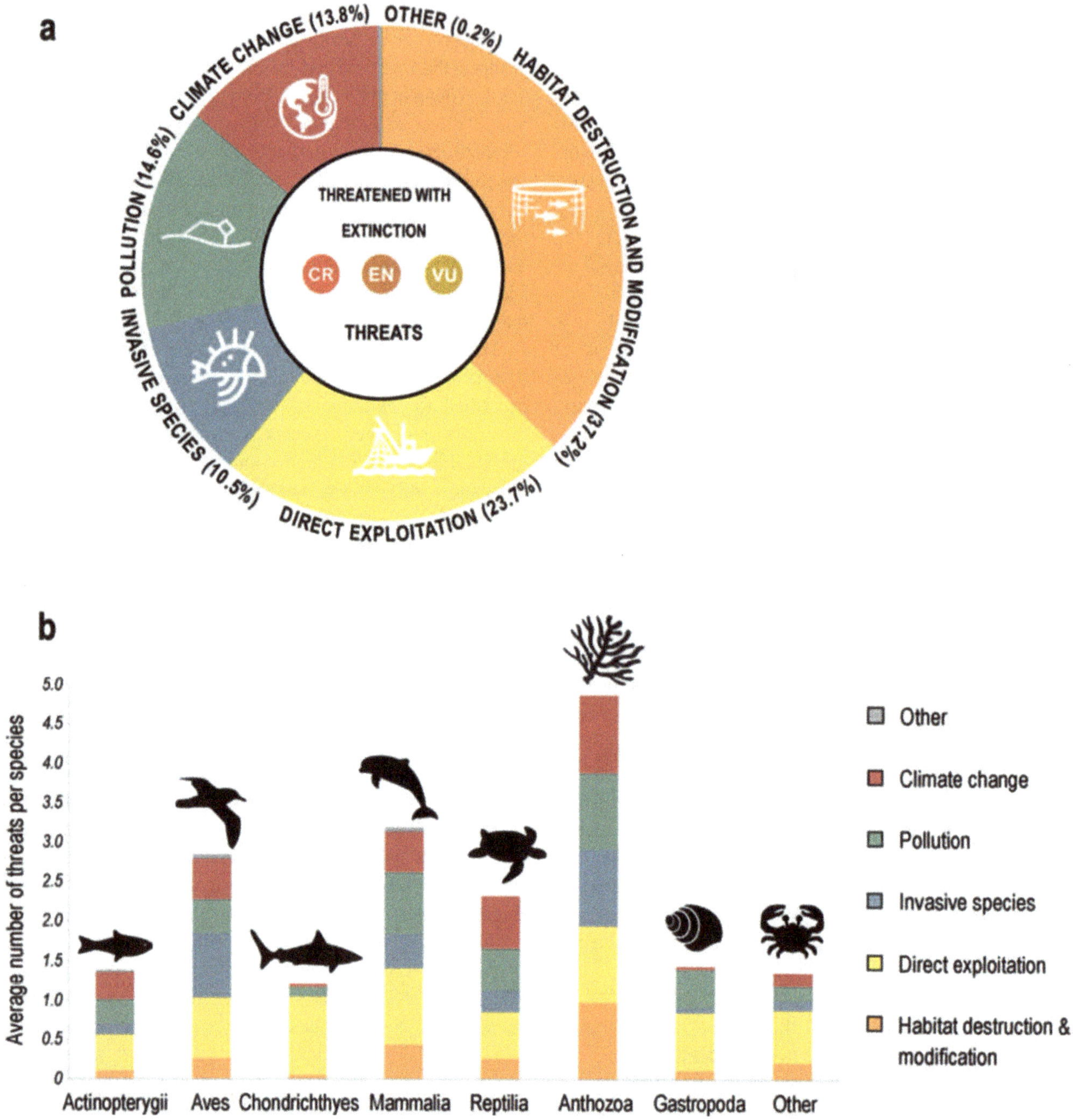

Fig. 4.3 (**a**) Relative importance of various anthropogenic stressors to species threatened with extinction (critically endangered (CR), endangered (EN) or vulnerable (VU)). (**b**) Average number of anthropogenic stressors affecting species threatened with extinction for various taxonomic groups. Several taxonomic groups with a low number of species threatened with extinction were pooled under "other" for visualization purposes ("Myxini," "Sarcopterygii," "Polychaeta," "Insecta," "Malacostraca," "Maxillopoda," "Merostomata," "Hydrozoa," and "Holothuroidea"). Colors indicate the percentage contribution of the different anthropogenic stressors. Data extracted from the IUCN Red List database of threats (IUCN 2018). Details on data compilation are provided in the Supplementary Material B

and destruction, which corresponds to large-scale habitat alterations driven by changes in sea ice extent (Walsh et al. 2016). Although Antarctica has low overall anthropogenic impacts (Halpern et al. 2015), the most common threat to threatened marine species in this region is overexploitation (57%). In summary, these results indicate that anthropogenic threats to threatened marine species differ considerably between marine regions, even if the overall average number of threats may be similar (Fig. 4.4; Halpern et al. 2015). This suggests that reducing anthropogenic impacts on threatened marine species requires targeted approaches specific to each marine region.

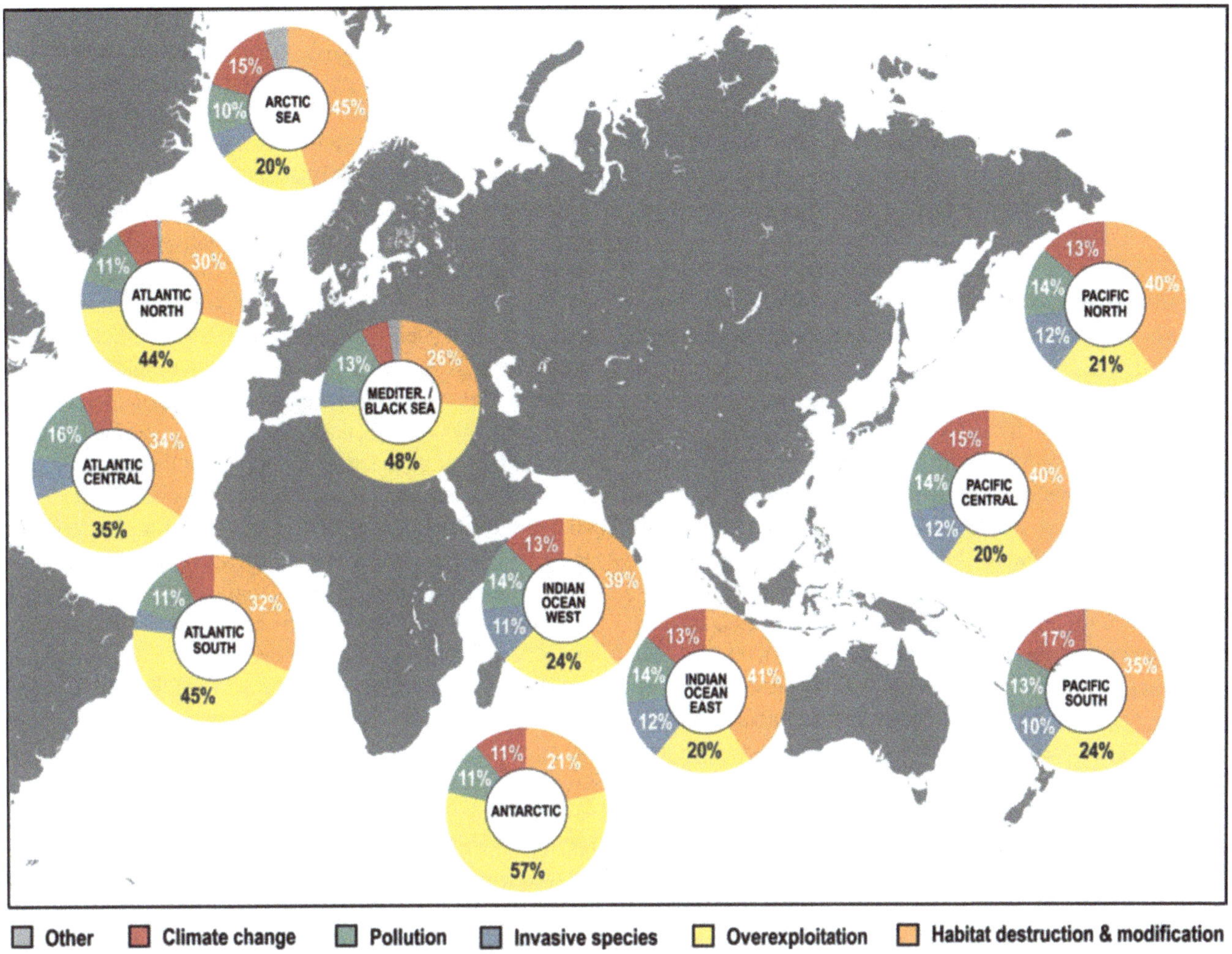

Fig. 4.4 The distribution of anthropogenic stressors faced by marine species threatened with extinction in various marine regions of the world. Numbers in the pie charts indicate the percentage contribution of an anthropogenic stressors' impact in a specific marine region. Percentages under 10% are not displayed for visualization purposes. Data extracted from IUCN Red List database of threats (IUCN 2018). Details on data compilation are provided in the Supplementary Material B

4.6 Mitigating Local-Scale Anthropogenic Stressors on Marine Biodiversity

Conservation interventions are a challenging task due to the large spatial scale and number of stakeholders involved in global threats to marine biodiversity, like climate change, interventions are challenging. The scale of these threats requires international cooperation and considerable changes in current pathways of production and consumption, which could take several decades to achieve, and current progress is slow (Anderson 2012). Nonetheless, emerging evidence from the literature suggests that also the mitigation of local-scale threats can aid in the conservation of marine populations. In a recent review, Lotze et al. (2011) found that 10–50% of surveyed species and ecosystems showed evidence of positive recovery, 95% of which were due to reduced local-scale threat impacts. Reducing these local-scale threat levels is, thus, particularly important given the persistent impact of more global-scale threats. Furthermore, while some small-scale threats can be abated by local governments and NGOs, other global threats such as biological invasions and climate change require international collaboration and the cooperation of all stakeholders. Discussing the mitigation of global-scale threats such as invasive species or climate change would be beyond the scope of this review. Thus, the following section will focus on mitigation strategies for local-scale threats such as overexploitation, habitat destruction and modification, and pollution.

Several regulations have been implemented to reduce overexploitation. The main regulatory tools employed in fisheries overharvesting are the use of closed areas, closed seasons, catch limits or bans, effort regulations, gear restrictions, size of fish length regulations, and quotas on certain species (Cooke and Cowx 2006). The use of marine protected areas (MPAs) such as marine reserves (MRs), where part of the marine habitat is set aside for population recovery and resource extraction is prohibited, has shown promising results. For instance, across 89 MRs, the density, biomass, and size of organisms, and the diversity of carnivorous fishes, herbivorous fishes, planktivorous fishes, and invertebrates increased in response to reserve establishment (Halpern 2003). Long-term datasets suggest this response can occur rapidly, detecting positive direct effects on target species and indirect effects on non-target species after 5 and 13 years, respectively (Babcock et al. 2010). Successful MPAs often share the key features of being no-take zones (no fishing), well-enforced (to prevent illegal harvesting), older than 10 years, larger than 100 km^2, and isolated by deep water or sand (Edgar et al. 2014). Additionally, it is critical to align the planning of the MPA with local societal considerations, as this determines the effectiveness of the MPA (Bennett et al. 2017). Future MPAs should be developed with these key features in mind to maximize their potential. More recently, dynamic modelling approaches encompassing population hindcasts, real-time data, seasonal forecasts, and climate projections have been proposed for dynamic closed zones that would alternate between periods open and closed to fishing (Hazen et al. 2018). Such management strategies based on near real-time data have the potential to adaptively protect both target and bycatch species while supporting fisheries. Non-target species may be under severe threat as their catch data might not be recorded and, thus, rates of decline in these species remain unknown (Lewison et al. 2014). Alterations to fishing gear along with new technological advances reduce the exploitation of non-target species. Increasing mesh size reduces non-target species' capture and prevents the mortality of smaller target species' size classes (Mahon and Hunte 2002). Furthermore, turtle exclusion devices, for example, have proven to be effective in protecting turtles from becoming entrapped in fishing nets (Gearhart et al. 2015). Illuminating gillnets with LED lights has also been shown to reduce turtle (Wang et al. 2013; Ortiz et al. 2016) and seabird bycatch (Mangel et al. 2018).

Reductions in overexploitation have led to rapid recoveries of several historically depleted marine populations, a pattern which is particularly evident for marine mammals (e.g., humpback whales, Northern elephant seals, bowhead whales). Following the ban on commercial whaling in the 1980s, several populations of Southern right whales, Western Arctic bowhead whales, and Northeast Pacific gray whales showed substantial recovery (Baker and Clapham 2004; Alter et al. 2007; Gerber et al. 2007). Similar population increases were observed after the ban on, or reduction of, pinniped hunting for fur, skin, blubber, and ivory (Lotze et al. 2006). Still, populations do not always recover from overexploitation. Many marine fish populations, for instance, have shown limited recovery despite fishing pressure reductions (Hutchings and Reynolds 2004). This might be partly explained by the low population growth rates at small population sizes (Allee effects), which can prevent the recovery of many marine populations (Courchamp et al. 2006). Nonetheless, these examples suggest that marine populations can show resilience to overexploitation and that appropriate management may facilitate population recoveries.

Marine habitats can be destroyed or modified either directly, through activities such as bottom trawling, coral harvesting, clearing of habitat for aquaculture and associated pollution, mining for fossil fuels/metals, and tourism; or indirectly as a consequence of ocean warming, acidification, and sea level rise (Rossi 2013). However, both ultimately lead to the physical destruction/modification or chemical alteration of the habitat which affects its suitability for marine life. The effects of habitat destruction and modification are often permanent or may require intense efforts to restore (Suding et al. 2004; van der Heide et al. 2007). Still, local-scale strategies can be adopted to protect marine habitats. In addition to the aforementioned positive effects on mitigating overexploitation, MRs have been shown to be a useful tool to counteract habitat destruction and modification. For instance, by excluding bottom trawling, MRs can safeguard seafloor habitats and the associated benthic organisms, which are often important ecosystem engineers (Rossi 2013). Furthermore, MRs ban harvesting activities of habitat-building organisms such as sponges, gorgonians, and reef-building corals. In the Bahamas, for example, both the coral cover and size distribution were significantly greater within a marine protected area compared to the surrounding unprotected area (Mumby and Harborne 2010). Additionally, development and resource extraction activities are prohibited within MPAs and MRs, thus, protecting marine habitats directly against destructive activities. Recent increases in habitat protection of coastal areas have contributed to ecosystem recovery in a variety of marine systems such as wetlands, mangroves, seagrass beds, kelp forests, and oyster and coral reefs (Lotze et al. 2011). Currently, approximately 4.8% of the oceans are within marine protected areas (Marine Conservation Institute 2019), and global goals are set on an expansion to at least 10% by 2020 under Aichi target 11 of the CBD (CBD, COP Decision X/2 2010), highlighting the value and broad applicability of MPAs as a tool to protect marine ecosystems.

Nonetheless, MPAs are not a perfect solution to mitigate marine threats at present. An MPA's ability to effectively protect species from overexploitation depends on the enforcement of harvesting regulations, as well as protected area size and connectivity. A review by Wood et al. (2008) found that the median size of MPAs was 4.6 km^2, and only half of the world's MPAs were part of a coherent MPA network. This suggests that the majority of MPAs might be insufficiently large to protect species with a large home-range or migrating behavior, such as marine megafauna (Agardy et al. 2011). Additionally, protected areas can only effectively negate habitat destruction and modification if their distribution is even and representative for the various habitats at risk, as well as the different biophysical, geographical, and political regions (Hoekstra et al. 2005; Wood et al. 2008). At present, MPAs are heavily biased toward coastal waters. The number of protected areas decreases exponentially with increasing distance from shore, and the pelagic region of the high seas is gravely underrepresented (Wood et al. 2008; Agardy et al. 2011; Thomas et al. 2014). Moreover, the majority of protected areas are situated either in the tropics (between 30° N and 30° S) or the upper northern hemisphere (> 50° N), and intermediate (30–50°) and polar (> 60°) latitudes are poorly covered.

Marine protected areas can also suffer unintended consequences of MPA establishment. For instance, the establishment of no-take zones might lead to a displacement of resource extraction to the area outside of the reserve. This concentrates the exploitation effort on a smaller area, increasing the pressure on an area which might already be heavily overexploited (Agardy et al. 2011). Additionally, there is increasing evidence the establishment of MPAs might promote the establishment and spread of invasive species (Byers 2005; Klinger et al. 2006; Francour et al. 2010). Finally, even when perfectly designed and managed, MPAs might still fail if the surrounding unprotected area is degraded. For example, the impact of pollution outside of the protected area might negatively affect the marine organisms within the MPA, thus, rendering it ineffective (Agardy et al. 2011). As such, the establishment of MPAs should go hand in hand with mitigation strategies for overexploitation and pollution in the surrounding matrix.

A common source of pollution in the marine environment is the deposition of excess nutrients, a process termed eutrophication. Strategies which are commonly adopted for the mitigation of eutrophication focus on reducing the amount of nitrogen and phosphorous runoff into waterways. Possible solutions to achieve this goal include reducing fertiliser use, more carefully handling manure, increasing soil conservation practices and continuing restoration of wetlands and other riparian buffers (Conley et al. 2009). Reductions in chemical pollution have facilitated population recovery in several marine species (Lotze et al. 2011). For instance, following population declines due to river damming and subsequent pollution, the Atlantic salmon in the St. Croix River (Canada) recovered strongly after efforts to reduce pollution (Lotze and Milewski 2004). In some instances, however, the anthropogenic addition of nutrients has been so great that it has caused ecosystem regime shifts (e.g., Österblom et al. 2007), in which case a reduced nutrient inflow may not be effective at facilitating population recovery.

Marine ecosystems are also commonly disturbed by noise pollution, which has been on the rise since the start of the industrial revolution. Common sources of noise include air and sea transportation, engines from large ocean tankers, icebreakers, marine dredging, construction activities of the oil and gas industry, offshore wind farms, and sound pulses emitted from seismic surveys (Scott 2004 and references therein). Nowacek et al. (2015) urge that an internationally signed agreement by member countries outline a protocol for seismic exploration. This agreement should include restrictions on the time and duration of seismic exploration in biologically important habitats, monitoring acoustic habitat ambient noise levels, developing methods to reduce the acoustic footprint of seismic surveys, creating an intergovernmental science organization, and developing environmental impact assessments. Other management strategies could include speed limits to reduce the noise in decibels emitted by ships, as well as noise reduction construction and design requirements for ships (Richardson et al. 1995, Scott 2004).

Finally, marine debris composed of plastic is one of the world's most pervasive pollution problems affecting our oceans—impacting marine organisms through entanglement and ingestion, use as a transport vector by invasive species, and the absorption of polychlorinated biphenyls (PCBs) from ingested plastics, among others (Derraik 2002; Sheavly and Register 2007). Other than global mitigation measures such as international legislation on garbage disposal at sea (MARPOL 1973), local-scale measures against marine debris pollution include education and conservation advocacy. The Ocean Conservancy organizes an annual International Coastal Cleanup (ICC), which collects information on the amounts and types of marine debris present globally. In 2009, 498,818 volunteers from 108 countries and locations collected 7.4 million pounds of marine debris from over 6,000 sites (Ocean Conservancy 2010). Public awareness campaigns have also been employed as a strategy raise awareness about plastic pollution. An alliance between conservation groups, government agencies, and consumers led a national (USA) ad campaign to help build awareness of boating and fishing groups about the impacts of fishing gear and packaging materials that enter the marine environment (Sheavly and Register 2007).

4.7 Conclusions

Overall, the status of marine biodiversity in the Anthropocene is complex. Globally, taxonomic assessments in the marine realm are highly incomplete. Similarly, the rate of assessment of marine species for extinction risk is also slow, at least compared to the terrestrial realm. This lack of assessment may have led to underestimations of the global extinction rate of marine species. Several authors have suggested that current extinction rates in the marine realm are low, but recent evidence suggests that this may be inaccurate due to low rates of assessment of extinction risk. Regardless, the loss of marine biodiversity is complex and extinctions or reductions in species richness at any scale do not adequately reflect the changes in marine biodiversity that are occurring. Directional changes in the composition of marine communities are occurring at local scales. These changes are nonrandom, as resident species are replaced by more widespread invaders, which may, over time, reduce diversity in space. The consequence of these changes is lower regional species richness. This, however, is infrequently quantified in the marine realm, and the consequences for ecosystem function and service delivery are poorly known.

In this context, we emphasize recent calls for careful quantification of trends in biodiversity loss at multiple spatial scales and the assessment of the possible effects of various forms of biodiversity loss on ecosystem functioning (e.g., McGill et al. 2015). In particular, population declines are ubiquitous in the marine realm and careful assessment of the effects on ecosystem functions and services are likely critical for understanding human impacts on the marine realm. The anthropogenic threats to marine biodiversity are diverse, cumulative and/or synergistic. Our analysis of the IUCN threats database shows that the main anthropogenic threats vary by both region and taxonomic group. Understanding trends in marine biodiversity thus requires assessing how different taxonomic groups in different regions respond to the various anthropogenic stressors, and how those affect marine biodiversity at different scales of space and time. Furthermore, while some small-scale threats can be abated by local governments and NGOs, other global threats require international collaboration and the cooperation of all stakeholders. This will be a challenge going forward, but will be necessary to fully support marine biodiversity in the Anthropocene.

Appendix

This article is related to the YOUMARES 9 conference session no. 5: "Species on the brink: navigating conservation in the Anthropocene." The original Call for Abstracts and the abstracts of the presentations within this session can be found in the Appendix "Conference Sessions and Abstracts", Chapter "4 Species on the brink: navigating conservation in the Anthropocene", of this book.

Supplementary Material A: Methods to Estimate Species Diversity

Early Attempts

Several early attempts were made to estimate global marine species diversity. Grassle and Maciolek (1992) used data from 233 box-core sediment samples (30 × 30 cm) taken at 2 km depth along a 176 km transect on the east coast of North America. In these samples, they found 460 (58%) undescribed species of marine macrofauna. Using standard rarefaction curves, they estimated that approximately 100 new species would be discovered per 100 km of seabed. Extrapolating over the entire ocean, they estimated that 10 million species occurred in marine benthic habitats. Shortly afterwards, May (1994) scaled back Grassle and Maciolek's (1992) estimates considerably by employing a more direct method of estimation. The method used by May (1994) works by increasing the number of globally described species by a factor related to the percentage of undiscovered species found in different sampling campaigns. In Grassle and Maciolek's (1992) samples, 58% of the species discovered were new. In 1994, there were approximately 200,000 marine macrofaunal species described. Thus, May (1994) estimated that there were likely to be 400,000–500,000 marine macrofaunal species.

Since these early estimates, several subsequent attempts have been made to estimate marine species diversity (reviewed in Caley et al. 2014). However, since 2010, five estimates of global marine diversity have been published (see Table 4.1). These estimates used variations on one of two methods: extrapolating based on past rates of species discovery or collating expert opinions of undiscovered species.

Extrapolating Based on Past Rates of Species Discovery

Statistical extrapolation of time-species description accumulation curves from biodiversity databases was used by Costello et al. (2012) and Appeltans et al. (2012) to estimate global marine diversity. Both fit non-homogenous renewable process models to the cumulative number of new species descriptions through time (based on description date) using records on the WoRMS database. These models (Costello and Wilson 2011) decompose time-series data into trend and variation components and allow for variation in the rate of discovery through time. For example, species description rates dropped markedly during the two World Wars (Costello et al. 2012). By fitting these models to the species description time-series, estimates of species that will be discovered at future time points can be made. However, certain analyses have shown that unless the time-species accumulation curves have approached asymptotic levels, estimates are highly uncertain (Bebber et al. 2007).

Higher taxonomic level accumulation curves have recently been used to overcome the asymptotic problem in estimating global species diversity (Mora et al. 2011). Higher taxonomic levels are more completely sampled than lower taxonomic levels (i.e., it is unlikely that many more discoveries at the family level will occur). Thus, Mora et al. (2011) used higher taxonomic (phylum to genus) description time-series from the WoRMS database and estimated their asymptotes. Using regression models, the authors related these higher taxonomic asymptotes to their taxonomic rank (1-5, phylum to genus) and used this model to extrapolate to the species level (taxonomic rank 6).

Estimates Based on Expert Opinion

The other method that has been applied to estimate global marine species diversity is to collate the estimates of experts of different marine groups. Appeltans et al. (2012) and Costello et al. (2010) used variations of this approach. Appeltans et al. (2012) collated estimates from 270 specialist taxonomists on the total species described, undescribed, and undiscovered for the different marine groups recognized in the WoRMS database. The WoRMS editors that contributed to these estimates are all experts in their taxonomic group and together represent ca. 5% of the active marine taxonomists. Despite their high level of expertise, all of the experts used markedly different methods to estimate the species number in their respective groups (see Table S2 in Appeltans et al. 2012).

In contrast, the expert opinion approach used by Costello et al. (2010) is a variation on the method used by May (1994). Over 360 scientists representing institutions in Antarctica, Atlantic Europe, Australia, the Baltic Sea, Canada, the Caribbean Sea, China, the Indian Ocean, Japan, the Mediterranean Sea, New Zealand, South Africa, South America, South Korea, and the USA were asked to rate the state of knowledge of different taxonomic groups on a 1-5 scale corresponding to different proportions of described species (e.g., 1 indicates that more than 80% of species are described). Based on these scores, Costello et al. (2010) estimated that 70-80% of marine species remain to be described. They then inflated the total described species at the time (~230,000) based on this estimate of uncertainty as per May (1994).

References

Appeltans W, Ahyong ST, Anderson G et al (2012) The magnitude of global marine species diversity. Curr Biol 22:2189–2202. doi:10.1016/j.cub.2012.09.03

Bebber DP, Marriott FH, Gaston KJ et al (2007) Predicting unknown species numbers using discovery curves. Proc R Soc Lond B Biol Sci 274:1651–1658. doi:10.1098/rspb.2007.0464

Caley MJ, Fisher R, Mengersen K (2014) Global species richness estimates have not converged. Trends Ecol Evol 29:187–188. doi:10.1016/j.tree.2014.02.002

Costello MJ, Coll M, Danovaro R et al (2010) A census of marine biodiversity knowledge, resources, and future challenges. PLoS ONE 5:e12110. doi:10.1371/journal.pone.0012110

Costello MJ, Wilson S, Houlding B (2012) Predicting total global species richness using rates of species description and estimates of taxonomic effort. Syst Biol 61:871–883. doi:10.1093/sysbio/syr080

Costello MJ, Wilson SP (2011) Predicting the number of known and unknown species in European seas using rates of description. Glob Ecol Biogeogr 20:319–330. doi:10.1111/j.1466-8238.2010.00603.x

Grassle JF, Maciolek NJ (1992) Deep-sea species richness: regional and local diversity estimates from quantitative bottom samples. Am Nat 139:313-341. doi:10.1086/285329

May RM (1994) Biological diversity: differences between land and sea. Philos Trans Royal Soc B 343:105–111. doi:10.1098/rstb.1994.0014

Mora C, Tittensor DP, Adl S et al (2011) How many species are there on Earth and in the ocean?. PLoS Biol 9:e1001127. doi:10.1371/journal.pbio.1001127

Supplementary Material B: Compilation of Threats Database

The threats database was compiled using data from the IUCN Red List of Endangered Species online "advanced search" tool (IUCN 2018; https://www.iucnredlist.org/). Marine species were selected, while species living at the interface between marine systems and freshwater or terrestrial systems were omitted from this study. Species listed as "threatened with extinction" (vulnerable – VU, endangered – EN, and critically endangered – CR) were then selected.

The IUCN database offers 12 types of threat categories. However, for visualization, and as the categories related to habitat disturbance are biased towards one type of ecosystem, we followed the approach of Young et al. (2016). The six categories related to disturbance (residential and commercial development, agriculture and aquaculture, energy production and mining, transportation and service corridors, human intrusions and disturbances, and natural system modifications) were pooled under the name "habitat destruction and modification." Additionally, the category "geological events" was pooled with "other options" under the common name "other." The final selection of threats categories was "habitat destruction and modification," "direct exploitation," "invasive species," "pollution," "climate change," and "other."

All species in the IUCN Red List database have a listing of threats affecting their populations at present. To assess the relative importance of the various threats to marine species threatened with extinction (Fig. 4.3a), we added the number of times a threat was listed for each marine species threatened with extinction. For the pooled threats (habitat destruction and modification and other), the number of threats per sub-category were added up to get the final number of threats.

We investigated the average number of threats and their relative importance for each marine taxonomic group threatened with extinction (Fig. 4.3b). We included all the taxonomic groups listed by the IUCN database as containing species which are threatened with extinction. In accordance with Fig. 4.1b, groups which contained few species were grouped under "other" for visualization purposes. Additionally, given the frequency by which marine mammals and reptiles are covered in the marine species conservation literature, these taxonomic groups were discussed separately (as opposed to grouping then under "other chordates," as was opted for in Fig. 4.1b). The number of times each threat was listed for a specific taxonomic group was added. For the "other" group, threats for the various subgroups were added up. Unlike the previous method, for the pooled threat categories, the average number of threats was calculated for each taxonomic group, and compared to total number of threats for that taxonomic group to estimate the relative importance of each threat. To calculate the average number of threats per species, we divided the total number threats by the number of species threatened with extinction for that taxonomic group. To construct the figure, the average number of threats per taxonomic group was multiplied by the relative importance of each threat within that group.

Finally, we assessed the relative importance of each threat in the different biogeographic marine regions. The IUCN database offers 19 types of marine regions, however, for visualization, and as certain regions had very similar threat distribution patterns, certain marine biogeographic regions were pooled. The final selection consisted of 11 marine regions: "Arctic sea," "Pacific North" (Pacific Northwest and Pacific Northeast), "Pacific Central" (Pacific Western Central and Pacific Eastern Central), "Pacific South" (Pacific Southwest and Pacific Southeast), "Mediterranean and Black Sea," "Indian Ocean Western," "Indian Ocean Eastern," "Atlantic North" (Atlantic Northwest and Atlantic Northeast), "Atlantic Central" (Atlantic Western Central and Atlantic Eastern Central), "Atlantic South" (Atlantic Southwest and Atlantic Southeast), and "Antarctic" (Pacific Antarctic, Indian Ocean Antarctic and Atlantic Antarctic). Here, we added up the number of times each threat was listed for species threatened with extinction within a certain marine biogeographic region. For the pooled marine regions, and the pooled threat categories, threat counts were added up.

References

IUCN (2018) The IUCN Red List of Threatened Species. Version 2018-1. Available: http://www.iucnredlist.org

Young HS, McCauley DJ, Galetti M et al (2016) Patterns, causes, and consequences of anthropocene defaunation. Annu Rev Ecol Evol Syst 47:333–358. doi:10.1146/annurev-ecolsys-112414-054142

References

Agardy T, di Sciara GN, Christie P (2011) Mind the gap: addressing the shortcomings of marine protected areas through large scale marine spatial planning. Mar Pol 35:226–232. https://doi.org/10.1016/j.marpol.2010.10.006

Airoldi L, Balata D, Beck MW (2008) The gray zone: relationships between habitat loss and marine diversity and their applications in conservation. J Exp Mar Biol Ecol 366:8–15. https://doi.org/10.1016/j.jembe.2008.07.034

Albins MA, Hixon MA (2013) Worst case scenario: potential long-term effects of invasive lionfish (*Pterois volitans*) on Atlantic and Caribbean coral-reef communities. Environ Biol Fishes 96:1151–1157. https://doi.org/10.1007/s10641-011-9795-1

Alter ES, Rynes E, Palumbi SR (2007) DNA evidence for historic population size and past ecosystem impacts of gray whales. Proc Natl Acad Sci U S A 104:15162–15167. https://doi.org/10.1073/pnas.0706056104

Altman S, Whitlatch RB (2007) Effects of small-scale disturbance on invasion success in marine communities. J Exp Mar Biol Ecol 342:15–29. https://doi.org/10.1016/j.jembe.2006.10.011

Anderson K (2012) Climate change going beyond dangerous–Brutal numbers and tenuous hope. Development Dialogue 61:16–40

Anderson SC, Flemming JM, Watson R et al (2010) Serial exploitation of global sea cucumber fisheries. Fish Fish 12:317–339. https://doi.org/10.1111/j.1467-2979.2010.00397.x

Appeltans W, Ahyong ST, Anderson G et al (2012) The magnitude of global marine species diversity. Curr Biol 22:2189–2202. https://doi.org/10.1016/j.cub.2012.09.03

Babcock RC, Shears NT, Alcala AC et al (2010) Decadal trends in marine reserves reveal differential rates of change in direct and indirect effects. Proc Natl Acad Sci U S A 107:18256–18261. https://doi.org/10.1073/pnas.0908012107

Baker SC, Clapham PJ (2004) Modelling the past and future of whales and whaling. Trends Ecol Evol 19:365–371. https://doi.org/10.1016/j.tree.2004.05.005

Baker AC, Glynn PW, Riegl B (2008) Climate change and coral reef bleaching: an ecological assessment of long-term impacts, recovery trends and future outlook. Estuar Coast Shelf Sci 80:435–471. https://doi.org/10.1016/j.ecss.2008.09.003

Balata D, Piazzi L, Benedetti-Cecchi L (2007a) Sediment disturbance and loss of beta diversity on subtidal rocky reefs. Ecology 88:2455–2461. https://doi.org/10.1890/07-0053.1

Balata D, Piazzi L, Cinelli F (2007b) Increase of sedimentation in a subtidal system: effects on the structure and diversity of macroalgal assemblages. J Exp Mar Biol Ecol 351:73–82. https://doi.org/10.1016/j.jembe.2007.06.019

Balmford A, Gaston KJ (1999) Why biodiversity surveys are good value. Nature 398:204–205. https://doi.org/10.1038/18339

Bando KJ (2006) The roles of competition and disturbance in marine invasion. Biol Invasions 8:755–763. https://doi.org/10.1007/s10530-005-3543-4

Barange M, Merino G, Blanchard JL et al (2014) Impacts of climate change on marine ecosystem production in societies dependent on fisheries. Nat Clim Chang 4:211. https://doi.org/10.1038/nclimate2119

Barbier EB, Hacker SD, Kennedy C et al (2011) The value of estuarine and coastal ecosystem services. Ecol Monogr 81:169–193. https://doi.org/10.1890/10-1510.1

Barnosky AD, Matzke N, Tomiya S et al (2011) Has the Earth's sixth mass extinction already arrived? Nature 471:51–57. https://doi.org/10.1038/nature09678

Baum JK, Worm B (2009) Cascading top-down effects of changing oceanic predator abundances. J Anim Ecol 78:699–714. https://doi.org/10.1111/j.1365-2656.2009.01531.x

Baum JK, Myers RA, Kehler DG et al (2003) Collapse and conservation of shark populations in the Northwest Atlantic. Science 299:389–392. https://doi.org/10.1126/science.1079777

Beck MW, Brumbaugh RD, Airoldi L et al (2011) Oyster reefs at risk and recommendations for conservation, restoration and management. BioScience 61:107–116. https://doi.org/10.1525/bio.2011.61.2.5

Bellwood DR, Hughes TP, Folke C et al (2004) Confronting the coral reef crisis. Nature 429:827. https://doi.org/10.1038/nature0269117

Bennett NJ, Roth R, Klain SC et al (2017) Conservation social science: understanding & integrating human dimensions to improve conservation. Biol Cons 205:93–108. https://doi.org/10.1016/j.biocon.2016.10.006

Bini LM, Diniz-Filho JA, Rangel TF et al (2006) Challenging Wallacean and Linnean shortfalls: knowledge gradients and conservation planning in a biodiversity hotspot. Divers Distrib 12:475–482. https://doi.org/10.1111/j.1366-9516.2006.00286.xs

Bouchet P, Lozouet P, Maestrati P et al (2002) Assessing the magnitude of species richness in tropical marine environments: exceptionally high numbers of molluscs at a New Caledonia site. Biol J Linn Soc 75:421–436. https://doi.org/10.1046/j.1095-8312.2002.00052.x

Brito D (2010) Overcoming the Linnean shortfall: data deficiency and biological survey priorities. Basic Appl Ecol 11:709–713. https://doi.org/10.1016/j.baae.2010.09.007

Brook BW, Sodhi NS, Bradshaw CJ (2008) Synergies among extinction drivers under global change. Trends Ecol Evol 23:453–460. https://doi.org/10.1016/j.tree.2008.03.011

Burke L, Reytar K, Spalding M et al (2011) Reefs at risk revisited. World Resources Institute, Washington, DC

Burkepile DE, Hay ME (2008) Herbivore species richness and feeding complementarity affect community structure and function on a coral reef. Proc Natl Acad Sci U S A 105:16201–16206. https://doi.org/10.1073/pnas.0801946105

Butchart SH, Walpole M, Collen B et al (2010) Global biodiversity: indicators of recent declines. Science 328:1164–1168. https://doi.org/10.1126/science.1187512

Byers JE (2005) Marine reserves enhance abundance but not competitive impacts of a harvested non-indigenous species. Ecol 86:487–500. https://doi.org/10.1890/03-0580

Caldeira K, Wickett ME (2003) Oceanography: anthropogenic carbon and ocean pH. Nature 425:365. https://doi.org/10.1038/425365a

Caley MJ, Munday PL (2003) Growth trades off with habitat specialization. Proc R Soc Lond B Biol Sci 270:175–177. https://doi.org/10.1098/rsbl.2003.0040

Cardinale BJ, Duffy JE, Gonzalez A et al (2012) Biodiversity loss and its impact on humanity. Nature 486:59–67

Cardinale BJ, Gonzalez A, Allington GR et al (2018) Is local biodiversity declining or not? A summary of the debate over analysis of species richness time trends. Biol Cons 219:175–183. https://doi.org/10.1016/j.biocon.2017.12.021

Carlton JT, Chapman JW, Geller JB et al (2017) Tsunami-driven rafting: Transoceanic species dispersal and implications for marine biogeography. Science 357:1402–1406. https://doi.org/10.1126/science.aao1498

CBD, COP Decision X/2 (2010) Strategic plan for biodiversity 2011–2020. Available: http://www.cbd.int/decision/cop/?id=12268

Ceballos G, Ehrlich PR, Barnosky AD et al (2015) Accelerated modern human-induced species losses: entering the sixth mass extinction. Sci Adv 1:e1400253. https://doi.org/10.1126/sciadv.1400253

Claudet J, Fraschetti S (2010) Human-driven impacts on marine habitats: a regional meta-analysis in the Mediterranean Sea. Biol Cons 143:2195–2206. https://doi.org/10.1016/j.biocon.2010.06.004

Clavel J, Julliard R, Devictor V (2011) Worldwide decline of specialist species: toward a global functional homogenization? Front Ecol Environ 9:222–228. https://doi.org/10.1890/080216

Coleman FC, Williams SL (2002) Overexploiting marine ecosystem engineers: potential consequences for biodiversity. Trends Ecol Evol 17:40–44. https://doi.org/10.1016/S0169-5347(01)02330-8

Colles A, Liow LH, Prinzing A (2009) Are specialists at risk under environmental change? Neoecological, paleoecological and phylogenetic approaches. Ecol Lett 12:849–863. https://doi.org/10.1111/j.1461-0248.2009.01336.x

Conley DJ, Paerl HW, Howarth RW et al (2009) Controlling eutrophication: nitrogen and phosphorus. Science 323:1014–1015. https://doi.org/10.1126/science.1167755

Cooke SJ, Cowx IG (2006) Contrasting recreational and commercial fishing: searching for common issues to promote unified conservation of fisheries resources and aquatic environments. Biol Cons 128:93–108. https://doi.org/10.1016/j.biocon.2005.09.019

Costanza R, d'Arge R, De Groot R et al (1997) The value of the world's ecosystem services and natural capital. Nature 6630:253–260

Costello MJ, Coll M, Danovaro R et al (2010) A census of marine biodiversity knowledge, resources, and future challenges. PLoS ONE 5:e12110. https://doi.org/10.1371/journal.pone.0012110

Costello MJ, Wilson S, Houlding B (2012) Predicting total global species richness using rates of species description and estimates of taxonomic effort. Syst Biol 61:871–883. https://doi.org/10.1093/sysbio/syr080

Costello MJ, Bouchet P, Boxshall G et al (2013) Global coordination and standardisation in marine biodiversity through the World Register of Marine Species (WoRMS) and related databases. PLoS ONE 8:e51629. https://doi.org/10.1371/journal.pone.0051629

Courchamp F, Angulo E, Rivalan P et al (2006) Rarity value and species extinction: the anthropogenic Allee effect. PLoS Biol 4:e415. https://doi.org/10.1371/journal.pbio.0040415

Crain CM, Kroeker K, Halpern BS (2008) Interactive and cumulative effects of multiple human stressors in marine systems. Ecol Lett 11:1304–1315. https://doi.org/10.1111/j.1461-0248.2008.01253.x

Crain CM, Halpern BS, Beck MW et al (2009) Understanding and managing human threats to the coastal marine environment. Ann NY Acad Sci 1162:39–62. https://doi.org/10.1111/j.1749-6632.2009.04496.x

Creel S, Christianson D (2008) Relationships between direct predation and risk effects. Trends Ecol Evol 23:194–201. https://doi.org/10.1016/J.TREE.2007.12.004

Creel S, Christianson D, Liley S et al (2007) Predation risk affects reproductive physiology and demography of elk. Science 315:960. https://doi.org/10.1126/science.1135918

Crowder LB, Norse E (2008) Essential ecological insights for marine ecosystem-based management and marine spatial planning. Mar Policy 32:772–778. https://doi.org/10.1016/J.MARPOL.2008.03.012

Crutzen PJ (2002) Geology of mankind. Nature 415:23. https://doi.org/10.1038/415023a

Davies RWD, Cripps SJ, Nickson A et al (2009) Defining and estimating global marine fisheries bycatch. Mar Policy 33:661–672. https://doi.org/10.1016/j.marpol.2009.01.003

de Vargas C, Audic S, Henry N et al (2015) Eukaryotic plankton diversity in the sunlit ocean. Science 348:1261605. https://doi.org/10.1126/science.1261605

De Vos JM, Joppa LN, Gittleman JL et al (2015) Estimating the normal background rate of species extinction. Conserv Biol 29:452–462. https://doi.org/10.1111/cobi.12380

Derraik JGB (2002) The pollution of the marine environment by plastic debris: a review. Mar Pollut Bull 44:842–852. https://doi.org/10.1016/S0025-326X(02)00220-5

Dewar WK, Bingham RJ, Iverson RL et al (2006) Does the marine biosphere mix the ocean? J Mar Res 64:541–561. https://doi.org/10.1357/002224006778715720

Dirzo R, Young HS, Galetti M et al (2014) Defaunation in the Anthropocene. Science 345:401–406. https://doi.org/10.1126/science.1251817

Doney SC, Ruckelshaus M, Duffy JE et al (2012) Climate change impacts on marine ecosystems. Annu Rev Mar Sci 4:11–37. https://doi.org/10.1146/annurev-marine-041911-111611

Dornelas M, Gotelli NJ, McGill B et al (2014) Assemblage time series reveal biodiversity change but not systematic loss. Science 344:296–299. https://doi.org/10.1126/science.1248484

Duffy JE (2002) Biodiversity and ecosystem function: the consumer connection. Oikos 99:201–219. https://doi.org/10.1034/j.1600-0706.2002.990201.x

Dulvy NK, Sadovy Y, Reynolds JD (2003) Extinction vulnerability in marine populations. Fish Fish 4:25–64. https://doi.org/10.1046/j.1467-2979.2003.00105.x

Edgar GJ, Stuart-Smith RD, Willis TJ et al (2014) Global conservation outcomes depend on marine protected areas with five key features. Nature 506:216–220

Elahi R, O'Connor MI, Byrnes JE et al (2015) Recent trends in local-scale marine biodiversity reflect community structure and human impacts. Curr Biol 25:1938–1943. https://doi.org/10.1016/j.cub.2015.05.030

Estes JA, Tinker MT, Williams TM et al (1998) Killer whale predation on sea otters linking oceanic and nearshore ecosystems. Science 282:473–476. https://doi.org/10.1126/science.282.5388.473

Estes JA, Terborgh J, Brashares JS (2011) Trophic downgrading of planet Earth. Science 333:301–306

Franco-Trecu V, Drago M, Katz H et al (2017) With the noose around the neck: marine debris entangling in otariid species. Environ Pollut 220:985–989. https://doi.org/10.1016/j.envpol.2016.11.057

Francour P, Mangialajo L, Pastor J (2010) Mediterranean marine protected areas and non-indigenous fish spreading. In: Golani D, Appelbaum-Golani B (eds) Fish invasions in the Mediterranean Sea: change and renewal. Pensoft Publishers, Sofia-Moscow, pp 127–144

Frank KT, Petrie B, Choi JS et al (2005) Trophic cascades in a formerly cod-dominated ecosystem. Science 308:1621–1623. https://doi.org/10.1126/science.1113075

Friedlander AM, DeMartini EE (2002) Contrast in density, size, and biomass of reef fishes between the northwestern and the main Hawaiian islands: the effects of fishing down apex predators. Mar Ecol Prog Ser 230:253–264. https://doi.org/10.3354/meps230253

Futuyama DJ, Moreno G (1988) The evolution of ecological specialization. Annu Rev Ecol Syst 19:207–233. https://doi.org/10.1146/annurev.es.19.110188.001231

Gamfeldt L, Lefcheck JS, Byrnes JE et al (2015) Marine biodiversity and ecosystem functioning: what's known and what's next? Oikos 124:252–265. https://doi.org/10.1111/oik.01549

García VB, Lucifora LO, Myers RA (2008) The importance of habitat and life history to extinction risk in sharks, skates, rays and chimaeras. Proc R Soc Lond B Biol Sci 275:83–89. https://doi.org/10.1098/rspb.2007.1295

Gardner TA, Côté IM, Gill JA et al (2003) Long-term region-wide declines in Caribbean corals. Science 301:958–960. https://doi.org/10.1126/science.1086050

Garrison LP, Link JS (2000) Dietary guild structure of the fish community in the Northeast United States continental shelf ecosystem. Mar Ecol Prog Ser 202:231–240. https://doi.org/10.3354/meps202231

Gaston K (1998) Species-range size distributions: products of speciation, extinction and transformation. Philos Trans R Soc B 353:219–230. https://doi.org/10.1098/rstb.1998.0204

Gaston KJ (2010) Biodiversity. In: Sodhi NS, Ehrlich PR (eds) Conservation biology for all. Oxford University Press, New York, pp 27–42

Gearhart JL, Hataway BD, Hopkins N et al (2015) 2012 Turtle excluder device (TED) testing and gear evaluations. NOAA Technical Memorandum NMFS-SEFSC-674, 0-29. https://doi.org/10.7289/V5JW8BTZ

Geburzi JC, McCarthy ML (2018) How do they do it? – Understanding the success of marine invasive species. In: Jungblut S, Liebich V, Bode M (eds) YOUMARES 8 – Oceans Across Boundaries: learning from each other. Springer, Cham, pp 109–124. https://doi.org/10.1007/978-3-319-93284-2_8

Gerber LR, Keller AC, DeMaster DP (2007) Ten thousand and increasing: is the western arctic population of bowhead whale endangered? Biol Conserv 137:577–583. https://doi.org/10.1016/j.biocon.2007.03.024

Grand J, Cummings MP, Rebelo TG et al (2007) Biased data reduce efficiency and effectiveness of conservation reserve networks. Ecol Lett 10:364–374. https://doi.org/10.1111/j.1461-0248.2007.01025.x

Gray JS (2002) Biomagnification in marine systems: the perspective of an ecologist. Mar Pollut Bull 45:46–52. https://doi.org/10.1016/S0025-326X(01)00323-X

Greene CM, Hall JE, Guilbault KR et al (2010) Improved viability of populations with diverse life-history portfolios. Biol Lett 6:382–386. https://doi.org/10.1098/rsbl.2009.0780

Griffiths HJ (2010) Antarctic marine biodiversity–what do we know about the distribution of life in the Southern Ocean? PLoS ONE 5:e11683. https://doi.org/10.1371/journal.pone.0011683

Guinotte JM, Fabry VJ (2008) Ocean acidification and its potential effects on marine ecosystems. Ann N Y Acad Sci 1134:320–342. https://doi.org/10.1196/annals.1439.013

Haberl H, Erb KH, Krausmann F et al (2007) Quantifying and mapping the human appropriation of net primary production in earth's terrestrial ecosystems. Proc Natl Acad Sci U S A 104:12942–12947. https://doi.org/10.1073/pnas.0704243104

Hairston NG, Smith FE, Slobodkin LB (1960) Community structure, population control, and competition. Am Nat 94:421–425. https://doi.org/10.1086/282146

Halpern BS (2003) The impact of marine reserves: do reserves work and does reserve size matter? Ecol Appl 13:117–137. https://doi.org/10.1890/1051-0761(2003)013[0117:TIOMRD]2.0.CO;2

Halpern BS, Frazier M, Potapenko J et al (2015) Spatial and temporal changes in cumulative human impacts on the world's ocean. Nat Commun 6:7615. https://doi.org/10.1038/ncomms8615

Harley CD, Randall Hughes A, Hultgren KM et al (2006) The impacts of climate change in coastal marine systems. Ecol Lett 9:228–241. https://doi.org/10.1111/j.1461-0248.2005.00871.x

Harnik PG, Lotze HK, Anderson SC et al (2012) Extinctions in ancient and modern seas. Trends Ecol Evol 27:608–617. https://doi.org/10.1016/j.tree.2012.07.010

Harrison PA, Berry PM, Simpson G et al (2014) Linkages between biodiversity attributes and ecosystem services: a systematic review. Ecosyst Serv 9:191–203. https://doi.org/10.1016/j.ecoser.2014.05.006

Harvell D, Altizer S, Cattadori IM et al (2009) Climate change and wildlife diseases: when does the host matter the most? Ecology 90:912–920. https://doi.org/10.1890/08-0616.1

Hazen EL, Scales KL, Maxwell SM et al (2018) A dynamic ocean management tool to reduce bycatch and support sustainable fisheries. Sci Adv 4:eaar3001. https://doi.org/10.1126/sciadv.aar3001

Heithaus MR, Frid A, Wirsing AJ et al (2008) Predicting ecological consequences of marine top predator declines. Trends Ecol Evol 23:202–210. https://doi.org/10.1016/J.TREE.2008.01.003

Hilborn R, Branch TA, Ernst B et al (2003) State of the World's Fisheries. Annu Rev Environ Resour 28:359–399. https://doi.org/10.1146/annurev.energy.28.050302.105509

Hillebrand H, Blasius B, Borer ET et al (2018) Biodiversity change is uncoupled from species richness trends: consequences for conservation and monitoring. J Appl Ecol 55:169–184. https://doi.org/10.1111/1365-2664.12959

Hobday AJ, Tegner MK, Haaker PL (2000) Over-exploitation of a broadcast spawning marine invertebrate: decline of the white abalone. Rev Fish Biol Fish 10:493–514. https://doi.org/10.1023/A:1012274101311

Hoegh-Guldberg O, Mumby PJ, Hooten AJ et al (2007) Coral reefs under rapid climate change and ocean acidification. Science 318:1737–1742. https://doi.org/10.1126/science.1152509

Hoekstra JM, Boucher TM, Ricketts TH et al (2005) Confronting a biome crisis: global disparities of habitat loss and protection. Ecol Lett 8:23–29. https://doi.org/10.1111/j.1461-0248.2004.00686.x

Hulme PE (2009) Trade, transport and trouble: managing invasive species pathways in an era of globalization. J Appl Ecol 46:10–18. https://doi.org/10.1111/j.1365-2664.2008.01600.x

Hutchings JA, Reynolds JD (2004) Marine fish population collapses: consequences for recovery and extinction. BioScience 54:297–309

Hutchings JA, Minto C, Ricard D et al (2010) Trends in the abundance of marine fishes. Can J Fish Aquat Sci 67:1205–1210. https://doi.org/10.1139/F10-081

IPCC (2013) Climate change 2013: the physical science basis. Contribution of working group I to the fifth assessment report of the Intergovernmental Panel on Climate Change. Cambridge University Press, Cambridge. https://doi.org/10.1017/CBO9781107415324

IUCN (2017) The IUCN Red List of Threatened Species. Version 2017-2. Available: http://www.iucnredlist.org

IUCN (2018) The IUCN Red List of Threatened Species. Version 2018-1. Available: http://www.iucnredlist.org

Jablonski D (1986) Background and mass extinctions: the alternation of macroevolutionary regimes. Science 231:129–133

Jablonski D (1994) Extinctions in the fossil record. Philos Trans Royal Soc B 344:11–17. https://doi.org/10.1098/rstb.1994.0045

Jackson JB, Kirby MX, Berger WH et al (2001) Historical overfishing and the recent collapse of coastal ecosystems. Science 293:629–637

Jennings S, Blanchard JL (2004) Fish abundance with no fishing: predictions based on macroecological theory. J Anim Ecol 73:632–642. https://doi.org/10.1111/j.0021-8790.2004.00839.x

Johannesen E, Ingvaldsen RB, Bogstad B et al (2012) Changes in Barents Sea ecosystem state, 1970–2009: climate fluctuations, human impact, and trophic interactions. ICES J Mar Sci 69:880–889. https://doi.org/10.1093/icesjms/fss046

Jones MC, Cheung WW (2014) Multi-model ensemble projections of climate change effects on global marine biodiversity. ICES J Mar Sci 72:741–752. https://doi.org/10.1093/icesjms/fsu172

Jones CG, Lawton JH, Shachak M (1994) Organisms as ecosystem engineers. In: Samson FB, Knopf FL (eds) Ecosystem management. Springer, New York, pp 130–147

Jones KR, Klein CJ, Halpern BS et al (2018) The location and protection status of Earth's diminishing marine wilderness. Curr Biol 28:2506–2512. https://doi.org/10.1016/j.cub.2018.06.010

Kaiser MJ, Collie JS, Hall SJ et al (2002) Modification of marine habitats by trawling activities: prognosis and solutions. Fish Fish 3:114–136. https://doi.org/10.1046/j.1467-2979.2002.00079.x

Kappel CV (2005) Losing pieces of the puzzle: threats to marine, estuarine, and diadromous species. Front Ecol Environ 3:275–282. https://doi.org/10.1890/1540-9295(2005)003[0275:LPOTPT]2.0.CO;2

Kinlan BP, Gaines SD (2003) Propagule dispersal in marine and terrestrial environments: a community perspective. Ecology 84:2007–2020. https://doi.org/10.1890/01-0622

Klinger T, Padilla DK, Britton-Simmons K (2006) Two invaders achieve high densities in reserves. Aquat Conserv 16:301–311. https://doi.org/10.1002/aqc.717

Knapp S, Schweiger O, Kraberg A et al (2017) Do drivers of biodiversity change differ in importance across marine and terrestrial systems – Or is it just different research communities' perspectives? Sci Total Environ 574:191–203. https://doi.org/10.1016/j.scitotenv.2016.09.002

Lafferty KD (2004) Fishing for lobsters indirectly increases epidemics in sea urchins. Ecol App 14:1566–1573. https://doi.org/10.1890/03-5088

Lafferty KD, Kuris AM (1993) Mass mortality of abalone *Haliotis cracherodii* on the California Channel Islands: Tests of epidemiological hypotheses. Mar Ecol Prog Ser 96:239–248

Lanyon JM (2019) Management of megafauna in estuaries and coastal waters: Moreton Bay as a case study. In: Wolanski E, Day JW, Elliot M et al (eds) Coasts and estuaries – the future, 1st edn. Elsevier, Amsterdam, pp 87–101. https://doi.org/10.1016/B978-0-12-814003-1.00006-X

Laundré JW, Hernández L, Ripple WJ (2010) The landscape of fear: ecological implications of being afraid. Open Ecol J 3:1–7

Lavery TJ, Roudnew B, Gill P et al (2010) Iron defecation by sperm whales stimulates carbon export in the Southern Ocean. Proc R Soc Lond B Biol Sci 277:3527–3531. https://doi.org/10.1098/rspb.2010.0863

Lawler JJ, Aukema JE, Grant JB et al (2006) Conservation science: a 20-year report card. Front Ecol Environ 4:473–480. https://doi.org/10.1890/15409295

Lefcheck JS, Byrnes JE, Isbell F et al (2015) Biodiversity enhances ecosystem multifunctionality across trophic levels and habitats. Nat Commun 6:6936. https://doi.org/10.1038/ncomms7936

Leray M, Knowlton N (2016) Censusing marine eukaryotic diversity in the twenty-first century. Philos Trans Royal Soc B 371:20150331. https://doi.org/10.1098/rstb.2015.0331

Lewison RL, Crowder LB, Wallace BP et al (2014) Global patterns of marine mammal, seabird, and sea turtle bycatch reveal taxa-specific and cumulative megafauna hotspots. Proc Natl Acad Sci U S A 111:5271–5276. https://doi.org/10.1073/pnas.1318960111

Ling SD, Scheibling RE, Rassweiler A et al (2015) Global regime shift dynamics of catastrophic sea urchin overgrazing. Philos Trans Royal Soc B 370:20130269. https://doi.org/10.1098/rstb.2013.0269

Lockwood JL, Russel GJ, Gittleman JL et al (2002) A metric for analysing taxonomic patterns of extinction risk. Conserv Biol 16:1137–1142. https://doi.org/10.1046/j.1523-1739.2002.01152.x

Lotze HK, Milewski I (2004) Two centuries of multiple human impacts and successive changes in a North Atlantic food web. Ecol Appl 14:1428–1447. https://doi.org/10.1890/03-5027

Lotze HK, Lenihan HS, Bourque BJ et al (2006) Depletion, degradation, and recovery potential of estuaries and coastal seas. Science 312:1806–1809. https://doi.org/10.1126/science.1128035

Lotze HK, Coll M, Magera AM et al (2011) Recovery of marine animal populations and ecosystems. Trends Ecol Evol 26:595–605. https://doi.org/10.1016/j.tree.2011.07.008

Magurran AE, Dornelas M, Moyes F et al (2015) Rapid biotic homogenization of marine fish assemblages. Nat Commun 6:8405

Mahon R, Hunte W (2002) Trap mesh selectivity and the management of reef fishes. Fish Fish 2:356–375. https://doi.org/10.1046/j.1467-2960.2001.00054.x

Mangel JC, Wang J, Alfaro-Shigueto J et al (2018) Illuminating gillnets to save seabirds and the potential for multi-taxa bycatch mitigation. R Soc Open Sci 5:180254. https://doi.org/10.1098/rsos.180254

Marine Conservation Institute (2019) MPAtlas [On-line]. Seattle, WA. Available at: www.mpatlas.org [Accessed 14 March 2019)]

Mark JC, Simon W, Brett H (2012) Predicting total global species richness using rates of species description and estimates of taxonomic effort. Systematic Biology 61 (5):871

MARPOL (1973) International convention for the prevention of pollution from ships: 1973 Convention, 1978 Protocol. Available: http://www.mar.ist.utl.pt/mventura/Projecto-Navios-I/IMO-Conventions%20%28copies%29/MARPOL.pdf

McCann KS (2000) The diversity–stability debate. Nature 405:228–233. https://doi.org/10.1038/35012234

McCauley DJ, Pinsky ML, Palumbi SR et al (2015) Marine defaunation: animal loss in the global ocean. Science 347:1255641. https://doi.org/10.1126/science.1255641

McGill BJ, Dornelas M, Gotelli NJ et al (2015) Fifteen forms of biodiversity trend in the Anthropocene. Trends Ecol Evol 30:104–113. https://doi.org/10.1016/j.tree.2014.11.006

McKinney ML (1997) Extinction vulnerability and selectivity: combining ecological and paleontological views. Annu Rev Ecol Syst 28:495–516. https://doi.org/10.1146/annurev.ecolsys.28.1.495

McKinney ML, Lockwood JL (1999) Biotic homogenization: a few winners replacing many losers in the next mass extinction. Trends Ecol Evol 14:450–453. https://doi.org/10.1016/S0169-5347(99)01679-1

McLeod E, Anthony KR, Andersson A et al (2013) Preparing to manage coral reefs for ocean acidification: lessons from coral bleaching. Front Ecol Environ 11:20–27. https://doi.org/10.1890/110240

Millennium Ecosystem Assessment (MEA) (2005) Ecosystems and human well-being. Island Press, Washington DC

Mills SL, Soule ME, Doak DF (1993) The keystone-species concept in ecology and conservation. BioScience 43:219–224. https://doi.org/10.2307/1312122

Molnar JL, Gamboa RL, Revenga C et al (2008) Assessing the global threat of invasive species to marine biodiversity. Front Ecol Environ 6:485–492. https://doi.org/10.1890/070064

Mora C, Tittensor DP, Adl S et al (2011) How many species are there on Earth and in the ocean? PLoS Biol 9:e1001127. https://doi.org/10.1371/journal.pbio.1001127

Mumby PJ, Harborne AR (2010) Marine reserves enhance the recovery of corals on Caribbean reefs. PLoS ONE 5:e8657. https://doi.org/10.1371/journal.pone.0008657

Mumby PJ, Hastings A, Edwards HJ (2007) Thresholds and the resilience of Caribbean coral reefs. Nature 450:98–101. https://doi.org/10.1038/nature06252

Myers RA, Worm B (2003) Rapid worldwide depletion of predatory fish communities. Nature 423:280–283. https://doi.org/10.1038/nature01610

Myers RA, Worm B (2005) Extinction, survival or recovery of large predatory fishes. Philos Trans R Soc Lond 360:13–20. https://doi.org/10.1098/rstb.2004.1573

Myers RA, Baum JK, Shepherd TD et al (2007) Cascading effects of the loss of apex predatory sharks from a coastal ocean. Science 315:1846–1850. https://doi.org/10.1126/science.1138657

Nowacek DP, Clark CW, Mann D et al (2015) Marine seismic surveys and ocean noise: time for coordinated and prudent planning. Front Ecol Environ 13:378–386. https://doi.org/10.1890/130286

O'Connor S, Ono R, Clarkson C (2011) Pelagic fishing at 42,000 years before the present and the maritime skills of modern humans. Science 334:1117–1121. https://doi.org/10.1126/science.1207703

O'Rourke KH, Williamson JG (2002) When did globalization begin? Eur Rev Econ Hist 6:23–50. https://doi.org/10.1017/S1361491602000023

Ocean Conservancy (2010) Trash travels. From our hands to the sea, around the globe, and through time. Available: https://oceanconservancy.org/wp-content/uploads/2017/04/2010-Ocean-Conservancy-ICC-Report.pdf

O'Connor MI, Bruno JF, Gaines SD et al (2007) Temperature control of larval dispersal and the implications for marine ecology, evolution, and conservation. Proc Natl Acad Sci U S A 104:1266–1271. https://doi.org/10.1073/pnas.0603422104

Olden JD (2006) Biotic homogenization: a new research agenda for conservation biogeography. J Biogeogr 33:2027–2039. https://doi.org/10.1111/j.1365-2699.2006.01572.x

Olden JD, LeRoy Poff N, Douglas MR et al (2004) Ecological and evolutionary consequences of biotic homogenization. Trends Ecol Evol 19:18–24. https://doi.org/10.1016/j.tree.2003.09.010

Olden JD, Hogan ZS, Vander Zanden MJ (2007) Small fish, big fish, red fish, blue fish: size-biased extinction risk of the world's freshwater and marine fishes. Glob Ecol Biogeogr 16:694–701. https://doi.org/10.1111/j.1466-8238.2007.00337.x

Orr JC, Fabry VJ, Aumont O et al (2005) Anthropogenic ocean acidification over the twenty-first century and its impact on calcifying organisms. Nature 437:681. https://doi.org/10.1038/nature04095

Ortiz N, Mangel JC, Wang J et al (2016) Reducing green turtle bycatch in small-scale fisheries using illuminated gillnets: the cost of saving a sea turtle. Mar Ecol Prog Ser 545:251–259. https://doi.org/10.3354/meps11610

Österblom H, Hansson S, Larsson U et al (2007) Human-induced trophic cascades and ecological regime shifts in the Baltic Sea. Ecosystems 10:877–889. https://doi.org/10.1007/s10021-007-9069-0

Pacifici M, Foden WB, Visconti P et al (2015) Assessing species vulnerability to climate change. Nat Clim Chang 5:215–224. https://doi.org/10.1038/nclimate2448

Packer C, Holt RD, Hudson PJ et al (2003) Keeping the herds healthy and alert: implications of predator control for infectious disease. Ecol Lett 6:797–802. https://doi.org/10.1046/j.1461-0248.2003.00500.x

Paine RT (1969) A note on trophic complexity and community stability. Am Nat 103:91–93. https://doi.org/10.2307/2459472

Paine RT (1980) Food webs: linkage, interaction strength and community infrastructure. J Anim Ecol 49:666–685. https://doi.org/10.2307/4220

Palumbi SR, Sandifer PA, Allan JD et al (2009) Managing for ocean biodiversity to sustain marine ecosystem services. Front Ecol Environ 7:204–211. https://doi.org/10.1890/070135

Pauly D, Christensen V, Dalsgaard J et al (1998) Fishing down marine food webs. Science 279:860–863. https://doi.org/10.1126/science.279.5352.860

Pershing AJ, Christensen LB, Record NR et al (2010) The impact of whaling on the ocean carbon cycle: why bigger was better. PLoS ONE 5:e12444. https://doi.org/10.1371/journal.pone.0012444

Petchey OL, Gaston KJ (2006) Functional diversity: back to basics and looking forward. Ecol Lett 9:741–758. https://doi.org/10.1111/j.1461-0248.2006.00924.x

Pimm SL, Russell GJ, Gittleman JL et al (1995) The future of biodiversity. Science 269:347–350. https://doi.org/10.1126/science.269.5222.347

Pimm SL, Jenkins CN, Abell R et al (2014) The biodiversity of species and their rates of extinction, distribution, and protection. Science 344:1246752. https://doi.org/10.1126/science.1246752

Poloczanska ES, Burrows MT, Brown CJ et al (2016) Responses of marine organisms to climate change across oceans. Front Mar Sci 3:1–21. https://doi.org/10.3389/fmars.2016.00062

Power ME, Tilman D, Estes JA et al (1996) Challenges in the quest for keystones: identifying keystone species is difficult – but essential to understanding how loss of species will affect ecosystems. BioScience 46:609–620. https://doi.org/10.2307/1312990

Purcell SW, Polidoro BA, Hamel JF et al (2014) The cost of being valuable: predictors of extinction risk in marine invertebrates exploited as luxury seafood. Proc R Soc Lond B Biol Sci 281:20133296. https://doi.org/10.1098/rspb.2013.3296

Reynolds JD, Dulvy NK, Goodwin NB et al (2005) Biology of extinction risk in marine fishes. Proc R Soc Lond B Biol Sci 272:2337–2344. https://doi.org/10.1098/rspb.2005.3281

Richardson WJ, Greene CR, Malme CJ et al (1995) Marine mammals and noise. Academic Press Inc, San Diego

Roberts CM, Hawkins JP (1999) Extinction risk at sea. Trends Ecol Evol 14:241–246. https://doi.org/10.1016/S0169-5347(98)01584-5

Roman J, McCarthy JJ (2010) The whale pump: marine mammals enhance primary productivity in a coastal basin. PLoS ONE 5:e13255. https://doi.org/10.1371/journal.pone.0013255

Roman J, Estes JA, Morissette L et al (2014) Whales as marine ecosystem engineers. Front Ecol Environ 12:377–385. https://doi.org/10.1890/130220

Rosenzweig ML (1995) Species diversity in space and time. Cambridge University Press, Cambridge

Rosenzweig C, Karoly D, Vicarelli M et al (2008) Attributing physical and biological impacts to anthropogenic climate change. Nature 453:353–357. https://doi.org/10.1038/nature06937

Ross ST (1986) Resource partitioning in fish assemblages: a review of field studies. Copeia:352–388. https://doi.org/10.2307/1444996

Rossi S (2013) The destruction of the "animal forests" in the oceans: towards an over-simplification of the benthic ecosystems. Ocean Coast Manage 84:77–85. https://doi.org/10.1016/j.ocecoaman.2013.07.004

Rudkin DM, Young GA, Nowlan GS (2008) The oldest horseshoe crab: a new xiphosurid from late ordovician konservat-lagerstätten deposits, Manitoba Canada. Palaeontol 51:1–9. https://doi.org/10.1111/j.1475-4983.2007.00746.x

Sala E, Ballesteros E, Dendrinos P et al (2012) The structure of Mediterranean rocky reef ecosystems across environmental and human gradients, and conservation implications. PLoS ONE 7:e32742. https://doi.org/10.1371/journal.pone.0032742

Sallan L, Galimberti AK (2015) Body-size reduction in vertebrates following the end-Devonian mass extinction. Science 350:812–815. https://doi.org/10.1126/science.aac7373

Sandel B, Arge L, Dalsgaard B et al (2011) The influence of late quaternary climate-change velocity on species endemism. Science 334:660–664. https://doi.org/10.1126/science.1210173

Säterberg T, Sellman S, Ebenman B (2013) High frequency of functional extinctions in ecological networks. Nature 499:468–470. https://doi.org/10.1038/nature12277

Sax DF, Gaines SD (2003) Species diversity: from global decreases to local increases. Trends Ecol Evol 18:561–566. https://doi.org/10.1016/S0169-5347(03)00224-6

Scavia D, Field JC, Boesch DF et al (2002) Climate change impacts on US coastal and marine ecosystems. Estuaries 25:149–164. https://doi.org/10.1007/BF02691304

Scheffer M, Carpenter S, Foley JA (2001) Catastrophic shifts in ecosystems. Nature 413:591–596

Schindler DE, Armstrong JB, Reed TE (2015) The portfolio concept in ecology and evolution. Front Ecol Environ 13:257–263. https://doi.org/10.1890/140275

Scott KN (2004) International regulation of undersea noise. Int Comp Law Q 53:287–324

Secretariat of the Convention on Biological Diversity and the Scientific and Technical Advisory Panel - GEF (2012) Impacts of marine debris on biodiversity: current status and potential solutions. CBD Technical Series 67:1–61

Severtsov AS, Shubkina AV (2015) Predator–prey interaction between individuals: 1. The role of predators in natural selection. Biol Bull 42:633–642. https://doi.org/10.1134/S1062359015070080

Sguotti C, Cormon X (2018) Regime shifts – a global challenge for the sustainable use of our marine resources. In: Jungblut S, Liebich V, Bode M (eds) YOUMARES 8 – Oceans Across Boundaries: learning from each other. Springer, Cham, pp 155–166. https://doi.org/10.1007/978-3-319-93284-2_11

Sheavly SB, Register KM (2007) Marine debris and plastics: environmental concerns, sources, impacts and solutions. J Polym Environ 15:301–305. https://doi.org/10.1007/s10924-007-0074-3

Shimada T, Limpus C, Jones R et al (2017) Aligning habitat use with management zoning to reduce vessel strike of sea turtles. Ocean Coast Manag 142:163–172. https://doi.org/10.1016/j.ocecoaman.2017.03.028

Smith FA, Lyons SK, Ernest SM et al (2003) Body mass of late quaternary mammals. Ecol 84:3403. https://doi.org/10.1890/02-9003

Sorte CJ, Williams SL, Carlton JT (2010) Marine range shifts and species introductions: comparative spread rates and community impacts. Global Ecol Biogeogr 19:303–316. https://doi.org/10.1111/j.1466-8238.2009.00519.x

Steffen W, Broadgate W, Deutsch L et al (2015) The trajectory of the Anthropocene: the great acceleration. Anthropocene Rev 2:81–98. https://doi.org/10.1177/2053019614564785

Steneck RS, Graham MH, Bourque BJ et al (2002) Kelp forest ecosystems: biodiversity, stability, resilience and future. Environ Conserv 29:436–459. https://doi.org/10.1017/S0376892902000322

Steneck RS, Leland A, McNaught DC et al (2013) Ecosystem flips, locks, and feedbacks: the lasting effects of fisheries on Maine's kelp forest ecosystem. Bull Mar Sci 89:31–55. https://doi.org/10.5343/bms.2011.1148

Suding KN, Gross KL, Houseman GR (2004) Alternative states and positive feedbacks in restoration ecology. Trends Ecol Evol 19:46–53. https://doi.org/10.1016/j.tree.2003.10.005

Thomas HL, Macsharry B, Morgan L et al (2014) Evaluating official marine protected area coverage for Aichi Target 11: appraising the data and methods that define our progress. Aquat Conserv 24:8–23. https://doi.org/10.1002/aqc.2511

Thrush SF, Gray JS, Hewitt JE et al (2006) Predicting the effects of habitat homogenization on marine biodiversity. Ecol Appl 16:1636–1642. https://doi.org/10.1890/1051-0761(2006)016[1636:PTEOHH]2.0.CO;2

United Nations General Assembly (2015) Transforming our world: the 2030 agenda for sustainable development. Available: http://www.un.org/ga/search/view_doc.asp?symbol=A/RES/70/1&Lang=E

Valiela I, Bowen JL, York JK (2001) Mangrove forests: One of the world's threatened major tropical environments: at least 35% of the area of mangrove forests has been lost in the past two decades, losses that exceed those for tropical rain forests and coral reefs, two other well-known threatened environments. BioScience 51:807–815. https://doi.org/10.1641/0006-3568(2001)051[0807:MFOOTW]2.0.CO;2

van der Heide T, van Nes EH, Geerling GW et al (2007) Positive feedbacks in seagrass ecosystems: implications for success in conservation and restoration. Ecosystems 10:1311–1322. https://doi.org/10.1007/s10021-007-9099-7

Villarrubia-Gómez P, Cornell SE, Fabres J (2018) Marine plastic pollution as a planetary boundary threat–the drifting piece in the sustainability puzzle. Mar Policy 96:213–220. https://doi.org/10.1016/j.marpol.2017.11.035

Walsh JE, Fetterer F, Scott Stewart J et al (2016) A database for depicting Arctic sea ice variations back to 1850. Geogr Rev 107:89–107. https://doi.org/10.1111/j.1931-0846.2016.12195.x

Wang J, Barkan J, Fisler S et al (2013) Developing ultraviolet illumination of gillnets as a method to reduce sea turtle bycatch. Biol Lett 9:20130383. https://doi.org/10.1098/rsbl.2013.0383
Waycott M, Duarte CM, Carruthers TJ et al (2009) Accelerating loss of seagrasses across the globe threatens coastal ecosystems. Proc Natl Acad Sci U S A 106:12377–12381. https://doi.org/10.1073/pnas.0905620106
Webb TJ, Mindel BL (2015) Global patterns of extinction risk in marine and non-marine systems. Curr Biol 25:506–511. https://doi.org/10.1016/j.cub.2014.12.023
Webb TJ, Berghe EV, O'Dor R (2010) Biodiversity's big wet secret: the global distribution of marine biological records reveals chronic under-exploration of the deep pelagic ocean. PLoS ONE 5:e10223. https://doi.org/10.1371/journal.pone.0010223
Wilcox C, Van Sebille E, Hardesty BD (2015) Threat of plastic pollution to seabirds is global, pervasive and increasing. Prod Natl Acad Sci U S A 112:11899–11904. https://doi.org/10.1073/pnas.1502108112
Wilmers CC, Estes JA, Edwards M et al (2012) Do trophic cascades affect the storage and flux of atmospheric carbon? An analysis of sea otters and kelp forests. Front Ecol Environ 10:409–415. https://doi.org/10.1890/110176
Winfree R, Fox JW, Williams NM et al (2015) Abundance of common species, not species richness, drives delivery of a real-world ecosystem service. Ecol Lett 18:626–635. https://doi.org/10.1111/ele.12424
Wood LJ, Fish L, Laughren J et al (2008) Assessing progress towards global marine protection targets: shortfalls in information and action. Oryx 42:340–351. https://doi.org/10.1017/s003060530800046x
Worm B, Tittensor DP (2011) Range contraction in large pelagic predators. Proc Natl Acad Sci U S A 108:11942–11947. https://doi.org/10.1073/pnas.1102353108
Worm B, Barbier EB, Beaumont N et al (2006) Impacts of biodiversity loss on ocean ecosystem services. Science 314:787–790. https://doi.org/10.1126/science.1132294
Worm B, Davis B, Kettemer L et al (2013) Global catches, exploitation rates, and rebuilding options for sharks. Mar Policy 40:194–204. https://doi.org/10.1016/j.marpol.2012.12.034
WoRMS Editorial Board (2018) World Register of Marine Species. http://www.marinespecies.org. Accessed 15 April 2018. doi:10.14284/170
WWF (2015) Living Blue Planet Report. Species, habitats and human well-being. WWF, Gland
Young HS, McCauley DJ, Galetti M et al (2016) Patterns, causes, and consequences of anthropocene defaunation. Annu Rev Ecol Evol Syst 47:333–358. https://doi.org/10.1146/annurev-ecolsys-112414-054142

Check for updates

5 Challenges in Marine Restoration Ecology: How Techniques, Assessment Metrics, and Ecosystem Valuation Can Lead to Improved Restoration Success

Laura Basconi, Charles Cadier, and Gustavo Guerrero-Limón

Abstract

Evaluating the effectiveness and success of coastal marine habitat restoration is often highly challenging and can vary substantially between different habitat types. The current article presents a state-of-the-art review of habitat-level restoration in the coastal marine environment. It sets out most successful techniques across habitats and suggestions of better metrics to assess their success. Improvements in restoration approach are outlined, with a particular focus on selective breeding, using recent advancements in genetics. Furthermore, the assessment of ecosystem services, as a metric to determine restoration success on a spatiotemporal scale, is addressed in this article. As the concept of ecosystem services is more tangible for a nonscientific audience, evaluating restoration success in this manner has the potential to greatly contribute to raising awareness of environmental issues and to implement socioeconomic policies. Moreover, habitat-based restoration has been proven to be an effective tool to address the issue of ecosystem service sustainability and poverty alleviation. Appropriate conservation management, prior to the implementation of restoration activities, is crucial to create an environment in which restoration efforts are likely to succeed.

The authors Laura Basconi, Charles Cadier and Gustavo Guerrero-Limón have been equally contributing to this chapter.

L. Basconi (✉)
Ca Foscari University, Venice, Italy
e-mail: 956409@stud.unive.it

C. Cadier (✉)
MER Consortium, UPV, Bilbao, Spain
e-mail: c.cadier@griffith.edu.au

G. Guerrero-Limón (✉)
MER Consortium, UPV, Bilbao, Spain

University of Liege, Ulg, Belgium
e-mail: g.guerrero@doct.uliege.be

5.1 Introduction

Restoration ecology is an emerging branch of environmental science which gained increased attention since the 1990s. The Society of Ecological Restoration, founded in 1983, states that ecological restoration is an intentional activity which has been initiated to accelerate the recovery of an ecosystem with respect to its health, integrity, and sustainability (SER 1998; Bullock et al. 2011). Frequently, the ecosystem that requires restoration has been degraded, damaged, transformed, or entirely destroyed as the direct or indirect result of human activities. Restoration ecology has the aim to restore the integrity of ecological systems, therefore restoring a critical range of variability in biodiversity, ecological processes and structures, regional and historical context, and sustainable cultural practices (SER 1998). In the last 20 years, restoration actions have been increasingly carried out all over the world (Swan et al. 2016; Zhang et al. 2018), and it is anticipated to become one of the most important fields within conservation science of the twenty-first century (Hobbs and Harris 2001).

At present, passive conservation aims to protect coastal marine habitats by removing or mitigating environmental stressors (e.g., removing polluting agents, increase water quality, and/or ban human uses at the damaged coastal site). Although these direct and indirect anthropogenic stressors once removed could allow for the natural recovery of these systems, in reality this does not always occur (Perrow and Davy 2002; Cox et al. 2017). For instance, in the case of seagrasses, even if there is an improvement in water quality and/or coastal tourism is banned, there could no longer be a population which can produce seeds nearby the damaged site. It impedes the natural recovery of the lost habitat (Nyström et al. 2012). In these conditions, restoration can help conservationists to reach their goal. Active conservation or restoration is the practice of rebuilding degraded, damaged, or destroyed ecosystems and habitats by active human intervention (Hobbs and Norton 1996; Palmer et al. 1997; Elliott

S. Jungblut et al. (eds.), *YOUMARES 9 - The Oceans: Our Research, Our Future*,
https://doi.org/10.1007/978-3-030-20389-4_5

et al. 2007). The implementation of appropriate conservation management, prior to any restoration attempt, is vital to reduce impacts making restoration feasible (Hobbs and Norton 1996; Benayas et al. 2009; McDonald et al. 2016).

Increased interest is not the result of scientific enthusiasm over a new research topic, but it is rather due to the urgent need to counteract the alarming decline in the cover of important habitats. Coastal marine ecosystems are being lost at alarming rates, and for them, passive protection could not be enough (Pandolfi et al. 2003; Hoekstra et al. 2005; Abelson et al. 2015; Doxa et al. 2017). Seagrass beds, mangroves, salt marshes, corals, and oyster reefs are important nursery habitats for species of economic interest (Robertson and Duke 1987; Hemminga and Duarte 2000). Active filtration in oyster reefs and sediment settlement facilitation in salt marshes, mangroves, and seagrass beds contribute to water purification (McLeod et al. 2011; Grabowsky et al. 2012). Moreover, salt marshes, mangroves, seagrass beds, macroalgal forests, corals, and oyster reefs are natural barriers against hydrodynamic forces and thus stabilize sediments (Moberg and Folke 1999; Hemminga and Duarte 2000; Lovelock et al. 2005; Gedan et al. 2011). Those habitats prevent coastal erosion (Callaway et al. 1997; Herkül and Kotta 2009). This sediment accumulation ultimately leads to the creation of organic-rich soils, acting as a carbon storage (*blue carbon*) and therefore a buffer against global warming (Nelleman et al. 2009; McLeod et al. 2011). Furthermore, leisure and recreational services are provided by coral reefs, for instance, generating scuba diving tourism which produces revenues to local communities (Moberg and Folke 1999). Some of these ecosystems also act as sites of social and cultural heritage such as sacred mangrove forests (Rönnbäck et al. 2007). The multitude of services and benefits derived from these coastal marine habitats highlight the need to restore them to have long-lasting ecosystem services provisioning.

Since the dawn of the industrial revolution, mankind has been inducing major changes at a global scale, being the primary cause of global warming. Humankind exploited nature lacking the consciousness of the long-term consequences. This unprecedented use brought to present enormous challenges to protect the environment. Ecosystem deterioration and loss of many important coastal habitats is one of the many repercussions of the multiple stresses that humans have caused over the past decades (Halpern et al. 2008; Waycott et al. 2009; Micheli et al. 2013). Cumulative impacts, such as overfishing, oil drilling, maritime transport, and coastal tourism, affect coastal marine habitats worldwide, leading to strong pressures to pristine environments, ultimately resulting in habitat loss (Airoldi and Beck 2007; Abelson et al. 2015; Doxa et al. 2017). Human activities have globally transformed or destroyed 30 to 50% of mangroves (Valiela et al. 2001; Duke et al. 2007; Giri et al. 2011; Richards and Friess 2017); 40% of global coral reefs, with a shocking prevision of almost complete disappearance by 2050 (Pandolfi et al. 2003; Eyre et al. 2018); 29% of seagrass beds (Waycott et al. 2009); and 85% of oyster reefs (Lotze et al. 2006; Beck et al. 2011). A comparable loss has been observed for macroalgal forests that, together with seagrass loss, has been considered a real "marine deforestation" (Airoldi and Beck 2007; Connell et al. 2008; Yu et al. 2012). Consequently, the decline of marine forests is leading to a reduction of structurally complex habitats, especially across temperate marine environments (Scheffer et al. 2001; Steneck et al. 2002; Sala 2004) with a significant reduction in species abundance and relative distributions (Novacek and Cleland 2001; Worm et al. 2006; Airoldi et al. 2008).

To counteract this loss, the most used mitigation action is the establishment of marine protected areas (MPAs). While recently O'Leary et al. (2017) stated that climate disturbance does not affect natural recovery, sea surface temperature (SST) increase and acidification due to climate change can present challenges for coastal marine habitat passively conserved (Harley et al. 2006; Hoegh-Guldberg and Bruno 2010). Furthermore, the stage of habitat degradation is often so pervasive that habitats require assisted recovery as well as active restoration (Young 2000; Perkol-Finkel and Airoldi 2010). Ecological restoration at the sea is, however, in its infancy, sometimes resulting in unsuccessful restoration attempts (Bayraktarov et al. 2016).

This paper aims to present marine restoration challenges through a review of published peer-reviewed studies with three main approaches: (i) habitat-based restoration techniques, (ii) restoration success among habitats and the importance of assessment metrics, and (iii) the importance of ecosystem service valuation.

5.2 Habitat-Level Restoration

Habitat restoration is an emerging field in marine ecology. A search on Google Scholar using the search terms "marine," "restoration," and "ecology" ended up with 191,000 articles displayed in the graph below (Fig. 5.1). Interest in marine restoration ecology worldwide arose in the 1990s, but just in the last 20 years, restoration actions have been increasingly carried out all over the world (a trend clearly identified also by Swan et al. (2016) and Zhang et al. (2018)).

Even though ecological restoration of marine ecosystems is quite young, some guidelines and rules have been identified, for instance, the need to define the historical baseline to which the system will be restored, prior to any restoration attempt (Seaman 2007). This has to be in a manner that restoration will be carried out in areas in which the species aimed to be restored was present before the disturbance. Moreover, the introduction of a species, such as a habitat

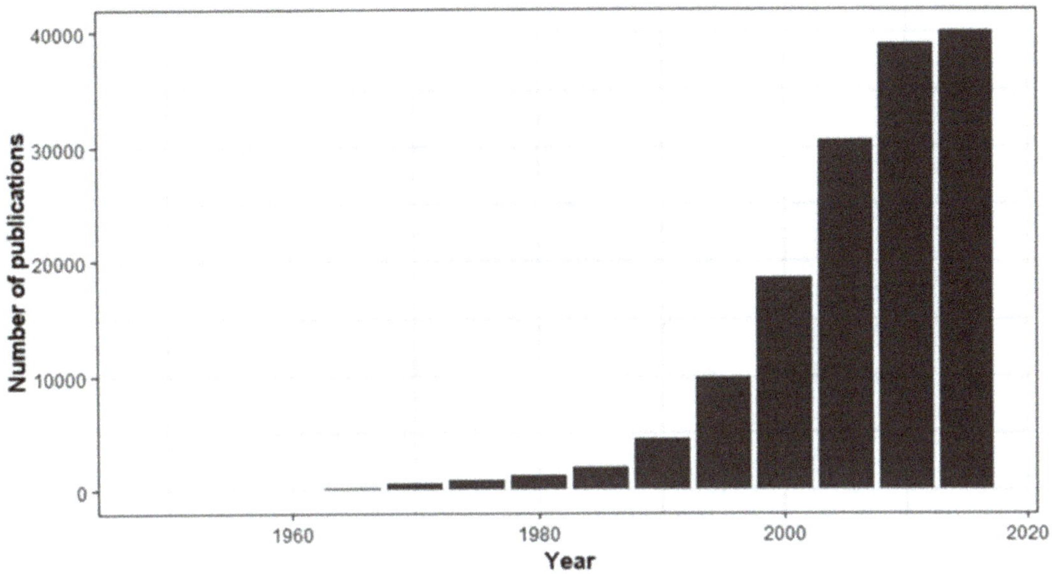

Fig. 5.1 The increase in published marine restoration articles from 1950 to the present day as results of Google Scholar search using the search terms "marine," "restoration," and "ecology" in one search. The past 20 years experienced a near exponential increase in scientific papers related to the topic of marine restoration

former, in a coastal area to enhance, for instance, productivity or merely as an afforestation action is not considered an ecological restoration (Elliott et al. 2007). In theory, any action of restoration without a historical reference point cannot be considered as such. But, in practice, only few habitats possess robust information on their historical baselines.

Finally, to allocate financial resources effectively, restoration should target areas in which its feasibility has been assessed (e.g., through ecological modelling) and especially where the environmental conditions are suitable for the survival of the target species. Abiotic conditions need to be checked in advance to see if loss of habitat is irreversible or not (Bellgrove et al. 2017). For instance, if the water conditions are detrimental to the introduction and persistence of the species, the water quality itself should be restored prior to implementing any restoration actions. Bayraktarov et al. (2016) stressed that low survivorship of restored corals was mainly due to inadequate site selection (e.g., sites with high sedimentation rates or strong currents) (Ammar et al. 2000; Fox et al. 2005). The same occurs in restoration efforts of seagrasses which often failed due to planting seagrass shoots at high wave energy locations without anchoring, inappropriate anchoring, or planting at sites with high levels of sediment movement and erosion (Bird et al. 1994; Ganassin and Gibbs 2008). Restoration actions can largely differ in terms of their applied strategies and approaches and whether they are focused at population or habitat levels or at an even broader landscape scale: the ecosystem level. To date, the most explored restoration approach is the habitat level. It focuses on one (or two) species that are known to act as habitat formers (i.e., targeted species are often ecosystem engineers that then act as foundation species) (Fig. 5.2). Once reintroduced, these species can provide fundamental ecosystem functions and processes that will ultimately benefit other associated organisms and lead to overall system recovery (Powers and Boyer 2014).

5.2.1 Habitat-Level Restoration Techniques

Habitat-level restoration in the marine environment concerns mainly coastal marine habitats, such as salt marshes, coral reefs, seagrasses, oyster reefs, macroalgae, and mangroves (Table 5.1). Attempts to restore these habitats encompass a variety of different techniques such as transplanting different stages of an organism life cycle, in the case of corals, seagrass, and macroalgae, the introduction of artificial substrata colonized by the target species, and planting of mangroves and salt marsh plants (Table 5.2). Restoration actions can in some cases be combined with measures to enhance water quality and improve the hydrodynamic conditions influencing these habitats. Based on a coarse literature review (considering even macroalgae) and information derived from Bayraktarov et al. (2016), the most explored technique, independently from the targeted habitat, is the transplantation approach. Transplantation consists of the movement of the species from a donor site where it is still present to another, where there is the need to restore the vanished habitat. Many of the examples reporting highly successful transplantations in coastal marine habitats are from transplants of entire coral colonies or fragments (branches of the coral) from donor colonies, as well as from coral farming techniques

Fig. 5.2 Coral farming of *Acroporacervicornis* and planting in Key Largo, Florida (USA), performed by the Coral Restoration Foundation (permission by Coral Restoration Foundation™)

(Bowden-Kerby 2001; Rinkevich 2005; Edwards and Gomez 2007; Bayraktarov et al. 2016).

Coral farming is the process of collecting fragments of corals from local reefs, raising them in nurseries until mature, and then installing them at the restoration site (Fig. 5.3) (Shaish et al. 2008; Tortolero-Langarica et al. 2014). For seagrasses, the most commonly used technique is the transplantation of seedlings, sprigs, shoots, or rhizomes (Bastyan and Cambridge 2008, Ganassin and Gibbs 2008; Balestri and Lardicci 2014). In particular, van Katwijk et al. (2016) pointed out that rhizome fragments, anchored using weights, are the most successful way to restore seagrass beds. For macroalgae, the most successful techniques involve the transplantation of both adults and early life stages of the organism (i.e., sporophytes or juveniles) which can be sourced from natural donor sites or aquaculture facilities (Terawaki et al. 2001, 2003; Falace et al. 2006; Yoon et al. 2014). Transplantation techniques have also been the most

Table 5.1 Frequently used species to restore coastal marine habitats. The first priority for a suitable species is the expectation of that species to survive at the site that is being restored. The species are listed randomly, not by proven restoration success. These species are habitat formers which once reintroduced can benefit other species, increase coastal productivity, and act as carbon sinks. These species have mostly resulted in successful restoration efforts (Bayraktarov et al. 2016)

Ecosystem	Species	References
Coral reefs	*Acropora* spp.	Edwards and Gomez (2007)
	Pocillopora spp.	Johnson et al. (2011)
	Porites spp.	Pizarro et al. (2012)
	Merulinascabricula	
	Milleporaalcicornis	
Seagrasses	*Cymodoceanodosa*	Ganassin and Gibbs (2008)
	Zostera spp.	Bastyan and Cambridge (2008)
	Posidonia spp.	Balestri and Lardicci (2014)
	Halodule spp.	
	Ruppiamaritima	
Mangrove forests	*Rhizophora* spp.	Goforth and Thomas (1979)
	Avicennia spp.	Ainodion et al. (2002)
	Laguncularia spp.	Primavera and Esteban (2008)
	Bruguiera spp.	Bosire et al. (2008)
	Laguncularia spp.	
Oyster reefs	*Crassostreavirginica*	Powers et al. (2009)
	Ostrea spp.	Rossi-Snook et al. (2010)
		Zarnoch and Schreibman (2012)
Macroalgae forests	*Phyllospora* spp.	Falace et al. (2006)
	Ulva spp.	Lee et al. (2008)
	Cystoseirabarbata	Perkol-Finkel and Airoldi (2010)
		Campbell et al. (2014)
Salt marshes	*Spartina* spp.	Armitage et al. (2006)
	Salicornia spp.	Zedler et al. (2003)
	Batismaritima	Castillo and Figueroa (2009)

utilized approach for mangroves, with many restoration programs carried out in countries with developing economies, where mangroves forests are located (Proffitt and Devlin 2005; Bayraktarov et al. 2016; Thorhauget al. 2017).

The fact that transplantation techniques are the most suitable for coastal marine habitats is also supported by the wide use of these techniques in terrestrial habitats, where restoration initiatives have been implemented for a longer time than the marine environment to counteract deforestation. Transplantation has also been used in the ecological restoration of estuaries and salt marshes (SER 1998; Zedler et al. 2003; Castillo and Figueroa 2009), which constitute good examples for coastal restoration activities due to the high number of reported successful efforts. Moreover, restoration programs, focused on these transitional water systems, often include an important involvement of private and public stakeholders who have funded long-lasting projects. For instance, the USA started the first restoration program of salt marshes as early as the 1970s (Tsihrintzis 1970).

It is often argued that restoration actions are focused on the present state of ecosystems, not considering the challenges they will face in the future (Harris et al. 2006). Ecological restoration needs to cope with a fast-changing environment which could lead to complications over projects and reduce the final success of restoration programs (Erwin 2009; Havens et al. 2015). Some solutions to this problem have, however, been advocated and utilized in many restoration initiatives. Coral reef restoration, for instance, benefits from advances in selective breeding techniques or the manipulation of the coral microbiome selective breeding (van Oppen et al. 2017).

5.2.2 Selective Breeding to Increase Restoration Technique Success

Climate change alters the physicochemical conditions of the ocean (IPCC 2014), causing organisms to adapt to new conditions that are different from the optimum. Since 1992, the idea of implementing genetic approaches to cope with global climate change has been proposed, particularly in terrestrial environments (Ledig and Kitzmiller 1992). Initially, non-native species that are more tolerant and, therefore, resistant to climate have been introduced to help enhance the resilience of species assemblages that are facing climate change (Ledig and Kitzmiller 1992; Harris et al. 2006). For instance, Ledig and Kitzmiller (1992) proposed to introduce non-native seeds to planting programs, artificially selecting those that are capable of surviving at higher temperatures. Humanity has been doing this since the domestication of plants and animals, always looking for desirable or specific traits (Hill and Caballero 1992). In the case of tree populations, they mostly rely on phenotypic plasticity to adapt to new conditions (Alfaro et al. 2014).

Climate change impacts in the marine environment become particularly severe when they affect coral reefs. Since climate change is related to ocean acidification, it poses a direct threat to coral reef health (Hoegh-Guldberg et al. 2007; Veron et al. 2009). Increased acidic conditions have detrimental effects on coral calcification and growth when the CO_2 concentration exceeds 560 ppm (Kleypas and Langdon 2006). It has, however, been demonstrated that some coral morphotypes or genera can survive extreme conditions of temperature and pH (Alcala et al. 1982; Lindahl 1998; Bowden-Kerby 2001). Latest global coral bleaching events have sharpened the focus on the use of assisted evolution (i.e., selective breeding and assisted gene flow (Aitken and Whitlock 2013), conditioning or epigenetic program-

Table 5.2 Most successful restoration techniques in different coastal marine habitats as summarized by Bayraktarov et al. (2016) and a coarse literature review with a particular focus on macroalgae. Macroalgae have not been considered by Bayraktarov et al. (2016), even though along the coasts they are an extremely important habitat formers, vital for coastal biodiversity and ecosystem functions. "Other techniques" do not always include human-mediated reintroduction; they can be facilitation measure

Ecosystem	Transplantation (adult life stages)	Planting (early life stages)	Other techniques
Coral reefs	Transplanting of the whole colony or fragmented corals; out-planting	Sexual propagation in aquaria; transplantation of juveniles	Coral farming facilitated by electrical field; deployment of artificial reef structures
Seagrasses	Transplanting seagrasses (cores or plugs)	Collecting or aquacultured seeds, seedlings, or rhizomes for transplantation	Deployment of hessian bags to stabilize the sandy bottom
Mangrove forests	Planting mangroves (saplings or small trees)	Planting mangroves (seeds, seedlings, or propagules)	Hydrological restoration (facilitation of a natural recovery)
Oyster reefs	–	Hatchery rearing of native oysters (and seeding)	Creating a no-harvest sanctuary with natural or artificial substrate
Macroalgae forests	Transplantation of adults	Transplanting of sporophyte, seedlings, spore, germlings, or juveniles	Cleaning the substrata from ephemeral algae or removal of grazers (facilitation to natural recovery); deployment of artificial substrata
Salt marshes	Planting of salt marsh plants containing plugs	Planting salt marsh seeds, seedlings, or sods	Construction; excavation and backfilling with clean soil

Fig. 5.3 Coral reef restoration in Quintana Roo, Mexico. Permission by Claudia Padilla Souza, Instituto Nacional de Pesca y Acuacultura (INAPESCA)

ming (Torda et al. 2017), the manipulation of the coral microbiome (Bourne et al. 2016), etc.) as a mean to enhance environmental stress tolerance of corals and increase the success of coral reef restoration efforts over longer periods (van Oppen et al. 2017). Genetic tolerance of host and zooxanthellae among heat-sensitive clones would provide enough evidence for change as the habitat moves to higher thermal regimes (Hoegh-Guldberg 1999). Another marine habitat used to attempt genetic selection are mangroves. Mangroves apparently have a better chance to adapt to quickly changing environmental conditions, as plants appear to have higher levels of genetic adaptation to current climate change (Rico et al. 2013). According to Xu et al. (2017), mangrove trees are highly adaptable organisms, although mangroves have the least diverse genome in comparison with other usually restored marine habitats, which may be due to the continual habitat turnover caused by the exposure to rising and falling sea levels in the geologically recent past. Mangroves are, therefore, thought to have better chances to thrive throughout the Climate Change Era. However, any optimism about their resilience at these times might be premature (Guo et al. 2018). Nevertheless, selecting individuals with the most appropriate genomes within the different mangrove species would constitute the best choice to use the most suitable organisms that could face future adverse and extreme conditions.

In the tropics, many macroalgal species live close to their thermal limits, and they will have to upregulate their response to tolerate sublethal temperature exposure. However, the effects of elevated CO_2 concentrations and thermal acclimation are not well-described for macroalgal forests (Koch et al. 2012). Cole et al. (2013) demonstrated that freshwater macroalgae of the genus *Oedogonium* are capable of assimilating higher amounts of CO_2 above present-day concentrations; thus, they are capable of improving their carbon storage features with consequences for their biomass. One of the most important ecosystem services is the sequestration and storage of CO_2 by algae, plants, and coral reefs, reducing the process of global warming (Moberg and Folke 1999; Corlett and Wescott 2013). With this in mind, the use of genetic tools, such as translocating genes from *Oedogonium* sp. to other algae species, could present an opportunity to modify organisms in a way to enhance their carbon uptake capacity and, therefore, remove larger amounts of CO_2 from the atmosphere and potentially smoothen the consequences of the impacts related to climate change.

Although enhancing specific traits on certain species using genetic technologies would sound like a good idea, specifically to an improvement to marine restoration success, in the end a rather philosophical question arises: Life always finds its way. No matter how bad conditions might get, there is always an alternative road. Thus, are we promoting/encouraging the next step of evolution or are we stopping it by selecting specific traits that are of interest to us? And, thus, how should we elaborate appropriate strategies to make restoration successful in a fast-changing world?

5.3 Measurements of Restoration Success

5.3.1 Survival

To date, restoration success has mainly been assessed as the average survival of the reintroduced species. However, even when survival of targeted species is high, assessment of regain in ecosystem functionality is often accounted for. Hence, it is difficult to see if the ecosystem is fully restored.

In fact, in the database of Bayraktarov et al. (2016), 61% of all observations on marine coastal restoration provided information on survival of restored organisms as an item-based success indicator. While Bayraktarov et al. (2016) did not include macroalgae records, they have been considered in this analysis, screening all peer-reviewed articles concerning attempts of macroalgae restoration. Macroalgae shall be included in the most restored habitat formers together with those already included in Bayraktarov et al. (2016): coral reefs, seagrasses, mangroves, oyster reefs, and salt marshes.

Macroalgal forests (i.e., kelp forests or *Cystoseira* spp., *Fucus* spp., *Sargassum* spp., *Phyllosphora* spp.) are disappearing worldwide, prompting many scientific groups all over the world to implement their restore (e.g., Terawaki et al. 2003; Falace et al. 2006; Perkol-Finkel et al. 2012). Therefore, the dataset of Bayraktarov et al. (2016) was supplied by other ten articles in which success of macroalgae restoration was assessed. Articles regarding macroalgae restoration available in the literature (up to 2016) were more than ten, but those are the only ones measuring success as survival of target species. To be consistent with the analysis of Bayrakratov et al. (2016), just those ten have been considered.

The average survival per habitat after restoration are 56.3% for coral reefs, 40.4% for seagrasses, 52.2% for mangroves, 55.1% for oyster reefs, 22.9% for macroalgae, and 57.2% for salt marshes. The number of studies in which the success of habitat restoration was assessed by survival (the only ones considered in this analysis) was different among habitats, revealing the fact that some habitats have been more explored in terms of restoration than others. This might be a cause of different success rates (Fig. 5.4). It can be claimed that the higher the number of restoration attempts, the more we succeed. On one hand, this agrees with low macroalgae restoration success (5.0% of median survival) (the least studied habitat in terms of restoration) and salt marshes and coral reef, the most restored and the most successful (with 64.8% and 64.5% of median survival, respectively). On the other hand, this disagrees with seagrass restoration, which is a well-explored habitat in terms of restoration, but that is not so successful (38.0% of median survival). Many studies, however, agreed on the specific challenges related to seagrass restoration (Ganassin and Gibbs 2008; van Katwijk et al. 2016; Cunha et al. 2012).

5.3.2 Ecosystem Services

Evaluating the success of restoration initiatives is extremely challenging. The use of one metric over another can result in profoundly different outcomes which in turn has important consequences for the management decisions and actions taken by stakeholders and decision makers regarding restoration projects. Therefore, it is of paramount importance to use the appropriated metrics and assessment techniques.

In theory, restoration activities can only be considered successful when ecosystem functioning and habitat resilience capacity are reverted to the state preceding the degradation (Peterson et al. 2003; Shackelford et al. 2013). These outcomes are difficult to evaluate. Therefore, assessment studies are looking for proxies. In the marine environment, most studies focus on rudimentary performance metrics (Ruiz-Jaen and Aide 2005). For instance, oyster reef restoration monitoring in the Gulf of Mexico used three success metrics based on sustainability of oysters: presence of vertical structure above the bottom, presence of live oysters, and evidence of recruitment (Powers et al. 2009; La Peyre et al.

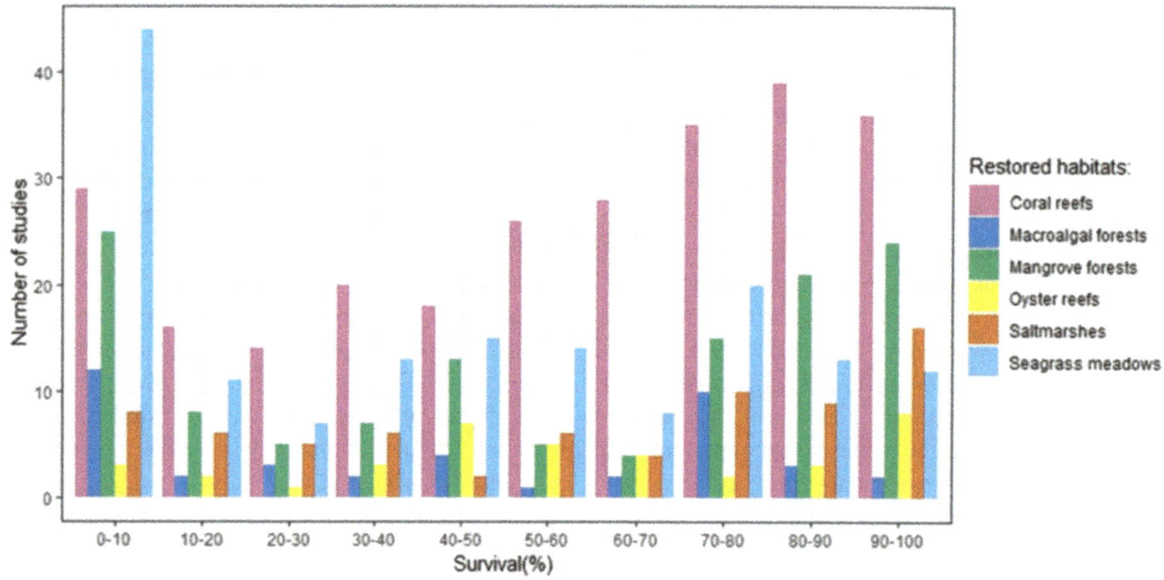

Fig. 5.4 Frequency of success in restoration, measured as percentage of survival reported by published articles until 2016 across the six mostly restored coastal marine habitats

2014). The most common metric used in all marine habitats is survival rate (Ruiz-Jaen and Aide 2005), whereas, in the terrestrial environment, assessment techniques are predominantly based on parameters such as biodiversity, vegetation structure, or ecological functions (Ruiz-Jaen and Aide 2005). Restoration science in marine coastal ecosystem is still in an early phase compared to terrestrial ecosystems (Suding 2011; Bayraktarov et al. 2016). Furthermore, monitoring programs base their evaluation of restoration success on short-term studies, even though previous research has repeatedly demonstrated that longer time scales provide better estimates of restoration success, since ecosystem services need longer time periods to recover (Bell et al. 2008; Bayraktarov et al. 2016). Better metrics need to be adopted in the marine environment to improve the assessment of restoration success, incorporating ecosystem services-based parameters (Fonseca et al. 2000; Paling et al. 2009). Ecosystem services-based parameters could facilitate the inclusion of economic and social interests in restoration projects while increasing biodiversity (Benayas et al. 2009; Adame et al. 2015).

Marine restoration monitoring can provide incorrect assessments of the outcome of restoration projects. They typically use survival of organisms transplanted (e.g., seagrass, mangroves, corals), whereas reviews advise to follow the path of terrestrial restoration monitoring that uses biodiversity, ecosystem structure, or ecological functions instead (Bayraktarov et al. 2016). It has been stated that restoration goals should focus on the provision of ecosystem services (Benayas et al. 2009). Therefore, the use of ecosystem services in restoration assessments might provide a better estimate of project success (Fonseca et al. 2000; Bayraktarov et al. 2016).

The use of ecosystem services-based metrics in restoration projects is quite new. To obtain the scale of the area to be restored to compensate for the habitats lost or degraded, a habitat equivalency analysis (HEA) was created by NOAA in 1997 (NOAA 1997). This analysis computes the interim of loss ecosystem services during the time the ecosystem was destroyed and the time where it reaches standards of equivalency after restoration. If an ecosystem is destroyed but its restoration takes place immediately after and reaches standards of ecological functions equivalent to those present previous to the perturbation in a short time, this interim would be relatively low. However, if the restoration takes place a long time after perturbation, and if the ecosystem takes a long time to reach the standards of equivalency, this interim would be higher. The novelty of this analysis consisted in integrating specific resource-based metrics as a proxy for the loss services. In an US federal court case, Fisher used seagrass shoot density as a proxy to provide compensation for the loss of sea grass within the Florida Keys National Marine Sanctuary (Fonseca et al. 2000). This protocol was demonstrated to be quite flexible and applicable to a wide range of habitats (Fonseca et al. 2000; Cabral et al. 2016; Grabowsky et al. 2012). Another tool to investigate the ecosystem services across a landscape is the InVEST (Integrated Valuation of Ecosystem Services and Trade-offs), which is based on cartographic representation of the ecological information within an area (Tallis and Polasky 2009). InVEST allows to estimate the change in ecosystem services in response to different management scenarios (Guerry et al. 2012). HEA and InVEST tools are utilized prior to the onset of restoration work, providing estimates of the required scale needed to compensate for the lost area and identifying the most appropriate location site of the restoration initiative. They set the outcomes of a project based on ecosystem services-based parameters, facilitating their use as metrics in assessment studies. The combined uses of these investigative tools have shown to provide valuable results as it delivers relevant information for decision makers in an integrated way using an ecosystem services framework as common currency and to be easily adapted to include more constraints and/or other ecosystem services (Cabral et al. 2016). However, evaluating restoration success using ecosystem services-based metrics is an emerging field in marine ecology, except for coastal wetlands (Zhao et al. 2016). Such metrics already exist for this habitat, such as the rapid assessment method (Galv-RAM) which uses a combination of biotic and abiotic parameters to obtain an ecosystem index score (Staszak and Armitage 2012). New metrics are being investigated for other habitats, such as the determination of food web structure through isotopic analysis for restored macroalgae beds (Kang et al. 2008), the use of fish assemblages to assess seagrass restoration success (Scapin et al. 2016), or the use of fish tracking and habitat use to assess the recovery of an estuary (Freedman et al. 2016). It is, however, a great challenge to determine the appropriate resource-based parameter to integrate those tools and then to estimate the recovery rate of ecosystems (Peterson et al. 2003; Carpenter et al. 2009). Moreover, the concept and metrics of marine coastal ecosystem services are still in their infancy and require further development (Liquete et al. 2013). The chosen metrics also have to match with the rate of ecosystem services recovery which in most case exceeds the time scale of most evaluation studies of restoration success (often <5 years) (Bell et al. 2008; De Groot et al. 2012; Bayraktarov et al. 2016). Recommendations to develop and use ecosystem-based metrics were formulated almost 20 years ago (Fonseca et al. 2000). Yet metrics used in marine restoration studies rarely focused on the recovery of ecosystem functions or services, and this might have led to restoration failure (Bayraktarov et al. 2016; Hein et al. 2017). Further research is needed to develop new ecosystem-based metrics, setting the appropriate goals for marine restoration to outweigh services loss after a perturbation.

5.4 Ecosystem Services as a Method to Link Restoration to Socioeconomic Sciences

The concept of ecosystem services plays an important role in the cost-benefit analysis of human activities impacting the natural capital. This concept has, therefore, been increasingly used in the context of socioeconomics. Associating a value with a service or resource is a powerful tool to highlight the ecological and also the financial loss because of anthropogenic impacts on habitats and ecosystems (Worm et al. 2006; Salomon and Dahms 2018). The Millennium Ecosystem Assessment (MEA 2005) has led to an exponential increase of scientific interest in this topic, from 2.5 papers per year during 1997–2006 to 25 papers per year since 2007 (Liquete et al. 2013). Marine coastal ecosystems can no longer be considered as inexhaustible, and their value to society and the costs associated with their loss and degradation need to be properly accounted for (Costanza et al. 1997; Carpenter et al. 2006; De Groot et al. 2012). Valuation of ecosystem services is nowadays considered as exceptionally important to contribute to the conservation and restoration of threatened and lost habitats. Economists demonstrated that restored, healthy habitats will generate value for both households and industry (Barbier et al. 2011). Recent studies confirm the general paradigm stating that ecosystem services would return through habitat restoration (Benayas et al. 2009) for mangroves (Rönnbäck et al. 2007), seagrasses (Reynolds et al. 2016), and even a whole estuarine system (salt marsh, mangrove, seagrass) (Russel and Greening 2015). New methods of ecosystem service valuation suggest that the economic benefits of restoration can outweigh their costs (Bullock et al. 2011; Grabowski et al. 2012; Speers et al. 2016; Adame et al. 2015). Cost-benefit computation derived from ecosystem services provided by restored habitats could incentivize managers and stakeholders to increase financial investment into marine restoration projects.

Valuation of ecosystem services has been a launch pad to raise interest and awareness of scientists and the general public to the socioeconomic importance of habitats and the services they provide. It has had an enormous impact on the perception of restoration as a way to improve human well-being (Costanza et al. 2014). However, it is challenging to estimate the value of many ecosystem services, as most of them do not have a market value. To compute the benefit to cost ratio of a restoration project, one needs to know which ecosystem services are provided by the habitat and how much financial value can be associated with them (Speers et al. 2016). Reference and meta-analysis studies estimating those parameters have been undertaken (Costanza et al. 1997, 2014; MEA 2005; Barbier et al. 2011; De Groot et al. 2012; Salomidi et al. 2012). The most commonly used estimation was obtained through the global assessment of The Economics of Ecosystems and Biodiversity (TEEB), reported by De Groot et al. (2012). This estimate was an international initiative undertaken by the UN Environment program. It collected information from 320 publications and published an overview of the value of ecosystem services provided by ten main biomes, from fishery stock to cultural and spiritual heritage (De Groot et al. 2012; Schröter et al. 2014). The estimates range from 490 international$/year[1] for a hectare of open ocean to 350,000 int$/year for a hectare of coral reefs. Seagrass and algae beds are estimated to be 28,916 int$/year per hectare and mangroves and tidal marshes up to 193,843 int$/year per hectare. These values include a mix of market and nonmarket values, with the latter being the most important (De Groot et al. 2012). Those nonmarket values are mainly estimated through existing studies of households' willingness to pay to protect the habitats (Mendelsohn and Olmstead 2009; Barbier et al. 2011). The TEEB report was picked up extensively by mass media and alongside the Millennium Assessment (MA) greatly influenced the communication to policy makers (Schröter et al. 2014). Costanza et al. (2014) used the TEEB report to determine the global economic loss for the whole marine environment in response to land use change between 1997 and 2011, obtaining a figure up to 10.9trillion$/year, mainly driven by coral reef degradation (Costanza et al. 2014). The interconnection between habitats may also greatly influence ecosystem services on a spatiotemporal scale (Barbier et al. 2011). A coral reef will not provide the same ecosystem services if mangroves or seagrass meadows are present or absent from nearby coastal area (Moberg and Folke 1999). More studies are required to assess the different ecosystem services delivered by "outsider" habitats (important but underrepresented in science) in a broader geographical range, taking into account the connectivity between habitats and spatiotemporal variability of the ecosystems.

Restoration projects have commonly been funded by governments, by private companies restoring a given ecosystem as compensatory measures of previous degradation and/or loss of habitat elsewhere, or through biobanking and biodiversity offset initiatives (Bullock et al. 2011). Valuation of ecosystem services has led to another way of funding which would address both ecological and social issues with ecosystem sustainability and poverty alleviation, respectively (Farley and Costanza 2010). Payment for ecosystem services (PES) schemes are emerging as new market-based approaches for restoration projects, based on the argument that people depend on ecosystem services and the way to ensure their continued provision is to pay for them (Redford and Adams 2009; Muradian et al. 2010). The previous statement has been increasingly used worldwide, especially in the terrestrial environment (Farley and Costanza 2010; Bullock et al. 2011). In the marine environment, the main market-based ecosystem service that could be integrated into a PES is carbon storage through the reducing emissions from deforestation and forest degradation (REDD+) set of international policies (Bullock et al. 2011; Locatelli et al. 2014). Blue forests (mangroves, seagrass, and macroalgae) are known to store a great quantity of carbon into their biomass or soil which works as a buffer for climate change (Nelleman et al. 2009; McLeod et al. 2011; Himes-Cornell et al. 2018). Seagrasses cover less than 0.2% of ocean bottom; yet they are estimated to account for 10% of the global carbon sequestered in marine sediment (Fourqurean et al. 2012), while mangrove accounts for 14% (Alongi 2012). Therefore, they are strong candidates for PES projects, especially mangroves due to a higher knowledge on ecosystem services and their financial valuation (Liquete et al. 2013; Locatelli et al. 2014). As mangroves are mainly present in developing countries, mangrove PES restoration schemes could contribute to alleviate poverty within the local coastal communities through restoration of ecosystem services, employment in restoration program, and benefit from the PES funds while tackling the issue of climate change at a global scale (Carpenter et al. 2006; Martinez-Alier 2014; Rodríguez 2018). A small-scale mangrove-based PES project already exists in Kenya, called "Mikoko Pamoja" (Fig. 5.5; see www.eafpes.org), but larger-scale projects are still difficult to implement due to the lack of local and regional institutional frameworks that could cope with the complexity of such schemes (Bullock et al. 2011). Even though much criticism has been voided against PES schemes, especially on their carbon-centric approach, neglecting other goods and services and their long-term viability, it is thought to be a vital tool to address the issue of the significance of restoration ecology to stakeholders and decision makers (Redford and Adams 2009; Bullock et al. 2011). Its implementation could be facilitated through the development of innovative tools such as ecosystem services mapping exercise.

The use of spatial analyses is extremely interesting in restoration ecology to select areas to restore with the highest cost-benefit ratio. A study by Adame et al. (2015) developed a novel restoration approach based on biodiversity and ecosystem services provided by a mangrove forest. The use of physical parameters to estimate ecosystem services might be tricky, but they assumed it to be more accurate than the common restoration area selection based on accessibility (Adame et al. 2015). These analyses do, however, have several limitations mainly due to a lack of knowledge in marine ecosystem boundaries, services, and connectivity (Liquete et al. 2013). Improvements in modelling and mapping technologies could, therefore, provide better information to advice and steer future ecosystem services policies, helping them to meet goals of sustainability of ecosystem services delivery.

[1]The international dollar, or the Geary-Khamis dollar, is a hypothetical unit of currency that is used to standardize monetary values across countries by correcting to the same purchasing power that the US dollar had in the USA at a given point in time (De Groot et al. 2012).

Fig. 5.5 Local woman planting mangrove juveniles (**a**) in Gazy Bay Boardwalk, Kenya (**b**). Permission of Mark Huxham (www.aces-org.co.uk/)

5.5 Discussion

Marine restoration ecology is an emerging field which could potentially become the main discipline in ecological science in the next decades. To reach this goal, reviews of success and failures are extremely important to address the main challenges faced. It will allow future projects to find solutions to these challenges and improve this field. Ecological restoration is critical for the persistence of marine habitats (Aronson and Van Andel 2012), and therefore this article provides important information on the relative success of habitat-based restoration and address the main challenges to improve it (Hobbs 2007; Zhang et al. 2018).

Different restoration methods have been utilized to restore habitats (Rinkevich 2005; Bastyan and Cambridge 2008; Ganassin and Gibbs 2008; Balestri and Lardicci 2014; Bowden-Kerby 2001; Tortolero-Langarica et al. 2014; Bayraktarov et al. 2016; Zhao et al. 2016). Transplantation has proven to be the most effective tool for most types of habitats (Bayraktarov et al. 2016). Restoration success is, however, highly dependent on the type of habitat. Highest restoration success is found within coral reefs and salt marshes, which are also the best studied habitats in terms of restoration techniques. Restoration success in the marine environment is more dependent of habitat-based research rather than financial input to the projects (Bayraktarov et al. 2016). Habitat-based research might include studies on top-down control of transplants by grazers and predators which is an important factor influencing marine restoration success (Zhang et al. 2018). Few studies on top-down interactions were directly employed and tested in restoration (four studies, 1%), but they were consistently found to have a significant effect on restoration success (Zhang et al. 2018). Also more studies on intra- and interspecies facilitation processes could lead to improved restoration success as recent research demonstrated that salt marsh restoration yield doubled simply by planting marsh grass plugs in aggregate (thus ameliorating abiotic stressors via increased intraspecific facilitation) (Silliman et al. 2015).

Global warming is an upcoming threat to marine habitats. Few are known about habitats' resilience to this threat. Each of them will be affected differently. Coral reefs will be highly affected by the increased seawater temperature and acidity (Veron et al. 2009). On the contrary, other habitat formers could benefit from an increased CO_2 concentration. The photosynthetic rates of seagrasses are CO_2-limited (Beer and Koch 1996). A study performed by Palacios and Zimmerman (2007) with the eelgrass *Zostera marina*, a temperate species, demonstrated that an increased CO_2 content in the atmosphere and oceans would lead to an increased area-specific productivity of seagrass meadows. A similar study performed by Ow et al. (2015) with three tropical seagrass species found the same result, yet explaining that responses were variable between species. These studies reinforce the emerging paradigm stating that seagrass meadows are likely to benefit significantly from a high-CO_2 world (Zimmerman et al. 2017). Acclimation and resilience capacity, alongside the genetic potential of marine habitats, need to be properly accounted for to address their vulnerability to global warming. While global warming might reduce the success of marine restoration, an inverse relationship also exists as marine habitats are known for their great capacity of carbon storage. Their restoration could be utilized as a strategy to buffer global warming (Ledig and Kitzmiller 1992; Harris et al. 2006). Restoring more and more degraded habitats would result in an increased carbon storage capacity of coastal marine environments (McLeod et al. 2011). This newly achieved capacity could then be maintained throughout future scenarios of global warming via the selection of adapted morphotypes and improved species plasticity.

Van Oppen et al. (2015) stressed the need for quick and reliable answers to the rapid changes. In the particular case of coral reefs, the authors revised the idea of assisted evolution as a good approach, agreeing, to some extent, with the ideas expressed in this work. Due to the lack of knowledge about how manipulation at the genetic level would interfere with natural processes, there is still uncertainty regarding the so-called genetically modified organisms (GMOs). Although in recent times, there has been an advancement in the knowledge regarding the evolutionary mechanisms, information is not enough to determine whether they will be, indeed, the best way to proceed in terms of artificial selection to increase the reach of ecological restoration programs.

These restoration strategies do not absolutely ensure the success of the projects, as it requires the development of reliable metrics. The use of parameters based on the return of ecosystem services provision is advocated as the most viable option (De Groot et al. 2012; Bayraktarov et al. 2016). Not all habitats have, however, been studied with the same attention, and there is still much to discover about their ecological functions leading to the provision of ecosystem services (Liquete et al. 2013). The use of the tools prior to restoration actions, such as the HEA protocol, combined with other tools such as InVEST, would provide clear objectives to marine restoration projects related to the provision of ecosystem services (Cabral et al. 2016). It would then be easier to develop and implement ecosystem services-based metrics to estimate restoration success during monitoring studies. Such metrics already exist for salt marshes and could be regarded as a foundation to develop other habitat-specific metrics (Staszak and Armitage 2012). They would also provide a common base to compare restoration projects in different habitats, leading to a better understanding of success and failure of marine ecosystem restoration. The need to improve communication between stakeholders and the general public has been repeatedly highlighted and would be aided by the use of ecosystem service metrics (Costanza et al. 2014; Nordlund et al. 2018). Reliable estimates of restoration success provided by these metrics would facilitate the involvement of stakeholders within these projects. Restoration has an underlying force residing within its socioeconomic aspect. Valuation of ecosystem services through global assessments, such as the TEEB and the MEA, has increased the awareness of the general public about the catastrophic consequences of habitat loss on the society as well as economy. They have greatly influenced stakeholders' perception of restoration (Costanza et al. 2014). An increased awareness of the ecological and financial gains obtained from ecological restoration through better valuation of ecosystem services will help decision makers to have the right judgment on feasibility and outcomes of restoration projects.

Many habitats in need to be restored are located in developing countries. Therefore, awareness of local communities' influence on ecosystem goods and services should involve and represent locals in restoring degraded habitats. Bottom-up management of conservation areas and restoration projects give more value and ownership to the local community which will then have an incentive to self-enforce any necessary policies. This will provide locals with the tools to establish sustainable living conditions and address poverty alleviation (Redford and Adams 2009; Farley and Costanza 2010; Locatelli et al. 2014). Moreover, if locals are taught to monitor the recovery of ecosystem services, these jobs can even be long-lasting. The Mikoko Pamoja project success is linked to the inclusion of Gazi and Makongeni communities (Fig. 5.5). They participate in the seeding of mangroves and benefit from the PES income with which they provide clean water and educational material to school children, mangrove conservation, and restoration and employ a full staff member to coordinate the project (UNEP and CIFOR 2014). The communities benefit from other mangrove-related income such as ecotourism with the "Gazi Bay Boardwalk" (Fig. 5.5) (Wylie et al. 2016). This project won the Equator Prize in 2017, rewarding outstanding community efforts to reduce poverty through the conservation and sustainable use of biodiversity (see also www.equatorinitiative.org). While this small-scale project was a success, it seems much less realistic to realize a similar one on a larger scale due to a lack of policy frameworks (Reford and Adams 2009; Bullock et al. 2011; Muradian and Rival 2012). The implementation of policies fostering restoration applications, such as the REDD+, is a necessary step to enhance projects on temporal and spatial scales.

Marine habitat restorations are win-win projects, in which all parts benefit from it. Locals regain lost ecosystem services and, if they are included in the management project, also benefit from long-lasting jobs. Stakeholders will see their investment refunded in the following years using PES schemes. The global population also benefits from these projects which, through the carbon storage capacity of coastal ecosystems, buffer global warming. Bottom-up management systems are now considered as the main method to implement long-term restoration projects. The standardization process and methods by the private sector could represent another way to accelerate the implementation of such projects. The inclusion of locals could guarantee a larger spatial and temporal scale for the projects, a higher investment fostered through PES schemes and/or similar approaches, and the implementation of protocols habitat by habitat. For instance, industrializing the process to maintain and keep enhanced coral stocks will benefit ecological restoration, but it will also be a great opportunity for aquarists, therefore, creating a branch of business with a potentially big money income and a chance to have organisms better adapted to greater ranges to environmental conditions (Van Oppen et al. 2015). Within this context, the mass production of organisms capable of tolerating higher stress conditions would be an option, but further investigation is needed.

5.6 Conclusion

In conclusion, restoration is not a better approach than passive conservation; both strategies work together to reach common goals to protect the natural capital (Elliott et al. 2007; Hobbs 2007; Zhang et al. 2018). We emphasize the fact that assisting natural recovery of ecosystems is no longer an option as their unassisted recovery rate is negligible and cannot cope with the pace of habitat loss. Restoration has to be seen as an integral part of our future ecosystem longevity and requires an urgent focus and implementation to address rapid changes and loss caused by both climate change and multiple direct human-related impacts. Marine restoration projects are not always successful. Failures are mainly due to a lack of habitat-based research in a broader geographical range and of reliable success metrics. Valuation of ecosystem services to increase public and stakeholders' awareness is an important step in marine habitat restoration. It would also improve the use of payment for ecosystem services schemes which are a useful tool to implement bottom-up management of marine restoration projects. Marine habitat restorations are win-win projects: they increase biodiversity, enforce local communities, and buffer climate change. We have the abilities to force such change; it is now time to unite our efforts and undertake the path of ecological restoration for the good of all.

Appendix

This article is related to the YOUMARES 9 conference session no. 6: "The challenge of marine restoration programs: habitats-based scientific research as a key to their success." The original Call for Abstracts and the abstracts of the presentations within this session can be found in the Appendix "Conference Sessions and Abstracts", Chapter "5 The challenge of marine restoration programs: habitats-based scientific research as a key to their success", of this book.

References

Abelson A, Halpern BS, Reed DC et al (2015) Upgrading marine ecosystem restoration using ecological-social concepts. Bio Science 66(2):156–163

Adame MF, Hermoso V, Perhans K et al (2015) Selecting cost-effective areas for restoration of ecosystem services. Conserv Biol 29(2):493–502

AinodionMJ, RobnettCR, AjoseTI (2002) Mangrove restoration by an operating company in the Niger Delta.Proceedings of the International Conference on Health, Safety and Environment in Oil and Gas Exploration and Production. Kuala Lumpur, pp 1001–1014

Airoldi L, Beck MW (2007) Loss, status and trends for coastal marine habitats of Europe. Oceanogr Mar Biol 45:345–405

Airoldi L, Balata D, Beck MW (2008) The gray zone: relationships between habitat loss and marine diversity and their applications in conservation. J Exp Mar Biol Ecol 366(1):8–15

Aitken SN, Whitlock MC (2013) Assisted gene flow to facilitate local adaptation to climate change. Annu Rev EcolEvolSyst 44:367–388

Alcala AC, Gomez ED, Alcala C (1982) Survival and growth of coral transplants in Central Philippines. Kalikasan 11(1):136–147

Alfaro RI, Fady B, Vendramin GG et al (2014) The role of forest genetic resources in responding to biotic and abiotic factors in the context of anthropogenic climate change. For Ecol Manag 333:76–87

Alongi DM (2012) Carbon sequestration in mangrove forests. Carbon Manag 3(3):313–322

Ammar MSA, Amin EM, Gundacker D et al (2000) One rational strategy for restoration of coral reefs: application of molecular biological tools to select sites for rehabilitation by asexual recruits. Mar Pollut Bull 40(7):618–627

Armitage AR, Boyer KE, Vance RR et al (2006) Restoring assemblages of salt marsh halophytes in the presence of a rapidly colonizing dominant species. Wetlands 26:667–676

Aronson J, Van Andel J (2012) Restoration ecology and the path to sustainability. In: Van Andel J, Aronson J (eds) Restoration ecology: the new frontier. Wiley, Chichester, pp 293–304

Balestri E, Lardicci C (2014) Effects of sediment fertilization and burial on *Cymodoceanodosa* transplants; implications for seagrass restoration under a changing climate. Restor Ecol 22:240–247

Barbier EB, Hacker SD, Kennedy C et al (2011) The value of estuarine and coastal ecosystem services. Ecol Monogr 81(2):169–193

Bastyan GR, Cambridge ML (2008) Transplantation as a method for restoring the seagrass *Posidoniaaustralis*. Estuar Coast Shelf Sci 79:289–299

Bayraktarov E, Saunders MI, Abdullah S et al (2016) The cost and feasibility of marine coastal restoration. Ecol Appl 26(4):1055–1074

Beck MW, Brumbaugh RD, Airoldi L et al (2011) Oyster reefs at risk and recommendations for conservation, restoration, and management. Bioscience 61(2):107–116

Beer S, Koch E (1996) Photosynthesis of marine macroalgae and seagrasses in globally changing CO_2 environments. Mar Ecol Prog Ser 141:199–204

Bell SS, Tewfik A, Hall MO et al (2008) Evaluation of seagrass planting and monitoring techniques: implications for assessing restoration success and habitat equivalency. Restor Ecol 16(3):407–416

Bellgrove A, McKenzie PF, Cameron H et al (2017) Restoring rocky intertidal communities: lessons from a benthic macroalgal ecosystem engineer. Mar Pollut Bull 117(1–2):17–27

Benayas JMR, Newton AC, Diaz A et al (2009) Enhancement of biodiversity and ecosystem services by ecological restoration: a meta-analysis. Science 325(5944):1121–1124

Bird KT, Jewett-Smith J, Fonseca MS (1994) Use of in-vitro propagated *Ruppiamaritima* for seagrass meadow restoration. J Coastal Res 10:732–737

Bosire JO, Dahdouh-Guebas F, Walton M et al (2008) Functionality of restored mangroves: a review. Aquat Bot 89(2):251–259

Bourne DG, Morrow KM, Webster NS (2016) Insights into coral microbiome: underpinning the health and resilience of coral reef systems. Annu Rev Ecol Evol 70:317–340

Bowden-Kerby A (2001) Low-tech coral reef restoration methods modeled after natural fragmentation processes. Bull Mar Sci 69(2):915–931

Bullock JM, Aronson J, Newton AC et al (2011) Restoration of ecosystem services and biodiversity: conflicts and opportunities. Trends Ecol Evol 26(10):541–549

Cabral P, Levrel H, Viard F et al (2016) Ecosystem services assessment and compensation costs for installing seaweed farms. Mar Policy 71:157–165

Callaway JC, DeLaune RD, Patrick WH Jr (1997) Sediment accretion rates from four coastal wetlands along the Gulf of Mexico. Coast Res 3(1):181–191

Campbell AH, Marzinelli EM, Vergés A et al (2014) Towards restoration of missing underwater forests. PLoS One 9(1):e84106

Carpenter SR, Bennett EM, Peterson GD (2006) Scenarios for ecosystem services: an overview. Ecol Soc 11(1):29

Carpenter SR, Mooney HA, Agard J et al (2009) Science for managing ecosystem services: beyond the millennium ecosystem assessment. Proc Natl Acad Sci U S A 106(5):1305–1312

Castillo JM, Figueroa E (2009) Restoring salt marshes using small cordgrass, *Spartinamaritima*. Restor Ecol 17:324–326

Cole AJ, Mata L, Paul NA et al (2013) Using CO_2 to enhance carbon capture and biomass applications of freshwater macroalgae. Glob Change Biol Bioenergy 6(6):637–645

Connell SD, Russell BD, Turner DJ et al (2008) Recovering a lost baseline: missing kelp forests from a metropolitan coast. Mar Ecol Prog Ser 360:63–72

Corlett RT, Wescott DA (2013) Will plants movement keeps up with climate change? Trends Ecol Evol 28(8):482–488

Costanza R, d'Arge R, De Groot R et al (1997) The value of the world's ecosystem services and natural capital. Nature 387:253–260

Costanza R, de Groot R, Sutton P et al (2014) Changes in the global value of ecosystem services. Glob Environ Change 26:152–158

Cox C, Valdivia A, McField M et al (2017) Establishment of marine protected areas alone does not restore coral reef communities in Belize. Mar Ecol Prog Ser 563:65–79

Cunha AH, Marbá NN, van Katwijk MM et al (2012) Changing paradigms in seagrass restoration. Restor Ecol 20(4):427–430

De Groot R, Brander L, Van Der Ploeg S et al (2012) Global estimates of the value of ecosystems and their services in monetary units. Ecosyst Serv 1(1):50–61

Doxa A, Albert CH, Leriche A et al (2017) Prioritizing conservation areas for coastal plant diversity under increasing urbanization. Environ Manag 201:425–434

Duke NC, Meynecke J-O, Dittman S et al (2007) A world without mangroves? Science 317(5834):41–42

EdwardsAJ, GomezED (2007) Reef restoration Concepts and Guidelines: making sensible management choices in the face of uncertainty. Coral Reef Targeted Research & Capacity Building for Management Programme. St Lucia, Australia

Elliott M, Burdon D, Hemingway KL et al (2007) Estuarine, coastal and marine ecosystem restoration: confusing management and science: a revision of concepts. Estuar Coast Shelf Sci 74(3):349–366

Erwin KL (2009) Wetlands and global climate change: the role of wetland restoration in a changing world. Wetl Ecol Manag 17:71–84

Eyre BD, Cyronak T, Drupp P et al (2018) Coral reefs will transition to net dissolving end of century. Science 359(6378):908–911

Falace A, Zanelli E, Bressan G (2006) Algal transplantation as a potential tool for artificial reef management and environmental mitigation. Bull Mar Sci 78(1):161–166

Farley J, Costanza R (2010) Payments for ecosystem services: from local to global. Ecol Econ 69(11):2060–2068

Fonseca MS, Julius BE, Kenworthy WJ (2000) Integrating biology and economics in seagrass restoration: How much is enough and why? Ecol Eng 15(3–4):227–237

Fourqurean JW, Duarte CM, Kennedy H et al (2012) Seagrass ecosystems as a globally significant carbon stock. Nat Geosci 5(7):505–509

Fox HE, Mous PJ, Pet JS et al (2005) Experimental assessment of coral reef rehabilitation following blast fishing. Conserv Biol 19(1):98–107

Freedman RM, Espasandin C, Holcombe EF et al (2016) Using movements and habitat utilization as a functional metric of restoration for estuarine juvenile fish habitat. Mar Coast Fish 8(1):361–373

GanassinC, GibbsP (2008) A review of seagrass planting as a means of habitat compensation following loss of seagrass meadow. In: NSW Department of Primary Industries – Fisheries Final Report Series No. 96. NSW Department of Primary Industries, Cronulla Fisheries Research Centre of Excellence. Cronulla, Australia

Gedan KB, Kirwan ML, Wolanski E et al (2011) The present and future role of coastal wetland vegetation in protecting shorelines: answering recent challenges to the paradigm. Clim Chang 106(1):7–29

Giri C, Ochieng E, Tieszen L et al (2011) Status and distribution of mangrove forests of the world using Earth observation satellite data. Glob Ecol Biogeogr 20(1):154–159

Goforth HJ, Thomas J (1979) Plantings of red mangroves (*Rhizophora mangle* L.) for stabilization of Marl shorelines in the Florida Keys. In: Cole D (ed) Proceedings of the sixth annual conference on wetlands restoration and creation, 19 May 1979. Hillsborough Community College, Tampa

Grabowski JH, Brumbaugh RD, Conrad RF et al (2012) Economic valuation of ecosystem services provided by oyster reefs. Bioscience 62(10):900–909

Guerry AD, Ruckelshaus MH, Arkema KK et al (2012) Modeling benefits from nature: using ecosystem services to inform coastal and marine spatial planning. Int J Biodivers Sci Ecosyst Serv Manage 8(1–2):107–121

Guo Z, Li X, He Z et al (2018) Extremely low genetic diversity across mangrove taxa reflects past sea level changes and hints at poor future responses. Glob Chang Biol 24(4):1741–1748

Halpern BS, Walbridge S, Selkoe KA et al (2008) A global map of human impact on marine ecosystems. Science 319(5865):948–952

Harley CDG, Randall Hughes A, Hultgren KM et al (2006) The impacts of climate change in coastal marine systems. Ecol Lett 9(2):228–241

Harris JA, Hobbs RJ, Higgs E et al (2006) Ecological restoration and global climate change. Restor Ecol 14(2):170–176

Havens K, Vitt P, Still S et al (2015) Seed sourcing for restoration in an era of climate change. Nat Area J 35(1):122–133

Hein MY, Willis BL, Beeden R et al (2017) The need for broader ecological and socioeconomic tools to evaluate the effectiveness of coral restoration programs. Restor Ecol 25(6):873–883

Hemminga MA, Duarte CM (2000) Seagrass ecology. Cambridge University Press, Cambridge

Herkül K, Kotta J (2009) Effects of eelgrass (*Zostera marina*) canopy removal and sediment addition on sediment characteristics and benthic communities in the Northern Baltic Sea. Mar Ecol 30:74–82

Hill WG, Caballero A (1992) Artificial selection experiments. Annu Rev Ecol Syst 23:287–310

Himes-Cornell A, Pendleton L, Atiyah P (2018) Valuing ecosystem services from blue forests: a systematic review of the valuation of salt marshes, sea grass beds and mangrove forests. Ecosyst Serv 30:36–48

Hobbs RJ (2007) Setting effective and realistic restoration goals: key directions for research. Restor Ecol 15(2):354–357

Hobbs RJ, Harris JA (2001) Restoration ecology: repairing the earth's ecosystems in the new millennium. Restor Ecol 9(2):239–246

Hobbs RJ, Norton DA (1996) Towards a conceptual framework for restoration ecology. Restor Ecol 4(2):93–110

Hoegh-Guldberg O (1999) Climate change, coral bleaching and future of the world's coral reefs. Mar Freshw Res 50:839–866

Hoegh-Guldberg O, Bruno JF (2010) The impact of climate change on the world's marine ecosystems. Science 328(5985):1523–1528

Hoegh-Guldberg O, Mumby PJ, Hooten AJ et al (2007) Coral reefs under rapid climate change and ocean acidification. Science 318(5857):1737–1742

Hoekstra JM, Boucher TM, Ricketts TH et al (2005) Confronting a biome crisis: global disparities of habitat loss and protection. Ecol Lett 8(1):23–29

IPCC (2014) Meyer LA, Pachauri, RK (eds) Synthesis report. Contribution of Working Groups I, II and III to the Fifth Assessment Report of the Intergovernmental Panel on Climate Change. IPCC, Geneva

Johnson M, Lustic C, Bartels E et al (2011) Caribbean Acropora restoration guide: best practices for propagation and population enhancement. The Nature Conservancy, Arlington

Kang CK, Choy EJ, Son Y et al (2008) Food web structure of a restored macroalgal bed in the eastern Korean peninsula determined by C and N stable isotope analyses. Mar Biol 153(6):1181–1198

KleypasJA, LangdonC (2006) coral reefs and changing seawater carbonate chemistry. In: Phinney JT, Hoegh-Guldberg O, Kleypas J et al (eds) Coral reefs and climate change: science and management, vol 61. American Geophysical Union, pp 73–110

Koch M, Bowes G, Ross C et al (2012) Climate change and ocean acidification effects on seagrasses and marine macroalgae. Glob Chang Biol 19(1):103–132

La Peyre M, Furlong J, Brown LA et al (2014) Oyster reef restoration in the Northern Gulf of Mexico: extent, methods and outcomes. Ocean Coast Manag 89:20–28

Ledig FT, Kitzmiller JH (1992) Genetic strategies for reforestation in the face of global climate change. For Ecol Manag 50(1–2):153–169

Lee JH, Sidharthan M, Jung SM et al (2008) Comparison of the effectiveness of four organic chemoattractants towards zoospores of *Ulva pertusa* and macrofouling. J Environ Biol 29:621–627

Lindahl U (1998) Low-tech rehabilitation of degraded coral reefs through transplantation of stag-horn coral. Ambio 27:645–650

Liquete C, Piroddi C, Drakou EG et al (2013) Current status and future prospects for the assessment of marine and coastal ecosystem services: a systematic review. PLoS One 8(7):e67737

Locatelli T, Binet T, Kairo JG et al (2014) Turning the tide: how blue carbon and payments for ecosystem services (PES) might help save mangrove forests. Ambio 43(8):981–995

Lotze HK, Lenihan HS, Bourque BJ et al (2006) Depletion, degradation, and recovery potential of estuaries and coastal seas. Science 312(5781):1806–1809

Lovelock CE, Feller IC, McKee KL et al (2005) Variation in mangrove forest structure and sediment characteristics in Bocas del Toro, Panama. Caribb J Sci 41(3):456–464

Martínez-Alier J (2014) Theenvironmentalism of the poor. Geoforum 54:239–241

McDonald T, Gann G, Jonson J et al (2016) International standards for the practice of ecological restoration–including principles and key concepts. Society for Ecological Restoration, Washington, DC

McLeod E, Chmura GL, Bouillon S et al (2011) A blueprint for blue carbon: toward an improved understanding of the role of vegetated coastal habitats in sequestering CO_2. Front Ecol Environ 9(10):552–560

MEA (2005) Millennium ecosystem assessment. World Resources Institute, Washington DC

Mendelsohn R, Olmstead S (2009) The economic valuation of environmental amenities and disamenities: methods and applications. Annu Rev Environ Resour 34:325–347

Micheli F, Halpern BS, Walbridge S et al (2013) Cumulative human impacts on Mediterranean and Black Sea marine ecosystems: assessing current pressures and opportunities. PLoS One 8(12):e79889

Moberg F, Folke C (1999) Ecological goods and services of coral reefs ecosystems. Ecol Econ 29:215–233

Muradian R, Rival L (2012) Between markets and hierarchies: the challenge of governing ecosystem services. Ecosyst Serv 1(1):93–100

Muradian R, Corbera E, Pascual U et al (2010) Reconciling theory and practice: an alternative conceptual framework for understanding payments for environmental services. Ecol Econ 69(6):1202–1208

Nelleman C, Corcoran E, Duarte CM et al (2009) Blue carbon: a rapid response assessment. United Nations Environment Programme, GRID-Arendal

NOAA (1997) National oceanic and atmospheric administration, Damage Assessment and Restoration Program. Habitat equivalency analysis: an overview. Policy and Technical Paper Series 95(1)

Norlund L, Jackson EL, Nakaoka M et al (2018) Seagrass ecosystem services -- What's next? Mar Pollut Bull 134:145–151

Novacek MJ, Cleland EE (2001) The current biodiversity extinction event: scenarios for mitigation and recovery. Proc Natl Acad Sci 98(10):5466–5470

Nyström M, Norström AV, Blenckner T et al (2012) Confronting feedbacks of degraded marine ecosystems. Ecosystems 15(5):695–710

O'Leary JK, Micheli F, Airoldi L et al (2017) The resilience of marine ecosystems to climatic disturbances. Bioscience 67(3):208–220

Ow YX, Collier CJ, Uthicke S (2015) Responses of three tropical seagrass species to CO_2 enrichment. Mar Biol 162(5):1005–1017

Palacios SL, Zimmerman RC (2007) Response of eelgrass *Zostera marina* to CO_2 enrichment: possible impacts of climate change and potential for remediation of coastal habitats. Mar Ecol Prog Ser 344:1–13

Paling EI, Fonseca M, van Katwijk MM et al (2009) Chapter 24: Seagrass restoration. In: Perillo W, Wolanski E, Brinson C et al (eds) Coastal wetlands: an integrated ecosystem approach, 1st edn. Elservier, Amsterdam, pp 687–713

Palmer MA, Ambrose RF, Poff NL (1997) Ecological theory and community restoration ecology. Restor Ecol 5(4):291–300

Pandolfi JM, Bradbury RH, Sala E et al (2003) Global trajectories of the long-term decline of coral reef ecosystems. Science 301(5635):955–958

Perkol-Finkel S, Airoldi L (2010) Loss and recovery potential of marine habitats: an experimental study of factors maintaining resilience in subtidal algal forests at the Adriatic sea. PLoS One 5(5):e10791

Perkol-Finkel S, Ferrario F, Nicotera V et al (2012) Conservation challenges in urban seascapes: promoting the growth of threatened species on coastal infrastructures. J Appl Ecol 49(6):1457–1466

Perrow MR, Davy AJ (2002) Handbook of ecological restoration, vol 2. Cambridge University Press, Cambridge

Peterson CH, Kneib RT, Manen CA (2003) Scaling restoration actions in the marine environment to meet quantitative targets of enhanced ecosystem services. Mar Ecol Prog Ser 264:73–176

PizarroV, CarrilloV, García-RuedaA(2012) Growing corals in line and floating nurseries at Tayrona Park. Abstracts of the 12th International Coral Reef Symposium, 2012, Cairns

Powers SP, Boyer KE (2014) Marine restoration ecology. In: Bertness MD, Bruno JP, Silliman BR, Stachowicz JJ (eds) Marine community ecology and conservation. Sinauer Publishing, Sunderland, pp 495–516

Powers S, Peterson C, Grabowski J et al (2009) Success of constructed oyster reefs in no-harvest sanctuaries: implications for restoration. Mar Ecol Prog Ser 389:159–170

Primavera J, Esteban J (2008) A review of mangrove rehabilitation in the Philippines: successes, failures and future prospects. Wetl Ecol Manag 16:345–358

Proffitt CE, Devlin DJ (2005) Long-term growth and succession in restored and natural mangrove forests in southwestern Florida. Wetl Ecol Manag 13:531–551

Redford KH, Adams WM (2009) Payment for ecosystem services and the challenge of saving nature. Conserv Biol 23(4):785–787

Reynolds LK, Waycott M, McGlathery KJ et al (2016) Ecosystem services returned through seagrass restoration. Restor Ecol 24(5):583–588

Richards DR, Friess DA (2017) Characterizing Coastal Ecosystem Service Trade-offs with Future Urban Development in a Tropical City. Environ Manag 60(5):961–973

Rico L, Ogaya R, Barbeta A et al (2013) Changes in DNA methylation fingerprint of *Quercus ilex* trees in response to experimental field drought simulating projected climate change. Plant Biol 16(2):419–427

Rinkevich B (2005) Conservation of coral reefs through active restoration measures: recent approaches and last decade progress. Environ Sci Technol 39(12):4333–4342

Robertson AI, Duke NC (1987) Mangroves as nursery sites: comparisons of the abundance and species composition of fish and crustaceans in mangroves and other nearshore habitats in tropical Australia. Mar Biol 96(2):193–205

Rodríguez FVL (2018) Mangrove concessions: an innovative strategy for community mangrove conservation in ecuador. In: Makowski C, Finkl CW (eds) Threats to mangrove forests, Hazards. Vulnerability and Management. Springer, Cham, pp 557–578

Rönnbäck P, Crona B, Ingwall L (2007) The return of ecosystem goods and services in replanted mangrove forests: perspectives from local communities in Kenya. Environ Conserv 34(4):313–324

Rossi-Snook K, Ozbay G, Marenghi F (2010) Oyster (*Crassostreavirginica*) gardening program for restoration in Delaware's Inland Bays, USA. AquacInt 18:61–67

Ruiz-Jaen MC, AideTM (2005) Restoration success: how is it being measured? Restor Ecol 13(3):569–577

Russell M, Greening H (2015) Estimating benefits in a recovering estuary: Tampa Bay, Florida. Estuar Coasts 38(1):9–18

Sala E (2004) The past and present topology and structure of Mediterranean subtidal rocky-shore food webs. Ecosystems 7:333–340

Salomidi M, Katsanevakis S, Borja Á et al (2012) Assessment of goods and services, vulnerability, and conservation status of European seabed biotopes: a stepping stone towards ecosystem-based marine spatial management. Medit Mar Sci 13(1):49–88

Salomon M, Dahms H (2018) Marine Ecosystem Services. In: Salomon M, Markus T (eds) Handbook on Marine environment protection. Springer, Cham, pp 67–75

Scapin L, Zucchetta M, Facca C et al (2016) Using fish assemblage to identify success criteria for seagrass habitat restoration. Web Ecol 16(1):33–36

Scheffer M, Carpenter S, Foley JA et al (2001) Catastrophic shifts in ecosystems. Nature 413:591–596

Schröter M, van der Zanden EH, van Oudenhoven AP (2014) Ecosystem services as a contested concept: a synthesis of critique and counterarguments. Conserv Lett 7(6):514–523

Seaman W (2007) Artificial habitats and the restoration of degraded marine ecosystems and fisheries. Hydrobiologia 580(1):143–155

SER (1998) The SER primer on ecological restoration.Society for Ecological Restoration, Science & Policy Working Group, Tucson

Shackelford N, Hobbs RJ, Burgar JM et al (2013) Primed for change: developing ecological restoration for the 21st century. Restor Ecol 21(3):297–304

Shaish L, Levy G, Gomez ED et al (2008) Fixed and suspended coral nurseries in the Philippines: establishing the first step in the "gardening concept" of reef restoration. J Exp Mar Biol Ecol 358:86–97

Silliman BR, Schrack E, He Q et al (2015) Facilitation shifts paradigms and can amplify coastal restoration efforts. Proc Natl Acad Sci 112(46):14295–14300

Speers AE, Besedin EY, Palardy JE et al (2016) Impacts of climate change and ocean acidification on coral reef fisheries: an integrated ecological-economic model. Ecol Econ 128:33–43

Staszak LA, Armitage AR (2012) Evaluating salt marsh restoration success with an index of ecosystem integrity. J Coast Res 29(2):410–418

Steneck RS, Graham MH, Bourque BJ et al (2002) Kelp forest ecosystems: biodiversity, stability, resilience and future. Environ Conserv 29(4):436–459

Suding KN (2011) Toward an era of restoration in ecology: successes, failures, and opportunities ahead. Annu Rev EcolEvolSyst 42:465–487

Swan KD, McPherson JM, Seddon PJ et al (2016) Managing marine biodiversity: the rising diversity and prevalence of marine conservation translocations. Conserv Lett 9(4):239–251

Tallis H, Polasky S (2009) Mapping and valuing ecosystem services as an approach for conservation and natural-resource management. Ann N Y Acad Sci 1162(1):265–283

Terawaki T, Hasegawa H, Arai S et al (2001) Management-free techniques for restoration of Eisenia and Ecklonia beds along the central Pacific coast of Japan. J Appl Phycol 13(1):13–17

Terawaki T, Yoshikawa K, Yoshida G et al (2003) Ecology and restoration techniques for Sargassum beds in the Seto Inland Sea, Japan. Mar Pollut Bull 47(1–6):198–201

Thorhaug A, Poulos HM, López-Portillo J et al (2017) Seagrass blue carbon dynamics in the Gulf of Mexico: stocks, losses from anthropogenic disturbance, and gains through seagrass restoration. Sci Total Environ 605:626–636

Torda G, Donelson JM, Aranda M et al (2017) Rapid adaptive responses to climate change in corals. Nat Clim Chang 7:627–636

Tortolero-Langarica JJA, Cupul-Magaña AL, Rodríguez-Troncoso AP (2014) Restoration of a degraded coral reef using a natural remediation process: a case study from a Central Mexican Pacific National Park. Ocean Coast Manag 96:12–19

Tsihrintzis VA (1970) Protection of wetlands from development impacts. WIT Trans Ecol Environ 34

UNEPand CIFOR (2014) Guiding principles for delivering coastal wetland carbon projects. United Nations Environment Programme, Nairobi Kenya and Center for International Forestry Research, Bogor

Valiela I, Bowen JL, York JK (2001) Mangrove forests: one of the world's threatened major tropical environments. Bioscience 51(10):807–815

Van Katwijk MM, Thorhaug A, Marbà N et al (2016) Global analysis of seagrass restoration: the importance of large-scale planting. J Appl Ecol 53(2):567–578

Van Oppen MJH, Oliver JK, Putnam HM et al (2015) Building coral reef resilience through assisted evolution. Proc Natl Acad Sci 112(8):2307–2313

Van Oppen MJH, Gates RD, Blackall LL et al (2017) Shifting paradigms in restoration of the world's coral reefs. Glob Chang Biol 23(9):3437–3448

Veron JEN, Hoegh-Guldberg O, Lenton TM et al (2009) The coral reef crisis: the critical importance of <350 ppm CO_2. Mar Pollut Bull 58:1428–1436

Waycott M, Duarte CM, Carruthers TJ et al (2009) Accelerating loss of seagrasses across the globe threatens coastal ecosystems. Proc Natl Acad Sci 106(30):12377–12381

Worm B, Barbier EB, Beaumont N et al (2006) Impacts of biodiversity loss on ocean ecosystem services. Science 314(5800):787–790

Wylie L, Sutton-Grier AE, Moore A (2016) Keys to successful blue carbon projects: lessons learned from global case studies. Mar Policy 65:76–84

Xu S, He Z, Zhang Z et al (2017) The origin, diversification and adaptation of a major mangrove clade (Rhizophorae) revealed by whole-genome sequencing. Natl Sci Rev 4(5):721–734

Yoon JT, Sun SM, Chung G (2014) Sargassum bed restoration by transplantation of germlings grown under protective mesh cage. J Appl Phycol 26(1):505–509

Young TP (2000) Restoration ecology and conservation biology. Biol Conserv 92(1):73–83

Yu YQ, Zhang QS, Tang YZ et al (2012) Establishment of intertidal seaweed beds of *Sargassumthunbergii* through habitat creation and germling seeding. Ecol Eng 44:10–17

ZarnochCB, SchreibmanMP (2012) Growth and reproduction of eastern oysters, *Crassostreavirginica*, in a New York City estuary: implications for restoration. In: Volume seven. Urban Habitats. Available via: www.urbanhabitats.org/v07n01/easternoysters_full.html. Accessed 12 April 2017

Zedler J, Morzaria-Luna H, Ward K (2003) The challenge of restoring vegetation on tidal, hypersaline substrates. Plant Soil 253:259–273

Zhang YS, Cioffi WR, Cope R et al (2018) Aglobal synthesis reveals gaps in coastal habitat restoration research. Sustainability 10:1040

Zhao Q, Bai J, Huang L et al (2016) A review of methodologies and success indicators for coastal wetland restoration. Ecol Indic 60:442–452

Zimmerman RC, Hill VJ, Jinuntuya M et al (2017) Experimental impacts of climate warming and ocean carbonation on eelgrass *Zostera marina*. Mar Ecol Prog Ser 566:1–15

6 Understanding How Microplastics Affect Marine Biota on the Cellular Level Is Important for Assessing Ecosystem Function: A Review

Natalie Prinz and Špela Korez

Abstract

Plastic has become indispensable for human life. When plastic debris is discarded into waterways, these items can interact with organisms. Of particular concern are microscopic plastic particles (microplastics) which are subject to ingestion by several taxa. This review summarizes the results of cutting-edge research about the interactions between a range of aquatic species and microplastics, including effects on biota physiology and secondary ingestion. Uptake pathways via digestive or ventilatory systems are discussed, including (1) the physical penetration of microplastic particles into cellular structures, (2) leaching of chemical additives or adsorbed persistent organic pollutants (POPs), and (3) consequences of bacterial or viral microbiota contamination associated with microplastic ingestion. Following uptake, a number of individual-level effects have been observed, including reduction of feeding activities, reduced growth and reproduction through cellular modifications, and oxidative stress. Microplastic-associated effects on marine biota have become increasingly investigated with growing concerns regarding human health through trophic transfer. We argue that research on the cellular interactions with microplastics provide an understanding of their impact to the organisms' fitness and, therefore, its ability to sustain their functional role in the ecosystem. The review summarizes information from 236 scientific publications. Of those, only 4.6% extrapolate their research of microplastic intake on individual species to the impact on ecosystem functioning. We emphasize the need for risk evaluation from organismal effects to an ecosystem level to effectively evaluate the effect of microplastic pollution on marine environments. Further studies are encouraged to investigate sublethal effects in the context of environmentally relevant microplastic pollution conditions.

Keywords

Plastics · Tissue level · Chemical contamination · Oxidative stress · Sublethal effects · Ecosystem function

6.1 Introduction

Plastic pollution is ubiquitous in the global environment. The Industrial Revolution paved the way for the rapid development in manufacturing of long-lasting plastic materials. Consequently, the volume of plastic waste produced has increased. Our reliance on this man-made material has led to what some call "the plastic age" (Thompson et al. 2009a). Worldwide ~348 million tons of plastics were produced in 2017, of which approximately 42% was used for single-use packaging (Geyer et al. 2017; Plastics Europe 2018). Littering, ineffective recycling management practices, weather events, etc. have all been linked to the release of plastics into the environment. It has been estimated that between 4.4 and 12.7 million tons of plastic enter the marine environment annually (Jambeck et al. 2015). Over time, in the environment and exposed to weathering, sunlight, and mechanical degradation, large plastics will become brittle and break down to secondary microplastics (<5 mm) and nanoplastics (<100 nm) (MSFD Technical Group on Marine Litter 2013). Secondary microplastics also include microfibers that are washed out of synthetic

N. Prinz
Faculty of Biology and Chemistry, University of Bremen, Bremen, Germany

Leibniz Centre for Tropical Marine Research (ZMT), Bremen, Germany

Š. Korez (✉)
Alfred Wegener Institute, Helmholtz Centre for Polar and Marine Research, Bremerhaven, Germany
e-mail: spela.korez@awi.de

S. Jungblut et al. (eds.), *YOUMARES 9 - The Oceans: Our Research, Our Future*,
https://doi.org/10.1007/978-3-030-20389-4_6

clothes (Browne et al. 2011). Primary microplastics are small particles designed to be used for manufacturing large plastic items, including virgin resin pellets and microbeads (from cosmetics and personal care products) (Andrady 2017). Many fibers and microbeads are too small to be removed by filters used in sewage systems and will be flushed into the sea (Carr et al. 2016; Lebreton et al. 2017). This makes the issue of marine plastics more pressing for the coming centuries due to a consistent increment in microplastic abundance (Browne et al. 2007). Nowadays, microplastics are omnipresent, in rivers, estuaries, on shorelines, the ocean surface or in the water column, and on the seafloor (GESAMP 2015). The ubiquitous nature of microplastics in the environment means that biota can, and will, interact with them from the surface waters of the ocean to the deep sea. The bioavailability of microplastics depends on their size, density, abundance, shape, and color (Wright et al. 2013a). Over 1401 marine species are known to interact with marine plastic debris in different ways (Ocean Plastics Lab 2018). However, entanglement and ingestion are the most common types of interaction between biota and plastics (Gregory 2009). Fouling of bacteria on plastic particles may promote the ingestion of plastic materials by biota (Zettler et al. 2013; Vroom et al. 2017). Microplastic ingestion has been described for many taxa of animals including plankton, invertebrates, fish, sea turtles, and marine mammals (Cole et al. 2013; Foekema et al. 2013; Schuyler et al. 2013; Hämer et al. 2014; Lusher et al. 2015; Scherer et al. 2018). Current research efforts focus on the effects of microplastics entering and being channeled up aquatic food chains. It is still being investigated which species are more susceptible to the encounter and uptake, and which mechanisms are simultaneously affected (Rochman et al. 2015). Many species have been observed to directly take up plastics, either by selective targeting of plastic items, or accidental ingestion by filtration or predation (Lusher 2015).

Most organisms are constantly confronted with inert particles of different sizes, shapes, and materials throughout their life. Seif et al. (2018) highlighted that, apart from plastic, metal, glass, and building materials were also found in the intestines of gulls. Microplastics are often similar in size to sediment particles or may resemble a grain of sand. Therefore, it is not surprising that animals in coastal areas, particularly filter feeders, consistently encounter natural particles as well as particles generated by human activity like microplastics (Van Cauwenberghe and Janssen 2014; Weber et al. 2018). Usually, if an animal is not able to digest an item, it egests it after some time (Garrett et al. 2012; Santana et al. 2017). Plastic particles represent foremost foreign bodies inside an organism; nevertheless, their charge, chemical composition, and contamination are of particular interest. In many cases, added chemicals in plastic manufacturing and persistent organic pollutants seem to be the actual threat. Increasingly, studies focus on physiological effects of microplastics on animals on an individual scale (Lusher 2015), as microplastics potentially cause cryptic sublethal effects that have to date rarely been investigated (Koelmans 2015). The effects include pathological stress, reproductive complications, changes in enzymes activities, reduced growth rate, and oxidative stress (Besseling et al. 2014; Sutton et al. 2016). Smaller particles (<100 nm) may have greater consequences upon ingestion, because they may end up in the tissues or even inside the cells (Lusher 2015). The time a particle spends inside the body (i.e., the retention time) is crucial for estimating chemical exchanges within the body. Many studies investigate the occurrence of plastic within the intestinal tract of an organism without discussing an impact on the animal itself (Boerger et al. 2010; Lusher et al. 2013; Battaglia et al. 2016; Rummel et al. 2016; Baalkhuyur et al. 2018). Yet, a wealth of studies identify effects of microplastic with artificial concentrations that are far beyond natural levels as currently encountered in the ocean (Pedà et al. 2016; Lusher et al. 2017; Critchell and Hoogenboom 2018). Nevertheless, findings provide evidence that plastic particles can cause internal wounds, lesions, or blockage of the digestive tract, which can promote a feeling of satiation that can lead to starvation, depletion of strength, and even death (Gregory 2009, Jovanović 2018).

It is important to disentangle the risks associated with ingested particles in an ecologically relevant context (Koelmans et al. 2017a). In a future of ever smaller particles, many organisms will be confronted with them, regardless of the size of the organism (Mattsson et al. 2017; Vendel et al. 2017; Critchell and Hoogenboom 2018).

This review evaluates the consequences of microplastic ingestion by summarizing the pathways of ingested microplastics and their subsequent effect on marine species, with some examples from freshwater species. The specific aims were to (i) collect results from current research of microplastic-derived impacts of organismal physiology and (ii) highlight the urgent need for embedding research on microbiological functioning of internal structures into the impact on ecosystem functioning. Further, this review aims to (iii) highlight the gaps of research that elaborate the sublethal effects of microplastics on an ecosystem function approach. An extensive literature review of 236 scientific publications resulted in this synthesized review. The percentage of articles discussing impacts on ecosystem function were calculated.

Three types of consequences of microplastics uptake through the digestive tract or the respiratory system have

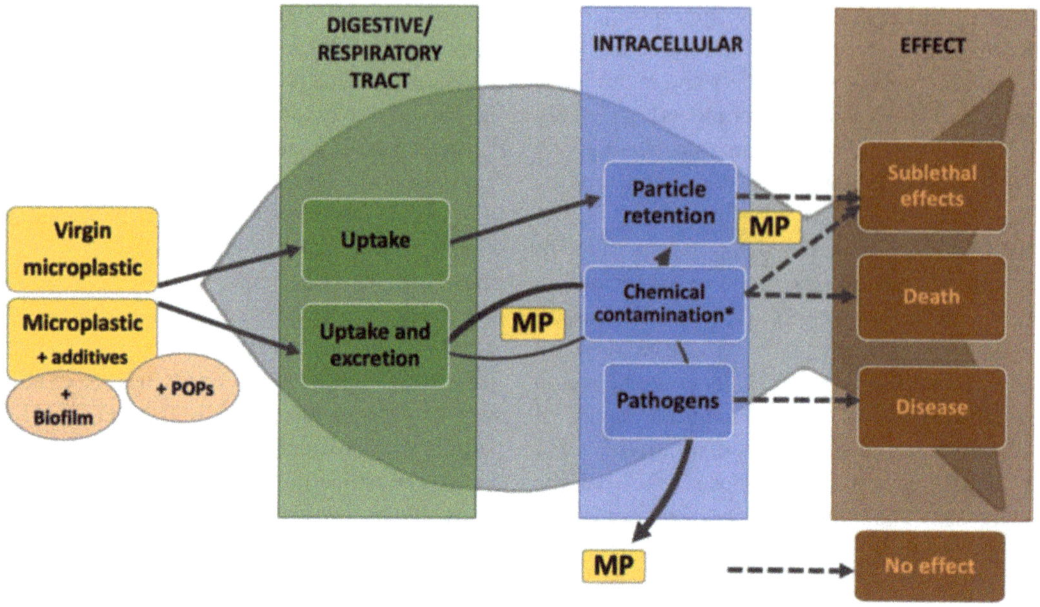

Fig. 6.1 Transfer pathways of microplastic particles and associated contaminants in the body of an organism (MP, microplastics; POPs, persistent organic pollutants). ∗Chemical impacts are graphically explored in more detail in Anbumani and Kakkar (2018)

been identified: (1) physical penetration of microplastic particles into cellular structures, (2) leaching of chemical additives or persistent organic pollutants (POPs) into the body, or (3) infecting eukaryotic and bacterial microbiota from the surface of ingested microplastics (Fig. 6.1). First, availability of microplastics to different biota will be discussed (Sect. 6.2). This entails the interactions of flora and fauna with microplastics. Further, known consequences of plastic particles in the tissues and cells are summarized (Sect. 6.3) with an evaluation on how cellular biomarkers are used (Sect. 6.4). Finally, the interactions between chemical pollutants and structures in the body are evaluated (Sect. 6.5) leading to a discussion about trophic cascading (Sect. 6.6) and human health (Sect. 6.7). Finally, this review discusses pathways of microplastic particle interaction with biota on the cellular level and concludes with suggestions for concrete research foci (Sect. 6.8).

6.2 Interactions of Different Organisms with Microplastics

6.2.1 Microplastic Interaction with Aquatic Primary Producers

Effects on algae are often neglected to be considered. Bhattacharya et al. (2010) reported that nanosized plastic beads can be adsorbed by a green algae (*Scenedesmus* spp.), hindering the photosynthetic activity. This occurrence was attributed to the physical chemistry of the particles when positively charged. Photosynthesis of a marine diatom (*Thalassiosira pseudonana*) and marine flagellate (*Dunaliella tertiolecta*) was not affected, although at high concentrations and decreasing particle size of uncharged polystyrene particles, growth was reduced (Sjollema et al. 2016). Microplastics can form aggregates with some phytoplankton species. The phytoplankton *Rhodomonas salina* has a tendency to incorporate more microplastic to the aggregate compared to *Chaetoceros neogracile* (Long et al. 2015). More concerning effects are addressed in a recent study by Kalčíková et al. (2017) with a freshwater species. Sharp polyethylene microplastics from exfoliating cosmetic products are reducing the viability of the root cells of the duckweed (*Lemna minor*), which detrimentally affects their growth. A similar phenomenon was observed in moss (*Sphagnum palustre*) where small aggregates of microplastics entered into the hyalocyte cells of the leaf. Bigger aggregates of microplastic adsorbed on the moss' surface (Capozzi et al. 2018). Adsorption was also observed in the colonial green algae *Scenedesmus* or seaweed *Fucus vesiculosus* (Bhattacharya et al. 2010; Gutow et al. 2016). Such results address the significance of primary producers interacting with microplastic (Yokota et al. 2017). Green et al. (2016) concluded that a reduction of macroalgal biomass can be responsible for the overall primary productivity of a sandy bottom ecosystem. This clearly alludes to further

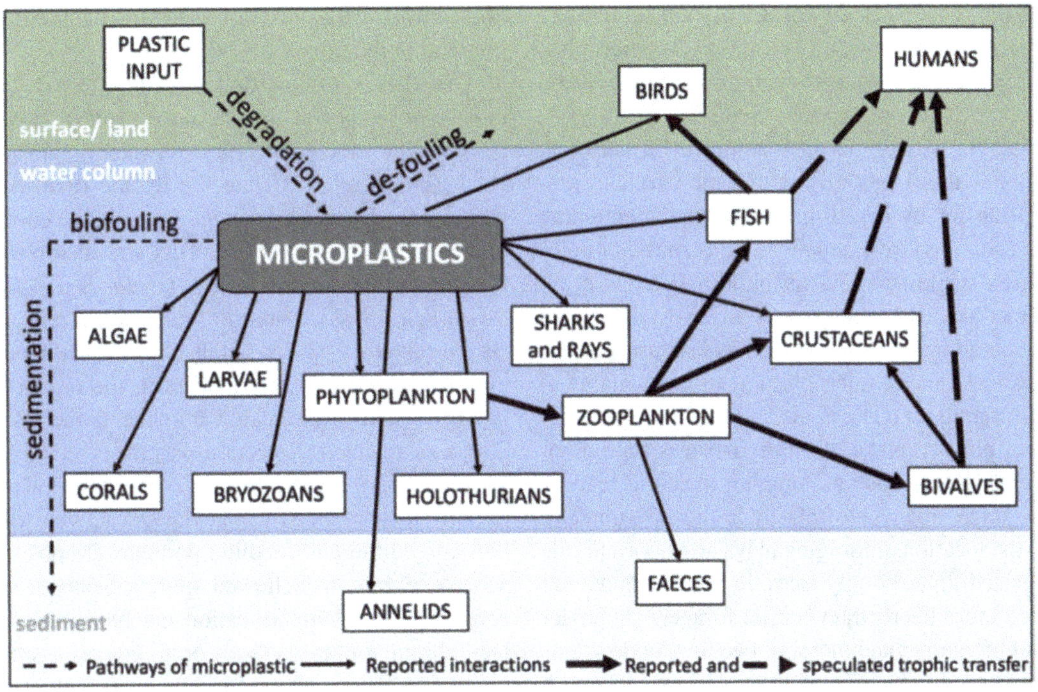

Fig. 6.2 Schematic presentation of microplastic interaction with different organisms in the food web. (Based on Wright et al. 2013a; Lusher 2015; Tosetto et al. 2017)

studies that quantify the effect of microplastics on the function that primary producers exhibit in the marine environment (Troost et al. 2018).

6.2.2 Microplastic Interactions with Invertebrates

The bioavailability of microplastic allows biological interactions with organisms of different feeding types (Fig. 6.2). Availability of microplastic is sometimes dependent on the organisms itself, as for Antarctic krill (*Euphausia superba*) which can biologically fragment microplastic into smaller nanoparticles upon ingestion (Dawson et al. 2018). Apart from the direct ingestion from the water, microplastic can be ingested through their prey (Watts et al. 2014; Green et al. 2015) or through adherence on the organs that are primarily not involved in digestion (Kolandhasamy et al. 2018). The latter was observed in blue mussels with microplastic presence in the gonad, mantle, adductor, visceral, and foot (Kolandhasamy et al. 2018). Here, the digestive gland contained the highest levels of microplastics; however, a clearance experiment showed the retention of microplastics also in other organs.

When microplastics aggregate with marine snow (Summers et al. 2018) or phytoplankton (Long et al. 2015) they are especially attainable for small and large filter feeders (Setälä et al. 2016, Besseling et al. 2015), and zooplankton (Cole et al. 2013). Over time, microplastic is introduced to the sediment habitat. Together with sediment particles or feces, it can be consumed by benthic suspension or deposit feeders and detritivores, such as annelids (Besseling et al. 2013). Cole et al. (2015) observed that microplastics encapsulated within the fecal pellets can be transferred between coprophagous copepod species. Furthermore, floating microplastics that wash onto the shore are available to invertebrates in the intertidal (Lourenço et al. 2017). Unsurprisingly, microplastic is not only ingested by marine invertebrates. Studies report about representative freshwater organisms such as zooplankton (*Daphnia magna*), amphipods (*Hyalella azteca*, *Gammarus pulex*), and sponges (*Hydra attenuate*) to be affected as well (Au et al. 2015; Rehse et al. 2016, 2018; Murphy and Quinn 2018; Weber et al. 2018).

6.2.3 Microplastic Interactions with Vertebrates

Predatory vertebrate species can ingest microplastic unintentionally, when misidentifying synthetic microparticles for prey. This is especially common when the actual prey is of distinctive color, like in the case of the family of fish

Gerreidae and blue copepods (Ory et al. 2017). In addition to fish (Ramos et al. 2012; Choy and Drazen 2013), microplastic was also reported in predators such as sea birds (Kühn and van Franeker 2012), sea turtles (Schuyler et al. 2013, Yaghmour et al. 2018), and marine mammals (Lusher et al. 2018). Vertebrates that ingested microplastic can also promote trophic transfer by ingesting microplastic-containing invertebrates (i.e., bivalves, amphipods, barnacles, polychaetes) or even while scraping on biofilm (Ramos et al. 2012; Reisser et al. 2014, Hodgson et al. 2018). Once the microplastic-containing organisms, such as fish, crustaceans or polychaetes egest feces, microplastic can be available to coprophagous organisms (Cole et al. 2016).

The level of plastic uptake of an organism is accommodated by several factors, such as foraging location, feeding strategies, life stage, and type of plastic in the environment. For example, the location of foraging plays an important role in what is ingested. Interestingly, oceanic juvenile and adult turtles ingested more debris than coastal foragers (Schuyler et al. 2013 and literature cited therein). Feeding mode seems to be correlated to the amount of plastic ingested by fish (Anastasopoulou et al. 2013; Romeo et al. 2015; Battaglia et al. 2016). Early life stages of fish are suggested to be increasingly confronted with microplastic, as they dwell close to the ocean surface where floating microplastic concentrate (N. Prinz, unpubl. data), or in the water column where particles become masked by microbial communities. Understanding differences in exposure conditions in the wild is of major importance to investigate how different species cope with exposure to microplastic in experimental set-ups (Rochman and Boxall 2014).

6.3 The Physical Aspect: Consequences of Microplastic Uptake

To quantify the interactions and effects of microplastic uptake in biota, laboratory exposure experiments are used on key species, resistant to versatile laboratory conditions (Devriese et al. 2015). To increase the probability of microplastic uptake, the concentrations often exceed environmental levels by several orders of magnitude. Studies provide us with the future scenario without appropriate current representation of the microplastic pollution (Rochman and Boxall 2014; Paul-Pont et al. 2018). Therefore, caution needs to be taken when interpreting the results. Furthermore, studies need to clearly disentangle consequences of exaggerated microplastic uptake from those likely encountered in the wild.

Physical effects of microplastics can be observed on individual or population level (Galloway and Lewis 2016). However, from 236 scientific publications reviewed herein, only 11 extrapolate results to the impact on ecosystem function (4.6%), with only three studies mentioning ecosystem function in the title (1.3%) (Table 6.1).

The effects of microplastics on specific specimen are mostly investigated in marine species. However, some already report the effects on freshwater organisms. Mattsson et al. (2017) reported that the uptake of microplastics by freshwater *Daphnia magna* positively correlated with microplastic concentrations. This was also observed in the marine species such as bivalves (*Macoma baltica*, *Mytilus trossulus*), mysids, and in the fiddler crab, *Uca rapax* (Brennecke et al. 2015; Setälä et al. 2016). Upon ingestion, microplastics are either retained in the organism, accumulated, egested, translocated into the tissue (Browne et al. 2008), or rejected. Rejection was observed in larvae of the sea urchin (*Tripneustes gratilla*). The ingestion of microplastic was thus reduced as the larvae actively discriminated between edible and inedible particles (Kaposi et al. 2014). Moreover, zebrafish showed spitting behavior in laboratory conditions as an identification mechanism of ingested but inedible microplastics (Kim et al. 2019). A similar mechanism of selection due to low nutritional content of the microplastic was observed also in blue mussels, where particles were excreted as pseudofeces (Wegner et al. 2012; Farrel and Nelson 2013).

Yet, possibly not all animals have the ability of rejection, and microplastics are likely retained. The effect of the retained microplastics depends on the particle size (Wright et al. 2013a) and seems to be affecting organisms in several ways. Some organisms like the Atlantic Sea scallop (*Placopecten magellanicus*) retain bigger beads longer, as they are probably transferred to the digestive gland for digestion. Smaller particles are trapped in the rejection grooves on the sorting tracts and egested (Brillant and MacDonald 2000). Wright et al. (2013b) attribute longer retention in lugworm to the low nutritional value of the particles and their extensive and energetically costly digestion. Similarly, in corals, the particles moved deep into their polyps, wrapped in their mesenterial tissue. Since the tissue is responsible for the digestion, this raises concerns of the ability to ingest natural food (Hall et al. 2015; Allen et al. 2017). Research on corals is still scarce, but some negative impacts on the health of stony corals were documented with the potential to be sublethal in the long term (Reichert et al. 2018; Tang et al. 2018). In addition to size, the shape of the microplastics is influential. Irregularly shaped microplastic can cause histopathological damages, as observed in the intestine of adult zebrafish (Duis and Coors 2016; Horton et al. 2017; Lei et al. 2018) and European sea bass (Pedà et al. 2016).

The ingestion, retention, and egestion can impair the nutritional health of the organisms. The lungworm (*Arenicola marina*) is used in several studies as an indicator species and ecosystem engineer (Green et al. 2016). Besseling et al.

Table 6.1 Selected studies investigating microplastic effects on marine and freshwater biota with regard to impacts on the organisms' function ordered after trophic level from algae to fish. Studies in bold mention ecosystem function in their title

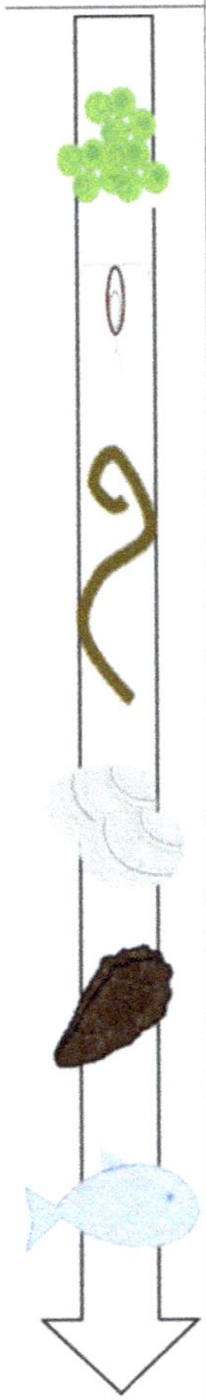

Common name	Species name	Plastics	Chemicals	Tissue effects	Ecologically relevant effects	References
Freshwater algae	*Chorella vulgaris*	Nano polystyrene (PS)	Positively charged	Adsorption of positively charged particles	Increased ROS, decreased chlorophyll concentration, photosynthesis rate of algae	Bhattacharya et al. (2010)
Marine and freshwater microalgae	*Dunaliella tertiolecta Thalassiosira pseudonana Chlorella vulgaris*	Micro Polysyrene (PS)	negatively charged carboxylated PS	Growth was negatively affected by uncharged particles	Altered growth of algae	Sjollema et al. (2016)
Copepod	***Calanus helgolandicus***	**Micro Polysyrene (PS)**	**n/a**	**Reductions in ingested carbon biomass owing to a subtle shift in the size of algal prey consumed**	**Reduced feeding on prey of copepods, energy depletion**	**Cole et al. (2015)**
Lugworm	***Arenicola marina***	**Micro polyvinyl chloride (PVC)**	**Nonylphenol, phenanthrene, Triclosan and PBDE-47**	**Higher susceptibility to oxidative stress**	**Diminished ability to engineer sediment burrows and remove pathogenic bacteria, mortality, ecophysiological function**	**Browne et al. (2013)**
Lugworm	*Arenicola marina*	Mirco un-plasticised polyvinylchloride (UPVC)	n/a	Long residence time in gut, inflammation, feeding apparatus affected	Reduced feeding activity, energy assimilation, seabed engineering and bioturbation affected	Wright et al. (2013b)
Lugworm	*Arenicola marina*	Micro polysytene (PS)	12 PCBs	Accumulation of PCBs in tissue	Weight loss, reduced feeding activity	Besseling et al. (2013)
Peppery furrow shell clam	*Scrobicularia plana*	Micro low-density polyethylene (LDPE)	benzo[a]pyrene (BaP) and perfluorooctane sulfonic acid (PFOS)	Neurotoxicity, mechanical injury to gills, MPs-adsorbed BaP and PFOS exerting a negative influence over the assessed biomarkers in this tissue	Bioindicator for evaluating the health status of coastal and estuarine ecosystems, playing a key role in their structure and functioning	O'Donovan et al. (2018)
European flat oyster, Blue mussel	***Ostrea edulis, Mytilus edulis***	**Micro polylactic acid (PLA), high density polyethylene (HDPE)**	**n/a**	**Reduced filtration rates**	**Altered filtration rates affecting concentrations and fluxes of benthic inorganic nitrogen**	**Green et al. (2017)**
Crucian carp, Bleak, Rudd Tench pike and Atlantic salmon	*Carassius carassius, Alburnus alburnus, Scardinius erythrophthalmus, Tinca tinca, Esox esox, Salmo salar*	Nano Virgin polystyrene (PS)	n/a	Triglycerides:cholesterol ratio in blood serum, nanoparticles bind to apolipoprotein A-I in fish serum in-vitro, restraining them from properly utilizing their fat reserves	Consumption of nanoparticle-containing zooplankton affects the feeding behavior of the fish	Cedervall et al. (2012)
Freshwater zooplankter, Crucian carp	*Daphnia magna, Carassius carassius*	Nano polystyrene (PS)	Positively charged	Direct interactions between plastic nanoparticles and brain tissue of fish	Reduce survival of aquatic zooplankter and penetrate the blood-to-brain barrier in a top consumer fish and cause behavioral disorders	Mattsson et al. (2017)
Beach hoppers, Frillfin goby	*Platorchestia smithi, Bathygobius krefftii*	Micro polyethylene (PE)	Sorbed PAHs	n/a	No effect on boldness and exploration variables in fish behavior	Tosetto et al. (2017)

(2013) and Wright et al. (2013b) observed reduced feeding rates and weight loss in the lugworm upon feeding on polystyrene microplastics. The authors observed reduced growth, maturity, reproduction, and somatic maintenance due to depleted energy reserves (Wright et al. 2013b). Langoustine (*Nephrops norvegicus*) lost body mass due to microplastic retention, which resulted in lower growth rates (Welden and Cowie 2016). Sea urchin larvae had reduced their body width, which was again related to reduced feeding efficiencies (Kaposi et al. 2014). Upon microplastic ingestion, Sussarellu et al. (2016) report that Pacific oysters (*Crassostrea gigas*) reallocated energy for reproduction to structural growth and maintenance. However, microplastic is differently affecting the nutritional health of the freshwater organisms. Weber et al. (2018) namely observed no significant effect on survival, development, metabolism (glycogen, lipid storage), or feeding activity of the freshwater amphipod *Gammarus pulex* upon microplastics ingestion. This is likely attributed to this species being a detritivore and adapted to non-digestible material.

Interestingly, the predatory performance of blue discus (*Symphysodon aequifasciatus*) juveniles and the common goby (*Pomatoschistus microps*) were negatively affected after microplastic exposure (de Sá et al. 2015; Fonte et al. 2016; Wen et al. 2018). Reduced performance raises concerns for survival of the organism as it diminishes the chances of capturing prey or escaping the predators. This might have subsequent effects on the population level if the levels of offspring are reduced on account of starvation, reduced growth, reproductive failure, and mortality (Ferreira et al. 2016; Galloway and Lewis 2016). In the studies by Lee et al. (2013) and Mazurais et al. (2015) mortality correlated with microplastic abundance in nauplii and copepodite stages of copepods and in the larvae of the European sea bass (*Dicentrarchus labrax*). Microplastic intake had lethal effects on fish larvae (Mazurais et al. 2015), forming a possible bottleneck in population dynamics which would lead into decrease of fish stocks (Steer et al. 2017; N. Prinz, unpubl. data).

Effects of microplastic were observed also on population level. Green and collaborators have shown a holistic effect of plastic on the function of bivalve-dominated sandy bottom ecosystems through measuring animal-mediated biogeochemical processes and abundance of different biota (Green et al. 2015, 2016, 2017). Only a few studies conclude with possible impacts on ecosystem function by microplastics that induced reduction of intracellular metabolic and endocrine functioning.

Galloway et al. (2017) described the potential impacts of microplastic exposure from the subcellular to the ecosystem level. This emphasizes again that effects on enzyme activity, oxidative damage, or gene expression can lead to sublethal pathological responses in the cells and organs, eventually harming entire populations through reduced fitness. The consequence of behavioral changes or community shifts can affect the ecosystem as we know it.

6.4 The Cellular Aspect: When Microplastic Particles Translocate into the Tissue

Current scientific efforts focus on more invasive effects of microplastics on organisms. Microplastic is not just affecting organisms when passing through the digestive system, but it can enter into the cells of the digestive tissue, be found in the blood and translocate between tissues (Volkheimer 1975, 1977). Browne et al. (2008) first showed the translocation of microplastic from the gut to the circulatory system of the blue mussels (*Mytilus edulis*) in 3 days. The particles stayed there for almost 50 days. The translocation to the hemocytes was not particle size-dependent, as both 3 μm and 9.6 μm small microspheres translocated. Nevertheless, the smaller particles showed a higher probability of entering into the hemolymph. Translocated microplastics were also found in the laboratory experiments with the shore crab (*Carcinus maenas*). After 1 h the 0.5 μm polystyrene microspheres were found in the stomach, hepatopancreas, ovary, gills, and hemolymph (Farrel and Nelson 2013). The experiments with bigger microparticles (10 μm) failed to show translocations to other organs (Watts et al. 2014), suggesting a size-dependent translocation in organisms. Similar observations of microplastic presence in the hepatopancreas, the stomach, and the gills were made in the laboratory experiments with the fiddler crab, *Uca repax* (Brennecke et al. 2015). Microplastics were observed in the endocytotic vacuoles of digestive epithelial cells of blue mussels, in their intestine and in the lumina of their primary and secondary ducts of the digestive gland. Epithelial cells of ducts and tubuli were eliminating microplastics, which were phagocytosed into the tissue, forming granulocytomas, an inflammatory response against the foreign particles (von Moos et al. 2012). The translocation of microplastic can sometimes be specific. In held mullet (*Mugil cephalus*), in zebrafish and in European Anchovies (*Engraulis encrasicolus*) microplastic translocated to their liver (Avio et al. 2015; Lu et al. 2016; Collard et al. 2017). Once translocated, microplastics can either cause oxidative stress (von Moos et al. 2012; Lu et al. 2016) or remain inert (Oliveira et al. 2013; Alomara et al. 2017).

6.4.1 Biomarkers Revealing the Effects of Microplastic on the Cellular Level

The direct impacts of microplastics on signaling pathways in the tissue are of interest to increase the knowledge on cellular effects. To investigate this, biomarkers are used as these biochemical tools measure an organisms' response to environmental contaminants (Monteiro et al. 2005). Many studies measured the activities of digestive enzymes as biomarkers. Microplastics affected, namely, digestive enzyme activities in the digestive system of isopods (*Idotea emarginata*), freshwater blue discus (*Symphysodon aequifasciatus*), silver barb (*Barbodes gonionotus*), and common carp (*Cyprinus carpio*) (Haghi and Banaee 2017; Romano et al. 2017; Wen et al. 2018; Š. Korez, unpubl. data). The affected activities of enzymes, such as lipase, esterase, trypsin, amylase, or alkaline phosphatase, show some kind of physiological challenge to the organism upon microplastic ingestion.

Microplastics enter into the cells through endocytosis or permeate through the lipid membrane when smaller than 50 nm (Fig. 6.3) (von Moos et al. 2012; Pinsino et al. 2015; Jeong et al. 2017). One biomarker used to estimate the health of an animal is the lysosomal membrane stability (LMS) which is sensitive to environmental pollutants (Moore et al. 2006). Lysosomes are single membrane organelles in the cell cytoplasm and are sensitive to environmental pollutants. Their function is cell-specific, however they are responsible for digesting the material taken into the cell (Martínez-Gómez et al. 2015). Microplastics were found in the lysosomes of blue mussels and caused the lysosomal membrane to destabilize, indicating that mussels were affected by the presence of these particles (von Moos et al. 2012; Avio et al. 2015).

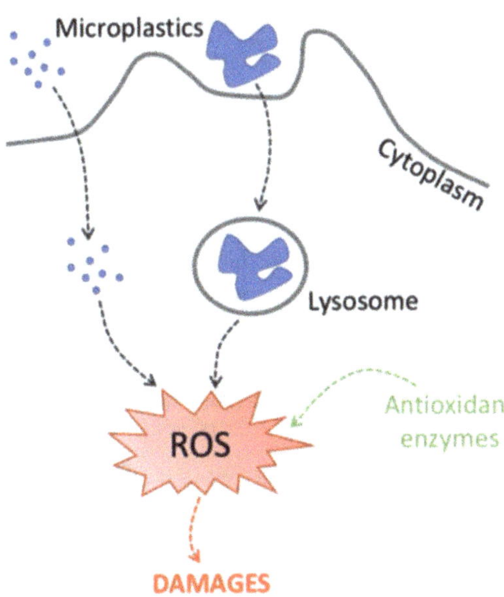

Fig. 6.3 Simplified schematic presentation of events upon microplastic entering the cell. Reactive oxygen species (ROS) are produced as a response to the foreign particle. The antioxidant enzymes are protecting the cell against ROS. In case of constant exposition to ROS, oxidative damages occur that target biomolecules. (Adapted after Wakamatsu et al. 2008)

Once in the cell, microplastic can induce oxidative stress due to a generation of reactive oxygen species (ROS). These are generated when particles are recognized as foreign particles by inflammatory cells, which generate an oxidative response (Miller et al. 2012). Through antioxidants, such as vitamins and enzymes, cells are usually appropriately protected (Lushchak 2011). Enzymes regulate the level of ROS in the cell but in the case of continuous exposure to microplastic can cause oxidative damage (Fig. 6.3) (Sureda et al. 2006).

There are a few studies concerning the biological effects of microplastics that use oxidative stress as a biomarker. Elevated ROS levels were observed in mussels (*Mytilus* spp.), monogonont rotifers (*Brachionus koreanus*), the labrid fish *Coris julis*, and the zebrafish *Danio rerio* after exposure to microplastic (Sureda et al. 2006; Paul-Pont et al. 2016; Jeong et al. 2016; Lu et al. 2016). Overall, microplastic toxicity generally increases with the decreasing particle size (Pan et al. 2007; Choi and Hu 2008; Jeong et al. 2016, 2017). Specifically, a negative correlation between ROS levels and decreasing microparticle sizes was shown in copepods, rotifers, and zebrafish (Jeong et al. 2016, 2017; Lu et al. 2016). The corresponding enzymatic defense mechanisms against elevated ROS follow the same trend of microplastic-size dependence. Rotifers (*B. koreanus*) and copepods (*Paracyclopina nana*) showed increased defense enzyme activities with decreasing microplastic size (Jeong et al. 2016, 2017). Endocytosis of nanoscale microplastics was not observed to induce oxidative stress responses in red mullet *Mullus surmuletus*. However, the increase in the activity of glutathione-S-transferase (GSF) was observed, suggesting activation of detoxification systems (Alomara et al. 2017). Above listed species experienced no cellular changes, increase in enzyme activity, or oxidative damage even though the organisms ingested microplastics.

ROS can have detrimental effects on biomolecules such as lipids when there are insufficient amounts of antioxidants present (Lushchak 2011). Lipid droplets in the liver of zebrafish confirmed that microplastics affect lipid metabolism (Lu et al. 2016). However, the trend was not universal in all organisms, as the lipid peroxidation levels remain unchanged in the labrid fish *Coris julis*, in the common goby and mussels (*Mytilus* spp.) after microplastics exposure (Sureda et al. 2006; Oliveira et al. 2013; Paul-Pont et al. 2016). The time of microplastic exposure plays a significant role, as short-term exposures showed no effect on lipid metabolism (von Moos et al. 2012; Avio et al. 2015). In addition to lipids, ROS can oxidate proteins and induce gene expression of specific metabolic pathways (Jeong et al. 2016, 2017). In copepods (*P. nana*) and rotifers (*B. koreanus*), kinase proteins were activated, indicating cell death (Jeong et al. 2016, 2017). In the copepod, *P. nana* (Jeong et al.

2017), the nematode, *Caenorhabditis elegans* (Lei et al. 2018), and *Mytilus galloprovicialis* microplastics up-regulated genes of cellular and immune defense pathways and enhanced the energy production (Avio et al. 2015; Détrée and Gallardo-Escárte 2017). The scleractinian coral (*Pocillopora damicornis*) showed induced antioxidant enzymes and detoxifying and immune enzyme activities were repressed (Tang et al. 2018). In *M. galloprovincialis* microplastic caused DNA damages (Avio et al. 2015).

6.5 The Chemical Aspect: Uptake of Leachates from Microplastics into the Body

Microplastics do not solely have consequences as a foreign body. If microplastics are ingested, they can act as a vector for the transfer of chemical contaminants to individuals. Given the diversity of contaminants in aquatic environments as well as the complex chemical structure of plastic polymers, a multitude of different chemical exchanges may occur inside the body of organisms upon ingestion (Karami et al. 2016a; Karami 2017). Also in this regard, the retention time of microplastic particles within the body is especially crucial for possible chemical exchanges into the cells (Welden and Cowie 2016). Some plastic polymers are considered biologically inert (Rist et al. 2018). Therefore, environmentally sorbed contaminants are of particular interest, as these chemicals can leach from the particle into the organism and affect metabolic pathways (Rochman 2015). An exact evaluation of the pollutants and their concentrations on the particle is needed to draw solid conclusions about the impact of microplastics on biota.

6.5.1 Leaching Additives and Persistent Organic Pollutants: The Real Threat?

Plastics are synthesized from monomers, which are polymerized to form macromolecular chains (Galloway 2015). Microplastics, in particular, can act as a vector for compounds that are added during plastic production and may be toxic to organisms (Browne et al. 2008; Hermabessiere et al. 2017). The final plastic polymers often include initiators, catalysts, solvents, stabilizers, plasticizers, flame retardants, pigments, and fillers (Crompton 2007; Galloway 2015). Because of their low molecular weight, toxic compounds, such as nonylphenol (NP) and bisphenol A (BPA), leach out of the plastic polymer, as they can naturally break down and release into the surrounding environment (Flint et al. 2012; Galloway 2015). Based on biodynamic modeling, microplastic-exposed animals, like lugworm and cod, are threatened by already low concentrations of NP and BPA (Koelmans et al. 2014, Bakir et al. 2016). Other evidence suggests BPA to cause reproductive toxicity in breeding zebrafish (*Danio rerio*) (Laing et al. 2016).

Alternatively, chemicals dissolved in the surrounding seawater can adsorb on the microplastic's surface. A multitude of factors influence the sorption-desorption of persistent pollutants (PPs) from the seawater onto microplastics, including shape, size, type of polymer, fouling, pH, temperature, PP concentration, and K_{ow} (n-Octanol/Water Partition Coefficient) of PPs (Teuten et al. 2007; Wang et al. 2016). Some persistent pollutants (PPs) that sorb onto microplastics are polycyclic aromatic hydrocarbons (PAHs), pesticides (dichlorodiphenyl trichloroethane, DDTs), polychlorinated biphenyls (PCBs), metals, and other endocrine disrupting chemicals (Ng and Obbard 2006; Cole et al. 2011; Bakir et al. 2014; Avio et al. 2015; Llorca et al. 2018). The importance of chemical exchange not only in the water column but in the sediment cannot be underestimated, as heavy metals from antifouling paints, fuel combustion, and industrial waste in sediments can sorb onto microplastics (Deheyn and Latz 2006; Holmes et al. 2012; Rochman et al. 2013; Khan et al. 2015; Brennecke et al. 2016). The global concentration of POPs in marine plastic pellets was estimated to be 1 – 10,000 ng g^{-1} (Ogata et al. 2009; Hirai et al. 2011).

Additives and pollutants sorbed onto microplastics are bioavailable to marine microorganisms which can metabolize them (Chua et al. 2014; Avio et al. 2015; Wardrop et al. 2016; Auta et al. 2017). Laboratory studies artificially spike microplastics to quantify in how far digestion is an important process in the so-called leaching or desorption of POPs. When particles containing adsorbed chemicals are ingested by an organism, the change in surrounding conditions can promote the release of pollutants (e.g., Besseling et al. 2013; Browne et al. 2013; Batel et al. 2016). Desorption rates of some contaminants in gut surfactants are up to 30 times faster than in the surrounding seawater (Bakir et al. 2014). These desorption rates are influenced by many factors such as pH and body temperature (Hollman et al. 2013; Bakir et al. 2014). For instance, PCBs may leach into fat tissue due to their hydrophobic properties (Hollman et al. 2013). In short-tailed shearwaters (*Puffinus tenuirostris*) from the field, chemical tracers were identified in the blubber tissue and the same tracers were isolated from plastics found in their stomachs (Tanaka et al. 2013). This is particularly interesting, as most studies up-to-date only investigate the digestive tract and draw conclusions from there. Some other important factors for leaching processes, like the constituent polymer, shapes, sizes, and buoyancy differences, are to be considered in bioassay protocols and microplastic toxicity testing (Karami et al. 2016b).

Biomarker responses in organisms like fish can provide insights in specific chemical interactions (Rudneva 2013). Measuring biomarkers such as the activity of enzymes is not only used for the effect of the inert particles on internal metabolism (Sect. 6.4.1) but also to elucidate the effect of chemical contaminants such as pesticides in the body (Ferreira et al. 2016). Plastic-associated chemicals can bind to specific cell receptors, which activate signaling pathways. In the common goby, virgin plastic particles did not induce acute toxicity of chromium (Luís et al. 2015). PAH, Benzo[a] pyrene, with which plastics were spiked, sorb into the intestine in adult zebrafish (Batel et al. 2016). The decrease in enzyme activity leads to a loss of energy (Oliveira et al. 2013), which can result in movement and vision difficulties and consequently influence predatory performance of the organisms (Ferreira et al. 2016; Fonte et al. 2016; Wen et al. 2018). This, in turn, could be investigated further to estimate the effect on the function of an organism in the ecosystem.

Chemical contaminants can have a wide range of harmful effects such as causing cancer and endocrine disruption, hepatic stress, birth defects, immune system problems, and early development issues (Teuten et al. 2009; GESAMP 2015; Rochman et al. 2013; Setälä et al. 2016; Auta et al. 2017). Bioaccumulation has been found in animal as well as in plant tissue with the consequence of ecotoxicity (Chua et al. 2014; Chae and An 2017; Smith 2018). Toxicity can already occur by simple attachment of contaminated microplastics on epithelia of zebrafish, with serious effects of waterborne toxic substances on early life stages (Batel et al. 2018). This shows that adherence rather than ingestion led to the accumulation of microplastics and associated toxicity (Batel et al. 2018). Furthermore, it is suggested that freshwater species suffer a higher risk, as the presence of salts in the water decrease the tendency of some chemicals to be sorbed onto plastic surfaces (Llorca et al. 2018).

Koelmans et al. (2016) suggested that microplastics ingestion by marine biota does not increase their exposure to hydrophobic organic compounds but could have a "cleaning effect", i.e., adsorption of bioaccumulated POPs onto microplastics, while being ingested. This theoretical explanation is supported by Rehse et al. (2018), who concluded that the presence of ingested microplastic particles can actually reduce the effects of BPA from surrounding water in freshwater zooplankton by a decreased body burden of the environmental pollutant. Kleinteich et al. (2018) found a similar result where a lower bioavailability of PAHs was found when they were sorbed to microplastics. As virgin particles not loaded with POPs did not cause any observable physical harm in zebrafish and clams (Batel et al. 2016; O'Donovan et al. 2018), there is evidence that chemical contamination is the key to understanding the exact impact of microplastic on marine biota (Hermabessiere et al. 2017).

Another line of evidence suggests that the combined effect of microplastics and sorbed contaminants altered organs homeostasis in a greater manner than the contaminants alone (Rainieri et al. 2018). This can only be further evaluated with controlled laboratory exposures to facilitate monitoring of the uptake, movement, and distribution of chemical compounds in whole organisms and excised tissues such as gills, intestinal tract, and liver (Lusher et al. 2017). Yet, little is known about the effects and influence of microplastic-associated toxins on the functionality of an organisms' body, and consequently associated altered ecosystem function (Table 1).

6.5.2 Microplastics as a Vector for Pathogens

A variety of biotic and abiotic particles can serve as vectors for pathogens, yet due to the persistence of plastic in the marine environment, microplastics are likely to travel farther and for longer periods of time than other types of foulable particles (Dobretsov 2010; Harrison et al. 2014). Contaminated microplastics within the marine environment may be transported between ocean basins and may contribute to the transfer of contaminants between ecosystems (Zarfl and Matthies 2010). This transfer is not limited to chemical contaminants, but also includes the transport of microbial communities consisting of "epiplastic" diatoms, coccolithophores, bryozoans, barnacles, dinoflagellates, invertebrate eggs, cyanobacteria, fungi, and bacteria (Zettler et al. 2013; Reisser et al. 2014; De Tender et al. 2015; Eich et al. 2015; Quero and Luna 2017). Bacterial communities associated with microplastics can potentially modify presently unpolluted habitats (Kleinteich et al. 2018).

Microplastics can serve as a substrate for microbiota as they offer a surface, the so-called *plastisphere* for attachment and settlement (Zettler et al. 2013). Microplastics can thus become a vector for non-ciliate pathogens, such as viruses (Masó et al. 2003; Pham et al. 2012) and pathogenic bacteria (Viršek et al. 2017). Studies in temperate and coral reef environments have investigated how pathogens on microplastic may trigger disease outbreaks in organisms. For example, Lamb et al. (2018) found that the likelihood of disease in corals increases from 4% to 89% when they are in contact with plastic, and, Goldstein et al. (2014) reported the transmission of the coral pathogen *Halofolliculina* spp. on plastic debris. Polypropylene marine debris is dominated by the genus *Vibrio* (Zettler et al. 2013), which are opportunistic pathogenic bacteria that can cause coral disease (Bourne et al. 2015). The microbial biofilm on microplastics, i.e., ecocorona (Lynch et al. 2014) can not only transport pathogens but influence the physical properties of the particle itself. A thick ecocorona reduces the ultraviolet (UV) light, reaching the surface of polyethylene particles by 90% (O'Brine and Thompson 2010) and makes the particle more hydrophilic (Lobelle and Cunliffe 2011). This increases a particle's sinking velocity (Li and Yuan 2002), which may influence their bioavailability by exposing organisms in other parts of the marine environment to microplastics and associated chemicals (Bråte et al. 2018).

Microplastic biofilms appear distinct compared to those on other marine substrata and are shaped by spatial and seasonal factors (Oberbeckmann et al. 2015). Foulon et al. (2016) summarize that the colonization of microplastics by the oyster-infecting *Vibrio crassostrea* is enhanced when the microplastic was already coated by a layer of primary marine aggregates. These secondary colonizers show a chemical attraction to the particle surface indicating a layering of colonizers in the ecocorona (Galloway et al. 2017).

These "camouflaged" plastic particles can be ingested by organisms such as zooplankton (Eich et al. 2015; Vroom et al. 2017) and even larger organisms. Some laboratory experiments concluded that bioavailability of plastics seems to be enhanced by particles that have been exposed to natural seawater for some time (Bråte et al. 2018). Yet, Allen et al. (2017) suggest that plastic contains phagostimulants that promote ingestion by corals. Interestingly, corals ingested more virgin plastic than plastics covered in microbial biofilm. Both lines of evidence highlight the likelihood of microplastic being ingested by different organisms for different reasons which needs to be better understood in a future with likely increasing amounts of microplastics in the ocean (Harrison et al. 2011; Allen et al. 2017).

Microorganisms in coastal sediments represent a key category of life with reference to understanding and mitigating the potential effects of microplastics, due to their role as drivers of the global functioning of the marine biosphere (Harrison et al. 2011). This is of particular interest with regards to their ability to biodegrade plastic-associated additives, contaminants, or even the plastics themselves (Harrison et al. 2011).

6.6 Trophic Cascade

It has been hypothesized that microplastics transfer within the marine food web from prey to predator (Fig. 6.2). The real extent to which trophic transfer occurs in the wild, however, remains largely unknown, although, laboratory studies have tried to investigate this (Nobre et al. 2015; Setälä et al. 2016). These studies demonstrated trophic transfer for low trophic level food chains, such as *Artemia* sp., crabs and fish (Murray and Cowie 2011; Farrell and Nelson 2013; Setälä et al. 2014; Watts et al. 2014; Batel et al. 2016). Observations of whole prey demonstrate trophic transfer from sand eels (*Ammodytes tobianus*) to plaice (*Pleuronectes plastessa*) in the wild. The lack of significant difference in microplastic abundance between predator and prey however suggests that microplastic is not retained by *P. platessa* (Welden et al. 2018). The likelihood of secondary ingestion is limited, as retention times and transit of particles through the gut of a prey organism can be relatively fast.

Interestingly, transfer of microplastics can occur from prey to predators, without evidences of microplastics persisting in their tissues after 10 days of exposure (Santana et al. 2017). Higher concentrations of microplastics were found in a predatory shellfish from the Persian Gulf, which lead the authors to suggest trophic transfer of microplastics in the food web without quantification in the prey (Naji et al. 2018). Seabird fecal pellets contained a similar composition of fibers to those which were identified in their macroinvertebrate prey which suggests that trophic transfer may be occurring (Lourenço et al. 2017). All predatory marine organisms are susceptible to ingest microplastic through their prey. Toothed marine mammals may be more likely to experience trophic transfer as primary route of microplastic ingestion than through direct intake (Lusher et al. 2016, Hocking et al. 2017). Feces of grey, harbor and fur seals or regurgitated fulmar remains of skuas suggest trophic transfer as these species are known to ingest whole prey (Eriksson and Burton 2003; Rebolledo et al. 2013; Hammer et al. 2016, Nelms et al. 2018). The contamination of microplastics appears to be transported into the deep ocean, not only by the change in density by fouling (Sect. 6.5.2) but through sinking of animal carcasses where it becomes available for scavengers (Clark et al. 2016).

A study by Mattsson et al. (2017) describes how plastic nanoparticles are transferred up through a freshwater algae-daphnia-fish food chain and enter the brain of the top consumer. The damaging effect on the brain leads to a disruption of the fish's natural behavior. In contrast, marine Kreffts's frill gobies (*Bathygobius kreffti*) (Tosetto et al. 2017) and an indo-pacific planktivore (*Acanthochromis polyacanthus*) (Critchell and Hoogenboom 2018) did not show altered behavior. Studies investigating animal's behavior are of extreme importance to draw conclusions about potential effects on ecosystem function. There are many relevant species for ecosystem function that need scientific attention (Rochman 2016; Wieczorek et al. 2018), such as different functional groups of fishes (Vendel et al. 2017).

An outdoor mesocosm experiment in sediment cores evaluated the potential effect of microplastics on the functioning of an ecosystem by quantifying the filtration rates of European flat oysters (*Ostrea edulis*) and blue mussels (*Mytilus edulis*) and the entire sedimentary community (Green et al. 2017). Filtration rates significantly decreased in *M. edulis* but increased in *O. edulis* when exposed to microplastics, affecting porewater ammonium. A decrease in biomass of benthic cyanobacteria and polychaetes emphasized the potential of microplastics to impact the functioning and structure of the sediment environment. Here, not only trophic transfer but the simultaneous effect of microplastic on the function biota in an ecosystem was stressed.

If trophic transfer occurs in the wild, this may also be a route for the transfer of any associated chemicals on the plastics. For example, laboratory experiments on simple artificial food chains, such as with nauplii and zebrafish, have estimated that a transfer of associated POPs occurs (Zhu et al. 2010). Bioaccumulation and biomagnification of chemical contaminants, such as PCBs and organochlorine pesticides (OCPs), are known to occur at higher trophic levels, particularly affecting marine top predators (Tsygankov et al. 2015; Jepson et al. 2016). Whether or not this chemical accumulation is connected to plastic-associated leaching remains unknown. It has been shown, however, that microplastic-associated chemicals can cause toxicity not only in marine animals (Choy and Drazen 2013; Rochman et al. 2015; Rummel et al. 2016; Karami et al. 2018) but also in humans (Hecht et al. 2010).

6.7 Microplastics and Human Health

Concerns of marine organism-derived microplastic and human health were extensively reviewed when microplastics began emerging as a potential threat to ecosystems (Thompson et al. 2009b; Talsness et al. 2009). Microplastic-induced toxicity and the evaluation of consequences for human health have been the focus of current literature (Revel et al. 2018). These concerns are magnified due to the presence of microplastic particles in food items worldwide. Research into the abundance of plastics in food has focused on seafood caught or cultured for human consumption (Van Cauwenberghe and Janssen 2014; Rochman et al. 2015; Naji et al. 2018). In dried fish the eviscerated flesh contained higher microplastic loads than the excised organs, which highlights that removing the digestive tract does not eliminate the risk of microplastic intake by consumers (Karami et al. 2017). When consuming an average portion of filter feeders like mussels, consumers can ingest up to 90 microplastic particles (Lusher et al. 2017). It was estimated that a European shellfish consumer annually ingests between 1800 and 11,000 microplastics (Van Cauwenberghe and Janssen 2014), with the potential for increased concentrations in farmed shellfish (Murphy 2018). Still, studies conclude that the low prevalence of often inert microplastics might indicate limited health risks as suggested by investigations of microplastic loads in canned fish (Karami et al. 2018). Particle uptake in the human body depends on the particle's size, surface charge and functionalization, hydrophobicity, and protein corona (Wright and Kelly 2017). The uptake of inert particles across the gut has been widely studied (O'Hagan 1996). Nanopolymers can be taken up across the gut into the circulation and be redistributed to the liver and spleen (Galloway 2015). In theory, all organs may be at risk following chronic exposure to nanopolymers. This includes the brain, testis, and reproductive organs, prior to their eventual excretion in urine and feces as evidenced in recent laboratory studies in invertebrates and fish (Jani et al. 1996; Garrett et al. 2012).

In fact, recent media has featured research on microplastics in other non-aquatic consumables such as bottled water, sugar, salt, beer, and honey (see EFSA 2016; Karami et al. 2017; Schymanski et al. 2017; Rist et al. 2018). Carbery et al. (2018) reviewed that there is no robust evidence for the transfer of microplastics and associated contaminants from seafood to humans and the implications for human health. Microplastic uptake through seafood consumption may be minimal when compared to other routes of human exposure, for example, fibers settling on consumables, or dust in the household (Catarino et al. 2018). Food items packaged in plastic may lead daily exposure to different plastic-associated chemicals up to 250 μg kg^{-1} body weight (EFSA 2011; Muncke 2011). Rist et al. (2018) describe that according to a comparison of two studies, exposure to microplastic ingestion from packaging is higher to a magnitude of 40 million compared to the exposure from shellfish. Prata (2018) summarized diseases originating from airborne microplastics and the consequences to human health; a person's lungs could be exposed to between 26 and 130 airborne microplastics per day. The continuous daily interaction with plastic items already leads to the presence of plastic and associated chemicals in the human body (Galloway 2015). Plastic additives, such as BPA, are a risk factor to human health (Srivastava and Godara 2017). Lithner et al. (2011) conducted a comprehensive ranking of plastic polymers, identifying physical, environmental, and health risks. The quantification of plastic particles in food is suggested to be included as one of the components of food safety management systems (Karami et al. 2018).

Given the long-term persistence of plastics within extensive variety of polymer types and additive composition, more research is required to adequately assess the risks that accumulation of micro- and nanoplastics in the body may pose and the true potential to induce pathology (Galloway 2015; Prata 2018). Furthermore, exposure to nanoplastics cannot be precisely estimated yet due to a lack of technological means (EFSA 2016). Despite the focus on human health being a major driving force to increase the investigation of marine biota and plastic interactions because of the economic value of marine protein, the diminished ecosystem service that some species might provide for humans should be highlighted.

6.8 Research Gaps and Future Work

In spite of almost a decade of research, microplastic research is still in its infancy, and it is still very difficult to estimate the cumulative risks of chronic exposure to plastics and their additives. This is due to the limited information available about rates of degradation and fragmentation, leaching of chemicals into environmental matrices, and entry into the food chain (Hermabessiere et al. 2017). Additionally, biological responses of microplastic on the molecular level are difficult to interpret, as the particles' chemical structure is complex and versatile. It can be concluded that current plastic use is not sustainable (Thompson et al. 2009a), which calls for an immediate change in plastic production, consumption, and human behavior, to reduce the amount of microplastics present in the environment. Mendenhall (2018) highlights the large-scale impacts of plastic debris on ecosystem function as a major knowledge-gap.

Although the informative review by Anbumani and Kakkar (2018) summarized different "ecological impacts" of microplastics on aquatic biota and the potential for ecological niche imbalance, an organism's role in ecosystem function is not discussed. Auta et al. (2017) also elaborate on the effects and fate of microplastic ingested by biota and suggest remedies such as microbial activity against microplastic contamination in the environment. Galloway et al. (2017) reviewed current literature and considered microplastic debris to become a planetary boundary threat through its effects on crucial processes exhibited by biota. Chae and An (2017) discuss different global concentrations in freshwater and marine environments, as well as the intrinsic and complex toxicological effects on biota. It is mentioned that research studying effects on generational and ecological effects is important, but no specific references are given.

There is a possibility that organisms may adapt to certain conditions, especially when they are exposed to low concentrations of contaminants for a longer period of time (Sureda et al. 2006). One could even propose that animals will evolutionarily adapt to microplastic concentrations in the environment, which in the future would not affect their fitness. Such suggestion could only apply to the organisms in water column or water surface habitats where the microplastic concentration is mostly stable. Organisms in the sediment or in the intertidal may however be exposed to an ever-increasing microplastic concentration in the near future (Lobelle and Cuncliffe 2011; Green et al. 2017). It is, therefore, critical to continuously evaluate removal rates from the water column towards the sediment or deep sea, as intended by analytical model approaches (Koelmans et al. 2017b). Reduced functionality is correlated to the disappearance of animals (Lusher et al. 2017). If biological processes at the base of ecosystems are altered because of the presence of microplastics, biologically mediated disruption to the long-term storage of carbon could occur (Villarrubia-Gómez et al. 2017). Despite attempts to model whether microplastics can affect the overall productivity of a marine ecosystem, no clear conclusion can be drawn yet (Troost et al. 2018).

Upon reviewing 222 journal articles, 9 book chapters, two reports, two dissertations and one exhibition, the following 9 research foci need to be especially considered in the future:

1. Laboratory studies should focus on experiments with environmentally relevant quantities and sizes of microplastic and contaminants to estimate actual impacts. Therefore, for instance, studies should include plastic particles, fouled in natural seawater to estimate the role of fouling and/or investigate the degree of chemical contamination from a certain area in the sea in the laboratory.
2. Studies suggest that toxicity of virgin microplastics, spiked microplastics, additives, or contaminants affect the organisms differently (Karami 2017). Further studies are needed to elucidate and distinguish these effects on different organisms and with regards to varying availability of plastic debris and POPs in different ecosystems.
3. Usually the digestive system is investigated for microplastic presence and their effects. Other tissues such as muscle tissue and fat (blubber) should be collected and analyzed for the presence of microplastic tracers and further compared to stomach analysis results (Tanaka et al. 2013; Lusher et al. 2015).
4. More studies on the base of the food chain and the subcellular level are necessary to conclude effects on the individual or population level. For this, we suggest microbiome studies and genetic tools.
5. Limited studies relate the effect on the ecological function of marine organisms after being influenced by microplastics and associated contaminants (Mattsson et al. 2017) (Table 1). Different feeding strategies need to be considered.
6. More research is needed to understand the potential impact of micro- and nanoplastics on primary production and food web interactions.
7. There is a necessity to develop techniques to identify bacterial communities on microplastics.
8. A special focus should be put on freshwater species as they may be at higher risk of some chemicals to be sorbed onto plastic surfaces (Llorca et al. 2018).
9. Many indigestible materials apart from plastic, such as wood, metal, glass and building materials, that are found in the nature that need to be considered. Therefore, other natural and anthropogenic materials should be considered as a comparison, when analyzing the effects of microplastics.

6.9 Summary

Along with ever-increasing plastic production, the amount of plastic waste that enters the oceans is also on the rise. The breakdown of larger debris into microplastic pieces is of high scientific concern as it can become bioavailable. In recent years, aquatic flora and fauna have been found to be affected in different ways when coming into contact with microplastics. This review summarizes that microplastics can attach or get physically ingested by almost all aquatic taxa or affect biota via leachates or pathogens from the microplastic surface. Some studies highlighted that under environmentally relevant levels, microplastic may not necessarily pose risks to the organisms, as particles are often inert. However, other lines of evidence found adverse physiological effects of microplastic in organisms, either through tissue damage, through cellular uptake, or through chemical contamination of leachates from the microplastics. In addition, microplastics can be a vector of pathogens into the tissue of organisms. Often, these effects do not cause death but a sublethal alteration of body functions. The consequences result in reduced primary productivity, compromised energy allocation, reduced growth, changed feeding efficiency, or altered predatory performance. Combined with other environmental stressors, this can lead to alterations of the ecological function of a species in the ecosystem. Only 4% of studies reviewed here investigated how reduced physiological processes, caused by microplastic, are linked with the ecological role, an organism and its population play. There is a general consensus that both the microplastic size and their concentration is critical to understand the impact on an organism. This review emphasized the importance that decreasing particle size and chemical contamination can affect organisms to the extent that critical body functions are impaired. This, in turn, can influence the functional role the organism fulfills in the ecosystem. Since microplastic is bioavailable to the smallest of organisms, secondary ingestion can occur, which may be channeled through the food web. Particular concern arises when microplastic is found in species for human consumption. Nevertheless, we argue that the uptake of plastic and plastic-associated chemicals occurs more through everyday sources in the urban environment, rather than seafood consumption and highlight the need to investigate the importance of impacted ecological functionality of species regarding ecosystem services for humans.

This review summarized cutting-edge research to understand some hazard potentials for different species and research gaps that still need to be examined. This particular field of science is necessary as reliable risk assessments are crucial, contributing to current environmental and societal discussions, and future perspectives concerning microplastic pollution. The focus should be set on the elucidation of microscopic impacts of plastics on biota for the sake of understanding the impact these small particles can have on populations and functionality of an entire ecosystem that needs to be protected.

Acknowledgments The authors would like to thank the two reviewers on their highly valuable comments that greatly improved the quality of this review, as well as Dr. Simon Jungblut, Dr. Sonia Bejarano, Hannah Earp, and Claire Shellem for constructive input through different stages of the manuscript.

Appendix

This article is related to the YOUMARES 9 conference session no. 7: "Submerged in Plastic: impacts of plastic pollution on marine biota". The original Call for Abstracts and the abstracts of the presentations within this session can be found in the Appendix "Conference Sessions and Abstracts", Chapter "6 Submerged in Plastic: Impacts of Plastic Pollution on Marine Biota", of this book.

References

Allen AS, Seymour AC, Rittschof D (2017) Chemoreception drives plastic consumption in a hard coral. Mar Pollut Bull 124(1):198–205

Alomara C, Sureda A, Capó X et al (2017) Microplastic ingestion by *Mullus surmuletus* Linnaeus, 1758 fish and its potential for causing oxidative stress. Environ Res 159:135–142

Anastasopoulou A, Mytilineou C, Smith CJ et al (2013) Plastic debris ingested by deep-water fish of the Ionian Sea (Eastern Mediterranean). Deep Sea Res I 74:11–13

Anbumani S, Kakkar P (2018) Ecotoxicological effects of microplastics on biota: a review. Environ Sci Pollut R 25(15):1473–1496

Andrady AL (2017) The plastic in microplastics: a review. Mar Pollut Bull 119(1):12–22

Au SY, Bruce TF, Bridges WC, Klaine SJ (2015) Responses of *Hyalella azteca* to acute and chronic microplastic exposures. Environ Toxicol Chem 34(11):2564–2572

Auta HS, Emenike CU, Fauziah SH (2017) Distribution and importance of microplastics in the marine environment: a review of the sources, fate, effects, and potential solutions. Environ Int 102:165–176

Avio CG, Gorbi S, Milan M et al (2015) Pollutants bioavailability and toxicological risk from microplastics to marine mussels. Environ Pollut 198:211–222

Baalkhuyur FM, Dohaish EJAB, Elhalwagy ME et al (2018) Microplastic in the gastrointestinal tract of fishes along the Saudi Arabian Red Sea coast. Mar Pollut Bull 131:407–415

Bakir A, Rowland SJ, Thompson RC (2014) Enhanced desorption of persistent organic pollutants from microplastics under simulated physiological conditions. Environ Pollut 185:16–23

Bakir A, O'Connor IA, Rowland SJ et al (2016) Relative importance of microplastics as a pathway for the transfer of hydrophobic organic chemicals to marine life. Environ Pollut 219:56–65

Batel A, Linti F, Scherer M et al (2016) The transfer of benzo [a]pyrene from microplastics to *Artemia* nauplii and further to zebrafish via a trophic food web experiment – CYP1A induction and visual tracking of persistent organic pollutants. Environ Toxicol Chem 35(7):1656–1666

Batel A, Borchert F, Reinwald H et al (2018) Microplastic accumulation patterns and transfer of benzo [a] pyrene to adult zebrafish (*Danio rerio*) gills and zebrafish embryos. Environ Pollut 235:918–930

Battaglia P, Pedà C, Musolino S et al (2016) Diet and first documented data on plastic ingestion of *Trachinotus ovatus* L. 1758 (Pisces: Carangidae) from the Strait of Messina (central Mediterranean Sea). Ital J Zool 83(1):121–129

Besseling E, Wegner A, Foekema EM et al (2013) Effects of microplastic on fitness and PCB bioaccumulation by the lugworm *Arenicola marina* (L.). Environ Sci Technol 47(1):593–600

Besseling E, Wang B, Lürling M et al (2014) Nanoplastic affects growth of *S. obliquus* and reproduction of *D. magna*. Environ Sci Technol 48(20):12336–12343

Besseling E, Foekema EM, Van Franeker JA et al (2015) Microplastic in a macro filter feeder: humpback whale *Megaptera novaeangliae*. Mar Pollut Bull 95(1):248–252

Bhattacharya P, Lin S, Turner JP et al (2010) Physical adsorption of charged plastic nanoparticles affects algal photosynthesis. J Phys Chem C 114(39):16556–16561

Boerger CM, Lattin GL, Moore SL et al (2010) Plastic ingestion by planktivorous fishes in the North Pacific Central Gyre. Mar Pollut Bull 60(12):2275–2278

Bourne DG, Ainsworth TD, Pollock FJ et al (2015) Towards a better understanding of white syndromes and their causes on Indo-Pacific coral reefs. Coral Reefs 34(1):233–242

Bråte ILN, Blázquez M, Brooks SJ et al (2018) Weathering impacts the uptake of polyethylene microparticles from toothpaste in Mediterranean mussels (*M. galloprovincialis*). Sci Total Environ 626:1310–1318

Brennecke D, Ferreira EC, Costa TM et al (2015) Ingested microplastics (> 100 μm) are translocated to organs of the tropical fiddler crab *Uca rapax*. Mar Pollut Bull 96(1-2):491–495

Brennecke D, Duarte B, Paiva F et al (2016) Microplastics as vectors for heavy metal contamination from the marine environment. Estuar Coast Shelf Sci 178:189–195

Brillant MG, MacDonald BA (2000) Postingestive selection in the sea scallop, *Placopecten magellanicus* (Gmelin): the role of particle size and density. J Exp Mar Biol Ecol 253(2):211–227

Browne MA, Galloway T, Thompson R (2007) Microplastic—an emerging contaminant of potential concern? Integr Environ Assess Manag 3(4):559–561

Browne MA, Dissanayake A, Galloway TS et al (2008) Ingested microscopic plastic translocates to the circulatory system of the mussel, *Mytilus edulis* (L.). Environ Sci Technol 42(13):5026–5031

Browne MA, Crump P, Niven SJ et al (2011) Accumulation of microplastic on shorelines worldwide: sources and sinks. Environ Sci Technol 45:9175–9179

Browne MA, Niven SJ, Galloway TS et al (2013) Microplastic moves pollutants and additives to worms, reducing functions linked to health and biodiversity. Curr Biol 23(23):2388–2392

Carbery M, O'Connor W, Palanisami T (2018) Trophic transfer of microplastics and mixed contaminants in the marine food web and implications for human health. Environ Int 115:400–409

Carr SA, Liu J, Tesoro AG (2016) Transport and fate of microplastic particles in wastewater treatment plants. Water Res 91:174–182

Capozzi F, Carotenuto R, Giordano S et al (2018) Evidence on the effectiveness of mosses for biomonitoring of microplastics in fresh water environment. Chemosphere 205:1–7

Catarino AI, Macchia V, Sanderson WG et al (2018) Low levels of microplastics (MP) in wild mussels indicate that MP ingestion by humans is minimal compared to exposure via household fibres fall-out during a meal. Environ Pollut 237:675–684

Cedervall T, Hansson LA, Lard M et al (2012) Food chain transport of nanoparticles affects behaviour and fat metabolism in fish. PloS One 7(2):e32254

Chae Y, An YJ (2017) Effects of micro-and nanoplastics on aquatic ecosystems: Current research trends and perspectives. Mar Pollut Bull 124(2):624–632

Choi O, Hu Z (2008) Size dependent and reactive oxygen species related nanosilver toxicity to nitrifying bacteria. Environ Sci Technol 42(12):4583–4588

Choy CA, Drazen JC (2013) Plastic for dinner? Observations of frequent debris ingestion by pelagic predatory fishes from the central North Pacific. Mar Ecol Prog Ser 485:155–163

Chua EM, Shimeta J, Nugegoda D et al (2014) Assimilation of polybrominated diphenyl ethers from microplastics by the marine amphipod, *Allorchestes compressa*. Environ Sci Technol 48:8127–8134

Clark JR, Cole M, Lindeque PK et al (2016) Marine microplastic debris: a targeted plan for understanding and quantifying interactions with marine life. Front Ecol Environ 14(6):317–324

Cole M, Lindeque P, Halsband C et al (2011) Microplastics as contaminants in the marine environment: A review. Mar Pollut Bull 62:2588–2597

Cole M, Lindeque P, Fileman E et al (2013) Microplastic ingestion by zooplankton. Environ Sci Technol 47(12):6646–6655

Cole M, Lindeque P, Fileman E et al (2015) The impact of polystyrene microplastics on feeding, function and fecundity in the marine copepod *Calanus helgolandicus*. Environ Sci Technol 49(2):1130–1137

Cole M, Lindeque PK, Fileman E et al (2016) Microplastics alter the properties and sinking rates of zooplankton faecal pellets. Environ Sci Technol 50(6):3239–3246

Collard F, Gilbert B, Compere P et al (2017) Microplastics in livers of European anchovies (*Engraulis encrasicolus*, L.). Environ Pollut 229:1000–1005

Critchell K, Hoogenboom MO (2018) Effects of microplastic exposure on the body condition and behaviour of planktivorous reef fish (*Acanthochromis polyacanthus*). PloS One 13(3):e0193308

Crompton T (2007) Additive migration from plastics into foods. A guide for analytical chemistry. Smithers Rapra Technology Limited, Shrewsbury

Dawson AL, Kawaguchi S, King CK et al (2018) Turning microplastics into nanoplastics through digestive fragmentation by Antarctic krill. Nat Commun 9(1):1001

Deheyn DD, Latz MA (2006) Bioavailability of metals along a contamination gradient in San Diego Bay (California, USA). Chemosphere 63:818–834

de Sá LC, Luís LG, Guilhermino L (2015) Effects of microplastics on juveniles of the common goby (*Pomatoschistus microps*): confusion with prey, reduction of the predatory performance and efficiency, and possible influence of developmental conditions. Environ Pollut 196:359–362

De Tender CA, Devriese LI, Haegeman A et al (2015) Bacterial community profiling of plastic litter in the Belgian part of the North sea. Environ Sci Technol 49:9629–9638

Détrée C, Gallardo-Escárate C (2017) Polyethylene microbeads induce transcriptional responses with tissue-dependent patterns in the mussel *Mytilus galloprovincialis*. J Molluscan Stud 83(2):220–225

Devriese LI, van der Meulen MD, Maes T et al (2015) Microplastic contamination in brown shrimp (*Crangon crangon*, Linnaeus 1758) from coastal waters of the Southern North Sea and Channel area. Mar Pollut Bull 98(1-2):179–187

Dobretsov S (2010) Marine biofilms. In: Dürr S, Thomason JC (eds) Biofouling. Blackwell, Oxford, pp 123–136

Duis K, Coors A (2016) Microplastics in the aquatic and terrestrial environment: sources (with a specific focus on personal care products), fate and effects. Environ Sci Eur 28(2):1–25

EFSA (2011) Scientific opinion: guidance on the risk assessment of the application of nanoscience and nanotechnologies in the food and feed chain. EFSA Journal 9:2140

EFSA (2016) Statement on the presence of microplastics and nanoplastics in food, with particular focus on seafood. EFSA Journal 14(6):4501

Eich A, Mildenberger T, Laforsch C et al (2015) Biofilm and diatom succession on polyethylene (PE) and biodegradable plastic bags in two marine habitats: early signs of degradation in the pelagic and benthic zone. PloS One 10(9):e0137201

Eriksson C, Burton H (2003) Origins and biological accumulation of plastic particles in fur seals from Macquarie Island. Ambio 32:380–384

Farrell P, Nelson K (2013) Trophic level transfer of microplastic: *Mytilus edulis* (L.) to *Carcinus maenas* (L.). Environ Pollut 177:1–3

Ferreira P, Fonte E, Soares ME et al (2016) Effects of multi-stressors on juveniles of the marine fish *Pomatoschistus microps*: gold nanoparticles, microplastics and temperature. Aquat Toxicol 170:89–103

Flint S, Markle T, Thompson S et al (2012) Bisphenol A exposure, effects, and policy: a wildlife perspective. J Environ Manage 104:19–34

Foekema EM, De Gruijter C, Mergia MT et al (2013) Plastic in North Sea fish. Environ Science Technol 47(15):8818–8824

Fonte E, Ferreira P, Guilhermino L (2016) Temperature rise and microplastics interact with the toxicity of the antibiotic cefalexin to juveniles of the common goby (*Pomatoschistus microps*): post-exposure predatory behaviour, acetylcholinesterase activity and lipid peroxidation. Aquat Toxicol 180:173–185

Foulon V, Le Roux F, Lambert C et al (2016) Colonization of polystyrene microparticles by Vibrio crassostreae: light and electron microscopic investigation. Environ Sci Technol 50(20):10988–10996

Galloway TS (2015) Micro- and Nanoplastics and human health. In: Bergmann M, Gutow L, Klages M (eds) Marine anthropogenic litter. Springer, Cham, pp 343–366

Galloway TS, Lewis CN (2016) Marine microplastics spell big problems for future generations. Proc Natl Acad Sci USA 113(9):2331–2333

Galloway TS, Cole M, Lewis C (2017) Interactions of microplastic debris throughout the marine ecosystem. Nat Ecol Evol 1(5):0116

Garrett NL, Lalatsa A, Uchegbu I et al (2012) Exploring uptake mechanisms of oral nanomedicines using multimodal nonlinear optical microscopy. J Biophotonics 5:458–468

GESAMP (2015) In: Kershaw PJ (ed) Sources, fate and effects of microplastics in the marine environment: a global assessment. IMO/FAO/UNESCO-IOC/ UNIDO/WMO/IAEA/UN/ UNEP/ UNDP Joint Group of Experts on the Scientific Aspects of Marine Environmental Protection. Rep. Stud. GESAMP No 90: p 1–96

Geyer R, Jambeck JR, Law KL (2017) Production, use, and fate of all plastics ever made. Sci Adv 3(7):e1700782

Goldstein MC, Carson HS, Eriksen M (2014) Relationship of diversity and habitat area in North Pacific plastic-associated rafting communities. Mar Biol 161(6):1441–1453

Green DS, Boots B, Blockley DJ et al (2015) Impacts of discarded plastic bags on marine assemblages and ecosystem functioning. Environ Sci Technol 49(9):5380–5389

Green DS, Boots B, Sigwart J et al (2016) Effects of conventional and biodegradable microplastics on a marine ecosystem engineer (*Arenicola marina*) and sediment nutrient cycling. Environ Pollut 208:426–434

Green DS, Boots B, O'Connor NE et al (2017) Microplastics affect the ecological functioning of an important biogenic habitat. Environ Sci Technol 51(1):68–77

Gregory MR (2009) Environmental implications of plastic debris in marine settings—entanglement, ingestion, smothering, hangers-on, hitch-hiking and alien invasions. Philos Trans Royal Soc B Biol Sci 364(1526):2013–3025

Gutow L, Eckerlebe A, Giménez L et al (2016) Experimental evaluation of seaweeds as a vector for microplastics into marine food webs. Environ Sci Technol 50(2):915–923

Haghi BN, Banaee M (2017) Effects of micro-plastic particles on paraquat toxicity to common carp (*Cyprinus carpio*): biochemical changes. Int J Environ Sci Technol 14(3):521–530

Hall NM, Berry KL, Rintoul L et al (2015) Microplastic ingestion by scleractinian corals. Mar Biol 162(3):725–732

Hämer J, Gutow L, Köhler A et al (2014) The fate of microplastics in the marine isopod *Idotea emarginata*. Environ Sci Technol 48:13451–13458

Hammer S, Nager RG, Johnson PCD et al (2016) Plastic debris in great skua (*Stercorarius skua*) pellets corresponds to seabird prey species. Mar Pollut Bull 103:206–210

Harrison JP, Sapp M, Schratzberger M et al (2011) Interactions between microorganisms and marine microplastics: a call for research. Mar Technol Soc J 45(2):12–20

Harrison JP, Schratzberger M, Sapp M et al (2014) Rapid bacterial colonization of low-density polyethylene microplastics in coastal sediment microcosms. BMC Microbiol 14(1):232

Hecht SS, Carmella SG, Villalta PW et al (2010) Analysis of phenanthrene and benzo[a]pyrene tetraol enantiomers in human urine: relevance to the bay region diol epoxide hypothesis of benzo[a] pyrene carcinogenesis and to biomarker studies. Chem Res Toxicol 23:900–908

Hermabessiere L, Dehaut A, Paul-Pont et al (2017) Occurrence and effects of plastic additives on marine environments and organisms: a review. Chemosphere 182:781-793

Hirai H, Takada H, Ogata Y et al (2011) Organic micropollutants in marine plastic debris from the open ocean and remote and urban beaches. Mar Pollut Bull 62:1683–1692

Hocking DP, Marx FG, Park T et al (2017) A behavioural framework for the evolution of feeding in predatory aquatic mammals. Proc Royal Soc B 284:1–10

Hodgson DJ, Bréchon A, Thompson RC (2018) Ingestion and fragmentation of plastic carrier bags by the amphipod *Orchestia gammarellus*: effects of plastic type and fouling load. Mar Pollut Bull 127:154–159

Hollman PC, Bouwmeester H, Peters RJ (2013) Microplastics in the aquatic food chain: sources, measurements, occurrence and potential health risks (Rikilt-Report 2013.003.). Rikilt-Institute of Food Safety, Wageningen UR (University & Research Centre), Netherlands, pp 1–27

Holmes LA, Turner A, Thompson RC (2012) Adsorption of trace metals to plastic resin pellets in the marine environment. Environ Pollut 160:42–48

Horton AA, Walton A, Spurgeon DJ et al (2017) Microplastics in freshwater and terrestrial environments: evaluating the current understanding to identify the knowledge gaps and future research priorities. Sci Total Environ 586:127–141

Jambeck JR, Geyer R, Wilcox C et al (2015) Plastic waste inputs from land into the ocean. Science 347:768–771

Jani PU, Nomura T, Yamashita F et al (1996) Biliary excretion of polystyrene microspheres with covalently linked FITC fluorescence after oral and parenteral administration to male wistar rats. J Drug Target 4:87–93

Jeong CB, Won EJ, Kang HM et al (2016) Microplastic size-dependent toxicity, oxidative stress induction, and p-JNK and p-P38 activation in the monogonont rotifer (*Brachionus koreanus*). Environ Sci Technol 50(16):8849–8857

Jeong CB, Kang HM, Lee MC et al (2017) Adverse effects of microplastics and oxidative stress-induced MAPK/Nrf2 pathway-mediated defense mechanisms in the marine copepod *Paracyclopina nana*. Sci Rep 7:41323

Jepson PD, Deaville R, Barber JL et al (2016) PCB pollution continues to impact populations of orcas and other dolphins in European waters. Sci Rep 6:1e17

Jovanović B (2018) Ingestion of microplastics by fish and its potential consequences from a physical perspective. Integr Environ Assess Manag 13(3):510–515

Kalčíková G, Gotvajn AŽ, Kladnik A et al (2017) Impact of polyethylene microbeads on the floating freshwater plant duckweed *Lemna minor*. Environ Pollut 230:1108–1115

Kaposi KL, Mos B, Kelaher BP et al (2014) Ingestion of microplastic has limited impact on a marine larva. Environ Sci Technol 48(3):1638–1645

Karami A (2017) Gaps in aquatic toxicological studies of microplastics. Chemosphere 184:841–848

Karami A, Romano N, Galloway T et al (2016a) Virgin microplastics cause toxicity and modulate the impacts of phenanthrene on biomarker responses in African catfish (*Clarias gariepinus*). Environ Res 151:58–70

Karami A, Golieskardi A, Choo CK et al (2016b) A high-performance protocol for extraction of microplastics in fish. Sci Total Environ 578:485–494

Karami A, Golieskardi A, Ho YB et al (2017) Microplastics in eviscerated flesh and excised organs of dried fish. Sci Rep 7(1):5473

Karami A, Golieskardi A, Choo CK et al (2018) Microplastic and mesoplastic contamination in canned sardines and sprats. Sci Total Environ 612:1380–1386

Khan FR, Syberg K, Shashoua Y et al (2015) Influence of polyethylene microplastic beads on the uptake and localization of silver in zebrafish (*Danio rerio*). Environ Pollut 206:73–79

Kim SW, Chae Y, Kim D et al (2019) Zebrafish can recognize microplastics as inedible materials: quantitative evidence of ingestion behavior. Sci Total Environ 649:156–162

Kleinteich J, Seidensticker S, Marggrander N et al (2018) Microplastics reduce short-term effects of environmental contaminants. Part II: Polyethylene particles decrease the effect of polycyclic aromatic hydrocarbons on microorganisms. I J Environ Res Public Health 15(2):287

Koelmans AA (2015) Modeling the role of microplastics in bioaccumulation of organic chemicals to marine aquatic organisms. Critical review. In: Bergmann M, Gutow L, Klages M (eds) Marine anthropogenic litter. Springer, Cham, pp 309–324

Koelmans AA, Gouin T, Thompson R et al (2014) Plastics in the marine environment. Environ Toxicol Chem 33(1):5–10

Koelmans AA, Bakir A, Burton GA et al (2016) Microplastic as a vector for chemicals in the aquatic environment: critical review and model-supported reinterpretation of empirical studies. Environ Sci Technol 50:3315–3326

Koelmans AA, Besseling E, Foekema E et al (2017a) Risks of plastic debris: unravelling fact, opinion, perception, and belief. Environ Sci Technol 51(20):11513–11519

Koelmans AA, Kooi M, Law KL et al (2017b) All is not lost: deriving a top-down mass budget of plastic at sea. Environ Res Lett 12(11):114028

Kolandhasamy P, Su L, Li J et al (2018) Adherence of microplastics to soft tissue of mussels: a novel way to uptake microplastics beyond ingestion. Sci Total Environ 610:635–640

Kühn S, van Franeker JA (2012) Plastic ingestion by the northern fulmar (*Fulmarus glacialis*) in Iceland. Mar Pollut Bull 64(6):1252–1254

Laing LV, Viana J, Dempster EL et al (2016) Bisphenol A causes reproductive toxicity, decreases dnmt1 transcription, and reduces global DNA methylation in breeding zebrafish (*Danio rerio*). Epigenetics 11:526–538

Lamb JB, Willis BL, Fiorenza EA et al (2018) Plastic waste associated with disease on coral reefs. Science 359(6374):460–462

Lebreton LC, Van der Zwet J, Damsteeg JW et al (2017) River plastic emissions to the world's oceans. Nature Commun 8:15611

Lee KW, Shim WJ, Kwon OY et al (2013) Size-dependent effects of micro polystyrene particles in the marine copepod *Tigriopus japonicus*. Environ Sci Technol 47(19):11278–11283

Lei L, Wu S, Lu S et al (2018) Microplastic particles cause intestinal damage and other adverse effects in zebrafish *Danio rerio* and nematode *Caenorhabditis elegans*. Sci Total Environ 619:1–8

Li X, Yuan Y (2002) Settling velocities and permeabilities of microbial aggregates. Water Res 36:3110–3120

Lithner D, Larsson Å, Dave G (2011) Environmental and health hazard ranking and assessment of plastic polymers based on chemical composition. Sci Total Environ 409:3309–3324

Llorca M, Schirinzi G, Martínez M et al (2018) Adsorption of perfluoroalkyl substances on microplastics under environmental conditions. Environ Pollut 235:680–691

Lobelle D, Cunliffe M (2011) Early microbial biofilm formation on marine plastic debris. Mar Pollut Bull 62(1):197–200

Long M, Moriceau B, Gallinari M et al (2015) Interactions between microplastics and phytoplankton aggregates: impact on their respective fates. Mar Chem 175:39–46

Lourenço PM, Serra-Gonçalves C, Ferreira JL et al (2017) Plastic and other microfibers in sediments, macroinvertebrates and shorebirds from three intertidal wetlands of southern Europe and west Africa. Environ Pollut 231:123–133

Lu Y, Zhang Y, Deng Y et al (2016) Uptake and accumulation of polystyrene microplastics in zebrafish (*Danio rerio*) and toxic effects in liver. Environ Sci Technol 50(7):4054–4060

Luís LG, Ferreira P, Fonte E et al (2015) Does the presence of microplastics influence the acute toxicity of chromium (VI) to early juveniles of the common goby (*Pomatoschistus microps*)? A study with juveniles from two wild estuarine populations. Aquat Toxicol 164:163–174

Lushchak VI (2011) Environmentally induced oxidative stress in aquatic animals. Aquat Toxicol 101(1):13–30

Lusher AL (2015) Microplastics in the marine environment: distribution, interactions and effects. In: Bergmann M, Gutow L, Klages M (eds) Marine anthropogenic litter. Springer, Cham, pp 245–307

Lusher AL, Mchugh M, Thompson RC (2013) Occurrence of microplastics in the gastrointestinal tract of pelagic and demersal fish from the English Channel. Mar Pollut Bull 67(1-2):94–99

Lusher AL, Hernandez-Milian G, O'Brien J et al (2015) Microplastic and macroplastic ingestion by a deep diving, oceanic cetacean: the True's beaked whale *Mesoplodon mirus*. Environ Pollut 199:185–191

Lusher AL, O'Donnell C, Officer R et al (2016) Microplastic interactions with North Atlantic mesopelagic fish. ICES J Mar Sci 73(4):1214–1225

Lusher AL, Welden NA, Sobral P et al (2017) Sampling, isolating and identifying microplastics ingested by fish and invertebrates. Anal Methods 9(9):1346–1360

Lusher AL, Hernandez-Milian G, Berrow S et al (2018) Incidence of marine debris in cetaceans stranded and bycaught in Ireland: recent findings and a review of historical knowledge. Environ Pollut 232:467–476

Lynch I, Dawson KA, Lead JR et al (2014) Nanoscience and the environment. In: Lead JR, Valsami-Jones E (eds) Frontiers of nanoscience, vol 7. Elsevier, Amsterdam, pp 127–156

Martínez-Gómez C, Bignell J, Lowe D (2015) Lysosomal membrane stability in mussels. ICES Tech Mar Environ Sci 56:1–41

Masó M, Garcés E, Pagès F et al (2003) Drifting plastic debris as a potential vector for dispersing Harmful Algal Bloom (HAB) species. Sci Mar 67:107–111

Mattsson K, Johnson EV, Malmendal A et al (2017) Brain damage and behavioural disorders in fish induced by plastic nanoparticles delivered through the food chain. Sci Rep 7(1):11452

Mazurais D, Ernande B, Quazuguel P et al (2015) Evaluation of the impact of polyethylene microbeads ingestion in European sea bass (*Dicentrarchus labrax*) larvae. Mar Environ Res 112:78–85

Mendenhall E (2018) Oceans of plastic: a research agenda to propel policy development. Mar Policy 96:291–298

Miller MR, Shaw CA, Langrish JP (2012) From particles to patients: oxidative stress and the cardiovascular effects of air pollution. Future Cardiol 8(4):577–602

Monteiro M, Quintaneiro C, Morgado F et al (2005) Characterization of the cholinesterases present in head tissues of the estuarine fish *Pomatoschistus microps*: application to biomonitoring. Ecotox Environ Safe 62(3):341–347

Moore MN, Allen JI, McVeigh A (2006) Environmental prognostics: an integrated model supporting lysosomal stress responses as predictive biomarkers of animal health status. Mar Environ Res 61(3):278–304

MSFD Technical Group on Marine Litter (2013) Guidance on monitoring of marine litter in European seas. JRC Scientific and Policy Reports, Publication Office of the European Union, pp 1–126

Muncke J (2011) Endocrine disrupting chemicals and other substances of concern in food contact materials: an updated review of exposure, effect and risk assessment. J Steroid Biochem Mol Biol 127:118–127

Murphy CL (2018) A Comparison of Microplastics in Farmed and Wild Shellfish near Vancouver Island and Potential Implications for Contaminant Transfer to Humans. Doctoral dissertation, Royal Roads University (Canada)

Murphy F, Quinn B (2018) The effects of microplastic on freshwater *Hydra attenuata* feeding, morphology & reproduction. Environ Pollut 234:487–494

Murray F, Cowie PR (2011) Plastic contamination in the decapod crustacean *Nephrops norvegicus* (Linnaeus, 1758). Mar Pollut Bull 62:1207–1217

Naji A, Nuri M, Vethaak AD (2018) Microplastics contamination in molluscs from the northern part of the Persian Gulf. Environ Pollut 235:113–120

Nelms SE, Galloway TS, Godley BJ et al (2018) Investigating microplastic trophic transfer in marine top predators. Environ Pollut 238:999–1007

Ng KL, Obbard JP (2006) Prevalence of microplastics in Singapore's coastal marine environment. Mar Pollut Bull 52:761–767

Nobre CR, Santana MF, Maluf A et al (2015) Assessment of microplastic toxicity to embryonic development of the sea urchin *Lytechinus variegatus* (Echinodermata: Echinoidea). Mar Pollut Bull 92(1–2):99–104

Oberbeckmann S, Löder MG, Labrenz M (2015) Marine microplastic-associated biofilms–a review. Environ Chem 12(5):551–562

O'Brine T, Thompson RC (2010) Degradation of plastic carrier bags in the marine environment. Mar Pollut Bull 60(12):2279–2283

Ocean Plastics Lab (2018) Exhibition: science vs. plastic waste. 9–19 April 2018, KDM, Brussels, Belgium

O'Donovan S, Mestre NC, Abel S et al (2018) Ecotoxicological effects of chemical contaminants adsorbed to microplastics in the clam *Scrobicularia plana*. Front Mar Sci 5:143

Ogata Y, Takada H, Mizukawa K et al (2009) International Pellet Watch: global monitoring of persistent organic pollutants (POPs) in coastal waters. 1. Initial phase data on PCBs, DDTs, and HCHs. Mar Pollut Bull 58(10):1437–1446

O'Hagan DT (1996) The intestinal uptake of particles and the implications for drug and antigen delivery. J Anat 189:477–482

Oliveira M, Ribeiro A, Hylland K et al (2013) Single and combined effects of microplastics and pyrene on juveniles (0+ group) of the common goby *Pomatoschistus microps* (Teleostei, Gobiidae). Ecol Indic 34:641–647

Ory NC, Sobral P, Ferreira JL et al (2017) Amberstripe scad *Decapterus muroadsi* (Carangidae) fish ingest blue microplastics resembling their copepod prey along the coast of Rapa Nui (Easter Island) in the South Pacific subtropical gyre. Sci Total Environ 586:430–437

Pan Y, Neuss S, Leifert A et al (2007) Size-dependent cytotoxicity of gold nanoparticles. Small 3(11):1941–1949

Paul-Pont I, Lacroix C, Fernández CG et al (2016) Exposure of marine mussels *Mytilus* spp. to polystyrene microplastics: toxicity and influence on fluoranthene bioaccumulation. Environ Pollut 216:724–737

Paul-Pont I, Tallec K, Gonzalez Fernandez C et al (2018) Constraints and priorities for conducting experimental exposures of marine organisms to microplastics. Front Mar Sci 5:252

Pedà C, Caccamo L, Fossi MC et al (2016) Intestinal alterations in European sea bass *Dicentrarchus labrax* (Linnaeus, 1758) exposed to microplastics: preliminary results. Environ Pollut 212:251–256

Pham PH, Jung J, Lumsden JS et al (2012) The potential of waste items in aquatic environments to act as fomites for viral haemorrhagic septicaemia virus. J Fish Dis 35(1):73–77

Pinsino A, Russo R, Bonaventura R et al (2015) Titanium dioxide nanoparticles stimulate sea urchin immune cell phagocytic activity involving TLR/p38 MAPK-mediated signalling pathway. Sci Rep 5:14492

Plastics Europe (2018) Plastics – the Facts 2018: an analysis of European plastics production, demand and waste data. Plastics Europe, Association of Plastics Manufacturers, Brussels, pp 1–44

Prata JC (2018) Airborne microplastics: consequences to human health? Environ Pollut 234:115–126

Quero GM, Luna GM (2017) Surfing and dining on the "plastisphere": microbial life on plastic marine debris. Adv Oceanogr Limnol 8(2):199–207

Rainieri S, Conlledo N, Larsen BK et al (2018) Combined effects of microplastics and chemical contaminants on the organ toxicity of zebrafish (*Danio rerio*). Environ Res 162:135–143

Ramos JA, Barletta M, Costa MF (2012) Ingestion of nylon threads by Gerreidae while using a tropical estuary as foraging grounds. Aquatic Biol 17(1):29–34

Rebolledo EL, Van Franeker JA, Jansen OE et al (2013) Plastic ingestion by harbour seals (*Phoca vitulina*) in The Netherlands. Mar Pollut Bull 67(1–2):200–202

Rehse S, Kloas W, Zarfl C (2016) Short-term exposure with high concentrations of pristine microplastic particles leads to immobilisation of *Daphnia magna*. Chemospehere 153:91–99

Rehse S, Kloas W, Zarfl C (2018) Microplastics reduce short-term effects of environmental contaminants. Part I: effects of Bisphenol A on freshwater zooplankton Are lower in presence of polyamide particles. Int J Environ Res Public Health 15:280

Reichert J, Schellenberg J, Schubert P et al (2018) Responses of reef building corals to microplastic exposure. Environ Pollut 237:955–960

Reisser J, Proietti M, Shaw J et al (2014) Ingestion of plastics at sea: does debris size really matter? Front Mar Sci 1:70

Revel M, Châtel A, Mouneyrac C (2018) Micro (nano) plastics: a threat to human health? Curr Opin Environ Sci Health 1:17–23

Rist S, Almroth BC, Hartmann NB et al (2018) A critical perspective on early communications concerning human health aspects of microplastics. Sci Total Environ 626:720–726

Rochman CM (2015) The complex mixture, fate and toxicity of chemicals associated with plastic debris in the marine environment. In: Bergmann M, Gutow L, Klages L (eds) Marine anthropogenic litter. Springer, Cham, pp 117–140

Rochman CM (2016) Ecologically relevant data are policy-relevant data. Science 352(6290):1172

Rochman CM, Boxall ABA (2014) Environmental relevance: a necessary component of experimental design to answer the question, "So what?". Integr Environ Assess Manag 10:311–312

Rochman CM, Hoh E, Kurobe T et al (2013) Ingested plastic transfers hazardous chemicals to fish and induces hepatic stress. Sci Rep 3:3263

Rochman CM, Tahir A, Williams SL et al (2015) Anthropogenic debris in seafood: plastic debris and fibers from textiles in fish and bivalves sold for human consumption. Sci Rep 5:14340

Romano N, Ashikin M, Teh JC et al (2017) Effects of pristine polyvinyl chloride fragments on whole body histology and protease activity in silver barb *Barbodes gonionotus* fry. Environ Pollut 237:1106–1111

Romeo T, Pietro B, Pedà C et al (2015) First evidence of presence of plastic debris in stomach of large pelagic fish in the Mediterranean Sea. Mar Pollut Bull 95(1):358–361

Rudneva I (2013) Biomarkers for stress in fish embryos and larvae. CRC Press, Boca Raton

Rummel CD, Löder MG, Fricke NF et al (2016) Plastic ingestion by pelagic and demersal fish from the North Sea and Baltic Sea. Mar Pollut Bull 102(1):134–141

Santana MFM, Moreira FT, Turra A (2017) Trophic transference of microplastics under a low exposure scenario: insights on the likelihood of particle cascading along marine food-webs. Mar Pollut Bull 121(1–2):154–159

Scherer C, Weber A, Lambert S et al (2018) Interactions of microplastics with freshwater biota. In: Wagner M, Lambert S (eds) Freshwater microplastics. Springer, Cham, pp 153–180

Schuyler Q, Hardesty BD, Wilcox C et al (2013) Global analysis of anthropogenic debris ingestion by sea turtles. Conserv Biol 28(1):129–139

Schymanski D, Goldbeck C, Humpf H et al (2017) Analysis of microplastics in water by micro-Raman spectroscopy: release of plastic particles from different packaging into mineral water. Water Res 129:154–162

Seif S, Provencher JF, Avery-Gomm S et al (2018) Plastic and non-plastic debris ingestion in three gull species feeding in an urban landfill environment. Arch Environ Contam Toxicol 74(3):349–360

Setälä O, Fleming-Lehtinen V, Lehtiniemi M (2014) Ingestion and transfer of microplastics in the planktonic food web. Environ Pollut 185:77–83

Setälä O, Norkko J, Lehtiniemi M (2016) Feeding type affects microplastic ingestion in a coastal invertebrate community. Mar Pollut Bull 102(1):95–101

Sjollema SB, Redondo-Hasselerharm P, Leslie HA et al (2016) Do plastic particles affect microalgal photosynthesis and growth? Aquat Toxicol 170:259–261

Smith M (2018) Do microplastic residuals in municipal compost bioaccumulate in plant tissue? Doctoral dissertation, Royal Roads University, Canada

Srivastava RK, Godara S (2017) Use of polycarbonate plastic products and human health. Int J Basic Clin Pharmacol 2(1):12–17

Steer M, Cole M, Thompson RC et al (2017) Microplastic ingestion in fish larvae in the western English Channel. Environ Pollut 226:250–259

Summers S, Henry T, Gutierrez T (2018) Agglomeration of nano-and microplastic particles in seawater by autochthonous and de novo-produced sources of exopolymeric substances. Mar Pollut Bull 130:258–267

Sureda A, Box A, Enseñat M et al (2006) Enzymatic antioxidant response of a labrid fish (*Coris julis*) liver to environmental caulerpenyne. Comp Biochem Physiol C Toxicol Pharmacol 144(2):191–196

Sussarellu R, Suquet M, Thomas Y et al (2016) Oyster reproduction is affected by exposure to polystyrene microplastics. Proc Natl Acad Sci USA 113(9):2430–2435

Sutton R, Mason SA, Stanek SK et al (2016) Microplastic contamination in the San Francisco Bay, California, USA. Mar Pollut Bull 109:230–235

Talsness CE, Andrade AJ, Kuriyama SN et al (2009) Components of plastic: experimental studies in animals and relevance for human health. Philos Trans Royal Soc B Biol Sci 364(1526):2079–2096

Tanaka K, Takada H, Yamashita R et al (2013) Accumulation of plastic-derived chemicals in tissues of seabirds ingesting marine plastics. Mar Pollut Bull 69:219–222

Tang J, Ni X, Zhou Z et al (2018) Acute microplastic exposure raises stress response and suppresses detoxification and immune capacities in the scleractinian coral *Pocillopora damicornis*. Environ Pollut 243:66–74

Teuten EL, Rowland SJ, Galloway TS et al (2007) Potential for plastics to transport hydrophobic contaminants. Environ Sci Technol 41:7759–7764

Teuten EL, Saquing JM, Knappe DR et al (2009) Transport and release of chemicals from plastics to the environment and to wildlife. Philos Trans Royal Soc B Biol Sci 364(1526):2027–2045

Thompson RC, Swan SH, Moore CJ et al (2009a) Our plastic age. Philos Trans Royal Soc B Biol Sci 364:1973–1976

Thompson RC, Moore CJ, Vom Saal FS et al (2009b) Plastics, the environment and human health: current consensus and future trends. Philos Trans Royal Soc B Biol Sci 364(1526):2153–2166

Tosetto L, Williamson JE, Brown C (2017) Trophic transfer of microplastics does not affect fish personality. Anim Behav 123:159–167

Troost TA, Desclaux T, Leslie HA et al (2018) Do microplastics affect marine ecosystem productivity? Mar Pollut Bull 135:17–29

Tsygankov VY, Boyarova MD, Lukyanova ON (2015) Bioaccumulation of persistent organochlorine pesticides (OCPs) by gray whale and Pacific walrus from the western part of the Bering Sea. Mar Pollut Bull 99:235–239

Van Cauwenberghe L, Janssen CR (2014) Microplastics in bivalves cultured for human consumption. Environ Pollut 193:65–70

Vendel AL, Bessa F, Alves VEN et al (2017) Widespread microplastic ingestion by fish assemblages in tropical estuaries subjected to anthropogenic pressures. Mar Pollut Bull 117(1-2):448–455

Villarrubia-Gómez P, Cornell SE, Fabres J (2017) Marine plastic pollution as a planetary boundary threat–The drifting piece in the sustainability puzzle. Mar Policy 96:213–220

Viršek MK, Lovšin MN, Koren Š et al (2017) Microplastics as a vector for the transport of the bacterial fish pathogen species *Aeromonas salmonicida*. Mar Pollut Bull 125(1–2):301–309

Volkheimer G (1975) Hematogenous dissemination of ingested polyvinyl chloride particles. Ann N Y Acad Sci 246(1):164–171

Volkheimer G (1977) Passage of particles through the wall of the gastrointenstinal tract. Environ Health Persp 9:215–225

von Moos N, Burkhardt-Holm P, Köhler A (2012) Uptake and effects of microplastics on cells and tissue of the blue mussel *Mytilus edulis* L. after an experimental exposure. Environ Sci Technol 46(20):11327–11335

Vroom RJ, Koelmans AA, Besseling E et al (2017) Aging of microplastics promotes their ingestion by marine zooplankton. Environ Pollut 231:987–996

Wakamatsu TH, Dogru M, Tsubota K (2008) Tearful relations: oxidative stress, inflammation and eye diseases. Arq Bras Oftalmol 71(6):72–79

Wang J, Tan Z, Peng J et al (2016) The behaviors of microplastics in the marine environment. Mar Environ Res 113:7–17

Wardrop P, Shimeta J, Nugegoda D et al (2016) Chemical pollutants sorbed to ingested microbeads from personal care products accumulate in fish. Environ Sci Technol 50:4037–4044

Watts AJ, Lewis C, Goodhead RM et al (2014) Uptake and retention of microplastics by the shore crab *Carcinus maenas*. Environ Sci Technol 48(15):8823–8830

Weber A, Scherer C, Brennholt N et al (2018) PET microplastics do not negatively affect the survival, development, metabolism and feeding activity of the freshwater invertebrate *Gammarus pulex*. Environ Pollut 234:181–189

Wegner A, Besseling E, Foekema EM et al (2012) Effects of nanopolystyrene on the feeding behavior of the blue mussel (*Mytilus edulis* L.). Environ Toxicol Chem 31(11):2490–2497

Welden NA, Cowie PR (2016) Long-term microplastic retention causes reduced body condition in the langoustine, *Nephrops norvegicus*. Environ Pollut 218:895–900

Welden NA, Abylkhani B, Howarth LM (2018) The effects of trophic transfer and environmental factors on microplastic uptake by plaice, *Pleuronectes plastessa*, and spider crab, *Maja squinado*. Environ Pollut 239:351–358

Wen B, Zhang N, Jin SR et al (2018) Microplastics have a more profound impact than elevated temperatures on the predatory performance, digestion and energy metabolism of an Amazonian cichlid. Aquat Toxicol 195:67–76

Wieczorek AM, Morrison L, Croot PL et al (2018) Frequency of microplastics in mesopelagic fishes from the Northwest Atlantic. Front Mar Sci 5:39

Wright SL, Kelly FJ (2017) Plastic and human health: a micro issue? Environ Sci Technol 51:6634–6647

Wright SL, Thompson RC, Galloway TS (2013a) The physical impacts of microplastics on marine organisms: a review. Environ Pollut 178:483–492

Wright SL, Rowe D, Thompson RC et al (2013b) Microplastic ingestion decreases energy reserves in marine worms. Curr Biol 23:R1031–R1033

Yaghmour F, Al Bousi M, Whittington-Jones B et al (2018) Marine debris ingestion of green sea turtles, *Chelonia mydas* (Linnaeus, 1758) from the eastern coast of the United Arab Emirates. Mar Pollut Bull 135:55–61

Yokota K, Waterfield H, Hastings C et al (2017) Finding the missing piece of the aquatic plastic pollution puzzle: interaction between primary producers and microplastics. Limnol Oceanogr Lett 2(4):91–104

Zarfl C, Matthies M (2010) Are marine plastic particles transport vectors for organic pollutants to the Arctic? Mar Pollut Bull 60:1810–1814

Zettler ER, Mincer TJ, Amaral-Zettler LA (2013) Life in the "plastisphere": microbial communities on plastic marine debris. Environ Sci Technol 47(13):7137–7146

Zhu XS, Wang JX, Zhang XZ et al (2010) Trophic transfer of TiO2 nanoparticles from *Daphnia* to zebra fish in a simplified freshwater food chain. Chemosphere 79:928–933

7 Chemical Biodiversity and Bioactivities of Saponins in Echinodermata with an Emphasis on Sea Cucumbers (Holothuroidea)

Elham Kamyab, Matthias Y. Kellermann, Andreas Kunzmann, and Peter J. Schupp

Abstract

Echinoderms are a source of a broad range of secondary metabolites with a large variety of bioactive properties. Although pigment and lipid derivatives are the major groups of bioactive compounds found in crinoids and ophiuroids, saponins represent the most abundant and diverse marine natural products (MNPs) in the phylum Echinodermata. This review is for researchers that are interested in MNPs derived from echinoderms, but with a particular focus on the structural diversity and biological function of saponins. Among the echinoderms, these steroidal compounds are mostly known for and structurally most diverse within sea cucumbers. Through compilation of extensive tables, this review provides a reference book, summarizing not only the major chemical classes of well-known secondary metabolites in the phylum Echinodermata but also further focusing on the presence of bioactive saponins in echinoderms in general and within different sea cucumber species in particular. The final compilation aims to correlate the vast structural diversity of saponins with known biological functions. The here presented data revealed that holothurians, holotoxins, cucumariosides, and echinosids are not only the most abundant saponin compounds in various genera of sea cucumbers but that these saponins can also be used as potential chemotaxonomic markers for different sea cucumber species. By studying the structure-function relationships of triterpene glycosides in echinoderms in general, or in particular within holothurians, the vast structural diversity, taxonomic distribution, and bioactivity of the molecules can be deciphered, which provides an opportunity to focus future research efforts on target species that contain MNPs with novel pharmacological activities.

Keywords

Secondary metabolites · Chemical diversity · Taxonomic markers · Structure-function analysis · Saponins · Echinoderms · Sea cucumber

7.1 Marine Natural Products (MNPs)

Compared to synthesized organic compounds, natural products (NPs) have long been used as efficient and often less harmful sources of drug molecules (Molinski et al. 2009). NPs refer to both primary and secondary metabolites; however, in the past, research on secondary metabolites mostly described ecological interactions of organisms with their environment, the pronounced biological and pharmacological activities, their great chemical diversity, and their higher tendency to interact with other biologically relevant molecules (Croteau et al. 2000).

The marine environment came into the focus of NPs right after technologies for studying marine ecosystems improved. Since the early 1900s, the idea of utilizing marine ecosystems as the potentially largest source for marine natural products (hereafter MNPs) was shaped. Although research on MNPs dates back more than 50 years and more than 32,000 studies related to MNPs have been published (MarineLit; http://pubs.rsc.org/marinlit/), only a few marine-derived compounds resulted in clinical trials (Mayer et al. 2017). That is, from 52 marine invertebrate-derived compounds that reached clinical trials, only seven compounds, isolated from sponges, mollusks, tunicates, and their associated bacteria, have so far

E. Kamyab (✉) · M. Y. Kellermann
Institute of Chemistry and Biology of the Marine Environment, University of Oldenburg, Oldenburg, Germany
e-mail: elham.kamyab@uni-oldenburg.de

A. Kunzmann
Leibniz Centre for Tropical Marine Research (ZMT) GmbH, Bremen, Germany

Faculty 02, University of Bremen, Bremen, Germany

P. J. Schupp (✉)
Institute of Chemistry and Biology of the Marine Environment, University of Oldenburg, Oldenburg, Germany

Helmholtz Institute for Functional Marine Biodiversity at the University of Oldenburg (HIFMB), Oldenburg, Germany

S. Jungblut et al. (eds.), *YOUMARES 9 - The Oceans: Our Research, Our Future*,
https://doi.org/10.1007/978-3-030-20389-4_7

been approved. Unfortunately, 45 of the total 52 MNPs have been discontinued from clinical trials (Fig. 7.1) due to low production yields and/or high costs.

In this review, we provide an overview on the MNPs reported from echinoderms with an emphasis on MNPs (i.e., particularly triterpene glycosides) reported from shallow water sea cucumbers. While there is extensive literature on the chemistry of MNPs from sessile marine organisms such as sponges, ascidians, and corals, MNP data on slow-moving invertebrates such as echinoderms are much more limited. Up to now, more than 7,000 living echinoderms species, divided into three subphyla and five different classes, have been described (Fig. 7.2). The evolutionary divergence of echinoderms with chordates rather than invertebrates makes their biochemistry and physiology rather similar with vertebrates. They can synthesize vertebrate-type steroids, which regulate their reproductive, growth, and developmental processes (Schoenmakers 1979). Therefore, it can be hypothesized that echinoderms can be promising substitution candidates of the synthetic compounds for producing efficient secondary metabolites helpful for human health. Although several defense mechanisms such as presence of spine, cuvierian tubules (CTs), evisceration, toxic secretion, and unpalatability are generally described for echinoderms and particularly for holothurians, they do not have a significant escape behavior and therefore likely depend on chemical defense strategies, such as triterpene glycosides, to protect themselves against predators (Iyengar and Harvell 2001; Bahrami et al. 2016). Saponins represent a diverse group of triterpene glycosides that have been mainly described from plants and are also one of the major secondary metabolite classes in Echinodermata including holothurians. Saponins are promising MNPs with the capacity to influence physiological and immunological processes and thus have been implicated as bioactive compounds in many ecological studies (Kalinin et al. 1996; Francis et al. 2002). In the following sections, we will discuss in more detail the role of saponins and other bioactive compounds in echinoderms in general, however, with a major focus on sea cucumbers.

7.2 MNPs in Echinoderms

From 28,609 MNPs that have been reported until 2016, more than 35% of the total compounds were isolated from echinoderms. However, the reported chemical diversity of MNPs from echinoderms, compared to other phyla, was not high (Blunt et al. 2018).

Typical reported MNPs derived from echinoderms are sulfated compounds that can be largely classified into two major groups: aromatics and saponins. Among the five classes of echinoderms (Fig. 7.2), aromatic sulfated compounds have only been reported in crinoids and ophiuroids as pigments derived from anthraquinones or naphthoquinones, whereas most of the saponins have been isolated from asteroids, echinoids, and holothuroids (Kornprobst et al. 1998) (Tables 7.1 and 7.2). Among various types of secondary metabolites that have been isolated

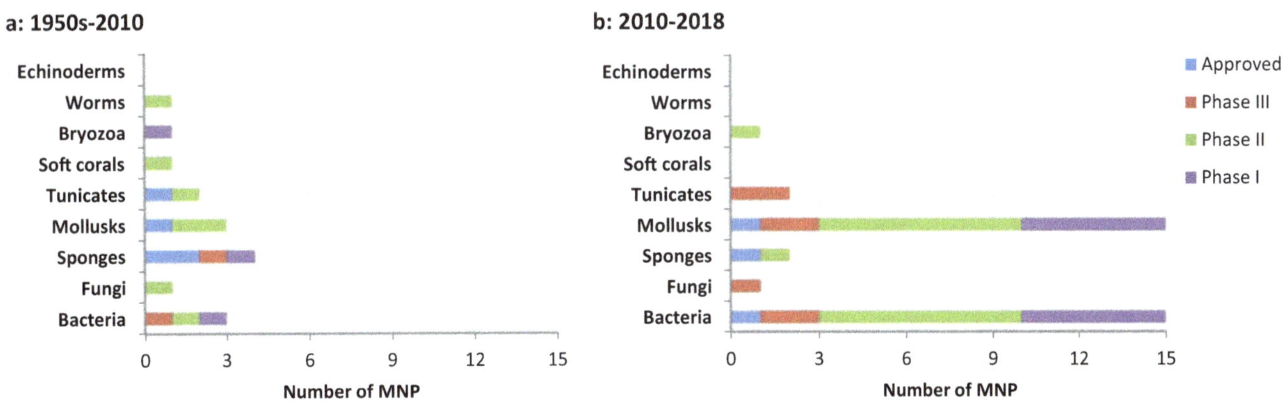

Fig. 7.1 Overview of marine organisms from which MNPs entered the pharmaceutical pipeline (**a**) from 1950s to 2010 and (**b**) from 2010 to 2018. (Compiled with data from Mayer and Hamann 2002; Mayer et al. 2017; http://marinepharmacology.midwestern.edu)

Fig. 7.2 Phylogenetic tree for the phylum Echinodermata (modified after Telford et al. 2014)

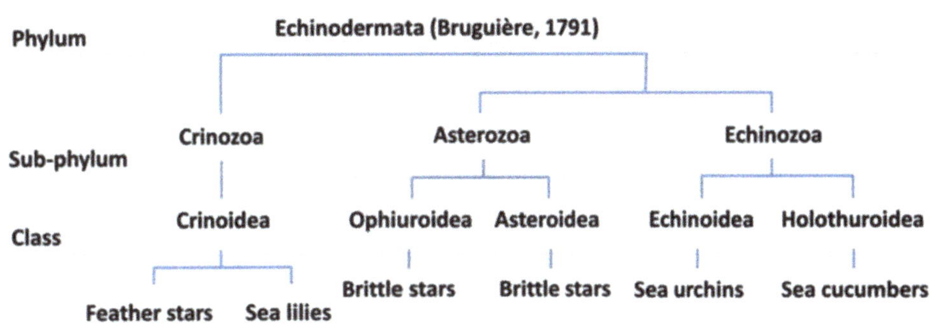

Table 7.1 Classes of echinoderms, major classes of secondary metabolites, examples of compounds, their bioactivity, and example species for which the compounds have been reported

	Major classes of secondary metabolites	Examples of bioactive compounds	Biological activity	Example of organisms	References
Crinoids: Lipids, pigments, polyketides	Polyketides	Rhodoptilometrin, crinemodin	Antipredatory	**Comanthus bennetti*	Rideout et al. (1979)
	Lipids	Ganglioside, cerebrosides	n.d.	**Comanthina schlegelii*	Inagaki et al. (2007)
	Naphthopyrones	Naphthopyrones comaparvin	Anti-inflammatory	*Comanthus parvicirrus*	Karin et al. (2004), Folmer et al. (2009), Chovolou et al. (2011), and Chen et al. (2014)
	Anthraquinoid pigments	Gymnochrome D	Antiviral	**Gymnocrinus richeri*	Laille et al. (1998)
Asteroids: Steroidal derivatives of cholesterol, fatty acids, ceramides, and few alkaloids and proteins	Lipids	Hexadecanoic acid	Antifouling	*Linckia laevigata*	Guenther et al. (2009)
	Lipids	Sphingolipids	n.d.	*Ophidiaster ophidianus*	Jin et al. (1994)
	Asterosaponins	Thornasteroside A	Antitumor	*Asteropsis carinifera*	Malyarenko et al. (2012)
	Fatty acids	Eicosanoic acid	n.d.	*Culcita novaeguineae*	Bruno et al. (1992) and Inagaki (2008)
	Protein	Ciguatoxins	n.d.	*Marthasterias glacialis*	Silva et al. (2015)
	Polyhydroxysteroids	Laeviusculosides	Hemolytic, cytotoxic activity	*Henricia leviuscula*	Ivanchina et al. (2006) and Fedorov et al. (2008)
Ophiuroids: Carotenoids, gangliosides, brominated indoles, phenyl propanoids, terpenes, steroids	Steroidal glycosides	Steroidal glycosides	Antiviral	*Ophiarachna incrassata*	D'Auria et al. (1993)
	Steroidal compound	Polyhydroxysterols	Antiviral	*Astrotoma agassizii*	Comin et al. (1999)
	Terpene	2,3-Dimethyl butenolide	Antitumor	*Ophiomastix mixta*	Lee et al. (2007)
	Carotenoid	Ophioxanthin	Antioxidant	*Ophioderma longicauda*	D'Auria et al. (1985)
Echinoids: Protein, polysaccharides, lipid, pigments	Naphthoquinoid pigment	Echinochrome A	Antioxidant, antimicrobial, anti-inflammatory, antitoxic agents	**Anthocidaris crassispina*	Berdyshev et al. (2007) and Jeong et al. (2014)
	Peptides	Strongylostatin	Anticancer	*Strongylocentrotus droebachiensis*	Pettit et al. (1981)
		Strongylocins	Antimicrobial	*Strongylocentrotus droebachiensis*	Li et al. (2008)
	Polysaccharide	Sulfated fucan	Anticoagulant	*Lytechinus variegatus*	Pereira et al. (1999)
	Steroidal compounds	n.d.	Anticancer	*Diadema savignyi*	Thao et al. (2015a)
	Ganglioside	DSG-A	Neuritogenic	*Diadema setosum*	Yamada et al. (2008)

(continued)

Table 7.1 (continued)

	Major classes of secondary metabolites	Examples of bioactive compounds	Biological activity	Example of organisms	References
Holothuroids: Triterpene glycosides, peptides, polysaccharides, lipids	Triterpene glycoside	Holothurins (A–B)	Antifungal, anticancer, ichthyotoxic	*Holothuria atra, Holothuria fuscocinerea*	Yamanouchi (1955), Kobayashi et al. (1991), Popov et al. (1994), and Zhang et al. (2006d)
	Triterpene glycoside	Echinoside A	Antifungal	*Actinopyga echinites*	Kitagawa et al. (1985)
	Triterpene glycoside	Holotoxin A–F	Anticancer, antifungal, antiprotozoa	*Apostichopus japonicus*	Kitagawa et al. (1976), Anisimov et al. (1983), Maltsev et al. (1985), and Wang et al. (2012)
	Triterpene glycosides	Holotoxin	Antifungal	*S. japonicus*	Yano et al. (2013)
	Polysaccharides	Glucosamine, Galactosamine	Antihyperlipidemic, antioxidant	*A. japonicus*	Liu et al. (2012)
	Sulfated polysaccharides	FucCS, GAGs	Anticoagulant, antithrombin, antiparasitic	*Ludwigothurea grisea*	Mourão et al. (1998), Borsig et al. (2007), and Marques et al. (2016)
	Sulfated polysaccharides	FucCS	Anticoagulant, antithrombin, antihyperglycemic, antiviral, insulin-sensitizing	*Thelenota ananas, Cucumaria frondosa*	Borsig et al. (2007), Huang et al. (2013), and Hu et al. (2014a)
	Sulfated polysaccharides	FucCS	Anticoagulant, antiparasitic	*Isostichopus badionotus*	Marques et al. (2016)
	Sulfated polysaccharides	GAGs	Antihyperlipidemic	**Metriatyla scabra*	Liu et al. (2002)
	Fatty acid	EPA-enriched PL, 12-MTA, ODAs	Antioxidant, antihyperglycemic, anticancer, antihyperlipidemic	*C. frondosa, Stichopus japonicus*	Yang et al. (2003), Nguyen et al. 2011, Hu et al. (2014b), Wu et al. (2014), and Ku et al. (2015)
	Lipid	Cerebrosides, galactocerebrosides, AMC-2	Anticancer, antihyperlipidemic	**Stichopus variegatus, Acaudina molpadioides, Bohadschia argus*	Sugawara et al. (2006), Ikeda et al. (2009), Zhang et al. (2012), and Du et al. (2015)
	Sphingolipid	Cerebroside	Antioxidant	*S. japonicus, Acaudina molpadioides*	Duan et al. (2016) and Xu et al. (2011)
	Lysophospholipid	LPC, L-PAF	Anti-inflammatory	*Holothuria atra*	Nishikawa et al. (2015)
	Peptide	Phenoloxidase, lysozyme	Antimicrobial	*C. frondosa*	Beauregard et al. (2001)
	Peptide	ACE inhibitory peptide	Antihypertension	*Acaudina molpadioides*	Zhao et al. (2009)
	Peptide	T-antigen-binding lectin	Antibacterial	*Holothuria scabra*	Gowda et al. (2008)
	Phenolic compounds	n.d.	Anti-inflammatory	*S. japonicus*	Song et al. (2016)

(continued)

Table 7.1 (continued)

	Major classes of secondary metabolites	Examples of bioactive compounds	Biological activity	Example of organisms	References
	Phenolic compounds	(Z)2,3-DPAN	Anticancer	*Holothuria parva*	Amidi et al. (2017)
	Pigments	Carotenoids	Antioxidant	*Holothuria atra*	Esmat et al. (2013)
	Pigments	β-carotene, echinenone, canthaxanthin, etc.	Antioxidant	*Plesiocolochirus minaeus*	Maoka et al. (2015)
	Sulfated alkene	2,6-DMHS, OS, DS	Antibacterial, antifungal	*A. japonicus*	La et al. (2012)
	Mucopolysaccharide	SJAMP	Antitumor, immunomodulatory effect	*S. japonicus*	Song et al. (2013)
	Glycolipid/ Sphingolipid	2,6-DMHS, OS, DS	Anticancer	*A. japonicus*	La et al. (2012)
	Saponin	Frondanol A_5	Anticancer	*C. frondosa*	Janakiram et al. (2010)and Jia et al. (2016)
	Saponin	n.d.	Antihyperlipidemic	*Pearsonothuria graeffei*	Hu et al. (2010) and Wu et al. (2015)
	Monosulfated triterpene glycosides	Cumaside	Radioprotective	*Cucumaria japonica*	Aminin et al. (2011)

n.d. not defined, *EPA-enriched PL* eicosapentaenoic acid-enriched phospholipids, *FucCS* fucosylated chondroitin sulfate, *GAGs* glycosaminoglycan, *2,6-DMHS* 2,6-dimethylheptyl sulfate, *OS* octyl sulfate, *DS* decyl sulfate, *ACE* angiotensin I-converting enzyme, *LPC* lysophosphatidylcholine, *L-PAF* lyso-platelet activating factor, *SCEA-F* ethyl acetate solvent fraction of sea cucumber, *EPA-enriched PC* eicosapentaenoic acid-enriched phosphatidylcholine lipids, *12-MTA* 12-methyltetradecanoic acid, *(Z)2,3-DPAN* (Z)-2,3-diphenylacrylonitrile, *SJAMP stichopus japonicus* acid mucopolysaccharide, *ODAs* 7(Z) octadecenoic acid, and 7(Z),10(Z)-octadecadienoic acid

*Based on WoRMS (2019), the accepted names changed from: *Metriatyla scabra to Holothuria scabra; Comanthus bennetti to Anneissia bennetti; Comanthina schlegelii to Comaster schlegelii; Gymnocrinus richeri to Neogymnocrinus richeri; Anthocidaris crassispina to Heliocidaris crassispina; Stichopus variegatus to Stichopus horrens*

from echinoderms, saponins are the most abundant. Compounds were derived from mainly two classes (i.e., Asteroidea and Holothuroidea) (Haug et al. 2002), which will be discussed in more detail in Sect. 7.3.

7.2.1 Crinoids (Feather Stars and Sea Lilies)

The most primitive form of current echinoderms are the crinoids (Karleskint et al. 2010). Sea lilies are, unlike feather stars, sessile and are found mainly in depths >100 m, whereas feather stars inhabit coral reefs from the intertidal to the deep-sea oceans. Moreover, feather stars are physically able to escape from predators by crawling, swimming, or hiding between corals or rocks (Ruppert et al. 2004; Karleskint et al. 2010). Furthermore, crinoids use other physical and chemical defense mechanisms to protect them against fish predators. For example, crinoids use spike-like pinnules as well as toxic chemical compounds such as polyketide derivatives and oxidized quinones that also give them their colorful appearance (Kenta et al. 2015; Feng et al. 2017). According to WoRMS[1] 2017, although they consist of nearly 700 species worldwide, until now only a few studies examined their bioactive compounds. According to the MarinLit database (2018), only 25 marine species from 16 different genera of crinoids have so far been screened for novel MNPs (Feng et al. 2017) (Table 7.1).

[1]World Register of Marine Species.

7.2.2 Asteroids (Sea Stars)

This class of echinoderms is, with over 1500 species, widely distributed and thus plays important ecological roles. Asteroids are opportunistic feeders, and species such as the temperate Ochre sea star *Pisaster ochraceus* and the tropical coral-eating crown of thorn sea star *Acanthaster planci* are keystone species (Paine 1969). Asteroids are known to use both physical and chemical defense mechanisms. Autotomy (i.e., found in *Evasterias troschelii* and *Pycnopodia helianthoides*), spines, modified tube feet called "pedicellaria," camouflage, quick locomotion, and shedding have been reported as physical defenses (Bryan et al. 1997; Candia Carnevali and Bonasoro 2001). However, some species such as the sea star *Pteraster tesselatus* rely to a great extent on their mucus as chemical defense (Nance and Braithwaite

Table 7.2 Steroidal compounds reported from echinoderms, except Holothuroids, and (if reported) their biological activities (Holothuroids see Table 7.3)

Class	Family	Species	Isolated compounds	Biological activity	References
Crinoids	Hemicrinidea	*Neogymnocrinus richeri*	Several steroids	n.d.	De Riccardis et al. (1991)
Asteroids	Asteriidae	*Asterias amurensis*	Thornasteroside A, versicosides A–C, and asteronylpentaglycoside sulfate, anasteroside B	n.d.	Hwang et al. (2011, 2014)
			Crude saponin	Insecticide and repellant activity	Park et al. (2009)
			Asterosides A–D, glycoside B, asterosaponins	n.d.	Riccio et al. (1988)
			Asterosaponin-4	Cytotoxic	Okano et al. (1985)
			Asterosaponin A, A_4	Antitumor	Ikegami et al. (1973)
		Asterias vulgaris	13 steroidal compounds	n.d.	Findlay and Agarwal (1983)
		Asterias forbesi	Forbeside D	n.d.	Findlay and He (1991)
			Forbesides A–B	Anti-inflammatory, Antiviral	Findlay et al. (1987)
			Forbesides C–E, E_1-E_3, F-H, L	n.d.	Findlay et al. (1989), Findlay and He (1991), D'Auria et al. (1993), and Jiang et al. (1993)
			Forbeside H	n.d.	Findlay et al. (1992)
		Asterias rubens	Ruberosides A–F	n.d.	Sandvoss et al. (2000, 2003)
		Asterias rathbuni	Rathbuniosides R_1-R_2	Cytotoxic	Prokof'eva et al. (2003)
		Anasterias minuta	Minutosides A, B	Antifungal	Chludil et al. (2002b)
			Anasterosides A–B, versicoside A	Antifungal	Chludil et al. (2002b)
		Asterias rollestoni	Amurensoside, forbeside		Zhang et al. (2013)
		Aphelasterias japonica	Aphelasteroside F	Inhibition of cell proliferation	Popov et al. (2016)
			Ophidianoside F	n.d.	Ivanchina et al. (2005)
			Aphelasteroside C (1), cheliferoside L_1 (2),3-O-sulfoasterone (3), forbeside E_3 (4), and 3-O-sulfothornasterol A (5) aphelaketotriol (6)	Hemolytic activity except compound (3)	Ivanchina et al. (2000)
		Leptasterias hylodes	Polyhydroxylated steroids	Antibacterial, hemolytic activity	Levina et al. (2010)
			Hylodoside A, novaeguinoside Y	Hemolytic activity	Levina et al. (2010)
		Leptasterias ochotensis	Leptasteriosides A–F	Anticancer	Malyarenko et al. (2014)
		Diplasterias brucei	Diplasteriosides A, B	Anticancer	Ivanchina et al. (2011)
		Coscinasterias tenuispina	Tenuispinosides A–C, coscinasteroside A–F	n.d.	Riccio et al. (1986d)
		Distolasterias nipon	Nipoglycosides A–D, versicoside A, and thornasteroside A	n.d.	Minale et al. (1995)
			Distolasterosides D_1–D_3	Neurogenic and neuroprotective effect	Palyanova et al. (2013)
		Distolasterias elegans	Pycnopodioside C	n.d.	Andriyashchenko et al. (1996)
		Lethasterias fusca	Lethasterioside A	Anticancer	Ivanchina et al. (2012)
		Lysastrosoma anthosticta	Lysaketotriol and iysaketodiol	Immunomodulatory activities	Levina et al. (2009)

(continued)

Table 7.2 (continued)

Class	Family	Species	Isolated compounds	Biological activity	References
			Luridosides A, marthasterone, marthasteroside, pyncopodioside C	n.d.	Levina et al. (2001)
		Marthasterias glacialis	Thornasteroside A, maculatoside A_1, A_2, B–C	n.d.	Bruno et al. (1984) and Minale et al. (1985)
	Oreasteridae	*Pentaceraster gracilis*	Pentacerosides A and B, maculatoside	Maculatoside: cytotoxic	Vien et al. (2017)
		[a]*Anthenea chinensis*	Anthenoside A, E, G, H, I, J, K	Antitumor	Ma et al. (2009a, 2010)
		Culcita novaeguineae	Culcinosides A–D	Cytotoxic	Lu et al. (2018)
			Novaeguinosides I,II, A–E, regularoside B	Antitumor	Tang et al. (2005) and Ngoan et al. (2015)
			Sodium (20R,24S)-6α-O-(4-O-sodiumsulfato-β-d-quinovopyranosyl)-5α-cholest-9(11)-en-3β,24-diol 3-sulfate	Anticancer	Ma et al. (2009b)
			Sodium (20R,24S)-6α-O-[3-O-methyl-β-d-quinovopyranosyl-(1→2)-β-d-xylopyranosyl-(1→3)-β-d-glucopyranosyl]-5α-cholest-9(11)-en-3β,24-diol 3-sulfate	Anticancer	Ma et al. (2009b)
			Galactocerebroside	n.d.	Inagaki et al. (2006)
			Polyhydroxylated steroids	Antibacterial, hemolytic activity	Levina et al. (2010)
			Hylodoside A, novaeguinoside Y	Hemolytic activity	Levina et al. (2010)
			Culcitoside C_2–C_3	Hemolytic activity, cytotoxic	Prokof'eva et al. (2003)
			Culcitoside C_1–C_8	n.d.	Kicha et al. (1985, 1986) and Iorizzi et al. (1991)
			Regularosides A–B, thornasteroside A, marthasteroside A_1	Cytotoxic	Tang et al. (2006)
			Asterosaponin 1, novaeguinosides I and II	Antitumor	Cheng et al. (2006) and Tang et al. (2009)
		Protoreaster nodosus	Nodososide	Anti-inflammatory, cytotoxic	Riccio et al. (1982b) and Thao et al. (2015b)
			Ganglioside, galactocerebroside, ganglioside PNG-2A	n.d.	Pan et al. (2010, 2012) and Kenta et al. (2015)
			Three steroids	n.d.	Riccio et al. (1982b) and Minale et al. (1984b)
			Protoreasteroside	n.d.	Riccio et al. (1985d)
		Pentaceraster alveolatus	Protoreasteroside	n.d.	Riccio et al. (1985d)
		Halityle regularis	Halityloside A–F, halityloside H	n.d.	Iorizzi et al. (1986)
			Regularosides A, B, thornasteroside A	n.d.	Riccio et al. (1986c)
		Oreaster reticulatus	Sulfated glycosides analog of nodososide	n.d.	De Correa et al. (1985)
			Reticulatosides A, B, ophidianoside F	n.d.	Iorizzi et al. (1995)
		Choriaster granulatus	Granulatosides A–E	D–E: Immunomodulatory effect	Pizza et al. (1985a) and Ivanchina et al. (2017, 2018)

(continued)

Table 7.2 (continued)

Class	Family	Species	Isolated compounds	Biological activity	References
	Ophidiasteridae	*Hacelia attenuata*	Nodososide, attenuatosides A-I, B-I, B-II, and C, polyhydroxysteroids	n.d.	Minale et al. (1983)
			Attenuatosides S-I–S-III, S-D, thornasteroid	n.d.	Minale et al. (1984a)
			Ophidianosides B, C, F	n.d.	Riccio et al. (1985c)
		Linckia laevigata	Thornasteroside A, marthasteroside A_1, ophidianoside F, maculatoside, laevigatoside	n.d.	Riccio et al. (1985b)
			Granulatoside A	Neuritogenic activity	Qi et al. (2006)
			Nodososide	n.d.	Minale et al. (1984c)
			Linckosides A–Q	Neuritogenic activity	Qi et al. (2002, 2004) and Han et al. (2006, 2007a)
			Linckosides L_1–L_7, echinasteroside C	Neuritogenic activity, cytotoxic	Kicha et al. (2007a, b, c)
		Ophidiaster ophidianus	Ophidianoside B–F	n.d.	Riccio et al. (1985c)
		Certonardoa semiregularis	Certonardoside A–J, halytoside D	Antiviral	Wang et al. (2002)
			Certonardoside K-N, culcitoside C_6	Cytotoxic, antibacterial	Wang et al. (2003)
			Certonardosterol Q_1–Q_7, B_2–B_4, A_2–A_4, D_2–D_5, H_3, H_4, E_2, E_3, P_1, O_1	Cytotoxic, antitumor	Wang et al. (2004a, b)
			Certonardoside B_2, B_3, P_1, P_2, O_1, J_2, J_3, I_2, I_3, H_2	Cytotoxic, antitumor	Wang et al. (2004a, 2005)
		Nardoa gomophia	Halityloside A, B, D, E, H, I, marthasteroside A_1, thornasteroide A, and 2 polyhydroxysteroids	n.d.	Riccio et al. (1986b)
		Nardoa novaecaledonia	Halityloside A, B, D	n.d.	Riccio et al. (1986b)
	Asterinidae	*Patiria pectinifera*	Polyhydroxysteroids	Cytotoxic.	Peng et al. (2010)
			Cucumarioside F_1, F_2	Indicative of trophic marker	Popov et al. (2014)
			Asterosaponin P_1,P_2, polyhydroxysteroids	Asterosaponin P1: neurogenic and neuroprotective effect	Kicha et al. (1983, 2000, 2004) and Palyanova et al. (2013)
			Pectinoside A	Immunological activity	Kawase et al. (2016)
		Asterina pectinifera	Pectiniosides A–J, acanthaglycciside C	Cytotoxic	Dubois et al. (1988), Honda et al. (1990), Jiang and Schmidt (1992), and Li et al. (2013)
		Asterina batheri	Astebatheriosides A–D	Astebatheriosides B-D: anti-inflammatory	Thao et al. (2013)
		Patiria miniata	Patiriosides A–G	Antitumor	Dubois et al. (1988), and D'Auria et al. (1990)

(continued)

Table 7.2 (continued)

Class	Family	Species	Isolated compounds	Biological activity	References
	Asteropectinidae	*Astropecten polyacanthus*	Astropectenols A, C, D	Antiparasitic	Thao et al. (2013, 2014)
		Astropecten monacanthus	Astrosteriosides A, D, C	Anti-inflammatory, anticancer	Thao et al. (2013, 2014) and Dai and Yu (2015)
		Craspidaster hesperus	Asterosaponin	n.d.	Wen et al. (2004)
		Psilaster cassiope	Psilasteroside	Cytotoxic	De Marino et al. (2003)
		Astropecten latespinosus	Latespinosides A–D	Weak-cytotoxic	Higuchi et al. (1996)
	Echinasteridae	*Henricia leviuscula*	Laevisculoside, laevisculoside A–J, H_2 sanguinosides A–B	Hemolytic activity	Kalinovskii et al. (2004), and Ivanchina et al. (2006)
			Laevisculoside G	Anticancer	Fedorov et al. (2008)
			Sanguinoside C	Cytotoxic	Levina et al. (2003)
		Henricia sanguinolenta	Laevisculoside, sanguinoside A–B	n.d.	Kalinovskii et al. (2004)
			Sanguinoside C	Cytotoxic	Levina et al. (2003)
		Henricia derjugini	Henricioside H_1–H_3, hexaol	n.d.	Ivanchina et al. (2004)
			Henricioside H_1, levisculoside G	Antifungal	Kaluzhskiy et al. (2017)
		Henricia sp.	Henriciosides H_1–H_3	n.d.	Kicha et al. (1993)
		Henricia downeyae	Asterosaponins	Antibacterial, antifungal, feeding deterrent	Palagiano et al. (1996)
		Echinaster brasiliensis	Brasilienoside, desulfated dihydro-echinasteroside A, echinasteroside B–G, marthasteroside A_1	n.d.	Iorizzi et al. (1993)
		Echinaster sepositus	22,23-epoxysteroidal (cyclic) glycosides	n.d.	Riccio et al. (1981), and Minale et al. (1997)
			Amurasterol, asterosterol	n.d.	De Simone et al. (1980)
			Sepositoside A	Cytotoxic	De Simone et al. (1981)
			Echinasterosides A, B_1, B_2, laeviusculosides C, I	n.d.	Zollo et al. (1985), Levina et al. (1987), and Iorizzi et al. (1993)
		Echinaster luzonicus	Sepositoside A, luzonicosides A, D	Cytotoxic, anticancer	De Simone et al. (1981), Riccio et al. (1982a), and Malyarenko et al. (2017)
	Stichasteridae	*Neosmilaster georgianus*	Santiagoside	n.d.	Vázquez et al. (1992)
		Cosmasterias lurida	Cosmasterosides A–D, forbeside H	n.d.	Roccatagliata et al. (1994)
			Luridosides A–B	n.d.	Maier et al. (1993)
	Asteropseidae	*Asteropsis carinifera*	Asteropsiside A, regularoside A, and thornasteroside A	Antitumor	Malyarenko et al. (2012)
			Cariniferosides A–F	No cytotoxicity	Malyarenko et al. (2011)
			Polyhydroxysteroids	n.d.	Malyarenko et al. (2010)

(continued)

Table 7.2 (continued)

Class	Family	Species	Isolated compounds	Biological activity	References
	Archasteridae	*Archaster typicus*	Five steroids	Anticancer	Yang et al. (2011)
			Archasterosides A–C	Anticancer	Kicha et al. (2010a, b)
	Luidiidae	*Luidia maculata*	Thornasteroside A, maculatosides A–C, A_2	Anticancer	Minale et al. (1985)
		Luidia quinaria	Thornasterol	n.d.	Andriyashchenko et al. (1996)
			Luidiaquinoside, psilasteroside	Cytotoxic	De Marino et al. (2003)
	Acanthasteridae	*Acanthaster planci*	Thornasterols A and B	Cytotoxic	Kitagawa and Kobayashi (1977, 1978)
			Acanthaglycoside B–F, marthasteroside A_1, and versicoside A–B	n.d.	Itakura and Komori (1986)
			5-Deoxyisonodososide, isonodososide	Cytotoxic	Pizza et al. (1985b)
			Nodososide	n.d.	Minale et al. (1984c)
	Goniopectinidae	*Goniopecten demonstrans*	Goniopectenosides A–C	Antifouling	De Marino et al. (2000)
		Hippasteria phrygiana	Hippasteriosides A–D	Hippasterioside D: anticancer	Kicha et al. (2011)
			Phrygiasterol (1), phrygioside B (2), borealoside C (3)	(1,2): Anticancer	Levina et al. (2004, 2005)
	Goniasteridae	*Mediaster murrayi*	Mediasteroside M_1	Anticancer	Prokof'eva et al. 2003
		Ceramaster patagonicus	Ceramasterosides C_1–C_3	Cytotoxic	Prokof'eva et al. (2003)
	Heliasteridae	*Heliaster helianthus*	Helianthoside	Cytotoxic	Vázquez et al. 1993
		Labidiaster annulatus	Labiasteroside A	n.d.	de Vivar et al. (1999)
	Solarestridae	[b]*Solaster borealis*	Solasteroside A, borealosides A–D, amurenoside B	Cytotoxic	Iorizzi et al. (1992)
	Zoroasteridae	*Myxoderma platyacanthum*	Myxodermoside A and 9 polyhydroxysteroids	n.d.	Finamore et al. (1991)
	Brisingidae	*Novodinia antillensis*	steroidal saponins: Sch 725737 and Sch 725739	Cytotoxic	Yang et al. (2007)
Ophiuroids	Ophiocomidae	[c]*Ophiocoma dentata*	Sulfated polyhydroxysterols	Antiviral	D'Auria et al. (1993)
		[d]*Ophiarthrum elegans*	Sulfated polyhydroxysterols	Antiviral	D'Auria et al. (1987, 1993)
		Ophiocoma erinaceus	n.d.	Hemolytic activity	Amini et al. (2014)
	Ophiopholidae	*Ophiopholis aculeata*	Sulfated polyhydroxysterols	Cytotoxic and hemolytic activity	Aminin et al. (1995)
	Ophiomyxidae	*Ophiarachna incrassata*	Sulfated polyhydroxysterols	Antiviral	D'Auria et al. (1987, 1993)
	Hemieuryalidae	*Ophioplocus januarii*	Sulfated steroids	Antiviral	Roccatagliata et al. (1996)
	Gorgonocephalidea	*Astrotoma agassizii*	Polyhydroxysterols	Antiviral	Comin et al. (1999)
	Ophiodermatidae	*Ophioderma longicauda*	Longicaudosides A–B	n.d.	Riccio et al. (1985a, 1986a)
Echinoids	Diadematidae	*Diadema savignyi*	Steroidal compounds	Anticancer	Thao et al. (2015a)
	Toxopneustidae	*Tripneustes gratilla*	Epidioxysterol	Cytotoxic	Liu et al. (2011)

[a]The accepted name changed from "*Anthenea chinesis*" to "*Anthenea pentagonula*"
[b]The accepted name changed from "*Solaster borealis*" to "*Crossaster borealis*"
[c]The accepted name changed from "*Ophiocoma dentata*" to "*Breviturma dentata*"
[d]The accepted name changed from "*Ophiarthrum elegans*" to "*Ophiomastix elegans*"

1979). Based on the hypothesis that saponins and saponin-like compounds produce various sugars upon hydrolysis (Fieser and Fieser 1956), Ward (1960) proposed that mucous-like compounds secreted from *Pteraster tessellates* have a saponin or saponin-like nature. Starfishes produce a wide range of MNPs (Table 7.2), which are largely described as lipid-like or lipid soluble molecules. Asteroids produce various steroidal derivatives, fatty acids, ceramides, and few alkaloids to either defend themselves or communicate (Table 7.1). Some of the latter compounds have been reported to possess pharmacological activities (Maier, 2008). After sea cucumbers, this group of echinoderms has also been reported to produce a large number of saponins, which have been isolated from different organs (i.e., stomach, arm, gonads, and digestive system) and possess various roles in digestion (Garneau et al. 1989; Demeyer et al. 2014), reproduction (Mackie et al. 1977) and the defense against potential predators (Harvey et al. 1987). Assessing the isolated steroidal glycosides from 1973 to 2016 revealed that most of the MNP studies on sea stars had focused on the families Asteroidea (26%), Echinasteridae (17%), Oreasteridae (16%), and Ophidiasteridae (13%; Table 7.2 and references therein).

The glycoside compounds of starfish are classified into three main groups of steroidal glycosides: asterosaponin, polyhydroxylated glycosides, and macrocyclic glycosides (Kicha et al. 2001; Maier 2008; Demeyer et al. 2014). Although steroidal glycosides are the characteristics of asteroids, triterpene glycosides have also been isolated from starfishes such as *Asterias rollestoni* (Zhan et al. 2006) and *Patiria pectinifera* (Popov et al. 2014). The isolated saponins from *A. rollestoni* (rollentosides A–B) have a similar aglycone and carbohydrate moiety than those observed in some sea cucumber species (Popov et al. 2014). Given the similar structures of rollentoside B (Zhan et al. 2006) and cucumarioside A_{15} that have been extracted from the sea cucumber *Eupentacta fraudatrix* (Silchenko et al. 2012a), it has been argued that the starfish fed on the sea cucumber (Popov et al. 2014; Fig. 7.3). Furthermore, it seems that *A. rollestoni* is

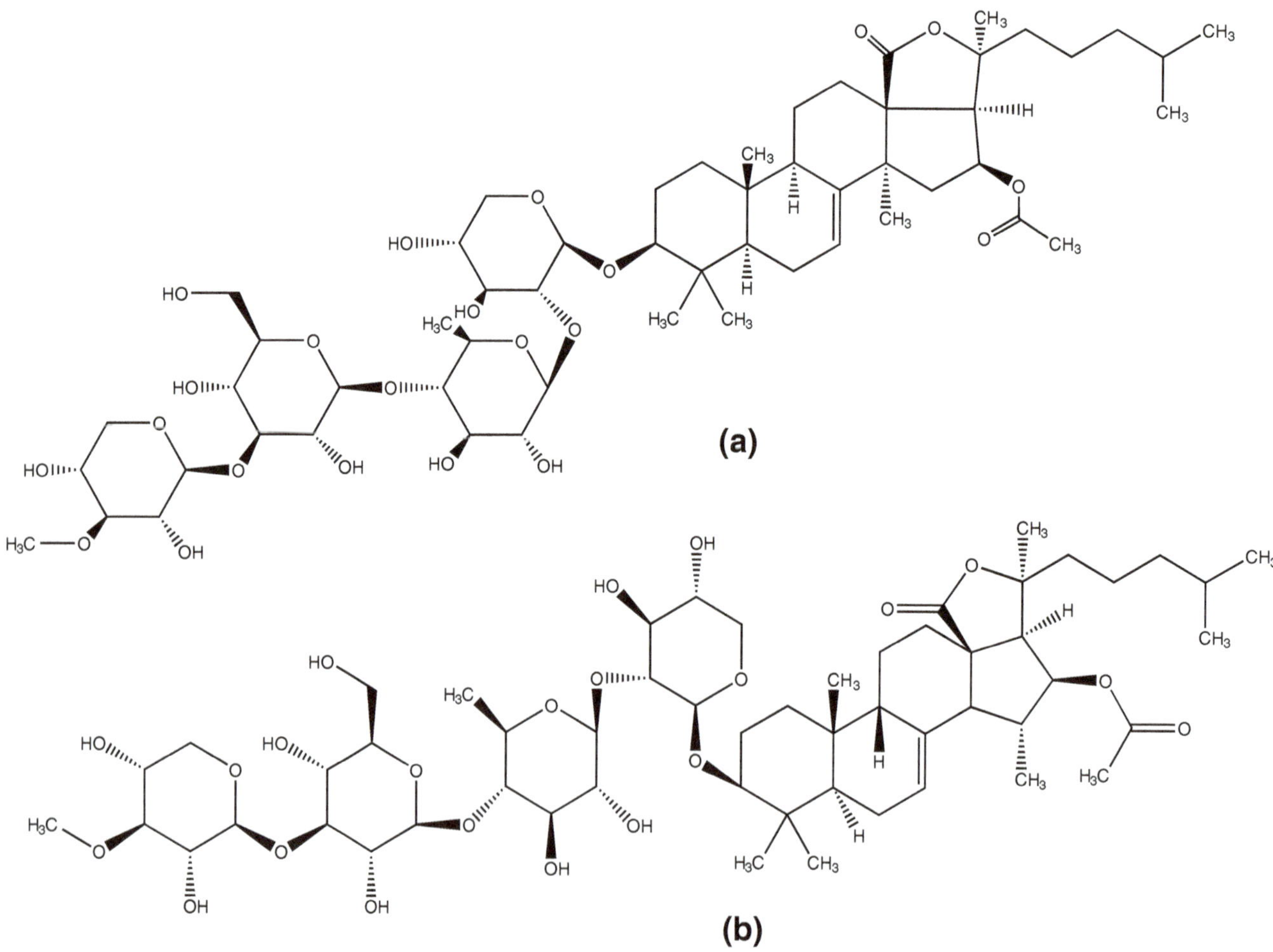

Fig. 7.3 (**a**) Rollentoside B isolated from *Asterias rollestoni* and (**b**) Cucumarioside A_{15} isolated from *Eupentacta fraudatrix* with similar chemical formula of $C_{55}H_{88}O_{22}$ (produced with ChemDraw, version 16.0.1.4 (77))

able to digest and also to accumulate the toxic triterpene glycosides that were originally derived from sea cucumbers.

7.2.3 Ophiuroids (Brittle Stars)

With over 2000 species, brittle stars are the largest group of echinoderms (Hickman et al. 2001). These organisms are widely distributed, and their feeding behavior can be suspension feeding, deposit feeding, and/or predation (Stöhr et al. 2012). Although brittle stars have numerous physical defense mechanisms such as fast locomotion, a quick removal of their extremities, and the ability to hide under rocks and crevices, some species still rely on chemical defenses. However, based on the MarineLit database, to this day only a few studies focused on ophiuroids. Nuzzo et al. (2017) mentioned that several classes of secondary metabolites such as carotenoids, gangliosides, brominated indoles, phenylpropanoids, several groups of terpenes, and steroids have been isolated from brittle stars (Table 7.1). The presence of sulfated steroids in starfish (see Sect. 7.3) and brittle stars is an indicator of the phylogenetically close relation between these two classes of echinoderms (Levina et al. 1996, 2007).

7.2.4 Echinoids (Sea Urchins)

Sea urchins, the living representative of echinoids, are free-moving echinoderms (Clemente et al. 2013). They typically have physical defense mechanisms such as fused skeleton plates, spines, and pedicellaria for pinching or capturing prey (Jangoux 1984). Some families such as Diadematidae, Echinothuriidae, and Toxopneustidae contain venoms (Thiel and Watling 2015). The main MNPs of sea urchins are proteins, polysaccharides, and pigments, which are located in the spines, testes, gonads, and/or pedicellaria (Shang et al. 2014; Jiao et al. 2015). Studies on their MNPs have mainly focused on proteins derived from naphthoquinone pigments that showed antibacterial, antioxidant, and anti-inflammatory activities. Few studies focused on steroidal components of sea urchins (Table 7.2), with the exception of *Tripneustes gratilla* (Liu et al., 2011) and *Diadema savignyi* (Thao et al. 2015a), from which several steroidal constituents had been described.

7.2.5 Holothuroids (Sea Cucumbers)

Sea cucumbers have been recognized as an interesting source of MNPs, since they are already used as traditional food and medicine source in Asian countries (i.e., healing wounds, eczema, arthritis, impotence; Ridzwan 2007; Althunibat et al. 2013). The enriched nutrition profile of sea cucumbers and their high protein, low sugar, and cholesterol-free content make holothurians a valuable food source, especially for people who suffer from hyperlipidemia (Wen et al. 2010; Bordbar et al. 2011). To date, antibacterial (Ghanbari et al. 2012; Soliman et al. 2016), antifungal (Ghannoum and Rice 1999; Soliman et al. 2016), antiviral (Mayer and Hamann 2002), antitumor and anticancer (Anisimov et al. 1973; Wu et al. 2007a; Janakiram et al. 2015; Fedorov et al. 2016), anti-schistosomal (Mona et al. 2012), and anti-inflammatory (Song et al. 2016) activities are the reported bioactive effects that were obtained from various classes of sea cucumber-derived secondary metabolites. Although a wide range of chemical classes from sea cucumbers such as peptides (Zhao et al. 2009; Song et al. 2016), polysaccharides (Liu et al. 2012; Marques et al. 2016), glycosphingolipids (Sugawara et al. 2006), polyunsaturated fatty acids (Yang et al. 2003; Hu et al. 2014b), and ceramides and gangliosides (Ikeda et al. 2009) were studied (Table 7.2), only a few products reached preclinical trials (Mayer et al. 2010).

7.3 Saponins in Echinoderms

The major group of bioactive compounds that are responsible for the biological activities of echinoderms are glycosides (Bhakuni and Rawat 2005; Dong et al. 2011). Saponins are common compounds that have been isolated from various terrestrial plants, but within the animal kingdom, they are reported only in few marine organism groups such as sponges (Kubanek et al. 2000), sea cucumbers (Yamanouchi 1955), and starfishes (Kitagawa and Kobayashi 1977). Echinoderms harbor in comparison to other marine invertebrates by far the most of the 350 reported saponin compounds.

Saponins are complex amphipathic glycosides composed of a steroid (largely found in sea stars) or triterpenoid aglycone (most commonly found in sea cucumbers) and a carbohydrate moiety (Minale et al. 1995). Saponins consist of hydrophilic (glycone) and hydrophobic (aglycone) components. The sugar moiety of saponins is mostly composed of glucose (Glc), xylose (Xyl), galactose (Gal), glucuronic acid (Glu), rhamnose (Rha), and/or methylpentose and is connected to the hydrophobic compartment (sapogenin) via glycosidic bonds. The nature of the side chains and the positions of various carbohydrate residues, or monosaccharide compositions, affect the membranotropic activities and functional properties of this chemical group.

Saponins show a broad range of bioactivities and ecological functions ranging from cytotoxic, hemolytic, antibacterial, antiviral, antifouling, antifungal, and anti-inflammatory activities, immunomodulatory effects, ichthyotoxicity, and deterrent/attractant properties for predators/symbionts (see Tables 7.2 and 7.3 for more details). Furthermore, the inter-

Table 7.3 Triterpene glycosides of different orders of holothurians and their bioactivity

Order	Family	Species	Saponin compounds	Biological activity	References
Apodida	Synaptidae	*Opheodesoma grisea*	n.d.	Hemolytic activity	Kalinin et al. (2008)
		Synapta maculata	Synaptoside A	Cytotoxic	Avilov et al. (2008)
			Synaptoside A_1	Antitumor	Avilov et al. (2008)
Elasipodida	Elpidiidae	*Kolga hyalina*	Kolgaosides A, B	Low cytotoxicity	Silchenko et al. (2014b)
		Rhipidothuria racovitzai	Achlioniceosides A_1–A_3	n.d.	Antonov et al. (2009)
Holothuriida	Holothuriidae	*Holothuria atra*	n.d.	Antifouling and antibacterial activities	Soliman et al. (2016)
			Holothurin A–B, Echinoside A–B	Antifungal	Kobayashi et al. 1991
			Ethanolic extracts	Antifungal and antibacterial	Abraham et al. (2002)
		Holothuria leucospilota	Holothurin A–B	Antimicrobial	Kitagawa et al. (1979, 1981d)
			Holothurin	Ichthyotoxic	Yamanouchi (1955)
			Leucospilotaside A–C, holothurin B, B_2	Leucospilotaside B: antitumor	Han et al. (2007b, 2008b, 2009a, 2010b)
			Holothurinoside E_1, holothurin A–B, B_3, desholothurin A, bivittoside D	Hemolytic activity	Van Dyck et al. (2010)
		n.d.	Holothurin A–B, A_2, holotoxin A_1	Immunomodulatory activity	Popov et al. (1994)
		Holothuria fuscocinerea	Holothurin	Hemolytic activity	Pocsidio (1983)
			Fuscocinerosides A–C, pervicoside C, holothurin A	Antifungal, cytotoxic	Zhang et al. 2006d
		Holothuria pulla	Holothurin	Hemolytic activity	Pocsidio (1983)
		Holothuria pervicax	Pervicosides A–C	Antifungal	Kitagawa et al. (1985, 1989)
		Holothuria mexicana	Holothurins A, B	Cytotoxic	Anisimov et al. (1980)
		Holothuria nobilis	Nobilisides A–D	Nobiliside A: antifungal, antitumor	Wu et al. (2006a, 2007a, 2009a), Zhang (2009), Guo and Xiong (2009) and Zhang and Zhu (2017)
			Echinoside A	Anticancer	Li et al. (2010)
		Holothuria scabra	24-dehydroechinoside A, echinoside A, holothurin A	Antifungal	Kobayashi et al. (1991)
			24-Dehydroechinoside A	Antitumor	Han et al. 2012
			Fusccosineroside C, echinoside A, holothurin A_1, A_4	Anticancer	Dang et al. (2007) and Han et al. (2012)
			Scabraside A,B, echinoside A, holothurin A_1	Antifungal	Han et al. (2008a, 2009b)
			Scabrasides A–D, fuscocineroside C	Antitumor	Han et al. (2009b, 2012)
			Holothurins A_3–A_4	Cytotoxic	Dang et al. (2007)
			Echinoside A and holothurin A_1–A_4	Cytotoxic	Han et al. (2009b)

(continued)

Table 7.3 (continued)

Order	Family	Species	Saponin compounds	Biological activity	References
			Crude extract	Antioxidant	Suwanmala et al. (2016)
			Ethanolic extracts	Antifungal and antibacterial	Abraham et al. (2002)
		Holothuria poli	Holothurins B_2–B_4, Holothurins A–B	n.d.	Silchenko et al. (2005c)
			methanol and aqueous extracts	Antifungal	Ismail et al. (2008)
			Bivittoside	Cytotoxic	Omran and Khedr (2015)
			Holothurinoside	Antifouling	Ozupek and Cavas (2017)
		Holothuria tubulosa	Holothurins A–B	n.d.	Silchenko et al. (2005c)
			Holothurinoside	Antifouling	Ozupek and Cavas (2017)
			Methanol and dichloromethane extracts	Anti-inflammatory	Herencia et al. (1998)
		**Holothuria fuscopunctata*	Impatienside B, arguside F, and pervicoside D	Antifungal	Yuan et al. (2009b)
			Axilogoside A, holothurin B	Antifungal	Yuan et al. (2008)
			Desulfated glycosides	Antifungal	Kobayashi et al. (1991)
		Holothuria forskali	Holothurinosides A–D	Antitumor, antiviral	Rodriguez et al. (1991)
		Holothuria floridana	n.d.	hypothermic, and hemolytic activities	Kaul (1986)
			Holothurins A_1–A_2, B_1	n.d.	Kuznetsova et al. (1982b), Oleinikova et al. (1982), and Oleinikova and Kuznetsova (1983)
			Holothurin A_1	Few inhibition of Na^+/K^+-ATPase activity	Gorshkova et al. (1989)
		Pearsonothuria graeffei	Disulfated Holothurin A	Hemolytic activity	Van Dyck et al. (2010)
			Desholothurin A, holothurinoside C	Hemolytic activity	Van Dyck et al. (2010)
			Holothurinoside J_1	Hemolytic activity	Van Dyck et al. (2010)
			Disulfated echinoside A	Stimulator of hepatic fatty acid β-oxidation and suppression of FA biosynthesis/anticancer	Zhao et al. (2011, 2012) and Wen et al. (2016)
			Disulfated echinoside A	Antimetastatic activity	Zhao et al. (2011)
			Echinoside A, disulfated echinoside A	Antitumor	Zhao et al. (2012)
			n.d.	Antihyperlipidemic activity	Hu et al. (2010)
		Holothuria grisea	Holothurin A_1	n.d.	Oleinikova et al. (1982)
			Griseaside A, 17-dehydroxyholothurinoside A	Cytotoxic	Yi et al. (2008)
		Holothuria edulis	Holothurin A	Anti-fungal	Kobayashi et al. (1991)
			Ethyl acetate fraction	Anti-inflammatory	Wijesinghe et al. (2015)
		Holothuria hilla	Hillasides A–C	Antitumor, cytotoxic	Wu et al. (2006b, 2007b, 2009b)
		Holothuria lessoni	Holothurinoside A_1, E_1	n.d.	Bahrami et al. (2014)
			Lessonioside A–D	Acetylated saponin	Bahrami et al. (2014) and Bahrami and Franco (2015)
			Lessonioside E–G, M	Nonacetylated saponin	Bahrami and Franco (2015)
			Holothurinoside X–Z	n.d.	Bahrami et al. (2014)

		Holothuria moebii	Sulfated and desulfated saponins	Cytotoxic	Yu et al. (2015)
		Holothuria sp.	n.d.	Antiviral	Farshadpour et al. (2014)
		Holothuria impatiens	Impatienside A[a], bivittoside D	Cytotoxic, Antitumor	Sun et al. (2007)
		Bohadschia argus	Bivittoside types	Antitumor	Kuznetsova et al. (1982a)
			Argusides A–E	Cytotoxic	Liu et al. (2007, 2008a, b)
			Holothurin C	Inhibition of Na^+/K^+-ATPase activity	Gorshkov et al. (1982)
		Bohadschia bivittata	Bivittosides A–D	Antifungal	Kitagawa et al. (1981c)
			Bivittosides A–B	Inhibition of Na^+/K^+-ATPase activity	Kitagawa et al. (1981c) and Gorshkova et al. (1989)
		Bohadschia vitiensis	Bivittoside D	Antiviral, anti-fungal, and spermicide	Lakshmi et al. (2008, 2012) and Maier (2008)
		Bohadschia marmorata	Impatiensides A[a]–B, marmortosides A–B, 25-acetoxybivittoside D, bivittoside D	Antifungal	Yuan et al. (2009a)
			Bivittosides	Antitumor	Kuznetsova et al. (1982a)
		Bohadschia cousteaui	Coustesides A–J	Antifungal	Elbandy et al. (2014)
		Bohadschia graeffei	Holothurin A, echinoside A	Antifungal	Kobayashi et al. (1991)
		Bohadschia subrubra	Impatienside A[a]; bivittoside C, D; araguside C; holothurinoside F, H, H_1, I, I_1, K_1	Hemolytic	Van Dyck et al. (2010)
		Actinopyga agassizi	Holothurin A	Ichthyotoxic	Chanley et al. (1959)
			24-Dehydroechinoside A	Ichthyotoxic	Kalinin et al. (2008)
			Holothurin	Antitumor	Sallivan et al. (1955)
			Holothurin	Antiparasitic (against *Trypanosoma lewisi*)	Styles (1970)
			Holothurin	Mitogenic activity	Nigrelli and Jakowska (1960)
			n.d.	Antibacterial, immunomodulatory effect	Kalinin et al. (2008)
		Actinopyga lecanora	n.d.	Hemolytic activity	Poscidio (1983)
			Holothurins A–B	Antifungal	Kumar et al. (2007)
			Holothurins A, A_1, B, lecanorosides A and B	Antitumor	Zhang et al. (2008)
			Holothurin A, holothurin B	Antiparasitic	Singh et al. (2008)
			n.d.	Antibacterial	Ghanbari et al. (2012)
		Actinopyga echinites	Echinosides A–B	Antifungal, antischistosomal	Kitagawa et al. (1980) and Melek et al. (2012)
			Ethanolic extracts	Antifungal and antibacterial	Abraham et al. (2002)

(continued)

Table 7.3 (continued)

Order	Family	Species	Saponin compounds	Biological activity	References
		Actinopyga mauritiana	Echinosides A–B, 24- ehydroechinosides A–B	Antifungal	Kobayashi et al. (1991)
			n.d.	Antineoplastic and cytotoxic	Pettit et al. (1976)
		Actinopyga flammea	Echinosides, Holothurinogenins	Antitumor, antifungal	Bhatnagar et al. (1985) and Mondol et al. (2017)
		Actinopyga miliaris	Ethanolic extracts	Antifungal and antibacterial	Abraham et al. (2002)
Dendrochirotida	Cucumariidae	*Cucumaria japonica*	Cucumarioside	Immunomodulatory effect	Polikarpova et al. (1990)
			Cucumarioside	Antifungal	Batrakov et al. (1980)
			Cucumarioside	mitogenic Antiproliferative activity	Turischev et al. (1991)
			Cucumarioside $A_{4\text{-}2}$	Hemolytic activity, immunomodulatory effect, antiviral	Aminin et al. (2001) and Kalinin et al. (2008)
			Cucumarioside A_2–2	Hemolytic, cytotoxic, inhibition of Na/K+-ATPase activity, immunomodulatory effect, antiviral, antitumor	Avilov et al. (1991b), Kalinin et al. (1996), Aminin et al. (2001), Agafonova et al. (2003), Menchinskaya et al. (2014), and Pislyagin et al. (2017)
			Cucumariosides $A_{1\text{-}2}$, $A_{2\text{-}3}$, $A_{2\text{-}4}$, $A_{4\text{-}2}$	n.d.	Avilov et al. (1991b)
			Cucumarioside A_3	Hemolytic, immunomodulatory effect	Aminin et al. (2001)
			Cucumariosides $A_{7\text{-}1}$, $A_{7\text{-}2}$, $A_{7\text{-}3}$	Hemolytic, cytotoxic, immunomodulatory effect	Drozdova et al. (1993), Kalinin et al. (1996), Aminin et al. (2001), and Agafonova et al. (2003)
			Cucumarioside A_6–2	Antitumor, cytotoxic, hemolytic, immunomodulatory effect	Kalinin et al. (1996), Drozdova et al. (1997), and Aminin et al. (2001)
			Cucumariosides $A_{0\text{-}1}$, $A_{0\text{-}2}$, $A_{0\text{-}3}$	n.d.	Drozdova et al. (1993)
			Cucumarioside A_3	Antitumor, hemolytic	Kalinin et al. (1996) and Drozdova et al. (1997)
			Cucumarioside G_1	Inhibition of Na/K+-ATPase activity	Anisimov et al. (1983)
			Cumaside	Immunomodulatory and hemolytic effect	Aminin et al. (2006)
			Cucumarioside	Immunomodulatory effect, antibacterial, Antiviral	Sedov et al. (1984, 1990), Grishin et al. (1990), and Aminin (2016)
		Cucumaria frondosa	Frondoside A	Antiproliferative effects, Antitumor, anticancer, Immunomodulatory effect	Al Shemaili et al. (2014), Girard et al. (1990), Al Marzouqi et al. (2011), Attoub et al. (2013), Ma et al. (2012), and Aminin et al. (2008)
			Frondoside D	n.d.	Yayli and Findlay (1999)
			Frondoside C	Antitumor	Avilov et al. (1998)
			Frondosides B, $A_{2\text{-}1}$–$A_{2\text{-}8}$	n.d.	Findlay et al. (1992) and Silchenko et al. (2005a, b)
			Frondoside $A_{7\text{-}2}$, $A_{7\text{-}3}$, $A_{7\text{-}4}$, isofrondoside C	n.d.	Silchenko et al. (2007a)

		**Cucumaria echinata*	Cucumechinosides A–F	Antifungal, anticancer, antiprotozoal	Miyamoto et al. (1990b)
			Cucumechinol A–C	n.d.	Miyamoto et al. (1990a)
			Disulfated Penaustrosides A–B	n.d.	Miyamoto et al. (1992)
			CEL-I	Hemolytic activity	Hatakeyama et al. (1999)
			CEL-III	Hemolytic activity	Oda et al. (1999)
		Cucumaria fallax	Fallaxosides B_1, C_1– C_2, D_1–D_7	Cytotoxic, hemolytic	Silchenko et al. (2016a)
		Cucumaria okhotensis	Okhotosides B_1–B_3	Antitumor, cytotoxic	Silchenko et al. (2008)
			Okhotosides $A_{2\text{-}1}$, $A_{1\text{-}1}$, B_1–B_3	Immunomodulatory activity, cytotoxic	Silchenko et al. (2007b, 2008) and Aminin et al. (2010)
			Frondosa A_1	Immunomodulatory activity	Aminin et al. (2010)
		Cucumaria Conicospermium	Cucumarioside $A_{2\text{-}5}$, $A_{3\text{-}2}$, $A_{3\text{-}3}$, Isokoreoside A, koreoside A	n.d.	Avilov et al. (2003)
		Cucumaria lefevrei	Lefevreosides A_1, A_2, C,D	n.d.	Rodriguez and Riguera (1989)
		Cucumaria koreaensis	Koreoside A	n.d.	Avilov et al. (1997)
		Cucumaria miniata	Cucumarioside $A_{7\text{-}3}$	n.d.	Drozdova et al. (1997)
		Mensamaria intercedens	Intercedensides A–C, D–I	Antitumor, cytotoxic	Zou et al. (2003, 2005)
		Hemioedema spectabilis	Hemoiedemosides A–B	Cytotoxic, antifungal	Chludil et al. (2002a)
		**Pentacta quadrangularis*	Phylinopside E (PE)	Antitumor, Antiangiogenesis, cytotoxic	Tian et al. (2005, 2007)
			Phylinopsides A,B, E, F	Cytotoxic	Yi et al. (2006) and Zhang et al. (2006a)
			Phylinopsides A–B, pentactaside I, II, and III	Cytotoxic	Han et al. (2010a)
			Pentactaside B, C	Antitumor	Han et al. (2010c)
		Pentacta australis	Desulfated penaustrosides A–D	n.d.	Miyamoto et al. (1992)
		Actinocucumis typica	Typicosides A_1, A_2, B_1	Immunomodulatory effect, cytotoxic	Pislyagin et al. (2014)
			Typicosides A_1, A_2, B_1, C_1, C_2, intercedenside A, holothurin B_3	Antifungal, cytotoxic	Silchenko et al. (2013b)
		Colochirus robustus	Colochiroside E	n.d.	Silchenko et al. (2016c)
			Colochirosides D, A_1–A_3, B_1–B_3	Cytotoxic, hemolytic activity	Silchenko et al. (2015a, 2016c, b)
		***Cercodemas anceps*	Colochiroside A	Antitumor	Cuong et al. (2015)
			Cercodemasoides A–E	Cytotoxic	Cuong et al. (2015)
		**Pseudocnus dubiosus leoninus*	Pseudocnoside A	Anticancer, antiproliferative	Careaga et al. (2014)

(continued)

Table 7.3 (continued)

Order	Family	Species	Saponin compounds	Biological activity	References
		Staurocucumis liouvillei	Liouvillosides A–B	Cytotoxic, antiviral	Maier et al. (2001)
			Liouvillosides A_1–A_5, B, B_2	n.d.	Antonov et al. (2008, 2011)
		***Staurocucumis turqueti*	Turquetoside A	n.d.	Silchenko et al. (2013d)
		Pseudocolochirus violaceus	Violaceusides A, B, C, E, I–III	Cytotoxic	Zhang et al. (2006b, c) and Silchenko et al. (2014a)
			Violaceuside D, G	Cytotoxic	Silchenko et al. (2014a)
		**Duasmodactyla kurilensis*	Kurilosides A, C	n.d.	Avilov et al. (1991a)
	Sclerodactylidae	*Eupentacta fraudatrix*	Cucumarioside G_1	Cytotoxic, hemolytic, inhibition of Na/K+-ATPase activity	Gorshkov et al. (1982), Afiyatullov et al. (1985), and Kalinin et al. (2008)
			Cucumarioside G_2	Hemolytic activity	Avilov et al. (1994)
			Cucumariosides H_1–H_8	Hemolytic activity, cytotoxic	Silchenko et al. (2012c)
			Cucumarioside C	Cytotoxic	Anisimov et al. (1974)
			Cucumariosides A_1–A_{10}, A_{14}, A_{15}, A_8	Antifungal, hemolytic activity	Melek et al. (2012) and Silchenko et al. (2012a, b)
			Cucumarioside B_2	Antifungal	Melek et al. (2012) and Silchenko et al. (2012a, b, d)
			Cucumariosides G_1, G_2, G_4, $G_{4\text{-}A}$	Hemolytic activity	Kalinin et al. (1992a, b, 2008)
			Cucumariosides F_1, F_2	n.d.	Popov et al. (2014)
			Cucumariosides I_1–I_3	Cytotoxic and immunostimulatory activities	Silchenko et al. (2013a, b, c)
			Cucumarioside B_1	Hemolytic activities, antifungal	Melek et al. (2012) and Silchenko et al. (2012a, b, d)
			Cucumariosides A_1–A_{15}	Cytotoxic	Silchenko et al. (2012a)
		**Cucumaria fraudatrix*	Cucumariosides G_1, G_3, $G_{3\text{-}A}$	Cytotoxic, hemolytic activity	Afiyatullov et al. (1985) and Popov (2002)
		Cladolabes schmeltzii	Cladolosides A_1–A_6, B_1–B_2, C, C_1–C_4, D, D_2, E_1, E_2, F_1, F_2, G, H_1, H_2, J_1, K_1, K_2, L_1, M, M_1, M_2, N, I_2, O, P, P_1, P_3, Q, R	Cytotoxic	Silchenko et al. (2013e, 2014c, 2015b, 2017a, 2018a, b)
	Psolidae	*Psolus fabricii*	Psolusoside A	Hemolytic activity	Kalinin et al. (1996)
			Psolusosides A, B	Inhibition of Na^+/K^+-ATPase activity	Kalinin et al. (1989a) and Gorshkova et al. (1999)
		Psolus eximius	Eximisoside A	n.d.	Kalinin et al. (1997)
		Psolus patagonicus	Patagonicosides A–C	Antifungal, cytotoxic	Murray et al. (2001), Muniain et al. (2008) and Careaga et al. (2011)
	Phyllophoridae	*Pentamera calcigera*	Calcigerosides B, C_1, C_2	Cytotoxic	Avilov et al. (2000b)
			Cucumarioside G_2	No cytotoxicity	Avilov et al. (2000b)
			Calcigerosides D_1–D_2, E	n.d.	Avilov et al. (2000a)
		Neothyonidium magnum	Magnumosides A_1–A_4, B_1–B_4, C_1–C_4	Cytotoxic, hemolytic	Silchenko et al. (2017b)

Synallactida	Stichopodidae	***Apostichopus japonicus*	Holotoxin A, B, C	Antifungal	Kitagawa et al. (1976) and Maltsev et al. (1984, 1985)
			Holotoxin A_1	Antiprotozoal, antitumor	Kitagawa et al. (1976) and Anisimov et al. (1983)
			Holotoxin A_1, cladoloside	Neurotoxic, cytotoxic	Yun et al. (2018)
			Holotoxin A_1	Cytotoxic, antiproliferative activity	Ishida et al. 1993 and Popov et al. 1994
			Holotoxins A_1, B_1	Inhibition of Na^+/K^+-ATPase activity	Maltsev et al. (1984) and Gorshkova et al. (1989)
			Apostichoposide C	Inhibition of Na^+/K^+-ATPase activity	Gorshkova et al. (1989)
			Holotoxin A_1, B_1 and holothurin A	Contraceptive effect	Mats et al. (1990)
			Cladoloside B	Antifungal	Wang et al. (2012)
			Holotoxins A_1, B_1, D_1, A	Antifungal	Wang et al. (2012)
			25,26-Dihydroxy-holotoxin A_1	Antifungal	Wang et al. (2012)
			Holotoxins D–G	Antifungal	Wang et al. (2012)
			(Nortriterpene glycoside) 26-nor-25-oxo-holotoxin A_1	Antifungal	Wang et al. (2012)
			Holotoxins F, G, H	Antifungal	Liu et al. (2012) and Wang et al. (2012)
		**Stichopus japonicus*	Stichopogenin A_4, A_2 (genuine aglycone holotoxin A)	Antifungal	Kitagawa et al. (1976)
			Crude saponin	Antioxidant	Husni et al. (2009)
		Stichopus chloronotus	Chloronoside A, C, D, E	Antimicrobial, antifungal, cytotoxic, antitumor	Anisimov et al. (1983) and Maltsev et al. (1985)
			Stichoposides C, D, E	Inhibition of Na/K+-ATPase activity	Gorshkova et al. (1989)
			Stichlorosides A_1–A_2, B_1, B_2, C_1, C_2	Antifungal	Kitagawa et al. (1981a, b)
			Stichoposides A, B	n.d.	Sharypov et al. (1981)
			n.d.	Antineoplastic and cytotoxic	Pettit et al. (1976)
		Stichopus horrens	Stichorrenosides A–D, stichoposide A	Cytotoxic	Cuong et al. (2017)
			Stichorrenosides E	Cytotoxic	Vien et al. (2018)
		**Stichopus variegatus*	Variegatusides A–F, holothurin B	Antifungal	Wang et al. (2014)
		Stichopus hermanni	Stichlorosides A_1, A_2, B_1, B_2, C_1, C_2	Antifungal	Kobayashi et al. (1991)
		**Stichopus parvimensis*	Parvimosides A, B	n.d.	De Moncerrat Iñiguez-Martinez et al. (2005)
		**Stichopus multifidus*	Astichoposide C	Inhibition of Na\K+-ATPase activity	Gorshkov et al. (1982)
		Thelenota ananas	Thelenotosides A–B	Antifungal	Stonik et al. (1982) and Maltsev et al. (1985)
			Saponin compounds	Antitumor	Pettit et al. (1976)
			Thelenotoside A	Immunomodulatory effect	Gorshkova et al. (1989)
			Stichlorosides A_1, A_2, B_1, B_2, C_1, C_2	Antifungal	Kobayashi et al. (1991)

(continued)

Table 7.3 (continued)

Order	Family	Species	Saponin compounds	Biological activity	References
			Telothurins A–B	Antitumor	Kuznetsova et al. (1982a)
			n.d.	Antineoplastic and cytotoxic	Pettit et al. (1976)
			n.d.	Antiviral	Hegde et al. (2002)
		Thelenota anax	Stichlorosides A_1, B_1, C_1	Antifungal	Kobayashi et al. (1991)
			Telothurin	Antitumor	Kuznetsova et al. (1982a)
			Stichoposides C–D	Antitumor, anticancer	Yun et al. (2012) and Park et al. (2014)
		Australostichopus mollis	Neothyonidioside	Antifungal	Yibmantasiri et al. (2012)
			Mollisosides A, B_1– B_2	n.d.	Moraes et al. (2005)
	Synallactida	*Synallactes nozawai*	Synallactosides A_1, A_2, B_1, B_2, C	n.d.	Silchenko et al. (2002)
Persiculida	Pseudostichopodidae	*Pseudostichopus trachus*	Pseudostichoposides A,B	n.d.	Kalinin et al. (1989b) and Silchenko et al. (2004)

[a]Marmoratoside A = impatienside A (Van Dyck et al. 2010)
*Based on WoRMS (2019), the accepted names changed from: *Stichopus multifidus* to *Astichopus multifidus; Neothyonidium magnum* to *Massinium magnum; Pseudocnus dubiosus leoninus* to *Pentactella leonina; Cucumaria fraudatrix* to *Eupentacta fraudatrix; Holothuria axiologa* to *Holothuria fuscopunctata; Penctata quadrangularis* to *Cholochirus quadrangularis; Stichopus variegatus* to *Stichopus horrens; Stichopus parvimensis* to *Apostichopus parvimensi; Duasmodactyla kurilensis* to *Thyonidium kurilensis; Pentacta quadrangularis* to *Colochirus quadrangularis; Bohadschia bivittata* to *Bohadschia vitiensis; Cucumaria echinata* to *Pseudocnus echinatus; Stichopus japonicus* to *Apostichopus japonicus*
**Synonymised names: *Staurocucumis turqueti = Cucumaria turqueti/Cucumaria spatha; Cercodemas anceps = Colochirus anceps*

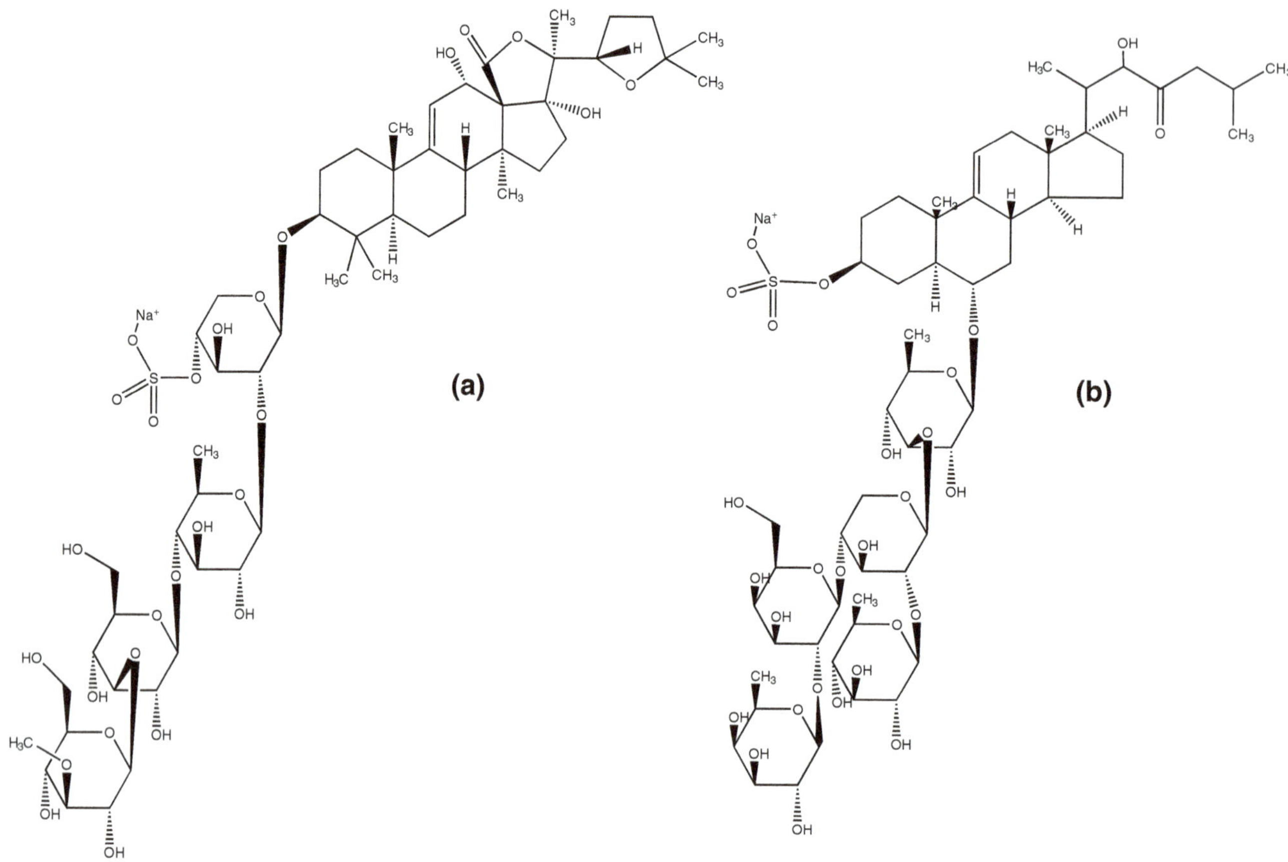

Fig. 7.4 Examples of (**a**) a triterpene glycoside structure: Holothurin A isolated from the sea cucumber *Holothuria leucospilota* (Kitagawa et al. 1981d) and (**b**) a steroidal glycoside structure: Thornasteroside A isolated from the sea star *Acanthaster planci* (Kitagawa and Kobayashi 1978) (produced with ChemDraw, version 16.0.1.4 (77))

actions between aglycone components (i.e., sapogenin) and sterols of the cell membranes can result in a saponification process that may lead to cell lysis (Bahrami et al. 2016).

The sulfate group seems to be one of the most essential groups in most saponins derived from ophiuroids, asteroids (Table 7.2), and holothuroids (Table 7.3). However, there is a basic difference in the position of this functional group between echinoderms (Fig. 7.4). For both sea stars and brittle stars, the sulfate group is located in the hydrophobic part (aglycone) of the molecule, whereas in holothurians the sulfate group is placed within the hydrophilic moiety (glycone) (Kornprobst et al. 1998). The structural differences of asterosaponin and triterpene glycosides showed that not only the presence but also the position of the sulfate groups may be important, resulting in potentially different biological activities of saponins (Maier 2008; Malyarenko et al. 2015).

As the sea cucumbers contain the highest variety of saponin species, we will next (see Sect. 7.3.1) focus on the distribution and function of triterpene glycosides that have been reported exclusively from holothurians.

7.3.1 Structural Diversity of Saponins in Holothuroids

The first report of polar and low volatile triterpene glycosides within the animal kingdom was in 1952 and originated from a sea cucumber extract (Nigrelli and Zahl 1952).The initial studies on the bioactive properties of compounds derived from sea cucumbers explained the ichthyotoxic activities of saponins, which were extracted from the body wall and the CTs of *Holothuria leucospilota* and *Actinopyga agassizi* (Nigrelli and Jakowska 1960; Yamanouchi, 1955). Most of the subsequently identified saponins were mainly isolated from three families of sea cucumbers: Holothuriidae, Stichopodiidae, and Cucumariidae (see Table 7.3).

The chemical structure of saponins in holothurians can be very complex in terms of the presence/absence and position of different functional groups (e.g., hydroxyl groups), which may differentiate them from other echinoderms as well as from each other marine invertebrates (Bahrami et al. 2014). The generic name of holothurian-derived saponins is Holothurin, which are nearly all 3β-glycosylated saponins

(Kornprobst et al. 1998). In most sea cucumbers, triterpene glycosides contain the aglycone lanosterol with an 18(20)-lactone (e.g., holostane 3β-ol; Kalinin 2000; Caulier et al. 2011) and an oligosaccharide chain that consists of D-Xyl, D-Quinov, D-Glc, D-3-O-methyl-Glc, and D-3-O-methyl-Xyl (Caulier et al. 2011; Bahrami et al. 2016).

Triterpene glycosides exhibit different bioactivities, which might aid the likelihood of survival for its producing organisms. This is also highlighted by their broad bioactivities as well as their broad ecological functions (e.g., anti-predatory defense). Although the structure of each unit affects the bioactivity of the compound, linear oligosaccharide structures (i.e., tetraosides) have shown to be the optimum quantity of monosaccharide units in the glycoside (Minale et al. 1995; Kalinin et al. 2008). Furthermore, allelopathic properties of saponins, as well as the presence of various functional groups like amides, hydroxyl groups, acetyl groups, and sulfate groups in different species of sea cucumber, can inhibit larval attachment of macroorganisms and also affect the growth of different strains of gram-positive and gram-negative bacteria (Soliman et al. 2016). By changing the hydrophobic-hydrophilic balance of bacterial cells, extracted saponins may affect permeability and stability of the bacterial cell wall, which in turn can ultimately lead to cellular death (Lawrence et al., 1957; Soliman et al. 2016). Additionally, due to their hydrophilic properties, saponins regulate oocyte maturation and can thus affect the reproduction cycle of organism by synchronizing the maturation process (Kalinin et al. 2008).

The vast chemical diversity of saponin in sea cucumbers makes them effective models for studying their biochemical evolution and applying these compounds as potential holothurian chemotaxonomic markers (Kalinin et al. 1996, 2008; Kalinin 2000). Depending on the taxonomic group of sea cucumbers, the number, composition, and location of monosaccharide units, and position of functional groups in the holostane skeleton (i.e., hydroxyl, acetylate, sulfate, double bonds, etc.) may affect the bioactivity of the compounds (Stonik 1986; Kalinin 2000). For example, the presence of trisulfated glycosides in members of the family Cucumariidae is unique for this taxonomic group (Bahrami et al. 2016). Recent chemotaxonomic analysis supported the evolution of saponins in both glycone and aglycone moieties.

The general trend of glycone evolution in Holothuroidea is from non-sulfated to sulfated compounds. Bondoc et al. (2013) studied saponins from three species of Holothuroidea by using MALDI[2]-FTICR[3] MS[4] and nano-HPLC[5]-chip Q-TOF[6]-MS, and by applying maximum likelihood analysis, molecular biology, and evolutionary software packages, they created mass chemical and genetic fingerprints of saponins. They concluded that evolution of saponins leads to glycone parts with higher membranolytic activities and hydrophilicity with lower metabolic cost (Kalinin and Stonik 1996; Bondoc et al. 2013; Kalinin et al. 2015). Therefore, the glycone evolution of Holothuroidea was likely in the following order (Kalinin et al. 2016):

1. Transition from non-sulfated to sulfated hexaoside or pentaosides
2. Changing from hexaoside and pentaosides to linear tetraosides and biosides:
   (a) Carbohydrate contains sulfate group at C-4 of first xylose unit.
   (b) Shifting sugars with C-6 Glc and 3-O-methyl-Glc to sulfated at C-4 of first xylose

Kalinin et al. (2015) mentioned that sulfated tetraosides are a common characteristic of Holothuria and Actinopyga; however, sea cucumbers of the genus *Bohadschia* contain both non-sulfated and sulfated carbohydrate units (i.e., hexaosides and tetraosides). Bivittoside D extracted from *Bohadschia vitiensis* is a hexaoside non-sulfated glycoside that evolved to a sulfated tetraoside (Holothurin A_2), which has been also found in *Holothuria scabra* (Dang et al. 2007) and *Pearsonothuria graeffei* (Zhao et al. 2011). Further structural modification leads to compounds with two monosaccharides (i.e., biosides such as echinoside B) from *Holothuria leucospilota* (Han et al. 2009a) and *Actinopyga echinites* (Kitagawa et al. 1985). The general direction of aglycone evolution is more complicated and depends on the presence or absence of lactone, keto, hydroxyl groups, as well as position of double bonds (Kalinin et al. 2015):

1. Presence/absence of lactone: It shifts from lanostane derivatives without lactone to lanostane with an 18(16)-lactone and holostane with an 18(20)-lactone.
2. Shifting the position of double bonds and the keto group. In general, transition of aglycones occurs from low oxidation to higher oxidized compounds.
   (a) Transition of aglycone compounds having a 7(8) double bond, and a carbonyl group at C-16, to compounds oxidized at C-22 or C-23 without the oxygen at C-16
   (b) Transition of aglycone molecule from 9(11) double bond and C-16 keto group to compounds having oxygen at C-16, and then to compounds without oxygen, but containing a 12α-hydroxyl group

[2] Matrix-assisted laser desorption/ionization.

[3] Fourier transformation cyclotron resonance.

[4] Mass spectrometry.

[5] Nano-high-performance liquid chromatography.

[6] Quadrupole time-of-flight.

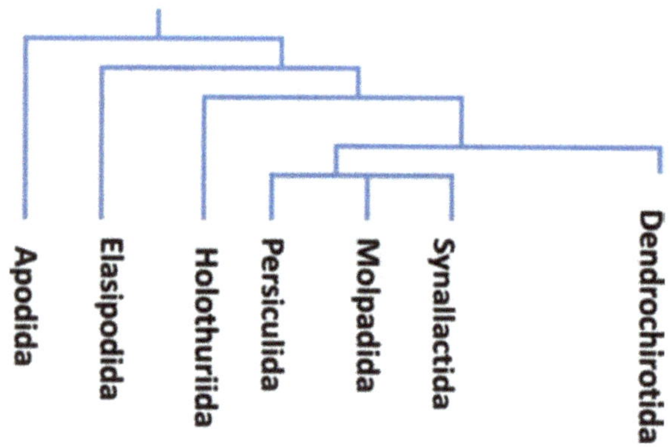

Fig. 7.5 Phylogeny of Holothuroids. Produced based on Miller et al. (2017). Holothuriida is the new accepted name for the order of Aspidochirotida

Overall, based on morphological, molecular, and paleontological analysis, there has been a clear evolutionary distance between Apodida and other species of the orders Dendrochirotida and Holothuriida (Fig. 7.5; Avilov et al. 2008). Several studies reported that the presence of the 3-O-methyl group in the terminal monosaccharide units of holothurians (*Psolus fabricii*, *Cucumaria japonica*, *Hemoidema spectabilis*, etc.) increased the membranolytic activities of the compound. Kalinin et al. (2008) described that during evolution of the terminal monosaccharide unit from glucoronic acid (GlcA) to Glc, the 3-O-methyl group was conserved due to the protective properties against predatory fish.

A unique group of sea cucumbers are the Synallactida. They are mostly epibenthic and their remarkable defense behavior is shedding (Kropp 1982). Their typical chemical defenses are holotoxins, stichoposides, and stichlorosides (Table 7.3). The common characteristics of stichoposides and holotoxins are the presence of a double bond at C-25 (C-26), while the presence of α-acetoxy group at C-23 and a 3-O-methyl-D-Glc in their polysaccharide chain are another feature of stichoposides. The presence of a keto-group at C-16 is observed for most holotoxins. Interestingly, there is a sulfate group present in stichoposides (Mondol et al. 2017). Thus, the presence of a particular aglycone or glycone glycoside can be a taxonomic marker for certain genera such as the genera *Bohadschia, Pearsonothuria*, and *Actinopyga* (Kalinin et al. 2016). The presence, expellability, and stickiness of CTs of Holothuriidae (i.e., *Bohadschia argus*, *Holothuria forskali*) affect the chemical diversity of triterpene glycosides of the sea cucumbers (Honey-Escandón et al. 2015). Among Holothuriidae, the genus *Bohadschia* is considered a more primitive group since it contains well-developed CTs with expellability and stickiness and possesses non-sulfated and less-oxidized glycosides in both the CT and body wall (Kalinin et al. 1996, 2008; Honey-Escandón et al. 2015). In contrast, more sulfated and oxidized glycosides have been reported within species without CTs or with dysfunctional CTs such as *Holothuria hilla* and *Actinopyga echinites* (Honey-Escandón et al., 2015). However, members of Dendrochirotida and Apodida also showed different patterns. Species of the order Apodida such as *Synapta maculata* are considered the most primitive group of Holothurians. They contain 3-O-methyl Glc-A in a carbohydrate chain and an 8(9) double bond in the aglycone moiety, which affects their membranolytic activity and hydrophilicity of the glycosides (Avilov et al. 2008).

7.4 Discussion and Conclusions

Predation, the biological interaction where a predator eats its prey, is a main driving force for community structure and ecosystem organization (Duffy and Hay 2001). It has been suggested that before the development of physical defenses, echinoderms used initially maternally derived chemical defenses from early larval stages to protect themselves against predators (Iyengar and Harvell 2001). Therefore, secondary metabolites play an important role in chemical defense of marine sessile and slow moving organisms and thus may affect and shape the community structure and increase the level of biodiversity of the ecosystem (Paul et al. 2007). Unfortunately, there is still a lack of information with regard to the ecological function of many MNPs, especially from echinoderms, while various pharmacological activities (e.g., antiviral, antitumor) have been widely reported. This represents a research opportunity for chemical ecologists who want to investigate how small modifications in molecules can affect ecological functions and community structure.

As summarized in Table 7.2, echinoderms have proven to be a rich source of bioactive compounds with most reported compounds in Asteroids and Holothuroids reported as saponins. Although various steroidal compounds of starfishes have been reported, only a few studies have investigated the biological activities of these compounds. Within ophiuroids, steroidal compounds, terpenes, and carotenoids have been isolated, and their mode of action has been summarized as antiviral and antitumor activities (Table 7.2).

The class Holothuria is a particularly rich source of MNPs with a multitude of reported activities. In the past decades, sea cucumbers have been increasingly harvested and consumed due to their nutritional values (high protein, low sugar, and no cholesterol (Liu et al. 2007, 2002; Wen et al. 2010) and their use in traditional medicine. Although a wide spectrum of bioactivities such as cytotoxic, hemolytic, antifungal, and immunomodulatory properties have been described for different sea cucumbers, in the extraction and compound purification process, often compounds with different chemical structures were combined, and thus the biological function of the individual compounds remain largely unknown. Therefore, their pharmaceutical potential has not yet been fully explored, which make them still promising candidates for the discovery of future MNPs with novel pharmaceutical applications. Furthermore, past studies focused largely on shallow-water holothurians, whereas deep-water specimens encounter particular harsh physicochemical conditions. Such conditions include strong hydrostatic pressure, low temperature, and possibly oxygen shortage, which could affect formation, structure, gene regulation, and biosynthesis of secondary metabolites, thus making deep-water specimens a potential interesting target for future MNP screening campaigns.

Saponins are highly diverse, common, and abundant MNPs in echinoderms. Among this group of the secondary metabolites, holothurins, holotoxins, cucumariosides, and echinosids are the most abundant compounds in various genera of sea cucumbers (Table 7.3). Most of the reported triterpene glycosides in sea cucumbers showed cytotoxicity as well as antifouling, antifungal, and antibacterial effects of saponins (Miyamoto et al. 1990b; Aminin et al. 2015; Soliman et al. 2016), providing sea cucumbers with an effective chemical defense mechanism against microbial attacks, fouling organisms, and potentially predators.

The principal mechanisms for the bioactivities of triterpene glycosides are most likely changing membranolytic effects and increased hydrophilicity of the compounds, which may not only affect their bioactivities but also make them potential trophic and taxonomic markers. Depending on the marine habitat and the defensive responses of holothurians, each group contains their own special mixture of saponins, which are often unique chemical signatures and thus can be used in chemotaxonomy to differentiate most holothurians at the family level. Furthermore, by studying structure-activity relationships (SAR), taxonomists may be able to predict physiological differences and their ecological role within the organisms.

Defense responses of holothurians vary at order or family levels, which is to some extend reflected in the stereochemistry of the saponins. The general evolution of aglycone is based on the presence/absence or position of lactone, keto, hydroxyl groups, and double bonds, which leads from low oxidized to more oxidized compounds. The direction of glycone evolution depends on the presence/absence or number and position of sulfate and acetoxy groups, type of sugar units and their (non)linear structure, as well as position of methyl group. For example, Apodida are considered as the most primitive sea cucumbers due to the presence of 3-O-methyl Glc-A in the glycone and 8(9) double bond in the aglycone moiety. Among Holothuriida, *Bohadschia* is the most primitive genus due to the presence of non-sulfated glycosides and functional CT.

In summary, studying the evolutionary pattern of structure-function relationships of holothurian's triterpene glycosides helps to understand their chemical-structural diversity, taxonomic distribution, ecological function, as well as bioactivity of the molecules, which can lead to a more targeted and efficient assessment of MNPs with novel pharmacological activities.

Acknowledgment The authors acknowledge funding by the BMBF via the Ginaico project, project number 16GW0106.We also thank the reviewers and editor for valuable comments which helped to improve the manuscript.

Appendix

This article is related to the YOUMARES 9 conference session no. 9: "Biodiversity of Benthic Holobionts: Chemical Ecology and Natural Products Chemistry in the Spotlight." The original Call for Abstracts and the abstracts of the presentations within this session can be found in the Appendix "Conference Sessions and Abstracts", Chapter "7 Biodiversity of Benthic Holobionts: Chemical Ecology and Natural Products Chemistry in the Spotlight", of this book.

References

Abraham TJ, Nagarajan J, Shanmugam SA (2002) Antimicrobial substances of potential biomedical importance from holothurian species. Indian J Mar Sci 31:161–164

Afiyatullov SS, Tishchenko LY, Stonik VA et al (1985) Structure of cucumarioside G_1 - A new triterpene glycoside from the holothurian *Cucumaria fraudatrix*. Chem Nat Compd 21:228–232. https://doi.org/10.1007/BF00714918

Agafonova IG, Aminin DL, Avilov SA et al (2003) Influence of Cucumariosides upon Intracellular $[Ca^{2+}]$i and Lysosomal Activity of Macrophages. J Agric Food Chem 51:6982–6986. https://doi.org/10.1021/jf034439x

Al Marzouqi N, Iratni R, Nemmar A et al (2011) Frondoside A inhibits human breast cancer cell survival, migration, invasion and the growth of breast tumor xenografts. Eur J Pharmacol 668:25–34. https://doi.org/10.1016/j.ejphar.2011.06.023

Al Shemaili J, Mensah-Brown E, Parekh K et al (2014) Frondoside A enhances the antiproliferative effects of gemcitabine in pancreatic cancer. Eur J Cancer 50:1391–1398. https://doi.org/10.1016/j.ejca.2014.01.002

Althunibat OY, Ridzwan BH, Taher M et al (2013) Antioxidant and cytotoxic properties of two sea cucumbers, *Holothuria edulis* Lesson and *Stichopus horrens* Selenka. Acta Biol Hung 64(1):10–20. https://doi.org/10.1556/ABiol.64.2013.1.2

Amidi S, Hashemi Z, Motallebi A et al (2017) Identification of (Z)-2,3-diphenylacrylonitrile as anti-cancer molecule in Persian gulf sea cucumber *Holothuria parva*. Mar Drugs 15:1–14. https://doi.org/10.3390/md15100314

Amini E, Nabiuni M, Baharara J et al (2014) Hemolytic and cytotoxic effects of saponin like compounds isolated from Persian Gulf brittle star (*Ophiocoma erinaceus*). J Coast Life Med 2:614–620. https://doi.org/10.12980/JCLM.2.2014JCLM-2014-0056

Aminin DL (2016) Immunomodulatory properties of sea cucumber triterpene glycosides. In: Gopalakrishnakone P, Haddad V Jr, Tubaro A et al (eds) Marine and Freshwater Toxins. Springer, Dordrecht, pp 382–397

Aminin DL, Agafonova IG, Fedorov SN (1995) Biological activity of disulphated polyhydroxysteroids from the pacific brittle star *Ophiopholis aculeata*. Comp Biochem Physiol Part C Pharmacol Toxicol Endocrinol 112C:201–204. https://doi.org/10.1016/0742-8413(95)02012-8

Aminin DL, Agafonova IG, Berdyshev EV et al (2001) Immunomodulatory properties of cucumariosides from the edible Far-Eastern holothurian *Cucumaria japonica*. J Med Food 4:127–135. https://doi.org/10.1089/109662001753165701

Aminin DL, Pinegin BV, Pichugina LV et al (2006) Immunomodulatory properties of Cumaside. Int Immunopharmacol 6:1070–1082. https://doi.org/10.1016/j.intimp.2006.01.017

Aminin DL, Agafonova IG, Kalinin VI et al (2008) Immunomodulatory properties of frondoside A, a major triterpene glycoside from the north Atlantic commercially harvested sea cucumber *Cucumaria frondosa*. J Med Food 11:443–453. https://doi.org/10.1089/jmf.2007.0530

Aminin DL, Silchenko AS, Avilov SA et al (2010) Immunomodulatory action of monosulfated triterpene glycosides from the sea cucumber *Cucumaria okhotensis*: stimulation of activity of mouse peritoneal macrophages. Nat Prod Commun 5:1877–1880

Aminin DL, Zaporozhets TA, Avilov S et al (2011) Radioprotective properties of cumaside, a complex of triterpene glycosides from the sea cucumber *Cucumaria japonica* and cholestrol. Nat Prod Commun 6:587–592

Aminin DL, Menchinskaya ES, Pisliagin EA et al (2015) Anticancer activity of sea cucumber triterpene glycosides. Mar Drugs 13:1202–1223. https://doi.org/10.3390/md13031202

Andriyashchenko PV, Levina EV, Kalinovskii AI (1996) Steroid compounds from the Pacific starfishes *Luidia quinaria* and *Distolasterias elegans* CA125:86975. Russ Chem Bull 45:455–458. https://doi.org/10.1007/BF01433994

Anisimov MM, Fronert EB, Kuznetsova TA et al (1973) The toxic effect of triterpene glycosides from *Stichopus japonicus* selenka on early embryogenesis of the sea urchin. Toxicon 11:109–111. https://doi.org/10.1016/0041-0101(73)90163-3

Anisimov MM, Shcheglov VV, Stonik VA et al (1974) The toxic effect of cucumarioside C from *Cucumaria fraudatrix* on early embryogenesis of the sea urchin. Toxicon 12:327–329

Anisimov MM, Prokofieva NG, Korotkikh LY et al (1980) Comparative study of cytotoxic activity of triterpene glycosides from marine organisms. Toxicon 18:221–223

Anisimov MM, Aminin DL, Rovin YG et al (1983) On the resistance of the cells of the sea cucumber *Stichopus japonicus* on the action of endotoxinstichoposide A. Dokl AN SSSR 270:991–993

Antonov A, Avilov S, Kalinovsky A (2008) Triterpene glycosides from Antarctic sea cucumbers. 1. Structure of Liouvillosides A_1, A_2, A_3, B_1, and B_2 from the sea cucumber *Staurocucumis liouvillei*: new procedure for separation of highly polar glycoside fractions and taxonomic revision. J Nat 71:1677–1685. https://doi.org/10.1021/np800173c

Antonov AS, Avilov SA, Kalinovsky AI et al (2009) Triterpene glycosides from antarctic sea cucumbers. 2. structure of Liouvillosides A_1, A_2, A_3 from the sea cucumber *Achlionice violaecuspidata* (=*Rhipidothuria racowitzai*). J Nat Prod 72:33–38

Antonov AS, Avilov SA, Kalinovsky AI et al (2011) Triterpene glycosides from Antarctic sea cucumbers III. Structures of liouvillosides A_4 and A_5, two minor disulphated tetraosides containing 3-O-methylquinovose as terminal monosaccharide units from the sea cucumber *Staurocucumis liouvillei* (Vaney). Nat Prod Res 25:1324–1333. https://doi.org/10.1080/14786419.2010.531017

Attoub S, Arafat K, Gélaude A et al (2013) Frondoside A Suppressive Effects on Lung Cancer Survival, Tumor Growth, Angiogenesis, Invasion, and Metastasis. PLoS One 8:1–10. https://doi.org/10.1371/journal.pone.0053087

Auria MVD, Riccio R, Minale L et al (1987) Novel marine steroid sulphates from Pacific Ophiuroids. Org Chem 52:3947–3952. https://doi.org/10.1021/jo00227a001

Avilov SA, Kalinovsky AI, Stonik VA (1991a) Two new triterpene glycosides from the holothurian *Duasmodactyla kurilensis*. Chem Nat Compd 27:188–192

Avilov SA, Stonik VA, Kalinovskii AI (1991b) Structures of four new triterpene glycosides from the holothurian *Cucumaria japonica*. Chem Nat Compd 26:670–675. https://doi.org/10.1007/BF00630079

Avilov SA, Kalinin VI, Makarieva TN et al (1994) Structure of cucumarioside G_2, a novel nonholostane glycoside from the sea cucumber *Eupentacta fraudatrix*. J Nat Prod 57:1166–1171. https://doi.org/10.1021/np50110a007

Avilov SA, Kalinovsky AI, Kalinin VI et al (1997) Koreoside A, a new nonholostane triterpene glycoside from the sea cucumber *Cucumaria koraiensis*. J Nat Prod 60:808–810. https://doi.org/10.1021/np970152g

Avilov SA, Drozdova OA, Kalinin VI et al (1998) Frondoside C, a new nonholostane triterpene glycoside from the sea cucumber *Cucumaria frondosa*: structure and cytotoxicity of its desulfated derivative. Sect Title Carbohydrates 76:137–141. https://doi.org/10.1139/cjc-76-2-137

Avilov SA, Antonov AS, Drozdova OA et al (2000a) Triterpene glycosides from the far eastern sea cucumber *Pentamera calcigera* II: Disulfated glycosides. J Nat Prod 63:1349–1355. https://doi.org/10.1021/np000002x

Avilov SA, Antonov AS, Drozdova OA et al (2000b) Triterpene glycosides from the Far-Eastern sea cucumber *Pentamera calcigera*. 1. Monosulphated glycosides and cytotoxicity of their unsulfated derivatives. J Nat Prod 63:65–71. https://doi.org/10.1021/np9903447

Avilov SA, Antonov AS, Silchenko AS et al (2003) Triterpene glycosides from the Far Eastern sea cucumber *Cucumaria conicospermium*. J Nat Prod 66:910–916. https://doi.org/10.1021/np030005k

Avilov SA, Silchenko AS, Antonov AS et al (2008) Synaptosides A and A1, triterpene glycosides from the sea cucumber *Synapta maculata* containing 3-O-methylglucuronic acid and their cytotoxic activity against tumor cells. J Nat Prod 71:525–531

Bahrami Y, Franco C (2015) Structure Elucidation of New Acetylated Saponins, Lessoniosides A, B, C, D, and E, and Non-Acetylated Saponins, Lessoniosides F and G, from the Viscera of the Sea Cucumber *Holothuria lessoni*. Mar Drugs 13:597–617. https://doi.org/10.3390/md13010597

Bahrami Y, Zhang W, Franco C (2014) Discovery of novel saponins from the viscera of the sea cucumber *Holothuria lessoni*. Mar Drugs 12:2633–2667. https://doi.org/10.3390/md12052633

Bahrami Y, Franco CMM, Benkendorff K (2016) Acetylated triterpene glycosides and their biological activity from holothuroidea reported in the past six decades. Mar Drugs 14:1–38. https://doi.org/10.3390/md14080147

Batrakov SG, Girshovich ES, Drozhzhina NS (1980) Triterpene glycosides with antifungal activity isolated from the sea cucumber, *Cucumaria japonica*. Antibiotiki 25:408–411

Beauregard KA, Truong NT, Zhang H et al (2001) The Detection and Isolation of a Novel Antimicrobial Peptide from an Echinoderm, *Cucumaria frondosa*. Adv Exp Med Biol 484:55–62. https://doi.org/10.1007/978-1-4615-1291-2

Berdyshev DV, Glazunov VP, Novikov VL (2007) 7-Ethyl-2,3,5,6,8-pentahydroxy-1,4-naphthoquinone (echinochrome A): A DFT study of the antioxidant mechanism. 1. Interaction of echinochrome A with hydroperoxyl radical. Russ Chem Bull 56:413–429. https://doi.org/10.1007/s11172-007-0067-3

Bhakuni DS, Rawat DS (2005) Bioactive Marine Natural Products. Springer, Dordrecht/Publishers, New Dehli

Bhatnagar S, Dudouet B, Ahond A et al (1985) Invertebres marins du lagon Neocaledonien IV. Saponines et sapogenines d'une holothurie, *Actinopyga flammea*. Bull Soc Chim Fr:124–129

Blunt JW, Carroll AR, Copp BR et al (2018) Marine natural products. Nat Prod Rep 35:8–53

Bondoc KGV, Lee H, Cruz LJ et al (2013) Chemical fingerprinting and phylogenetic mapping of saponin congeners from three tropical holothurian sea cucumbers. Comp Biochem Physiol B Biochem Mol Biol 166:182–193. https://doi.org/10.1016/j.cbpb.2013.09.002

Bordbar S, Anwar F, Saari N (2011) High-value components and bioactives from sea cucumbers for functional foods - A review. Mar Drugs 9:1761–1805. https://doi.org/10.3390/md9101761

Borsig L, Wang L, Cavalcante MCM et al (2007) Selectin blocking activity of a fucosylated chondroitin sulfate glycosaminoglycan from sea cucumber: Effect on tumor metastasis and neutrophil recruitment. J Biol Chem 282:14984–14991. https://doi.org/10.1074/jbc.M610560200

Bruno I, Minale L, Pizza C et al (1984) Starfish saponins. Part 14. Structures of the steroidal glycoside sulphated from the starfish *Marthasterias glacialis*. J Chem Soc Perkin Trans I 0: 1875-1883

Bruno I, D'Auria MV, Iorizzi M et al (1992) Marine eicosanoids: Occurrence of 8,11,12-trihydroxylated eicosanoic acids in starfishes. Experientia 48:114–115. https://doi.org/10.1007/BF01923622

Bryan PJ, Mcclintock JB, Hopkins TS (1997) Structural and chemical defenses of echinoderms from the northern Gulf of Mexico. Exp Mar Biol Ecol 210:173–186

Candia Carnevali MD, Bonasoro F (2001) Introduction to the biology of regeneration in echinoderms. Microsc Res Tech 55:365–368. https://doi.org/10.1002/jemt.1184

Careaga VP, Muniain C, Maier MS (2011) Patagonicosides B and C, two antifungal sulphated triterpene glycosides from the sea cucumber *Psolus patagonicus*. Chem Biodivers 8:467–475

Careaga VP, Bueno C, Muniain C et al (2014) Pseudocnoside A, a new cytotoxic and antiproliferative triterpene glycoside from the sea cucumber *Pseudocnus dubiosus leoninus*. Nat Prod Res 28:213–220. https://doi.org/10.1080/14786419.2012.751596

Caulier G, Van Dyck S, Gerbaux P et al (2011) Review of saponin diversity in sea cucumbers belonging to the family Holothuriidae. SPC Beche-de-mer Inf Bull 31:48–54

Chanley JD, Ledeen R, Wax J et al (1959) Holothurin. I. The Isolation, Properties and Sugar Components of Holothurin A. Am Chem Soc 81:5180–5183. https://doi.org/10.1021/ja01528a040

Chen LC, Lin YY, Jean YH et al (2014) Anti-inflammatory and analgesic effects of the marine-derived compound comaparvin isolated from the crinoid: *Comanthus bennetti*. Molecules 19:14667–14686. https://doi.org/10.3390/molecules190914667

Cheng G, Zhang X, Tang H-F et al (2006) Asterosaponin 1, a cytostatic compound from the starfish *Culcita novaeguineae*, functions by inducing apoptosis in human glioblastoma U87MG cells. J Neurooncol 79:235–241. https://doi.org/10.1007/s11060-006-9136-y

Chludil HD, Muniain CC, Seldes AM et al (2002a) Cytotoxic and antifungal triterpene glycosides from the Patagonian sea cucumber *Hemoiedema spectabilis*. J Nat Prod 65:860–865. https://doi.org/10.1021/np0106236

Chludil HD, Seldes AM, Maier MS (2002b) Antifungal steroidal glycosides from the Patagonian starfish *Anasterias minuta*: Structure - activity correlations. J Nat Prod 65:153–157. https://doi.org/10.1021/np010332x

Chovolou Y, Ebada SS, Wätjen W et al (2011) Identification of angular naphthopyrones from the Philippine echinoderm Comanthus species as inhibitors of the NF-κB signaling pathway. Eur J Pharmacol 657:26–34. https://doi.org/10.1016/j.ejphar.2011.01.039

Clemente S, Hernández JC, Montaño-Moctezuma G et al (2013) Predators of juvenile sea urchins and the effect of habitat refuges. Mar Biol 160:579–590. https://doi.org/10.1007/s00227-012-2114-3

Comin MJ, Maier MS, Roccatagliata AJ et al (1999) Evaluation of the antiviral activity of natural sulphated polyhydroxysteroids and their synthetic derivatives and analogs. Steroids 64:335–340. https://doi.org/10.1016/S0039-128X(99)00016-1

Croteau R, Kutchan TM, Lewis NG (2000) Natural products (Secondary Metabolites). Biochem Mol Biol Plants 7:1250–1318. https://doi.org/10.1016/j.phytochem.2011.10.011

Cuong NX, Vien LT, Hanh TTH et al (2015) Cytotoxic triterpene saponins from Cercodemas anceps. Bioorg Med Chem Lett 25:3151–3156. https://doi.org/10.1016/j.bmcl.2015.06.005

Cuong NX, Vien LT, Hoang L et al (2017) Cytotoxic triterpene diglycosides from the sea cucumber *Stichopus horrens*. Bioorg Med Chem Lett 27:2939–2942. https://doi.org/10.1016/j.bmcl.2017.05.003

D'Auria MV, Riccio R, Minale L (1985) Ophioxanthin, a new marine carotonoid sulphate from the Ophiuroid *Ophioderma longicaudum*. Tetrahedron Lett 26:1871–1872

D'Auria MV, Riccio R, Minale L et al (1987) Novel marine steroid sulphates from pacific ophiuroids. J Org Chem 52(18):3947–3952. https://doi.org/10.1021/jo00227a001

D'Auria MV, Maria I, Minale L et al (1990) Starfish saponins part 40. Structures of two new Asterosaponins from the starfish *Patritia miniata*: Patirioside A, and Patirioside B. J Chem Soc Perkin Trans I 1:1019–1023

D'Auria MV, Minale L, Riccio R (1993) Polyoxygenated Steroids of Marine Origin. Chem Rev 93:1839–1895. https://doi.org/10.1021/cr00021a010

Dai Y, Yu B (2015) Total synthesis of astrosterioside A, an anti-inflammatory asterosaponin. Chem Commun 51:13826–13829. https://doi.org/10.1039/C5CC04734J

Dang NH, Van Thanh N, Van Kiem P et al (2007) Two New Triterpene Glycosides from the Vietnamese Sea Cucumber *Holothuria scabra*. Arch Pharm Res 30:1387–1391. https://doi.org/10.1007/BF02977361

De Correa RS, Duque C, Riccio R et al (1985) Starfish saponins, Part 21. Steroidal glycosides from the starfish *Oreaster Reticulatus*. J Nat Prod 48:751–755. https://doi.org/10.1021/np50041a006

De Marino S, Iorizzi M, Zollo F et al (2000) Three new asterosaponins from the starfish *Goniopecten demonstrans*. European J Org Chem 2000:4093–4098. https://doi.org/10.1002/1099-0690(200012)2000:24<4093::AID-EJOC4093>3.0.CO;2-M

De Marino S, Borbone N, Iorizzi M et al (2003) Bioactive asterosaponins from the starfish *Luidia quinaria* and *Psilaster cassiope*. Isolation and structure characterization by two-dimensional NMR spectroscopy. J Nat Prod 66:515–519. https://doi.org/10.1021/np0205046

De Moncerrat Iñiguez-Martinez AM, Guerra-Rivas G, Rios T et al (2005) Triterpenoid oligoglycosides from the sea cucumber *Stichopus parvimensis*. J Nat Prod 68:1669–1673. https://doi.org/10.1021/np050196m

De Riccardis F, Giovannitti B, Iorizzi M, Minale L, Riccio R, Debitus C, De FBR (1991) Sterol Composition of the " Living Fossil " Crinoid *Gymnocrinus richeri*. Comp Biochem Physiol B Comp Biochem 100:647–651

De Simone F, Dini A, Minale L et al (1980) The Sterols of the Asteroid *Echinaster sepositus*. Comp Biochem Physiol B Biochem Mol Biol 66:351–357

De Simone F, Dini A, Finamore E et al (1981) Starfish saponins. Part 5. Structure of sepositoside A, a novel steroidal cyclic glycoside from the starfish *Echinaster sepositus*. J Chem Soc Perkin Trans I 1:1855–1862. https://doi.org/10.1039/P19810001855

de Vivar M, Maier M, Seldes AM (1999) Polar metabolites from the Antarctic starfish *Labidiaster annulatus*. An des la Asoc Quim Argentina 87:247–253

Demeyer M, De WJ, Caulier G et al (2014) Molecular diversity and body distribution of saponins in the sea star *Asterias rubens* by mass spectrometry. Comp Biochem Physiol B Biochem Mol Biol 168:1–11. https://doi.org/10.1016/j.cbpb.2013.10.004

Dong G, Xu T, Yang B et al (2011) Chemical constituents and bioactivities of starfish. Chem Biodivers 8:740–791. https://doi.org/10.1002/cbdv.200900344

Drozdova OA, Avilov SA, Kalinovskii AI et al (1993) New glycosides from the holothurian *Cucumaria japonica*. Chem Nat Compd 29:200–205

Drozdova OA, Avilov SA, Kalinin VI et al (1997) Cytotoxic triterpene glycosides from far-eastern sea cucumbers belonging to the genus *Cucumaria*. Liebigs Ann 1997:2351–2356. https://doi.org/10.1002/jlac.199719971125

Du L, Xu J, Xue Y et al (2015) Cerebrosides from sea cucumber ameliorates cancer-associated cachexia in mice by attenuating adipose atrophy. J Funct Foods 17:352–363. https://doi.org/10.1016/j.jff.2015.05.040

Duan J, Ishida M, Aida K et al (2016) Dietary cerebroside from sea cucumber (*Stichopus japonicus*): absorption and effects on skin barrier and cecal short-chain fatty acids. J Agric Food Chem 64:7014–7021. https://doi.org/10.1021/acs.jafc.6b02564

Dubois M-A, Noguchi Y, Higuchi R et al (1988) Structures of two new oligoglycoside sulfates: Pectinioside C and pectinioside D. Liebigs Ann 1988:495–500. https://doi.org/10.1002/jlac.198819880603

Duffy JE, Hay ME (2001) The ecology and evolution of marine consumer-prey interactions. In: Bertness MD, Hay ME, Gaines SD (eds) Marine community ecology. Sinauer Associates, Sunderland, pp 131–157

Elbandy M, Rho JR, Afifi R (2014) Analysis of saponins as bioactive zoochemicals from the marine functional food sea cucumber *Bohadschia cousteaui*. Eur Food Res Technol 238:937–955. https://doi.org/10.1007/s00217-014-2171-6

Esmat AY, Said MM, Soliman AA et al (2013) Bioactive compounds, antioxidant potential, and hepatoprotective activity of sea cucumber (*Holothuria atra*) against thioacetamide intoxication in rats. Nutrition 29:258–267. https://doi.org/10.1016/j.nut.2012.06.004

Farshadpour F, Gharibi S, Taherzadeh M et al (2014) Antiviral activity of *Holothuria* sp. a sea cucumber against herpes simplex virus type 1 (HSV-1). Eur Rev Med Pharmacol Sci 18:333–337

Fedorov SN, Shubina LK, Kicha AA et al (2008) Proapoptotic and Anticarcinogenic Activities of Leviusculoside G from the Starfish *Henricia leviuscula* and Probable Molecular Mechanism. Nat Prod Commun 3:1575–1580

Fedorov SN, Dyshlovoy SA, Kuzmich AS et al (2016) In vitro anticancer activities of some triterpene glycosides from holothurians of Cucumariidae, Stichopod, Psolidae, Holothuriidae, and Synaptidae families. Nat Prod Commun 11:1239–1242

Feng Y, Khokhar S, Davis RA (2017) Crinoids: ancient organisms, modern chemistry. Nat Prod Rep 34:571–584. https://doi.org/10.1039/C6NP00093B

Fieser LF, Fieser M (1956) Organic chemistry. Reinhold, New York

Finamore E, Minale L, Riccio R et al (1991) Novel marine polyhydroxylated steroids from the starfish *Myxoderma platyacanthurn*. J Org Chem 56:1146–1153. https://doi.org/10.1021/jo00003a043

Findlay JA, Agarwal VK (1983) Aglycones from the saponin of the starfish *Asterias vulgaris*. J Nat Prod 46:876–880. https://doi.org/10.1021/np50030a008

Findlay JA, He ZQ (1991) Polyhydroxylated steroidal glycosides from the starfish *Asterias forbesi*. J Nat Prod 54:428–435. https://doi.org/10.1021/np50074a013

Findlay JA, Jaseja M, Burnell DJ (1987) Major saponins from the starfish *Asterias forbesi*. Complete structures by nuclear magnetic resonance methods. Can J Chem 65:1384–1391. https://doi.org/10.1139/v87-234

Findlay JA, Findlay A, Findlay A et al (1989) Forbeside E: a novel sulphated sterol glycoside from *Asterias forbesi*. Can J Chem 67:2078–2080

Findlay JA, He Z-Q, Blackwell B (1990) Minor saponins from the starfish *Asterias forbesi*. Can J Chem 68:1215–1217. https://doi.org/10.1139/v90-188

Findlay JA, Yayli N, Radics L (1992) Novel sulfated oligosaccharides from the sea cucumber *cucumaria frondosa*. J Nat Prod 55:93–101. https://doi.org/10.1021/np50079a014

Folmer F, Jaspars M, Solano G et al (2009) The inhibition of TNF-α-induced NF-κB activation by marine natural products. Biochem Pharmacol 78:592–606. https://doi.org/10.1016/j.bcp.2009.05.009

Francis G, Kerem Z, Makkar HPS et al (2002) The biological action of saponins in animal systems: a review. Br J Nutr 88:587. https://doi.org/10.1079/BJN2002725

Garneau FX, Harvey C, Simard JL et al (1989) The distribution of asterosaponins in various body components of the starfish *Leptasterias polaris*. Comp Biochem Physiol B Biochem Mol Biol 92:411–416. https://doi.org/10.1016/0305-0491(89)90302-7

Ghanbari R, Ebrahimpour A, Abdul-Hamid A et al (2012) *Actinopyga lecanora* hydrolysates as natural antibacterial agents. Int J Mol Sci 13:16796–16811. https://doi.org/10.3390/ijms131216796

Ghannoum MA, Rice LB (1999) Antifungal agents: Mode of action, mechanisms of resistance, and correlation of these mechanisms with bacterial resistance. Clin Microbiol Rev 12:501-517. doi:10.1.1.322-6182

Girard M, Bélanger J, ApSimon JW et al (1990) Frondoside A. A novel triterpene glycoside from the holothurian *Cucumaria frondosa*. Can J Chem 68:11–18. https://doi.org/10.1139/v90-003

Gorshkov BA, Gorshkova IA, Stonik VA et al (1982) Effect of marine glycosides on adenosinetriphosphatase activity. Toxicon 20:655–658. https://doi.org/10.1016/0041-0101(82)90059-9

Gorshkova IA, Gorshkov BA, Stonik VA (1989) Inhibition of rat brain Na^+-K^+-ATPase by triterpene glycosides from holothurians. Toxicon 27:927–936. https://doi.org/10.1016/0041-0101(89)90104-9

Gorshkova IA, Kalinin VI, Gorshkov BA et al (1999) Two different modes of inhibition of the rat brain Na^+, K^+-ATPase by triterpene glycosides, psolusosides A and B from the Holothurian *Psolus fabricii*. Comp Biochem Physiol C Pharmacol Toxicol Endocrinol 122:101–108. https://doi.org/10.1016/S0742-8413(98)10085-3

Gowda NM, Goswami U, Khan MI (2008) T-antigen binding lectin with antibacterial activity from marine invertebrate, sea cucumber (*Holothuria scabra*): Possible involvement in differential recognition of bacteria. J Invertebr Pathol 99:141–145. https://doi.org/10.1016/j.jip.2008.04.003

Grishin Y, Besednova NN, Stonik VA et al (1990) The regulation of hemopoesis and immunogenesis by triterpene glycosides from holothurians. Radiobiologija 30:556

Guenther J, Wright AD, Burns K et al (2009) Chemical antifouling defences of sea stars: Effects of the natural products hexadecanoic acid, cholesterol, lathosterol and sitosterol. Mar Ecol Prog Ser 385:137–149. https://doi.org/10.3354/meps08034

Guo D, Xiong Y (2009) Preparation and characterization of *Holothuria nobilis* saponins nobiliside A freeze-dried liposome CA152:19119. Dier Junyi Daxue Xuebao 30:202–207

Han C, Qi J, Ojika M (2006) Structure–activity relationships of novel neuritogenic steroid glycosides from the Okinawan starfish *Linckia laevigata*. Bioorg Med Chem 14:4458–4465. https://doi.org/10.1016/j.bmc.2006.02.032

Han C, Qi J, Ojika M (2007a) Linckosides M-Q: Neuritogenic steroid glycosides from the Okinawan starfish *Linckia laevigata*. J Nat Med 61:138–145. https://doi.org/10.1007/s11418-006-0107-6

Han H, Yi YH, Li L et al (2007b) A new triterpene glycoside from sea cucumber *Holothuria leucospilota*. Chinese Chem Lett 18:161–164. https://doi.org/10.1016/j.cclet.2006.12.027

Han H, Yi Y, La M et al (2008a) Studies on antifungal and antitumor activities of scabraside A, B from *Holothuria scabra* Jaeger. Zhongguo Yaolixue Tongbao 24:1111–1112

Han H, Yi YH, Liu BS et al (2008b) Leucospilotaside C, a new sulphated triterpene glycoside from sea cucumber *Holothuria leucospilota*. Chinese Chem Lett 19:1462–1464. https://doi.org/10.1016/j.cclet.2008.09.051

Han H, Yi Y-H, Li L et al (2009a) Triterpene Glycosides from Sea Cucumber *Holothuria leucospilota*. Chin J Nat Med 7:346–350. https://doi.org/10.3724/SP.J.1009.2009.00346

Han H, Yi Y, Li L et al (2009b) Antifungal active triterpene glycosides from sea cucumber *Holothuria scabra*. Acta Pharmaceutica Sinica 44(6):620–624

Han H, Xu Q-Z, Tang H-F et al (2010a) Cytotoxic Holostane-Type Triterpene Glycosides from the Sea Cucumber *Pentacta quadrangularis*. Planta Med 76:1900–1904. https://doi.org/10.1055/s-0030-1249854

Han H, Zhang W, Yi YH et al (2010b) A novel sulphated holostane glycoside from sea cucumber *Holothuria leucospilota*. Chem Biodivers 7:1764–1769. https://doi.org/10.1002/cbdv.200900094

Han H, Xu QZ, Yi YH et al (2010c) Two new cytotoxic disulphated holostane glycosides from the sea cucumber *Pentacta quadrangularis*. Chem Biodivers 7:158–167. https://doi.org/10.1002/cbdv.200800324

Han H, Li L, Yi Y et al (2012) Triterpene Glycosides from Sea Cucumber *Holothuria scabra* with Cytotoxic Activity. Chinese Herb Med 4:183–188. https://doi.org/10.3969/j.issn.1674-6384.2012.03.002

Harvey C, Garneau FX, Himmelman JH (1987) Chemodetection of the predatory seastar *Leptasterias polaris* by the whelk *Buccinum undatum*. Mar Ecol Prog Ser 40:79–86

Hatakeyama T, Kamine T, Konishi Y et al (1999) Carbohydrate-dependent hemolytic activity of the conjugate composed of a C-type lectin, CEL-I, and an amphiphilic α-helical peptide, 43-βAla2. Biosci Biotechnol Biochem 63:1312–1314. https://doi.org/10.1271/bbb.63.1312

Haug T, Kjuul AK, Styrvold OB et al (2002) Antibacterial activity in *Strongylocentrotus droebachiensis* (Echinoidea), *Cucumaria frondosa* (Holothuroidea), and *Asterias rubens* (Asteroidea). J Invertebr Pathol 81:94.102. https://doi.org/10.1016/S0022-2011(02)00153-2

Hegde VR, Chan TM, Pu H et al (2002) Two selective novel triterpene glycosides from sea cucumber, *Theleonata ananas*: inhibitors of chemokine receptor-5. Bioorganic Med Chem Lett 12:3203–3205

Herencia F, Ubeda A, Ferrandiz ML et al (1998) Anti-inflammatory activity in mice of extracts from mediterranean marine invertebrates. Pharmacol Lett 62:115–120

Hickman CP, Roberts L, Larson A (2001) Integrated principles of zoology. Mosby Publishers, St. Louis

Higuchi R, Fujita M, Matsumoto S et al (1996) Isolation and structure of four new steroid glycoside Di-0-sulphates from the starfish *Asteropecten latespinosus*. Liebigs Ann 1996:837–840

Honda M, Igarashi T, Komori T (1990) Structure of Pectinioside C: determination of the stereochemistry of the C-17 side chain of the steroidal aglycone. Liebigs Ann 1990:547–553

Honey-Escandón M, Arreguín-Espinosa R, Solís-Marín FA et al (2015) Biological and taxonomic perspective of triterpenoid glycosides of sea cucumbers of the family Holothuriidae (Echinodermata, Holothuroidea). Comp Biochem Physiol Part B Biochem Mol Biol 180:16–39. https://doi.org/10.1016/j.cbpb.2014.09.007

Hu X-Q, Wang YM, Wang JF et al (2010) Dietary saponins of sea cucumber alleviate orotic acid-induced fatty liver in rats via PPARα and SREBP-1c signaling. Lipids Health Dis 9:25. https://doi.org/10.1186/1476-511X-9-25

Hu S, Chang Y, He M et al (2014a) Fucosylated chondroitin sulfate from sea cucumber improves insulin sensitivity via activation of PI3K/PKB pathway. J Food Sci 79:H1424–H1427. https://doi.org/10.1111/1750-3841.12465

Hu S, Xu L, Shi D et al (2014b) Eicosapentaenoic acid-enriched phosphatidylcholine isolated from *Cucumaria frondosa* exhibits anti-hyperglycemic effects via activating phosphoinositide 3-kinase/protein kinase B signal pathway. J Biosci Bioeng 117:457–463. https://doi.org/10.1016/j.jbiosc.2013.09.005

Huang N, Wu MY, Zheng CB et al (2013) The depolymerized fucosylated chondroitin sulfate from sea cucumber potently inhibits HIV replication via interfering with virus entry. Carbohydr Res 380:64–69. https://doi.org/10.1016/j.carres.2013.07.010

Husni A, Shin IS, You S et al (2009) Antioxidant Properties of Water and Aqueous Ethanol Extracts and Their Crude Saponin Fractions from a Far-eastern Sea Cucumber, *Stichopus japonicus*. Food Sci Technol 18:419–424

Hwang IH, Kim WD, Kim SJ et al (2011) Asterosaponins Isolated from the Starfish *Asterias amurensis*. Chem Pharm Bull 59:78–83

Hwang IH, Kulkarni R, Yang MH et al (2014) Complete NMR assignments of undegraded asterosaponins from *Asterias amurensis*. Arch Pharm Res 37:1252–1263. https://doi.org/10.1007/s12272-014-0374-9

Ikeda Y, Inagaki M, Yamada K et al (2009) Isolation and structure of a galactocerebroside from the sea cucumber *Bohadschia argus*. Chem Pharm Bull (Tokyo) 57:315–317. https://doi.org/10.1248/cpb.57.315

Ikegami S, Kamiya Y, Tamura S (1973) Studies on Asterosaponins-V: A novel steroid conjugate, 5 alfa-PREGN-9(11)-ENE-3beta,6alfa-DIOL-20-1-3-Sulfate, from a starfish saponin. Asterosaponin A. Tetrahedron 29:1807–1810

Inagaki M (2008) Structure and biological activity of glycosphingolipids from starfish and feather stars. Yakugaku zasshi 128(8):1187–1194. https://doi.org/10.1248/yakushi.128

Inagaki M, Nakata T, Higuchi R (2006) Isolation and structure of a galactocerebroside molecular species from the starfish *Culcita novaeguineae*. Chem Pharm Bull 54:260–261. https://doi.org/10.1248/cpb.54.260

Inagaki M, Shiizaki M, Hiwatashi T et al (2007) Constituents of Crinoidea. 5. Isolation and structure of a new glycosyl inositolphosphoceramide-type ganglioside from the feather star *Comanthina schlegeli*. Chem Pharm Bull (Tokyo) 55:1649–1651. https://doi.org/10.1248/cpb.55.1649

Iorizzi M, Minale L, Riccio R et al (1986) Starfish saponins, part 23. Steroidal glycosides from the starfish *Halityle regularis*. J Nat Prod 49:67–78. https://doi.org/10.1021/np50043a007

Iorizzi M, Minale L, Riccio R et al (1991) Starfish Saponins, Part 46. Steroidal Glycosides and Polyhydroxysteroids from the Starfish *Culcita novaeguineae*. J Nat Prod 54:1254–1264. https://doi.org/10.1021/np50077a003

Iorizzi M, Minale L, Riccio R et al (1992) Starfish saponins, 48. isolation of fifteen sterol constituents (six glycosides and nine polyhydroxysteroids) from the starfish *Solaster borealis*. J Nat Prod 55:866–877. https://doi.org/10.1021/np50085a005

Iorizzi M, De Riccardis F, Minale L et al (1993) Starfish saponins, 52. Chemical constituents from the starfish *Echinaster brasiliensis*. J Nat Prod 56:2149–2162. https://doi.org/10.1021/np50102a018

Iorizzi M, Bifulco G, De Riccardis F et al (1995) Starfish saponins, part 53. A reinvestigation of the polar steroids from the starfish *Oreaster reticulatus*: Isolation of sixteen steroidal oligoglycosides and six polyhydroxysteroids. J Nat Prod 58:10–26. https://doi.org/10.1021/np50115a002

Ishida H, Hirota Y, Nakazawa H (1993) Effect of sub-skinning concentrations of saponin on intracellular Ca^{2+} and plasma membrane fluidity in cultured cardiac cells. BBA - Biomembr 1145:58–62. https://doi.org/10.1016/0005-2736(93)90381-9

Ismail H, Lemriss S, Ben Aoun Z et al (2008) Antifungal activity of aqueous and methanolic extracts from the Mediterranean sea cucumber, *Holothuria polii*. J Mycol Med 18:23–26. https://doi.org/10.1016/j.mycmed.2008.01.002

Itakura Y, Komori T (1986) Biologically Active Glycosides from Asteroidea, X. Steroid Oligoglycosides from the Starfish *Acanthaster planci L.*, 3. Structures of Four New Oligoglycoside Sulfates. Liebigs Ann 1986:499–508. https://doi.org/10.1002/jlac.198619860308

Ivanchina NV, Kich A, Kalinovsky A et al (2000) Hemolytic polar steroidal constitutents of the starfish *Aphelasterias japonica*. J Nat Prod 63(8):1178–1181. https://doi.org/10.1021/np000030f

Ivanchina NV, Kicha AA, Kalinovsky AI et al (2004) Absolute configuration of side chains of polyhydroxylated steroidal compounds from the starfish *Henricia derjugini*. Russ Chem Bull 53:2639–2642. https://doi.org/10.1007/s11172-005-0166-y

Ivanchina NV, Malyarenko TV, Kicha AA et al (2005) Asterosaponin ophidianoside F from gonads of the Far-Eastern starfish *Aphelasterias japonica*. Chem Nat Compd 41:481–482. https://doi.org/10.1007/s10600-005-0187-7

Ivanchina NV, Kicha AA, Kalinovsky AI et al (2006) Polar steroidal compounds from the Far Eastern starfish *Henricia leviuscula*. J Nat Prod 69:224–228. https://doi.org/10.1021/np050373j

Ivanchina NV, Malyarenko TV, Kicha AA et al (2011) Structures and cytotoxic activities of two new asterosaponins from the antarctic starfish *Diplasterias brucei*. Russ J Bioorganic Chem 37:499–506. https://doi.org/10.1134/S1068162011030083

Ivanchina NV, Kalinovsky AI, Kicha AA et al (2012) Two New Asterosaponins from the Far Eastern Starfish *Lethasterias fusca*. Nat Prod Commun 7(7):853–858

Ivanchina NV, Malyarenko TV, Kicha AA et al (2017) A new steroidal glycoside granulatoside C from the starfish *Choriaster granulatus*, unexpectedly combining structural features of polar steroids from several different marine invertebrate phyla. Nat Prod Commun 12:1585–1588

Ivanchina NV, Kicha AA, Malyarenko TV et al (2018) Granulatosides D, E and other polar steroid compounds from the starfish *Choriaster granulatus*. Their immunomodulatory activity and cytotoxicity. Nat Prod Res:1–8. https://doi.org/10.1080/14786419.2018.1463223

Iyengar EV, Harvell CD (2001) Predator deterrence of early developmental stages of temperate lecithotrophic asteroids and holothuroids. J Exp Mar Bio Ecol 264:171–188. https://doi.org/10.1016/S0022-0981(01)00314-8

Janakiram NB, Mohammed A, Zhang Y et al (2010) Chemopreventive effects of Frondanol A5, a *Cucumaria frondosa* extract, against rat colon carcinogenesis and inhibition of human colon cancer cell growth. Cancer Prev Res 3:82–91. https://doi.org/10.1158/1940-6207.CAPR-09-0112

Janakiram NB, Mohammed A, Rao CV (2015) Sea cucumbers metabolites as potent anti-cancer agents. Mar Drugs 13:2909–2923. https://doi.org/10.3390/md13052909

Jangoux M (1984) Diseases of echinoderms. Helgoländer Meeresun 37:207–216. https://doi.org/10.1007/BF01989305

Jeong SH, Kim HK, Song IS et al (2014) Echinochrome a protects mitochondrial function in cardiomyocytes against cardiotoxic drugs. Mar Drugs 12:2922–2936. https://doi.org/10.3390/md12052922

Jia Z, Song Y, Tao S et al (2016) Structure of sphingolipids from sea cucumber *Cucumaria frondosa* and structure-specific cytotoxicity against human hepg2 cells. Lipids 51:321–334. https://doi.org/10.1007/s11745-016-4128-y

Jiang Z-H, Schmidt RR (1992) The hexasaccharide moiety of pectinioside. Liebigs Ann 1992:75–982

Jiang Z-H, Han X-B, Schmidt RR (1993) Synthesis of the sulfated steroidal glycosides Forbeside E3 and E1. Liebigs Ann 1993:1179–1184

Jiao H, Shang X, Dong Q et al (2015) Polysaccharide constituents of three types of sea urchin shells and their anti-inflammatory activities. Mar Drugs 13:5882–5900. https://doi.org/10.3390/md13095882

Jin W, Rinehart KL, Jares-Erijman EA (1994) Ophidiacerebrosides: cytotoxic glycosphingolipids containing a novel sphingosine from a sea star. J Org Chem 59:144–147. https://doi.org/10.1021/jo00080a023

Kalinin VI (2000) System-theoretical (Holistic) approach to the modelling of structural-functional relationships of biomolecules and their evolution: an example of triterpene glycosides from sea cucumbers (Echinodermata, Holothurioidea). J Theor Biol 206:151–168. https://doi.org/10.1006/jtbi.2000.2110

Kalinin VI, Stonik VA (1996) Application of morphological trends of evolution to phylogenetic interpretation of chemotaxonomic data. J Theor Biol 180:1–10. https://doi.org/10.1006/jtbi.1996.0073

Kalinin VI, Kalinovskii AI, Stonik VA et al (1989a) Structure of psolusoside B- A nonholostane triterpene glycoside of the holothurian genus *Psolus*. Chem Nat Compd 25:311–317

Kalinin VI, Stonik VA, Kalinovskii AI et al (1989b) Structure of pseudostichoposide A- The main triterpene glycoside from the holothurian *Pseudostichopus trachus*. Chem Nat Compd 25:577–582

Kalinin VI, Avilov SA, Kalinovskii AI et al (1992a) Cucumarioside G_3- A minor triterpene glycoside from the holothurian *Eupentacta fraudatrix*. Chem Nat Compd 28:635–636

Kalinin VI, Avilov SA, Kalinovskii AI et al (1992b) Cucumarioside G_4 - A new triterpenglycoside from the holothurian *Eupentacta fraudatrix*. Chem Nat Compd 28:600–603

Kalinin VI, Prokofieva NG, Likhatskaya GN et al (1996) Hemolytic activities of triterpene glycosides from the holothurian order Dendrochirotida: Some trends in the evolution of this group of toxins. Toxicon 34:475–483. https://doi.org/10.1016/0041-0101(95)00142-5

Kalinin VI, Avilov SA, Kalinina EY et al (1997) Structure of eximisoside A, a novel triterpene glycoside from the Far-Eastern sea cucumber *Psolus eximius*. J Nat Prod 60:817–819. https://doi.org/10.1021/np9701541

Kalinin VI, Aminin DL, Avilov SA et al (2008) Triterpene glycosides from sea cucucmbers (Holothurioidea, Echinodermata). Biological activities and functions. In: Atta-Ur-Rahman (ed) Studies in natural products chemistry (Bioactive natural products) Elsevier Science Publisher 35:135–196

Kalinin VI, Avilov SA, Silchenko AS et al (2015) Triterpene glycosides of sea cucumbers (Holothuroidea, Echinodermata) as taxonomic markers. Nat Prod Commun 10:21–26

Kalinin VI, Silchenko AS, Avilov SA (2016) Taxonomic Significance and Ecological Role of Triterpene Glycosides from Holothurians. Biol Bull 43:616–624. https://doi.org/10.1134/S1062359016060108

Kalinovskii AI, Levina EV, Stonik VA et al (2004) Steroid polyols from the far eastern starfish *Henricia sanguinolenta* and *H. leviuscula leviuscula*. Russ J Bioorganic Chem 30:191–195. https://doi.org/10.1023/B:RUBI.0000023107.90150.09

Kaluzhskiy LA, Shkel TV, Ivanchina NV et al (2017) Structural Analogues of Lanosterol from Marine Organisms of the Class Asteroidea as potential inhibitors of human and *Candida albicans* lanosterol 14α-demethylases. Nat Prod Commun 12:1843–1846

Karin M, Yamamoto Y, Wang QM (2004) The IKK NF-κB system: A treasure trove for drug development. Nat Rev Drug Discov 3:17–26. https://doi.org/10.1038/nrd1279

Karleskint G, Turner R, Small JW (2010) In: Brooks/Cole, Belmont (ed) Introduction to marine biology, 3rd edn

Kaul P (1986) Marine pharmacology: bioactive molecules from the sea. Annu Rev Pharmacol Toxicol 26:117–142. https://doi.org/10.1146/annurev.pharmtox.26.1.117

Kawase O, Ohno O, Suenaga K et al (2016) Immunological Adjuvant Activity of Pectinioside A, the Steroidal Saponin from the Starfish *Patiria pectinifera*. Nat Prod Commun 11:605–606

Kenta G, Tatsuya S, Hideki T et al (2015) Total Synthesis and Neuritogenic Activity Evaluation of Ganglioside PNG-2A from the Starfish *Protoreaster nodosus*. Asian J Org Chem 4:1160–1171. https://doi.org/10.1002/ajoc.201500282

Kicha AA, Kallnovsky AI, Levina E et al (1983) Asterosaponin P_1 from the starfish *Patria pectinifera*. Tetrahedron Lett 24:3893–3896

Kicha AA, Kalinovsky AI, Levina EV et al (1985) Culcitoside C_1 from starfishes *Culcita novaeguineae* and *Linckia guildingi*. Chem Nat Compd 21:760–762

Kicha AA, Kalinovskii AI, Andrishchenko PV (1986) Culcitosides C_2 and C_3 from the starfish *Culcita novaeguineae*. Chem Nat Compd 22:557–560. https://doi.org/10.1007/BF00599260

Kicha AA, Kalinovsky AI, Gorbach NV et al (1993) New polyhydroxysteroids from the far-eastern starfish *Henricia* sp. Chem Nat Compd 29:206–210

Kicha AA, Ivanchina NV, Kalinovsky AI et al (2000) Asterosaponin P2 from the Far-Eastern starfish *Patiria (asterina) pectinifera*. Russ Chem Bull 49:1794–1795

Kicha AA, Ivanchina NV, Kalinovsky AI et al (2001) Sulphated steroid compounds from the starfish *Aphelasterias japonica* of the Kuril population. Russ Chem Bull 50:724–727

Kicha AA, Ivanchina NV, Stonik VA (2004) Seasonal variations in polyhydroxysteroids and related glycosides from digestive tissues of the starfish *Patiria (=Asterina) pectinifera*. Comp Biochem Physiol Part B Biochem Mol Biol 139:581–585. https://doi.org/10.1016/j.cbpc.2004.06.011

Kicha AA, Ivanchina NV, Kalinovsky A et al (2007a) Sulfated steroid glycosides from the Viet Namese starfish *Linckia laevigata*. Chem Nat Compd 43:76–80. https://doi.org/10.1007/s10600-007-0036-y

Kicha AA, Ivanchina NV, Kalinovsky A et al (2007b) New neuritogenic steroid glycosides from the Vietnamese starfish *Linckia laevigata*. Nat Prod Commun 2:41–46

Kicha AA, Ivanchina NV, Kalinovsky AI et al (2007c) Four new steroid glycosides from the Vietnamese starfish *Linckia laevigata*. Russ Chem Bull 56:823–830. https://doi.org/10.1007/s11172-007-0123-z

Kicha AA, Ivanchina NV, Huong TTT et al (2010a) Two new asterosaponins, archasterosides A and B, from the Vietnamese starfish *Archaster typicus* and their anticancer properties. Bioorganic Med Chem Lett 20:3826–3830. https://doi.org/10.1016/j.bmcl.2010.04.005

Kicha AA, Ivanchina NV, Huong TTT et al (2010b) Minor asterosaponin archasteroside C from the starfish *Archaster typicus*. Russ Chem Bull 59:2133–2136. https://doi.org/10.1007/s11172-010-0368-9

Kicha AA, Kalinovsky AI, Ivanchina NV et al (2011) Four new asterosaponins, hippasteriosides A - D, from the Far Eastern starfish *Hippasteria kurilensis*. Chem Biodivers 8:166–175. https://doi.org/10.1002/cbdv.200900402

Kitagawa I, Kobayashi M (1977) On the structure of the major saponin from *Acanthaster planci*. Tetrahedron Lett 2:859–862

Kitagawa I, Kobayashi M (1978) Saponin and Sapogenol. XXVI. Steroidal saponins from the starfish *Acanthaster planci L.* (Crown of Thorns). (2). Structure of the major saponin Thornasteroside A. Chem Pharm Bull 26:1864–1873. https://doi.org/10.1248/cpb.37.3229

Kitagawa I, Sugawara T, Yosioka I et al (1976) Saponin and sapogenol. XIV. Antifungal glycosides from the sea cucumber *Stichopus japonicus* Selenka: 1. Structure of Stichopogenin A_4, the genuine aglycone of holotoxin A. Chem Pharm Bull 24:266–274. https://doi.org/10.1248/cpb.37.3229

Kitagawa I, Nishino T, Kyogoku Y (1979) Structure of holothurin A a biologically active triterpene-oligoglycoside from the sea cucumber *Holothuria leucospilota* Brandt. Tetrahedron Lett:1419–1422. https://doi.org/10.1016/S0040-4039(01)86166-9

Kitagawa I, Inamoto T, Fuchida M et al (1980) Structures of Echinoside A and B, two antifungal oligoglycosides from the sea cucumber *Actinopyga echinites* (JAEGER). Chem Pharm Bull 28:1651–1653. https://doi.org/10.1248/cpb.37.3229

Kitagawa I, Kobayashi K, Inamoto T et al (1981a) The structure of six antifungal oligoglycosides, Stichlorosides A_1,A_2,B_1,B_2,C, and C_2, from the sea cucumber *Stichopus chloronotus* (Brandt). Chem Pharm Bull 29:2387–2391

Kitagawa I, Kobayashi K, Inamoto T et al (1981b) Stichlorogenol and Dehydrostichlorogenol, Genuine Aglycones of Stichlorosides A_1, B_1, C_1 and A_2, B_2, C_2, from the Sea Cucumber *Stichopus Chloronotus* (BRANDT). Biosci Biotechnol Biochem 29:1189–1192. https://doi.org/10.1248/cpb.37.3229

Kitagawa I, Kobayashi M, Hori M et al (1981c) Structures of four new triterpenoidal oligoglycosides, Bivittoside A, B, C, and D, from the sea cucumber *Bohadschia bivittata* MITSUKURI. Chem Pharm Bull 29:282–285. https://doi.org/10.1093/jxb/erl177

Kitagawa I, Nishino T, Kobayashi M et al (1981d) Marine Natural Products. VIII. Bioactive triterpene- oligoglycosides from the sea cucumber *Holothuria leucospilota* (Brandt). Structure of holothurin A. Chem Pharm Bull 29:1951–1956. https://doi.org/10.1248/cpb.37.3229

Kitagawa I, Kobayashi M, Inamoto T et al (1985) Marine Natural Products. XIV. Structures of echinosides A and B, antifungal lanostane oligosides from the sea cucumber *Actinopyga echinites* (Jaeger). Chem Pharm Bull (Tokyo) 33:5214–5224

Kitagawa I, Kobayashi M, Hori M et al (1989) Marine Natural Producs. XVIII. Four lanostane- type triterpene oligoglycosides, bivittosides A,B,C, and D from the Okinawan sea cucumber *Bohadschia bivittata* (Mitsukuri). Chem Pharm Bull 37:61–67

Kobayashi M, Hori M, Kan K et al (1991) Marine Natural Products. XXVII Distribution of Lanostane-type triterpene oligoglycosides in ten kind of Okinawan sea cucumbers. Chem Pharm Bull 39:2282–2287. https://doi.org/10.1248/cpb.37.3229

Kornprobst J-M, Sallenave C, Barnathan G (1998) Sulfated compounds from marine organisms. Comp Biochem Physiol B Biochem Mol Biol 119:1–51. https://doi.org/10.1016/S0305-0491(97)00168-5

Kropp RK (1982) Responses of Five Holothurian Species to Attacks by a Predatory Gastropod *Tonna perdix*. Pacific Sci 36:445–452

Kubanek J, Pawlik JR, Eve TM et al (2000) Triterpene glycosides defend the Caribbean reef sponge *Erylus formosus* from predatory fishes. Mar Ecol Prog Ser 207:69–77. https://doi.org/10.3354/meps207069

Kumar R, Chaturvedi AK, Shukla PK et al (2007) Antifungal activity in triterpene glycosides from the sea cucumber *Actinopyga lecanora*. Bioorg Med Chem Lett 17:4387–4391. https://doi.org/10.1016/j.bmcl.2006.12.052

Kuznetsova TA, Anisimov MM, Popov AM et al (1982a) A comparative study in vitro of physiological activity of triterpene glycosides of marine invertebrates of echinoderm type. Comp Biochem Physiol Part C Pharmacol Toxicol Endocrinol 73:41–43. https://doi.org/10.1016/0306-4492(82)90165-4

Kuznetsova TA, Kalinovskaya NI, Kalinovskii AI et al (1982b) Glycosides of marine invertebrates. XIV. Structure of holothurin B_1 from the holothurian *Holothuria floridana*. Chem Nat Compd 18:449–451. https://doi.org/10.1007/BF00579642

La M-P, Li C, Li L et al (2012) New bioactive sulfated alkenes from the sea cucumber *Apostichopus japonicus*. Chem Biodivers 9:1166–1171. https://doi.org/10.1002/cbdv.201100324

Laille M, Gerald F, Debitus C (1998) In vitro antiviral activity on dengue virus of marine natural products. C Cell Mol life Sci 54:167–170. https://doi.org/10.1007/s000180050138

Lakshmi V, Saxena A, Mishra SK et al (2008) Spermicidal Activity of Bivittoside D from *Bohadschia vitiensis*. Arch Med Res 39:631–638. https://doi.org/10.1016/j.arcmed.2008.06.007

Lakshmi V, Srivastava S, Mishra SK, Shukla PK (2012) Antifungal activity of bivittoside-D 14 from *Bohadschia vitiensis* (Semper). Nat Prod Res 26(10):913–918

Lawrence PG, Harold PL, Francis OG (1957) Antibiotics and Chemotherapy 4(1):1980–1989

Lee J, Wang W, Hong J et al (2007) A new 2,3-dimethyl butenolide from the brittle star *Ophiomastix mixta*. Chem Pharm Bull (Tokyo) 55:459–461. https://doi.org/10.1248/cpb.55.459

Levina EV, Kalinovskii AI, Andriyaschenko PV et al (1987) Steroid Glycosides from the starfish *Echinaster sepositus*. Chem Nat Compd 23:206–209

Levina EV, Andriyaschenko PV, Stonik VA et al (1996) Ophiuroid-type steroids in starfish of the genus *Pteraster*. Comp Biochem Physiol B Biochem Mol Biol 114:49–52. https://doi.org/10.1016/0305-0491(95)02121-3

Levina EV, Andriyashchenko PV, Kalinovskii AI et al (2001) Steroid compounds from the Pacific starfish *Lysastrosoma anthosticta*. Russ Chem Bull 50:313–315. https://doi.org/10.1023/A:1009503006894

Levina EV, Kalinovskii AI, Stonik VA et al (2003) Steroidal polyols from Far-Eastern starfishes *Henricia sanguinolenta* and *H. leviuscula leviuscula*. Russ Chem Bull 52:1623–1628. https://doi.org/10.1023/A:1025613714119

Levina EV, Kalinovskii AI, Andriyashchenko PV et al (2004) A new steroidal glycoside phrygioside A and its aglycone from the starfish *Hippasteria phrygiana*. Russ Chem Bull 53:2634–2638. https://doi.org/10.1007/s11172-005-0165z

Levina EV, Kalinovsky AI, Andriyashenko PV et al (2005) Phrygiasterol, a cytotoxic cyclopropane-containing polyhydroxysteroid, and related compounds from the Pacific starfish *Hippasteria phrygiana*. J Nat Prod 68:1541–1544. https://doi.org/10.1021/np049610t

Levina EV, Kalinovskii AI, Dmitrenok PS (2007) Steroid compounds from the Far East starfish *Pteraster obscurus* and the ophiura *Asteronyx loveni*. Bioorg Khim 33:365–370. https://doi.org/10.1134/S1068162007030119

Levina EV, Kalinovsky A, Dmitrenok PS (2009) Bioactive Steroidal Sulphates from the Ambulakrums of the Pacific Starfish *Lysastrosoma anthosticta*. Nat Prod Commun 4:1041–1046

Levina EV, Kalinovsky A, Dmitrenok PS et al (2010) Two new steroidal saponins, Hylodoside A and Novaeguinoside Y, from the starfish *Leptasterias hylodes reticulata* and *Culcita novaeguineae* (Juvenile). Nat Prod Commun 5:1737–1742

Li C, Haug T, Styrvold OB et al (2008) Strongylocins, novel antimicrobial peptides from the green sea urchin, *Strongylocentrotus droebachiensis*. Dev Comp Immunol 32:1430–1440. https://doi.org/10.1016/j.dci.2008.06.013

Li M, Miao ZH, Chen Z et al (2010) Echinoside A, a new marine-derived anticancer saponin, targets topoisomerase2α by unique interference with its DNA binding and catalytic cycle. Ann Oncol 21:597–607. https://doi.org/10.1093/annonc/mdp335

Li Z, Chen G, Lu X et al (2013) Three new steroid glycosides from the starfish *Asterina pectinifera*. Nat Prod Res 27:1816–1822. https://doi.org/10.1080/14786419.2012.761621

Liu HH, Ko WC, Hu ML (2002) Hypolipidemic effect of glycosaminoglycans from the sea cucumber *Metriatyla scabra* in rats fed a cholesterol-supplemented diet. J Agric Food Chem 50:3602–3606. https://doi.org/10.1021/jf020070k

Liu BS, Yi YH, Li L et al (2007) Arguside A: A new cytotoxic triterpene glycoside from the sea cucumber *Bohadschia argus* Jaeger. Chem Biodivers 4:2845–2851. https://doi.org/10.1002/cbdv.200790234

Liu BS, Yi YH, Li L et al (2008a) Argusides D and E, two new cytotoxic triterpene glycosides from the sea cucumber *Bohadschia argus* Jaeger. Chem Biodivers 5:1425–1433. https://doi.org/10.1002/cbdv.200890131

Liu BS, Yi YH, Li L et al (2008b) Argusides B and C, two new cytotoxic triterpene glycosides from the sea cucumber *Bohadschia argus* Jaeger. Chem Biodivers 5:1288–1297. https://doi.org/10.1002/cbdv.200890115

Liu Y, Yan H, Wen K et al (2011) Identification of epidioxysterol from south China sea urchin *Tripneustes gratilla* Linnaeus and its cytotoxic activity. J Food Biochem 35:932–938. https://doi.org/10.1111/j.1745-4514.2010.00426.x

Liu X, Sun Z, Zhang M et al (2012) Antioxidant and antihyperlipidemic activities of polysaccharides from sea cucumber *Apostichopus japonicus*. Carbohydr Polym 90:1664–1670. https://doi.org/10.1016/j.carbpol.2012.07.047

Lu Y, Li H, Wang M et al (2018) Cytotoxic polyhydroxysteroidal glycosides from starfish *Culcita novaeguineae*. Mar Drugs 16:92. https://doi.org/10.3390/md16030092

Ma N, Tang HF, Qiu F et al (2009a) A new polyhydroxysteroidal glycoside from the starfish *Anthenea chinensis*. Chinese Chem Lett 20:1231–1234. https://doi.org/10.1016/j.cclet.2009.05.012

Ma XG, Tang HF, Zhao CH et al (2009b) Two new 24-hydroxylated asterosaponins from *Culcita novaeguineae*. Chinese Chem Lett 20:1227–1230. https://doi.org/10.1016/J.CCLET.2009.05.031

Ma N, Tang HF, Qiu F et al (2010) Polyhydroxysteroidal glycosides from the starfish *Anthenea chinensis*. J Nat Prod 73:590–597. https://doi.org/10.1021/np9007188

Ma X, Kundu N, Collin PD et al (2012) Frondoside A inhibits breast cancer metastasis and antagonizes prostaglandin E receptors EP_4 and EP_2. Breast Cancer Res Treat 132:1001–1008. https://doi.org/10.1007/s10549-011-1675-z

Mackie AM, Singh HT, Owen JM (1977) Studies on the distribution, biosynthesis and function of steroidal saponins in echinoderms. Comp Biochem Physiol B Comp Biochem 56:9–14. https://doi.org/10.1016/0305-0491(77)90214-0

Maier MS (2008) Biological activities of sulfated glycosides from echinoderms. In: Atta-Ur-Rahman (ed) Studies in natural products Chemistry. Elsevier Science Publisher 35:311–354

Maier MS, Roccatagliata A, Seldes AM (1993) Two Novel Steroidal Glycoside Sulphates from the Starfish *Cosmasterias lurida*. J Nat Prod 56:939–942. https://doi.org/10.1021/np50096a020

Maier MS, Roccatagliata AJ, Kuriss A et al (2001) Two new cytotoxic and virucidal trisulphated triterpene glycosides from the antarctic sea cucumber *Staurocucumis liouvillei*. J Nat Prod 64:732–736. https://doi.org/10.1021/np000584i

Maltsev II, Stonik VA, Kalinovsky AI (1984) Triterpene glycosides from sea cucumber *Stichopus japonicus* Selenka. Comp Biochem Physiol B Comp Biochem 78:421–426

Maltsev II, Stekhova SI, Schentsova EB et al (1985) Antimicrobial activities of glycosides from the sea cucumbers of family Stichopodidae. Khim-Pharm Zhurn 19:54–56

Malyarenko TV, Kicha AA, Ivanchina NV et al (2010) Three new polyhydroxysteroids from the tropical starfish *Asteropsis carinifera*. Russ J Bioorganic Chem 36:755–761. https://doi.org/10.1134/S1068162010060129

Malyarenko TV, Kicha AA, Ivanchina NV et al (2011) Cariniferosides A–F and other steroidal biglycosides from the starfish *Asteropsis carinifera*. Steroids 76:1280–1287. https://doi.org/10.1016/J.STEROIDS.2011.06.006

Malyarenko TV, Kicha AA, Ivanchina NV (2012) Asteropsiside A and other asterosaponins from the starfish *Asteropsis carinifera*. Russ Chem Bull 61:1986–1991. https://doi.org/10.1007/s11172-012-0275-3

Malyarenko TV, Kicha AA, Ivanchina NV et al (2014) Asterosaponins from the Far Eastern starfish *Leptasterias ochotensis* and their anticancer activity. Steroids 87:119–127. https://doi.org/10.1016/j.steroids.2014.05.027

Malyarenko TV, Malyarenko OS, Ivanchina NV et al (2015) Four new sulfated polar steroids from the Far Eastern starfish *Leptasterias ochotensis*: Structures and activities. Mar Drugs 13:4418–4435. https://doi.org/10.3390/md13074418

Malyarenko OS, Dyshlovoy SA, Kicha AA et al (2017) The inhibitory activity of Luzonicosides from the starfish *Echinaster luzonicus* against human melanoma cells. Mar Drugs 15:1–11. https://doi.org/10.3390/md15070227

Maoka T, Nakachi S, Kobayashi R et al (2015) A new carotenoid, 9Z,9′Z-tetrahydroastaxanthin, from the sea cucumber *Plesiocolochirus minutus*. Tetrahedron Lett 56:5954–5955. https://doi.org/10.1016/j.tetlet.2015.09.060

Marques J, Vilanova E, Mourão PAS et al (2016) Marine organism sulphated polysaccharides exhibiting significant antimalarial activity and inhibition of red blood cell invasion by Plasmodium. Sci Rep 6:1–14. https://doi.org/10.1038/srep24368

Mats MN, Korkhov VV, Stepanov VR et al (1990) The contraceptive activity of triterpene glycosides-the total sum of holotoxins A1 and B1 and holothurin A in an experiment. Farmakol Toksikol 53:45–47

Mayer AMS, Hamann MT (2002) Marine pharmacology in 1999: compounds with antibacterial, anticoagulant, antifungal, anthelmintic, anti-inflammatory, antiplatelet, antiprotozoal and antiviral activities affecting the cardiovascular, endocrine, immune and nervous systems, and other misc. Comp Biochem Physiol 132:315–339. https://doi.org/10.1007/s10126-003-0007-7

Mayer AMS, Glaser KB, Cuevas C et al (2010) The odyssey of marine pharmaceuticals: a current pipeline perspective. Trends Pharmacol Sci 31:255–265. https://doi.org/10.1016/j.tips.2010.02.005

Mayer AMS, Rodr AD, Taglialatela-Scafati O et al (2017) Marine pharmacology in 2012 – 2013: marine the immune and nervous systems, and other miscellaneous mechanisms of action. Mar Drugs 15:1–61. https://doi.org/10.3390/md15090273

Melek FR, Tadros MM, Yousif F et al (2012) Screening of marine extracts for schistosomicidal activity in vitro. Isolation of the triterpene glycosides echinosides A and B with potential activity from the Sea Cucumbers *Actinopyga echinites* and *Holothuria polii*. Pharm Biol 50:490–496. https://doi.org/10.3109/13880209.2011.615842

Menchinskaya ES, Pislyagin EA, Kovalchyk SN et al (2014) Antitumor activity of cucumarioside A2-2. Chemotherapy 59:181–191. https://doi.org/10.1159/000354156

Miller A, Kerr A, Paulay G et al (2017) Moleculra phylogeny of extant Holothuroidae (Echinodermata). Mol Phylogenetics Evol 111:110–131. https://doi.org/10.1016/j.ympev.2017.02.014

Minale L, Pizza C, Zollo F (1983) Starfish saponins. Part 9. A novel 24-O-glycosidated steroid from the starfish *Hacelia attenuata*. Experientia 39:567–569

Minale L, Pizza C, Plomitallo A et al (1984a) Starfish saponins. XII. Sulphated steroid glycosides from the starfish *Hacelia attenuata*. Gazz Chim Ital 114:151–158

Minale L, Pizza C, Riccio R et al (1984b) Minor Polyhydroxylated Sterols from the Starfish *Protoreaster nodosus*. J Nat Prod 47:790–795. https://doi.org/10.1021/np50035a006

Minale L, Pizza C, Riccio R et al (1984c) Starfish Saponins, XIII. Occurrence of Nodososide in the Starfish *Acanthaster Planci* and *Linckia Laevigata*. J Nat Prod 47:558. https://doi.org/10.1021/np50033a037

Minale L, Riccio R, Squillace Greco O et al (1985) Starfish saponins-XVI. Composition of the steroidal glycoside sulphates from the starfish *Luidia maculata*. Comp Biochem Physiol B Biochem 80:113–118. https://doi.org/10.1016/0305-0491(85)90431-6

Minale L, Riccio R, Zollo F (1995) Structural Studies on Chemical Constituents of Echinoderms. Stud Nat Prod Chem 15:43–110

Minale L, Riccio R, De Simone F et al (1997) Starfish saponin II: 22,23-Epoxysteroids, minor genins from the starfish *Echinaster sepositus*. Tetrahedron Lett 20:645–648. https://doi.org/10.1038/sj.onc.1209954

Miyamoto T, Togawa K, Higuchi R et al (1990a) Constituents of holothuroidea, Isolation and structures of three triterpenoid aglycones, cucumechinol A, B, and C, from the sea cucumber *Cucumaria echinata*. Liebigs Ann 1990:39–42. https://doi.org/10.1002/jlac.199019900106

Miyamoto T, Togawa K, Higuchi R et al (1990b) Six newly identified biologically active triterpenoid glycoside sulfates from the sea cucumber *Cucumaria echinata*. Liebigs Ann 1990:453–460

Miyamoto T, Togawa K, Higuchi R (1992) Structures of four new triterpenoid oligoglycosides: DS-penaustrosides A, B, C, and D from the sea cucumber *Pentacta australis*. J Nat Prod 55:940–946. https://doi.org/10.1021/np50085a014

Molinski TF, Dalisay DS, Lievens SL et al (2009) Drug development from marine natural products. Nat Rev Drug Discov 8:69–85. https://doi.org/10.1038/nrd2487

Mona MH, Omran NEE, Mansoor MA (2012) Antischistosomal effect of holothurin extracted from some Egyptian sea cucumbers. Pharm Biol 50:1144–1150. https://doi.org/10.3109/13880209.2012.661741

Mondol MAM, Shin HJ, Rahman MA et al (2017) Sea cucumber glycosides: Chemical structures, producing species and important biological properties. Mar Drugs 15:317. https://doi.org/10.3390/md15100317

Moraes G, Northcote PT, Silchenko AS et al (2005) Mollisosides A, B_1, and B_2: Minor triterpene glycosides from the New Zealand and south Australian sea cucumber *Australostichopus mollis*. J Nat Prod 68:842–847. https://doi.org/10.1021/np050049o

Mourão PAS, Guimarães MAM, Mulloy B (1998) Antithrombotic activity of a fucosylated chondroitin sulphate from echinoderm: Sulphated fucose branches on the polysaccharide account for its antithrombotic action. Br J Haematol 101:647–652. https://doi.org/10.1046/j.1365-2141.1998.00769.x

Muniain C, Centurion R, Careaga C et al (2008) Chemical ecology and bioactivity of triterpene glycosides from the sea cucumber *Psolus patagonicus (Dendrochirotida: Psolidae)*. J Mar Biol Assoc UK 88(4):817–823

Murray AP, Muniaín C, Seldes AM et al (2001) Patagonicoside A: A novel antifungal disulfated triterpene glycoside from the sea cucumber *Psolus patagonicus*. Tetrahedron 57:9563–9568. https://doi.org/10.1016/S0040-4020(01)00970-X

Nance JM, Braithwaite LF (1979) The function of mucous secretions in the cushion star *Pteraster tesselatus* Ives. J Exp Mar Biol Ecol 40:259–266. https://doi.org/10.1016/0022-0981(79)90055-8

Ngoan BT, Hanh TTH, Vien LT et al (2015) Asterosaponins and glycosylated polyhydroxysteroids from the starfish *Culcita novaeguineae* and their cytotoxic activities. J Asian Nat Prod Res 17:1010–1017. https://doi.org/10.1080/10286020.2015.1041930

Nguyen TH, Um BH, Kim SM (2011) Two unsaturated fatty acids with Potent α-Glucosidase inhibitory Activity purified from the body wall of sea cucumber (*Stichopus japonicus*). J Food Sci 76:208–214. https://doi.org/10.1111/j.1750-3841.2011.02391.x

Nigrelli RF, Jakowska S (1960) Effects of Holothurin, a steroid saponin from the Bahamian sea cucumber (*Actinopyga Agassizi*), on various biological systems. Ann N Y Acad Sci 17:884–892. https://doi.org/10.1111/j.1749-6632.1960.tb26431.x

Nigrelli R, Zahl P (1952) Some biological characteristics of Holothurin. Exp Biol Med 81(2):379–380. https://doi.org/10.3181/00379727-81-19882

Nishikawa Y, Furukawa A, Shiga I et al (2015) Cytoprotective effects of lysophospholipids from sea cucumber *Holothuria atra*. PLoS One 10:1–14. https://doi.org/10.1371/journal.pone.0135701

Nuzzo G, Gomes BA, Amodeo P et al (2017) Isolation of chemigrene sesquiterpenes and absolute configuration of isoobtusadiene from the brittle star *Ophionereis reticulata*. J Nat Prod 80:3049–3053

Oda T, Shinmura N, Nishioka Y et al (1999) Effect of the Hemolytic Lectin CEL-III from Holothuroidea *Cucumaria echinata* on the ANS Fluorescence Responses in Sensitive MDCK and Resistant CHO Cells. J Biochem 125:713–720

Okano K, Ohkawa N, Ikegami S (1985) Structure of ovarian Asterosaponin-4, an inhibitor of spontaneous oocyte maturation from the Starfish *Asterias amurensis*. Agri Biol Chemi 49:2823–2826

Oleinikova GK, Kuznetsova TA (1983) Two-stage smith degradation of holothurin B_1 from the holothurian *Holothuria floridana*. Chem Nat Compd 19:508–509. https://doi.org/10.1007/BF00575731

Oleinikova GK, Kuznetsova TA, Rovnykh NV et al (1982) Glycosides of marine invertebrates. XVIII. Holothurin A_2 from the Caribbean holothurian *Holothuria floridana*. Chem Nat Compd 18:501–502

Omran NE, Khedr AM (2015) Structure elucidation, protein profile and the antitumor effect of the biological active substance extracted from sea cucumber *Holothuria polii*. Toxicol Ind Health 31:1–8. https://doi.org/10.1177/0748233712466135

Ozupek NM, Cavas L (2017) Triterpene glycosides associated antifouling activity from *Holothuria tubulosa* and *H. polii*. Reg Stud Mar Sci 13:32–41. https://doi.org/10.1016/j.rsma.2017.04.003

Paine RT (1969) A Note on Trophic Complexity and Community Stability. Am Nat 103:91–93

Palagiano E, Zollo F, Minale L et al (1996) Isolation of 20 glycosides from the starfish *Henricia downeyae*, collected in the Gulf of Mexico. J Nat Prod 59:348–354. https://doi.org/10.1021/np9601014

Palyanova NV, Pankova TM, Starostina MV et al (2013) Neuritogenic and neuroprotective effects of polar steroids from the far east starfishes *Patiria pectinifera* and *Distolasterias nipon*. Mar Drugs 11:1440–1455. https://doi.org/10.3390/md11051440

Pan K, Inagaki M, Ohno N et al (2010) Identification of sixteen new galactocerebrosides from the starfish *Protoreaster nodosus*. Chem Pharm Bull (Tokyo) 58:470–474. https://doi.org/10.1248/cpb.58.470

Pan K, Tanaka C, Inagaki M et al (2012) Isolation and structure elucidation of GM4-type gangliosides from the Okinawan starfish *Protoreaster nodosus*. Mar Drugs 10:2467–2480. https://doi.org/10.3390/md10112467

Park HY, Kim JY, Kim HJ et al (2009) Insecticidal and repelenet activities of crude saponin from the starfish *Asterias amuerensis*. Fish Sci Technol 12:1–5

Park J-I, Bae H-R, Kim CG et al (2014) Relationships between chemical structures and functions of triterpene glycosides isolated from sea cucumbers. Front Chem 2:1–14. https://doi.org/10.3389/fchem.2014.00077

Paul VJ, Arthur KE, Ritson-Williams R et al (2007) Chemical Defenses: From Compounds to Communities Linked. Biol Bull 213:226–251

Peng Y, Zheng J, Huang R et al (2010) Polyhydroxy steroids and saponins from China sea starfish *Asterina pectinifera*. Chem Pharm Bull 58:856–858. https://doi.org/10.1248/cpb.58.856

Pereira MS, Mulloy B, Moura PAS (1999) Structure and Anticoagulant Activity of Sulphated Fucans. J Biol Chem 274:7656–7667. https://doi.org/10.1074/jbc.M002422200

Pettit GR, Herald CL, Herald DL (1976) Antineoplastic agents XLV: Sea cucumber cytotoxic saponins. J Pharm Sci 65:1975–1976

Pettit GR, Hasler JA, Paull KD et al (1981) Antineoplastic Agents. 76. The Sea Urchin *Strongylocentrotus droebachiensis*. J Nat Prod 44:701–704. https://doi.org/10.1021/np50018a015

Pislyagin EA, Aminin DL, Silchenko AS et al (2014) Immunomodulatory action of triterpene glycosides isolated from the sea cucumber *Actinocucumis typica*. Structure-activity relationships. Nat Prod Commun Immunomodul 9:771–772

Pislyagin EA, Manzhulo IV, Gorpenchenko TY et al (2017) Cucumarioside $A_{2\text{-}2}$ causes macrophage activation in mouse spleen. Mar Drugs 15:1–15. https://doi.org/10.3390/md15110341

Pizza C, Minale L, Laurent D (1985a) Starfish saponins: XXVII. Steroidal glycosides from the starfish *Choriaster granulatus*. Gazz Chim Ital 115:585–589

Pizza C, Pezzullo P, Minale L et al (1985b) Starfish saponins. Part 20. Two novel steroidal glycosides from the starfish *Acanthaster planci* (L). J Chem Res Synop 1985:76–77

Pocsidio GN (1983) The Mutagenicity Potential of Holothurin of Some Philippine Holothurin. Philipp J Sci 112:1–12

Polikarpova SI, Volkova ON, Sedov AM et al (1990) Cytogenetic study of the mutagenicity of cucumarioside. Genetika 26:1682–1685

Popov AM (2002) A comparative study of the hemolytic and cytotoxic activities of triterpenoids isolated from ginseng and sea cucumbers. Biol Bull 29:120–128. https://doi.org/10.1023/A:1014398714718

Popov A, Atopkina L, Samoshina NF et al (1994) Immunomodulating activity of tetracyclic triterpene glycosides of the dammarane and holostane series. Antibiot Khimioter 39:19–25

Popov RS, Avilov SA, Silchenko AS et al (2014) Cucumariosides F_1 and F_2, two new triterpene glycosides from the sea cucumber *Eupentacta fraudatrix* and their LC-ESI MS / MS identification in the star fish *Patiria pectinifera*, a predator of the sea cucumber. Biochem Syst Ecol 57:191-197. doi:https://doi.org/10.1016/j.bse.2014.08.009

Popov R, Ivanchina N, Kalinovsky A et al (2016) Aphelasteroside F, a new asterosaponin from the far eastern starfish *Aphelasterias japonica*. Nat Prod Commun 11:1247–1250

Prokof'eva NG, Chaikina EL, Kicha AA et al (2003) Biological activities of steroid glycosides from starfish. Comp Biochem Physiol Part B Biochem Mol Biol 134:695–701. https://doi.org/10.1016/S1096-4959(03)00029-0

Qi J, Ojika M, Sakagami Y (2002) linckosides A and B, two new neuritogenic steroid glycosides from the okinawan starfish *Linckia laevigata*. Bioorg Med Chem 10:1961–1966. https://doi.org/10.1016/S0968-0896(02)00006-8

Qi J, Ojika M, Sakagami Y (2004) Linckosides C–E, three new neuritogenic steroid glycosides from the Okinawan starfish *Linckia laevigata*. Bioorg Med Chem 12:4259–4265. https://doi.org/10.1016/j.bmc.2004.04.049

Qi J, Han C, Sasayama Y et al (2006) Granulatoside A, a starfish steroid glycoside, enhances PC12 cell neuritogenesis induced by nerve growth factor through an activation of MAP kinase. ChemMedChem 1:1351–1354. https://doi.org/10.1002/cmdc.200600190

Riccio R, De Simone E, Dini A et al (1981) Starfish saponins VI - unique 22,23-epoxysteroidal cyclic glycosides, minor constituents from *Echinaster sepositus*. Tetrahedron Lett 22:1557–1560. https://doi.org/10.1016/S0040-4039(01)90377-6

Riccio R, Dini A, Minale L et al (1982a) Starfish saponins VII. Structure of Luzonicoside, a further steroidal cyclic glycoside from the pacific starfish *Echinaster Luzonicus*. Experientia 38:68–70

Riccio R, Minale L, Pagonis S et al (1982b) A novel group of highly hydroxylated steroids from the starfish *Protoreaster nodosus*. Tetrahedron 38:3615–3622. https://doi.org/10.1016/0040-4020(82)80069-0

Riccio R, D'Auria MV, Minale L (1985a) Unusual sulfated marine steroids from the ophiuroid *Ophioderma longicaudum*. Tetrahedron 41:6041–6046. https://doi.org/10.1016/S0040-4020(01)91445-0

Riccio R, Greco OS, Minale L et al (1985b) Starfish saponins, part 18. steroidal glycoside sulfates from the starfish *Linckia laevigata*. J Nat Prod 48:97–101. https://doi.org/10.1021/np50037a017

Riccio R, Pizza C, Squillace-Greco O et al (1985c) Starfish saponins. Part 17. steroidal glycoside sulfates from the starfish *Ophidiaster ophidianus* (Lamarck), and *Hacelia attenuata* (Gray). J Chem Soc Perkin Trans I 1:655–660

Riccio R, Zollo F, Finamore E et al (1985d) Starfish saponins, 19. A novel steroidal glycoside sulphate from the starfishes *Protoreaster nodosus* and *Pentaceraster alveolatus*. J Nat Prod 48:266–272. https://doi.org/10.1021/np50038a011

Riccio R, D'Auria MV, Minale L (1986a) Two New Steroidal Glycoside Sulphates, Longicaudoside-A and -B, from the Mediterranean Ophiuroid *Ophioderma longicaudum*. J Org Chem 51:533–536. https://doi.org/10.1021/jo00354a025

Riccio R, Greco OS, Minale L et al (1986b) Starfish saponins, part 28. steroidal glycosides from pacific starfishes of the genus *Nardoa*. J Nat Prod 49:1141–1143. https://doi.org/10.1021/np50048a036

Riccio R, Iorizzi M, Greco OS et al (1986c) Starfish Saponins, Part 22. Asterosaponins from the Starfish *Halityle Regularis*: A Novel 22,23-Epoxysteroidal Glycoside Sulfate. J Nat Prod 48:756–765. https://doi.org/10.1021/np50041a007

Riccio R, Iorizzi M, Minale L (1986d) Starfish Saponins. Isolation of Sixteen Steroidal Glycosides and Three Polyhydroxysteroids from the Mediterranean Starfish *Coscinasterias Tenuispina*. Bull des Sociétés Chim Belges 95:869–893. https://doi.org/10.1002/bscb.19860950912

Riccio R, Iorizzi M, Minale L et al (1988) Starfish saponins. Part 34. Novel steroidal glycoside sulphates from the starfish *Asterias amurensis*. J Chem Soc Perkin Trans 1:1337–1347. https://doi.org/10.1039/P19880001337

Rideout JA, Smith NB, Sutherland MD (1979) Chemical defense of crinoids by polyketide sulphates. Experientia 35:1273–1274. https://doi.org/10.1007/BF01963951

Ridzwan BH (2007) Sea cucumber: the Malaysian heritage. Research Centre, IULM, Kuala Lumpur

Roccatagliata AJ, Maier MS, Seldes AM (1994) Starfish saponins, part 2. Steroidal oligoglycosides from the starfish *Cosmasterias lurida*. J Nat Prod 57:747–754. https://doi.org/10.1021/np50108a010

Roccatagliata AJ, Maier MS, Seldes AM et al (1996) Antiviral sulphated steroids from the ophiuroid *Ophioplocus januarii*. J Nat Prod 59:887–889. https://doi.org/10.1021/np960171a

Rodriguez J, Riguera R (1989) Lefevreiosides: four novel triterpenoid glycosides from the sea cucumber *Cucumaria lefevrei*. ChemInform 21:2620–2636. https://doi.org/10.1002/chin.199014299

Rodriguez J, Castro R, Riguera R (1991) Holothurinosides: new antitumor non sulphated triterpenoid glycosides from the sea cucumber *Holothuria forskali*. Tetrahedron 47:4753–4762

Ruppert EE, Fox RS, Barnes RD (2004) Invertebrate zoology: a functional evolutionary approach. Thomson Brook/Cole, Belmont

Sallivan TD, Ladue KT, Nigrelli RF (1955) The effect of holothurin, a steroid saponin of animal origin, on Krebs-2 ascites tumors in Swiss mice. Zoologica 40:49–52

Sandvoss M, Pham H, Levsen K et al (2000) Isolation and structural elucidation of Steroid Oligoglycosides from the starfish *Asterias rubens* by means of direct online LC-NMR-MS hyphenation and One- and two-dimensional NMR investigations. Eur J Org Chem 2000:1253–1262

Sandvoss M, Preiss A, Levsen K et al (2003) Two new asterosaponins from the starfish Asterias rubens: Application of a cryogenic NMR probe head. Magn Reson Chem 41:949–954

Schoenmakers HJN (1979) In vitro biosynthesis of steroids from cholestrol by the ovaries and pyloric caeca of the starfish *Asterias rubens*. Comp Bochem Physiol B 63:179–184

Sedov AM, Shepeleva IB, Zakharova NS et al (1984) Effect of cucumarioside (a triterpene glycoside from the holothurian *Cucumaria japonica*) on the development of an immune response in mice to corpuscular pertussis vaccine. Zhurnal mikrobiologii, epidemiologii, i immunobiologii 9:100–104

Sedov AM, Apollonin AV, Sevast'ianova EK et al (1990) Stimulation of nonspecific antibacterial resistance of mice to opportunistic gram-negative microorganisms with triterpene glycosides from Holothuroidea. Antibiot Khimioter 35:23–26

Shang X, Liu X, Zhang J et al (2014) Traditional chinese medicine -Sea urchin. Mini-Rev Med Chem 14:537–542

Sharypov VF, Chumak AD, Stonik VA et al (1981) Glycosides of marine invertebrates. X. The structure of stichoposides A and B from the holothurians *Stichopus cloronotus*. Chem Nat Compd 17:139–142

Silchenko AS, Avilov SA, Antonov AA et al (2002) Triterpene glycosides from the deep-water North-Pacific sea cucumber *Synallactes nozawai* Mitsukuri. J Nat Prod 65:1802–1808. https://doi.org/10.1021/np0202881

Silchenko AS, Avilov SA, Kalinin VI et al (2004) Pseudostichoposide B – new triterpene glycoside with unprecedent type of sulfatation from the deep-water North-Pacific sea cucumber *pseudostichopus trachus*. Nat Prod Res 18:565–570. https://doi.org/10.1080/14786410310001630591

Silchenko AS, Avilov SA, Antonov AS et al (2005a) Glycosides from the sea cucumber *Cucumaria frondosa*. III. Structure of frondosides A_{2-1}, A_{2-2}, A_{2-3}, and A_{2-6}, four new minor monosulphated triterpene glycosides. Can J Chem 83:21–27. https://doi.org/10.1139/v05-243

Silchenko AS, Avilov SA, Antonov AS et al (2005b) Glycosides from the sea cucumber *Cucumaria frondosa*. IV. Structure of frondosides A_{2-4}, A_{2-7}, and A_{2-8}, three new minor monosulphated triterpene glycosides. Can J Chem 83:2120–2126. https://doi.org/10.1139/v05-243

Silchenko AS, Stonik VA, Avilov SA et al (2005c) Holothurins B_2, B_3, and B_4, new triterpene glycosides from Mediterranean sea cucumbers of the genus Holothuria. J Nat Prod 68:564–567. https://doi.org/10.1021/np049631n

Silchenko AS, Avilov SA, Antonov AS et al (2007a) Glycosides from the North Atlantic sea cucumber *Cucumaria frondosa* — Structures of five new minor trisulfated triterpene oligoglycosides, frondosides A_{7-1}, A_{7-2}, A_{7-3}, A_{7-4}, and isofrondoside C. Can J Chem 85:626–636. https://doi.org/10.1139/v04-163

Silchenko AS, Avilov SA, Kalinin VI et al (2007b) Monosulfated triterpene glycosides from *Cucumaria okhotensis* Levin et Stepanov, a new species of sea cucumbers from Sea of Okhotsk. Bioorg Khim 33:81–90. https://doi.org/10.1134/S1068162007010098

Silchenko AS, Avilov SA, Kalinin VI et al (2008) Constituents of the sea cucumber *Cucumaria okhotensis*. Structures of okhotosides B_1-B_3 and cytotoxic activities of some glycosides from this species. Nat Prod 71:351–356. https://doi.org/10.1021/np0705413

Silchenko AS, Kalinovsky AI, Avilov SA et al (2012a) Triterpene glycosides from the sea cucumber *Eupentacta fraudatrix*. Structure and biological action of Cucumariosides A_1, A_3, A_4, A_5, A_6, A_{12} and A_{15}, seven new minor non-sulfated tetraosides and unprecedented 25-keto, 27-norholostane aglycone. Nat Prod Commun 7:517–525

Silchenko AS, Kalinovsky AI, Avilov SA et al (2012b) Triterpene Glycosides from the Sea Cucumber *Eupentacta fraudatrix*. Structure and Cytotoxic Action of Cucumariosides A_2, A_7, A_9, A_{10}, A_{11}, A_{13} and A_{14}, Seven New Minor Non-Sulfated Tetraosides and an Aglycone with an Uncommon 18-Hydroxy Group. Nat Prod Commun Triterpene 7:845–852

Silchenko AS, Kalinovsky AL, Avilov SA et al (2012c) Structures and cytotoxic properties of cucumariosides H_2,H_3 and H_4 from the see cucumber *Eupentacta fraudatrix*. Nat Prod Res 26(19):1765–1774. https://doi.org/10.1080/14786419.2011.602637

Silchenko AS, Kalinovsky AL, Avilov SA et al (2012d) Triterpene glycosides from the sea cucumber *Eupentacta fraudatrix*. Structure and biological action of Cucumariosides B_1, and B_2,two new minor non-sulfated tetraosides and unprecedented 25-keto, 27-norholostane aglycone. Nat Prod Commun 7:517–525

Silchenko AS, Kalinovsky AI, Avilov SA et al (2013a) Triterpene glycosides from the sea cucumber *Eupentacta fraudatrix*. Structure and biological action of cucumariosides I_1, I_3, I_4, three new minor disulfated pentaosides. Nat Prod Commun 8:1053–1058

Silchenko AS, Kalinovsky AI, Avilov SA et al (2013b) Structures and biological activities of typicosides A_1, A_2, B_1, C_1 and C_2, triterpene glycosides from the sea cucumber *Actinocucumis typica*. Nat Prod Commun 8:301–310

Silchenko AS, Kalinovsky AI, Avilov SA et al (2013c) Structure of cucumarioside I_2 from the sea cucumber *Eupentacta fraudatrix* (Djakonov et Baranova) and cytotoxic and immunostimulatory activities of this saponin and relative compounds. Nat Prod Res 27:1776–1783. https://doi.org/10.1080/14786419.2013.778851

Silchenko AS, Kalinovsky AI, Avilov SA et al (2013d) Triterpene glycosides from Antarctic sea cucumbers IV. Turquetoside A, a 3-O-methylquinovose containing disulfated tetraoside from the sea cucumber *Staurocucumis turqueti* (Vaney, 1906) (=*Cucumaria spatha*). Biochem Syst Ecol 51:45–49. https://doi.org/10.1016/j.bse.2013.08.012

Silchenko AS, Kalinovsky AI, Avilov SA et al (2013e) Structure and biological action of Cladolosides B_1, B_2, C, C_1, C_2 and D, six new triterpene glycosides from the sea cucumber *Cladolabes schmeltzii*. Nat Prod Commun 8:1527–1534

Silchenko AS, Kalinovsky AI, Avilov SA et al (2014a) Structures of Violaceusosides C, D, E and G, Sulfated Triterpene Glycosides from the Sea Cucumber *Pseudocolochirus violaceus* (Cucumariidae, Dendrochirotida). Nat Prod Rep 9:391–399

Silchenko AS, Kalinovsky AI, Avilov SI et al (2014b) Kolgaosides A and B, Two New Triterpene Glycosides from the Arctic Deep Water Sea Cucumber *Kolga hyalina* (Elasipodida: Elpidiidae). Nat Prod Commun 9:1259–1264

Silchenko AS, Kalinovsky AI, Avilov et al (2014c) Triterpene glycosides from the sea cucumber *Cladolabes schmeltzii*. II. Structure and biological action of cladolosides A1-A6. Nat Prod Commun 9:1421-1429

Silchenko AS, Kalinovsky AI, Avilov SA et al (2015a) Colochirosides B_1, B_2, B_3 and C, novel sulfated triterpene glycosides from the sea cucumber *Colochirus robustus* (Cucumariidae, Dendrochirotida). NPC. Nat Prod Commun 10:1687–1694. https://doi.org/10.1080/13531040802284544

Silchenko AS, Kalinovsky AI, Avilov SA et al (2015b) Structures and biological activities of cladolosides C_3, E_1, E_2, F_1, F_2, G, H_1 and H_2, eight triterpene glycosides from the sea cucumber *Cladolabes schmeltzii* with one known and four new carbohydrate chains. Carbohydr Res 414:22–31. https://doi.org/10.1016/j.carres.2015.06.005

Silchenko AS, Kalinovsky AI, Avilov SA et al (2016a) Structures and biogenesis of fallaxosides D_4, D_5, D_6 and D_7, trisulfated non-holostane triterpene glycosides from the sea cucumber Cucumaria fallax. Molecules 21:2–13. https://doi.org/10.3390/molecules21070939

Silchenko AS, Kalinovsky AI, Avilov SA et al (2016b) Colochirosides A_1, A_2, A_3, and D, Four Novel Sulfated Triterpene Glycosides from the Sea Cucumber *Colochirus robustus* (Cucumariidae, Dendrochirotida). Nat Prod Commun 11:381–387

Silchenko AS, Kalinovsky AI, Avilov SA et al (2016c) Colochiroside E, an unusual non-holostane triterpene sulfated trioside from the sea cucumber *Colochirus robustus* and evidence of the impossibility of a 7(8)-double bond Migration in lanostane derivatives having an 18(16)-lactone. Nat Prod Commun 11:741–746

Silchenko AS, Kalinovsky AI, Avilov SA et al (2017a) Cladolosides I_1, I_2, J_1, K_1, K_2 and L_1, monosulfated triterpene glycosides with new carbohydrate chains from the sea cucumber *Cladolabes schmeltzii*. Carbohydr Res 445:80–87. https://doi.org/10.1016/j.carres.2017.04.016

Silchenko AS, Kalinovsky AI, Avilov SA et al (2017b) Nine new triterpene glycosides, magnumosides A_1–A_4, B_1, B_2, C_1, C_2 and C_4, from the Vietnamese sea cucumber *Neothyonidium* (=*Massinium*) *magnum*: Structures and activities against tumor cells independently and in synergy with radioactive irradiation. Mar Drugs 15:1–22. https://doi.org/10.3390/md15080256

Silchenko AS, Kalinovsky AI, Avilov SA et al (2018a) Cladolosides O, P, P_1-P3 and R, triterpene glycosides with two novel types of carbohydrate chains from the sea cucumber *Cladolabes schmeltzii*. Carbohydr Res 468:73–79

Silchenko AS, Kalinovsky AI, Avilov SA et al (2018b) Cladolosides C_4,D_1,D_2, M, M_1, M^2, N, and Q, new triterpene glycosides with diverse carbohydrate chains from the sea cucumber *Cladolabes schmeltzii*. An uncommon 20,21,22,23,24,25,26,27-okta-nor-lanostane aglycone. The synergism of inhibitory action of non-toxic dose of the glycosides and radioactive irradiation on colony formation of HT-29 cancer cells. Carbohydr Res 468:36–44

Silva M, Rodriguez I, Barreiro A et al (2015) First report of ciguatoxins in two starfish species: *Ophidiaster ophidianus* and *Marthasterias glacialis*. Toxins (Basel) 7:3740–3757. https://doi.org/10.3390/toxins7093740

Singh N, Kumar R, Gupta S et al (2008) Antileishmanial activity in vitro and in vivo of constituents of sea cucumber *Actinopyga lecanora*. Parasitol Res 103:351–354. https://doi.org/10.1007/s00436-008-0979-3

Soliman YA, Ibrahim AM, Tadros HRZ et al (2016) Antifouling and Antibacterial Activities of Marine Bioactive Compounds Extracted from some Red Sea Cucumber. Contemp Appl Sci 3:83–103

Song Y, Jin SJ, Cui LH et al (2013) Immunomodulatory effect of *Stichopus japonicus* acid mucopolysaccharide on experimental hepatocellular carcinoma in rats. Molecules 18:7179–7193. https://doi.org/10.3390/molecules18067179

Song J, Li T, Cheng X et al (2016) Sea cucumber peptides exert anti-inflammatory activity through suppressing NF-κB and MAPK and inducing HO-1 in RAW264.7 macrophages. Food Funct 7:2773–2779. https://doi.org/10.1039/c5fo01622c

Stöhr S, O'Hara TD, Thuy B (2012) Global diversity of brittle stars (Echinodermata: Ophiuroidea). PLoS One 7:e31940. https://doi.org/10.1371/journal.pone.0031940

Stonik VA (1986) Some terpenoid and steroid derivatives from echinoderms and sponges. Pure Appl Chem 58:423–436. https://doi.org/10.1351/pac198658030423

Stonik VA, Mal'tsev II, Elyakov GB (1982) The structure of Thelenotosides A and B from the Holothurian *Theleonata ananas*. Chem Nat Compd 18:590–593

Styles TJ (1970) Effect of Holothurin on *Trypanosoma zewisi* Infections in Rats. J Protozool 17:196–198

Sugawara T, Zaima N, Yamamoto A et al (2006) Isolation of Sphingoid Bases of Sea Cucumber Cerebrosides and Their Cytotoxicity against Human Colon Cancer Cells. Biosci Biotechnol Biochem 70:2906–2912. https://doi.org/10.1271/bbb.60318

Sun P, Liu BS, Yi YH et al (2007) A new cytotoxic lanostane-type triterpene glycoside from the sea cucumber *Holothuria impatiens*. Chem Biodivers 4:450–457. https://doi.org/10.1002/cbdv.200790037

Suwanmala J, Lu S, Tang Q et al (2016) Comparison of Antifatigue Activity of Five Sea Cucumber Species in a Mouse Model of Intense Exercise. J Food Nutr Res 4:12–19. https://doi.org/10.12691/jfnr-4-1-3

Tang HF, Yi Y, Li L et al (2005) Three new asterosaponins from the starfish *Culcita novaeguineae* and their bioactivity. Planta Med 71(5):458–463. https://doi.org/10.1055/s-2005-871215

Tang HF, Yi YH, Li L et al (2006) Asterosaponins from the starfish *Culcita novaeguineae* and their bioactivities. Fitoterapia 77:28–34. https://doi.org/10.1016/J.FITOTE.2005.07.009

Tang HF, Yi YH, Li L et al (2009) Bioactive Asterosaponins from the Starfish *Culcita novaeguineae*. J Nat Prod 68:337–341. https://doi.org/10.1021/np0401617

Telford MJ, Lowe CJ, Cameron CB et al (2014) Phylogenomic analysis of echinoderm class relationships supports Asterozoa. Proc R Soc B Biol Sci 281:20140479–20140479. https://doi.org/10.1098/rspb.2014.0479

Thao NP, Cuong NX, Luyen BTT et al (2013) Anti-inflammatory asterosaponins from the starfish *Astropecten monacanthus*. J Nat Prod 76:1764–1770. https://doi.org/10.1021/np400492a

Thao NP, No JH, Luyen BTT et al (2014) Secondary metabolites from Vietnamese marine invertebrates with activity against *Trypanosoma brucei* and *T. cruzi*. Molecules 19:7869–7880. https://doi.org/10.3390/molecules19067869

Thao NP, Luyen BTT, Kim EJ et al (2015a) Steroidal constituents from the edible sea urchin *Diadema savignyi* Michelin induce apoptosis in human cancer cells. J Med Food 18:45–53. https://doi.org/10.1089/jmf.2013.3105

Thao NP, Luyen BTT, Koo JE et al (2015b) Anti-inflammatory components of the Vietnamese starfish *Protoreaster nodosus*. Biol Res 48:12. https://doi.org/10.1186/s40659-015-0002-2

Thiel M, Watling L (2015) Lifestyles and feeding biology. the natural history of the Crustacea, vol 2. Oxford Universiry Press, Oxford

Tian F, Zhang X, Tong Y et al (2005) PE, a new sulphated saponin from sea cucumber, exhibits anti-angiogenic and anti-tumor activities in vitro and in vivo. Cancer Biol Ther 4:874–882. https://doi.org/10.4161/cbt.4.8.1917

Tian F, Zhu C, Zhang X et al (2007) Philinopside E , a new sulfated saponin from sea cucumber, blocks the interaction between kinase insert domain-containing receptor (KDR) and avb3 integrin via binding to the extracellular domain of KDR. Mol Pharmacol 72:545–552. https://doi.org/10.1124/mol.107.036350.receptor

Turischev SN, Bolshakova GB, Sakandelidze OG et al (1991) Influence of complexes of holothurian triterpene glycosides on liver generation. Izv Akad Nauk SSSR, Ser Biol 2:306–310

Van Dyck S, Gerbaux P, Flammang P (2010) Qualitative and quantitative saponin contents in five sea cucumbers from the Indian ocean. Mar Drugs 8:173–189. https://doi.org/10.3390/md8010173

Vázquez MJ, Quiñoá E, Riguera R et al (1992) Santiagoside, the first asterosaponin from an antarctic starfish (*Neosmilaster georgianus*). Tetrahedron 48:6739–6746. https://doi.org/10.1016/S0040-4020(01)80019-3

Vázquez MJ, Quindo E, Riguera R (1993) Helianthoside from *Heliaster helianthus*, an asterosaponin with a C3'- sulfated pyranose. Can J Chem 71:1174–1151

Vien LT, Ngoan BT, Hanh TTH et al (2017) Steroid glycosides from the starfish *Pentaceraster gracilis*. J Asian Nat Prod Res 19:474–480. https://doi.org/10.1080/10286020.2016.1235038

Vien LT, Hoang L, Hanh TTH et al (2018) Triterpene tetraglycosides from the sea cucumber *Stichopus horrens*. Nat Prod Res 32:1039–1043. https://doi.org/10.1080/14786419.2017.1378206

Wang W, Li F, Alam N et al (2002) New saponins from the starfish *Certonardoa semiregularis*. J Nat Prod 65:1649–1656. https://doi.org/10.1021/np020234r

Wang W, Li F, Hong J et al (2003) Four New Saponins from the Starfish *Certonardoa semiregularis*. Chem Pharm Bull 51:435–439. https://doi.org/10.1248/cpb.51.435

Wang W, Hong J, Lee C-O et al (2004a) Cytotoxic Sterols and Saponins from the Starfish *Certonardoa semiregularis*. J Nat Prod 67:584–591. https://doi.org/10.1021/np030427u

Wang W, Jang H, Hong J et al (2004b) Additional cytotoxic sterols and saponins from the starfish *Certonardoa semiregularis*. J Nat Prod 67:1654–1660. https://doi.org/10.1021/np049869b

Wang W, Jang H, Hong J et al (2005) New Cytotoxic Sulphated Saponins from the Starfish *Certonardoa semiregularis*. Arch Pharmacal Res 28:285–289

Wang Z, Zhang H, Yuan W et al (2012) Antifungal nortriterpene and triterpene glycosides from the sea cucumber *Apostichopus japonicus* Selenka. Food Chem 132:295–300. https://doi.org/10.1016/J.FOODCHEM.2011.10.080

Wang XH, Zou ZR, Yi YH et al (2014) Variegatusides: New non-sulphated triterpene glycosides from the sea cucumber *Stichopus variegates* Semper. Mar Drugs 12:2004–2018. https://doi.org/10.3390/md12042004

Ward JA (1960) A further investigation on the swimming reaction of *Stomphia coccinea*. Master thesis. University of Washington

Wen Z, Zhu Z, Shen D et al (2004) Determination of asterosaponins in *Asterias amurensis* and *Craspidaster hesperus* with ultraviolet spectrophotometry. Fenxi Kexue Xuebao 20:592–594

Wen J, Hu C, Fan S (2010) Chemical composition and nutritional quality of sea cucumbers. J Sci Food Agric 90:2469–2474. https://doi.org/10.1002/jsfa.4108

Wen M, Fu X, Han X et al (2016) Sea cucumber saponin echinoside A (EA) stimulates hepatic fatty acid β-oxidation and suppresses fatty acid biosynthesis coupling in a diurnal pattern. J Nutr Sci Vitaminol (Tokyo) 62:170–177. https://doi.org/10.3177/jnsv.62.170

Wijesinghe WAJP, Vairappan CS, Jeo YJ (2015) Exploitation of Health Promoting Potentials of Edible Sea Cucumber (*Holothuria edulis*): Search of New Bioactive Components as Functional Ingredients. Int Proc Chem Biol Environ Eng 86:36–41. https://doi.org/10.7763/IPCBEE

Wu J, Yi Y-H, Tang H-F et al (2006a) Nobilisides A - C, Three New Triterpene Glycosides from the Sea Cucumber *Holothuria nobilis*. Planta Med 72:932–935. https://doi.org/10.1055/s-2006-931603

Wu J, Yi Y, Tang H et al (2006b) Structure and Cytotoxicity of a New Lanostane-Type Triterpene Glycoside from the Sea Cucumber *Holothuria hilla* holothurians. Chem Biodivers 3:1249–1254

Wu J, Yi Y, Wu H et al (2007a) Studies on the in vitro antifungal and antitumor activities of nobiliside A from the sea cucumber *holothuria nobilis* Selenka. Zhongguo Yaolixue Tongbao 23:139–140

Wu J, Yi YH, Tang HF et al (2007b) Hillasides A and B, two new cytotoxic triterpene glycosides from the sea cucumber *Holothuria hilla* Lesson. J Asian Nat Prod Res 9:609–615. https://doi.org/10.1080/10286020600882676

Wu J, Zhang J, Ding P et al (2009a) Nobiliside C purified from *Holothuria nobilis* for use as antitumor agent. marinlit ID: A21616 (21787). Faming Zhuanli Shenqing Gongkai Shuomin

Wu J, Zhang J, Ding P et al (2009b) Anti-tumor compound hillaside a separated from *Holothuria hilla* CA151:181594. Marinlit ID: A22103 (22279) Faming Zhuanli Shenqing Gongkai Shuomin

Wu FJ, Xue Y, Liu XF et al (2014) The protective effect of eicosapentaenoic acid-enriched phospholipids from sea cucumber *Cucumaria frondosa* on oxidative stress in PC12 cells and SAMP8 mice. Neurochem Int 64:9–17. https://doi.org/10.1016/j.neuint.2013.10.015

Wu M, Xu L, Zhao L et al (2015) Structural analysis and anticoagulant activities of the novel sulphated fucan possessing a regular well-defined repeating unit from sea cucumber. Mar Drugs 13:2063–2084. https://doi.org/10.3390/md13042063

Xu J, Wang Y-M, Feng T-Y et al (2011) Isolation and anti-fatty liver activity of a novel cerebroside from the sea cucumber *Acaudina molpadioides*. Biosci Biotechnol Biochem 75:1466–1471. https://doi.org/10.1271/bbb.110126

Yamada K, Tanabe K, Miyamoto T et al (2008) Isolation and Structure of a Monomethylated Ganglioside Possessing Neuritogenic Activity from the Ovary of the Sea Urchin *Diadema setosum*. Chem Pharm Bull 56:734–737. https://doi.org/10.1248/cpb.56.734

Yamanouchi T (1955) On the poisonous substance contained in Holothurians. Mar Biol Lab 4:183–203

Yang P, Collin P, Madden T et al (2003) Inhibition of proliferation of PC3 cells by the branched-chain fatty acid, 12-methyltetradecanoic acid, is associated with inhibition of 5-lipoxygenase. Prostate 55:281–291. https://doi.org/10.1002/pros.10243

Yang S-W, Chan T-M, Buevich A et al (2007) Novel steroidal saponins, Sch 725737 and Sch 725739, from a marine starfish, *Novodinia antillensis*. Bioorg Med Chem Lett 17:5543–5547. https://doi.org/10.1016/j.bmcl.2007.08.025

Yang X-W, Chen X-Q, Dong G et al (2011) Isolation and structural characterisation of five new and 14 known metabolites from the commercial starfish *Archaster typicus*. Food Chem 124:1634–1638. https://doi.org/10.1016/j.foodchem.2010.08.033

Yano A, Abe A, Aizawa F et al (2013) The effect of eating sea cucumber jelly on Candida load in the oral cavity of elderly individuals in a nursing home. Mar Drugs 11:4993–5007. https://doi.org/10.3390/md11124993

Yayli N, Findlay JA (1999) A triterpenoid saponin from *Cucumaria frondosa*. Phytochemistry 50:135–138. https://doi.org/10.1016/S0031-9422(98)00463-4

Yi Y, Xu Q, Li L et al (2006) Philinopsides A and B , two new sulfated triterpene glycosides from the sea cucumber *Pentacta quadrangularis*. Helv Acta 89:54–63

Yi Y, Sun G, Li L et al (2008) Purification of triterpene saponin compound griseaside A from *Holothuria grisea* Selenka for cancer therapy and antitumor agent development. MarinLit ID: A20342 (20506). Faming Zhuanli Shenqing Gongkai Shuomin

Yibmantasiri P, Leahy DC, Busby BP et al (2012) Molecular basis for fungicidal action of neothyonidioside, a triterpene glycoside from the sea cucumber, *Australostichopus mollis*. Mol Biosyst 8:902. https://doi.org/10.1039/c2mb05426d

Yu S, Ye X, Huang H et al (2015) Bioactive Sulfated Saponins from Sea Cucumber *Holothuria moebii*. Planta Med 81:152–159. https://doi.org/10.1055/s-0034-1383404

Yuan W-H, Yi Y-H, Xue M et al (2008) Two Antifungal Active Triterpene Glycosides from Sea Cucumber *Holothuria* (*Microthele*) *axiloga*. Chin J Nat Med 6:105–108. https://doi.org/10.1016/S1875-5364(09)60010-8

Yuan W-H, Yi Y, Tang H-F et al (2009a) Antifungal Triterpene Glycosides from the Sea Cucumber *Bohadschia marmorata*. Planta Med 75:168–173. https://doi.org/10.1055/s-0028-1088348

Yuan WH, Yi YH, Tan RX et al (2009b) Antifungal triterpene glycosides from the sea cucumber *Holothuria* (*Microthele*) *axiloga*. Planta Med 75:647–653. https://doi.org/10.1055/s-0029-1185381

Yun SH, Park ES, Shin SW et al (2012) Stichoposide C induces apoptosis through the generation of ceramide in leukemia and colorectal cancer cells and shows in vivo antitumor activity. Clin Cancer Res 18:5934–5948. https://doi.org/10.1158/1078-0432.CCR-12-0655

Yun SH, Sim EH, Han SH et al (2018) Holotoxin A1 induces apoptosis by activating acid sphyngomyelinase and neutral sphyngomyelinase in K562 and human primary leukemia cells. Mar Drugs 16:123

Zhan Y-C, Sun Y, Li W et al (2006) A new triterpene glycoside from *Asterias rollentoni*. J Asian Nat Prod Res 8:631–636. https://doi.org/10.1080/10286020500208626

Zhang J (2009) Antitumor effects of nobiliside B from sea cucumber *Holothuria nobilis* and its acetoxy compounds CA151:279085. Zhongguo Haiyang Yaowu 28:41–43

Zhang J-J, Zhu K-Q (2017) A novel antitumor compound nobiliside D isolated from sea cucumber (*Holothuria nobilis* Selenka). Exp Ther Med 14:1653–1658. https://doi.org/10.3892/etm.2017.4656

Zhang S-L, Li L, Yi Y-H et al (2006a) Philinopsides E and F, two new sulfated triterpene glycosides from the sea cucumber *Pentacta quadrangularis*. Nat Prod Res 20:399–407. https://doi.org/10.1080/14786410500185584

Zhang S-Y, Yi Y-H, Tang H-F (2006b) Cytotoxic sulfated triterpene glycosides from the sea cucumber *Pseudocolochirus violaceus*. Chem Biodivers 3:807–817. https://doi.org/10.1002/cbdv.200690083

Zhang S-Y, Yi Y-H, Tang H-F et al (2006c) Two new bioactive triterpene glycosides from the sea cucumber *Pseudocolochirus violaceus*. J Asian Nat Prod Res 8:1–8. https://doi.org/10.1080/10286020500034972

Zhang SY, Yi YH, Tang HF (2006d) Bioactive triterpene glycosides from the sea cucumber *Holothuria fuscocinerea*. J Nat Prod 69:1492–1495. https://doi.org/10.1021/np060106t

Zhang S-L, Li L, Sun P et al (2008) Lecanorosides A and B, two new triterpene glycosides from the sea cucumber *Actinopyga lecanora*. J Asian Nat Prod Res 10:1097-1103. doi:https://doi.org/10.1080/10286020701604813

Zhang B, Xue C, Hu X et al (2012) Dietary sea cucumber cerebroside alleviates orotic acid-induced excess hepatic adipopexis in rats. Lipids Health Dis 11(1). https://doi.org/10.1186/1476-511X-11-48

Zhang G, Ren H-H, Zhang Y-B et al (2013) Chemical constituents of the starfish *Asterias rollestoni* Bell. Biochem Syst Ecol 51:203–206. https://doi.org/10.1016/J.BSE.2013.08.031

Zhao Y, Li B, Dong S et al (2009) A novel ACE inhibitory peptide isolated from *Acaudina molpadioidea* hydrolysate. Peptides 30:1028–1033. https://doi.org/10.1016/j.peptides.2009.03.002

Zhao Q, Liu Z, Xue Y et al (2011) Ds-echinoside A, a new triterpene glycoside derived from sea cucumber, exhibits antimetastatic activity via the inhibition of NF-κB-dependent MMP-9 and VEGF expressions. J Zhejiang Univ Sci B 12:534–544. https://doi.org/10.1631/jzus.B1000217

Zhao Q, Xue Y, Wang J et al (2012) In vitro and in vivo anti-tumour activities of echinoside A and ds-echinoside A from *Pearsonothuria graeffei*. J Sci Food Agric 92:965–974. https://doi.org/10.1002/jsfa.4678

Zollo F, Finamore E, Minale L (1985) Starfish saponins XXIV. Two novel steroidal glycoside sulphates from the starfish *Echinaster sepositus*. Gazz Chim Ital 115:303–306

Zou Z-R, Yi Y-H, Wu H-M et al (2003) Intercedensides A–C, three new cytotoxic triterpene glycosides from the sea cucumber *Mensamaria intercedens* (Lampert). J Nat Prod 66:1055–1060. https://doi.org/10.1021/np030064y

Zou Z, Yi Y, Wu H et al (2005) Intercedensides D-I, cytotoxic triterpene glycosides from the sea cucumber *Mensamaria intercedens* Lampert. J Nat Prod 68:540–546. https://doi.org/10.1021/np040205b

8 Secondary Metabolites of Marine Microbes: From Natural Products Chemistry to Chemical Ecology

Lars-Erik Petersen, Matthias Y. Kellermann, and Peter J. Schupp

Abstract

Marine natural products (MNPs) exhibit a wide range of pharmaceutically relevant bioactivities, including antibiotic, antiviral, anticancer, or anti-inflammatory properties. Besides marine macroorganisms such as sponges, algae, or corals, specifically marine bacteria and fungi have shown to produce novel secondary metabolites (SMs) with unique and diverse chemical structures that may hold the key for the development of novel drugs or drug leads. Apart from highlighting their potential benefit to humankind, this review is focusing on the manifold functions of SMs in the marine ecosystem. For example, potent MNPs have the ability to exile predators and competing organisms, act as attractants for mating purposes, or serve as dye for the expulsion or attraction of other organisms. A large compilation of literature on the role of MNPs in marine ecology is available, and several reviews evaluated the function of MNPs for the aforementioned topics. Therefore, we focused the second part of this review on the importance of bioactive compounds from crustose coralline algae (CCA) and their role during coral settlement, a topic that has received less attention. It has been shown that certain SMs derived from CCA and their associated bacteria are able to induce attachment and/or metamorphosis of many benthic invertebrate larvae, including globally threatened reef-building scleractinian corals. This review provides an overview on bioactivities of MNPs from marine microbes and their potential use in medicine as well as on the latest findings of the chemical ecology and settlement process of scleractinian corals and other invertebrate larvae.

Keywords

Marine natural products · Secondary metabolites · Marine bacteria · Marine fungi · Crustose coralline algae · Settlement · Coral larvae

8.1 Introduction: Definition of Secondary Metabolism

Over millions of years, evolution has created a multitude of diverse organisms and biocoenosis. Besides individual differences within their appearance and way of life, the ability of absorbing, processing, and secreting substances from and into the environment can be found in all living organisms (Madigan et al. 2003). The biosynthesis and breakdown of these substances, including proteins, fats, or nucleic acids, is commonly known as primary metabolism with the compounds involved known as "primary metabolites" (Dewick 2002; Dias et al. 2012). The primary metabolism of plants, animals, humans, and prokaryotic microorganisms shows great similarity and displays the essential uniformity of all living matters; it thus serves as a driving force for the survival and reproduction of all life (Kreis 2007). In contrast, the mechanism by which an organism synthesizes "secondary metabolites" (SMs), frequently associated with the term "natural products" (NPs), is known as secondary metabolism (Dias et al. 2012). SMs are defined as molecules with a molecular weight ranging between 100 and 1000 Da (Breinbauer et al. 2002) and, unlike primary metabolites, are

L.-E. Petersen (✉) · M. Y. Kellermann
Institute of Chemistry and Biology of the Marine Environment, University of Oldenburg, Oldenburg, Germany
e-mail: lars-erik.petersen1@uni-oldenburg.de

P. J. Schupp (✉)
Institute of Chemistry and Biology of the Marine Environment, University of Oldenburg, Oldenburg, Germany

Helmholtz Institute for Functional Marine Biodiversity at the University of Oldenburg (HIFMB), Oldenburg, Germany
e-mail: peter.schupp@uni-oldenburg.de

S. Jungblut et al. (eds.), *YOUMARES 9 - The Oceans: Our Research, Our Future*,
https://doi.org/10.1007/978-3-030-20389-4_8

often found to be unique to an organism or a specific taxonomic group. They do not directly contribute to the basal metabolism of its producing organism but rather act as crucial factors to either attract, deter, or kill other organisms and thus increase their likelihood of survival (Kreysa and Grabley 2007). For example, SMs have been found in both prokaryotic and eukaryotic microorganisms, with unicellular bacteria (e.g., *Bacillus* spp., *Pseudomonas* spp.), eukaryotic fungi (e.g., *Penicillium* spp., *Aspergillus* spp.), filamentous actinomyces (e.g., *Streptomyces* spp.), and terrestrial plants being the most frequently studied and versatile producers (Bérdy 2005). Many SMs are only produced under specific circumstances to serve different purposes: they can exile predators or competing organisms because of their toxic nature (Dewick 2002), act as attractants toward the same species for mating purposes (Gurnani et al. 2014), or serve as dyes for the expulsion and attraction of other creatures (Pichersky and Gang 2000). A possible explanation why organisms produce a high variety of bioactive SMs is that these molecules provide producers with a selective advantage against competing organisms and, furthermore, act as an adaptation to environmental conditions (Jensen et al. 2005; O'Brien and Wright 2011; Letzel et al. 2013; Macheleidt et al. 2016). Moreover, several natural products (NPs) have the ability to protect against nonbiological impacts, such as high light intensities or elevated temperatures, and to obtain reproduction advantages for their producers (Ludwig-Müller and Gutzeit 2014). In the marine environment, SMs fulfill manifold tasks for their producers as they, for instance, act as a chemical defense against predators (Rohde et al. 2015; Helber et al. 2017; Rohde and Schupp 2018) or have antimicrobial effects against pathogenic microbes (Goecke et al. 2010; Rohde et al. 2015; Helber et al. 2018). Furthermore, MNPs are important for inducing larval settlement of benthic invertebrates (Yvin et al. 1985; Morse et al. 1988; Tebben et al. 2011, 2015; Harder et al. 2018), thereby maintaining and controlling community functioning and population dynamics. Besides their ecological impact, many NPs have been reported to exhibit a wide range of medically relevant bioactivities (Keller et al. 2005; Blunt et al. 2018), thus serving as promising molecules for the development of new drugs or drug leads (Heilmann 2007).

8.2 Marine Natural Products Chemistry: The Ocean as a Rich and Versatile Habitat

The ocean covers more than 70% of our planet's surface and likely represents the origin of Earth's life. In terms of species diversity, certain marine ecosystems, such as coral reefs, are thought to outnumber even tropical rain forests (Haefner 2003). Until today, the number of marine species that inhabit the world's oceans is not truly known; however, experts estimated a number approaching 1–2 million species (Simmons et al. 2005; Das et al. 2006). In the past, marine sponges were an interesting source for novel NPs; these sessile organisms can produce bioactive substances for chemical defense against natural predators, such as fishes (Rohde et al. 2015), as well as prevent overgrowth by competing organisms (Proksch 1994; Ortlepp et al. 2006). Furthermore, sponges serve as incubators for particular associated microorganisms like bacteria and fungi that also can contribute to the production of bioactive compounds (Radjasa et al. 2011; Wiese et al. 2011). Sponges being sessile, soft-bodied organisms, which mostly lack morphological defenses like biological armature or spines, depend to a large extend on bioactive metabolites for their survival and the survival of their associated microbial symbionts (Proksch et al. 2006). Accordingly, marine NP research has its origin in the discovery of the two nucleosides spongothymidine and spongouridine by Bergmann and coworkers in the 1950s, who isolated both active compounds from the Caribbean sponge *Cryptotethya crypta* (Bergmann and Feeneyz 1951; Bergmann and Burke 1955). These two SMs served as lead structures for the development of the synthetic antivirals cytarabine (Fig. 8.1a) and vidarabine (Fig. 8.1b) (Mayer et al. 2010) and, therefore, display exemplarily the tremendous potential of MNPs for the development of new drugs (Gulder and Moore 2009). Although promising and still relevant, sponges and their associated microorganisms are not the only marine source producing bioactive compounds. Marine NP research has expanded its efforts in exploring worldwide oceans and their inhabitants from macro- to microorganisms as rich sources for novel SMs. Until today, this resulted in new MNPs being continuously described (Table 8.1) (Martins et al. 2014). For instance, 1163 novel compounds derived from marine organisms were described only in 2013 (Blunt et al. 2015).

Over the past decades, it has been obvious that unknown NPs are more likely found when high quality materials from novel sources are examined (Goodfellow and Fiedler 2010). Unfortunately, the acquisition of marine organisms, compared to that of terrestrial organisms, is often more difficult and thus making the exploration and collection of marine samples (i.e., deep-sea organisms) very expensive (Molinski et al. 2009). However, past progress in marine technologies, such as easy accessible scuba diving equipment as well as remotely operated vehicles (ROVs), facilitated the investigations beyond the intertidal areas and led to the exploration of new marine organisms that can potentially produce a huge range of novel chemical compounds with unique bioactivities (Gerwick and Moore 2012). Since many MNPs are released into the water, the concentration of bioactive compounds is rapidly diluted via diffusion processes, and thus MNPs must be highly potent to have a long-reaching effect (Haefner 2003). Past studies have

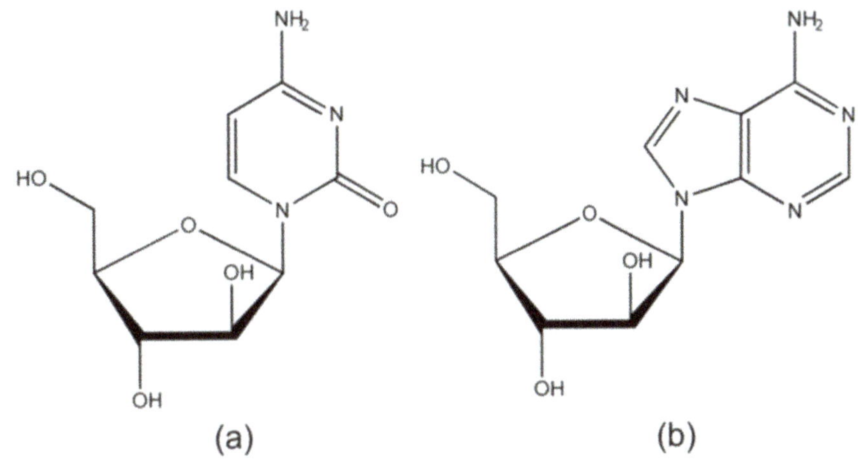

Fig. 8.1 (**a**) Cytarabine and (**b**) vidarabine (Mayer et al. 2010) (created with ChemDraw, v. 16.0.1.4)

Table 8.1 Pipeline of marine pharmaceuticals until 2018 (according to http://marinepharmacology.midwestern.edu/clinical_pipeline.html, accessed 28 January 2018)

Compound	Chemical class	Source org.	Therapeutic area	Status 2018
Cytarabine	Nucleoside	Sponge	Cancer	FDA-approved
Vidarabine	Nucleoside	Sponge	Antiviral	FDA-approved
Ziconotide	Peptide	Cone snail	Chronic pain	FDA-approved
Trabectedin	Alkaloid	Tunicate	Cancer	FDA-approved
Brentuximab vedotin	Antibody drug conjugate	Mollusk	Cancer	FDA-approved
Eribulin mesylate	Macrolide	Sponge	Cancer	FDA-approved
Omega-3-acid ethyl ester	Omega-3 fatty acid	Fish	Hypertriglyceridemia	FDA-approved
Plinabulin	Diketopiperazine	Fungus	Cancer	Phase III
Plitidepsin	Depsipeptide	Tunicate	Cancer	Phase III
Bryostatin	Macrolide lactone	Bryozoan	Alzheimer's	Phase II
Plocabulin	Polyketide	Sponge	Cancer	Phase II
Marizomib	Beta-lactone-gamma-lactam	Bacterium	Cancer	Phase I

Chemical structures of all compounds listed in this table can be found in Figs. 8.1 and 8.2

shown that MNPs cover a wide variety of biological activities (Fig. 8.3), such as anticancer (Nastrucci et al. 2012), antibacterial (Hughes and Fenical 2010), antifungal, and antiviral effects (Mayer et al. 2013), making them a promising source for novel drugs. Figure 8.3 shows that different chemical classes of MNPs are showing equal proportions among a vast set of bioactivities, leading to the assumption that most chemical structures could either be developed or serve as scaffolds for the development of new drugs against various diseases (Hu et al. 2015).

Besides the investigation on marine invertebrates or algae, modern marine biotechnology expanded its interests onto the exploration of marine bacteria and fungi, since the latter have been recognized as renewable producers of SMs (i.e., under controlled laboratory conditions) in the drug discovery process (Waters et al. 2010). Both, bacteria and fungi associated with marine macroorganisms have shown to be potent producers of bioactive substances, in some cases with prominent activities against several pathogenic germs, viruses, and tumor cells (Imhoff et al. 2011 and references therein).

8.2.1 Marine Bacteria: Widely Distributed Producers of Promising Natural Products

Marine microorganisms managed to conquer every marine habitat ranging from shallow and deep marine waters, polar regions, and deep-sea hydrothermal vents to diverse coral reef ecosystems. Particularly, the surface of macroorganisms, such as algae, sponges, and corals, is a favorable ecological niche for marine microorganisms. In many cases, bacteria live in close association with higher organisms and form symbiotic or mutualistic relationships (Lee et al. 2009; Kazamia et al. 2012; Cooper and Smith 2015). There is growing evidence that the microbial community composition on marine macroorganisms is habitat and even species specific. Examples include differences in communities found on the surface of different algae (Lachnit et al. 2009), between different parts of the rhizoid and phylloid of the brown alga *Saccharina latissima* (Staufenberger et al. 2008), between different sponge species (Thomas et al. 2016; Moitinho-Silva et al. 2017), as well as between outer

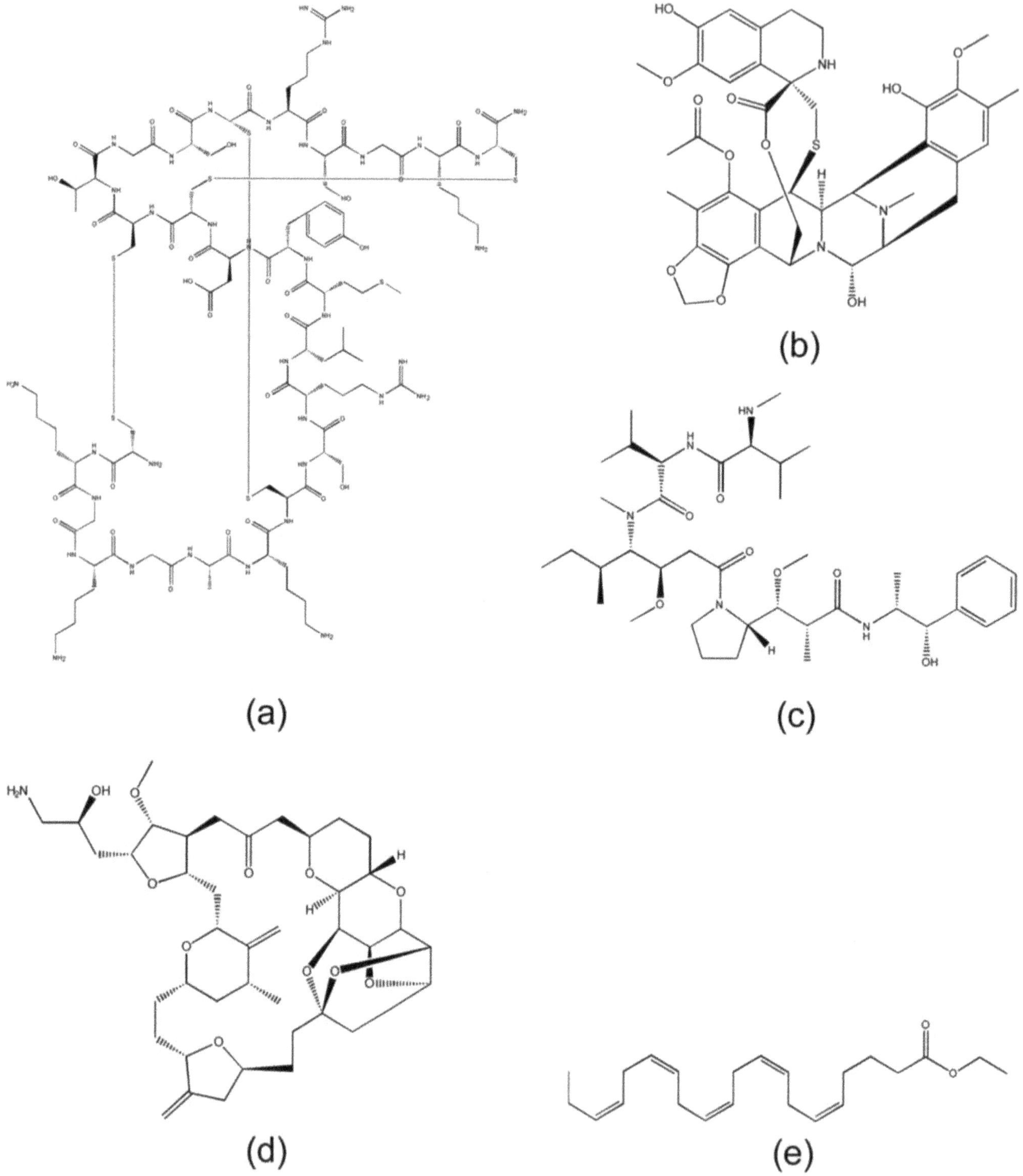

Fig. 8.2 Selected marine pharmaceuticals. (**a**) Ziconotide, (**b**) trabectedin, (**c**) monomethyl auristatin E, (**d**) eribulin, (**e**) omega-3 fatty acid, (**f**) plinabulin, (**g**) plitidepsin, (**h**) bryostatin, (**i**) plocabulin, and (**j**) marizomib (Mayer et al. 2010; modified from Lee et al. 2015; Pantazopoulou et al. 2018) (created with ChemDraw, v. 16.0.1.4)

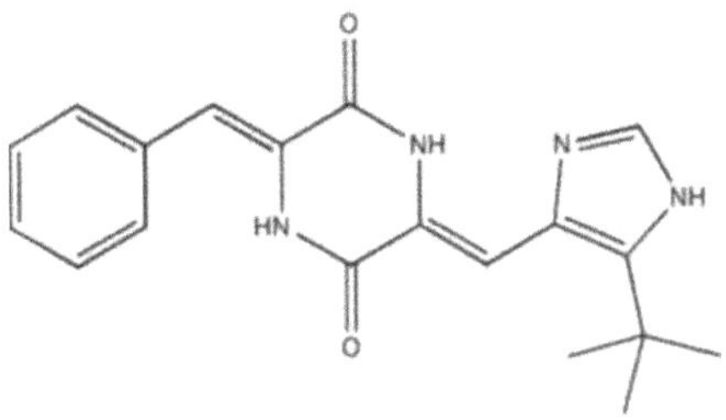

(f)

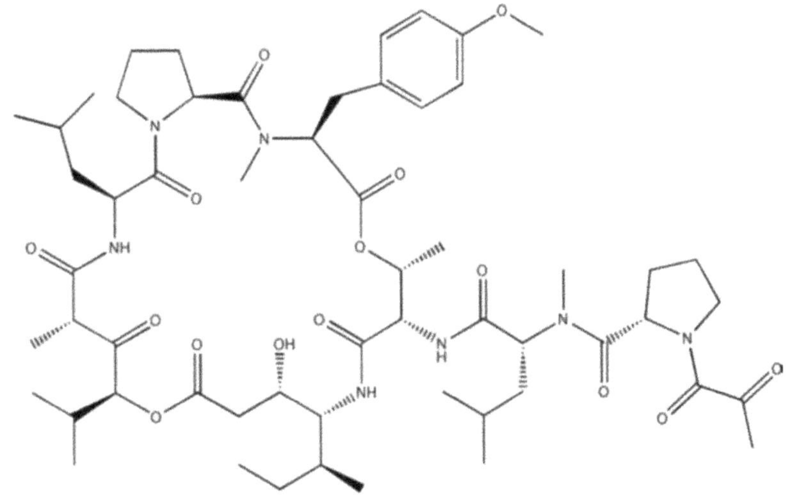

(g)

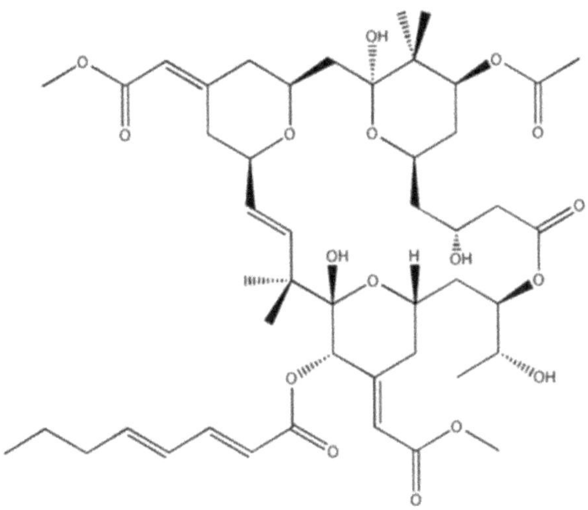

(h)

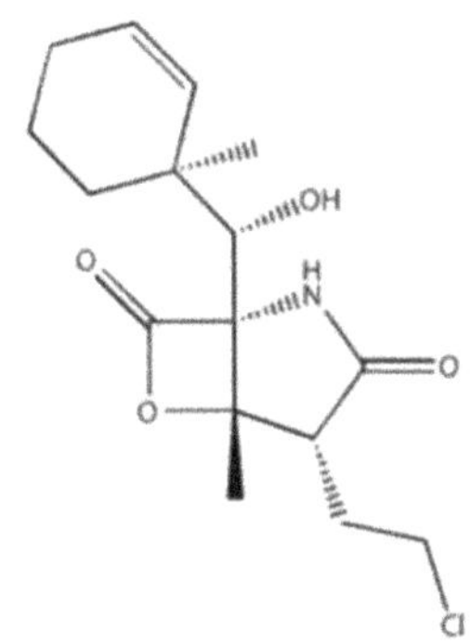

(j)

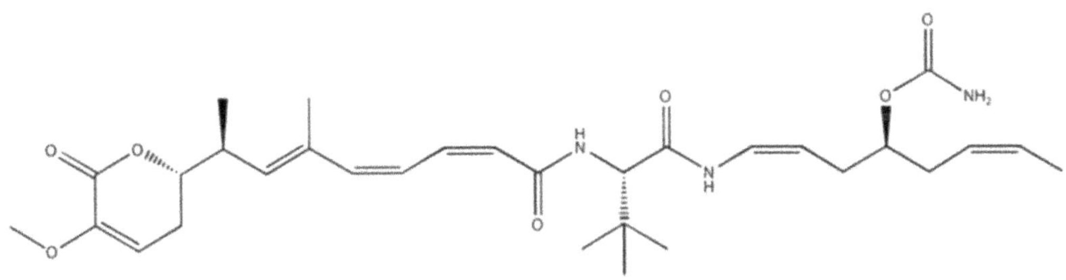

(i)

Fig. 8.2 (continued)

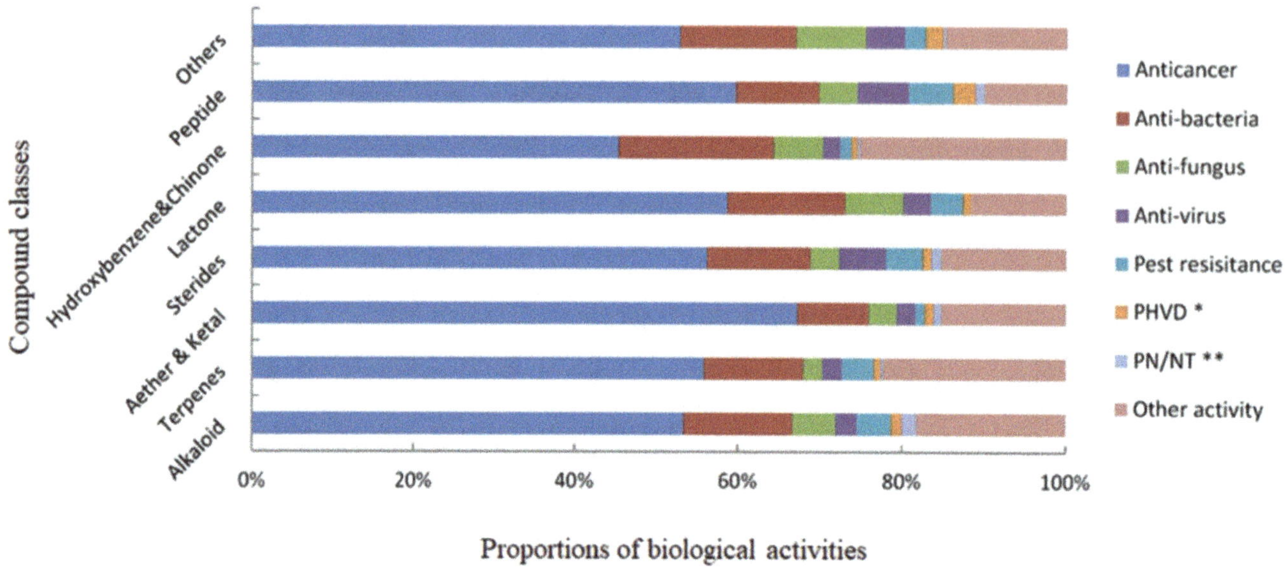

Fig. 8.3 Analysis of new marine-derived compounds from 1985 to 2012 according to chemical classes and biological activities (*PHVD, prevention of heart and vascular disease; **PN/NT, protection of neurons/neurotoxicity) (modified from Hu et al. 2015)

and inner parts of the sponge *Tethya aurantium* (Thiel et al. 2007). However, the great microbial diversity of marine environments remains nearly untapped. Simon and Daniel (2011) estimated that less than 0.1%, probably solely 0.01%, of all microbes in the oceans have been characterized. Molecular analysis of marine metagenomes revealed a great number of phylogenetic lines of so far uncultured groups of bacteria and archaea (DeLong et al. 2006; Simon and Daniel 2009; Hug et al. 2016). Besides their important roles in shaping community structures and in mediating microbe-microbe as well as microbe-host interactions, marine bacteria are suggested to represent a treasure box of new compounds for biotechnology. This assumption is due to their high biodiversity and the gap of knowledge regarding their potential of NP biosynthesis (Imhoff et al. 2011). Yet, much evidence is given that marine bacteria produce new compounds useful for the discovery of novel pharmaceuticals (Rahman et al. 2010; Waters et al. 2010; Blunt et al. 2018). From 1997 to 2008, about 660 new marine bacterial NPs were identified. Most of them originated from the classes *Actinobacteria* (40%) and *Cyanobacteria* (33%), followed by *Proteobacteria* (12%) and members of the *Bacteroidetes* and *Firmicutes* (5%) (Williams 2009). In comparison, 179 novel NPs have been isolated from marine bacteria in 2016. This is only a moderate increase compared to the average number of new marine bacterial compounds in the last 3 years, but a significant increase from the average for the period of 2010 to 2012 (Blunt et al. 2018). Members of the *Actinobacteria* are a rich source of NPs and hold an unmatched capacity for the generation of new drugs (Bull et al. 2005; Bull and Stach 2007; Fenical and Jensen 2006). The first bioactive compound extracted from a marine actinomycete was the antibiotic SS-228Y, showing antibacterial activity to gram-positive bacteria. This biomolecule was proposed to be a peri-hydroxyquinone derivative produced by *Streptomyces purpurogeneiscleroticus* (*Chainia purpurogena*) collected from sea mud (Okazaki et al. 1975). Until today, the genus *Streptomyces* continues to be a prolific source of new and interesting chemistry; numerous compounds showed exciting bioactivities. For example, *S. spinoverrucosus*, isolated from a sand sample from the Bahamian tidal flats, produced the dibohemamines A–C, three new dimeric bohemamines. These compounds were shown to be formed via a nonenzymatic process with formaldehyde, which was also detectable in the growth media. Both metabolites dibohemamines B (Fig. 8.4a) and C exhibited potent activity against lung cancer cells, with IC_{50} values of 140 nM and 145 nM, respectively (Fu et al. 2016). Another *Streptomyces* sp. isolated from a marine sediment sample collected off Oceanside, California, USA, produced the ansalactams A–D. The novel ansalactam derivatives (B–D) represent three new carbon skeletons and, therefore, display the plasticity within the ansamycin biosynthetic pathway. The latter three novel metabolites showed moderate antibacterial activity against MRSA (methicillin-resistant *Staphylococcus aureus*) (Wilson et al. 2011; Le et al. 2016). Apart from *Streptomyces*, species of the genera *Salinispora* and *Marinispora* were found to produce structurally novel bioactive compounds. A *Salinispora tropica* strain was isolated from a sediment sample, collected from a mangrove environment in Chub Cay, Bahamas. This strain produced several β-lactone-gamma-lactams, the salinospo-

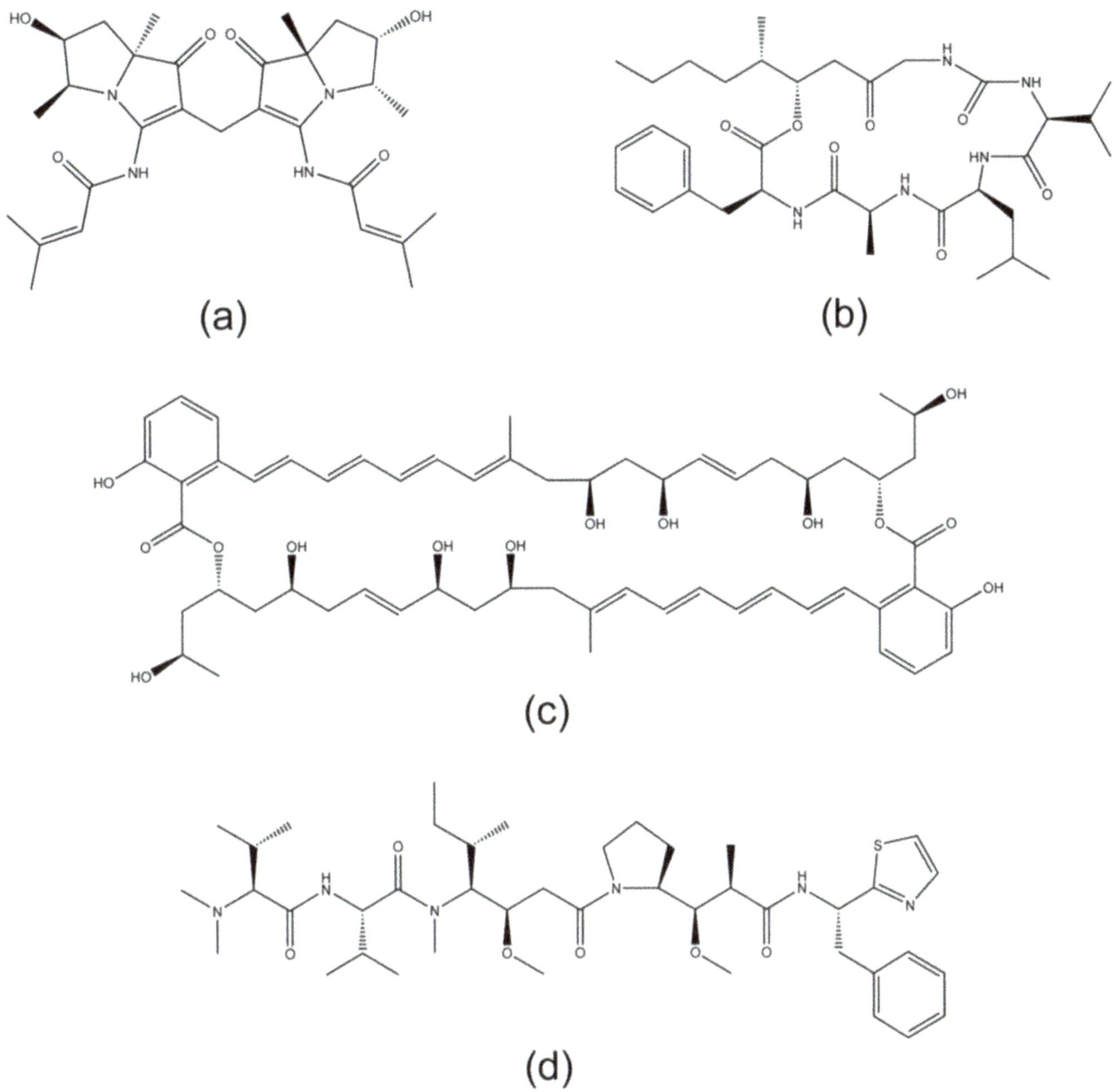

Fig. 8.4 (**a**) Dibohemamine B (Fu et al. 2016), (**b**) arenamide B (Asolkar et al. 2009), (**c**) marinomycin A (Kwon et al. 2006), and (**d**) dolastatin 10 (Bai et al. 1990) (created with ChemDraw, v. 16.0.1.4)

ramides, which represent a new family of SMs (Feling et al. 2003; Williams et al. 2005). Specifically, salinosporamide A (marizomib; see Fig. 8.2j) displayed potent in vitro cytotoxicity against HCT-116 human colon carcinoma with an IC_{50} value of only 11 ng mL^{-1}. Furthermore, this compound showed great potency against NCI-H226 non-small cell lung cancer, SK-MEL-28 melanoma, MDA-MB-435 breast cancer, and SF-539 CNS cancer, all with LC_{50} values less than 10 nm (Feling et al. 2003). As displayed in Table 8.1, marizomib has entered Phase I human clinical trials for the treatment of multiple myeloma (Martins et al. 2014; http://marinepharmacology.midwestern.edu/clinical_pipeline.html, accessed 28 January 2018). Asolkar et al. (2009) found three new cyclohexadepsipeptides, namely, arenamides A–C, produced in the fermentation broth of a marine *Salinispora arenicola*, isolated from a sediment sample. Arenamide A and B (Fig. 8.4b) blocked TNF (tumor necrosis factor) induced activation in transfected 293/NFκB-Luc human embryonic kidney cells in a time- and dose-dependent manner with IC_{50} values of 3.7 μM and 1.7 μM, respectively. The compounds also inhibited nitric oxide (NO) and prostaglandin E_2 (PGE_2) production with lipopolysaccharide (LPS)-induced RAW 264.7 macrophages. The authors suggest that the anti-inflammatory and chemoprevention characteristics of arenamides A and B are worth further investigation (Asolkar et al. 2009). Other examples for antibiotics with antitumor activity from marine *Actinobacteria* are the marinomycins. A *Marinispora* strain, isolated from an offshore sediment sample, produced the marinomycins A–D. The most promising compound within this novel class of polyketides is marinomycin A (Fig. 8.4c). It shows selectivity against several human melanoma cell lines with an IC_{50} value of 5 nM for SK-MEL5 melanoma cells (Kwon et al. 2006). Besides *Actinobacteria*, members of marine *Cyanobacteria* are known to produce bioactive SMs too. For example, the peptide dolastatin 10 (Fig. 8.4d) was originally isolated from the sea hare *Dolabella auricularia* (Bai et al. 1990) but was then shown to be produced by the cyanobacterium *Symploca* sp. (Luesch et al. 2001). This natural product (NP) was used as a model for the synthetic development of soblidotin, which has entered Phase III clinical trials (Mayer et al. 2010). The cyclic depsipeptide largazole is produced by another marine *Symploca* sp. and inhibited the growth of highly invasive transformed human mammary epithelial cells in a dose-dependent manner (GI_{50} 7.7 nM). It induced cytotoxicity at higher concentrations (117 nM) (Taori et al. 2008). All these examples show the potential of marine bacteria, specifically *Actinobacteria* and *Cyanobacteria*, to produce chemicals that cover a broad range of bioactivities and might be used for the generation of novel drug candidates.

8.2.2 Marine Fungi: Bioprospecting the Future

Compared to bacteria, the basic knowledge on marine fungi, hereinafter referring to obligate and facultative marine fungi, is still deficient in matters of diversity and ecological importance (Imhoff et al. 2011). The term "marine fungi" applies rather to an ecological background than to a distinct taxonomy or a physiological approach (Kohlmeyer and Kohlmeyer 1979). Within biology, marine fungi are mainly separated into two groups, namely, obligate marine fungi, which grow and sporulate exclusively in marine habitats, and facultative marine fungi, which originate from freshwater or terrestrial milieus and are capable to grow also in the marine environment (Kohlmeyer and Kohlmeyer 1979). By 1996, mycologists estimated the number of marine fungi to be approximately 1500 species, and by 2011, biodiversity estimations of marine fungi were placed to be more than 10,000 species (Jones 2011).

According to Overy et al. (2014), the examination of new substrata and geological locations will greatly increase the number of total species through the rapid discovery of new fungal species. However, marine fungal strains have been isolated from nearly every possible marine habitat until today, including soil and sediment (Wang et al. 2013; Simões et al. 2015), marine invertebrates (e.g., sponges and corals) (Wiese et al. 2011; Amend et al. 2012), marine plants (e.g., algae) (Loque et al. 2010), and marine vertebrates (fishes) (Rateb and Ebel 2011). Algae have been used primarily as a source for bioprospecting fungal diversity, closely followed by sponges and mangrove habitats (Fig. 8.5). Efforts to isolate these symbionts within new and sometimes extreme habitats are still being made. A study on the fungal community by a culture-dependent approach revealed that several Antarctic sponges of the phylum *Ascomycota* were a rich source of associated fungi and novel SMs, with some of them showing antimicrobial, antitumoral, and antioxidant potential (Henríquez et al. 2014). Furthermore, due to the development of deep-sea instrumentation and new techniques used for sampling, the deep-sea habitat emerged as a new and highly promising source for marine fungal biodiversity, and thus an excessive number of novel fungal specimen have been retrieved (Wang et al. 2015). For example, Burgaud et al. (2009) investigated the biodiversity of culturable filamentous fungi and uncovered the presence of both *Ascomycota* and *Basidiomycota* associated with different deep-sea samples, including sediment, mussels, shrimps, and smoker rock scrapings. However, as an outcome of the recent bioprospecting efforts, biotechnological interests have mostly turned to marine microorganisms and notably fungi as a likely source for MNPs (Fig. 8.6) (Bhadury et al. 2006).

By 1992, only 15 marine fungal metabolites had been described (Fenical and Jensen 1993), and this number rose to 270 until 2002 (Bugni and Ireland 2004). Within the period from 2006 until mid-2010, Rateb and Ebel (2011) summarized 690 NPs from fungi isolated from marine habitats. With *Penicillium* spp. and *Aspergillus* spp. being the most potent producers, their study revealed that nearly 50% of the compounds are polyketides and their prenylated forms, whereas alkaloids, terpenoids, and peptides contributed 15%–20% (Rateb and Ebel 2011). A famous example of a NP from a marine fungus is the diketopiperazine halimide (Fig. 8.6a), an aromatic alkaloid of a marine *Aspergillus* sp. isolated from the green alga *Halimeda copiosa* (Fenical 1999). Its synthetic analog, plinabulin (Fig. 8.2f), is showing antitumor activity by causing tubulin depolymerization, thereby leading to the disruption of tumor cells followed by necrosis of the tumor itself (Gullo et al. 2006). Up to today, plinabulin is the only marine fungal synthetic analog that has entered clinical trials and successfully passed the first and second phase (Gomes et al. 2015; www.beyondspringpharma.com/en/pipeline/, accessed 28

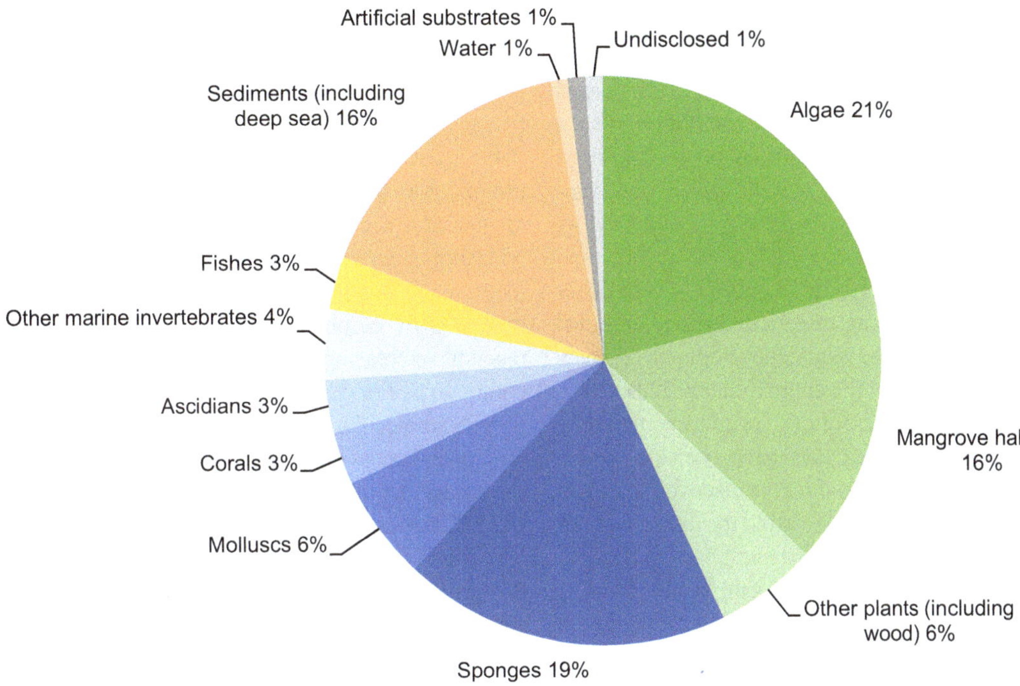

Fig. 8.5 Sources of marine fungi-producing MNPs until 2010 (reproduced from Rateb and Ebel 2011)

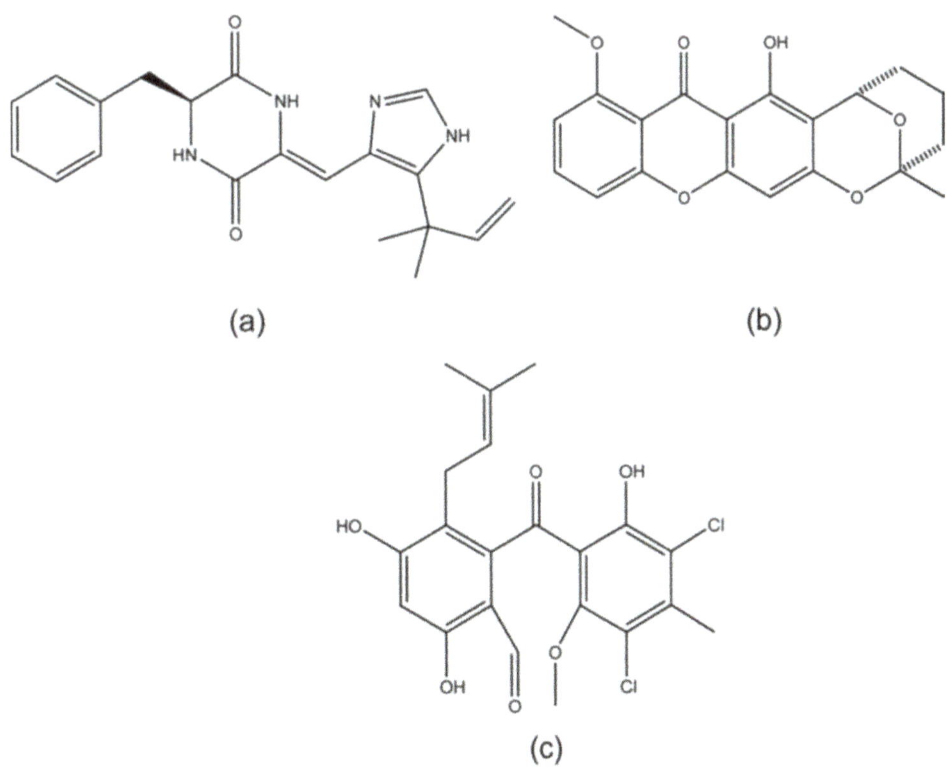

Fig. 8.6 (**a**) Halimide (Fenical 1999), (**b**) chaetoxanthone B (Gademann and Kobylinska 2009), and (c) pestalone (Cueto et al. 2001) (created with ChemDraw, v. 16.0.1.4)

January 2018). A further example was given by Pontius et al. (2008), who isolated chaetoxanthone B (Fig. 8.6b) from a marine *Chaetomium* sp., showing selective antimalarial activity against *Plasmodium falciparum* (IC_{50} = 0.5 μg mL^{-1}) and moderate activity against *Trypanosoma cruzi* (IC_{50} = 1.5 μg mL^{-1}) without or with only minimal cytotoxicity toward cultured eukaryotic cells. Another promising marine NP is the chlorinated benzophenone pestalone (Fig. 8.6c), which has been isolated from the fungus *Pestalotia* sp., which is associated with the brown alga *Rosenvingea* sp. and was collected near the Bahamas Islands. Although pestalone was only produced when *Pestalotia* sp. was cocultured with a marine bacterium, this compound showed potent activities against methicillin-resistant *Staphylococcus aureus* and vancomycin-resistant *Enterococcus faecium* strains, indicated by minimum inhibitory concentrations (MIC) of 37 ng mL^{-1} and 78 ng mL^{-1}, respectively (Cueto et al. 2001). These latter examples encourage the ongoing research activities on novel marine fungal species for the future development of new drugs. Considering that 38% of the approximately 22,000 bioactive microbial metabolites are of fungal origin, and that only about 5% of the world's fungal taxa have been described, fungi exhibit a tremendous potential for the discovery of novel bioactive SMs (Schulz et al. 2008). To avoid the rediscovery of already known compounds, specialized and effective dereplication strategies need to be constantly employed (Martins et al. 2014). For this purpose, the most common techniques are a combination of chemical compound separation hyphenated to various spectroscopic or mass-selective detection methods such as high-performance liquid chromatography (HPLC) coupled to either a diode array detector (HPLC-DAD) or a mass spectrometer (HPLC-MS) (Wolfender et al. 2010). Besides nuclear magnetic resonance (NMR) spectroscopy, HPLC-MS is another predominant analytical technique for the fast detection and identification of SMs and other small molecules. A major advantage of MS over NMR is that MS-based methods are far more sensitive, making it the method of choice when it comes to first-pass compound detection and identification in high-throughput screening applications (Carrano and Marinelli 2015). Moreover, it provides accurate mass even within the nanogram range, which can be used as a search criterion or query in nearly all NP databases (Nielsen et al. 2011). On the contrary, NMR is by far the most efficient method to unambiguously elucidate complex structures of small molecules (Hubert et al. 2017). One of its advantages compared to MS strategies is that it serves as a quantitative analysis without the need of a suitable reference material (Kurita and Linington 2015). The ^{1}H-NMR is also useful for evaluating the purity of a given sample. For example, impurities such as lipids are somewhat invisible in HPLC-DAD-MS techniques due to their low UV absorption, hydrophobicity, and contumaciousness to ionization, but they can easily be seen in ^{1}H-NMR (Carrano and Marinelli 2015). After the collection of UV/VIS absorption spectra, molecular mass, and further structure data, the gained information needs to be compared with database entries. Over the decades, many different databases covering a wide range of compounds have been established (Mohamed et al. 2016; Guijas et al. 2018), including general compound libraries like SciFinder (www. scifinder.cas.org, accessed 28 January 2018), NP libraries such as AntiBase (www.wiley-vch.de/stmdata/antibase.php, accessed 28 January 2018) or Dictionary of Natural Products (dnp.chemnetbase.com, accessed 28 January 2018), and even some free-to-use databases like ChemSpider (www.chemspider.com, accessed 28 January 2018), PubChem (pubchem.ncbi.nlm.nih.gov, accessed 28 January 2018), or Metlin (metlin.scripps.edu, accessed 28 January 2018). In addition to this widespread dereplication approach, fragmentation-based MS methods, also referred to as MS/MS or tandem mass spectrometry, in combination with molecular networking are receiving increasing attention for the identification of unknown compounds. For example, the Global Natural Products Social (GNPS) molecular networking website (http://gnps.ucsd.edu) is an open-access knowledge base that aims to let NP chemists work together and share their raw, processed, or identified MS/MS spectrometry data. We believe that crowdsourced curation of freely available reference MS libraries as well as a fast-growing database of MS/MS spectra will rapidly accelerate the annotation and thus the search of prior unknown compounds (Wang et al. 2016; Kind et al. 2017; Quinn et al. 2017).

8.3 Marine Chemical Ecology: Predator-Prey Interactions and Competition

During the last decades, marine chemical ecology has evolved from a young science with mostly NP chemists finding new SMs with potentially obscure ecological functions into a matured research field that simultaneously combines chemical and biological aspects. Besides their side effect of exhibiting utilizable bioactivities for humankind, chemical cues possess major influences on every organizational level in the marine system. Several reviews highlight the importance of chemical communication between benthic and pelagic organisms for a better understanding of marine ecosystem functioning (i.e., Hay 1996, 2009; Sieg et al. 2011; Paul et al. 2011; Puglisi et al. 2014). However, most marine organisms are rather organized in highly biodiverse and productive communities occurring in ocean fringes, such as coral reefs or offshore zones, than being distributed all over the ocean (Simmons et al. 2005; Das et al. 2006). Many of these biological communities are characterized by the presence of extremely harsh conditions in matters of UV radiation (light stress at water surface), temperature, pressure, and salinity. In addition to these environmental stressors, sessile benthic organisms are often in strong competition for available resources such as space (to settle and grow) and nutrients. As a result, survival and reproduction between the competing organisms can strongly depend on their ability to produce bioactive SMs (de Carvalho and Fernandes 2010). These bioactive substances can perform various tasks for their producers and associated organisms; for instance, SMs work as a chemical defense against predators

(Pohnert 2004; Kubicek et al. 2011; Rasher et al. 2013; Rohde et al. 2015; Helber et al. 2017), function as attractants toward consumers (Sakata 1989), have antimicrobial effects against pathogenic microbes (Goecke et al. 2010; Puglisi et al. 2014; Helber et al. 2018), guide the opposing sex by letting individuals find and evaluate potential mating partners through chemical cues (Lonsdale et al. 1998; Li et al. 2002), or act as settlement cues for invertebrate larvae to initiate the transformation into a sessile, juvenile form (Morse et al. 1988; Heyward and Negri 1999; Negri et al. 2001; Kitamura et al. 2009; Tebben et al. 2011, 2015; Sneed et al. 2014). For example, different classes of macroalgae defend themselves chemically against herbivores and produce SMs with antimicrobial and antifouling activity (Schupp and Paul 1994; Paul et al. 2014; Schwartz et al. 2016). Specifically, brown algae of the family Dictyotaceae produce several classes of diterpenes that defend their producers against herbivores but have also shown activity against other competitors. It has been reported that natural concentrations of a diterpene of the dolastane class (Fig. 8.7a), originally isolated from the brown alga *Canistrocarpus cervicornis*, reduce feeding activity by the sea urchin *Lytechinus variegatus* (Bianco et al. 2010). In a study of Craft et al. (2013), lipophilic extracts of nine subtropical algae were offered to four subtropical and three cold-temperate sea urchins at two concentrations. While the extracts of the subtropical marine algae *Caulerpa sertularioides*, *Dictyota pulchella*, and *D. ciliolate* deterred all urchins, the other macroalgae extracts from the cold-adapted areas led to different feeding resistance patterns. Apart from anti-herbivore activity, many macroalgae are known for their antimicrobial and antifouling activity (Goecke et al. 2010, 2012; Pérez et al. 2016; Schwartz et al. 2017). A review by Harder et al. (2012) highlights the crucial role of halogenated furanones (Fig. 8.7b, c) within the red alga *Delisea pulchra* and how these compounds interact with the associated bacteria. The halogenated furanones deter fouling by bacterial pathogens and epiphytic bacteria through interference with bacterial quorum sensing. By imitating quorum sensing-mediating acyl homoserine lactones to block the same receptor sites, halogenated furanones can manipulate bacterial colonization and biofilm formation as well as bleaching and diseases caused by pathogenic bacteria. Besides macroalgae, sponges and their associated microbes are another prolific source of potentially novel NPs with promising bioactivities. Although the ecological role of sponge crude extracts has been evaluated for numerous sponge species, assignment of activities to specific NPs is lacking behind. Investigated bioactivities included antipredatory, antifouling, antimicrobial, and allelopathic functions (Rohde et al. 2015; Helber et al. 2017, 2018). Several studies are providing evidence that sponges are chemically defended from predation and pathogens by compounds that either the host or other associated microorganisms had produced (Pawlik 2011; Hentschel et al. 2012). The Mediterranean sponge *Axinella verrucosa*, collected from the Gulf of Naples, Italy, produces hymenidin (Fig. 8.7d) and debromo-carteramine A (Fig. 8.7e), two bromopyrroles that are also known from other sponges. The *n*-butanol part of the *A. verrucosa* extract, containing the two bromopyrroles as well as the pure hymenidin, showed activity against microbial fouling and deterred feeding by the generalist shrimp *Palaemon elegans* at naturally occurring concentrations (Haber et al. 2011).

Several studies have shown that sponges of the same genus and even of the same species can produce different SMs. This circumstance raises the question to what extend SMs have evolutionary advantages for the survival of sponges. A study of Noyer et al. (2011) showed that several populations of *Spongia lamella*, collected in the Mediterranean Sea, spanning a region of 1200 km, had an extremely high intraspecific chemical diversity. While nitenin (Fig. 8.7f) and ergosteryl myristate (Fig. 8.7g) were the major metabolites, the number of compounds as well as their concentrations changed among populations collected from different geographic locations. The authors suggested that these variations may have been due to both genetic and environmental factors. A further study on *S. lamella* revealed that the populations from the five regions (Portugal, Gibraltar, Baleares, Catalonia, and South France) significantly differed within their genetic and chemical diversity as well as their associated bacteria (Noyer and Becerro 2012). Similarly, Rohde et al. (2012) found different metabolites and compound concentrations in the tropical sponge *Stylissa massa* across different ocean basins and within sites. Compound concentration varied among individuals, and no correlation between compound concentrations and factors such as depth, UV, predation, and microbial growth could be identified. The authors concluded that concentrations could be affected by other selective pressures such as water temperature, water quality, light conditions, and food availability or that the observed variations reflected population-specific constitutive defenses. Another activity that has received increased attention recently are allelopathic actions of sponges by which they can outcompete scleractinian corals. Sponges have become an increasingly dominant species in the Caribbean reefs (Maliao et al. 2008; Colvard and Edmunds 2011; Perry et al. 2013; Villamizar et al. 2013; Loh and Pawlik 2014) and to a lesser extend in the Indo-Pacific (Bell and Smith 2004; Bell et al. 2013; Helber et al. 2017) as coral reef systems are permanently threatened by multiple decades of loss of reef-building corals due to climate change, disease, and pollution. In contrast to the calcium carbonate skeleton of corals, sponge skeletons are made of silica or protein, making them less sensitive to ocean acidification and temperature shifts (Pawlik 2011; Bell et al. 2013). Apart from being environmentally more robust, sponges can also outcompete corals

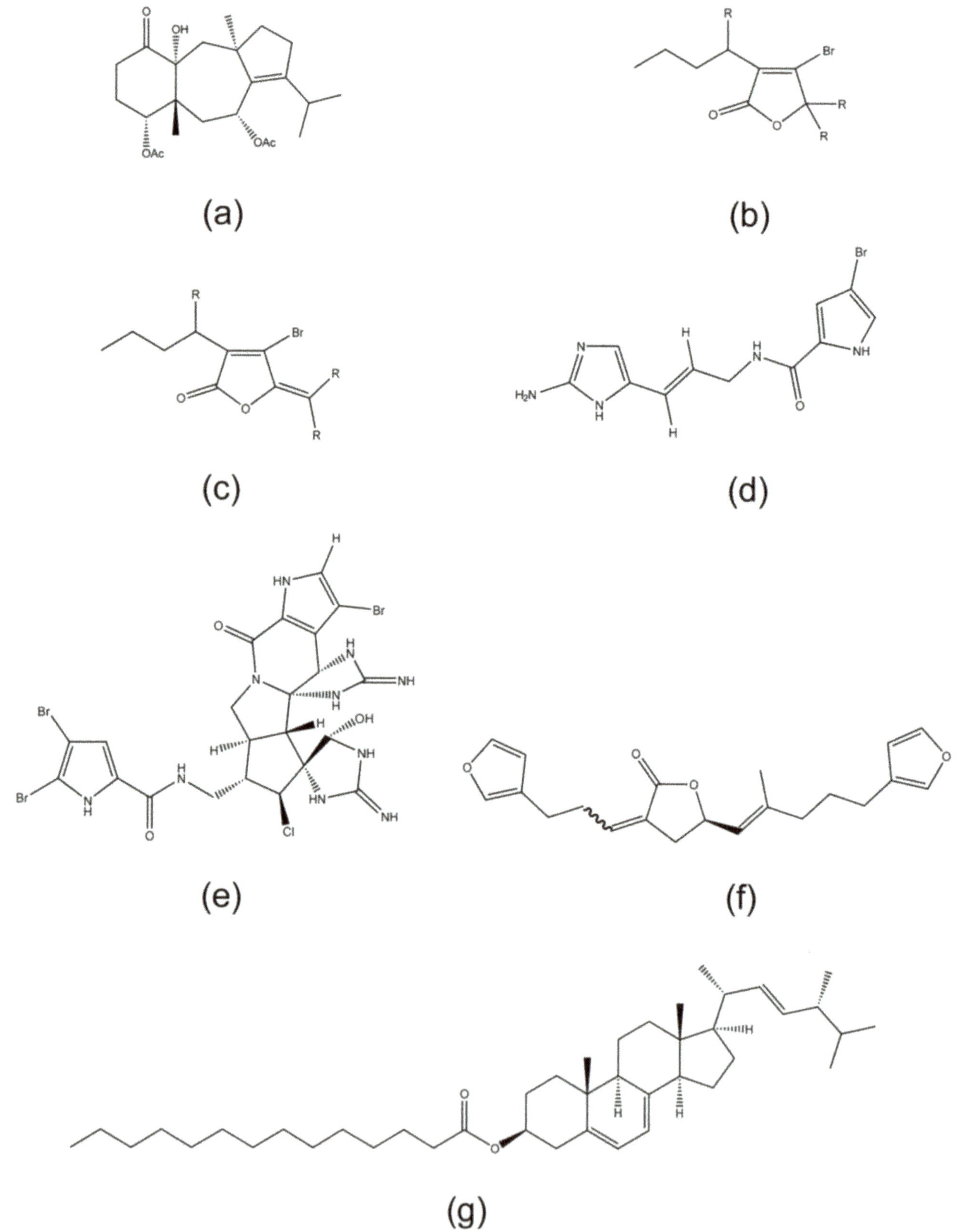

Fig. 8.7 (**a**) A diterpene (Bianco et al. 2010), (**b**, **c**) two halogenated furanones (Harder et al. 2012), (**d**) hymenidin, (**e**) debromo-carteramine A (Haber et al. 2011), (**f**) nintenin, and (**g**) ergosteryl myristate (Noyer et al. 2011) (created with ChemDraw, v. 16.0.1.4)

by chemically affecting the coral symbionts through allelopathy. Crude extracts of several sponges collected from the Caribbean reefs were embedded in stable gels at natural concentrations and caused a decrease in the photosynthetic potential of the symbiotic zooxanthellae from the brain coral *Diploria labyrinthiformis*. Interestingly, sponge extracts influenced the symbiotic zooxanthellae in two ways: impairing photosynthesis with bleaching and with only little or no bleaching at all (Pawlik and Steindler 2007). Similarly, organic extracts of three sponges collected from Zanzibar reduced the photosynthetic performance of symbionts in the scleractinian coral *Porites* sp. (Helber et al. 2018). In addition to allelopathy on adults, it has been reported that sponge-derived SMs can negatively affect invertebrate larvae settlement too (Thompson 1985; Thompson et al. 1985; Bingham and Young 1991; Hellio et al. 2005). Since there have been several reviews in recent years on the role of SMs in chemical ecology and specifically chemical defense (Paul et al. 2011; Pawlik 2011; Rohde and Schupp 2018), we decided to focus in the remaining part of this review on the role of SMs during the settlement process of invertebrates (a role which has to this point received less attention).

8.3.1 Marine Invertebrate Larvae Settlement: Role of Secondary Metabolites

Many benthic marine invertebrates such as corals, sponges, mussels, or worms have a planktonic phase followed by a metamorphic event that transforms them into a less mobile or immobile, sessile benthic form. Since the process of attachment and metamorphosis for most organisms is generally irreversible (Thorson 1950), the choice of a suitable location for settlement is crucial for invertebrate larvae regarding survival, population dynamics, and community functioning. In the past, two models have been developed to explain the settlement of marine invertebrate larvae: (1) the stochastic model postulated that the settlement process happens randomly as soon as suitable substrate becomes available and that postmetamorphic events arrange the final distribution of juveniles, and (2) the deterministic model suggested that specific environmental factors determine the attachment and metamorphosis of larvae as well as their final distribution. Nowadays, there is great evidence that the settlement process of invertebrate larvae is mainly biologically and chemically driven, although environmental parameters may also influence settlement behavior (Sebens 1983; Morse et al. 1988; Mundy and Babcock 1998; Lau and Qian 2001; Lau et al. 2005; Tebben et al. 2015; Da-Anoy et al. 2017). Chemical settlement cues are produced by a variety of marine organisms. Some invertebrate larvae like to settle among individuals of their own species, while others preferably settle upon other species, resulting in gregarious or associative settlement, respectively. Gregarious settlement has been reported for many phyla including polychaete worms and barnacles (Hadfield and Paul 2001). Live adults of the polychaete *Hydroides dianthus* were capable of eliciting gregarious settlement responses in conspecific larvae. Interestingly, settlement in response to live adults with or without their tubes as well as to their amputated tentacular crowns was significantly greater compared to dead worms, empty tubes, or biofilm covered slides. Moreover, after extraction of aggregations of adult worms with organic solvents, the inductive capacity of the remaining tissue was lost, and the activity went into both the nonpolar and polar fractions of the crude extract (Toonen and Pawlik 1996). Gregarious settlement of invertebrate larvae has long been assumed to be induced by contact with adult conspecifics (Crisp and Meadows 1962, 1963; Clare and Matsumura 2000). It has been shown that a glycoprotein with high molecular weight isolated from the adult barnacle *Amphibalanus amphitrite*, termed the settlement-inducing protein complex (SIPC), induced settlement of cypris larvae (Matsumura et al. 1998; Dreanno et al. 2006). Nevertheless, there are reports showing that waterborne cues are able to induce gregarious settlement as well. Endo et al. (2009) isolated a previously undescribed ~32-kDa water-soluble protein from extracts of *A. amphitrite* adults that is distinct from SIPC and induced settlement of cyprids. This protein quickly induced searching behavior of conspecific larvae and was therefore proposed to act as a waterborne settlement pheromone. Elbourne and Clare (2010) provided evidence that settlement of *A. amphitrite* larvae can be induced by an unknown waterborne cue produced by conspecific adults both in the field and in the laboratory. These authors suggest that the ecological role of water-soluble settlement cues might be to facilitate the transition of invertebrate larvae out of the plankton by stimulating searching behavior, rather than attachment and metamorphosis caused by surface-bound settlement cues. Besides gregarious settlement, associative settlement is another form and can be divided into several subcategories, including herbivorous/predatory relationships, parasitic relationships, and nonparasitic or symbiotic relationships. There are already some fully and partially characterized chemical compounds described; however, their ecological relevance often remains obscure (Pawlik 1992; Hadfield and Paul 2001). The quinol jacaranone (Fig. 8.8a), isolated from the red alga *Delesseria sanguinea*, induces larval settlement in *Pecten maximus* (Yvin et al. 1985), although this scallop has previously not been described to settle on this kind of red alga with any specificity (Chevolot et al. 1991; Nicolas et al. 1998). Another example is given by Williamson et al. (2000), who at first isolated a water-soluble complex of the sugar floridoside and isethionic acid in a 1:1 ratio from *Delesseria pulchra*. This floridoside-isethionic acid complex induced metamorphosis and reversible settlement in the sea urchin *Holopneustes pur-*

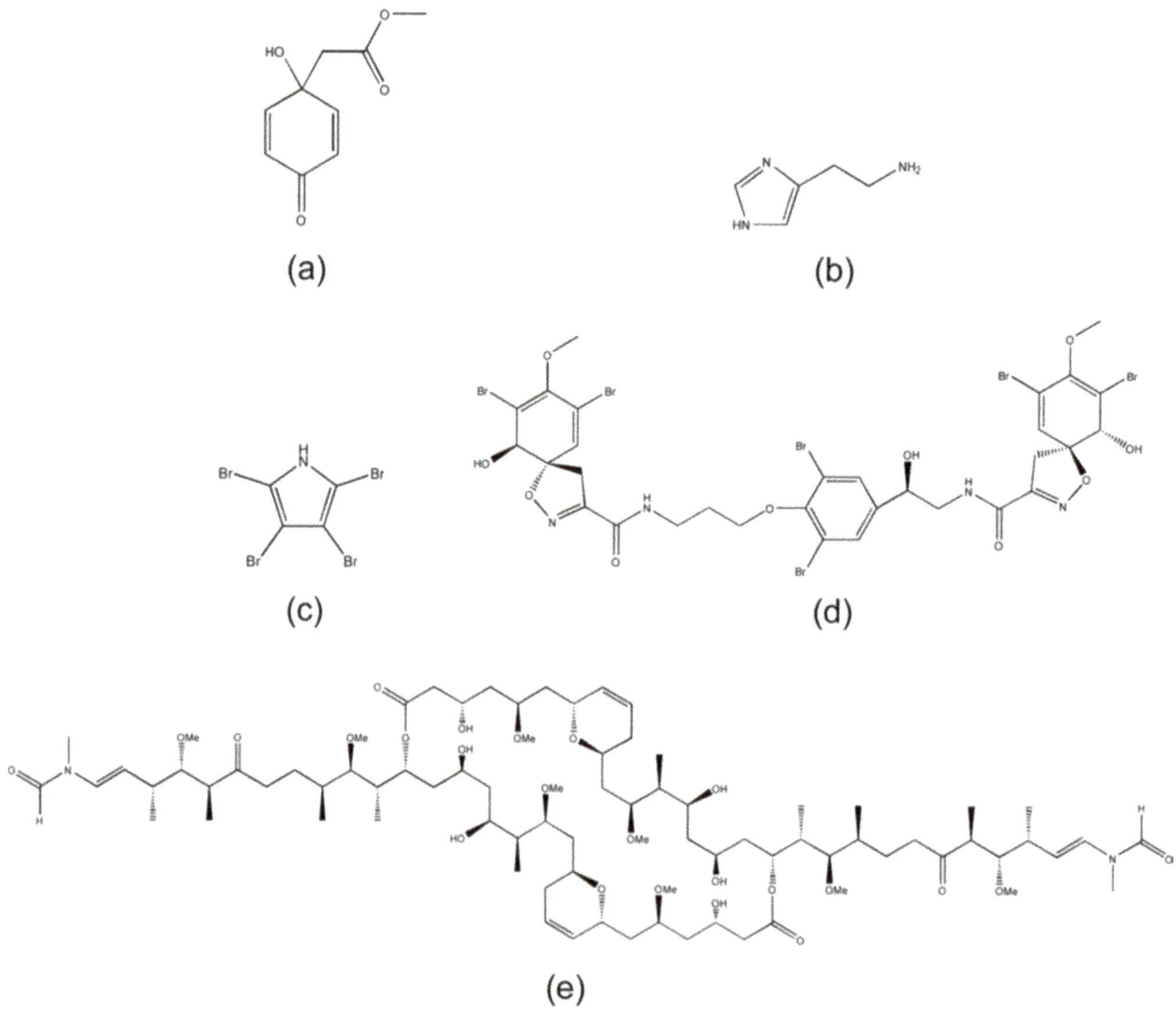

Fig. 8.8 (**a**) Jacaranone (Yvin et al. 1985), (**b**) histamine (Williamson et al. 2000), (**c**) tetrabromopyrrole (Tebben et al. 2011), (**d**) 11-deoxyfistularin-3 (Kitamura et al. 2007), and (**e**) luminaolide (Maru et al. 2013) (created with ChemDraw, v. 16.0.1.4)

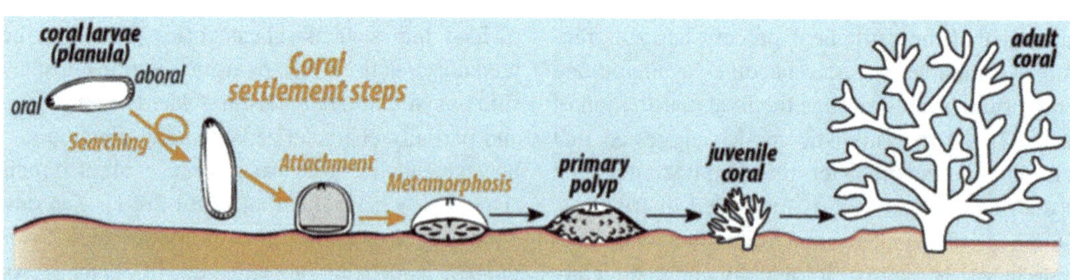

Fig. 8.9 Settlement and early life stages of scleractinian corals. This figure highlights on the first steps of a coral larvae (searching, attachment, and metamorphosis) toward an adult coral

purascens. In a following study, Swanson et al. (2004) were unable to reproduce these results. Instead, they found that histamine (Fig. 8.8b), isolated from the polar extract of *D. pulchra*, induced rapid settlement in 80–100% of the larvae of *H. purpurascens*. As larval settlement can be distinguished between searching, attachment, and metamorphosis (see Fig. 8.9; exemplarily shown for coral larvae), it is questionable if a single cue can induce the settlement chain or if several different cues are sequentially involved. Several studies on the settlement of different marine invertebrate larvae have indeed proven that single molecules can induce the whole chain of settlement, although only a few catalysts have been fully chemically characterized (Yvin et al. 1985; Pawlik 1986; Pawlik et al. 1991; Tsukamoto et al. 1999; Swanson et al. 2004; Dreanno et al. 2006). Only three of the latter studies have supported the ecological role of their investigated signaling molecules by also applying ecologically realistic concentrations (Pawlik 1986; Swanson et al. 2004; Dreanno et al. 2006). Meanwhile, tetrabromopyrrole (Fig. 8.8c), a tetrabrominated pyrrole, has been isolated from a marine *Pseudoalteromonas* sp. associated to the crustose coralline algae (CCA) *Neogoniolithon fosliei*, showing settlement activity in larvae of the branching stony coral *Acropora millepora*. Interestingly, coral larvae directly underwent metamorphosis by developing into primary polyps within a few hours, but only a small amount of them conducted attachment to the substratum, a process normally administered before metamorphosis is initiated (Tebben et al. 2011). Somewhat the same applies to 11-deoxyfistularin-3 (Fig. 8.8d), a bromotyrosine derivative that has been isolated from an unnamed CCA overgrowing coral rubble collected in Okinawa, Japan. This secondary metabolite induced solely metamorphosis in larvae of the scleractinian coral *Pseudosiderastrea tayamai*. Metamorphosis activity was further enhanced by the addition of fucoxanthinol and fucoxanthin, which are two carotenoids that had been isolated from the same CCA as well (Kitamura et al. 2007). Given the high number of studies on marine invertebrate settlement, it is very likely that additional examples of larvae relying on waterborne or surface-bound cues for gregarious and associative settlement will be described in the future. Furthermore, we are convinced that future studies will not only focus on the discovery of novel chemical settlement cues but also provide more information on their role in the mechanism of the settlement cascade and on their broader ecological functions.

8.3.2 Coral Larvae Settlement: Search for Novel Settlement Cues

Coral reefs are among the world's most diverse ecosystems and serve as nursery grounds and feeding areas for many reef-dependent animal species. Due to their relative complex physical structure, coral reefs shape the otherwise flat sea floor into a three-dimensional structure that provides a combination of food and shelter for a high biomass of commercially important fish species and other associated fauna (Moberg and Folke 1999). Besides their manifold ecosystem services, coral reefs affect humankind by having a major impact on economy and politics. Coral reefs provide food via commercial fisheries (Pauly et al. 2002), protect coastlines from destruction by waves (Barbier et al. 2011), and generate income from food and tourism (Bellwood et al. 2004). Unfortunately, corals reefs are also highly threatened ecosystems with some local and global factors being responsible. Local factors include declining water quality, destructive fishing, and increased pollution from urban areas. Global factors are global warming and ocean acidification, due to a dramatic rise of carbon dioxide levels in the atmosphere over the past century (Hoegh-Guldberg et al. 2007). Stony "reef-building" corals (Scleractinia) live in symbiosis with microalgae named zooxanthellae, which provide their coral hosts with up to 90% of their energy through photosynthesis (Stanley 2006). This relationship can be disrupted by environmental stressors such as long-lasting temperature increases together with intense periods of high sun irradiance. As a response to the latter stressor, corals expel their algae and thus lose their photosynthetic pigments at the same time, leading to the phenomenon of skeletal-looking bright white corals, a process better known as "coral bleaching" (Ainsworth et al. 2016; Heron et al. 2016). Such bleaching events have been increasing in the last two decades, thereby affecting reefs on a global scale. The severity of events caused coral mortalities of over 60% in some locations (Eakin et al. 2010; Hughes et al. 2017). Predicted impacts of persistent bleaching events include a reduction of reef biodiversity and coral cover, up to the total extinction of local coral species (Brainard et al. 2011, 2013). Furthermore, ocean acidification decelerates the calcification of corals by reducing the concentration of carbonate (CO_3^{2-}), and thus making it even less available for marine calcifiers (Hoegh-Guldberg et al. 2007). To counteract the fast and massive coral decline, a better understanding of the recovery and population dynamics of stony corals needs to be developed. because threatened coral reef systems depend on the recruitment of new individuals (Mumby and Steneck 2008). The recruitment process can be divided into (1) the development of competent larvae in the water column (spawners) or within the corals itself (brooders), (2) the settlement (searching, attachment, and larval metamorphosis) onto suitable substrata (Fig. 8.9), and (3) the survival of juvenile corals (Ritson-Williams et al. 2009). Since the survival rate of juvenile corals is likely influenced by the type of substratum chosen for settlement (Harrington et al. 2004; Ritson-Williams et al. 2010), finding a suitable settlement ground may be a critical step within the recruitment process.

Although chemical cues are believed to serve as the primary determinants of coral settlement, to some extent, physical properties have shown to influence coral larvae settlement as well. In a field study, larvae of five different species of scleractinian corals, including *Goniastrea favulus*, *G. aspera*,

Acropora tenuis, *Oxypora lacera*, and *Montipora peltiformis*, have shown to favor locations with lower light intensity (Mundy and Babcock 1998). Also, a study by Mason et al. (2011) demonstrated that both larvae of *Porites astreoides* and *Acropora palmata* consistently settled on different red or red-orange plastic materials while, at the same time, disdaining other colors such as green, blue, or white. It was suggested that this consistent response to red or reddish surfaces is related to long-wavelength photosensitivity and thus might be a potential strategy to artificially promote coral larvae settlement. Over the past decades, many studies have shown that coral larvae settle in response to either live CCA or organic extracts of CCA. For instance, larvae of the agariciid corals *Agaricia tenuifolia*, *A. humilis*, and *A. agaricites* have reported to settle to different degrees of stringency and specificity on Caribbean CCA, specifically *Hydrolithon boergesenii*. The responsible morphogenic inducer was fractionated by ultrafiltration and shown to be a water-insoluble, ether-insoluble, and acetone-insoluble unstable biochemical, which is apparently associated with the cell walls of the inducing CCA (Morse et al. 1988). Further studies by Morse and Morse (1991) have shown that the inductive molecule is a sulfated glycosaminoglycan. A field study of Price (2010) showed that larvae of several scleractinian corals, including *Pocillopora* spp., *Acropora* spp., and *Porites* spp., do indeed prefer specific CCA species for in situ settlement, as they recruited more frequently on *Titanoderma prototypum* than on other CCA. A similar settlement specificity has been demonstrated for larvae of the scleractinian corals *A. palmata* and *Montastraea faveolata*. The latter two species have been tested for their rates of settlement on ten different species of red algae, including eight different CCA species, resulting in strong settlement preferences of larvae from both corals to different CCA. *A. palmata* settled on surfaces of *H. boergesenii*, *Lithoporella atlantica*, *Neogoniolithon affine*, and *Titanoderma prototypum*, but showed no settlement on *N. mamillare*. Larvae of *M. faveolata* settled on surfaces of *Amphiroa tribulus*, *H. boergesenii*, *N. affine*, *N. munitum*, and *T. prototypum*, but no settlement occurred on *N. mamillare*, *Porolithon pachydermum*, and a noncoralline *Peyssonnelia* sp. The authors of this study suggested that patterns of coral distribution might be dependent on the red algae distribution (Ritson-Williams et al. 2014). However, in many cases of coral larvae settlement, the chemical identity of the presumed settlement-inducing molecule is just poorly described or remains largely unknown. In the past, the identity of these CCA-associated chemical cues was presumed to be cell wall-bound and thought to be some kind of high molecular mass polysaccharides (Morse and Morse 1991; Morse et al. 1994, 1996). Other studies chemically fully described the chemical signaling molecules; however, the detected cue often did not initiate the entire settlement cascade (e.g., 11-deoxyfistularin-3, Fig. 8.8d) (Kitamura et al. 2007) or just function as a settlement enhancer, such as luminaolide (Fig. 8.8e). The macrodiolide luminaolide was originally isolated from the CCA *H. reinboldii* and greatly enhanced the metamorphosis activity in *Leptastrea purpurea* when combined with another fraction that eluted at 80% aqueous methanol by octadecyl silica gel column chromatography (Kitamura et al. 2009; Maru et al. 2013). Interestingly, chemical inducers for larval settlement were also discovered in coral rubble and the skeleton of the massive coral *Goniastrea* sp. (Heyward and Negri 1999), indicating that coral larvae settlement can be either induced by a variety of chemical cues or by specific cues from multiple sources. In the past, bacterial biofilms have received notable attention as suitable settlement ground for many marine invertebrate larvae (Johnson et al. 1991; Pawlik 1992; Huang and Hadfield 2003; Huggett et al. 2006; Hadfield 2011). It was shown that coral reef biofilms, which were more than 2 weeks old, are able to induce settlement in the scleractinian *A. microphthalma*. FISH (fluorescence in situ hybridization) analysis revealed that the overall community composition of these biofilms was dominated by classes of *Alphaproteobacteria*, *Betaproteobacteria*, *Gammaproteobacteria*, and *Cytophagia-Flavobacteria* of *Bacteroidetes* (Webster et al. 2004). Apart from bacterial multispecies biofilms, a specific strain belonging to the genus *Pseudoalteromonas* (*Pseudoalteromonas* A3) isolated from the CCA *H. onkodes* was able to induce full settlement, including attachment and metamorphosis, in the reef-building corals *A. willisae* and *A. millepora*

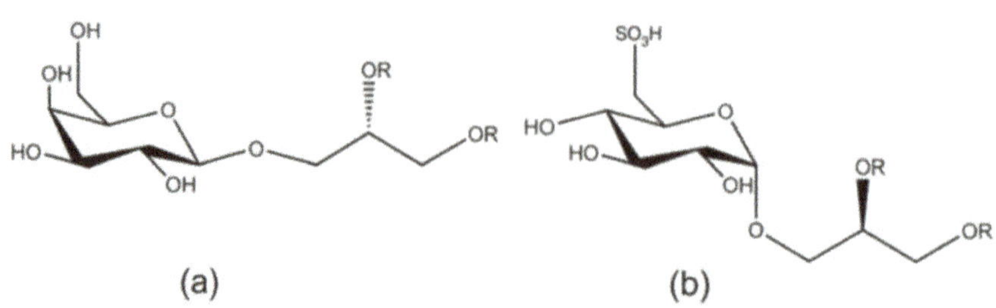

Fig. 8.10 Chemical structures of (**a**) (2S)-1-O-(7Z,10Z,13Z-hexadecatrienoyl)-3-O-β-D-galactopyranosyl-sn-glycerol and (**b**) (2R)-1-O-(palmitoyl)-3-O-α-D-(6′-sulfoquinovosyl)-sn-glycerol (Tebben et al. 2015) (created with ChemDraw, v. 16.0.1.4)

(Negri et al. 2001). Another metamorphosis-inducing cue, named tetrabromopyrrole, was also isolated from a *Pseudoalteromonas* sp. (Tebben et al. 2011). This compound might have widespread importance among Caribbean corals as it induced settlement in the brooder *Porites astreoides* as well as in the spawning species *Orbicella franksi* and *A. palmata* (Sneed et al. 2014). Further studies on the ecological relevance of Pseudoalteromonads and tetrabromopyrrole in the coral settlement process revealed that the respective bacteria and its compound did not elicit the same rates of coral larvae settlement as CCA and instead introduced morphogenic processes that are often fatal to the larvae. Instead it was found that CCA-derived molecules, belonging to the chemical classes of glycoglycerolipids (Fig. 8.10a, b) and high molecular weight polysaccharides, are the major contributors of the mixed fractions and caused larval settlement at equivalent concentrations present in live CCA (Tebben et al. 2015).

8.4 Conclusions

Secondary metabolites are investigated for their outstanding pharmaceutical applications as well as for their ecological relevance. Many MNPs have been found to elicit a broad range of bioactivities and, therefore, continue to be a prolific source for the generation of new drugs or drug leads. We believe that the exploration of new and extreme habitats will advance the discovery of novel macro- and microorganisms and, thus, might lead to the detection and isolation of novel NPs. Specifically, marine fungi still represent an underestimated but rich source for new SMs, although their distribution and ecological role often remains scarce. The hyphenation of state-of-the-art techniques such as chromatographic separation, mass spectrometry, and nuclear magnetic resonance spectroscopy is a suitable way to facilitate NP screening effort. Particularly, the use of multiple secondary metabolite databases as well as MS/MS approaches in combination with molecular networking makes the search for novel NPs more efficient and, at the same time, lowers the risk of rediscovery. In terms of chemical ecology, SMs fulfill manifold roles for their producers. Besides predator-prey and algae-herbivore interactions, marine chemical ecology has also shifted its focus on marine invertebrate settlement behavior. Specifically, the role of SMs as signaling molecules for coral larvae settlement has gained interest during the last decades. CCA and their associated microorganisms are the best-known sources for coral larvae settlement cues, but until today, only few settlement compounds have been chemically fully described. Furthermore, the knowledge of the interplay between coral larva, settlement cue, settlement cue-producing organism (may it be the CCA or its associated microbes), and other environmental factors such as light intensity is still limited and needs to be improved for a deeper comprehension of coral reef functioning. We are only at the beginning of understanding the role of SMs in the marine environment and many fascinating discoveries are yet to come.

Acknowledgments This study was carried out in the framework of the PhD research training group "The Ecology of Molecules" (EcoMol) supported by the Lower Saxony Ministry for Science and Culture. We also thank the reviewers for valuable comments which helped to improve the manuscript.

Appendix

This article is related to the YOUMARES 9 conference session no. 9: "Biodiversity of Benthic Holobionts: Chemical Ecology and Natural Products Chemistry in the Spotlight." The original Call for Abstracts and the abstracts of the presentations within this session can be found in the Appendix "Conference Sessions and Abstracts", Chapter "7 Biodiversity of Benthic Holobionts: Chemical Ecology and Natural Products Chemistry in the Spotlight", of this book.

References

Ainsworth TD, Heron SF, Ortiz JC et al (2016) Climate change disables coral bleaching protection on the Great Barrier Reef. Science 352:338–342

Amend AS, Barshis DJ, Oliver TA (2012) Coral-associated marine fungi form novel lineages and heterogeneous assemblages. ISME J 6:1291–1301

Asolkar RN, Freel KC, Jensen PR et al (2009) Arenamides A–C, cytotoxic NFκB inhibitors from the marine actinomycete *Salinispora Arenicola*. J Nat Prod 72:396–402

Bai R, Pettit GR, Hamel E (1990) Dolastatin 10, a powerful cytostatic peptide derived from a marine animal. Inhibition of tubulin polymerization mediated through the vinca alkaloid binding domain. Biochem Pharmacol 39:1941–1949

Barbier EB, Hacker SD, Kennedy C et al (2011) The value of estuarine and coastal ecosystem services. Ecol Monogr 81:169–193

Bell JJ, Smith D (2004) Ecology of sponge assemblages (Porifera) in the Wakatobi region, south-east Sulawesi, Indonesia: richness and abundance. J Mar Biol Assoc UK 84:581–591

Bell JJ, Davy SK, Jones T et al (2013) Could some coral reefs become sponge reefs as our climate changes? Glob Change Biol 19:2613–2624

Bellwood DR, Hughes TP, Folke C et al (2004) Confronting the coral reed crisis. Nature 429:827–833

Bérdy J (2005) Bioactive microbial metabolites. A personal view. J Antibiot 58:1–26

Bergmann W, Burke DC (1955) Contributions to the study of marine products. XXXIX. The nucleosides of sponges. III. Spongothymidine and spongouridine. J Org Chem 20:1501–1507

Bergmann W, Feeneyz RJ (1951) Contributions to the study of marine products. XXXII. The nucleosides of sponges. I. J Org Chem 16:981–987

Bhadury P, Mohammad BT, Wright PC (2006) The current status of natural products from marine fungi and their potential as anti-infective agents. J Ind Microbiol Biotechnol 33:325–337

Bianco ÉM, Teixeira VL, Pereira RC (2010) Chemical defenses of the tropical marine seaweed *Canistrocarpus cervicornis* against herbivory by sea urchin. Braz J Oceanogr 58:213–218

Bingham BL, Young CM (1991) Influence of sponges on invertebrate recruitment: a field test of allelopathy. Mar Biol 109:19–26

Blunt JW, Copp BR, Keyzers RA et al (2015) Marine natural products. Nat Prod Rep 32:116–211

Blunt JW, Carroll AR, Copp BR et al (2018) Marine natural products. Nat Prod Rep 35:8–53

Brainard RE, Birkeland C, Eakin CM et al (2011) Status review report of 82 candidate coral species petitioned under the U.S. endangered species act. U.S. Department of Commerce, NOAA Technical Memorandum, NOAA-TM-NMFS-PIFSC-27, 530 p + 1 appendix

Brainard RE, Weijerman M, Eakin CM et al (2013) Incorporating climate and ocean change into extinction risk assessments for 82 coral species. Conserv Biol 27:1169–1178

Breinbauer R, Vetter IR, Waldmann H (2002) Von Proteindomänen zu Wirkstoffkandidaten – Naturstoffe als Leitstrukturen für das Design und die Synthese von Substanzbibliotheken. Angew Chem 114:3002–3015

Bugni TS, Ireland CM (2004) Marine-derived fungi: a chemically and biologically diverse group of microorganisms. Nat Prod Rep 21:143–163

Bull AT, Stach JEM (2007) Marine actinobacteria: new opportunities for natural product search and discovery. Trends Microbiol 15:491–499

Bull AT, Stach JEM, Ward AC et al (2005) Marine actinobacteria: perspectives, challenges, future directions. Antonie Van Leeuwenhoek 87:65–79

Burgaud G, Le Calvez T, Arzur D et al (2009) Diversity of culturable marine filamentous fungi from deep-sea hydrothermal vents. Environ Microbiol 11:1588–1600

Carrano L, Marinelli F (2015) The relevance of chemical dereplication in microbial natural product screening. J Appl Bioanal 1:55–67

Chevolot L, Cochard J-C, Yvin J-C (1991) Chemical induction of larval metamorphosis of *Pecten maximus* with a note on the nature of naturally occurring triggering substances. Mar Ecol Prog Ser 74:83–89

Clare AS, Matsumura K (2000) Nature and perception of barnacle settlement pheromones. Biofouling 15:57–71

Colvard NB, Edmunds PJ (2011) Decadal-scale changes in abundance of non-scleractinian invertebrates on a Caribbean coral reef. J Exp Mar Biol Ecol 397:153–160

Cooper MB, Smith AG (2015) Exploring mutualistic interactions between microalgae and bacteria in the omics age. Curr Opin Plant Biol 26:147–153

Craft JD, Paul VJ, Sotka EE (2013) Biogeographic and phylogenetic effects on feeding resistance of generalist herbivores toward plant chemical defenses. Ecology 94:18–24

Crisp DJ, Meadows PS (1962) The chemical basis of gregariousness in cirripedes. Proc R Soc Lond B Biol Sci 156:500–520

Crisp DJ, Meadows PS (1963) Adsorbed layers: the stimulus to settlement in barnacles. Proc R Soc Lond B Biol Sci 158:364–387

Cueto M, Jensen PR, Kauffman C et al (2001) Pestalone, a new antibiotic produced by a marine fungus in response to bacterial challenge. J Nat Prod 64:1444–1446

Da-Anoy JP, Villanueva RD, Cabaitan PC et al (2017) Effects of coral extracts on survivorship, swimming behavior, and settlement of *Pocillopora damicornis* larvae. J Exp Mar Biol Ecol 486:93–97

Das S, Lyla PS, Ajmal Khan S (2006) Marine microbial diversity and ecology: importance and future perspectives. Curr Sci 90:1325–1335

de Carvalho CCCR, Fernandes P (2010) Production of metabolites as bacterial responses to the marine environment. Mar Drugs 8:705–727

DeLong EF, Preston CM, Mincer T et al (2006) Community genomics among stratified microbial assemblages in the ocean's interior. Science 311:496–503

Dewick PM (2002) Medicinal natural products. A biosynthetic approach. Wiley, Chichester

Dias DA, Urban S, Roessner U (2012) A historical overview of natural products in drug discovery. Meta 2:303–336

Dreanno C, Matsumura K, Dohmae N et al (2006) An α2-macroglobulin-like protein is the cue to gregarious settlement of the barnacle *Balanus amphitrite*. Proc Natl Acad Sci U S A 103:14396–14401

Eakin CM, Morgan JA, Heron SF et al (2010) Caribbean corals in crisis: record thermal stress, bleaching, and mortality in 2005. PLoS One 5:e13969. https://doi.org/10.1371/journal.pone.0013969

Elbourne PD, Clare AS (2010) Ecological relevance of a conspecific, waterborne settlement cue in *Balanus amphitrite* (Cirripedia). J Exp Mar Biol Ecol 392:99–106

Endo N, Nogata Y, Yoshimura E et al (2009) Purification and partial amino acid sequence analysis of the larval settlement-inducing pheromone from adult extracts of the barnacle, *Balanus amphitrite* (= *Amphibalanus amphitrite*). Biofouling 25:429–434

Feling RH, Buchanan GO, Mincer TJ et al (2003) Salinosporamide A: a highly cytotoxic proteasome inhibitor from a novel microbial source, a marine bacterium of the new genus *Salinospora*. Angew Chem Int Ed 42:355–357

Fenical W (1999) Halimide, a cytotoxic marine natural product, and derivatives thereof. US Patent WO 1999048889 A1

Fenical W, Jensen PR (1993) Marine microorganisms: a new biomedical resource. In: Attaway DH, Zaborsky OR (eds) Marine biotechnology. Plenum Press, New York, pp 419–457

Fenical W, Jensen PR (2006) Developing a new resource for drug discovery: marine actinomycete bacteria. Nat Chem Biol 2:666–673

Fu P, Legako A, La S et al (2016) Discovery, characterization, and analogue synthesis of bohemamine dimers generated by non-enzymatic biosynthesis. Chem Eur J 22:3491–3495

Gademann K, Kobylinska J (2009) Antimalarial natural products of marine and freshwater origin. Chem Rec 9:187–198

Gerwick WH, Moore BS (2012) Lessons from the past and charting the future of marine natural products drug discovery and chemical biology. Chem Biol 19:85–98

Goecke F, Labes A, Wiese J et al (2010) Chemical interactions between marine macroalgae and bacteria. Mar Ecol Prog Ser 409:267–300

Goecke F, Labes A, Wiese J et al (2012) Dual effect of macroalgal extracts on growth of bacteria in Western Baltic Sea. Rev Biol Mar Oceanogr 47:75–86

Gomes NGM, Lefranc F, Kijjoa A et al (2015) Can some marine-derived fungal metabolites become actual anticancer agents? Mar Drugs 13:3950–3991

Goodfellow M, Fiedler H-P (2010) A guide to successful bioprospecting: informed by actinobacterial systematics. Antonie Van Leeuwenhoek 98:119–142

Guijas C, Montenegro-Burke JR, Domingo-Almenara X et al (2018) Metlin: a technology platform for identifying knowns and unknowns. Anal Chem 90:3156–3164

Gulder TAM, Moore BS (2009) Chasing the treasures of the sea – bacterial marine natural products. Curr Opin Microbiol 12:252–260

Gullo VP, McAlpine J, Lam KS et al (2006) Drug discovery from natural products. J Ind Microbiol Biotechnol 33:523–531

Gurnani N, Mehta D, Gupta M et al (2014) Natural products: source of potential drugs. Afr J Basic Appl Sci 6:171–186

Haber M, Carbone M, Mollo E et al (2011) Chemical defense against predators and bacterial fouling in the Mediterranean sponges *Axinella polypoides* and *A. verrucosa*. Mar Ecol Prog Ser 422:113–122

Hadfield MG (2011) Biofilms and marine invertebrate larvae: what bacteria produce that larvae use to choose settlement sites. Annu Rev Mar Sci 3:453–470

Hadfield MG, Paul VJ (2001) Natural chemical cues for settlement and metamorphosis of marine-invertebrate larvae. In: McClintock JB, Baker BJ (eds) Marine chemical ecology. CRC Press, Boca Raton, pp 431–461

Haefner B (2003) Drugs from the deep: marine natural products as drug candidates. Drug Discov Today 8:536–544

Harder T, Campbell AH, Egan S et al (2012) Chemical mediation of ternary interactions between marine holobionts and their environment as exemplified by the red alga *Delisea pulchra*. J Chem Ecol 38:442–450

Harder T, Tebben J, Möller M et al (2018) Chemical ecology of marine invertebrate larval settlement. In: Puglisi M, Becerro MA (eds) Chemical ecology: the ecological impact of marine natural products. CRC Press, Boca Raton, pp 328–355

Harrington L, Fabricius K, De'ath G et al (2004) Recognition and selection of settlement substrata determine post-settlement survival in corals. Ecology 85:3428–3437

Hay ME (1996) Marine chemical ecology: what's known and what's next? J Exp Mar Bio Ecol 200:103–134

Hay ME (2009) Marine chemical ecology: chemical signals and cues structure marine populations, communities, and ecosystems. Annu Rev Mar Sci 1:193–212

Heilmann J (2007) Wirkstoffe auf Basis biologisch aktiver Naturstoffe. Chem unserer Zeit 41:376–389

Helber SB, de Voogd NJ, Muhando CA et al (2017) Anti-predatory effects of organic extracts of 10 common reef sponges from Zanzibar. Hydrobiologia 790:247–258

Helber SB, Hoeijmakers DJJ, Muhando CA et al (2018) Sponge chemical defenses are a possible mechanism for increasing sponge abundance on reefs in Zanzibar. PLoS One 13:e0197617. https://doi.org/10.1371/journal.pone.0197617

Hellio C, Tsoukatou M, Maréchal J-P et al (2005) Inhibitory effects of Mediterranean sponge extracts and metabolites on larval settlement of the barnacle *Balanus amphitrite*. Mar Biotechnol 7:297–305

Henríquez M, Vergara K, Norambuena J et al (2014) Diversity of cultivable fungi associated with Antarctic marine sponges and screening for their antimicrobial, antitumoral and antioxidant potential. World J Microbiol Biotechnol 30:65–76

Hentschel U, Piel J, Degnan SM et al (2012) Genomic insights into the marine sponge microbiome. Nat Rev Microbiol 10:641–654

Heron SF, Maynard JA, van Hooidonk R et al (2016) Warming trends and bleaching stress of the world's coral reefs 1985–2012. Sci Rep 6:38402. https://doi.org/10.1038/srep38402

Heyward AJ, Negri AP (1999) Natural inducers for coral larval metamorphosis. Coral Reefs 18:273–279

Hoegh-Guldberg O, Mumby PJ, Hooten AJ et al (2007) Coral reefs under rapid climate change and ocean acidification. Science 318:1737–1742

Hu Y, Chen J, Hu G et al (2015) Statistical research on the bioactivity of new marine natural products discovered during the 28 years from 1985 to 2012. Mar Drugs 13:202–221

Huang S, Hadfield MG (2003) Composition and density of bacterial biofilms determine larval settlement of the polychaete *Hydroides elegans*. Mar Ecol Prog Ser 260:161–172

Hubert J, Nuzillard J-M, Renault J-H (2017) Dereplication strategies in natural product research: how many tools and methodologies behind the same concept? Phytochem Rev 16:55–95

Hug LA, Baker BJ, Anantharaman K et al (2016) A new view of the tree of life. Nat Commun 7:11870. https://doi.org/10.1038/nmicrobiol.2016.48

Huggett MJ, Williamson JE, de Nys R et al (2006) Larval settlement of the common Australian sea urchin *Heliocidaris erythrogramma* in response to bacteria from the surface of coralline algae. Oecologia 149:604–619

Hughes CC, Fenical W (2010) Antibacterials from the sea. Chem Eur J 16:12512–12525

Hughes TP, Kerry JT, Álvarez-Noriega M et al (2017) Global warming and recurrent mass bleaching of corals. Nature 543:373–377

Imhoff JF, Labes A, Wiese J (2011) Bio-mining the microbial treasures of the ocean: new natural products. Biotechnol Adv 29:468–482

Jensen PR, Mincer TJ, Williams PG et al (2005) Marine actinomycete diversity and natural product discovery. Antonie Van Leeuwenhoek 87:43–48

Johnson CR, Sutton DC, Olson RR et al (1991) Settlement of crown-of-thorns starfish: role of bacteria on surfaces of coralline algae and a hypothesis for deepwater recruitment. Mar Ecol Prog Ser 71:143–162

Jones EBG (2011) Fifty years of marine mycology. Fungal Divers 50:73–112

Kazamia E, Czesnick H, Van Nguyen TT et al (2012) Mutualistic interactions between vitamin B_{12}-dependent algae and heterotrophic bacteria exhibit regulation. Environ Microbiol 14:1466–1476

Keller NP, Turner G, Bennett JW (2005) Fungal secondary metabolism. From biochemistry to genomics. Nat Rev Microbiol 3:937–947

Kind T, Tsugawa H, Cajka T et al (2017) Identification of small molecules using accurate mass MS/MS search. Mass Spec Rev 37:513–532

Kitamura M, Koyama T, Nakano Y et al (2007) Characterization of a natural inducer of coral larval metamorphosis. J Exp Mar Bio Ecol 340:96–102

Kitamura M, Schupp PJ, Nakano Y et al (2009) Luminaolide, a novel metamorphosis-enhancing macrodiolide for scleractinian coral larvae from crustose coralline algae. Tetrahedron Lett 50:6606–6609

Kohlmeyer J, Kohlmeyer E (1979) Marine mycology. The higher fungi. Academic, New York

Kreis W (2007) Prinzipien des Sekundärstoffwechsels. In: Hänsel R, Sticher O (eds) Pharmakognosie – Phytopharmazie. Springer, Heidelberg, pp 4–30

Kreysa G, Grabley S (2007) Vorbild Natur. Stand und Perspektiven der Naturstoff-Forschung in Deutschland. DECHEMA e.V., Frankfurt am Main

Kubicek A, Bessho K, Nakaoka M et al (2011) Inducible defence and its modulation by environmental stress in the red alga *Chondrus yendoi* (Yamada and Mikami in Mikami, 1965) from Honshu Island, Japan. J Exp Mar Bio Ecol 397:208–213

Kurita KL, Linington RG (2015) Connecting phenotype and chemotype: high-content discovery strategies for natural products research. J Nat Prod 78:587–596

Kwon HC, Kauffman CA, Jensen PR et al (2006) Marinomycins A-D, antitumor-antibiotics of a new structure class from a marine actinomycete of the recently discovered genus "*Marinispora*". J Am Chem Soc 128:1622–1632

Lachnit T, Blümel M, Imhoff JF et al (2009) Specific epibacterial communities on macroalgae: phylogeny matters more than habitat. Aquat Biol 5:181–186

Lau SCK, Qian P-Y (2001) Larval settlement in the serpulid polychaete *Hydroides elegans* in response to bacterial films: an investigation of the nature of putative larval settlement cue. Mar Biol 138:321–328

Lau SCK, Thiyagarajan V, Cheung SCK et al (2005) Roles of bacterial community composition in biofilms as a mediator for larval settlement of three marine invertebrates. Aquat Microb Ecol 38:41–51

Le TC, Yang I, Yoon YJ (2016) Ansalactams B–D illustrate further biosynthetic plasticity within the ansamycin pathway. Org Lett 18:2256–2259

Lee OO, Wong YH, Qian P-Y (2009) Inter- and intraspecific variations of bacterial communities associated with marine sponges from San Juan Island, Washington. Appl Environ Microbiol 75:3513–3521

Lee J-Y, Orlikova B, Diederich M (2015) Signal transducers and activators of transcription (STAT) regulatory networks in marine organisms: from physiological observations towards marine drug discovery. Mar Drugs 13:4967–4984

Letzel A-C, Pidot SJ, Hertweck C (2013) A genomic approach to the cryptic secondary metabolome of the anaerobic world. Nat Prod Rep 30:392–428

Li W, Scott AP, Siefkes MJ et al (2002) Bile acid secreted by male sea lamprey that acts as a sex pheromone. Science 296:138–141

Loh T-L, Pawlik JR (2014) Chemical defenses and resource trade-offs structure sponge communities on Caribbean coral reefs. Proc Natl Acad Sci U S A 111:4151–4156

Lonsdale DJ, Frey MA, Snell TW (1998) The role of chemical signals in copepod reproduction. J Mar Syst 15:1–12

Loque CP, Medeiros AO, Pellizzari FM et al (2010) Fungal community associated with marine macroalgae from Antarctica. Polar Biol 33:641–648

Ludwig-Müller J, Gutzeit H (2014) Biologie von Naturstoffen. Synthese, biologische Funktionen und Bedeutung für die Gesundheit. UTB, Stuttgart

Luesch H, Moore RE, Paul VJ et al (2001) Isolation of dolastatin 10 from the marine cyanobacterium *Symploca* species VP642 and total stereochemistry and biological evaluation of its analogue symplostatin 1. J Nat Prod 64:907–910

Macheleidt J, Mattern DJ, Fischer J et al (2016) Regulation and role of fungal secondary metabolites. Annu Rev Genet 50:371–392

Madigan MT, Martinko JM, Parker J (2003) Brock biology of microorganisms. Prentice Hall, Upper Saddle River

Maliao RJ, Turingan RG, Lin J (2008) Phase-shift in coral reef communities in the Florida keys National Marine Sanctuary (FKNMS), USA. Mar Biol 154:841–853

Martins A, Vieira H, Gaspar H et al (2014) Marketed marine natural products in the pharmaceutical and cosmeceutical industries: tips for success. Mar Drugs 12:1066–1101

Maru N, Inuzuka T, Yamamoto K et al (2013) Relative configuration of luminaolide. Tetrahedron Lett 54:4385–4387

Mason B, Beard M, Miller MW (2011) Coral larvae settle at a higher frequency on red surfaces. Coral Reefs 30:667–676

Matsumura K, Nagano M, Fusetani N (1998) Purification of a larval settlement-inducing protein complex (SIPC) of the barnacle, *Balanus amphitrite*. J Exp Zool 281:12–20

Mayer AMS, Glaser KB, Cuevas C (2010) The odyssey of marine pharmaceuticals: a current pipeline perspective. Trends Pharmacol Sci 31:255–265

Mayer AMS, Rodríguez AD, Taglialatela-Scafati O et al (2013) Marine pharmacology in 2009–2011: marine compounds with antibacterial, antidiabetic, antifungal, anti-inflammatory, antiprotozoal, antituberculosis, and antiviral activities; affecting the immune and nervous systems, and other miscellaneous mechanisms of action. Mar Drugs 11:2510–2573

Moberg F, Folke C (1999) Ecological goods and services of coral reef ecosystems. Ecol Econ 29:215–233

Mohamed A, Nguyen CH, Mamitsuka H (2016) Current status and prospects of computational resources for natural product dereplication: a review. Brief Bioinform 17:309–321

Moitinho-Silva L, Nielsen S, Amir A et al (2017) The sponge microbiome project. Gigascience 6:1–7

Molinski TF, Dalisay DS, Lievens SL et al (2009) Drug development from marine natural products. Nat Rev Drug Discov 8:69–85

Morse DE, Morse ANC (1991) Enzymatic characterization of the morphogen recognized by *Agaricia humilis* (scleractinian coral) larvae. Biol Bull 181:104–122

Morse DE, Hooker N, Morse ANC et al (1988) Control of larval metamorphosis and recruitment in sympatric agariciid coral. J Exp Mar Bio Ecol 116:193–217

Morse DE, Morse ANC, Raimondi PT et al (1994) Morphogen-based chemical flypaper for *Agaricia humilis* coral larvae. Biol Bull 186:172–181

Morse ANC, Iwao K, Baba M et al (1996) An ancient chemosensory mechanism brings new life to coral reefs. Biol Bull 191:149–154

Mumby PJ, Steneck RS (2008) Coral reef management and conservation in light of rapidly evolving ecological paradigms. Trends Ecol Evol 23:555–563

Mundy CN, Babcock RC (1998) Role of light intensity and spectral quality in coral settlement: implications for depth-dependent settlement? J Exp Mar Bio Ecol 223:235–255

Nastrucci C, Cesario A, Russo P (2012) Anticancer drug discovery from the marine environment. Recent Pat Anticancer Drug Discov 7:218–232

Negri AP, Webster NS, Hill RT (2001) Metamorphosis of broadcast spawning corals in response to bacteria isolated from crustose algae. Mar Ecol Prog Ser 223:121–131

Nicolas L, Robert R, Chevolot L (1998) Comparative effects of inducers on metamorphosis of the Japanese oyster *Crassostrea gigas* and the great scallop *Pecten maximus*. Biofouling 12:189–203

Nielsen KF, Månsson M, Rank C et al (2011) Dereplication of microbial natural products by LC-DAD-TOFMS. J Nat Prod 74:2338–2348

Noyer C, Becerro MA (2012) Relationship between genetic, chemical, and bacterial diversity in the Atlanto-Mediterranean bath sponge *Spongia lamella*. Hydrobiologia 687:85–99

Noyer C, Thomas OP, Becerro MA (2011) Patterns of chemical diversity in the Mediterranean sponge *Spongia lamella*. PLoS One 6:e20844. https://doi.org/10.1371/journal.pone.0020844

O'Brien J, Wright GD (2011) An ecological perspective of microbial secondary metabolism. Curr Opin Biotechnol 22:552–558

Okazaki T, Kitahara T, Okami Y (1975) Studies on marine microorganisms. IV. A new antibiotic SS-228 Y produced by *Chainia* isolated from shallow sea mud. J Antibiot 28:176–184

Ortlepp S, Sjögren M, Dahlström M et al (2006) Antifouling activity of bromotyrosine-derived sponge metabolites and synthetic analogues. Mar Biotechnol 9:776–785

Overy DP, Bayman P, Kerr RG et al (2014) An assessment of natural product discovery from marine (*sensu strictu*) and marine-derived fungi. Mycology 5:145–167

Pantazopoulou A, Galmarini CM, Penalva MA (2018) Molecular basis of resistance to the microtubule-depolymerizing antitumor compound plocabulin. Sci Rep 8:8616. https://doi.org/10.1038/s41598-018-26736-3

Paul VJ, Ritson-Williams R, Sharp K (2011) Marine chemical ecology in benthic environments. Nat Prod Rep 28:345–387

Paul VJ, Ritson-Williams R, Campbell J et al (2014) Algal chemical ecology in a changing ocean. Planta Med 80:IL11. https://doi.org/10.1055/s-0034-1382302

Pauly D, Christensen V, Guénette S et al (2002) Towards sustainability in world fisheries. Nature 418:689–695

Pawlik JR (1986) Chemical induction of larval settlement and metamorphosis in the reef-building tube worm *Phragnmtopoma californica* (Sabellariidae: Polychaeta). Mar Biol 91:59–68

Pawlik JR (1992) Chemical ecology of the settlement of benthic marine invertebrates. Oceanogr Mar Biol 30:273–335

Pawlik JR (2011) The chemical ecology of sponges on Caribbean reefs: natural products shape natural systems. Bioscience 61:888–898

Pawlik JR, Steindler L (2007) Chemical warfare on coral reefs: sponge metabolites differentially affect coral symbiosis *in situ*. Limnol Oceanogr 52:907–911

Pawlik JR, Butman CA, Starczak VR (1991) Hydrodynamic facilitation of gregarious settlement of a reef-building tube worm. Science 251:421–424

Pérez MJ, Falqué E, Domínguez H (2016) Antimicrobial action of compounds from marine seaweed. Mar Drugs 14:52. https://doi.org/10.3390/md14030052

Perry CT, Murphy GN, Kench PS et al (2013) Caribbean-wide decline in carbonate production threatens coral reef growth. Nat Commun 4:1402. https://doi.org/10.1038/ncomms2409

Pichersky E, Gang DR (2000) Genetics and biochemistry of secondary metabolites in plants: an evolutionary perspective. Trends Plant Sci 5:439–445

Pohnert G (2004) Chemical defense strategies of marine organisms. Top Curr Chem 239:179–219

Pontius A, Krick A, Kehraus S et al (2008) Antiprotozoal activities of heterocyclic-substituted xanthones from the marine-derived fungus *Chaetomium* sp. J Nat Prod 71:1579–1584

Price N (2010) Habitat selection, facilitation, and biotic settlement cues affect distribution and performance of coral recruits in French Polynesia. Oecologia 163:747–758

Proksch P (1994) Defensive roles for secondary metabolites from marine sponges and sponge-feeding nudibranchs. Toxicon 32:639–655

Proksch P, Edrada-Ebel R, Ebel R (2006) Bioaktive Naturstoffe aus marinen Schwämmen: Apotheke am Meeresgrund. Biol unserer Zeit 36:150–159

Puglisi MP, Sneed JM, Sharp K et al (2014) Marine chemical ecology in benthic environments. Nat Prod Rep 31:1510–1553

Quinn RA, Nothias L-F, Vining O (2017) Molecular networking as a drug discovery, drug metabolism, and precision medicine strategy. Trends Pharmacol Sci 38:143–154

Radjasa OK, Vaske YM, Navarro G et al (2011) Highlights of marine invertebrate-derived biosynthetic products: their biomedical potential and possible production by microbial associants. Bioorg Med Chem 19:6658–6674

Rahman H, Austin B, Mitchell WJ et al (2010) Novel anti-infective compounds from marine bacteria. Mar Drugs 8:498–518

Rasher DB, Hoey AS, Hay M (2013) Consumer diversity interacts with prey defenses to drive ecosystem function. Ecology 94:1347–1358

Rateb ME, Ebel R (2011) Secondary metabolites of fungi from marine habitats. Nat Prod Rep 28:290–344

Ritson-Williams R, Arnold SN, Fogarty ND et al (2009) New perspectives on ecological mechanisms affecting coral recruitment on reefs. Smithson Contrib Mar Sci (38):437–457

Ritson-Williams R, Paul VJ, Arnold SN et al (2010) Larval settlement preferences and post-settlement survival of the threatened Caribbean corals *Acropora palmata* and *A. cervicornis*. Coral Reefs 29:71–81

Ritson-Williams R, Arnold SN, Paul VJ et al (2014) Larval settlement preferences of *Acropora palmata* and *Montastraea faveolata* in response to diverse red algae. Coral Reefs 33:59–66

Rohde S, Schupp PJ (2018) Spatial and temporal variability in sponge chemical defense. In: Puglisi M, Becerro MA (eds) Chemical ecology: the ecological impact of marine natural products. CRC press, Boca Raton, pp 372–397

Rohde S, Gochfeld DJ, Ankisetty S et al (2012) Spatial variability in secondary metabolites of the Indo-Pacific sponge *Stylissa massa*. J Chem Ecol 38:463–475

Rohde S, Nietzer S, Schupp PJ (2015) Prevalence and mechanisms of dynamic chemical defenses in tropical sponges. PLoS One 10:e0132236. https://doi.org/10.1371/journal.pone.0132236

Sakata K (1989) Feeding attractants and stimulants for marine gastropods. In: Scheuer PJ (ed) Bioorganic marine chemistry. Springer, Heidelberg, pp 115–129

Schulz B, Draeger S, de la Cruz TE et al (2008) Screening strategies for obtaining novel, biologically active, fungal secondary metabolites from marine habitats. Bot Mar 51:219–234

Schupp PJ, Paul VJ (1994) Calcium carbonate and secondary metabolites in tropical seaweeds: variable effects on herbivorous fishes. Ecology 75:1172–1185

Schwartz N, Rohde S, Hiromori S et al (2016) Understanding the invasion success of *Sargassum muticum*: herbivore preferences for native and invasive *Sargassum* spp. Mar Biol 163:181. https://doi.org/10.1007/s00227-016-2953-4

Schwartz N, Rohde S, Dobretsov S et al (2017) The role of chemical antifouling defence in the invasion success of *Sargassum muticum*: a comparison of native and invasive brown algae. PLoS One 12:e0189761. https://doi.org/10.1371/journal.pone.0189761

Sebens KP (1983) Settlement and metamorphosis of a temperate soft-coral larva (Alcyonium siderium Verrill): induction by crustose algae. Biol Bull 165:286–304

Sieg RD, Poulson-Ellestad KL, Kubanek J (2011) Chemical ecology of the marine plankton. Nat Prod Rep 28:388–399

Simmons TL, Andrianasolo E, McPhail K et al (2005) Marine natural products as anticancer drugs. Mol Cancer Ther 4:333–342

Simões MF, Antunes A, Ottoni CA et al (2015) Soil and rhizosphere associated fungi in gray mangroves (*Avicennia marina*) from the Red Sea – a metagenomic approach. Genomics Proteomics Bioinformatics 13:310–320

Simon C, Daniel R (2009) Achievements and new knowledge unraveled by metagenomic approaches. Appl Microbiol Biotechnol 85:265–276

Simon C, Daniel R (2011) Metagenomic analyses: past and future trends. Appl Environ Microbiol 77:1153–1161

Sneed JM, Sharp KH, Ritchie KB et al (2014) The chemical cue tetrabromopyrrole from a biofilm bacterium induces settlement of multiple Caribbean corals. Proc R Soc Lond B Biol Sci 281:20133086. https://doi.org/10.1098/rspb.2013.3086

Stanley GD Jr (2006) Photosymbiosis and the evolution of modern coral reefs. Science 312:857–858

Staufenberger T, Thiel V, Wiese J et al (2008) Phylogenetic analysis of bacteria associated with *Laminaria saccharina*. FEMS Microbiol Ecol 64:65–77

Swanson RL, Williamson JE, De Nys R et al (2004) Induction of settlement of larvae of the sea urchin *Holopneustes purpurascens* by histamine from a host alga. Biol Bull 206:161–172

Taori K, Paul VJ, Luesch H (2008) Structure and activity of largazole, a potent antiproliferative agent from the Floridian marine cyanobacterium *Symploca* sp. J Am Chem Soc 130:1806–1807

Tebben J, Tapiolas DM, Motti CA et al (2011) Induction of larval metamorphosis of the coral *Acropora millepora* by tetrabromopyrrole isolated from a *Pseudoalteromonas* bacterium. PLoS One 6:e19082. https://doi.org/10.1371/journal.pone.0019082

Tebben J, Motti CA, Siboni N et al (2015) Chemical mediation of coral larval settlement by crustose coralline algae. Sci Rep 5:10803. https://doi.org/10.1038/srep10803

Thiel V, Neulinger SC, Staufenberger T et al (2007) Spatial distribution of sponge-associated bacteria in the Mediterranean sponge *Tethya aurantium*. FEMS Microbiol Ecol 59:47–63

Thomas T, Moitinho-Silva L, Lurgi M et al (2016) Diversity, structure and convergent evolution of the global sponge microbiome. Nat Commun 7:11870. https://doi.org/10.1038/ncomms11870

Thompson JE (1985) Exudation of biologically-active metabolites in the sponge *Aplysina fistularis*. I. Biological evidence. Mar Biol 88:23–26

Thompson JE, Walker RP, Faulkner DJ (1985) Screening and bioassays for biologically-active substances from forty marine sponge species from San Diego, California, USA. Mar Biol 88:11–21

Thorson G (1950) Reproduction and larval ecology of marine bottom invertebrates. Biol Rev Camb Philos Soc 25:1–45

Toonen RJ, Pawlik JR (1996) Settlement of the tube worm *Hydroides dianthus* (Polychaeta: Serpulidae): cues for gregarious settlement. Mar Biol 126:725–733

Tsukamoto S, Kato H, Hirota H et al (1999) Lumichrome. A larval metamorphosis-inducing substance in the ascidian *Halocynthia roretzi*. Eur J Biochem 264:785–789

Villamizar E, Díaz MC, Rützler K et al (2013) Biodiversity, ecological structure, and change in the sponge community of different geomorphological zones of the barrier fore reef at Carrie Bow Cay, Belize. Mar Ecol 35:425–435

Wang M-H, Li X-M, Li C-S et al (2013) Secondary metabolites from *Penicillium pinophilum* SD-272, a marine sediment-derived fungus. Mar Drugs 11:2230–2238

Wang Y-Z, Xue Y-R, Liu C-H (2015) A brief review of bioactive metabolites derived from deep-sea fungi. Mar Drugs 13:4594–4616

Wang M, Carver JJ, Phelan VV et al (2016) Sharing and community curation of mass spectrometry data with GNPS. Nat Biotechnol 34:828–837

Waters AL, Hill RT, Place AR et al (2010) The expanding role of marine microbes in pharmaceutical development. Curr Opin Biotechnol 21:780–786

Webster NS, Smith LD, Heyward AJ et al (2004) Metamorphosis of a scleractinian coral in response to microbial biofilms. Appl Environ Microbiol 70:1213–1221

Wiese J, Ohlendorf B, Blümel M et al (2011) Phylogenetic identification of fungi isolated from the marine sponge *Tethya aurantium* and identification of their secondary metabolites. Mar Drugs 9:561–585

Williams PG (2009) Panning for chemical gold: marine bacteria as a source of new therapeutics. Trends Biotechnol 27:45–52

Williams PG, Buchanan GO, Feling RH et al (2005) New cytotoxic salinosporamides from the marine actinomycete *Salinispora tropica*. J Org Chem 70:6196–6203

Williamson JE, De Nys R, Kumar N et al (2000) Induction of metamorphosis in the sea urchin *Holopneustes purpurascens* by a metabolite complex from the algal host *Delisea pulchra*. Biol Bull 198:332–345

Wilson MC, Nam S-J, Gulder TAM (2011) Structure and biosynthesis of the marine streptomycete ansamycin ansalactam A and its distinctive branched chain polyketide extender unit. J Am Chem Soc 133:1971–1977

Wolfender J-L, Marti G, Queiroz EF (2010) Advances in techniques for profiling crude extracts and for the rapid identification of natural products: dereplication, quality control and metabolomics. Curr Org Chem 14:1808–1832

Yvin JC, Chevolot L, Chevolot-Magueur AM et al (1985) First isolation of jacaranone from an alga, *Delesseria sanguinea*, a metamorphosis inducer of *Pecten* larvae. J Nat Prod 48:814–816

9 Sponges Revealed: A Synthesis of Their Overlooked Ecological Functions Within Aquatic Ecosystems

Mainah Folkers and Titus Rombouts

Abstract

While sponges are the oldest still living multicellular animals on this planet and omnipresent within aquatic ecosystems, they have not been studied nearly as much compared to the recognized ecosystem drivers in coral reefs: corals, algae, and fish. We therefore want to take this opportunity to illustrate the diversity, functionality, and sheer survivability of these ancient animals. Beyond its multitude of external shapes and colors, sponges hold a unique internal aquiferous system. This system of afferent and efferent canals is intricately linked to supply its key function as a filter feeder. By filtering both particulate and dissolved material, sponges fill a niche in nutrient cycling. Moreover, the survivability of sponges is demonstrated in the variety of habitats it resides in; from freshwater canals to polar deep seas. In formerly uninhabitable environments, sponges can potentially create biodiversity hotspots by providing habitat complexity and shelter from predators. This review will give insight into the early life history, morphology, diet, and reproduction of sponges. Furthermore, it is imperative to consider their function as habitat facilitator, nutrient cycler, and, last but not the least, their potential for future pharmaceuticals. The emphasis in the proceedings has been specifically put on the role of sponges as nutrient cycler as they play a role in the three essential elements: carbon, nitrogen, and phosphorous. With all this in mind, it should be clear that even though sponges are relatively overlooked marine invertebrates, they should be studied similarly to corals and respected as a key ecosystem driver in novel and established environments.

Keywords

Porifera · Filter feeders · Nutrient cycling · Habitat facilitation · Marine natural products

9.1 Introduction

Ever since marine research has been documented several hundreds of years ago, we usually consider coral reefs as iconic examples of biological hotspots. These reefs have provided potential services to the ecosystem, among them are their services for early humans to maintain nutritional uptake. We have typically considered coral reefs to consist of three big ecosystem drivers: corals, algae and fish. Yet, within these coral reefs lie a less familiar but equally important builder and energy conveyer: sponges. Slowly, recognition develops that sponges are key ecosystem engineers. They have the capacity to retain nutrients and transform them into a bio-available form back into their surrounding habitat while providing protection for motile fauna. In fact, on Caribbean coral reefs, sponges usually show a higher diversity and higher abundance compared to corals (Diaz and Rützler 2001).

However, apart from these iconic coral reefs, there is a multitude of habitats in which sponges thrive (Gili and Coma 1998). Habitats can range from the deep sea to turbulent freshwater canals among cities. Within these various habitats, sponges perform important so-called benthic-pelagic coupling, which is a crucial ecosystem service to recycle pelagic nutrients toward the benthos that would otherwise be unavailable to higher trophic levels (Griffiths et al. 2017). For example, the Caribbean giant barrel sponge *Xestospongia muta* shows a large role in the carbon transfer from the water column to the benthos (McMurray et al. 2017). Moreover, it could be suggested that, in previously uncolonized marine environments, sponges are among the first settlers creating a three-dimensional habitat, allowing benthic ecosystem

Mainah Folkers and Titus Rombouts contributed equally to this chapter.

M. Folkers (✉) · T. Rombouts (✉)
Institute for Biodiversity and Ecosystem Dynamics, University of Amsterdam, Amsterdam, The Netherlands

S. Jungblut et al. (eds.), *YOUMARES 9 - The Oceans: Our Research, Our Future*,
https://doi.org/10.1007/978-3-030-20389-4_9

hotspots to develop. However, research is needed to elucidate the potential functional role of sponges as ecosystem engineers.

Apart from the potential functional role sponges fulfill, there is also a wide debate on the phylogenetic relationship among major animal lineages; yet recent research has shown through genomic data that sponges rather than the proposed comb jellies (Dunn et al. 2008) can be interpreted as the sister group to the remaining animals (Nakanishi et al. 2014; Pisani et al. 2015). Considering sponges could be among the first multicellular animals, they are remarkable study objects with respect to evolution. Unlike cnidarians and ctenophores, sponges lack a nervous system but do allow cells to move through layers and accordingly change function (Nakanishi et al. 2014). These characteristics make sponges a unique animal filling niches within the aquatic environment. Yet research regarding sponges until the year 2017 is substantially lower (637 publications) compared to corals (1590 publications) (Fig. 9.1). However, the research regarding sponges is on the rise with projects such as the "EU horizon 2020"-funded SponGES. We hope to appeal to more funding projects in the future to be able to investigate sponges at similar levels to corals.

This review is arranged in four sections. The first section will draw the attention to sponges' life history, morphology, diet, and reproduction. All these components add to the special position of sponges related to other marine animals. The second section focuses on how sponge morphology plays a role in facilitating habitats for other life forms such as fish, crustaceans, and other invertebrates. Within these habitats, sponges play an important role in the cycling of nutrients to make carbon, nitrogen, and phosphorous bioavailable, which is argued in the third section. Finally, the fourth section will dive into the human-related prospects of sponges, in both their physical form and at a molecular level as marine natural products valuable to the pharmaceutical industry.

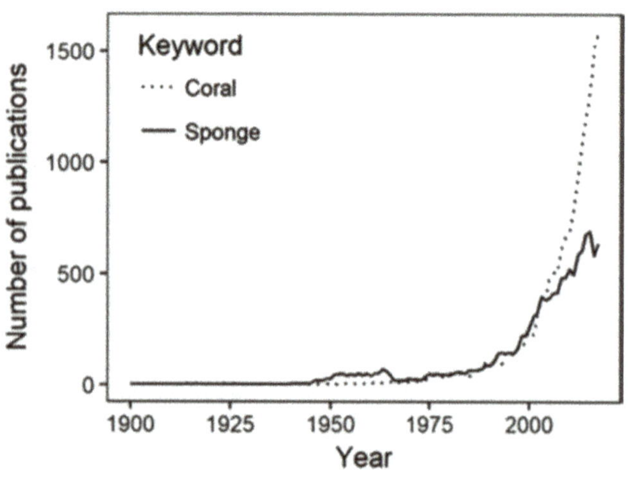

Fig. 9.1 Annual number of scientific publications on the PubMed database including the words "coral" (dotted line) or "sponge" (solid line) in the publication title or as a keyword from 1900 to 2017

9.2 Sponge Characteristics

9.2.1 History and Phylogeny

Before we can understand the potentially important ecosystem functions of sponges as filter feeders, we need to establish a rudimentary familiarity with the history and morphology of sponges. Sponges (Porifera) have diverged earliest from within the metazoans around 600 million years ago and are one of the most diversified invertebrate phyla present in both the marine (~8000 species) and freshwater (~150 species) environment (van Soest et al. 2018). Yet discussion remains if sponges, rather than ctenophores, are considered the sister group to all the remaining animals (Pisani et al. 2015; Adamska 2016; Simion et al. 2017). There are three classes of sponges which, in general, display bathymetric differences in abundance: Calcarea, Demospongiae, and Hexactinellida (Fig. 9.2) (Reid 1968). Furthermore, later research has shown sponges differ over depth in body size and shape in shallow (Bell and Barnes 2000) and deep areas (Maldonado and Young 1996).

Calcarea or calcareous sponges are restricted to shallow environments where it is least demanding to produce calcium carbonate (Vacelet 1988). Demospongiae constitute to about 90% of all sponge species and live in the widest range of habitats (Zenkevich et al. 1960). From the epipelagic until the bathypelagic zone, they thrive in both freshwater and marine environments, under various shapes and sizes (van Soest et al. 2018). Finally, Hexactinellida or glass sponges are the least flexible species. They possess a net of amoebocytes where the epidermal cells would be in other sponge classes. Their cells are interspersed with glass spicules protruding on the outside, which makes them a very rigid class (Barnes 1982). This class is present in polar regions and ocean depths of the abyssal pelagic zone. All three classes have hard skeletal elements called spicules to support their body. The spicules of Calcarea contain calcium carbonate, while the latter two are made up of hydrated silicon dioxide. Recently, the fourth class of sponges has been recognized to be phylogenetically well distinct from their closest relative Demospongiae: Homoscleromorpha (Gazave et al. 2012). They display a relatively simple body structure and with 184 species constitute to only 2% of the sponge species recorded (Hooper and van Soest 2002; van Soest et al. 2018). Therefore, our focus will remain on the three aforementioned classes.

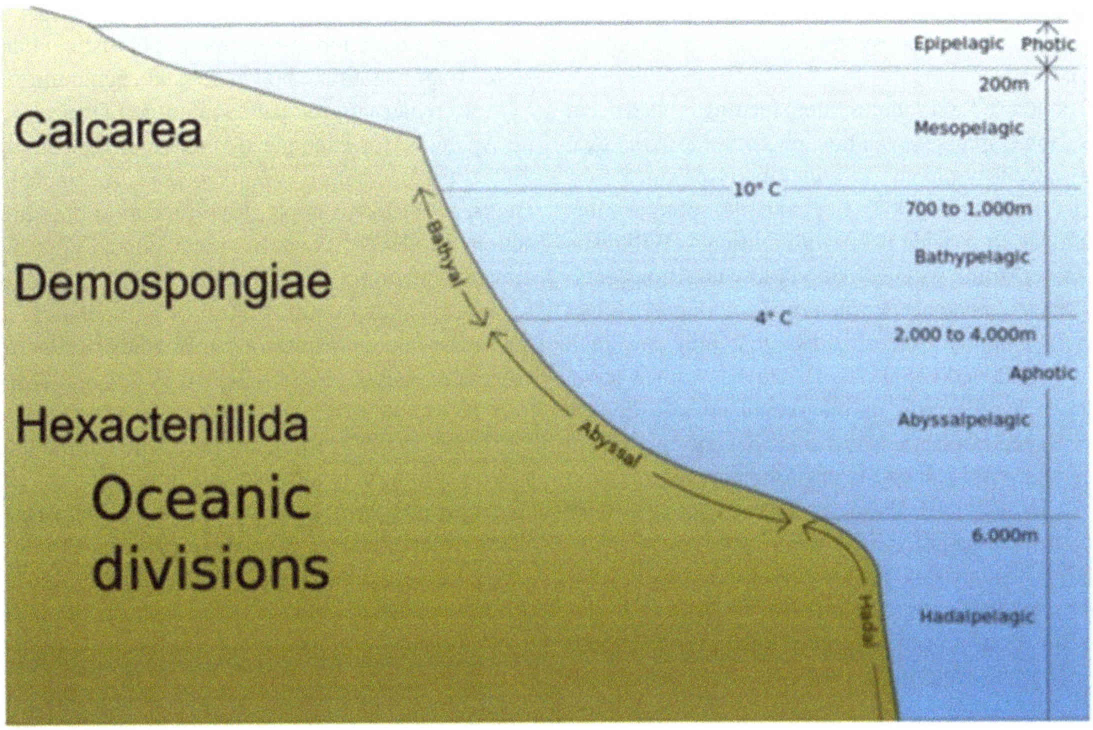

Fig. 9.2 General trend in the bathymetric distribution of three main sponge classes

9.2.2 Morphology

Sponges are radially symmetrical sessile filter feeders with a unique aquiferous system, which is an arrangement of afferent and efferent canals conducting water through chambers lined with flagellate choanocyte cells (Simpson 1984). These choanocyte cells propel water actively through the sponge's aquiferous system. Three conditions of this system exist with increasing size and complexity. Demospongiae are known to display the most complex and folded leuconoid condition, with many inhalant canals (ostia) collared by choanocyte cells together with one or more converging exhalant canals (oscula) (Boury-Esnault and Rützler 1997). This folding increases surface area of cells in contact with the surrounding seawater. A large number of the sponge cells (archaeocytes) are totipotent (Müller 2006), meaning they can change form and function. This is especially useful when perturbations occur surrounding a sessile filter feeder. This totipotency allows cells to migrate over the three different cell types present: pinacocytes, mesohyl cells, and choanocytes (Müller 2006). Sponges unlike more complex multicellular organisms do not have nervous, digestive or circulatory systems. In its place, sponges rely on water flowing through their bodies to nourish them with food and oxygen while simultaneously excreting waste. Furthermore, they do not show bilateral symmetry. Instead, sponges show radial symmetry, which allows for maximal efficiency in water flow around the central cavity of the sponge.

9.2.3 Diet

Traditionally, researchers thought most sponge species relied on particulate food sources such as bacteria and plankton (Kahn et al. 2015). However, a century ago already, suggestions were made that the traditional view of sponges only feeding upon particulate food sources (Reiswig 1971; Pile et al. 1996, 1997), was insufficient to sustain their nutritional requirements (von Putter 1914).

Nowadays, sponges have been suggested to play an important role in dissolved organic matter (DOM) cycling, thereby fueling "benthic-pelagic coupling" (de Goeij et al. 2013; Lesser 2006). This will be discussed in more details in the section on nutrient cycling (see Sect. 9.4). The question remains, however, if this food source is taken up by the endosymbionts abundantly present in the sponge's tissue or by the sponge itself. The presence of microorganisms in marine sponges has been identified already 80 years ago (Dosse 1939). We know now that sponges may host large amounts of microbes (Gloeckner et al. 2014; Taylor et al. 2007), within some cases up to 40% of their body mass (Taylor et al. 2007). High microbial abundance (HMA) sponges harbor as many as 10^8 to 10^{10} cells · g^{-1} of sponge wet weight, being two to four orders of magnitude higher than microbial abundance in seawater. Low microbial abundance (LMA) sponges contain $<10^6$ cells · g^{-1} of sponge wet weight. This distinction in microbial abundance could have a considerable effect on their capacity to feed on dissolved compared to particulate food sources. Sometimes, 48–80% of sponge's energy supply comes from these endosymbiotic microbes (Ruppert et al. 2004).

Apart from these food sources, more feeding modes occur in sponges. Some sponges even host photosynthesizing cyanobacteria as endosymbionts to additionally produce food and oxygen (Taylor et al. 2007). For example, sponges often host green algae to provide them with nutrients (Wilkinson 1992). However, some species living in low nutritious environments have become carnivorous sponges (class Demospongiae; family Cladorhizidae) that prey on small crustaceans (Maldonado et al. 2015). Little is known about their ability to capture prey as they count up to a diverse group of 328 species (van Soest et al. 2018) only present in challenging and remote deep-sea environments (Maldonado et al. 2015). Interestingly, these carnivorous sponges have lost most of their aquiferous system and choanocytes. Therefore, it is unsurprising that they are opportunistic feeders together with their endosymbiotic methanotrophic bacteria, which can act as a complementary food source to these deep-sea sponges (Vacelet et al. 1995).

9.2.4 Life History

Very little is known about the life cycle of sponges with respect to population dynamics, which is very important for conservation (Maldonado et al. 2015). We do know that, similar to other metazoans, sponges can use sexual reproduction through both viviparous and oviparous species. Even though it is more difficult to study viviparous species, due to internal maturation, we know more about their life cycle compared to oviparous species (Leys and Ereskovsky 2006). Moreover, sightings of egg spawning are rare compared to corals, which suggests a more viviparous lifestyle in sponges, but numbers are still largely unknown. Hexactinellida and Calcarea are viviparous, while most oviparous sponges are found in the Demospongiae (Leys and Ereskovsky 2006).

Similar to corals, sponges are hermaphrodites, in which case they release both sperm and eggs. Due to the absence of organs, gametes are, respectively, transformed from the choanocytes and archeocytes. After the capture of sperm by a host, fertilization and hatching usually occur internally after which the larvae swim out until they find a place to settle. In the case of deepwater Hexactinellida, it is difficult to determine the early life history. Yet some studies have found that two of those species are productive year-round (Ijima and Okada1901; Okada 1928) and one was only seasonally active in early summer (Boury-Esnault and Vacelet 1994). This shows that deepwater environments can be influenced by seasonal fluctuations in some cases.

In contrast, the totipotency of sponge cells also allows for asexual reproduction, by means of four distinct methods: fission, fragmentation, budding, and gemmule formation. Fission creates large clonal populations of especially encrusting sponges, for example *Crambe crambe* (Calderón et al. 2007). Fragmentation usually occurs in turbulent environments, with, for example, high predation pressure or wave action. Similar to fission, fragmentation allows nearby recolonization within a single habitat in the case of some coral reef species where almost 30% of a population consists of the same genome (Wulff 1986). Budding occurs in a limited number of species, such as *Tethya citrina* and *Tethya aurantium* (Gaino et al. 2006). Finally, one special adaptation is the formation of gemmules predominantly by freshwater species during unfavorable conditions (Kilian 1952). These survival pods form due to considerable temperature differences experienced in freshwater environments in comparison to the ocean (Manconi and Pronzato 2016). These pods of unspecialized cells remain dormant until conditions improve, and they either recolonize their parental skeletons or start a new colony.

Depending on where sponges live, they can grow from a few years in temperate regions to hundreds of years in both tropical and deep-sea environments. Some sponges grow only 0.2 mm per year which makes specimens of over 1 m in diameter over 5000 years old (Ruppert et al. 2004).

9.3 Sponges as Habitat Providers

Aggregations of sponges are observed in many different environments: coral reefs, mangrove forests, deep sea regions and polar regions. Sponge aggregations have been shown to increase habitat complexity and consequently increase the abundance and biodiversity of benthic associated species (Maldonado et al. 2015). Sponges provide associated species with various services such as shelter from predation, food availability, breeding grounds, and substratum to settle on.

9.3.1 Tropical Habitat Providers

On tropical coral reefs, much habitat complexity is provided by corals. However, sponges may also contribute to habitat facilitation either directly or indirectly. For example, on Caribbean reefs, 39 sponge-dwelling fish species were found (Tyler and Böhlke 1972). Different degrees of sponge associations were described by Tyler and Böhlke (1972). Some goby species are classified as obligate sponge-dwellers with some even showing morphologically specialized features for living exclusively inside sponges (Tyler and Böhlke 1972). Other fish species of various families are simply fortuitous sponge-dwellers. They only use sponges for the deposit or brooding of eggs and usually live outside of sponges.

Tube- and vase-shaped sponges on coral reefs off Key Largo, Florida, offer a physical barrier that lowers fish predation pressure on brittle stars (Henkel and Pawlik 2005). A chemical defense to deter fish predators is lacking in these sponges, which might make them a preferred surface for deposit feeders next to a predation refuge (Henkel and Pawlik 2005). Sponge-associated brittle stars are known to consume detrital particles adhering to the sponges' surface (Hendler 1984).

Indirectly, sponges in Bahamian caves contribute to increased herbivore abundance on coral reefs. The cave sponges provide corals and algae with enhanced nutrient levels. Coral cover and diversity was higher close to cave openings compared to similar sites further away (Slattery et al. 2013). The increased habitat complexity (through corals) and increased food availability (through algae) result in increased herbivory.

9.3.2 Deep-Sea Habitat Providers

In the deep sea, scarcity of complex structural habitat makes sponge grounds one of the most important hotspots for biodiversity (Hogg et al. 2010). Demosponges (Klitgaard 1995; Maldonado et al. 2015) and glass sponges (Beaulieu 2001) have been described as abundant deep-sea habitat providers for associated fauna. Klitgaard (1995) found 242 species associated with deep-sea demosponge aggregations in the North Atlantic. The majority of the sponge-associated fauna used the sponges as substratum.

Biohermal glass sponge reefs increase habitat complexity through biohermal growth. Glass sponges are able to fuse their spicules by a process called secondary silicification. Young sponges settle on the silica skeletons left behind by their ancestors. Fish, crustaceans, nudibranchs, and infaunal polychaetes were found to be more abundant in biohermal glass sponge reefs compared to surrounding areas (Maldonado et al. 2015). This could be due to the improved hydrodynamics of the boundary layer and the shelter the glass sponge reef topography provides to the benthic fauna. Additionally, the energy and nutrient cycling which increases benthic-pelagic coupling (see Sect. 9.4) could also help increase local benthic biodiversity.

Lithistid sponges are known to form rigid massive silica skeletons, which do not easily dissolve. New lithistid sponge recruits can settle on these skeletons, which results in biohermal growth much like glass sponge reef growth. Lithistid sponge mounds on the seabed attract fish and various macroinvertebrates (Maldonado et al. 2015).

In the deep sea off California, USA, 139 taxa of marine organisms were found to be associated with glass sponge stalks (Beaulieu 2001). These micro cryptic habitats were dominated by zoanthids and polychaetes that used the stalks as a hard substratum to grow on.

Carnivorous cladorhizid sponges on the Macquarie Ridge live among many other invertebrates (Maldonado et al. 2015). However, cladorhizid sponges might not increase biodiversity by increasing habitat complexity. Because of their carnivorous characteristics, cladorhizid sponges might prevent larvae from settling.

9.3.3 Arctic Habitat Providers

Seabed gouging is a process that is described as drifting ice going through the benthos with the keel when passing through shallower waters. Seabed gouging is known to disturb glass sponge aggregations in the Weddell Sea, Antarctica, leaving behind sponge spicule mats, which serve as a substratum for other organisms (Maldonado et al. 2015). Muddy, soft seabeds were linked to species-poor communities, while solid sponge spicule mats were linked to species-rich communities (Hogg et al. 2010).

9.3.4 Habitat for Commercially Important Species

Sponges provide structural habitat which harbors food and/or provides shelter from predators for fishes and crustaceans (Butler et al. 1995; Ryer et al. 2004; Miller et al. 2012). Some sponge-associated species are also commercially important for fisheries.

High mortality of sponges after cyanobacterial blooms in Florida Bay had consequences for juvenile Caribbean spiny lobster living around these sponges. These lobsters are important to commercial fisheries. The spiny lobsters were exposed to increased predation due to the lack of shelter that was previously provided by sponges (Butler et al. 1995).

Other sponge-associated species are threatened by fishing activities (such as trawling and dredging). When fishing gear (such as nets and long lines) is towed across the seabed, it removes and damages large epibenthic organisms, including sponges. Groundfish (such as cod and ling) are often caught in trawl nets along with sponges (Hogg et al. 2010). Cod, ling, halibut, and Pacific Ocean perch are commercially important fish species that might face negative consequences of sponge habitat loss. Regulation is needed to protect sponge aggregations and the species living in close proximity to them.

Juvenile halibut showed a strong preference for habitats with a 16% sponge coverage compared to habitats with bare sand in laboratory experiments (Ryer et al. 2004). This strong preference could be due to less predation vulnerability for halibuts. Ryer et al. (2004) observed predators being impeded on their prey pursuit by the sponges.

Rockfishes, including the Pacific Ocean perch, are associated with sponges in the southeastern Bering Sea (Miller et al. 2012). The sponges are thought to support diverse and abundant macroinvertebrate communities that serve as prey for rockfish next to providing shelter from predators (Miller et al. 2012).

9.4 Nutrient Cycling by Sponges

9.4.1 Sponge Loop

Nutrient cycling is essential to maintain a balance between food and waste for all species in an ecosystem food web. In marine environments, primary producers on reefs, such as corals and algae, release 50% of their mucus, of which 80% is dissolved directly into the adjacent seawater (Wild et al. 2004, 2008). DOM, consisting of, e.g., carbohydrates, lipids, and proteins, is an abundant potential food source on reefs for microbes (Azam et al. 1983), yet largely unavailable to most heterotrophic reef inhabitants (de Goeij et al. 2013).

Conventionally, microbial degradation of DOM in the water column and sediment has been considered the primary pathway in DOM cycling (Harvey 2006; Wild et al. 2004, 2009). However, decades ago already, suggestions were made that the traditional view of sponges only feeding upon particulate food sources, such as bacteria and plankton (Reiswig 1971; Pile et al. 1996, 1997), was insufficient to sustain their nutritional requirements (von Putter 1914).

Recently, sponges have been discovered to take up DOM, thereby providing an important role in benthic-pelagic coupling (de Goeij et al. 2013; Lesser 2006). Only 42% of the dissolved carbon taken up from the ambient water is respired by sponges. Therefore, the remainder is most likely either assimilated, used for reproduction, or converted into particulate organic matter (POM) through cell shedding (Alexander et al. 2014; de Goeij et al. 2009). Assimilation, in the form of growth, takes place very little in encrusting sponges; thus, conversion to POM is considered the preferred route. This shedding takes place mainly among the choanocytes which, unsurprisingly, also show high proliferation rates (Alexander et al. 2015) 2900 times faster compared to their normal growth rate of other cells (Ayling 1983). A recent study has revealed that these other cells, specifically from the mesohyl, contribute additionally to the production of POM (Maldonado 2016). This high cell turnover can be a clever mechanism to prevent permanent damage to the sponge caused by environmental stress (de Goeij et al. 2009) and creates the opportunity for higher trophic levels to feed on these cells (de Goeij et al. 2013).

The turnover of DOM by sponges is faster than by microbes (van Duyl et al. 2008) and equals the order of magnitude of the gross primary production in the Caribbean reef ecosystem (de Goeij et al. 2013). Thus, apart from the conventional microbial loop (Azam et al. 1983), accounting for only 10% of nutrient cycling, a sponge loop (de Goeij et al. 2013), accounting for 90%, now supports a major role in the DOM reincorporation pathway. The produced POM is, thereafter, most likely consumed by detritivores, which can be present as associated sponge fauna in so-called consumer-resource interactions (de Goeij et al. 2013; Rix et al. 2016).

Most food web models do not include the role sponges have in cycling resources in their environments, which makes the many current models incomplete. Future food web models can be improved by adding sponge energy and nutrient cycling. Moreover, sponges carry out benthic-pelagic coupling, which is crucial to retain nutrients within an environment. Within this section, we will focus on the three important cycles of: carbon, nitrogen and phosphorous, as key components to sustaining life.

9.4.2 Carbon Cycling by Sponges

Carbon is one of the main components of biological life forms, and sponges play an important role in carbon cycling in aquatic ecosystems. They take up and release carbon to their environment in several ways (Rix et al. 2017).

Organic matter dissolves in water after extracellular release or cell lysis by primary producers such as phytoplankton, macrophytes, and coral symbionts. Dissolved organic carbon (DOC) concentration can differ in space and time in aquatic ecosystems: 0.7 to 45 mg $\cdot$ L^{-1} in rivers, 0.7 to 330 mg $\cdot$ L^{-1} in lakes, and 0.5 to 3.0 mg $\cdot$ L^{-1} in the ocean (Mulholland 2003). DOC is a large energy resource in aquatic environments and makes up a large part of DOM (Wild et al. 2004). For example, on coral reefs, more than 90% of the total organic matter consists of DOM (Carlson 2002). However, the carbon fraction of DOM is not readily available to most organisms. Mostly sponges and microbes utilize DOC as an energy source. The DOC uptake of Caribbean coral reef sponges is estimated to be 90 to 350 mmol C $\cdot$ m^{-2} $\cdot$ day^{-1} (de Goeij and Van Duyl 2007), which is comparable to the gross primary production of 200 to 600 mm C $\cdot$ m^{-2} $\cdot$ day^{-1} on coral reefs (Hatcher 1990). DOC removal on coral reefs is mostly accounted for by sponges compared to only 5 to 50 mmol C $\cdot$ m^{-2} $\cdot$ day^{-1} of microbial DOC uptake (de Goeij and Van Duyl 2007; Haas et al. 2011). Total organic carbon (TOC) uptake of coral reef sponges consists predominantly (56 to 97%) of DOC (de Goeij et al. 2017). Next to filter feeding on carbon sources, some sponge species host photoautotrophic symbionts, which photosynthesize and transfer carbon into the sponge's tissue (Wilkinson 1983; Fiore et al. 2013). Wilkinson (1983) found that the carbon fixation rate in coral reef sponges containing symbiotic cyanobacteria was only 2.4 to 6.6% in dark conditions compared to light conditions. However, sponges usually do not rely chiefly on the symbionts for nutrition. Additionally, apart from dissolved food sources, particulate sources such as bacteria can act as a food source (Kahn et al. 2015).

After DOC uptake, 3.7 to 14.7 μmol DOC · mmol $C_{sponge}^{-1} \cdot 12\ h^{-1}$ is assimilated inside sponge cells and sponge-associated microbe cells (Rix et al. 2017). Carbon fixation in sponges is not restricted to the abundance of associated microbes because LMA sponges also take up DOC (de Goeij et al. 2017). Part of the carbon taken up by sponges is respired; another part is used to grow. De Goeij et al. (2017) estimated a daily net biomass increase of 38% for *Halisarca caerulea* if all assimilated carbon (61% of the TOC uptake) would be used for growth. However, sponges do not grow as fast as expected. Only 2.2% of TOC uptake was used for a daily biomass increase of 1.3% (Alexander et al. 2015). Instead, sponges show a rapid cell turnover. An average cell cycle of only 6 hours was determined for *H. caerulea* (de Goeij et al. 2009). This cell turnover is the result of rapid cell proliferation and shedding of old cells. Fifteen to 24% of the carbon assimilated by sponge holobionts is released as particulate organic carbon (POC) (Rix et al. 2017). This is how sponges transform energy in the form of DOC to POC. Carbon becomes available to detritivores (such as ophiuroids, crustaceans, snails, and polychaetes) that consume sponge-derived POC (de Goeij et al. 2013; Rix et al. 2017).

Alternatively, sponges make carbon available to their environment through bio-erosion. Excavating sponges break down calcium carbonate chemically by dissolution and mechanically by chip production (Zundelevich et al. 2007).

9.4.3 Nitrogen Cycling by Sponges

Apart from carbon, nitrogen is important in marine ecosystems as it is essential to produce amino acids which in turn make proteins and DNA. Moreover, it is often a limiting nutrient to meet energy requirements in tropical reefs (Muscatine and Porter 1977; Delgado and Lapointe 1994; Fiore et al. 2013). Paradoxically, the abundance of nitrogen in air (78%) remains unavailable for animals unless nitrogen-fixing bacteria, e.g., cyanobacteria, or to a lesser extent, lightning converts nitrogen to a biologically available form. The (re)cycling of inorganic nitrogen is therefore imperative and occurs via different pathways both in surface waters and the deep sea. Nitrification is considered a source of bioavailable nitrogen, whereas denitrification is considered a sink. These processes are of importance to species-specific ecosystem services to the surrounding environment. For example, benthic microbial nitrifiers provide nitrate to the root nodular system of macrophytes.

Former studies have shown that inorganic nitrogen cycling in sponges takes place and is mediated by the microbial biofilm present in the sponge's tissue (Taylor et al. 2007; Hoffmann et al. 2005; Fiore et al. 2010; Schläppy et al. 2010; Gloeckner et al. 2014). For example, Mediterranean sponges can actively switch between aerobic and anaerobic metabolism by inhibiting water flow over time (Hoffmann et al. 2008). This could induce anaerobic environments to trigger supposed "coupled nitrification-denitrification," meaning a part of the nitrified nitrate is subsequently transformed into nitrogen gas.

Unsurprisingly, the two most present processes to consider in microbial nitrogen cycling in sponges are nitrification and denitrification (Southwell et al. 2008) (Fig. 9.3). Nitrification is a two-tiered process where ammonia-oxidizing bacteria (AOB) perform the first, often rate-limiting, step from ammonium to nitrite, while the nitrite-oxidizing bacteria (NOB) oxidize the latter to nitrate. AOBs are usually beta- and gamma-proteobacteria, and NOB belong to *Nitrobacter* and *Nitrococcus* family (Bayer et al. 2008). Factors that affect nitrification include metabolic interactions in the microbial community and respiration rates but most importantly oxygen concentrations (Müller et al. 2004). If oxygen is lacking, nitrification cannot take place. Sponges, like many other marine invertebrates, usually excrete ammonium as a waste product (Brusca and Brusca 1990). Therefore, the unexpected excretion of nitrate was the first evidence of microbial nitrification within the sponge (Diaz and Ward 1997; Jiménez and Ribes 2007).

Apart from aerobic nitrification, anaerobic processes such as denitrification or anaerobic ammonium oxidation (anammox) also occur regularly in sponges and have, for example, been shown in the tropical sponge *Xestospongia muta* (Fiore

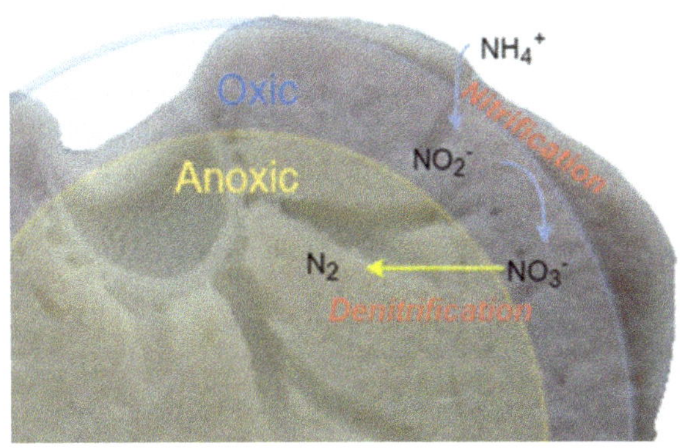

Fig. 9.3 Simplified diagram of nitrification and denitrification superimposed on a vertical slice of a sponge *Geodia barretti*

et al. 2013). First, to confirm the presence of anaerobic zones within sponge tissue, Hoffmann et al. (2005) have investigated bacterial metabolic activity ex situ within the HMA sponge *Geodia barretti*. Their research revealed a steep oxygen profile within intact sponges in which anoxic microenvironments could allow denitrifying bacteria to reduce nitrate in nitrogen gas. Furthermore, Fiore et al. (2013) found a negative efflux of nitrate, indicating that either denitrification or anammox was taking place inside the sponge. However, the Caribbean sponge *X. muta* was actively pumping during the study contradicting the earlier hypothesis of Fiore et al. (2010) where lack of pumping would equal denitrification. Finally, Hoffmann et al. (2009) discovered both nitrification and denitrification in *G. barretti* with rates of 566 and 92 nmol $N \cdot cm^{-3}$ sponge $\cdot$ day^{-1}, respectively. However, this research was performed with explants (3 cm^3 radial cylinders of cut sponge tissue). Even though sponges are known for their totipotent cells and quick regeneration after damaging (Alexander et al. 2015), using explants alters the aquiferous system to such extent that pumping is most likely inhibited.

Although the deep sea is often viewed as an uninhabitable environment, particular areas like hydrothermal vents, cold seeps, and sponge grounds actually harbor complex ecosystems (Klitgaard 1995; Roberts et al. 2006; Cathalot et al. 2015). Their dependence on nitrogen is influenced by the influx of inorganic and organic nitrogen from surface waters through vertical mixing (Romera-Castillo et al. 2016) and bottom water advection (Davies et al. 2009). Regarding nitrogen cycling, deep-sea sponges (DSS) can be of particular interest, as in some areas they make up 90% of the benthic biomass (Klitgaard and Tendal 2004; Murillo et al. 2016) and are important filter feeders (Kutti et al. 2013). However, the relevance of DSS is only currently emerging with studies showing that sponges have the potential to recycle essential elements like carbon and nitrogen (Witte et al. 1997; van Oevelen et al. 2009; Rix et al. 2016). As mentioned before in Sect. 4.1, in shallow tropical reefs, sponges are known to retain carbon through the sponge loop, whereas they release nitrogen (de Goeij et al. 2013). Rix et al. (2016) were the first to examine that DSS might have a similar potential in fueling the ecosystem with nitrogen resembling their warm water counterparts. However, she only performed this in ex situ aquarium experiments. Whether a potential cold water sponge loop also takes place in the deep sea is yet to be confirmed, especially regarding the cycling of inorganic nutrients.

9.4.4 Phosphorous Cycling by Sponges

Finally, phosphorous is essential for the biological synthesis and for the transfer of energy (Tyrreell 1999). In aquatic environments, phosphorous budgets consist of three components: particulate phosphorous, dissolved inorganic phosphate, and dissolved organic phosphate (Maldonado et al. 2012). We know that of these three, the latter is present most abundantly (Dyhrman et al. 2007). However, phosphorous cycling has not undergone comprehensive research as the two nutrients aforementioned. Therefore, data is limited to a few studies which have only concluded that both low and high microbial abundant sponges act as a net source of phosphate reviewed in Maldonado et al. (2012). Further research into phosphorous cycling by sponges would give us more insight into the potential limiting factors of sponge habitats.

9.5 Marine Natural Products from Sponges

9.5.1 Introduction

Since the early Egyptian times, sponge skeletons have been harvested for its cleaning properties and as hygienic tools (Pronzato 2003). Likewise, the earliest medicinal feature, in Greek times, a cold wet sponge placed on the heart, would resurrect the fainted (Jesionowski et al. 2018). However, these days, the capacities of sponges have shifted from a biomaterial to a biomolecular source (Jesionowski et al. 2018). Within the last 20 years, the detection of marine natural products (MNP) has increased, with an estimate of 15,000 MNPs discovered until 2010 (Hu et al. 2011). In the search for bioactive compounds in the marine environment, vertebrate animals such as fish, sharks, and snakes have been examined. Among the invertebrates, more groups have been examined including tunicates, echinoderms, algae, mollusks, corals, and sponges. Finally, microorganisms were examined, and of those several bacteria, fungi, and cyanobacteria showed potential (Alonso et al. 2003). The fact that sponges can harbor high densities of microorganisms in the mesohyl layer makes them very potent study objects for novel bioactive compounds (Alexander 2015). Moreover, sponges lack an immune system, protective armor, and mobility which pressed on the evolution to synthesize compounds for defensive purposes. For example, instead of an immune system, when invaded with foreign material, sponges produce a range of chemicals, such as 3-alkyl-pyridinium, that inhibits movement of surrounding cells preventing the use of the sponge's internal transport system (Sepčić et al. 1999).

Interestingly, there is considerable debate about whether sponges are the true producer of these compounds and not just hosts to the true producers: microorganisms (Jensen and Fenical 1994; Bultel-Poncé et al. 1997; Hentschel et al. 2006 reviewed in Mehbub et al. 2016b). It would not be a surprise if the majority of these compounds are a result of the symbiotic microorganisms rather than the host, considering in HMA sponges the body mass could be accounted for by bacteria to up to 40% (Taylor et al. 2007). Either way, it is important to investigate the possibilities for MNPs within the holobiont as they ultimately provide the compound as a whole organism. This holobiont approach does lead to some discussion about the true producer and if they can do so without each other's presence. Several reviews have discussed this. In Jensen and Fenical (1994), the problem is mentioned that the sponge does contain a microbial community distinct of the surrounding water, implying that bacteria need the sponge host to be initially present. A prime example of misjudgments of the true producer was found after flow cytometric separation of sponge and microbial fraction localizing the true producer: a prokaryotic cell (Unson and Faulkner 1993).

9.5.2 Potential for Exploitation

The potential to exploit marine sources for pharmaceuticals has been of major importance in recent times since we cannot only rely on terrestrial sources alone. Moreover, it is of importance to investigate these potential pharmaceuticals, because infectious microorganisms constantly evolve a resistance against current pharmaceuticals. Several reviews have focused on MNPs in general (Faulkner 2000; Blunt et al. 2017, 2018); yet some have also focused only on sponge-specific MNPs (Mehbub et al. 2016a).

The reason for sponges to entail such a vast majority of the MNPs found in the marine environment could well be caused by their survival over the past 580 million years. During this time, they have undergone huge environmental changes which induced specialization and formation of vastly different species over the entire aquatic environment. Sponges are at present divided into 4 distinct classes, 25 orders, 128 families, and 680 genera (Abad et al. 2011). These specializations over all the different groups together with their capacity to hold endosymbionts led to the production of a large range of (secondary) metabolites.

Among all marine species, sponges are the most investigated with nearly 30% of all MNPs discovered (Mehbub et al. 2016a). This accounts for a total of 4851 compounds of which 1499 isolated only between 2008 and 2012. The compounds were classified into 18 chemical classes among them: acids, alkaloids, esters, fatty acids, and further less relevant groups (Mehbub et al. 2016a). These compounds contain a wide variety of bioactivities: anticancer, antiviral, antibacterial, anti-inflammatory, and many more neural activities (Chakraborty et al. 2009). The latter is of importance because of the vast presence of patients with neurodegenerative diseases especially in high-income countries (Global Health Estimates 2016). Moreover, many studies have investigated the neuroprotective capacities of the MNPs in sponges. Compounds were found with activities such as modulation of the neurotransmitters acetylcholinesterase and glutamate, decreasing oxidative stress, enrichment of serotonin, and neurite growth (Alghazwi et al. 2016). However, none of these compounds have yet been developed as a finalized marine pharmaceutical.

9.5.3 Culturing of Sponges

Since the eighteenth century, reconstructive growth of sponges has been recorded, with major advances made at the beginning of the twentieth century by Henry Moore in Florida (Jesionowski et al. 2018). Later research has investigated the potential of secondary metabolites regarding their antimicrobial activity (Thompson et al. 1985) acting as therapeutic drugs, collagen, and optical equipment (Munro et al. 1999). However, this has so far been held back by what is referred to as "the supply problem" (Osinga et al. 2003). The supply problem states that we are limited by the small amount of chemical present within the sponge compared to its biomass. Moreover, the dilution effect of the ocean requires compounds to be very stable and highly active (Abad et al. 2011) which would explain the low quantities found. To overcome this, we are required to grow enormous sponge biomass to perform acceptable preclinical and clinical trials. As harvesting the enormous sponge biomass from the environment would not be sustainable, therefore we have to look at alternatives. Thus, opportunities in biotechnological methods are progressively favored to avoid the supply problem. One possible solution is to have the sponge-associated microorganisms to flourish independently of their host to produce larger quantities of MNPs. Another growing area is the use of sponge cell cultures, which avoids the complex environment necessary for whole organisms (Müller et al. 2000). Yet some advance has been made to express these biosynthetic pathways of interest in more easily cultivatable hosts to overcome the supply problem (Wilson et al. 2014). Nevertheless, large-scale production of sponge biomass for MNPs remains unsuccessful.

As mentioned afore, sponges can grow in vastly different natural environments, from the deep sea to tropical reefs. At present, in situ culture has been the only successful approach at harvesting considerable biomass of sponges. The usual method is to use asexual reproduction by fragmentation which creates explants hanging from buoy lines in the water column (Osinga et al. 2003). However, in situ culturing has serious drawbacks such as incoming diseases, abrupt weather changes, or habitat disruption. This has impeded large-scale in situ production. Therefore, Osinga et al. (2003) started to experiment with growing sponges in vivo in bioreactors. They succeeded to grow a tropical sponge *Pseudosuberites andrewsi* on a single marine diatom species either intact or as a filtered crude extract. However, unlike freshwater sponges

Ephydatia fluviatilis and *Spongilla alba*, which have shown to grow successfully on a stock of *Escherichia coli* (Francis et al. 1990), the lack in further culture literature might explain that proper food source or mixture to grow marine sponges is still elusive. Though discussion remains even on the selection of one food source, we know marine sponges feed on several food sources simultaneously, such as: pico-plankton, bacteria, viruses and dissolved organic substrates (de Goeij et al. 2013). It could well be that the quality of the particulate food sources and the composition of these dissolved organic substrates are decisive. Yet apart from food sources, there are abiotic variables (such as temperature, salinity, and light) that can determine growth. Therefore, more research should be undertaken finding the balance in food sources while breeding marine sponges under various laboratory conditions.

9.6 Conclusions

In this review, we have tried to outline the importance of sponges to their environments as well as to humans. Sponges are a diverse phylum (consisting of more than 8000 species). In this phylum, a broad range of morphologies, feeding habits, and reproduction strategies are present. Much about these various characteristics of sponges is yet to be discovered and understood. These future discoveries are not only interesting for sponge-specific knowledge but also of importance to a wider understanding of evolution and aquatic ecosystems in general.

Sponges are important to their environment for multiple reasons. Sponges provide habitat and food for fish, crustaceans, and many other animals in a variety of ecosystems ranging from tropical coral reefs to deep-sea sponge grounds. Among these sponge-dwelling organisms are even commercially important species.

Additionally, sponges shape the communities surrounding them by playing a significant role in the cycling of energy and nutrients. Sponges make carbon, nitrogen, and potentially phosphorus available to higher trophic levels. By doing so, these elements are retained within the ecosystems sponges live in and would otherwise not be available to many organisms.

Lastly, sponges are also important to humans in upcoming pharmaceutical research. The high abundance of symbiotic microorganisms living within sponges makes sponges good candidates for novel bioactive compound discovery. It is expected that sponge-associated marine natural products have potential as bioactive compounds in drugs. Therefore, it is important to invest time in uncovering the importance of this phylum within all aquatic ecosystems.

Appendix

This article is related to the YOUMARES 9 conference session no. 11: "Sponges (Porifera): Fantastic filter feeders." The original Call for Abstracts and the abstracts of the presentations within this session can be found in the Appendix "Conference Sessions and Abstracts", Chapter "9 Sponges (Porifera): Fantastic filter feeders", of this book.

References

Abad M, Bedoya L, Bermejo P (2011) Marine compounds and their antimicrobial activities. In: Méndez-Vilas A (ed) Science against microbial pathogens: communicating current research and technological advances, vol 2, 3rd edn. Formatex, Badajoz, pp 1293–1306

Adamska M (2016) Sponges as models to study emergence of complex animals. Curr Opin Genet Dev 39:21–28. https://doi.org/10.1016/j.gde.2016.05.026

Alexander B (2015) Cell turnover in marine sponges: insight into poriferan physiology and nutrient cycling in benthic ecosystems. Dissertation, University of Amsterdam

Alexander B, Liebrand K, Osinga R et al (2014) Cell turnover and detritus production in marine sponges from tropical and temperate benthic ecosystems. PLoS One 9(10):e109486. https://doi.org/10.1371/journal.pone.0109486

Alexander B, Achlatis M, Osinga R et al (2015) Cell kinetics during regeneration in the sponge *Halisarca caerulea*: how local is the response to tissue damage? PeerJ 3:e820. https://doi.org/10.7717/peerj.820

Alghazwi M, Qi Kan Y, Zhang W et al (2016) Neuroprotective activities of marine natural products from marine sponges. Curr Med Chem 23(4):360–382

Alonso D, Khalil Z, Satkunanthan N et al (2003) Drugs from the sea: conotoxins as drug leads for neuropathic pain and other neurological conditions. Mini-Rev Med Chem 3(7):785–787

Ayling AL (1983) Growth and regeneration rates in thinly encrusting demospongiae from temperate waters. Biol Bull 165(2):343–352. https://doi.org/10.2307/1541200

Azam F, Fenchel T, Field J et al (1983) The ecological role of water-column microbes in the sea. Mar Ecol Prog Ser 10:257 263. https://doi.org/10.3354/meps010257

Barnes RD (1982) Invertebrate zoology. Saunders College Publishing, Philadelphia

Bayer K, Schmitt S, Hentschel U (2008) Physiology, phylogeny and in situ evidence for bacterial and archaeal nitrifiers in the marine sponge *Aplysina aerophoba*. Environ Microbiol 10(11):2942–2955

Beaulieu SE (2001) Life on glass houses: sponge stalk communities in the deep sea. Mar Biol 138(4):803–817

Bell JJ, Barnes DKA (2000) The influences of bathymetry and flow regime upon the morphology of sublittoral sponge communities. J Mar Biol Assoc UK 80:707–718

Blunt WJ, Copp RB, Keyzers AR et al (2017) Marine natural products. Nat Prod Rep 34:235–294. https://doi.org/10.1039/C6NP00124F

Blunt WJ, Carroll RA, Copp RB et al (2018) Marine natural products. Nat Prod Rep 35:8–53. https://doi.org/10.1039/C7NP00052A

Boury-Esnault N, Rützler K (1997) Thesaurus of sponge morphology. Smithson Contrib Zool 596:1–55. https://doi.org/10.5479/si.00810282.596

Boury-Esnault N, Vacelet J (1994) Preliminary studies on the organization and development of a hexactinellid sponge from a Mediterranean cave, *Oopsacas minuta*. In: vanSoestRWM, vanKempenTMG, BraekmanJet al (eds) Sponges in time and space: biology, chemistry, paleontology. 4th international Porifera Congress, Amsterdam, April1993. Balkema, Rotterdam, pp407–416

Brusca R, Brusca G (1990) Phylum Porifera: the sponges. In: Brusca R, Brusca G (eds) The invertebrates, 2nd edn. Sinaeur Press, Sunderland, pp 181–210

Bultel-Poncé V, Debitus C, Blond A et al (1997) Lutoside: an acyl-1-(acyl-6′-mannobiosyl)-3-glycerol isolated from the sponge-associated bacterium *Micrococcus luteus*. Tetrahedron Lett 38:5805–5808. https://doi.org/10.1016/S0040-4039(97)01283-5

Butler MJ, Hunt JH, Herrnkind WF et al (1995) Cascading disturbances in Florida bay, USA: cyanobacteria blooms, sponge mortality, and implications for juvenile spiny lobsters Panulirus argus. Mar Ecol Prog Ser 129:119–125

Calderón I, Ortega N, Duran S et al (2007) Finding the relevant scale: clonality and genetic structure in a marine invertebrate (*Crambe crambe*, Porifera). Mol Ecol 16:1799–1810. https://doi.org/10.1111/j.1365-294X.2007.03276

Carlson CA (2002) Production and removal processes. In: Hansell DA, Carlson CA (eds) Biogeochemistry of marine dissolved organic matter, 2nd edn. Academic Press, London, pp 91–151

Cathalot C, Van Oevelen D, Cox TJS et al (2015) Cold-water coral reefs and adjacent sponge grounds: hotspots of benthic respiration and organic carbon cycling in the deep sea. Front Mar Sci 2:37. https://doi.org/10.3389/fmars.2015.00037

Chakraborty C, Hsu CH, Wen ZH et al (2009) Anticancer drugs discovery and development from marine organisms. Curr Top Med Chem 9(16):1536–1545

Davies AJ, Duineveld GCA, Lavaleye MSS et al (2009) Downwelling and deep-water bottom currents as food supply mechanisms to the cold-water coral *Lophelia pertusa* (Scleractinia) at the Mingulay reef complex. Limnol Oceanogr 54(2):620–629. https://doi.org/10.4319/lo.2009.54.2.0620

de Goeij JM, van Duyl FC (2007) Coral cavities are sinks of dissolved organic carbon (DOC). Limnol Oceanogr 52(6):2608–2617

de Goeij JM, de Kluijver A, van Duyl FC et al (2009) Cell kinetics of the marine sponge *Halisarca caerulea* reveal rapid cell turnover and shedding. J Exp Biol 212(23):3892–3900

de Goeij JM, van Oevelen D, Vermeij MJA et al (2013) Surviving in a Marine Desert: the sponge loop retains resources within coral reefs. Science 342(6154):108–110

de Goeij JM, Lesser MP, Pawlik JR (2017) Nutrient fluxes and ecological functions of coral reef sponges in a changing ocean. In: Carballo JL, Bell JJ (eds) Climate change, ocean acidification and sponges, 1st edn. Springer, Cham, pp 373–410

Delgado O, Lapointe BE (1994) Nutrient-limited productivity of calcareous versus fleshy macroalgae in a eutrophic, carbonate-rich tropical marine environment. Coral Reefs 13(3):151–159. https://doi.org/10.1007/BF00301191

Diaz MC, Rützler K (2001) Sponges: an essential component of Caribbean coral reefs. Bull Mar Sci 69(2):535–546

Diaz MC, Ward BB (1997) Sponge-mediated nitrification in tropical benthic communities. Mar Ecol Prog Ser 156:97–107

Dosse G (1939) Bakterien-und Pilzbefunde sowie pathologische und Fäulnisvorgänge in Meeres-und Süßwasserschwämmen. Z Parasitenkd 11(2–3):331–356

Dunn CW, Hejnol A, Matus DQ et al (2008) Broad phylogenomic sampling improves resolution of the animal tree of life. Nature 452:745–749. https://doi.org/10.1038/nature06614

Dyhrman ST, Ammerman JW, van Mooy BAS (2007) Microbes and the marine phosphorus cycle. Oceanography 20:110–116

Faulkner DJ (2000) Marine natural products. Nat Prod Rep 17:7–55. https://doi.org/10.1039/A809395D

Fiore CL, Jarett JK, Olson ND et al (2010) Nitrogen fixation and nitrogen transformations in marine symbioses. Trends Microbiol 18(10):455–463. https://doi.org/10.1016/j.tim.2010.07.001

Fiore CL, Baker DM, Lesser MP (2013) Nitrogen biogeochemistry in the Caribbean sponge, *Xestospongia muta*: a source or sink of dissolved inorganic nitrogen? PLoS One 8(8):e72961. https://doi.org/10.1371/journal.pone.0072961

Francis JC, Bart L, Poirrier MA (1990) Effect of medium pH on the growth rate of *Ephydatia fluviatilis* in laboratory culture. In: RützlerK, MacintyreVV, SmithKT (eds) New perspectives in sponge biology. 3rd international sponge conference, Woods Hole, November 1985. Smithsonian Institution Press, Washington, DC, pp 485–490

Gaino E, Scalera Liaci L, Sciscioli M et al (2006) Investigation of the budding process in *Tethya citrina* and *Tethya aurantium* (Porifera, Demospongiae). Zoomorphology 125:87. https://doi.org/10.1007/s00435-006-0015-z

Gazave E, Lapébie P, Ereskovsky AV et al (2012) No longer Demospongiae: Homoscleromorpha formal nomination as a fourth class of Porifera. In: Maldonado M, Turon X, Becerro M et al (eds) Ancient animals, New challenges, Developments in hydrobiology, vol 219. Springer, Dordrecht, pp 3–10

Gili JM, Coma R (1998) Benthic suspension feeders: their paramount role in littoral marine food webs. Trends Ecol Evol 13:316–321. https://doi.org/10.1016/S0169-5347(98)01365-2

Global Health Estimates (2016) Deaths by Cause, Age, Sex, by Country and by Region, 2000–2016. World Health Organization, Geneva. http://www.who.int/healthinfo/global_burden_disease/estimates/en/. Accessed 31 Oct 2018

Gloeckner V, Wehrl M, Moitinho-Silva L et al (2014) The HMA-LMA dichotomy revisited: an electron microscopical survey of 56 sponge species. Biol Bull 227(1):78–88. https://doi.org/10.1086/BBLv227n1p78

Griffiths JR, Kadin M, Nascimento FJ et al (2017) The importance of benthic–pelagic coupling for marine ecosystem functioning in a changing world. Glob Chang Biol 23(6):2179–2196

Haas AF, Nelson CE, Wegley Kelly Let al (2011) Effects of coral reef benthic primary producers on dissolved organic carbon and microbial activity. PLOS ONE 6:e27973. doi:https://doi.org/10.1371/journal.pone.0027973 pmid:22125645

Harvey HR (2006) Sources and cycling of organic matter in the marine water column. In: Volkman JK (ed) Marine organic matter: biomarkers, isotopes and DNA, The handbook of environmental chemistry, vol 2N. Springer, Heidelberg, pp 1–25

Hatcher BG (1990) Coral reef primary productivity. A hierarchy of pattern and process. Trends Ecol Evol 5(5):149–155. doi:https://doi.org/10.1016/0169-5347(90)90221-X pmid:21232343

Hendler G (1984) The association of *Ophiothrix lineata* and *Callyspongia vaginalis*: a Brittlestar-sponge cleaning symbiosis? Mar Ecol 5(1):9–27

Henkel TP, Pawlik JR (2005) Habitat use by sponge-dwelling brittle stars. Mar Biol 146(2):301–313

Hentschel U, Usher KM, Taylor MW (2006) Marine sponges as microbial fermenters. FEMS Microbiol Ecol 55(2):167–177. https://doi.org/10.1111/j.1574-6941.2005.00046.x

Hoffmann F, Larsen O, Thiel V et al (2005) An anaerobic world in sponges. Geomicrobiol J 22(1–2):1–10. https://doi.org/10.1080/01490450590922505

Hoffmann F, Røy H, Bayer K et al (2008) Oxygen dynamics and transport in the Mediterranean sponge *Aplysina aerophoba*. Mar Biol 153(6):1257–1264. https://doi.org/10.1007/s00227-008-0905-3

Hoffmann F, Radax R, Woebken D et al (2009) Complex nitrogen cycling in the sponge *Geodia barretti*. Environ Microbiol 11(9):2228–2243. https://doi.org/10.1111/j.1462-2920.2009.01944.x

Hogg MM, Tendal OS, Conway KWet al (2010) Deep-sea sponge grounds: reservoirs of biodiversity. UNEP-WCMC biodiversity series, vol 32. UNEP-WCMC, Cambridge

Hooper JNA, van Soest RWM (2002) Systema Porifera: a guide to the classification of sponges. Kluwer Academic/Plenum Publishers, New York

Hu GP, Yuan J, Sun L et al (2011) Statistical research on marine natural products based on data obtained between 1985 and 2008. Mar Drugs 9(4):514–525

Ijima I, Okada Y (1901) Studies on the Hexactinellida: contribution I. (Euplectellidae). Imperial University of Tokyo, Tokyo. https://doi.org/10.5962/bhl.title.16267

Jensen PR, Fenical W (1994) Strategies for the discovery of secondary metabolites from marine bacteria: ecological perspectives. Annu Rev Microbiol 48:559–584

Jesionowski T, Norman M, Żółtowska-Aksamitowska S et al (2018) Marine spongin: naturally prefabricated 3D scaffold-based biomaterial. Mar Drugs 16(3):88

Jiménez E, Ribes M (2007) Sponges as a source of dissolved inorganic nitrogen: nitrification mediated by temperate sponges. Limnol Oceanogr 52(3):948–958. https://doi.org/10.4319/lo.2007.52.3.0948

Kahn AS, Yahel G, Chu JW et al (2015) Benthic grazing and carbon sequestration by deep-water glass sponge reefs. Limnol Oceanogr 60(1):78–88

Kilian EF (1952) Wasserströmung und Nahrungsaufnahme beim Süsswasserschwamm *Ephydatia fluviatilis*. Z Vgl Physiol 34:407–447

Klitgaard AB (1995) The fauna associated with outer shelf and upper slope sponges (Porifera, Demospongiae) at the Faroe Islands, northeastern Atlantic. Sarsia 80(1):1–22

Klitgaard AB, Tendal OS (2004) Distribution and species composition of mass occurrences of large-sized sponges in the Northeast Atlantic. Prog Oceanogr 61(1):57–98. https://doi.org/10.1016/j.pocean.2004.06.002

Kutti T, Bannister RJ, Fosså JH (2013) Community structure and ecological function of deep-water sponge grounds in the Traenadypet MPA—Northern Norwegian continental shelf. Cont Shelf Res 69:21–30. https://doi.org/10.1016/j.csr.2013.09.011

Lesser MP (2006) Benthic-pelagic coupling on coral reefs: feeding and growth of Caribbean sponges. J Exp Mar Biol Ecol 328(2):277–288

Leys SP, Ereskovsky AV (2006) Embryogenesis and larval differentiation in sponges. Canadian J Zool 84(2):262–287

Maldonado M (2016) Sponge waste that fuels marine oligotrophic food webs: a re-assessment of its origin and nature. Mar Ecol 37(3):477–491. https://doi.org/10.1111/maec.12256

Maldonado M, Young CM (1996) Bathymetric patterns of sponge distribution on the Bahamian slope. Deep-Sea Res I 43:897–915. https://doi.org/10.1016/0967-0637(96)00042-8

Maldonado M, Ribes M, van Duyl FC (2012) Nutrient fluxes through sponges: biology, budgets, and ecological implications. Adv Mar Biol 62:113–182

Maldonado M, Aguilar R, Bannister RJ et al (2015) Sponge grounds as key marine habitats: a synthetic review of types, structure, functional roles, and conservation concerns. In: Rossi S, Bramanti L, Gori A et al (eds) Marine animal forests: the ecology of benthic biodiversity hotspots, 1st edn. Springer, Switzerland, pp 1–39

Manconi R, Pronzato R (2016) How to survive and persist in temporary freshwater? Adaptive traits of sponges (Porifera: Spongillida): a review. Hydrobiologia 782:11–22. https://doi.org/10.1007/s10750-016-2714-x

McMurray SE, Pawlik JR, Finelli CM (2017) Demography alters carbon flux for a dominant benthic suspension feeder, the giant barrel sponge, on Conch reef, Florida keys. Funct Ecol 31(11):2188–2198. https://doi.org/10.1111/1365-2435.12908

Mehbub MF, Perkins MV, Zhang W et al (2016a) New marine natural products from sponges (Porifera) of the order Dictyoceratida (2001 to 2012); a promising source for drug discovery, exploration and future prospects. Biotechnol Adv 34:473–491. https://doi.org/10.1016/j.biotechadv.2015.12.008

Mehbub MF, Tanner JE, Barnett SJ et al (2016b) The role of sponge-bacteria interactions: the sponge *Aplysilla rosea* challenged by its associated bacterium *Streptomyces* ACT-52A in a controlled aquarium system. Appl Microbiol Biotechnol 100:10609–10626. https://doi.org/10.1007/s00253-016-7878-9

Miller RJ, Hocevar J, Stone RP et al (2012) Structure-forming corals and sponges and their use as fish habitat in Bering Sea submarine canyons. PLoS One 7(3):e33885. https://doi.org/10.1371/journal.pone.0033885

Mulholland PJ (2003) Large-scale patterns in dissolved organic carbon concentration, flux, and sources. In: Findlay SEG, Sinsabaugh RL (eds) Aquatic ecosystems, 2nd edn. Academic, London, pp 139–159

Müller WEG (2006) The stem cell concept in sponges (Porifera): metazoan traits. Semin Cell Dev Biol 17:481–491. https://doi.org/10.1016/j.semcdb.2006.05.006

Müller WEG, Böhm M, Batel R et al (2000) Application of cell culture for the production of bioactive compounds from sponges: synthesis of Avarol by primmorphs from *Dysidea avara*. J Nat Prod 63:1077–1081

Müller WEG, Grebenjuk VA, Thakur NL et al (2004) Oxygen-controlled bacterial growth in the sponge *Suberites domuncula*: toward a molecular understanding of the symbiotic relationships between sponge and bacteria. Appl Environ Microbiol 70(4):2332–2341

Munro MHG, Blunt JW, Dumdei EJ et al (1999) The discovery and development of marine compounds with pharmaceutical potential. J Biotechnol 70(1–3):15–25. https://doi.org/10.1016/S0168-1656(99)00052-8

Murillo FJ, Kenchington E, Lawson JM et al (2016) Ancient deep-sea sponge grounds on the Flemish Cap and Grand Bank, Northwest Atlantic. Mar Biol 163(3):63. https://doi.org/10.1007/s00227-016-2839-5

Muscatine L, Porter JW (1977) Reef corals: mutualistic symbioses adapted to nutrient-poor environments. Bioscience 27(7):454–460. https://doi.org/10.2307/1297526

Nakanishi N, Sogabe S, Degnan BM (2014) Evolutionary origin of gastrulation: insights from sponge development. BMC Biol 12:26. https://doi.org/10.1186/1741-7007-12-26

Okada Y (1928) On the development of a hexactinellid sponge, *Farrea sollasii*. J Fac Sci Imper Univ Tok Sect IV Zool 2:1–27

Osinga R, Belarbi EH, Grima EM et al (2003) Progress towards a controlled culture of the marine sponge *Pseudosuberitesandrewsi* in a bioreactor. J Biotechnol 100(2):141–146

Pile AJ, Patterson M, Witman J et al (1996) In situ grazing on plankton <10 μm by the boreal sponge *Mycale lingua*. Mar Ecol Prog Ser 141:95–102. https://doi.org/10.3354/meps141095

Pile AJ, Patterson MR, Savarese M et al (1997) Trophic effects of sponge feeding within Lake Baikal's littoral zone. 2. Sponge abundance, diet, feeding efficiency, and carbon flux. Limnol Oceanogr 42(1):178–184. https://doi.org/10.4319/lo.1997.42.1.0178

Pisani D, Pett W, Dohrmann M et al (2015) Genomic data do not support comb jellies as the sister group to all other animals. Proc Natl Acad Sci U S A 112(50):15402–15407. https://doi.org/10.1073/pnas.1518127112

Pronzato R (2003) Mediterranean Sponge fauna: a biological, historical and cultural heritage. Biogeographia 24:91–99. https://doi.org/10.21426/B6110118

Reid REH (1968) Bathymetric distributions of Calcarea and Hexactinellida in the present and the past. Geol Mag 105:546–559. https://doi.org/10.1017/S0016756800055904

Reiswig HM (1971) Particle feeding in natural populations of three marine demosponges. Biol Bull 141:568–591. https://doi.org/10.2307/1540270

Rix L, de Goeij JM, Mueller CE et al (2016) Coral mucus fuels the sponge loop in warm- and cold-water coral reef ecosystems. Sci Rep 6(1):18715. https://doi.org/10.1038/srep18715

Rix L, Goeij JM, van Oevelen D et al (2017) Differential recycling of coral and algal dissolved organic matter via the sponge loop. Funct Ecol 31(3):778–789

Roberts JM, Wheeler AJ, Freiwald A (2006) Reefs of the deep: the biology and geology of cold-water coral ecosystems. Science 312(5773):543–547

Romera-Castillo C, Letscher RT, Hansell DA (2016) New nutrients exert fundamental control on dissolved organic carbon accumulation in the surface Atlantic Ocean. Proc Natl Acad Sci U S A 113(38):10497–10502. https://doi.org/10.1073/pnas.1605344113

Ruppert EE, Fox RS, Barnes RD (2004) Invertebrate zoology, 7th edn. Cengage Learning US, Fort Worth

Ryer CH, Stoner AW, Titgen RH (2004) Behavioral mechanisms underlying the refuge value of benthic habitat structure for two flatfishes with differing anti-predator strategies. Mar Ecol Prog Ser 268:231–243

Schläppy ML, Schöttner SI, Lavik G et al (2010) Evidence of nitrification and denitrification in high and low microbial abundance sponges. Mar Biol 157(3):593–602. https://doi.org/10.1007/s00227-009-1344-5

Sepčić K, Poklar N, Vesnaver G et al (1999) Interaction of 3-Alkylpyridinium polymers from the sea sponge *Renierasarai* with insect acetylcholinesterase. J Protein Chem 18:251–257. https://doi.org/10.1023/A:1021096726288

Simion P, Philippe H, Baurain D et al (2017) A large and consistent Phylogenomic dataset supports sponges as the sister group to all other animals. Curr Biol 27:958–967. https://doi.org/10.1016/j.cub.2017.02.031

Simpson TL (1984) The cell biology of sponges. Springer, New York

Slattery M, Gochfeld DJ, Easson CG et al (2013) Facilitation of coral reef biodiversity and health by cave sponge communities. Mar Ecol Prog Ser 476:71–86

Southwell MW, Popp BN, Martens CS (2008) Nitrification controls on fluxes and isotopic composition of nitrate from Florida keys sponges. Mar Chem 108(1):96–108. https://doi.org/10.1016/j.marchem.2007.10.005

Taylor MW, Radax R, Steger D et al (2007) Sponge-associated microorganisms: evolution, ecology, and biotechnological potential. Microbiol Mol Biol Rev 71(2):295–347. https://doi.org/10.1128/MMBR.00040-06

Thompson JE, Walker RP, Faulkner DJ (1985) Screening and bioassays for biologically-active substances from forty marine sponge species from San Diego, California, USA. Mar Biol 88(1):11–21. https://doi.org/10.1007/BF00393038

Tyler JC, Böhlke JE (1972) Records of sponge-dwelling fishes, primarily of the Caribbean. Bull Mar Sci 22(3):601–642

Tyrreell T (1999) The relative influences of nitrogen and phosphorus on oceanic primary production. Nature 400:525–531

Unson MD, Faulkner DJ (1993) Cyanobacterial symbiont biosynthesis of chlorinated metabolites from *Dysideaherbacea* (Porifera). Experientia 49:349–353. https://doi.org/10.1007/BF01923420

Vacelet J (1988) Indications de profondeur données par les Spongiaires dans les milieux benthiques actuels. Géol Médit 15:13–26. https://doi.org/10.3406/geolm.1988.1392

Vacelet J, Boury-Esnault N, Fiala-Medioni A et al (1995) A methanotrophic carnivorous sponge. Nature 377:296. https://doi.org/10.1038/377296a0

van Duyl F, Hegeman J, Hoogstraten A et al (2008) Dissolved carbon fixation by sponge–microbe consortia of deep water coral mounds in the northeastern Atlantic Ocean. Mar Ecol Prog Ser 358:137–150. https://doi.org/10.3354/meps07370

van Oevelen D, Duineveld G, Lavaleye Met al (2009) The cold-water coral community as a hot spot for carbon cycling on continental margins: a food-web analysis from Rockall Bank (Northeast Atlantic). Limnol Oceanogr54(6):1829–1844

van Soest RWM, Boury-Esnault N, HooperJNAet al (2018) World Porifera database. http://www.marinespecies.org/porifera. Accessed 30 Oct 2018

Von Putter A (1914) Der Stoffwechsel der Kieselschwämme. Z Allg Physiol 16:65–114

Wild C, Huettel M, Klueter A et al (2004) Coral mucus functions as an energy carrier and particle trap in the reef ecosystem. Nature 428(6978):66–70. https://doi.org/10.1038/nature02344

Wild C, Mayr C, Wehrmann L et al (2008) Organic matter release by cold water corals and its implication for fauna–microbe interaction. Mar Ecol Prog Ser 372:67–75

Wild C, Wehrmann L, Mayr C et al (2009) Microbial degradation of cold-water coral-derived organic matter: potential implication for organic C cycling in the water column above Tisler reef. Aquatic Biol 7(1–2):71–80. https://doi.org/10.3354/ab00185

Wilkinson CR (1983) Net primary productivity in coral reef sponges. Science 219(4583):410–412

Wilkinson CR (1992) Symbiotic interactions between marine sponges and algae. In: Reisser W (ed) Algae and symbiosis: plants, animals, fungi, viruses. Interactions explored. Biopress, Bristol, pp 112–151

Wilson MC, Mori T, Rückert C et al (2014) An environmental bacterial taxon with a large and distinct metabolic repertoire. Nature 506:58–62

Witte U, Brattegard T, Graf G et al (1997) Particle capture and deposition by deep-sea sponges from the Norwegian-Greenland Sea. Mar Ecol Prog Ser 154:241–252

Wulff JL (1986) Variation in clone structure of fragmenting coral reef sponges. Biol J Linnean Soc 27:311–330. https://doi.org/10.1111/j.1095-8312.1986.tb01740.x

Zenkevich LA, Barsanova NG, Beliaev G (1960) Quantitative distribution of bottom fauna in the abyssal area of the World Ocean. Dokl Akad Nauk SSSR 130:183–186

Zundelevich A, Lazar B, Ilan M (2007) Chemical versus mechanical bioerosion of coral reefs by boring sponges – lessons from *Pionecf. vastifica*. J Exp Biol 210(1):91–96

10 Theories, Vectors, and Computer Models: Marine Invasion Science in the Anthropocene

Philipp Laeseke, Jessica Schiller, Jonas Letschert, and Sara Doolittle Llanos

Abstract

Marine invasions are well-recognized as a worldwide threat to biodiversity and cause for tremendous economic damage. Fundamental aspects in invasion ecology are not yet fully understood, as there is neither a clear definition of invasive species nor their characteristics. Likewise, regulations to tackle marine invasions are fragmentary and either restricted to specific regions or certain aspects of the invasion process. Nonetheless, marine anthropogenic vectors (e.g., vessel fouling, ballast water, aquaculture, marine static structures, floating debris, and human-mediated climate change) are well described. The most important distribution vector for marine non-indigenous species is the shipping sector, composed by vessel fouling and ballast water discharge. Ship traffic is a constantly growing sector, as not only ship sizes are increasing, but also remote environments such as the polar regions are becoming accessible for commercial use. To mitigate invasions, it is necessary to evaluate species' capability to invade a certain habitat, as well as the risk of a region of becoming invaded. On an ecological level, this may be achieved by Ecological Niche Modelling based on environmental data. In combination with quantitative vector data, sophisticated species distribution models may be developed. Especially the ever-increasing amount of available data allows for comprehensive modelling approaches to predict marine invasions and provide valuable information for policy makers. For this article, we reviewed available literature to provide brief insights into the backgrounds and regulations of major marine vectors, as well as species distribution modelling. Finally, we present some state-of-the-art modelling approaches based on ecological and vector data, beneficial for realistic risk assessments.

Keywords

Non-indigenous species · Marine vectors · Species distribution modelling · Regulations · Anthropogenic debris

10.1 Non-indigenous and Invasive Species

Non-indigenous species (NIS) can have negative effects on receiving ecosystems and are considered one of the major global threats to biodiversity (Ruiz et al. 1997; Casas et al. 2004; Raffo et al. 2009). Apart from ecological consequences, substantial economic damage can be caused by overly abundant introduced species or harmful species such as pathogens (e.g., Pimentel et al. 2000, 2001). The effects of introductions and establishments of new species in a community are unpredictable, as a multitude of biotic and abiotic factors determine the onset and further development of an invasion. Depending on the receiving habitat and the observed parameter, the same species can have negative but also positive effects (McLaughlan et al. 2014). Because of the variety of factors of each invasion, understanding them on the species-, pathway-, and ecosystem level is essential for adequate evaluation and possible management.

Despite their ecological and economic relevance, not even the basic terminology of introduced or invasive species is clearly determined among scientists and regulations. Over time, several definitions have been proposed for biological invasions. The most basic one is being a non-indigenous

All authors equally contributed to this chapter.

P. Laeseke (✉) · J. Schiller
Marine Botany, University of Bremen, Bremen, Germany
e-mail: philipp.laeseke@uni-bremen.de

J. Letschert
Leibniz Centre for Tropical Marine Research (ZMT), Bremen, Germany

S. D. Llanos
Groningen Institute for Evolutionary Life-Sciences GELIFES, University of Groningen, Groningen, The Netherlands

S. Jungblut et al. (eds.), *YOUMARES 9 - The Oceans: Our Research, Our Future*,
https://doi.org/10.1007/978-3-030-20389-4_10

species (NIS), namely, a species introduced after the discovery of America and the onset of large-scale transatlantic ship traffic (Leppäkoski et al. 2013; Ricciardi et al. 2013). More specifically, Richardson and Pyšek (2006) defined invasion ecology as the study of human-mediated introductions of species to areas beyond their native range without considering the impact on the invaded habitat. Alpert et al. (2000) included effects of NIS and described an invasive species as "one that both spreads in space and has negative effects on species already in the space that it enters." According to Boudouresque and Verlaque (2002) introduced and invasive species can be differentiated by the conspicuous role the latter play in the recipient ecosystems, which is characterized by becoming dominant and potentially taking the place of keystone species. The previous examples show how much definitions can vary in only a few studies – with more being considered, they even begin to contradict each other, both in terminology and procedure (e.g., Blackburn et al. 2011; Guy-Haim et al. 2018).

Although not clearly defined, bioinvasions are a topic of public interest (García-Llorente et al. 2008) and there are several national eradication programs and policies established (see New Zealand, USA; Myers et al. 2000, Wotton et al. 2004). However, on a global scale, overarching regulations to mitigate marine invasions are missing. This is reflected in the EU legislative 1143/2014, which only deals with anthropogenically introduced species, but does not consider naturally introduced species. Moreover, international conventions for marine traffic are not binding across the globe or only concern certain aspects of dispersal mechanisms (see Sect. 10.2). One reason for this fragmentation among marine NIS regulations might be the influence of economic interests, which dilute scientific expertise (Margolis et al. 2005).

To develop efficient regulations, it is essential to gain an in-depth understanding of human-mediated vectors and factors influencing invasion success. Ecological Niche Models (ENM) can be powerful in evaluating invasion potential and are currently implemented at the frontier of invasion science (see Sect. 10.3). Figure 10.1 sets the framework for this article, in which we summarize knowledge on anthropogenic vectors and give insights into methods and developments of ENM as a potential forecasting tool. We intend to contribute to the understanding of bioinvasions at a broader scale and shine a light on necessary future efforts to develop efficient regulations.

10.2 Anthropogenic Vectors

Defining which vector has the highest impact in terms of the number of introductions, establishment rate, and effects on the new habitat is challenging because their effectiveness and frequency vary with time and geographical region (Williams et al. 2013). In general, failed introductions and invasions pose a problem in cross-vector analysis, because they mostly remain hidden, leading to strong biases in introduction rates per vector (Zenni and Nuñez 2013). About four decades ago, ship traffic and aquaculture were identified as the major vectors for marine human-mediated introductions (Carlton 1979). Recent studies suggest that this assumption has not changed much and efforts have been undertaken to rank vectors regarding their potential of dispersing NIS. On a global level, a positive correlation between cargo ship traffic and marine introductions reveals the vast contribution of marine traffic to create connectivity across distant geographic regions (Seebens et al. 2016). Ship traffic can be divided into two NIS pathways: the colonization of vessel hulls with sessile or small motile species (hereafter fouling species), and the transportation of organisms and their early-life stages (eggs, larvae) in ballast water tanks (Ruiz et al. 1997; Cohen and Carlton 1998; Godwin 2003). On a regional level, a cross-vector comparison in California revealed vessel fouling as the most important vector followed by ballast water and aquaculture (Williams et al. 2013). However, the authors claim that results cannot be extrapolated and are case-specific with respect to area, time, and vector composition.

This review examines the following marine vectors: vessel fouling, ballast water, mariculture, marine static structures, floating anthropogenic litter, and human-mediated climate change. This selection encompasses the major vectors, affecting most marine ecosystems worldwide. Live species trade with ornamental (Weigle et al. 2005) and bait species (Weigle et al. 2005; Fowler et al. 2016) represent minor vectors and therefore will not be elaborated in this article. Canals play an important role in the distribution of marine species on regional scales (see Gollasch 2006 for the influence of the Suez Canal on Mediterranean species composition). However, they represent the removal of physical barriers between adjacent regions and allow migration in a variety of ways (e.g., shipping related or natural dispersal), which are covered in the sections mentioned above. Therefore, we do not include an individual chapter on this vector.

10.2.1 Vessel Fouling

The importance of hull fouling for marine invasions is unquestionable. A convenient parameter to quantify the marine invasion risk through hull fouling is the wetted surface area (WSA) of ships (Miller et al. 2018) and an approach of calculating the WSA of the world fleet of commercial vessels resulted in 325×10^6 m^2(Moser et al. 2016). Marine traffic is continuously increasing and even remote areas, such as

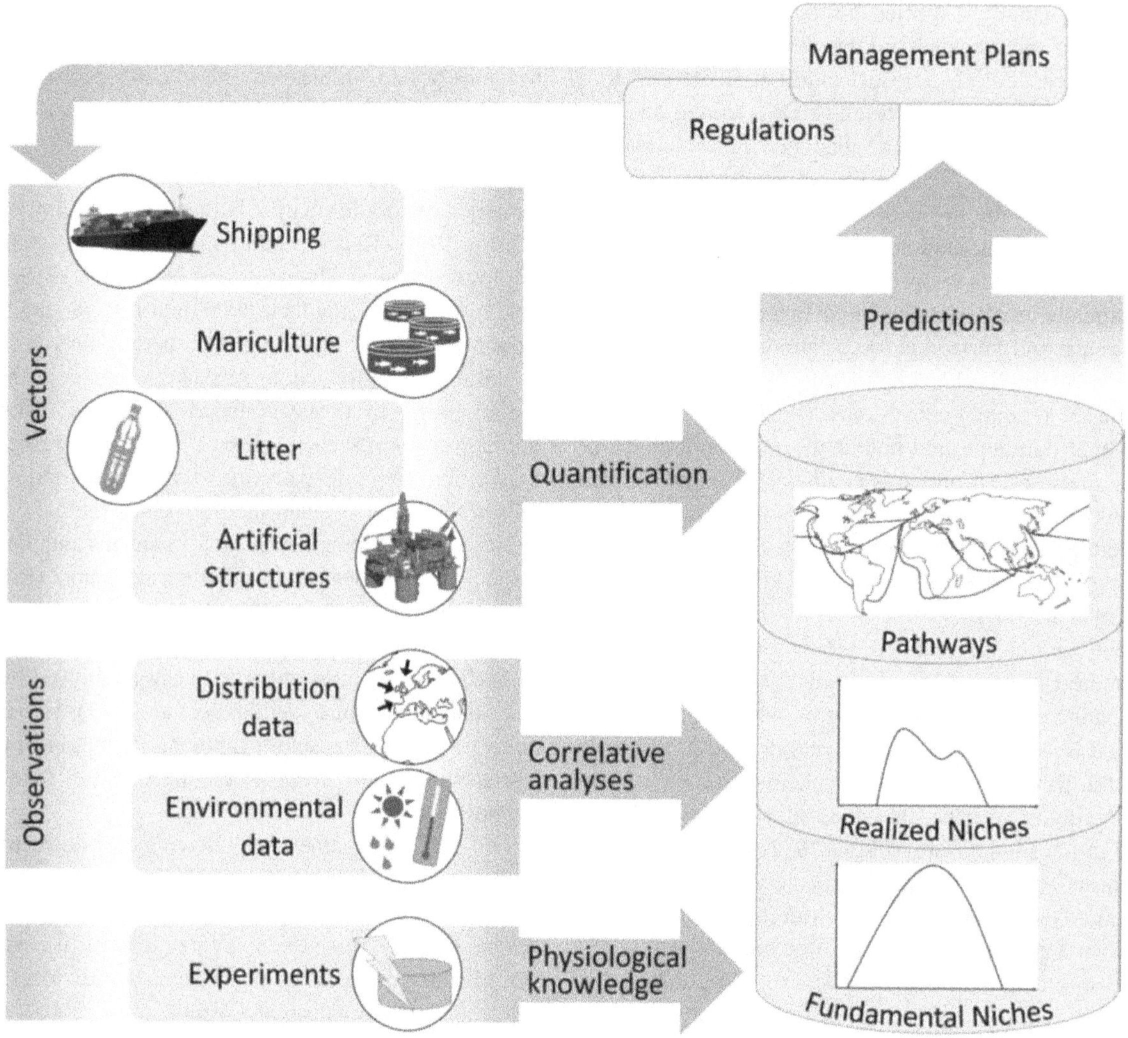

Fig. 10.1 Marine Bioinvasions in the Anthropocene: the most important vectors for alien invasive species across geographic regions are anthropogenic transportation means, such as shipping- and mariculture-related transfers. Also passively drifting litter and stable structures contribute to the transport and introduction of species. Quantification of introductions along these vectors allows for identification of major pathways across the globe. Ecological Niche Modelling can help to identify suitable environmental conditions for species in question. While correlative approaches are well established for the investigation of realized niches, laboratory studies can yield important additional information about the species' fundamental niche and hence contribute to the understanding of ecological mechanisms which influence a species' distribution potential. Transportation data and Ecological Niche Models can be combined to evaluate invasion risk. Identification of areas with high introduction pressure and understanding of the species being transported along are an important step prior to the development of regulations, management plans, and mitigation strategies. However, to date, only few international regulations are effective which successfully control the spread of species

the Arctic, become available for commercial shipping due to melting sea ice (Miller and Ruiz 2014).

Antifouling coatings are applied to vessel hulls and repel many fouling species that would normally settle on submerged vessel areas (Williams et al. 2013). Yet, there are certain organisms that are immune to antifouling components such as the bryozoan *Watersipora subtorquata*, which may serve as a foundation species providing settlement space for subsequent epibionts (Floerl et al. 2004). Small disruptions of 1–2 cm in antifouling coatings may enable the settlement of a wide range of sessile marine species, which can easily be overseen in cryptic spots like keels or propeller shafts (Piola and Johnston 2008). Godwin (2003) observed weaknesses of antifouling coatings at weld seams and spots where smaller boats were placed on wooden blocks while painted. He also assumed that slow velocities and long port stays increase the potential of sessile species to settle and survive on vessel hulls (Godwin 2003). Kauano et al. (2017) tested persistence of fouling species after being dragged with 5, 15, and 20 knots for 20 min. Although the overall trend shows a negative correlation between velocity and persistence, 90% of the species were present with at least 20% of their original abundance after being dragged with 20 knots. Some limitation to vessel fouling is provided by desiccation. Kauano et al. (2017) found that most soft-bodied sessile species died after being outside of the water for 24 hours, whereas barnacles survived 120 hours. Another example for desiccation resistance are sporophytes of the invasive kelp *Undaria pinnatifida* that released viable spores even after 3 days outside of the water (Bollen et al. 2017).

Large cargo vessels such as bulk carriers, tankers, and container ships are usually equipped with slow-speed engines (Endresen et al. 2003), meaning that they rarely travel faster than 20 knots, and are only put on dry dock every 5 years when their hulls are cleaned and repainted. Yet, these vessels represent 79% of the WSA of the commercial world fleet and substantially contribute to geographical connectivity (Moser et al. 2016; Seebens et al. 2016). In combination with the knowledge mentioned above, this may explain why vessel fouling is still a major pathway for NIS on a global scale.

Trends in the marine traffic industry favor larger container ships and hub-ports (Shenkar and Rosen 2018), from which smaller transport vessels carry goods to smaller ports, representing one example of secondary spread. Small-scale boating may contribute to secondary spread of NIS, especially in areas with intense tourism or recreational activities (Anderson et al. 2015). Many marine invertebrates, such as ascidians and bryozoans, have very short natural dispersal ranges and hence marine traffic or rafting debris is likely to enable their long-range dispersal (Petersen and Svane 1995). This is underlined by a case study in the great barrier reef where sessile NIS were found about 80 km offshore at an isolated coral reef that is frequently visited by boats (Piola and Johnston 2008).

In 2011, the International Marine Organization (IMO) published a resolution for the responsible management of vessel fouling to reduce the risk of NIS introduction (IMO 2011). However, these are mere voluntary guidelines and despite the global significance of vessel fouling for NIS dispersal, there is no enforced regulation on an international level yet. There are some examples for implemented hull fouling standards on a national and regional level represented by New Zealand (Ministry for Primary Industries 2014), and the National Park of Galapagos, Ecuador (Campbell et al. 2015). Both regulations require clean vessel hulls and antifouling coatings prior to the arrival of vessels.

10.2.2 Ballast Water

Ballast water discharge is the vector with the most management rules among the important anthropogenic dispersal mechanisms. The International Convention for the Control and Management of Ships' Ballast Water and Sediments (hereafter the BW Convention) was adopted by the IMO in 2004 and came into force in September 2017 (IMO 2004). According to requirements regulating the behavior of ballast water discharge, the BW Convention can be split into two major parts.

The first part obliges incoming vessels to exchange their ballast water at least 200 nautical miles offshore in a minimum depth of 200 m. The USA, not a signatory to the BW Convention, implemented a similar requirement. A study assessing ballast water exchanges in the USA from 2005 to 2007 found that most vessels abide with this rule, however, especially vessels that journeyed along the South and North American coasts still exchange their ballast water in coastal areas frequently (Miller et al. 2011). Similar results were obtained by a study targeting the Taiwanese maritime cargo sector showing that up to 30% of the surveyed ships exchange ballast water closer to shore than 200 nautical miles (Liu et al. 2014).

The second part of the BW Convention restricts the total amount of viable organisms in discharged ballast water to up to ten with a size of >50 μm m^{-3} plus up to ten with a size of <50 μm ml^{-1} (IMO 2004). To meet these restrictions, vessels are obliged to install ballast water treatment plants (e.g., electro-chlorination, UV treatment, and filtration). Given those conditions, Reusser et al. (2013) developed a model to predict the invasion rate per year through foreign ballast water discharge in the US Pacific Coast. Based on existing invasive species records and assuming a linear relationship between discharged organisms and successful invasions, they calculated that a new invasion would only occur every 10–100 years.

Shipping routes and source regions of ballast water affect the survivability of organisms at the ship's destination (Verling et al. 2005). For example, do transport routes through the Panama Canal expose attached specimens to tropical and partially freshwater conditions leading to temperature and osmotic stress (Miller and Ruiz 2014). The BW Convention requires ships to keep records of ballast water activities, so that uptake areas can be compared to discharge areas on demand and high risks of introductions can be avoided. Additionally, port states are empowered to conduct ballast water controls on incoming foreign ships and, if necessary, impose sanctions.

Still, a minimum risk of biointroduction remains and is positively correlated to the amount of ballast water discharged in an area (Reusser et al. 2013). This is important to consider in major ports serving as hubs for international maritime trade such as Shanghai, Singapore, or Rotterdam. Moreover, a study of the Chinese ballast water capacity confirmed the rising amount of ballast water in line with the growing maritime transport sector (Zhang et al. 2017).

10.2.3 Mariculture

Many marine species have been intentionally transported across broad geographical distances to be husbanded in aquacultures. The largest contributors to the global mariculture industry are Asian countries with China being by far the most important among them (FAO 2016). Other countries, such as Norway, Chile, and Indonesia have fast-growing mariculture industries as well (Buschmann et al. 2009). The most important cultured organisms worldwide are finfish, mollusks, crustaceans, and seaweed species. In 2014, 580 aquatic species have been registered with the FAO as husbanded species (FAO 2016). These species are often non-indigenous in the place where they are kept, meaning an escape would directly lead to an introduction into the new habitat. Examples for intentionally introduced species are the Pacific oyster *Crassostrea gigas*, domesticated salmon, and many seaweed species (Naylor et al. 2001). In contrast, accidental introductions may occur due to associated hitchhikers such as parasites, algae (e.g., *Codium fragile*), and various fouling species that live on or in aquaculture gear and husbanded species (Naylor et al. 2001).

Focusing on introductions to urban areas, Padayachee et al. (2017) investigated the taxa composition introduced by several marine vectors and found a significant difference between the categories Mariculture and Fisheries. Vertebrates were almost exclusively introduced for cultivation, while plants dominated, and were exclusive to, the equipment-facilitated arrivals. The continuous transfer of equipment and seed stock between maricultures has an especially high potential of species introduction (Forrest and Blakemore 2006). One striking example for this is the kelp *Undaria pinnatifida*, used for mariculture. It arrived to Europe alongside the Pacific oyster and has since been spread independently of oyster cultivation for farming or as a fouling species and recently reached German waters (Schiller et al. 2018). This was largely enabled by its tolerance to various conditions, including surviving overland transport on boat hulls or ropes (Bollen et al. 2017).

After vessel fouling, shellfish farming is considered the second most important vector for the 277 registered non-indigenous seaweeds worldwide. Especially red corticated algae, but also a variety of other taxa, live in association with farmed shellfish (Williams and Smith 2007). Seaweed mariculture itself is only a minor but efficient way of seaweed introductions, because farmed algae are specifically chosen for their competitiveness (Williams and Smith 2007). Interestingly, seaweed mariculture is the fastest-growing sector of aquaculture posing one-quarter of the global volume produced by aquaculture (FAO 2016). This growth is mainly due to seaweed farms in Indonesia and China that were established during the last 20 years. Between 2004 and 2014, the global aquaculture industry has grown rapidly and the percentage share of total worldwide fish harvest increased from 31.1% to 44.1% (wild catches and aquaculture products including non-food uses; FAO 2016). While regulations on an international level are missing, there are some examples for guidelines of the treatment of aquaculture organisms and gear, proposing sterilization prior to moving it to a new location. An example is the Australian National Biofouling Management Guidelines for the Aquaculture Industry that proposes different treatment methods such as exposures to air, fresh water, heat, or chemicals (NSPMMPI 2013).

10.2.4 Static Maritime Structures

There is a growing number of various static maritime structures (SMS), which are occasionally relocated and thus pose a risk to transport marine NIS or serve as stepping stones (i.e., oil and gas platforms, offshore wind farms, navigational buoys, non-cargo barges, and dry docks; Iacarella et al. 2018). Most SMS are characterized by their large and complex wetted surface area (WSA), providing space for fouling organisms, which, in turn, may attract predators (Friedlander et al. 2014; Todd et al. 2018). These artificial communities often differ from surrounding species assemblages (Stachowicz et al. 2002). Oil and gas platforms represent a major part of SMS and will therefore be the main focus of this section.

After being stationary for years, oil and gas platforms may be moved to a new service location, for repair, or decommission. To be able to navigate, they are either equipped with engines, towed by tug vessels (wet-tow), or carried on heavy lifting ships (dry-tow; Robertson et al. 2018). The former two options pose a risk for NIS dispersal, because platforms stay in the water during transport and are transported at very low speed (<8 knots), allowing associated organisms to travel along. In contrast to vessel fouling, translocated oil and gas platforms may introduce entire ecosystems to new geographical areas, including large sessile and mobile species across all trophic levels from algae to vertebrates (Ferreira et al. 2006; Yeo et al. 2009). Incidences of stranded or intentionally moved oil and gas platforms prove the introduction of a range of invertebrate species (Foster and Willan 1979; Ferreira et al. 2006; Page et al. 2006; Yeo et al. 2009), as well as fish species (Yeo et al. 2009; Wanless et al. 2010; Pajuelo et al. 2016).

Abandoned oil and gas platforms are frequently transformed into artificial reefs instead of being decommissioned ("rigs to reefs"; reviewed by Bull and Love 2019), because they foster entire marine ecosystems and due to high demolishment costs. This practice is largely unregulated with respect to its biological implications, an issue in need of addressing, considering that a large number of the roughly

7000 oil platforms worldwide were already reaching the end of their service time in 2003 (Hamzah 2003).

Iacarella et al. (2018) emphasized that regulations concerning marine NIS introductions through SMS are still missing. This is especially worrying considering that the Arctic might become more available for commercial use, including drilling operations, with decreasing sea ice.

10.2.5 Marine Litter

We have long known about how ocean currents can transport a wide variety of structures, which may then serve as a raft for fouling species (Guppy 1917; Thiel and Gutow 2005; Wichmann et al. 2012). The presence of floating plastic debris in the oceans has increased tremendously in recent decades and continues to grow (PlasticsEurope 2013). Due to this increment of potential vectors, we very well might be on the brink of a new era for marine invasions.

The exact sources of anthropogenic debris are often unknown, since trajectories of floating objects are hard to track, being influenced by seasonal variations in wind and current conditions (Kiessling et al. 2015). The United Nations Joint Group of Experts on the Scientific Aspects of Marine Pollution (GESAMP) have estimated that land-based sources account for up to 80% of the world's marine litter, 60–95% of the waste being plastic debris (Sheavly 2005). However, shipping activities have also been a major source of marine litter (Scott 1972). Despite agreements to forbid ship waste dumping (London Dumping Convention, promulgated in 1972; Lentz 1987), compliance and enforcement still pose significant challenges (Carpenter and Macgill 2005). In fact, in some regions up to 95% of all litter items are shipping-related (Van Franeker et al. 2011), and debris composition in the Baltic Sea and North Pacific Ocean leaves little doubt that ocean-based sources are major contributors to marine debris (Moore and Allen 2000; Fleet et al. 2009; Keller et al. 2010; Watters et al. 2010; Schlining et al. 2013).

The predominance of plastic as floating litter and as accumulated debris on shorelines is not due to the amounts in which it is produced relative to other types of waste, but to its remarkable persistence and durability (Andrady 2015). The long life expectancy of a piece of plastic contrasts to the natural processes of consumption and decomposition that organic flotsam eventually undergoes (Vandendriessche et al. 2007). It is because of this persistence that today we are facing the possibility of human litter more than doubling rafting opportunities, particularly at high latitudes (Barnes 2002), and potentially propagating fauna outside of their native ranges (Barnes et al. 2009; Gregory 2009) and up to the most remote polar marine environments (Barnes et al. 2010; Lusher et al. 2015). Because of its overall high numbers, plastic debris offers rafting opportunities that quantitatively surpass other floating substrata in the oceans. As Goldstein et al. (2012) suggest, many species may no longer be limited by the availability of suitable substrata to adhere to. On top of enhancing transport of rafting communities, the availability of plastic may favor the transport of certain species over others. This is because rafting communities on litter and, e.g., macroalgae are described as similar, but less species rich in the former (Winston et al. 1997; Gregory 2009).

Over 1200 taxa have been associated with natural and anthropogenic flotsam (Thiel and Gutow 2005), and many organisms and potential invaders were first described on marine litter (Jara and Jaramillo 1979; Stevens et al. 1996; Winston et al. 1997; Cadée 2003). One most notable event was the record of a 188-ton piece of a former dock, dislodged during a tsunami in Japan in 2011, stranded in Oregon and accounting for the first record of over 100 species non-native to the west coast of the USA (Choong and Calder 2013). While samples taken from beach litter collections show a bias towards sessile organisms with hard calcified structures (Winston et al. 1997; Gregory 2009), debris collected afloat include a higher diversity of soft-bodied and/or motile species (Astudillo et al. 2009; Goldstein et al. 2014). Overall, cnidarians, bryozoans, mollusks, and crustaceans seem to be the most abundant taxa registered. Today, we know plastic can host a variety of pathogens: the ciliate *Halofolliculina* sp., which targets coral skeletal structures (Goldstein et al. 2014), potential human and animal pathogens of the genus *Vibrio* (Zettler et al. 2013), and dinoflagellates known to cause harmful algal blooms (Masó et al. 2003).

What ensures colonization and survival during transport on a plastic raft? Kiessling et al. (2015) reviewed 82 publications with the aim of characterizing marine debris rafters, their biological traits, and identifying the specific conditions rafters face in order to survive their voyages. Their results suggest that a majority of species act as facultative rafters (77%), as fully sessile (59%), and as suspension feeders (72%). This can easily be compared to communities of algae rafts, which are more complex at the structural level, and more capable of hosting mobile species with different feeding patterns (Thiel and Gutow 2005).

Colonization might influence certain characteristics of a plastic raft. Floating behavior might be altered, as the added weight of rafters may stabilize an otherwise highly buoyant and unbalanced object. This would increase colonization probability (Bravo et al. 2011) and the succession of the rafting community, but heavy fouling on a plastic item may increase the raft's weight and cause it to sink (Ye and Andrady 1991; Barnes et al. 2009). If this causes death and loss of rafters, it may result in decolonization and resurfacing of the item (Ye and Andrady 1991), extending the life of plastic as a vector. The size of a particular piece of debris can also play a part in influencing the species richness and density of organisms rafting on it. Studies have shown a positive correlation between higher taxonomic richness and a larger surface area of plastic debris (Carson et al. 2013; Goldstein et al. 2014). However, this may be due to stochastic effects, biased sampling efforts (smaller items sink faster when colonized by fewer organisms) or other characteristics of the raft such as stability (Goldstein et al. 2014).

Although it is not expected that marine litter opens up novel pathways that are not available for other rafting materials (Lewis et al. 2005), it is more durable, more pervasive, and travels slower in comparison with vessel hulls, factors that might favor the survival of rafters (Barnes 2002). Therefore, the presence of plastic debris in the ocean might be adding another dimension to rafting and dispersal opportunities.

Today, we are familiar with calls to consider plastic as hazardous materials (Rochman et al. 2013), investing in better controls for waste management (European Commission 2018), and seeing strong lobbying in certain sectors of social media. As Rech et al. (2016) state, our main research priorities should center around estimating the impact of marine litter on NIS dispersal, and identifying sources and sinks by better understanding behavior of debris in ocean currents. Future research should consider unifying sampling methods to obtain comparable results and including base knowledge of local communities to better monitor arrivals of NIS while continuing our advance in taxonomic and genetic identification methods to be able to better identify species that might be cryptic or yet unknown to us (Carlton and Fowler 2018).

Finally, recognizing that the plastic problem is theoretically an avoidable one, research should be accompanied by management that aims in the direction of education and public awareness, the surveillance and protection of sink zones, and the reduction of production through taxation and banning.

10.2.6 Climate Change

Hellmann et al. (2008) identified possible ways in which climate change may affect NIS either directly or by influencing their competitors or dispersal: Firstly, climate change alters traits of habitats such as temperatures and CO_2 concentrations, which may reduce environmental constraints for marine invaders and diminish native species' competitiveness. Ultimately, this would increase the establishment rate of NIS in a new habitat. Secondly, climate change alters human-induced propagule pressure by affecting maritime tourism, cargo, and recreational activities. Finally, Hellmann et al. (2008) argued that climate change may lead to range-shifts of species, a trend that has been documented multiple times in the scientific literature, and which does not only affect NIS, but also native species (Sorte et al. 2010b; Carlton 2011; Wernberg et al. 2011; Canning-Clode and Carlton 2017; Martínez et al. 2018).

Although marine range-shifts occurs at a slower rate than marine introductions through anthropogenic vectors, the impacts on ecological communities in both scenarios can be very similar (Sorte et al. 2010a) and thus range-shifts due to human-induced climate change may be considered a type of anthropogenic introduction.

Climate change predictions include not only a change in the overall temperature but also the increasing climate variability (Rhein et al. 2013). Aperiodic cold snaps have been observed to reduce the number of invasive species (Canning-Clode et al. 2011). In this particular example, a cold snap in January 2010 in Florida, USA caused high mortalities of many marine organisms, among them the invasive porcelain crab *Petrolisthes armatus* (Firth et al. 2011; Kemp et al. 2011). Testing the survivability of *P. armatus* in cold water treatments, Canning-Clode et al. (2011) found that abnormal cold temperatures decrease the population of the invasive crab. Cold snaps limiting NIS might be relevant worldwide, but do not balance out climate change-induced range-shifts of NIS (Canning-Clode and Carlton 2017). In fact, individual examples show that NIS may expand to a broader distribution range after its population got reduced by a cold snap (Crickenberger and Moran 2013). Canning-Clode and Carlton (2017) assumed that NIS surges will eventually outnumber NIS setbacks along with predicted warming climate. This is underlined by several studies showing the beneficial impact of warmer water on NIS (Stachowicz et al. 2002; Sorte et al. 2010b; Kersting et al. 2015).

Stachowicz et al. (2002) found several benefits for non-native fouling species in warmer water temperatures. During a 10-year monitoring campaign, starting in 1991, they found a positive correlation between mean temperature and total recruitment of NIS, whereas the opposite trend was observed for native species. Additionally, non-native fouling species started their recruitment earlier in warmer waters, a remarkable advantage over native species. Stachowicz et al. (2002) also tested the growth of two non-native and one native ascidian species under different water temperatures resulting in faster growth of the former in warm water conditions. Sorte et al. (2010b) conducted mortality experiments with four native and seven non-native sessile species (bryozoans, colonial and solitary tunicates, and hydroids) in increased temperature treatments. They observed that the temperature at which only 50% of the species were alive is 3 °C higher for NIS than for native species, suggesting that NIS are more resistant to abnormally high temperatures.

Overall, there seems to be a trend of species shifting their ranges polewards along the continental coasts with proceeding climate change (Müller et al. 2009; Sorte et al. 2010a; Wernberg et al. 2011; Morley et al. 2018).

10.3 Forecasting

Of all transferred species, only a small number become truly invasive (see the "tens rule"; Williamson and Fitter 1996). Identifying the potential of an introduced species for dispersal and establishment can be useful in risk assessment. In this cause, Ecological Niche Modelling (ENM) and Species Distribution Models (SDM) can be of great help when predictions of species' potential distributions are needed. Conservation biology can, besides other applications (Guisan and Thuiller 2005; Gavin et al. 2014), profit from SDMs for risk assessment of invasions (Peterson 2003; Thuiller et al. 2005; Seebens et al. 2016). An ecological niche represents an *n*-dimensional (e.g., food-availability and temperature gradient) space in which a species can thrive (Hutchinson 1957). For distribution modelling, a model is usually calibrated on a species' niche and then projected onto the geographic space of interest. Here, the calibration process is conducted on available information of a species' known distribution and/or biological traits and the projection area is compared with the needs of a species. Like that, the suitability of an area can be evaluated and visualized. Calibration and projection can be done on historical and present-day data and allow predictions for simulated environmental conditions as, e.g., for future or past climate scenarios. The importance of invasion-risk assessments is underlined by Leung et al. (2002) who developed a bio-economic model as a framework to assess costs and benefits of invasions and their prevention efforts. Leung et al. (2002) demonstrated that investment in prevention over damage repair is to be preferred for society. For risk assessments, the recognition of suitable habitat of a species is of central interest. Hence, ENM is an important tool for policy makers to evaluate and to react to possible invasions before they can get economically or ecologically out of hand. Although ENM/SDM-related publications have become more and more abundant and yield valuable information for a diverse array of interests, there is a huge gap in the number of available publications between the terrestrial and the aquatic realm and between organizational organism levels. While a lot of studies are accessible for especially terrestrial higher plants, mammals, and birds, aquatic (small) taxa are still underrepresented (Soininen and Luoto 2014). Hence, methodological aspects in the following section are partly explained based on terrestrial studies. To understand the underlying concept of ecological niche modelling, Soberon and Peterson (2005) elaborated the work of Pulliam (2000) and presented the BAM-diagram. The BAM diagram consists of a set of suitable biotic, abiotic and accessible (movement) spaces. Thus, A represents the fundamental niche and the intersection of B and A represents the realized niche of a species. The fundamental niche is the space which can theoretically be inhabited by a species. Contrary, the realized niche represents the fundamental niche which is actually inhabited but truncated due to abiotic or biotic factors. M can contain naturally accessible regions as well as regions which are reachable through anthropogenic influence. Restrictions of M can be inherent (dispersal capacity of a species) or external and either of natural (e.g., land bridges) or artificial character (e.g., dams; Watters 1996, Ovidio and Philippart 2002). In the context of this paper, M (with respect to dispersal vectors) and A (with respect to changing climate) play major roles. Implementation of B (as biotic interactions) into models is still an area of investigation and rather case-specific than following established concepts.

10.3.1 Limitations of Models Through Knowledge Gaps

Distribution modelling is the projection of an identified niche from one geographic range to another under the presumption that species occupy the same conditions in both regions (Peterson and Vieglais 2001). Therefore, environmental data (predictor variables) in a species' distributional range is correlated with occurrence data (response variable). Nowadays, more and more databases are becoming available to provide researchers with valuable data for predictors (e.g., bio-oracle, worldclim, MerraClim) as well as distributional data (e.g., gbif, iobis) in addition to available primary sources (e.g., herbaria, museum collections, scientific reports, field guides, citizen science projects). Although correlative models have great predictive power, they can only identify the realized niche based on available distributional and environmental data. However, species are not necessarily in equilibrium with their environment and not all suitable environmental combinations might be represented in the distributional training range (Jackson and Overpeck 2000). Hence, these models may underestimate the fundamental niche of a species, leading to narrowed projections of suitable habitat (Kearney and Porter 2009; Martínez et al. 2015). Additionally, even in native distributional ranges parts of populations are in fact sink populations (Soberon and Peterson 2005) and might, therefore, reflect unsuitable environmental conditions for reproduction but suitable for survival. There are many other possible cases, in which the observed distribution of a species does not cover all suitable environmental conditions (e.g., sampling bias, seasonality, anthropogenic influences, and recent introduction). In any case, models based on unfilled niches could lead to erroneous assumptions on suitability of habitat for a given species (Peterson 2005). Likewise, projections into niche space beyond the identified realized niche can only be speculative. Therefore, models based on physiological knowledge are an important addition to classic correlative models (Kearney and Porter 2009). These mechanistic models make use of physiological knowledge (e.g., survival thresholds or performance over environmental gradients) to identify abiotically suitable spaces (Kearney and Porter 2009; Buckley et al. 2010; Diamond et al. 2012; Martínez et al. 2015). Identification of physiological limits is more laborious than correlative modelling, but these models are not subject to incomplete distribution data. In 2017, the GlobTherm database has been launched (Bennett et al. 2018), which includes experimentally determined thermal tolerances for more than 2000 aquatic and terrestrial species, providing a promising tool for future more holistic modelling approaches. While correlative models might assume too narrow niches, mechanistic models might, in ignorance of biological influences, assume too wide niches. The resulting discrepancies might bear the potential to investigate factors influencing the prevalence of a species and be useful in invasion risk assessment.

10.3.2 Invasions and Niche Shifts

It is a central assumption of SDM that species do occupy the same niche in their novel range as in their native range and across time periods. However, this has been subject of debates and evaluation studies in the past under the term "niche shift." This term implies changes in the realized niche of a species with respect to the centroid of the niche, the margins, and/or frequency of occupied environmental conditions (Guisan et al. 2014). Although studies have been published, which suggest niche shifts in invaded territories (e.g., Maron et al. 2004; Fitzpatrick et al. 2006; Broennimann et al. 2007), ecological niches have also been shown to be a rather conservative feature of species' and can be transferred to other than the native regions (Prinzing et al. 2001; Broennimann et al. 2007; Tingley et al. 2009; Petitpierre et al. 2012). Petitpierre et al. (2012) investigated 50 holarctic terrestrial plant species from herbs to trees and found niche expansion of more than 10% in the invaded range for only 14% of the studied species. Furthermore, they stated that genetic admixing (repeated introductions or hybridization) or reduced competition in the novel range do not automatically lead to substantial niche expansions. Ecological niches are even conserved over evolutionary time scales (i.e., several million years), as has been shown by Peterson et al. (1999). The authors built ecological niche models for 37 sister taxa of birds, mammals, and butterflies, and were able to reciprocally predict the geographical distribution of the respective sister taxon with high accuracy. Larger niche dissimilarity was found only on the higher taxonomic family level. Naturally, due to shared ancestry, niches of sister taxa tend to be highly similar. This was demonstrated by comparison to more distant taxa and to what can be expected from their environmental background alone (Warren et al. 2008). However, comparison of ecological niches of sister taxa and respective outlier groups does not necessarily indicate close phylogenetic relationship (Warren et al. 2008). Hence, not only phylogenetic relationship but also the environmental framework defines the species' niches. This is in line with Ackerly (2003): Species' niches are maintained throughout space and time and adaptation in specific traits seems to have a more stabilizing function in maintaining this niche. Adaptive evolution still may occur under the following scenarios: When a species colonizes islands in environmental space (not necessarily equivalent with geographical space), in trailing edge populations during migrations, or adaptation within the occupied niche space (due to environmental changes) (Ackerly 2003). To identify true niche shifts, Guisan et al. (2014) propose to build ENMs with gradually trimmed environmental data from the native and novel range and to investigate the effects of rare climatic conditions on resulting niche overlap metrics.

Large niche shifts can erroneously be assumed when species niches are derived from unfilled niches. For example, if a species occupies environmental conditions in a new geographic region (e.g., an invaded site) which are not found in its native range. In fact, populations rather persist at the edges than in the center of their historical distributional range (Lomolino and Channell 1995), and building a niche model on occurrence data from a certain time span can only yield a snapshot of the actual niche and might result in biased projections (Faurby and Araújo 2018). Including historical distributional and environmental data may be important to prevent modelling of biased niches. Also the findings of Peterson et al. (1999) might allow to considerably enhance the available information on tolerated environmental conditions by carefully including the realized niches of sister taxa. Distributional ranges and concomitant realized niches are massively narrowed through anthropogenic influences (e.g., extinctions or displacement; Lomolino and Channell 1995) but, in contrast, can become enormously enhanced through dispersal events (anthropogenically through increasing global trade and aquaculture, Ruiz et al. 1997, or naturally through drift, Waters 2008). An example of how unfilled niches may pose problems for accurate ecological niche modelling is given by Peterson (2005): He explained how non-equilibrium distributional data may lead to biased niche assumptions and hence underestimate a species' niche as in the case in Ganeshaiah et al. (2003). Ganeshaiah et al. (2003) modelled the ecological niche of the terrestrial sugarcane woolly aphid (*Ceratovacuna lanigera*) to predict its invasion potential across India. However, they used distributional records which were collected during the process of migration, and therefore could not cover the whole range of suitable environmental conditions. Thus, the suitable range was underestimated and Peterson (2005) suggested using native distributional data to train an ENM in order to capture the whole ecological niche for more realistic information on potential habitat.

While biotic interactions play a central role in ecology, they are only recently being integrated into niche modelling. Their former omission could be due to either the difficulties of implementing highly complex and dynamic biotic interactions, population and dispersal dynamics or because of the general assumption that rather abiotic (climatic) factors are the main drivers of species distributions (Woodward and Williams 1987). Models based on climatic variables alone have good predictive power, but biotic interactions can be of major importance (Araújo and Luoto 2007; Soininen et al. 2013). However, including biotic interactions does not necessarily lead to an increase in predictive power (Raath et al. 2018), but, in contrast, might even lead to a decrease (Silva et al. 2014). Nevertheless, they can be of great importance when it comes to predicting invasion success, as has been shown by Silva et al. (2014) for the crown-of-thorns sea star.

10.3.3 Assessing Invasions

Predicted high suitability of a certain range does not imply a risk of invasion. Species distribution models give information on how well a certain area matches the requirements of a certain species. Intact environments and species communities as well as geographic obstacles may prevent the invasion of a species. In fact, successful invasions often follow several failed introductions (Sax and Brown 2000) and although a potentially suitable habitat might be available, intrinsic and extrinsic factors are able to prevent a successful invasion. Firn et al. (2011) did not find evidence for a general pattern of higher abundance in non-native ranges of introduced species ("invasion paradox," Sax and Brown 2000). Only a small fraction of the 26 investigated species did show a higher abundance in the new range. The other species were as abundant as in their native ranges or less abundant.

Simple ecological niche models can give good estimations of the extent of range shifts under projected climate conditions. However, at finer scales, models can be improved by including high-detail data for dispersal capacity and land-use data when it comes to accurate local predictions (Fordham et al. 2018). Even if data on dispersal capacity cannot be included in a model, general predictions of invasion risk are possible. Thuiller et al. (2005) showed that invasion prediction based on climatic variables in combination with economic data such as tourism and trade intensity is a usable and important tool in identifying invasion risk between regions at a global scale. Tourism and trade were used as proxies for propagule pressure from source regions (namely, South Africa) to target regions. For risk assessment of marine bioinvasions via ports and main shipping routes, an intermediate distance between origin and recipient port of 8000 to 10,000 km seems to be significant for high-risk assignment (Seebens et al. 2013). Generally, main traffic highways across the oceans exist, of which some have higher invasion probabilities than others (e.g., between Asia & Europe and Asia & North-America). Seebens et al. (2016) successfully developed a model to predict migrations of marine algae-based only on occupied environmental conditions and marine traffic data. Furthermore, they used historical invasion data to identify the invasion risk for ecoregions around the world and were able to identify the respective invasion probability.

Predictive models are not equally good among taxa. Soininen et al. (2013) found that predictive power of species distribution models decreases with decreasing body size of the organism under investigation to exceptionally low values when compared to taxonomic groups of larger body size. This might be due to the fact that especially small planktonic taxa might be drifted to unsuitable habitats and therefore exhibit source and sink populations alike within their distributional range. Soininen and Luoto (2014) further investigated how species-specific traits can influence the predictive power of distribution models. They investigated 4911 AUC values ("Area under curve," an indicator of predictive power of a model) of 50 publications on taxonomically widespread organisms. One conclusion was that predictability increases with body size, which might be due to the fact that smaller organisms are more prone to colonization-extinction dynamics, fine-scale environmental fluctuations, and have less niche plasticity than larger taxa. Interestingly, they did not find a trend in predictability over dispersal mode (i.e., passive, non-flying active, and flying) and thereby underlined the findings of Kharouba et al. (2013). Furthermore, niches of organisms from lower trophic levels might be more reliably predicted by abiotic predictors alone than from higher trophic levels (Soininen and Luoto 2014). However, this notion could not be verified in an earlier comparative study by Huntley et al. (2004) on 306 higher plant, insect, and bird taxa.

10.4 Conclusions

Under increasing globalization and blue growth (i.e., marine cargo shipping, mariculture, oil and gas drilling, and deep-sea mining), human-mediated vectors will continuously homogenize marine species assemblages across biogeographical regions. Although ballast water is the only vector with a worldwide binding regulation, guidelines to face the threat of invasions are constantly improved and official frameworks are beginning to be implemented on a broader scale. For example, the European Union is now considering plastic pollution as a threat and carrying out mitigation strategies such as banning certain one-use products. An efficient starting point for the prevention of bionvasions could be to control the number of organisms attached to vessel hulls and in ballast water tanks prior the vessel's arrival in hub-ports.

Species Distribution Models (SDMs) have the potential to become powerful and valuable tools in identifying high-risk areas and species and developing mitigation strategies. To achieve this, we conclude that two things are necessary: quantifying vectors (e.g., wetted surface area (WSA) of ships), and gathering non-indigenous species records and making these publicly available. Moreover, quantifying human-mediated vectors may also facilitate the opportunity of performing holistic cross-vector comparisons. One example would be to compute WSA values and complexity degrees, not only for ships but also for floating plastic and oil platforms, which would enable realistic comparisons between these three vectors. At the onset of increased economic exploration in polar regions, it is imperative to push the understanding of bioinvasions and lower the risk of potential ecosystem shifts due to unintended species introductions.

Appendix

This article is related to the YOUMARES 9 conference session no. 13: "Higher temperatures and higher speed – Marine Bioinvasions in a changing world." The original Call for Abstracts and the abstracts of the presentations within this session can be found in the Appendix "Conference Sessions and Abstracts", Chapter "9 Higher temperatures and higher speed – Marine Bioinvasions in a changing world", of this book.

References

Ackerly DD (2003) Community assembly, niche conservatism, and adaptive evolution in changing environments. Int J Plant Sci 164:S165–S184

Alpert P, Bone E, Holzapfel C (2000) Invasiveness, invasibility and the role of environmental stress in the spread of non-native plants. Perspecti Plant Ecol Evol Syst 3:52–66

Anderson LG, Rocliffe S, Haddaway NR et al (2015) The role of tourism and recreation in the spread of non-native species: a systematic review and meta-analysis. PLoS One 10:e0140833

Andrady AL (2015) Persistence of plastic litter in the oceans. In: Bergmann M, Gutow L, Klages M (eds) Marine anthropogenic litter. Springer, Cham, pp 57–72

Araújo MB, Luoto M (2007) The importance of biotic interactions for modelling species distributions under climate change. Glob Ecol Biogeogr 16:743–753

Astudillo JC, Bravo M, Dumont CP et al (2009) Detached aquaculture buoys in the SE Pacific: potential dispersal vehicles for associated organisms. Aquat Biol 5:219–231

Barnes DK (2002) Biodiversity: invasions by marine life on plastic debris. Nature 416:808–809

Barnes DK, Galgani F, Thompson RC et al (2009) Accumulation and fragmentation of plastic debris in global environments. Philos Trans R Soc B Biol Sci 364:1985–1998

Barnes DK, Walters A, Gonçalves L (2010) Macroplastics at sea around Antarctica. Mar Environ Res 70:250–252

Bennett JM, Calosi P, Clusella-Trullas S et al (2018) GlobTherm, a global database on thermal tolerances for aquatic and terrestrial organisms. Sci Data 5:180022

Blackburn TM, Pyšek P, Bacher S et al (2011) A proposed unified framework for biological invasions. Trends Ecol Evol 26:333–339

Bollen M, Battershill CN, Pilditch CA et al (2017) Desiccation tolerance of different life stages of the invasive marine kelp *Undaria pinnatifida*: potential for overland transport as invasion vector. J Exp Mar Biol Ecol 496:1–8

Boudouresque CF, Verlaque M (2002) Biological pollution in the Mediterranean Sea: invasive versus introduced macrophytes. Mar Pollut Bull 44:32–38

Bravo M, Astudillo JC, Lancellotti D et al (2011) Rafting on abiotic substrata: properties of floating items and their influence on community succession. Mar Ecol Prog Ser 439:1–17

Broennimann O, Treier UA, Muller-Scharer H et al (2007) Evidence of climatic niche shift during biological invasion. Ecol Lett 10:701–709

Buckley LB, Urban MC, Angilletta MJ et al (2010) Can mechanism inform species' distribution models? Ecol Lett 13:1041–1054

Bull AS, Love MS (2019) Worldwide oil and gas platform decommissioning: a review of practices and reefing options. Ocean Coast Manage 168:274–306

Buschmann AH, Cabello F, Young K et al (2009) Salmon aquaculture and coastal ecosystem health in Chile: analysis of regulations, environmental impacts and bioremediation systems. Ocean Coast Manage 52:243–249

Cadée M (2003) Een vondst van de Atlantische Pareloester *Pinctada imbracata* (Röding, 1789) in een plastic fles op het Noordwijkse strand. Het Zeepard 63:76–78

Campbell ML, Keith I, Hewitt CL et al (2015) Evolving marine biosecurity in the Galapagos Islands. Manag Biol Invasion 6:227–230

Canning-Clode J, Carlton JT (2017) Refining and expanding global climate change scenarios in the sea: poleward creep complexities, range termini, and setbacks and surges. Divers Distrib 23:463–473

Canning-Clode J, Fowler AE, Byers JE et al (2011) 'Caribbean creep' chills out: climate change and marine invasive species. PLoS One 6:e29657

Carlton JT (1979) Introduced invertebrates of San Francisco Bay. In: Conomos TJ (ed) San Francisco, the urbanized estuary. American Association for the Advancement of Science, San Francisco, pp 427–444

Carlton JT (2011) Invertebrates, marine. In: Simberloff D, Rejmánek M (eds) Encyclopedia of biological invasions, vol 3. University of California Press, Berkeley

Carlton JT, Fowler AE (2018) Ocean rafting and marine debris: a broader vector menu requires a greater appetite for invasion biology research support. Aquat Invasions 13:11–15

Carpenter A, Macgill SM (2005) The EU Directive on port reception facilities for ship-generated waste and cargo residues: the results of a second survey on the provision and uptake of facilities in North Sea ports. Mar Poll Bullut 50:1541–1547

Carson HS, Nerheim MS, Carroll KA et al (2013) The plastic-associated microorganisms of the North Pacific gyre. Mar Pollut Bull 75:126–132

Casas G, Scrosati R, Piriz ML (2004) The invasive kelp *Undaria pinnatifida* (Phaeophyceae, Laminariales) reduces native seaweed diversity in Nuevo Gulf (Patagonia, Argentina). Biol Invasions 6:411–416

Choong HH, Calder DR (2013) *Sertularella mutsuensis* Stechow, 1931 (Cnidaria: Hydrozoa: Sertulariidae) from Japanese tsunami debris:

systematics and evidence for transoceanic dispersal. BioInvasions Rec 2:33.38

Cohen AN, Carlton JT (1998) Accelerating invasion rate in a highly invaded estuary. Science 279:555–558

Crickenberger S, Moran A (2013) Rapid range shift in an introduced tropical marine invertebrate. PLoS One 8:e78008

Diamond SE, Nichols LM, McCoy N et al (2012) A physiological trait-based approach to predicting the responses of species to experimental climate warming. Ecology 93:2305–2312

Endresen Ø, Sørgård E, Sundet JK et al (2003) Emission from international sea transportation and environmental impact. J Geophys Res 108:4560

European Commission (2018) A European strategy for plastics in a circular economy. European Commission, Brussels

FAO (2016) The state of world fisheries and aquaculture 2016. Contributing to food security and nutrition for all. FAO, Rome

Faurby S, Araújo MB (2018) Anthropogenic range contractions bias species climate change forecasts. Nat Clim Chang 8:252

Ferreira C, Gonçalves J, Coutinho R (2006) Ship hulls and oil platforms as potential vectors to marine species introduction. J Coas Res SI39:1340–1345

Firn J, Moore JL, MacDougall AS et al (2011) Abundance of introduced species at home predicts abundance away in herbaceous communities. Ecol Lett 14:274–281

Firth LB, Knights AM, Bell SS (2011) Air temperature and winter mortality: implications for the persistence of the invasive mussel *Perna viridis* in the intertidal zone of the South-Eastern United States. J Exp Mar Biol Ecol 400:250–256

Fitzpatrick MC, Weltzin JF, Sanders NJ et al (2006) The biogeography of prediction error: why does the introduced range of the fire ant over-predict its native range? Glob Ecol Biogeogr 16:24–33

Fleet D, Van Franeker JA, Dagevos Jet al (2009) Marine litter. Thematic report no. 3.8. In: Marencic H, CWSS (eds) Quality status report 2009. Wadden Sea Ecosystem no. 25. Common Wadden Sea Secretariat (CWSS), Wilhelmshaven, Germany

Floerl O, Pool TK, Inglis GJ (2004) Positive interactions between non-indigenous species facilitate transport by human vectors. Ecol Appl 14:1724–1736

Fordham DA, Bertelsmeier C, Brook BW et al (2018) How complex should models be? Comparing correlative and mechanistic range dynamics models. Glob Chang Biol 24:1357–1370

Forrest BM, Blakemore KA (2006) Evaluation of treatments to reduce the spread of a marine plant pest with aquaculture transfers. Aquaculture 257:333–345

Foster B, Willan R (1979) Foreign barnacles transported to New Zealand on an oil platform. New Zeal J Mar Fresh 13:143–149

Fowler AE, Blakeslee AM, Canning-Clode J et al (2016) Opening Pandora's bait box: a potent vector for biological invasions of live marine species. Divers Distrib 22:30–42

Friedlander AM, Ballesteros E, Fay M et al (2014) Marine communities on oil platforms in Gabon, West Africa: high biodiversity oases in a low biodiversity environment. PLoS One 9:e103709

Ganeshaiah K, Barve N, Nath N et al (2003) Predicting the potential geographical distribution of the sugarcane woolly aphid using GARP and DIVA-GIS. Curr Sci 85:1526–1528

García-Llorente M, Martín-López B, González JA et al (2008) Social perceptions of the impacts and benefits of invasive alien species: implications for management. Biol Conserv 141:2969–2983

Gavin DG, Fitzpatrick MC, Gugger PF et al (2014) Climate refugia: joint inference from fossil records, species distribution models and phylogeography. New Phytol 204:37–54

Godwin LS (2003) Hull fouling of maritime vessels as a pathway for marine species invasions to the Hawaiian islands. Biofouling 19:123–131

Goldstein MC, Rosenberg M, Cheng L (2012) Increased oceanic microplastic debris enhances oviposition in an endemic pelagic insect. Biol Lett 8:817–820

Goldstein MC, Carson HS, Eriksen M (2014) Relationship of diversity and habitat area in North Pacific plastic-associated rafting communities. Mar Biol 161:1441–1453

Gollasch S (2006) Overview on introduced aquatic species in European navigational and adjacent waters. Helgol Mar Res 60:84–89

Gregory MR (2009) Environmental implications of plastic debris in marine settings – entanglement, ingestion, smothering, hangers-on, hitch-hiking and alien invasions. Philos Trans R Soc B Biol Sci 364:2013–2025

Guisan A, Thuiller W (2005) Predicting species distribution: offering more than simple habitat models. Ecol Lett 8:993–1009

Guisan A, Petitpierre B, Broennimann O et al (2014) Unifying niche shift studies: insights from biological invasions. Trends Ecol Evol 29:260–269

Guppy HB (1917) Plants, seeds, and currents in the West Indies and Azores. Williams & Norgate, London

Guy-Haim T, Lyons DA, Kotta J et al (2018) Diverse effects of invasive ecosystem engineers on marine biodiversity and ecosystem functions: a global review and meta-analysis. Glob Chang Biol 24:906–924

Hamzah B (2003) International rules on decommissioning of offshore installations: some observations. Mar Policy 27:339–348

Hellmann JJ, Byers JE, Bierwagen BG et al (2008) Five potential consequences of climate change for invasive species. Conserv Biol 22:534–543

Huntley B, Green RE, Collingham YC et al (2004) The performance of models relating species geographical distributions to climate is independent of trophic level. Ecol Lett 7:417–426

Hutchinson GE (1957) Concluding remarks. Cold Spring Harb Symp Quant Biol 22:415–427

Iacarella JC, Davidson IC, Dunham A (2018) Biotic exchange from movement of 'static'maritime structures. Biol Invasions:1–11

IMO (2004) International convention for the control and management of ships' ballast water and sediments. International Maritime Organization, London

IMO (2011) Guidelines for the control and Management of Ship's biofouling to minimize the transfer of invasive aquatic species. International Maritime Organization, London

Jackson ST, Overpeck JT (2000) Responses of plant populations and communities to environmental changes of the late quaternary. Paleobiology 26:194–220

Jara C, Jaramillo E (1979) Hallazgo de Planes marinus Rathbun, 1914, sobre boya a la deriva en Bahía de Maiquillahue, Chile. (Crustacea. Decapoda, Grapsidae). Medio Ambiente (Chile) 4:108–113

Kauano RV, Roper JJ, Rocha RM (2017) Small boats as vectors of marine invasion: experimental test of velocity and desiccation as limits. Mar Biol 164:27

Kearney M, Porter W (2009) Mechanistic niche modelling: combining physiological and spatial data to predict species' ranges. Ecol Lett 12:334–350

Keller AA, Fruh EL, Johnson MM et al (2010) Distribution and abundance of anthropogenic marine debris along the shelf and slope of the US west coast. Mar Pollut Bull 60:692–700

Kemp DW, Oakley CA, Thornhill DJ et al (2011) Catastrophic mortality on inshore coral reefs of the Florida keys due to severe low-temperature stress. Glob Chang Biol 17:3468–3477

Kersting DK, Cebrian E, Casado Cet al (2015) Experimental evidence of the synergistic effects of warming and invasive algae on a temperate reef-builder coral. Sci Rep5:18635

Kharouba HM, McCune JL, Thuiller W et al (2013) Do ecological differences between taxonomic groups influence the relationship between species' distributions and climate? A global meta-analysis using species distribution models. Ecography 36:657–664

Kiessling T, Gutow L, Thiel M (2015) Marine litter as habitat and dispersal vector. In: Bergmann M, Gutow L, Klages M (eds) Marine anthropogenic litter. Springer, Cham, pp 141–181

Lentz SA (1987) Plastics in the marine environment: legal approaches for international action. Mar Pollut Bull 18:361–365
Leppäkoski E, Gollasch S, Olenin S (2013) Invasive aquatic species of Europe. Distribution, impacts and management. Springer, Dordrecht
Leung B, Lodge DM, Finnoff D et al (2002) An ounce of prevention or a pound of cure: bioeconomic risk analysis of invasive species. Proc R Soc Lond B Biol Sci 269:2407–2413
Lewis PN, Riddle MJ, Smith SD (2005) Assisted passage or passive drift: a comparison of alternative transport mechanisms for non-indigenous coastal species into the Southern Ocean. Antarct Sci 17:183–191
Liu T-K, Chang C-H, Chou M-L (2014) Management strategies to prevent the introduction of non-indigenous aquatic species in response to the ballast water convention in Taiwan. Mar Policy 44:187–195
Lomolino MV, Channell R (1995) Splendid isolation: patterns of geographic range collapse in endangered mammals. J Mammal 76:335–347
Lusher AL, Tirelli V, O'Connor I et al (2015) Microplastics in Arctic polar waters: the first reported values of particles in surface and sub-surface samples. Sci Rep 5:14947
Margolis M, Shogren JF, Fischer C (2005) How trade politics affect invasive species control. Ecol Econ 52:305–313
Maron JL, Vila M, Bommarco R et al (2004) Rapid evolution of an invasive plant. Ecol Monogr 74:261–280
Martínez B, Arenas F, Trilla A et al (2015) Combining physiological threshold knowledge to species distribution models is key to improving forecasts of the future niche for macroalgae. Glob Chang Biol 21:1422–1433
Martínez B, Radford B, Thomsen MS et al (2018) Distribution models predict large contractions of habitat-forming seaweeds in response to ocean warming. Diver Distrib 24:1350–1366
Masó M, Garcés E, Pagès F et al (2003) Drifting plastic debris as a potential vector for dispersing Harmful Algal Bloom (HAB) species. Sci Mar 67:107–111
McLaughlan C, Gallardo B, Aldridge D (2014) How complete is our knowledge of the ecosystem services impacts of Europe's top 10 invasive species? Acta Oecol 54:119–130
Miller AW, Ruiz GM (2014) Arctic shipping and marine invaders. Nat Clim Chang 4:413
Miller AW, Minton MS, Ruiz GM (2011) Geographic limitations and regional differences in ships' ballast water management to reduce marine invasions in the contiguous United States. Bioscience 61:880–887
Miller AW, Davidson IC, Minton MS et al (2018) Evaluation of wetted surface area of commercial ships as biofouling habitat flux to the United States. Biol Invasions 20:1977–1990
Ministry for Primary Industries (2014) Craft risk management standard: biofouling on vessels arriving to New Zealand. Ministry for Primary Industries, Wellington
Moore SL, Allen MJ (2000) Distribution of anthropogenic and natural debris on the mainland shelf of the Southern California bight. Mar Pollut Bull 40:83–88
Morley JW, Selden RL, Latour RJ et al (2018) Projecting shifts in thermal habitat for 686 species on the north American continental shelf. PLoS One 13:e0196127
Moser CS, Wier TP, Grant JF et al (2016) Quantifying the total wetted surface area of the world fleet: a first step in determining the potential extent of ships' biofouling. Biol Invasions 18:265–277
Müller R, Laepple T, Bartsch I et al (2009) Impact of oceanic warming on the distribution of seaweeds in polar and cold-temperate waters. Bot Mar 52:617–638
Myers JH, Simberloff D, Kuris AM et al (2000) Eradication revisited: dealing with exotic species. Trends Ecol Evol 15:316–320
Naylor RL, Williams SL, Strong DR (2001) Aquaculture – a gateway for exotic species. Science 294:1655–1656
NSPMMPI (2013) National biofouling management guidelines for the aquaculture industry. Department of Agriculture and Water Resources, Canberra
Ovidio M, Philippart J-C (2002) The impact of small physical obstacles on upstream movements of six species of fish. Hydrobiologia 483:55–69
Padayachee AL, Irlich UM, Faulkner KT et al (2017) How do invasive species travel to and through urban environments? Biol Invasions 19:3557–3570
Page HM, Dugan JE, Culver CS et al (2006) Exotic invertebrate species on offshore oil platforms. Mar Ecol Prog Ser 325:101–107
Pajuelo JG, González JA, Triay-Portella R et al (2016) Introduction of non-native marine fish species to the Canary Islands waters through oil platforms as vectors. J Mar Sys 163:23–30
Petersen JK, Svane I (1995) Larval dispersal in the ascidian *Ciona intestinalis* (L.). evidence for a closed population. J Exp Mar Biol Ecol 186:89–102
Peterson AT (2003) Predicting the geography of species' invasions via ecological niche modeling. Q R Biol 78:419–433
Peterson AT (2005) Predicting potential geographic distributions of invading species. Curr Sci 89:9
Peterson AT, Vieglais DA (2001) Predicting species invasions using ecological niche modeling: new approaches from bioinformatics attack a pressing problem: a new approach to ecological niche modeling, based on new tools drawn from biodiversity informatics, is applied to the challenge of predicting potential species' invasions. Bioscience 51:363–371
Peterson AT, Soberón J, Sánchez-Cordero V (1999) Conservatism of ecological niches in evolutionary time. Science 285:1265–1267
Petitpierre B, Kueffer C, Broennimann O et al (2012) Climatic niche shifts are rare among terrestrial plant invaders. Science 335:1344–1348
Pimentel D, Lach L, Zuniga R et al (2000) Environmental and economic costs of nonindigenous species in the United States. Bioscience 50:53–65
Pimentel D, McNair S, Janecka J et al (2001) Economic and environmental threats of alien plant, animal, and microbe invasions. Agric Ecosyst Environ 84:1–20
Piola RF, Johnston EL (2008) The potential for translocation of marine species via small-scale disruptions to antifouling surfaces. Biofouling 24:145–155
PlasticsEurope (2013) Plastics – the facts 2013: an analysis of European latest plastics production. Available at:www.plasticseurope.org
Prinzing A, Durka W, Klotz S et al (2001) The niche of higher plants: evidence for phylogenetic conservatism. Proc R Soc Lond B Biol Sci 268:2383–2389
Pulliam HR (2000) On the relationship between niche and distribution. Ecol Lett 3:349–361
Raath MJ, le Roux PC, Veldtman R et al (2018) Incorporating biotic interactions in the distribution models of African wild silk moths (*Gonometa* species, Lasiocampidae) using different representations of modelled host tree distributions. Austral Ecol 43:316–327
Raffo MP, Eyras MC, Iribarne OO (2009) The invasion of *Undaria pinnatifida* to a *Macrocystis pyrifera* kelp in Patagonia (Argentina, south-west Atlantic). J Mar Biol Assoc UK 89:1571–1580
Rech S, Borrell Y, García-Vazquez E (2016) Marine litter as a vector for non-native species: what we need to know. Mar Pollut Bull 113:40–43
Reusser DA, Lee I, Frazier M et al (2013) Per capita invasion probabilities: an empirical model to predict rates of invasion via ballast water. Ecol Appl 23:321–330
Rhein M, Rintoul SR, Aoki S, Campos E et al (2013) Observations: ocean pages. In: Stocker TF, Quin D, Plattner GK et al (eds) Climate change 2013: the physical science basis. Contribution of working group I to the Fifth Assessment Report of the Intergovernmental Panel on Climate Change. Cambridge University Press, Cambridge

Ricciardi A, Hoopes MF, Marchetti MP et al (2013) Progress toward understanding the ecological impacts of nonnative species. Ecol Monogr 83:263–282

Richardson DM, Pyšek P (2006) Plant invasions: merging the concepts of species invasiveness and community invasibility. Prog Phys Geogr 30:409–431

Robertson DR, Dominguez-Dominguez O, Victor B et al (2018) An Indo-Pacific damselfish (*Neopomacentrus cyanomos*) in the Gulf of Mexico: origin and mode of introduction. PeerJ 6:e4328

Rochman CM, Browne MA, Halpern BS et al (2013) Policy: classify plastic waste as hazardous. Nature 494:169

Ruiz GM, Carlton JT, Grosholz ED et al (1997) Global invasions of marine and estuarine habitats by non-indigenous species: mechanisms, extent, and consequences. Am Zool 37:621–632

Sax DF, Brown JH (2000) The paradox of invasion. Glob Ecol Biogeogr 9:363–371

Schiller J, Lackschewitz D, Buschbaum C et al (2018) Heading northward to Scandinavia: *Undaria pinnatifida* in the northern Wadden Sea. Bot Mar 61:365–371

Schlining K, Von Thun S, Kuhnz L et al (2013) Debris in the deep: using a 22-year video annotation database to survey marine litter in Monterey canyon, Central California, USA. Deep Sea Res 79:96–105

Scott G (1972) Plastics packaging and coastal pollution. Int J Environ Stud 3:35–36

Seebens H, Gastner MT, Blasius B (2013) The risk of marine bioinvasion caused by global shipping. Ecol Lett 16:782–790

Seebens H, Schwartz N, Schupp PJ et al (2016) Predicting the spread of marine species introduced by global shipping. Proc Natl Acad Sci U S A 113:5646–5651

Sheavly SB (2005) Sixth meeting of the UN open-ended informal consultative processes on oceans & the law of the sea marine debris-an overview of a critical issue for our oceans.The Ocean Conservancy, June 6–10 2005

Shenkar N, Rosen D (2018) How has the invention of the shipping container influenced marine bioinvasion? Manag Biol Invasion 9:187–194

Silva DP, Gonzalez VH, Melo GAR et al (2014) Seeking the flowers for the bees: integrating biotic interactions into niche models to assess the distribution of the exotic bee species *Lithurgus huberi* in South America. Ecol Model 273:200–209

Soberon J, Peterson AT (2005) Interpretation of models of fundamental ecological niches and species' distributional areas. Biodivers Inform 2:1–10

Soininen J, Luoto M (2014) Predictability in species distributions: a global analysis across organisms and ecosystems. Glob Ecol Biogeogr 23:1264–1274

Soininen J, Korhonen JJ, Luoto M (2013) Stochastic species distributions are driven by organism size. Ecology 94:660–670

Sorte CJ, Williams SL, Carlton JT (2010a) Marine range shifts and species introductions: comparative spread rates and community impacts. Glob Ecol Biogeogr 19:303–316

Sorte CJ, Williams SL, Zerebecki RA (2010b) Ocean warming increases threat of invasive species in a marine fouling community. Ecology 91:2198–2204

Stachowicz JJ, Terwin JR, Whitlatch RB et al (2002) Linking climate change and biological invasions: ocean warming facilitates nonindigenous species invasions. Proc Nat Acad Sci USA 99:15497–15500

Stevens LM, Gregory MR, Foster BA (1996) Fouling Bryozoa on pelagic and moored plastics from northern New Zealand bryozoans in space and time. In: Gordon DP, Smith AM, Grant-Mackie JA (eds) Bryozoans in space and time. In: Proceedings of the 10th international bryozoology conference. NIWA, Wellington, p 321

Thiel M, Gutow L (2005) The ecology of rafting in the marine environment. II. The rafting organisms and community. Oceanogr Mar Biol Annu Rev 43:279–418

Thuiller W, Richardson DM, Pyšek P et al (2005) Niche-based modelling as a tool for predicting the risk of alien plant invasions at a global scale. Glob Chang Biol 11:2234–2250

Tingley MW, Monahan WB, Beissinger SR et al (2009) Birds track their Grinnellian niche through a century of climate change. Proc Nat Acad Sci USA 106:19637–19643

Todd VL, Lavallin EW, Macreadie PI (2018) Quantitative analysis of fish and invertebrate assemblage dynamics in association with a North Sea oil and gas installation complex. Mar Environ Res 142:69–79

Van Franeker JA, Blaize C, Danielsen J et al (2011) Monitoring plastic ingestion by the northern fulmar *Fulmarus glacialis* in the North Sea. Environ Pollut 159:2609–2615

Vandendriessche S, Vincx M, Degraer S (2007) Floating seaweed and the influences of temperature, grazing and clump size on raft longevity—a microcosm study. J Exp Mar Biol Ecol 343:64–73

Verling E, Ruiz GM, Smith LD et al (2005) Supply-side invasion ecology: characterizing propagule pressure in coastal ecosystems. Proc R Soc Lond B Biol Sci 272:1249–1257

Wanless RM, Scott S, Sauer WH et al (2010) Semi-submersible rigs: a vector transporting entire marine communities around the world. Biol Invasions 12:2573–2583

Warren DL, Glor RE, Turelli M (2008) Environmental niche equivalency versus conservatism: quantitative approaches to niche evolution. Evolution 62:2868–2883

Waters JM (2008) Driven by the westwind drift? A synthesis of southern temperate marine biogeography, with new directions for dispersalism. J Biogeogr 35:417–427

Watters GT (1996) Small dams as barriers to freshwater mussels (Bivalvia, Unionoida) and their hosts. Biol Conserv 75:79–85

Watters DL, Yoklavich MM, Love MS et al (2010) Assessing marine debris in deep seafloor habitats of California. Mar Pollut Bull 60:131–138

Weigle SM, Smith LD, Carlton JT et al (2005) Assessing the risk of introducing exotic species via the live marine species trade. Conserv Biol 19:213–223

Wernberg T, Russell Bayden D, Thomsen Mads S et al (2011) Seaweed communities in retreat from ocean warming. Curr Biol 21:1828–1832

Wichmann C-S, Hinojosa IA, Thiel M (2012) Floating kelps in Patagonian Fjords: an important vehicle for rafting invertebrates and its relevance for biogeography. Mar Biol 159:2035–2049

Williams SL, Smith JE (2007) A global review of the distribution, taxonomy, and impacts of introduced seaweeds. Annu Rev Ecol Evol Syst 38:327–359

Williams SL, Davidson IC, Pasari JR et al (2013) Managing multiple vectors for marine invasions in an increasingly connected world. Bioscience 63:952–966

Williamson M, Fitter A (1996) The varying success of invaders. Ecology 77:1661–1666

Winston JE, Gregory MR, Stevens LM (1997) Encrusters, epibionts, and other biota associated with pelagic plastics: a review of biogeographical, environmental, and conservation issues. In: Coe JM, Rogers DB (eds) Marine Debris. Springer, New York, pp 81–97

Woodward FI, Williams B (1987) Climate and plant distribution at global and local scales. Vegetatio 69:189–197

Wotton DM, O'Brien C, Stuart MD et al (2004) Eradication success down under: heat treatment of a sunken trawler to kill the invasive seaweed *Undaria pinnatifida*. Mar Pollut Bull 49:844–849

Ye S, Andrady AL (1991) Fouling of floating plastic debris under Biscayne Bay exposure conditions. Mar Pollut Bull 22:608–613

Yeo DC, Ahyong ST, Lodge DM et al (2009) Semisubmersible oil platforms: understudied and potentially major vectors of biofouling-mediated invasions. Biofouling 26:179–186

Zenni RD, Nuñez MA (2013) The elephant in the room: the role of failed invasions in understanding invasion biology. Oikos 122:801–815

Zettler ER, Mincer TJ, Amaral-Zettler LA (2013) Life in the "plastisphere": microbial communities on plastic marine debris. Environ Sci Technol 47:7137–7146

Zhang X, Bai M, Tian Y et al (2017) The estimation for ballast water discharged to China from 2007 to 2014. Mar Pollut Bull 124:89–93

Benthos-Pelagos Interconnectivity: Antarctic Shelf Examples

11

Santiago E. A. Pineda-Metz

Abstract

This review focuses on studies dealing with the coupling between the benthic and pelagic realms on Antarctic shelves and on factors that regulate these processes. Such studies in Antarctic waters are scarce, especially on the shelves, where flux studies via moorings are highly endangered by drifting icebergs. Nevertheless, such studies are essential to understand these processes and functioning of the cold water ecosystem and how energy is transported through its different compartments. Different abiotic (e.g., currents, sea ice, water depth, topography of the seafloor, seasonality) and biotic (e.g., composition and structure of the benthic and pelagic flora and fauna, primary production, vertical migrations) factors are presented as parameters regulating the coupling between benthos and pelagos, here defined as benthos-pelagos interconnectivity. Regional variability in these parameters may result in delayed or even different coupling and/or decoupling of these realms. This is exemplarily discussed comparing the West Antarctic Peninsula (WAP) and Eastern Weddell Sea Shelf (EWSS). While in the WAP both compartments appear decoupled, on the EWSS both compartments appear tightly connected. The development of the benthos in the Larsen embayments after the shelf ice disintegration is described as an example of how changes in the pelagic realm affect and modify also the benthic realm.

Keywords

Bentho-pelagic coupling · Pelago-benthic coupling · Carbon flux · Weddell Sea Shelf · Antarctic Peninsula shelf

S. E. A. Pineda-Metz (✉)
Alfred-Wegener-Institut Helmholtz-Zentrum für Polar- und Meeresforschung, Bremerhaven, Germany

Universität Bremen (Fachbereich 2 Biologie/Chemie), Bremen, Germany
e-mail: santiago.pineda.metz@awi.de

S. Jungblut et al. (eds.), *YOUMARES 9 - The Oceans: Our Research, Our Future*,
https://doi.org/10.1007/978-3-030-20389-4_11

11.1 Bentho-Pelagic or Pelago-Benthic Coupling? A Short Introduction

When thinking of biotic (e.g., diversity, abundance, biomass) and abiotic (e.g., particle concentration, sediment grain size) parameters of both, benthic and pelagic realms, we start noticing lines or processes connecting them. One of the first studies on this connectivity was that of Hargrave (1973). He pointed out that both realms are connected by the flow of matter, especially that of carbon. Since that study, this interconnection between benthos and pelagos has been referred to as bentho-pelagic or pelago-benthic coupling. While the terms bentho-pelagic and pelago-benthic appear exchangeable, each one alludes to the predominant or driving component and direction in the coupling (Renaud et al. 2008). In bentho-pelagic coupling, it is the benthos which modifies or influences the pelagos. Contrastingly, in pelago-benthic coupling it is the pelagos which influences or modifies the benthos. In some literature bentho-pelagic coupling is referred to as "upward" coupling, while pelago-benthic coupling is referred to as "downward" coupling (e.g., Smith et al. 2006).

With this review, I aim to exemplify in a concise and simple way how benthos-pelagos interconnectivity, i.e., upward and downward coupling, works in the Southern Ocean with special focus on Antarctic shelf ecosystems (Fig. 11.1). My second aim is to enable non-experts to get a rough picture of the Antarctic benthos-pelagos interconnectivity.

11.1.1 Pelago-benthic Coupling

The first approaches used to describe the coupling between pelagos and benthos included measurements of carbon input from the water column to calculate how much of this carbon was assimilated in the sediment (Hargrave 1973). Currently, studies of downward mass flux are still the most common type of coupling studies (e.g., Cattaneo-Vietti et al. 1999; Smith et al. 2006, 2008; Isla et al. 2006a, b, 2011). Other

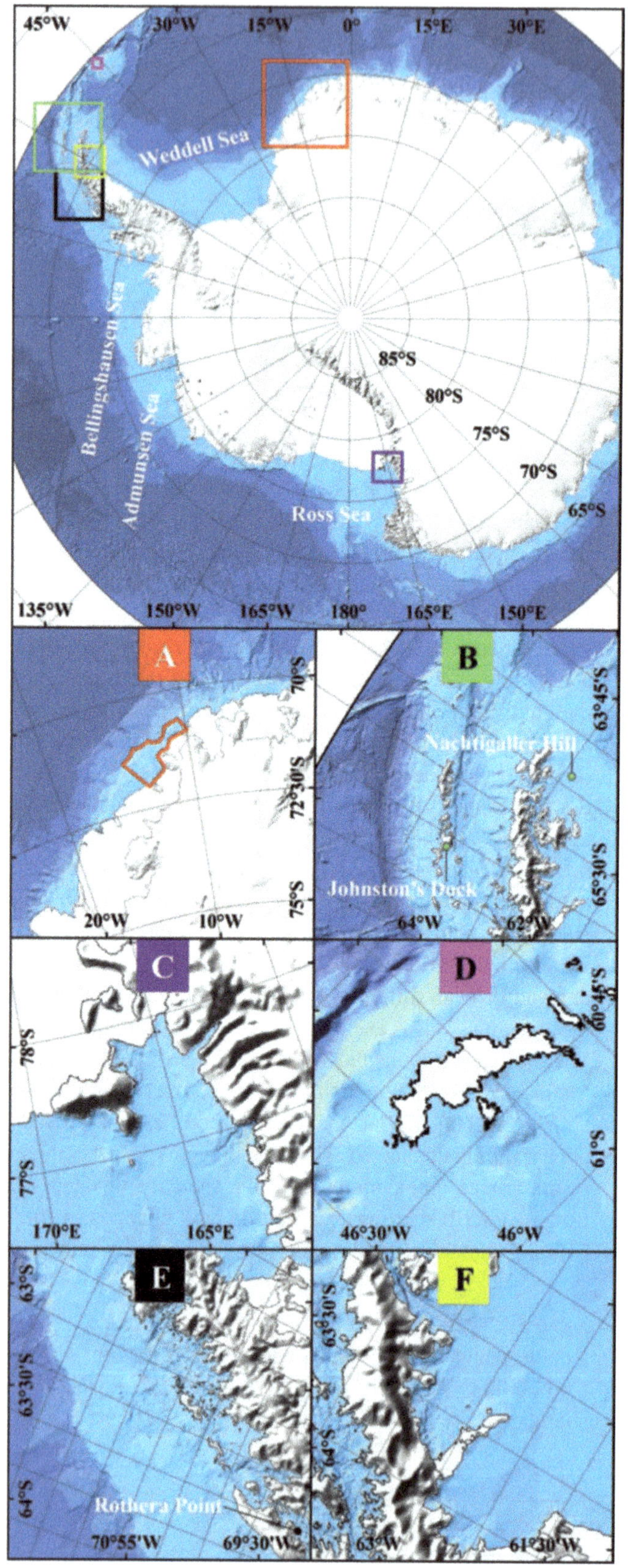

Fig. 11.1 Map of the Antarctic continent including locations mentioned in the review. (**a**) Austasen and Kapp Norvegia, EWSS; (**b**) Bransfield Strait and tip of the Antarctic Peninsula; (**c**) McMurdo Sound, Ross Sea; (**d**) Signy and Orcadas Islands; (**e**) Rothera Point and area studied within the frame of the Food for Benthos on the Antarctic Continental Shelf (FOODBANCS) project in the West Aantarctic Peninsula; and (**f**) Larsen embayments, east coast of the Antarctic Peninsula. (Modified after Arndt et al. (2013))

approaches to study pelago-benthic coupling include recruitment of benthic organisms via meroplanktonic larvae (Bowden 2005), change of sediment characteristics (Collier et al. 2000; Hauck et al. 2012; Isla 2016b), pelagic characteristics and seasonal patterns and how these affect benthic processes such as feeding activity (Barnes and Clarke 1995; McClintic et al. 2008; Souster et al. 2018), reproduction (Pearse et al. 1991; Stanwell-Smith et al. 1999; Brockington et al. 2001; Galley et al. 2005), growth rates and carbon fixed by benthos (Dayton 1989; Brey and Clarke 1993; Clarke 2003; Barnes et al. 2006, 2016, 2018; Barnes 2015), and benthic distribution patterns (Barry 1988; Barry and Dayton 1988; Graf 1989; Bathmann et al. 1991; Gutt et al. 1998; Sumida et al. 2008; Segelken-Voigt et al. 2016; Jansen et al. 2018).

11.1.2 Bentho-pelagic Coupling

Less common than pelago-benthic coupling studies are those that show an effect from the benthos to the pelagos, i.e., a bentho-pelagic coupling. One clear example of this "upward" coupling is the regulation of particulate matter flow in the benthic boundary layer by means of benthic structures (Graf and Rosenberg 1997; Mercuri et al. 2008; Tatián et al. 2008); another example of these processes is the increase of abundance and diversity of plankton by the release of meroplanktonic larvae from benthic organisms into the water (Bowden 2005; Schnack-Schiel and Isla 2005). Benthic processes also create feeding grounds for birds, seals, and zooplankton (Arntz et al. 1994; Ligowski 2000; Schmidt et al. 2011), they enhance primary production through export of micronutrients from remineralization and consumption/excretion processes of pelagic communities (Doering 1989; Smith et al. 2006; Schmidt et al. 2011), and can regulate the chemical characteristics of the water column (Doering 1989; Sedwick et al. 2000; Tatián et al. 2008).

11.2 Regulating Factors of Benthic and Pelagic Processes

In general terms, the interconnectivity between benthos and pelagos could be regarded as "weak" or "strong." This alludes to how directly changes in pelagos are reflected in benthos and vice versa. When seen as a correlation, it would be how strong the correlation between compartments is. The strength of the coupling between benthos and pelagos depends on seasonality in both compartments, the ecology and structure of benthic and pelagic communities, water depth, seafloor topography, water circulation (e.g., tides and currents), and wind, all affecting the transport of particles and thus carbon flux from one compartment to the other. Around the Antarctic continent, another factor playing a major role for the regulation of this coupling between benthos and pelagos is the influence of ice in any of its forms (e.g., sea ice and disintegrated shelf ice, i.e., icebergs).

11.2.1 Sea Ice

The Southern Ocean is characterized by its large extension of sea ice, which covers up to 20×10^6 km^2 during Austral winter and 4×10^6 km^2 during summer (Fig. 11.2), making sea ice-associated ecosystems one of the most dynamic and largest ecosystems on Earth (Arrigo et al. 1997; Thomas and Dieckmann 2002; Michels et al. 2008). The retreat of sea ice during summer increases the water column stability, seeds summer phytoplankton blooms, and works as a source for micronutrients such as iron (as well as other particles), favoring phytoplankton blooms and explaining the higher productivity near sea ice edges as compared to open waters (Clarke 1988; Sedwick and DiTullio 1997; Sedwick et al. 2000; Kang et al. 2001; Donnelly et al. 2006). It has been shown that reduction of the sea ice duration also contributes to an increase of carbon drawdown by benthic organisms (Barnes 2015).

Sea ice starts growing during March to its enormous extension in Austral winter. The high coverage of sea ice and snow during winter time diminishes the light entering the water column, thus causing a drastic decrease in local productivity and particle flux (Scharek et al. 1994; Isla et al. 2006a). However, autotrophic plankton entrapped by sea ice during its formation (along with nutrients and consumers) continues primary production in winter time, which can be four to five times higher than water column production (Garrison and Close 1993). While lower than summer production, sea ice primary production has been pointed out to serve as a possible food source for meroplanktonic larvae (Bowden 2005) and various krill life stages (Nicol 2006; Kohlbach et al. 2017; Schaafsma et al. 2017). These few examples show how the sea ice summer/winter cycle regulates primary and secondary production in the water column and the particle flux, thus directly influencing the benthos-pelagos interconnectivity.

11.2.2 Depth, Topography, Currents, and Wind

One conspicuous aspect of the Antarctic shelf is its depth. While other shelf ecosystems in the world are shallower (down to around 200 m depth), the isostatic pressure generated by the ice cap on the Antarctic continent deepens the surrounding shelf down to 400–600 m and even down to 800–1000 m in some regions (Gallardo 1987; Smith et al. 2006; Sumida et al. 2008). Smith et al. (2006) pointed out

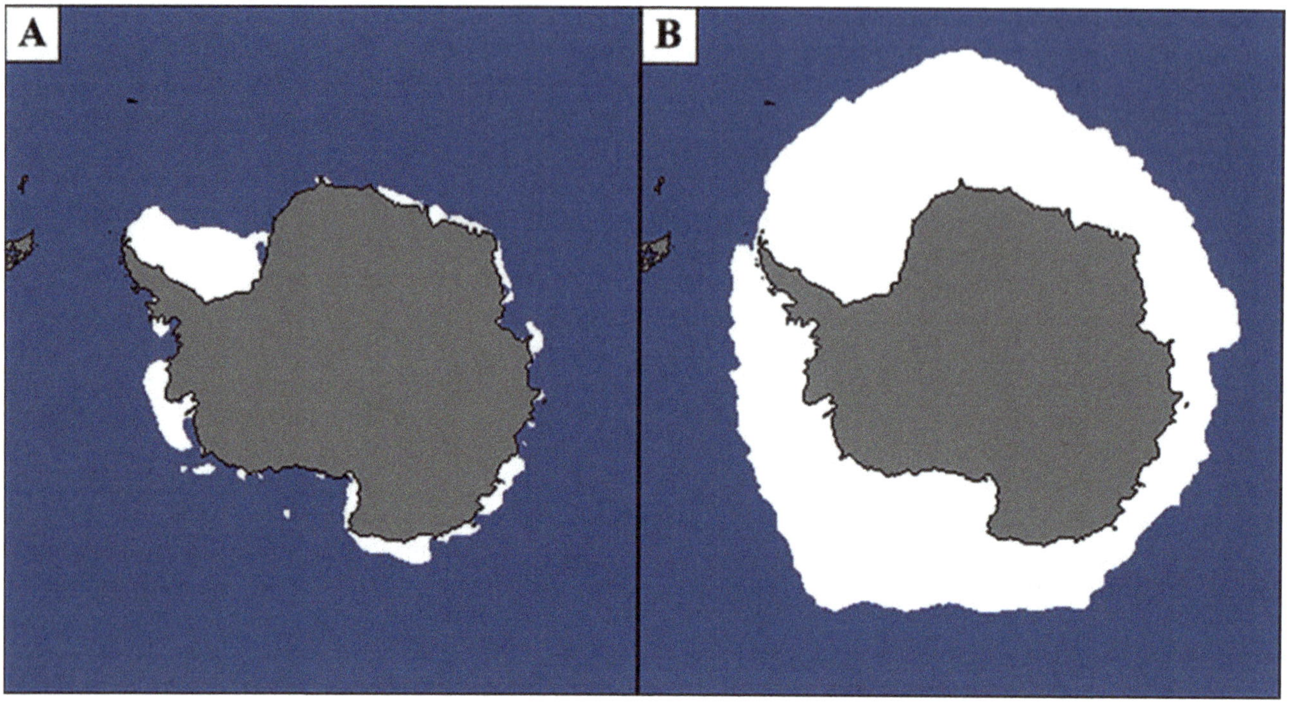

Fig. 11.2 Examples of sea ice extension during (**a**) summer (February 2018) and (**b**) winter (August 2018). (Modified after Fetterer et al. (2018))

that the increased depth of the Antarctic shelf with its complex topography and current systems may reduce the strength of the coupling by increasing the time particles spent in the water column, allowing local characteristics of the benthic habitat to mask the pelagic signals on the seafloor. However, the effect of depth on particle receding time in the water column will depend on the nature of the particles, e.g., on their flocculation ability, and other environmental factors such as wind forcing, which regulates deposition or advection of particles (biological factors are treated later). For the Eastern Weddell Sea Shelf (EWSS; Fig. 11.1a), it has been described that particle flux is rather fast. Total mass fluxes measured at mid-water and near the seafloor with sediment traps appeared to be similar, and it has been noted that particles can reach the seafloor within days despite the long 400–600 m depth trip from the euphotic zone to the seafloor (Bathmann et al. 1991; Isla et al. 2006a, 2009). For the Ross Sea, while Dunbar et al. (1998) recorded mean settling velocities of 176–245 m d^{-1} for different types of fecal pellets, DiTullio et al. (2000) found aggregates of *Phaeocystis antarctica* to sink at speeds >200 m d^{-1}, i.e., it could take 1–3 days for pellets or *Phaeocystis* aggregates to reach the seafloor.

The topography of the shelf influences the benthos-pelagos interconnectivity as well. Topography affects benthic distribution patterns and the transport and deposition of particles suspended in the water column alike. Dorschel et al. (2014) pointed out that topographic features such as range hills, mounds, and seamounts modify water current pathways and their strength. Their study of the benthos at Nachtigaller Hill (Fig. 11.1b) at the tip of the Antarctic Peninsula described depth as one main factor explaining benthic distribution patterns. They related this to food availability for the benthos, which could have been enhanced by the topography of Nachtigaller Hill. Another topographic feature affecting water currents is the width of the shelf. Along wider shelves the currents tend to be weaker; stronger currents are more usual when the shelf is narrow. Gutt et al. (1998) found relatively weaker current regimes on wider shelves of the EWSS to be beneficial for particle settling, which in turn benefits deposit-feeding organisms. Conversely, the narrower areas off Austasen and Kapp Norvegia (Fig. 11.1a) on the EWSS generate relatively stronger currents promoting resuspension of particles and thus being favorable for suspension feeder-dominated community types (Gutt et al. 1998).

Currents, tides, and advection of water parcels on the shelf also play a role in the benthos-pelagos interconnectivity. In some cases they weaken; in others they mask coupling processes between the compartments. An example can be drawn from the study of Isla et al. (2006b) at Johnston's Dock (Fig. 11.1b), where water current induced transport and advection of particles from shallower shelf areas enhance particle flux to deeper parts (Fig. 11.3). Other studies conducted in waters of the West Antarctic Peninsula (WAP) found particle flux on the deeper shelf to be enhanced by advected material originating from shallower shelves. This allochthonous input weakens the connection between benthic distribution patterns and metabolism of benthic organ-

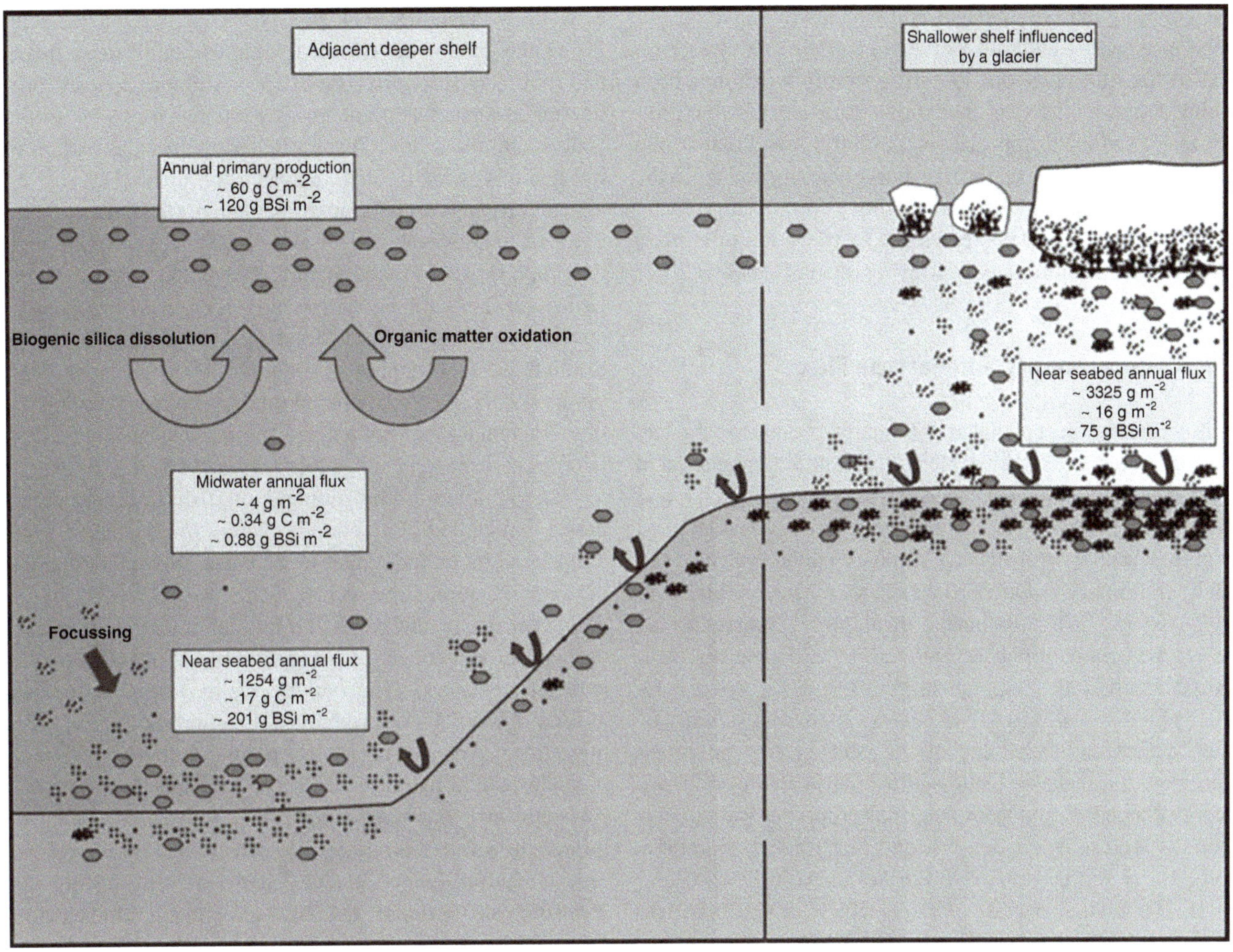

Fig. 11.3 Main particle fluxes at mooring sites around Johnston's Glacier (Johnston's Dock) studied by Isla et al. (2006b). Approximate annual total mass (g m^{-2}), organic carbon (g C m^{-2}), and biogenic silica (g BSi m^{-2}) values are given. The sketch shows that most particles produced offshore over the deep shelf (polygons) do not reach mid-water; the material settling in the shallower shelf feeds the deeper shelf via advection. Glacier and floating icebergs deliver coarse and fine sediments (dense clusters and circles, respectively) onto shallow areas, but mostly the latter reaches the deeper shelf. Near the seabed, resuspension of sediments is represented by curved arrows. (Modified after Isla et al. (2006b) with permission from Springer)

isms with primary production and local input of particles (McClintic et al. 2008; Sumida et al. 2008). Another clear example of the role of currents in the benthos-pelagos interconnectivity can be found in McMurdo Sound (Fig. 11.1c). Barry (1988) and Barry and Dayton (1988) found benthic distribution patterns to be coupled with primary production regimes and water circulation patterns. Circulation on the eastern side of McMurdo has a southward direction toward the Ross Ice Shelf and transports productive waters, which fuel rich benthic communities, whereas on the western side of the Sound, where less productive waters arrive from the ice shelf, a poorer benthic community is found.

Wind affects directly the benthos-pelagos interconnectivity by partly regulating sea ice and polynya formation, sea ice displacement, and mixed layer depth. While during winter periods, cold winds absorb heat from the water surface enhancing sea ice formation, in summer periods strong winds push away sea ice forming coastal polynyas (Isla 2016a). Wind-driven dispersal of the sea ice prior to its melting can prevent local release of algae trapped in the ice which would normally seed a local bloom (Riebesell et al. 1991). Furthermore, the strength of wind can also regulate the depth of the mixing layer in both a beneficial and prejudicial way. Where winds are relatively weaker, a shallower mix layer is formed (especially close to the ice edge). This shallower mix layer can foster larger blooms than deeper mixed layers (Ducklow et al. 2006). Conversely, in areas where winds are relatively stronger, a deeper mix layer is found. Deeper mixed layers can abruptly interrupt phytoplankton blooms, thus inhibiting primary production (Gleitz et al. 1994; Dunbar et al. 1998; Ducklow et al. 2006). While the deepening of the mix layer by wind action appears prejudicial for the coupling between pelagos and benthos by reducing primary production and thus its related particle flux, a deepening of the mix layer due to strong stormy winds has been pointed out to increase total downward particle flux. By means of sediment traps, Isla et al. (2009) found that strong stormy winds enhanced the transport of organic matter to the seabed. In their study, the flux resulting from a storm event which lasted a few days represented 53% of the total mass flux collected at mid-water during a period of 30 days.

11.2.3 Seasonality and Particle Flux

It is commonly accepted that the Antarctic benthic realm can be considered as a rather stable system with little variation in environmental parameters such as temperature, salinity, and water currents, whereas the pelagic realm is considered as highly seasonal with distinct summer/winter cycles, especially in primary production and sea ice extension (Gallardo 1987; Clarke 1988; Bathmann et al. 1991; Scharek et al. 1994; Arntz et al. 1994; Arrigo et al. 1998; Palanques et al. 2002; Smith et al. 2006; Isla et al. 2009, 2011; Rossi et al. 2013; Flores et al. 2014; Isla 2016b). While the stability of the benthos and instability of the pelagos are commonly accepted, the intrinsic biotic and abiotic factors of both are highly dependent on local water mass properties and circulation, as well as wind, sea ice, and topographic conditions (e.g., Barry and Dayton 1988; Barthel and Gutt 1992; Gleitz et al. 1994; Dunbar et al. 1998; Ducklow et al. 2006; Isla et al. 2009; Hauck et al. 2010; Barnes 2015).

11.2.3.1 Pelagic Realm

Primary production in the water column is key in regulating the flux of particles. Most of the primary production is proposed to be generated within the seasonal sea ice zone, especially in waters close to the retreating sea ice edge, where water column stability and nutrient concentrations are high. Driven by melting of sea ice, these locations also act as seeding grounds for primary production in the euphotic zone, enabled by released sea ice algae and enhanced input of nutrients (Scharek et al. 1994; Sedwick and DiTullio 1997; Sedwick et al. 2000; Arrigo et al. 2008; Bertolin and Schloss 2009; Isla et al. 2009; Isla 2016b). The primary production in the seasonal sea ice zone was estimated to be 1300 Tg C y^{-1}, of which 420 Tg C y^{-1} are generated in the marginal sea ice zone and roughly 5% of production of the seasonal sea ice zone is produced by sea ice algae (Lizotte 2001). The importance of primary production regulating particle fluxes matches with zooplankton activities, because zooplankton quickly reacts to phytoplankton blooms (Flores et al. 2014). Grazing pressure is one of the main regulators of phytoplankton blooms. Fecal pellets resulting from this grazing largely contribute and regulate particle fluxes (Bathmann et al. 1991; Palanques et al. 2002; Isla et al. 2009; Rossi et al. 2013) and change the chemical composition of these fluxes and their size structure (Isla 2016b). Summer primary production and zooplanktonic grazing amount for >95% of the yearly total mass flux. This particle flux provides carbon to the benthos, which equals between <1 up to 18% of the annual primary production of a region (Bathmann et al. 1991; Palanques et al. 2002; Isla et al. 2006a, 2009). Although the proportion of carbon reaching the seafloor appears negligible to low, it is still enough to support biomass-rich benthic communities and to form "food banks" (Gutt et al. 1998; Smith et al. 2006; Isla et al. 2009, 2011), as observed, e.g., on the EWSS, where benthic biomass is high and communities are mainly constituted by sessile suspension feeders (Gerdes et al. 1992; Gutt and Starmans 1998).

Vertical migration by zooplankton, fish, or diving vertebrates is regarded as a common feature of aquatic environments, and on an individual level, these provide a trade-off between nutrition and survival (Schmidt et al. 2011). In the context of this review, vertical migration refers to any causal vertical movement (e.g., foraging expeditions and avoidance of predators). The benthic realm works as feeding ground for various vertebrates, thus promoting vertical migrations. Arntz et al. (1994) pointed out that seals and penguins often dive deep to feed on benthic invertebrates. Antarctic krill *Euphausia superba* has also been found to migrate down to 3000 m depth either to feed on the seabed or as a result of being satiated (Ligowski 2000; Tarling and Johnson 2006; Schmidt et al. 2011). While migrating, swimming organisms release carbon and nutrients in form of feces. Release of feces near the benthos could mean an extra input of available food for benthic organisms. Conversely, excretion of a mix of benthic organic material and lithogenic particles in the upper water column would increase the concentration of labile iron which could enhance primary production (Schmidt et al. 2011).

11.2.3.2 Deposition and Resuspension

Specific particle composition and flux rates in a region are not just a question of primary production and associated zooplanktonic activity. They also are affected by local deposition and resuspension processes. Water currents, especially near the seabed, are one key environmental factor regulating deposition and resuspension. Another key environmental factor are icebergs. Iceberg scours change the seabed topography, affect the near seabed current regime, and modify the deposition regime in the area by trapping particles in the scours mark (working as a sort of "sediment trap"). Iceberg scour marks can be 10s to 100s meters wide, several meters deep, and 10s of meters or even kilometers long (Gutt 2001; Gerdes et al. 2003). On the other hand, iceberg scours can also enhance resuspension by generating an upward particle flux (Gutt 2001; Barnes et al. 2018).

A recent study on the effect of icebergs and sea ice on "blue carbon" (carbon in organisms) pointed out that in March 2017, 47 giant icebergs larger than 30 km^2 occurred in Antarctica, 6 of which exceeded 1000 km^2 in area (Barnes et al. 2018). Initially, any iceberg scour would resuspend already fixed blue carbon and increase the open water area by breaking and displacing sea ice. The combination of additional resuspended material and open water area would result in an increase of primary production, which in turn would promote benthic growth. As a result, deposition would be increased not only by the enhanced primary production but also by the proportional increase of benthic suspension feeder biomass (Barnes et al. 2018).

The studies of Mercuri et al. (2008), Tatián et al. (2008), and Barnes et al. (2016, 2018) are examples of how benthic organisms affect deposition and resuspension. Micro-, macro-, and megafauna as well as marine flora directly affect the sediment erodibility and regulate sediment mixing, which greatly affects the benthos-pelagos interconnectivity (Orvain et al. 2012; Queirós et al. 2015). Benthic organisms may decrease sediment roughness by mucus, bacterial mats, or diatom film production, thus reducing the resuspension ability of sediments (de Jonge and van den Bergs 1987; Grant and Bathmann 1987; Paterson 1989; Self et al. 1989; Delgado et al. 1991; Dade et al. 1992; de Jonge and van Beusekom 1995). In Antarctic benthos, hexactinellid sponges exemplify how organisms can reduce resuspension and enhance deposition. These sponges cement and consolidate sediments, enhance biodiversity by promoting the immigration of other sponge species, provide refuges to other taxa, and generate spicule mats (Fig. 11.4), which work as silicon traps (Barthel 1992; Barthel and Gutt 1992; Gutt et al. 2013a). Sponges and other filter feeders collect particles from the water column, thus enhancing the downward flux of particles and their deposition (Barthel 1992; Mercuri et al. 2008; Tatián et al. 2008). This biodeposition effect is enhanced by the increase of biodiversity provided by sponges. Furthermore, spicule mats reduce resuspension by covering the sediment, thus reducing its erodibility. Other structures that enhance deposition are tube formations (Fig. 11.4). High density of polychaete tubes could generate an attracting effect equal to that of baffles in sediment traps, albeit in a reduced area (Frithsen and Doering 1986). Contrastingly, other activities of benthic organisms such as pellet production and bioturbation with formation of mounds, pits, tubes, and tracks can change the sediment structure and enhance particle resuspension (Eckman et al. 1981; Eckman and Nowell 1984; Luckenbach 1986; Davis 1993). Resuspended material tends to be rich in nutrients and contains also micronutrients such as iron, which could, in shallower shelf areas with upwelling or those shelf areas where deep mixing occur, enhance summer primary production (Doering 1989; Sedwick et al. 2000).

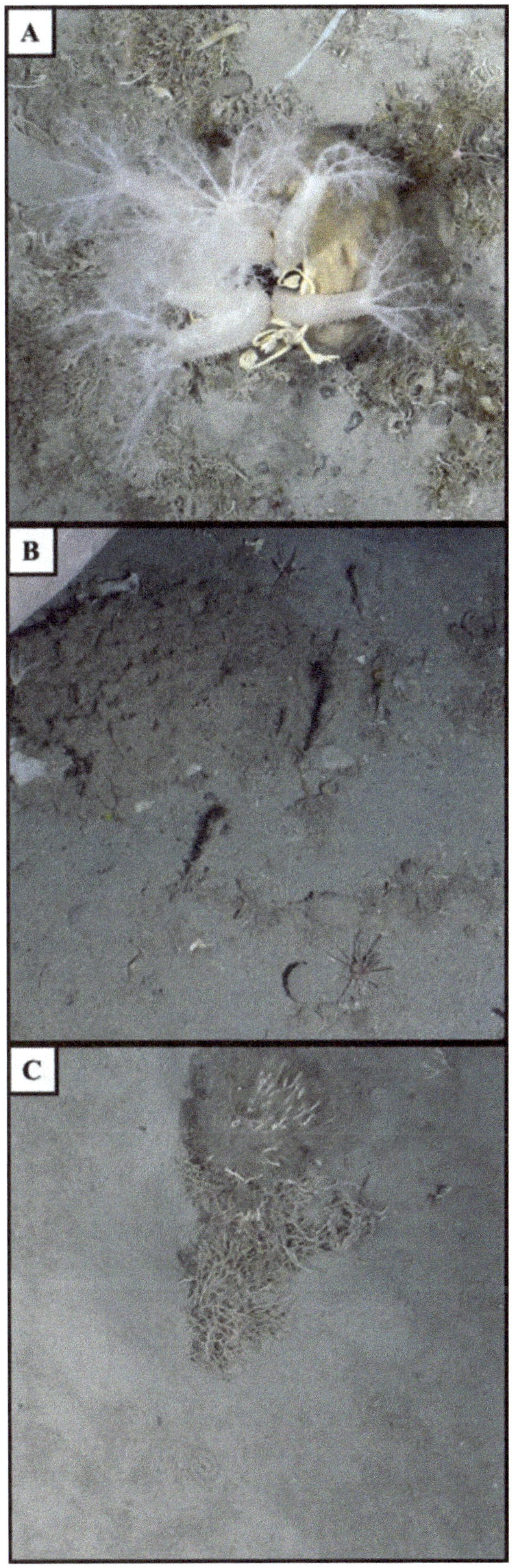

Fig. 11.4 Examples of benthic structures which modify particle resuspension and deposition: (**a**) a three-dimensional structure provided by sponges and associated organisms; (**b**) a spicule mat covering part of the seabed; and **C**) a cluster of polychaete tubes. Images (**a**) and (**c**) were modified after Piepenburg (2016). Image (**b**) was kindly provided by D. Gerdes and modified

11.2.3.3 Benthic Realm

The marked seasonal differences in the pelagic realm, especially the reduction of carbon flux in winter (see Sects. 11.2.1 and 11.2.3.1), has been thought to directly regulate benthic processes such as reproduction, growth, feeding activity, sexual development, recruitment of juveniles, and also benthic distribution patterns. However, studies on bentho-pelagic and pelago-benthic coupling in Antarctic waters have shown differences between benthic and pelagic seasonality to be less important in regulating benthic processes and that both compartments could be less coupled than thought, partly due to the effect of currents, lateral advection, and tides (see Sect. 11.2.2). Stanwell-Smith et al. (1999) studied meroplanktonic larvae released by benthic organisms and described these larvae to be present throughout the year. In some cases, the larval peak was clearly decoupled from the summer bloom, and the recruitment of benthic organisms was described to occur year-round or with a tendency to happen during winter months (Bowden 2005; Galley et al. 2005). Similarly, Sumida et al. (2008) found recruitment of holothurians to occur during winter, but these deposit feeders were actively feeding throughout the whole year. Measurements of metabolic activity via thorium (Th) isotopes made by McClintic et al. (2008) confirmed benthos to be metabolically active year-round. Results from the studies of Sumida et al. (2008) and McClintic et al. (2008) conducted in the WAP agreed with earlier findings made by Barnes and Clarke (1995), who recorded feeding activities of bryozoans, holothurians, polychaetes, and hydroids at Signy Island (Fig. 11.1d). However, Barnes and Clarke (1995) did not find any feeding activity during short periods of time during winter. Similarly, a study conducted at Rothera Point (Fig. 11.1e) by Brockington et al. (2001) on the feeding activity and nutritional status of the sea urchin, *Sterechinus neumayeri*, found this species to completely stop feeding during winter. In a recent study, Souster et al. (2018) measured the seasonality of oxygen consumption of five benthic invertebrates and found the oxygen consumption of suspension and deposit feeders to be independent from the input provided by the local summer flux. It has been proposed that benthic organisms can feed or be metabolically active year-round by changing their feeding mechanism, as is known for some sponges, polychaetes, bivalves, and cnidarians (Cattaneo-Vietti et al. 1999; Orejas et al. 2001).

11.3 Regional Patterns in Coupling Processes

The interaction between biotic and abiotic factors regulating the benthos-pelagos interconnectivity will have direct implications on how "strong" or "weak" the coupling between realms is and how changes in one of the compartments may affect its counterpart. When comparing different Antarctic regions, we observe differences in flux regulators and in the structure of the respective benthic communities. These differences reflect how variable the strength of the coupling between benthos and pelagos is. To exemplify how coupled or decoupled systems appear, I compared data obtained in WAP and EWSS waters. Furthermore, I include the example of the Larsen area (Fig. 11.1f) to exemplify how changes in the pelagos affect and modify the benthos.

11.3.1 West Antarctic Peninsula

To describe the benthos-pelagos interconnectivity on the WAP shelf, I focused on studies from the Bransfield Strait (Palanques et al. 2002; Isla et al. 2006b), Rothera Point (Souster et al. 2018), and those conducted within the frame of the "Food for Benthos on the Antarctic Continental Shelf" project (FOODBANCS; e.g., Smith et al. 2006; McClintic et al. 2008; Sumida et al. 2008). All locations are marked in Fig. 11.1b, e. According to these studies, the coupling between the pelagic primary production and benthic biological processes in these areas appears "weak." As already stated, the study of McClintic et al. (2008) with Th isotopes not only showed benthos to be metabolically active the whole year; it also showed that the delivery of this isotope to the sediment was not related to local downward flux, suggesting more influence from advected material than from local production. Investigation of the shelf fauna via video recordings (Sumida et al. 2008, 2014) also shows proof of a "weak" local coupling. They found holothurians to recruit during winter, i.e., independently from local food input. Sumida et al. (2008, 2014) also recorded feces of holothurian year-round, but with hints to higher feeding rates during summer, which appears to be the result of better food quality in this season (Sumida et al. 2014). The study of Souster et al. (2018) found results partly different to those of Sumida et al. (2008, 2014). Souster et al. (2018) described primary consumers (suspension and deposit feeders) to maintain a rather stable metabolic activity year-round, regardless of food input, while secondary consumers (scavengers and predators) showed higher metabolic activity during summer than winter. These authors attributed the seasonal metabolic differences of secondary consumers to be related to better quality of food items rather than to their quantity.

Studies conducted in the WAP evidence advection of material to be more important than locally produced particle fluxes. Palanques et al. (2002) found a high amount of the sediments captured by their traps located in the deeper Bransfield Strait (BS; Fig. 11.1b) to originate from shallower areas of the BS. The sediment fluxes near the bottom accounted for 18% of the annual primary production, and these fluxes included benthic organisms and particles resuspended and laterally transported from shallower adjacent areas. The study of Isla et al. (2006b) found that sedimentation generated by the Johnson's Glacier (Johnson's Dock, Fig. 11.1b) was comprised mostly of fine sediment. These particles were rich in organic matter, and near-bottom lateral transport of this resuspended matter was the main source of carbon flux into deeper basins (Isla et al. 2006b). These evidences suggest the shallow coastal areas of the WAP to be highly nutritive. Via advection from these shallower areas, the adjacent deeper basins are provided with organic matter. This material is accumulated and forms green mats or "food banks." These green mats ensure the presence of food for benthos during the low production autumn and winter seasons (Smith et al. 2006). The formation of these "food banks" via advected material and a dominance of deposit feeders might explain the restricted meaning of locally generated particle fluxes between pelagic and benthic realms in the WAP (McClintic et al. 2008; Sumida et al. 2008; Souster et al. 2018).

11.3.2 Eastern Weddell Sea Shelf

The "weak" interconnectivity in the WAP appears to be connected to how particle fluxes are mainly regulated by advection processes from shallower shelves to deeper basins, where "food banks" are formed (Isla et al. 2006b; Smith et al. 2006; McClintic et al. 2008; Sumida et al. 2008). On the EWSS, downward particle transport off Austasen and Kapp Norvegia (Fig. 11.1a) has been described to be fast (Bathmann et al. 1991; Isla et al. 2009), despite the relatively stronger currents caused by the narrow shelf. This "fast" downward flux is evidenced by (a) how sediments quickly reflect the local bloom and its associated characteristics (Bathmann et al. 1991; Isla et al. 2009) and (b) how bottom sediments are especially nutritive during summer/autumn (Isla et al. 2011). The efficient transport of carbon from the pelagic to the benthic realm in combination with the resuspension of particles could explain the benthic community characteristic on the EWSS. Benthic communities in this region have been described as rich in sessile suspension feeders, especially glass sponges, which not only increase diversity by creating three-dimensional structures with space for many other species but also explain the high biomass of the EWSS benthos, which is higher than that of other subregions in the Weddell Sea including the tip of the Antarctic Peninsula (Table 11.1; Barthel 1992; Barthel and Gutt 1992; Gerdes et al. 1992; Arntz et al. 1994; Gutt and Starmans 1998; Sañé et al. 2012; Gerdes 2014a, b; S.E.A. Pineda-Metz unpublished data). This high biomass of suspension feeders also influences deposition and sediment chemistry. It seems feasible that suspended particles are largely consumed by suspension feeders, thus transforming the chemical composition of these particles and reducing the amount of organic carbon remaining for incorporation into the sediment. The efficient local flux patterns in combination with particle resuspension and high biomass of suspension feeders which benefit from these conditions might explain the "stronger" coupling between benthic and pelagic realms on the EWSS contrarily to what was found in the WAP region.

Table 11.1 Depth ranges and wet weight biomass data (g_{ww} m^{-2}) from multi-box corer samples collected in four subregions of the Weddell Sea: Tip of the Antarctic Peninsula (TAP), Larsen embayments (LA), Filchner Region (FR), and Eastern Weddell Sea Shelf (EWSS) (S.E.A. Pineda-Metz, unpublished data)

Subregion	Depth range (m)	Biomass (g_{ww} m^{-2})		
		Range	Mean	Median
TAP	187–934	30–3485	423	223
LA	202–850	2–786	78	16
FR	254–1217	1–335	51	24
EWSS	248–1486	1–103,235	4811	134

11.3.3 The Changing Situation of Larsen

The Larsen embayments on the eastern coast of the Antarctic Peninsula (Fig. 11.1f) may serve as an example of how changes in the pelagic system influence benthos. Studies in the embayments formerly covered by the Larsen A and B ice shelves reflected a shift from an oligotrophic system to one with enhanced production and flux rates (Sañé et al. 2011). Before the disintegration of the shelf ice in 1995 and 2002, respectively, the shelf benthos appeared impoverished and in an early developmental stage as compared to the EWSS. Sessile suspension feeders showed low biomasses, and several deep-sea species on the shelf reflected the oligotrophic conditions resembling the deep sea (Gutt et al. 2011; Sañé et al. 2012; Gerdes 2014a, b). The disintegration of shelf ice created new space offshore for enhanced local primary production, shifting toward a more eutrophic and productive pelagic realm (Bertolin and Schloss 2009). Within a relatively short time, this enhanced pelagic production led to a shift also in the composition of the benthos (Fillinger et al. 2013; Gutt et al. 2013b). Benthos shifted from an ascidian dominated to a sponge- and ophiuroid-dominated fauna. Suspension-feeding ophiuroids were replaced by a more abundant deposit-feeding ophiuroid fauna, and sponges increased two- to threefold in terms of abundance and biomass (Fillinger et al. 2013; Gutt et al. 2013b).

11.4 Outlook

Studies on the coupling between the benthic and pelagic realms are difficult approaches with complex sampling programs, which require similar temporal and spatial scales for drawing accurate conclusions about coupling processes and their meaning for both compartments (Raffaelli et al. 2003; Renaud et al. 2008). This review on benthos-pelagos interconnectivity includes attempts to describe regulating factors that connect the benthic and pelagic both realms.

Based on "real data", I draw assumptions to distinguish between specific coupling processes in different Antarctic regions. These assumptions are made on only few studies, which were not all intended to study the benthos-pelagos interconnectivity per se but aimed to study processes individually. This implies that my hypothetical assumptions need further testing. This shows also that many gaps remain and filling them will be of paramount importance to better understand how both realms are connected and how carbon cycling works on Antarctic shelves.

There have been a series of attempts to connect the Antarctic benthic and pelagic realms, reflected (but not restricted) to the works of Barry (1988) Barry and Dayton (1988), Dayton (1989), Ligowski (2000), Schnack-Schiel and Isla (2005), Barnes et al. (2006, 2016, 2018), Isla et al. (2006a; b), Smith et al. (2006, 2008), McClintic et al. (2008), Mercuri et al. (2008), Tatián et al. (2008), Schimdt et al. (2011), Sañé et al. (2011, 2012), Barnes (2015), Jansen et al. (2018), and Souster et al. (2018). Promising attempts to fill regional gaps have also been made. The FOODBANCS project (Smith et al. 2006, 2008) gives a clear hint of how the coupling (or decoupling) between benthos and pelagos works in shelves of the WAP. In this modern age, modelling has gained great importance. Models on how pelagic particles are distributed and are related to benthic distribution patterns are starting to be developed (e.g., Jansen et al. 2018). While promising, attempts on modelling and correlating benthic and pelagic processes are still in early stages. Other Antarctic areas with a long history of studies such as the Weddell Sea need the available data to be reviewed, sorted, and used to start drawing lines between benthic and pelagic realms, as attempted in this review. This first step will help to set the course of future studies and point out a red line on how benthos-pelagos interactions could be investigated in different Antarctic regions, which in turn will provide an excellent tool to understand how the ongoing and predicted climate change will affect the Antarctic shelves.

Acknowledgments I would like to thank the organizers of the YOUMARES 9 conference for giving me the opportunity not only to work on this manuscript but also to organize the session "Connecting the bentho-pelagic dots." Also thanks to Chester Sands and another anonymous reviewer who took the job of greatly improving this review. I also feel deeply grateful to Dieter Gerdes, without whom this manuscript would look a lot messier and harder to read as it is. Many thanks go also to L. Metz, M. Pineda, and H. Costa, who founded part of my work. Last, but not least, thanks to the staff of the CPT N°704, who helped me to start paying attention to the full picture some years ago.

Appendix

This article is related to the YOUMARES 9 conference session no. 14: "Connecting the bentho-pelagic dots." The original Call for Abstracts and the abstracts of the presentations within this session can be found in the Appendix "Conference Sessions and Abstracts", Chapter "10 Connecting the bentho-pelagic dots", of this book.

References

Arndt JE, Schenke HW, Jakobsson M et al (2013) The international bathymetric chart of the Southern Ocean (IBCSO) version 1.0 – a new bathymetric compilation covering circum-Antarctic water. Geophys Res Lett 40:3111–3117. https://doi.org/10.1002/grl.50413

Arntz WE, Brey T, Gallardo VA (1994) Antarctic zoobenthos. Oceanogr Mar Biol 32:241–304

Arrigo KR, Worthen DL, Lizotte MP et al (1997) Primary production in Antarctic Sea ice. Science 276(5311):394–397. https://doi.org/10.1126/science.276.5311.394

Arrigo KR, Worthen DL, Schnell A et al (1998) Primary production in Southern ocean waters. J Geophys Res 103:15587–11560. https://doi.org/10.1028/1998JC000289

Arrigo KR, van Dijken GL, Bushinsky S (2008) Primary production in the Southern Ocean, 1997-2006. J Geophys Res 113:C08004. https://doi.org/10.1029/2007JC004551

Barnes DKA (2015) Antarctic Sea ice losses drive gains in benthic carbon drawdown. Curr Biol 25:R775–R792

Barnes KA, Clarke A (1995) Feeding activity in Antarctic suspension feeders. Polar Biol 15:335–340

Barnes DKA, Webb K, Linse K (2006) Slow growth of Antarctic bryozoans increases over 20 years and is anomalously high in 2003. Mar Ecol Prog Ser 314:187–195

Barnes DKA, Ireland L, Hogg OT et al (2016) Why is the South Orkney Island shelf (the world's first high seas marine protected area) a carbon immobilization hotspot? Glob Change Biol 22:1110–1120. https://doi.org/10.1111/gcb.13157

Barnes DKA, Fleming A, Sands CJ et al (2018) Icebergs, sea ice, blue carbon and Antarctic climate feedbacks. Phil Trans R Soc A 376:2017176. https://doi.org/10.1098/rsta.2017.0176

Barry JP (1988) Hydrographic patterns in McMurdo Sound, Antarctica and their relationship to local benthic communities. Polar Biol 8:377–391

Barry JP, Dayton PK (1988) Current patterns in McMurdo Sound, Antarctica and their relationship to local biotic communities. Polar Biol 8:367–376

Barthel D (1992) Do hexactinellids structure Antarctic sponge associations? Ophelia 36:111–118

Barthel D, Gutt J (1992) Sponge associations in the eastern Weddell Sea. Antarct Sci 4:157–150

Bathmann E, Fischer G, Müller PJ et al (1991) Short-term variations in particulate matter sedimentation off Kapp Norvegia, Weddell Sea,

Antarctica: relation to water mass advection, ice cover, plankton biomass and feeding activity. Polar Biol 11:185–195

Bertolin ML, Schloss IR (2009) Phytoplankton production after the collapse of the Larsen a ice shelf. Antarct Polar Biol 32:1435–1446. https://doi.org/10.1007/s00300-009-638-x

Bowden DA (2005) Seasonality of recruitment in Antarctic sessile marine benthos. Mar Ecol Prog Ser 297:101–118

Brey T, Clarke A (1993) Population dynamics of marine benthic invertebrates in Antarctic and subantarctic environments: are there unique adaptations? Antarct Sci 5(3):253–266. https://doi.org/10.1017/S0954102093000343

Brockington S, Clarke A, Chapman ALG (2001) Seasonality of feeding and nutritional status during the austral winter in the Antarctic Sea urchin *Sterechinus neumayeri*. Mar Biol 139:127–138

Cattaneo-Vietti R, Chiantore MC, Misic C et al (1999) The role of pelagic-benthic coupling in structuring littoral benthic communities at Terra Nova Bay (Ross Sea) and in the Straits of Magellan. Sci Mar 63(1):113–121

Clarke A (1988) Seasonality in the Antarctic marine environment. Comp Biochem Physiol B 90(3):461–473

Clarke A (2003) Costs and consequences of evolutionary temperature adaptation. Trends Ecol Evol 18(11):573–581

Collier R, Dymond J, Honjo S et al (2000) The vertical flux of biogenic and lithogenic material in the Ross Sea: Moored sediment trap observations 1996–1998. Deep-Sea Res II 47:3491–3520

Dade WB, Nowell ARM, Jumars PA (1992) Predicting erosion resistance of muds. Mar Geol 105:285–297

Davis WR (1993) The role of bioturbation in sediment resuspension and its interaction with physical shearing. J Exp Mar Biol Ecol 171:187–200

Dayton PK (1989) Interdecadal variation in an Antarctic sponge and its predators from oceanographic climate shifts. Science 245:1484–1486

de Jonge VN, van Beusekom JEE (1995) Wind- and tide-induced resuspension of sediment and microphytobenthos from tidal flats in the Ems estuary. Limnol Oceanogr 40:766–778

de Jonge VN, van den Bergs J (1987) Experiments on the resuspension of estuarine sediments containing benthic diatoms. Estuar Coastal Shelf Sci 24:725–740

Delgado M, de Jonge VN, Peletier H (1991) Experiments on resuspension of natural microphytobenthos populations. Mar Biol 108:321–328

DiTullio GR, Grebmeier JM, Arrigo KR et al (2000) Rapid and early export of *Phaeocystis Antarctica* blooms in the Ross Sea, Antarctica. Nature 404:595–598

Doering P (1989) On the contribution of the benthos to pelagic production. J Mar Res 47:371–383

Donnelly J, Sutton TT, Torres JJ (2006) Distribution and abundance of micronekton and microzooplankton in the NW Weddell Sea: relation to a spring ice-edge bloom. Polar Biol 29:280–293. https://doi.org/10.1007/s00300-005-0051-z

Dorschel B, Gutt J, Piepenburg D et al (2014) The influence of the geomorphological and sedimentological settings on the distribution of epibenthic assemblages on a flat topped hill on the over-deepened shelf of the western Weddell Sea (Southern Ocean). Biogeosciences 11:3797–3817. https://doi.org/10.5194/bg-11-3797-2014

Ducklow HW, Frase W, Karl DM et al (2006) Water-column processes in the West Antarctic Peninsula and the Ross Sea: interannual variations and foodweb structure. Deep-Sea Res II 53:834–852

Dunbar RB, Leventer AR, Mucciarone DA (1998) Water column sediment fluxes in the Ross Sea, Antarctica: atmospheric and sea ice forcing. J Geophys Res 103:30741–30759

Eckman JE, Nowell ARM (1984) Boundary skin friction and sediment transport about an animal-tube mimic. Sedimentology 31:851–862

Eckman JE, Nowell ARM, Jumars PA (1981) Sediment destabilization by animal tubes. J Mar Res 39:361–374

Fetterer F, Knowles K, Meier W et al (2018) Sea Ice Index, Version3 [February and August 2018]. NSIDC: National Snow and Ice Data Center, Boulder. https://doi.org/10.7265/N5K072F8. Accessed 11 Oct 2018

Fillinger L, Janussen D, Lundäalv T et al (2013) Rapid glass sponge expansion after climate-induced Antarctic ice shelf collapse. Curr Biol 23:1330–1334. https://doi.org/10.1016/j.cub.2013.05.051

Flores H, Hunt BPV, Kruse S et al (2014) Seasonal changes in the vertical distribution and community structure of Antarctic macrozooplankton and micronekton. Deep-Sea Res I 84:127–141

Frithsen JB, Doering PH (1986) Active enhancement of particle removal from the water column by tentaculate benthic polychaetes. Ophelia 25:169–182

Gallardo VA (1987) The sublittoral macrofaunal benthos of the Antarctic shelf. Environ Int 13:71–81

Galley EA, Tyler PA, Clarke A et al (2005) Reproductive biology and biochemical composition of the brooding echinoid *Amphipneustes lorioli* on the Antarctic continental shelf. Mar Biol 148:59–71. https://doi.org/10.1007/s00227-005-0069-3

Garrison DL, Close AR (1993) Winter ecology of the sea ice biota in Weddell Sea pack ice. Mar Ecol Prog Ser 96:17–31

Gerdes D (2014a) Biomass of macrozoobenthos in surface sediments sampled during POLARSTERN cruise ANT-XXIII/8. Alfred Wegener Institute, Helmholtz Center for Polar and Marine Research, Bremerhaven, PANGAEA. https://doi.org/10.1594/PANGAEA.834054. Accessed 20 Aug 2015

Gerdes D (2014b) Biomass of macrozoobenthos in surface sediments sampled during POLARSTERN cruise ANT-XXVII/3. Alfred Wegener Institute, Helmholtz Center for Polar and Marine Research, Bremerhaven, PANGAEA. https://doi.org/10.1594/PANGAEA.834058. Accessed 20 Aug 2015

Gerdes D, Klages M, Arntz WE et al (1992) Quantitative investigations on macrobenthos communities of the southeastern Weddell Sea shelf based on multibox corer samples. Polar Biol 12:291–301

Gerdes D, Hilbig B, Montiel A (2003) Impact of iceberg scouring on macrobenthic communities in the high-Antarctic Weddell Sea. Polar Biol 26:295–301

Gleitz M, Bathmann EV, Lochte K (1994) Build-up and decline of summer phytoplankton biomass in the eastern Weddell Sea, Antarctica. Polar Biol 14:413–422

Graf G (1989) Benthic-pelagic coupling in a deep-sea benthic community. Nature 341:437–439

Graf G, Rosenberg R (1997) Bioresuspension and biodeposition: a review. J Mar Sys 11:269–278

Grant J, Bathmann EV (1987) Swept away: resuspension of bacterial mats regulates benthic-pelagic exchange of sulfur. Science 236:1472–1474. https://doi.org/10.1126/science.236.4807.1472

Gutt J (2001) On the direct impact of ice on marine benthic communities, a review. Polar Biol 24:553–564. https://doi.org/10.1007/s003000100262

Gutt J, Starmans A (1998) Structure and biodiversity of megabenthos in the Weddell and Lazarev Seas (Antarctica): ecological role of physical parameters and biological interactions. Polar Biol 20:229–247

Gutt J, Starmans A, Dieckmann G (1998) Phytodetritus deposited on the Antarctic shelf and upper slope: its relevance for the benthic system. J Mar Syst 17:435–444

Gutt J, Barrat I, Domack E et al (2011) Biodiversity change after climate-induced ice-shelf collapse in the Antarctic. Deep-Sea Res II 58:74–83

Gutt J, Böhmer A, Dimmler W (2013a) Antarctic sponge spicule mats shape microbenthic diversity and act as a silicon trap. Mar Ecol Prog Ser 480:57–71. https://doi.org/10.3354/meps10226

Gutt J, Cape M, Dimmler W et al (2013b) Shifts in Antarctic megabenthic structure after ice-shelf disintegration in the Larsen area east of the Antarctic peninsula. Polar Biol 36:895–906. https://doi.org/10.1007/s00300-013-1315-7

Hargrave BT (1973) Coupling carbon flow through some pelagic and benthic communities. J Fish Res Board Can 30:1317–1326

Hauck J, Hoppema M, Bellerby RGJ et al (2010) Data-based estimation of antropogenic carbon and acidification in the Weddell Sea on a decadal timescale. J Geophys Res 115:C03004. https://doi.org/10.1029/2009JC005479

Hauck J, Gerdes D, Hillenbrand C-D et al (2012) Distribution and mineralogy of carbonate sediments on Antarctic shelves. J Mar Syst 90:77–87. https://doi.org/10.1016/j.marsys.2011.09.005

Isla E (2016a) Environmental controls on sediment composition and particle fluxes over the Antarctic continental shelf. In: Beylich A, Dixon J, Zwoliński Z (eds) Source-to-sink fluxes in undisturbed cold environments. Cambridge University Press, Cambridge, pp 199–212. https://doi.org/10.1017/CBO9781107705791.017

Isla E (2016b) Organic carbon and biogenic silica in marine sediments in the vicinities of the Antarctic Peninsula: spatial patterns across a climatic gradient. Polar Biol 39:819–828. https://doi.org/10.1007/s00300-015-1833-6

Isla E, Gerdes D, Palanques A et al (2006a) Particle fluxes and tides near the continental ice edge on the eastern Weddell Sea shelf. Deep-Sea Res II 53:866–874

Isla E, Gerdes D, Palanques A et al (2006b) Relationships between Antarctic coastal and Deep-sea particle fluxes: implications for the deep-sea benthos. Polar Biol 29:249–256

Isla E, Gerdes D, Palanques A et al (2009) Downward particle flux, wind and a phytoplankton bloom over a polar continental shelf: a stormy impulse for the biological pump. Mar Geol 259:59–72

Isla E, Gerdes D, Rossi S et al (2011) Biochemical characteristics of surface sediments on the Eastern Weddell Sea continental shelf, Antarctica: is there any evidence of seasonal patterns? Polar Biol 34:1125–1133

Jansen J, Hill NA, Dunstan PK et al (2018) Abundance and richness of key Antarctic seafloor fauna correlates with modelled food availability. Nat Ecol Evol 2:71–80. https://doi.org/10.1038/s41559-017-0392-3

Kang S-H, Kang J-S, Lee S et al (2001) Antarctic phytoplankton assemblages in the marginal ice zone of the northwestern Weddell Sea. J Plankton Res 23(4):333–352

Kohlbach D, Lange BA, Schaafsma FL et al (2017) Ice algae-produced carbon is critical for overwintering of Antarctic krill *Euphausia superba*. Front Mar Sci 4:310. https://doi.org/10.3389/fmars.2017.00310

Ligowski R (2000) Benthic feeding by krill, *Euphausia superba* Dana, in coastal waters off West Antarctica and in Admiralty Bay, South Shetland Islands. Polar Biol 23:619–625

Lizotte MP (2001) The contribution of sea ice algae to Antarctic marine primary production. Am Zool 41(1):57–73. https://doi.org/10.1668/0003-1569(2001)041[0057:TCOSIA]2.0.CO;2

Luckenbach MR (1986) Sediment stability around animal tubes: the roles of hydrodynamic processes and biotic activity. Limnol Oceanogr 31:779–787

McClintic MA, DeMaster DJ, Thomas CJ et al (2008) Testing the FOODBANCS hypothesis: seasonal variations in near-bottom particle flux, bioturbation intensity, and deposit feeding based on ^{234}Th measurements. Deep-Sea Res II 55:2425–2437. https://doi.org/10.1016/j.dsr2.2008.06.003

Mercuri G, Tatián M, Momo F et al (2008) Massive input of terrigenous sediment into potter cove during austral summer and the effects on the bivalve *Laternula elliptica*: a laboratory experiment. Ber Polar Meeresforsch 571:111–117

Michels J, Dieckmann GS, Thomas DN et al (2008) Short-term biogenic particle flux under late spring sea ice in the western Weddell Sea. Deep-Sea Res II 55:1024–1039

Nicol S (2006) Krill, currents, and sea ice: *Euphausia superba* and its changing environment. Bioscience 56(2):111–120

Orejas C, Gile JM, López-Gonzélez J et al (2001) Feeding strategies and diet composition of four Antarctic cnidarian species. Polar Biol 24:620–627. https://doi.org/10.1007/s03000100272

Orvain F, Le Hir P, Sauriau P-G et al (2012) Modelling the effects of macrofauna on sediment transport and bed elevation: application over a cross-shore mudflat profile and model variation. Estuar Coastal Shelf Sci 108:64–75. https://doi.org/10.1016/j.ecss.2011.12.036

Palanques A, Isla E, Puig P et al (2002) Annual evolution of downward particle fluxes in the Western Bransfield Strait (Antarctica) during the FRUELA project. Deep-Sea Res II 49:903–920

Paterson DM (1989) Short-term changes in the erodibility of intertidal cohesive sediments related to the migratory behaviour of epipelic diatoms. Limnol Oceanogr 34:223–234

Pearse JS, McClintock JB, Bosch I (1991) Reproduction of Antarctic benthic marine invertebrates: tempos, modes, and timing. Am Zool 31(1):65–80

Piepenburg D (2016) Seabed photographs taken along OFOS profiles during Polarstern cruise PS96 (ANT-XXXI/2 FROSN). Alfred Wegener Institute, Helmholtz Center for Polar and Marine Research, Bremerhaven, PANGAEA. https://doi.org/10.1594/PANGAEA.862097. Accessed 31 Oct 2016

Queirós AN, Stephens N, Cook R et al (2015) Can benthic community structure be used to predict the process of bioturbation in real ecosystems. Prog Oceanogr 137:559–569

Raffaelli D, Bell E, Weithoff G et al (2003) The ups and downs of benthic ecology: considerations of scale, heterogeneity and surveillance for benthic-pelagic coupling. J Exp Mar Biol Ecol 285-286:191–203. https://doi.org/10.1016/S0022-0981(02)00527-0

Renaud PE, Morata N, Carroll ML et al (2008) Pelagic-benthic coupling in the western Barents Sea: processes and time scales. Deep-Sea Res II 55:2372–2380. https://doi.org/10.1016/j.dsr2.2008.05.017

Riebesell U, Schloss I, Smetack V (1991) Aggregation of algae released from melting sea ice: implications for seeding and sedimentation. Polar Biol 11:239–248

Rossi S, Isla E, Martínez-García A et al (2013) Transfer of seston lipids during a flagellate bloom from the surface to the benthic community in the Weddell Sea. Sci Mar 77(3):397–407. https://doi.org/10.3989/scimar.03835.30A

Sañé E, Isla E, Grémare A et al (2011) Pigments in sediments beneath recently collapsed ice shelves: the case of Larsen A and B shelves, Antarctic peninsula. J Sea Res 65:94–102

Sañé E, Isla E, Gerdes D et al (2012) Benthic macrofauna assemblages and biochemical properties of sediments in two Antarctic regions differently affected by climate change. Cont Shelf Res 35:53–63

Schaafsma FL, Kohlbach D, David C et al (2017) Spatio-temporal variability in the winter diet of larval and juvenile Antarctic krill, *Euphausia superba*, in ice-covered waters. Mar Ecol Prog Ser 580:101–115. https://doi.org/10.3354/meps12309

Scharek R, Smetacek V, Fahrbach E et al (1994) The transition from winter to early spring in the eastern Weddell Sea, Antarctica: plankton biomass and composition in relation to hydrography and nutrients. Deep-Sea Res I 41(8):1231–1250

Schmidt K, Atkinson A, Steigenberger S et al (2011) Seabed foraging by Antarctic krill: implications for stock assessment, bentho-pelagic coupling, and the vertical transfer of iron. Limnol Oceanogr 56(4):1411–1428. https://doi.org/10.4317/lo.2011.56.4.1411

Schnack-Schiel SB, Isla E (2005) The role of zooplankton in the pelagic-benthic coupling of the Southern Ocean. Sci Mar 69(2):39–55

Sedwick PN, DiTullio G (1997) Regulation of algal blooms in Antarctic shelf waters by the release of iron from melting sea ice. Geophys Res Lett 24(20):2515–2518

Sedwick P, DiTullio GR, Mackey DJ (2000) Iron and manganese in the Ross Sea, Antarctica: seasonal iron limitation in Antarctic shelf waters. J Geophys Res 105(C5):11321–11336

Segelken-Voigt A, Bracher A, Dorschel B et al (2016) Spatial distribution patterns of ascidians (Ascidiacea: Tunicata) on the continental shelves off the northern Antarctic Peninsula. Polar Biol 39:863–879. https://doi.org/10.1007/s00300-016-1909-y

Self RFL, Nowell ARM, Jumars PA (1989) Factors controlling critical shears for deposition and erosion of individual grains. Mar Geol 86:181–199

Smith CR, Minks S, DeMaster DJ (2006) A synthesis of bentho-pelagic coupling on the Antarctic shelf: food banks, ecosystem inertia and global climate change. Deep-Sea Res II 53:875–894. https://doi.org/10.16/j.dsr2.2006.02.001

Smith CR, Mincks S, DeMaster DJ (2008) The FOODBANCS project: introduction and sinking fluxes of organic carbon, chlorophyll-*a* and phytodetritus on the western Antarctic Peninsula continental shelf. Deep-Sea Res II 55:2404–2414

Souster TA, Morley SA, Peck LS (2018) Seasonality of oxygen consumption in five common Antarctic benthic marine invertebrates. Polar Biol 41(5):897–908. https://doi.org/10.1007/s00300-018-2251-3

Stanwell-Smith D, Peck LS, Clarke A et al (1999) The distribution, abundance and seasonality of pelagic marine invertebrate larvae in the maritime Antarctic. Philos Trans R Soc B 354:471–484

Sumida PYG, Bernardino AF, Stedall VP et al (2008) Temporal changes in benthic megafaunal abundance and composition across the West Antarctic Peninsula shelf: results from video surveys. Deep-Sea Res II 55:2465–2477

Sumida PYG, Smith CR, Bernardino AF et al (2014) Seasonal dynamics of megafauna on the deep West Antarctic Peninsula shelf in response to variable phytodetrital influx. R Soc Open Sci 1:140294. https://doi.org/10.1098/rsos.140294

Tarling GA, Johnson ML (2006) Satiation gives krill that sinking feeling. Curr Biol 16:R83–R84. https://doi.org/10.1016/j.cub.2006.01.044

Tatián M, Mercuri G, Fuentes VL et al (2008) Role of benthic filter feeders in pelagic-benthic coupling: assimilation, biodeposition and particle flux. Ber Polar Meeresforsch 571:118–127

Thomas DN, Dieckmann GS (2002) Antarctic Sea ice – a habitat for extremophiles. Science 295(5555):641–644. https://doi.org/10.1126/science.1063391

12 Investigating the Land-Sea Transition Zone

Stephan Ludger Seibert, Julius Degenhardt, Janis Ahrens, Anja Reckhardt, Kai Schwalfenberg, and Hannelore Waska

Abstract

Terrestrial and marine environments merge at the land-sea transition zone. This zone is important as ~38% of the world's population live by and depend on the coastal regions, and oceans are considerably affected by it. Furthermore, terrestrial and marine groundwater and seawater mix in the subterranean estuary (STE), where submarine groundwater discharge (SGD), i.e., discharging fresh groundwater and recirculated seawater, results in significant solute fluxes to the sea. With this article, we focus on advances of geochemical, microbiological, and technological aspects related to fresh groundwater, SGD, and STE in sandy coastal areas, using the barrier island Spiekeroog as a case study area. Previous studies showed that the fresh groundwater composition in sandy coastal aquifers is governed by calcareous shell dissolution, cation exchange, and organic matter degradation. Biogeochemical reactions in the STE further modify the water composition of SGD, with a dependence on residence time. Microbial communities, which are present in coastal sediments and usually follow salinity and redox gradients, are the driver for the degradation of organic matter. Regarding organic matter sources in the STE, it is evident that dissolved organic matter is primarily of marine origin and that SGD delivers degraded dissolved organic matter back into the ocean. Furthermore, recent studies used radiotracers, such as radium and radon, and seepage meters as reliable tools to quantify rates and fluxes associated with SGD. We conclude that, despite the advances being made, the complexity and interactions of the different processes at land-sea transition zones require multidisciplinary scientific approaches.

Keywords

Biogeochemistry · Freshwater lens · Marine sensors · Radiotracers · Spiekeroog Island · Submarine groundwater discharge · Subterranean estuary

12.1 Introduction

The coastal area is the interface between terrestrial and marine environments. From both environmental and (socio-)economic perspectives (Barbier et al. 2011), these are areas of enormous relevance, since ~38% of the world's population lives within 100 km distance from the coastline (UNEP 2014). In environmental sciences, the importance of this land-sea transition zone arises from a number of processes: (i) Land-sea interactions significantly affect global ocean material inventories due to export of terrestrially derived compounds to the ocean (Jeandel and Oelkers 2015); (ii) extensive land use and growing populations in coastal areas have a high impact on coastal ecosystems and aquifers, for example, high population densities in coastal areas present a risk for fresh-

S. L. Seibert (✉)
Hydrogeology and Landscape Hydrology Group, Institute for Biology and Environmental Sciences, Carl von Ossietzky University of Oldenburg, Oldenburg, Germany
e-mail: stephan.seibert@uol.de

J. Degenhardt (✉)
Paleomicrobiology Group, Institute for Chemistry and Biology of the Marine Environment (ICBM), Carl von Ossietzky University of Oldenburg, Oldenburg, Germany
e-mail: julius.degenhardt@uol.de

J. Ahrens · A. Reckhardt
Microbiogeochemistry Group, Institute for Chemistry and Biology of the Marine Environment (ICBM), Carl von Ossietzky University of Oldenburg, Oldenburg, Germany

K. Schwalfenberg
Marine Sensor Systems Group, Institute for Chemistry and Biology of the Marine Environment (ICBM), Carl von Ossietzky University of Oldenburg, Wilhelmshaven, Germany

H. Waska
Research Group for Marine Geochemistry (ICBM-MPI Bridging Group), Institute for Chemistry and Biology of the Marine Environment (ICBM), Carl von Ossietzky University of Oldenburg, Oldenburg, Germany

S. Jungblut et al. (eds.), *YOUMARES 9 - The Oceans: Our Research, Our Future*,
https://doi.org/10.1007/978-3-030-20389-4_12

water resources as well as for coastal ecosystems due to overexploitation, potentially leading to saltwater intrusion, and the disposal of waste and release of contaminants (Post 2005); and (iii) climate change and rising sea levels affect the coastal areas disproportionally relative to their size, giving them the description as the "frontlines of climate change" (Barbier 2015). For instance, it is expected that rising sea levels will increase land loss, storm intensities, floodings, and saltwater intrusion in coastal areas (e.g., McGranahan et al. 2007; Werner and Simmons 2009; Nicholls and Cazenave 2010). Continental fluxes to the ocean include riverine and atmospheric inputs, as well as submarine groundwater discharge (SGD). SGD was defined as "[…] any and all flow of water on continental margins from the seabed to the coastal ocean […]" by Burnett et al. (2003) and comprises terrestrial groundwater as well as recirculated seawater. The great importance of SGD has become widely accepted with emerging applications of radiotracer balances, and it could be demonstrated that SGD contributes significant fluxes of nutrients (Hays and Ullman 2007; Tait et al. 2014) and metals (Basu et al. 2001; Windom et al. 2006) to coastal oceans. In some cases, groundwater-derived nutrient and metal inputs may even rival river inputs (Moore 1996). The zone where terrestrial groundwater and seawater mix was consequently termed "subterranean estuary" (STE, Moore 1999) and shown to be a hot spot of biogeochemical reactions, further affecting the composition of SGD (Charette and Sholkovitz 2002; Roy et al. 2011; McAllister et al. 2015).

The objective of this article is to give an overview of recent advances made in scientific fields related to the land-sea transition zone, with a focus on STE and SGD in sandy coastal areas. Sandy beaches are an important part of the global land-sea transition zone, as they account for one third of the world's ice-free coastline (Luijendijk et al. 2018). Furthermore, the high permeability of sandy beach sediments permits advective pore water flow and allows for efficient transport of water constituents, resulting in steep environmental gradients and dynamic interfaces (Huettel et al. 1998; Riedel et al. 2010). We use the barrier island Spiekeroog located in the Southern North Sea, Germany, as a case study area in some of the following sections. Spiekeroog Island is characterized by mesotidal sandy beaches impacted by strong hydro-morphodynamic variations (Flemming and Davis 1994) and steep pore water salinity gradients due to the evolution of freshwater lenses below the dune areas. Therefore, Spiekeroog Island presents an excellent study site for the investigation of land-sea transition zones, SGD, and chemical reactions in sandy coastal STE.

As the freshwater composition is not only critical for drinking water supply in coastal areas but also for the chemical processes happening in the STE, the evolution of coastal freshwater aquifers is reviewed in Sect. 12.2, using freshwater lenses below (barrier) islands as an example. The effect of biogeochemical processes, including microbially mediated redox processes, and salinity gradients on the pore water chemistry of sandy STE is examined in Sect. 12.3. Moreover, the role and sources of labile organic matter, which fuels sedimentary redox processes in sandy STE, is reviewed in Sect. 12.4. Here, we focus on the concentration and reactivity of individual (dissolved) organic compounds controlling the rate and efficiency of organic matter degradation in the STE. Furthermore, the composition of microbial communities in tidal sands and their ability to utilize different electron donors and acceptors are explored with respect to the biogeochemical processes in, and signature of, sandy STE in Sect. 12.5. As biological processes often are kinetically driven, water infiltration rates and residence times are of great relevance for the chemical composition of pore waters and constituent transport in the STE and, thus, for the efficiency of the biogeochemical reactor (Anschutz et al. 2009; Tamborski et al. 2017). The presence of natural radioisotopes, such as radium (Ra) and radon (Rn) isotopes, provides the basis for determining groundwater residence times and quantifying SGD, which is reviewed in Sect. 12.6. Lastly, substantial progress in sensor technologies has been made since the discovery of SGD. Today, new methods based on the state-of-the-art sensor technologies can produce continuous datasets, which allow for detailed insights into physicochemical processes at the land-sea transition zone and provide a basis for modeling approaches, as examined in Sect. 12.7.

12.2 The Hydrochemical Evolution of Coastal Fresh Groundwater: Using Barrier Island Freshwater Lenses as an Example

Coastal freshwater aquifers are important for the land-sea transition zone, as fresh groundwater eventually discharges into the subterranean estuary (STE) and finally the ocean (e.g., Burnett et al. 2003; Röper et al. 2015). Here, the freshwater chemistry is of importance for further biogeochemical reactions and potentially plays a significant role for fluxes of nutrients and reduced species (Moore 1999, 2010; Slomp and Van Cappellen 2004; Spiteri et al. 2008). Freshwater lenses (FWL) on barrier islands present a special case of coastal freshwater aquifers and allow for hydrochemical investigations of groundwater at the land-sea transition zone. This section provides an overview of advances which were made regarding the hydrochemical evolution of FWL, as an example for coastal freshwater aquifers.

FWL on barrier islands evolve in dune areas that are not prone to inundation events (Röper et al. 2013; Holt et al. 2017). The formation process is mainly driven by density differences between infiltrating freshwater and saline groundwater (Drabbe and Badon Ghijben 1889; Herzberg 1901; Fetter 1972; Vacher 1988), while factors such as tides, groundwater recharge rates, land propagation, or climate change may play an additional role (Röper et al. 2013; Stuyfzand 2017; Holt et al. 2019). Being the only natural freshwater resource on barrier islands, FWL are precious for both drinking water supply and freshwater ecosystems. However, storm surges, sea level rise, and overexploitation pose serious threats to these sensitive aquifers (Anderson 2002; Post 2005; Post and Houben 2017).

A major process that determines the water composition of FWL below barrier islands is the dissolution of calcareous shell debris (Röper et al. 2012; Houben et al. 2014; Seibert et al. 2018; Fig. 12.1a). Shell dissolution results from the infiltration of acidic rain into dune sediments, with further acidity being supplied to the subsurface due to carbon dioxide production from respiration in the soil by vegetation and microorganisms (Stuyfzand 1993, 1998). Marine shells, including those in the North Sea region, are predominantly made up of biogenic carbonates (e.g., Blackmon and Todd 1959; Hild 1997; Milliman et al. 2012; Seibert et al. 2018; Winde and Böttcher, unpubl. data), and the dissolution of these typically leads to an enrichment of bicarbonate and calcium in FWL (e.g., Röper et al. 2012; Houben et al. 2014). The dissolution of calcite may take place in the water-unsaturated zone, i.e., an open system with respect to a CO_2-bearing gas phase, or in the water-saturated zone, i.e., a closed system (Appelo and Postma 2005). To distinguish between open or closed system conditions during calcite dissolution and to study further carbonate mineral reactions within the aquifer, carbon isotopes are widely used as an additional tool to bicarbonate and calcium concentrations

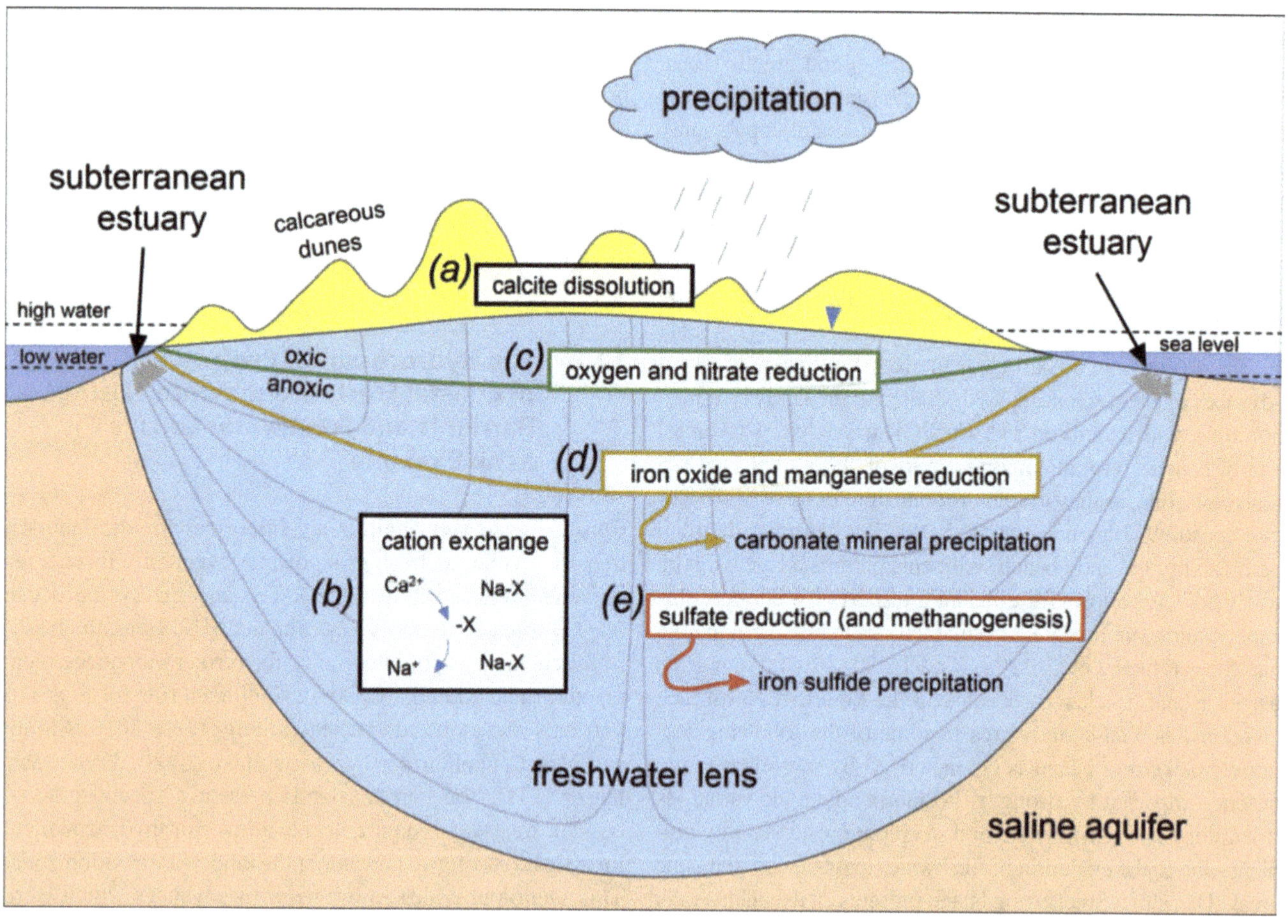

Fig. 12.1 Conceptual model of a freshwater lens below a barrier island and important hydrogeochemical processes. The hydrologic cycle of the freshwater lens starts with precipitation infiltrating into the dune sediments, followed by groundwater recharge, flow through the freshwater lens (conceptual flowlines are indicated), and discharge in the STE. (**a**) calcite dissolution, typically proceeding in the unsaturated dune sediments, (**b**) cation exchange in a freshening aquifer, and (**c**) to (**e**) redox processes related to the degradation of organic matter, i.e., zones of oxygen and nitrate reduction (green colors), iron and manganese oxide reduction (brown colors), and sulfate reduction as well as methanogenesis (red colors), respectively, are indicated. (**d**) brown and (**e**) red arrows highlight that reduced species (i.e., Fe^{2+}, Mn^{2+}, and H_2S) may precipitate as carbonate minerals (e.g., siderite and rhodochrosite) and iron sulfides, respectively, in the anoxic zone of the freshwater lens. Note that the figure is not to scale. The vertical freshwater lens extent is exaggerated in comparison to the horizontal extent. Typically, horizontal and vertical extents of FWL below barrier islands range between several meters to kilometers and several meters to tens of meters, respectively

(Deines et al. 1974; Chapelle and Knobel 1985; Böttcher 1999). This is applicable because ^{13}C signatures of dissolved inorganic carbon (DIC) produced in the unsaturated zone are a function of the ^{13}C signature of the soil CO_2 gas phase and the pH, while the ^{13}C signature of DIC produced in the saturated zone is a function of both the signature of the dissolved, acidity-providing CO_2 and the dissolving carbonate (Deines et al. 1974; Appelo and Postma 2005). With regard to FWL, Bryan et al. (2017) and Seibert et al. (2018) used ^{13}C signatures of DIC to identify carbonate recrystallization processes within the aquifer. By applying ^{13}C signatures of DIC, Seibert et al. (2018) could further infer that degrading organic matter in the investigated unsaturated dune sediments presumably had a terrestrial origin.

Cation exchange is another prominent process at the land-sea transition zone. The governing principle is that the exchange of cations along a flow path is retarded compared to the pore water velocity, depending on the aquifer cation exchange capacity (CEC) and the composition of the initial (e.g., saline groundwater) and the infiltrating solutions (e.g., fresh groundwater) (Appelo and Geirnaert 1991; Stuyfzand 1993, 1999; Appelo and Postma 2005). Aquifer freshening accompanied by cation exchange is a commonly observed process in coastal aquifers and proceeds via the exchange of dissolved calcium (the dominant cation in freshwater of calcareous aquifers) for sodium, potassium, and magnesium (the dominant cations in seawater) bound to the aquifer solid phase (Fig. 12.1b) until equilibrium with fresh groundwater is reached (Appelo and Postma 2005). Therefore, a simultaneous decrease of calcium and an increase of sodium, potassium, and magnesium are typically observed along the flow path of a freshening aquifer (Chapelle and Knobel 1983; Beekman and Appelo 1991; Appelo and Postma 2005). For the case of evolving FWL below barrier islands, previous studies have shown that water compositions usually reflect the above-described trends for major ions in a freshening aquifer (Röper et al. 2012; Houben et al. 2014; Seibert et al. 2018). Cation exchange can further trigger calcite dissolution in a freshening aquifer due to the removal of solute calcium, subsequently leading to a subsaturation with respect to calcite. This was, for example, observed by Andersen et al. (2005) for a coastal sand aquifer prone to storm floodings, and Seibert et al. (2018) found indications for calcite dissolution triggered by cation exchange in a FWL of Spiekeroog Island, which was especially relevant at the early stages of the lens formation. Freshening times of FWL below sandy barrier islands typically range between several hundreds and thousands of years, after a hydrodynamic steady-state has been reached. As an example, Seibert et al. (2018) calculated a total freshening time of ~600 years for the freshwater aquifer below Spiekeroog Island, Germany. Freshening times may, however, greatly vary depending on aquifer properties (e.g., CEC), recharge rates, and end-member compositions.

The water composition of FWL below barrier islands is further modified by the degradation of organic matter via the consumption of different electron acceptors (Fig. 12.1c–e). Following the classical redox cascade based on overall energy yields and competitive exclusion, oxygen is consumed first followed by nitrate, manganese and iron oxide, and sulfate reduction and methanogenesis (e.g., Lovley and Phillips 1987; Chapelle and Lovley 1992; Appelo and Postma 2005), and reduced species (e.g., ammonium, sulfide, ferrous iron) are released to the groundwater under increasingly anoxic conditions. However, the redox sequences may not be as strictly separated in nature, because zones of iron oxide and sulfate reduction and methanogenesis can overlap, with a dependence on organic matter reactivity, iron oxide stability, and sulfate concentrations (Postma and Jakobsen 1996; Jakobsen and Postma 1999). While oxic to metal oxide reducing conditions were reported for near-surface groundwater of a FWL below a barrier island in Georgia, USA, by Snyder et al. (2004) and oxic conditions for a FWL in a limestone aquifer at Rottnest Island, Australia, by Bryan et al. (2017), Seibert et al. (2018) could show that also sulfate reduction and localized methanogenesis can be important pathways for the decomposition of organic matter in deeper and older groundwater of a barrier island FWL. The reduced species which are produced under anoxic conditions, such as dissolved ferrous iron and sulfide, may further undergo secondary reactions (Fig. 12.1d, e). For instance, dissolved sulfide may precipitate as iron sulfide (e.g., mackinawite and/or pyrite) if reactive iron oxides are present (Schoonen 2004; Rickard and Morse 2005), which was observed for the brackish transition zone of a young FWL at Spiekeroog Island (Seibert et al. 2019). Manganese and ferrous iron may precipitate as metal carbonates (e.g., rhodochrosite and siderite) if hydrochemical conditions are favorable (e.g., Seibert et al. 2018).

In conclusion, previous research has shown that the hydrochemical evolution of FWL in sandy calcareous barrier islands is dominated by calcite dissolution, typically leading to increased bicarbonate and calcium concentrations in young groundwater. Furthermore, cation exchange can affect the groundwater cation composition, with sodium and calcium concentrations commonly increasing and decreasing, respectively, along the flow path in a freshening aquifer. Finally, redox processes related to the oxidation of organic matter result in the consumption of different electron acceptors and the release of reduced species to the groundwater, which may cause secondary reactions such as the precipitation of iron sulfides or metal carbonates. Fresh submarine groundwater discharge (SGD) at barrier islands may, thus, supply anoxic groundwater (iron oxide- to sulfate-(methanogenic-)reducing conditions) to the STE with elevated concentrations of nutrients from degrading organic matter (e.g., ammonium and phosphate) and further reduced species, such as dissolved sulfide. Although the governing processes for the evolution of fresh groundwater in sandy coastal aquifers could be identified, future research could benefit from the following motivations: (i) Fresh groundwater end-member compositions should be better characterized to evaluate the role of fresh SGD for the chemical processes occurring in the STE and to quantify material fluxes from land to ocean; (ii) the effect of changing boundary conditions (e.g., changing amounts of groundwater recharge, variations of anthropogenic pollution inputs, effects of groundwater pumping and/or inundation events) as well as local effects (e.g., heterogeneities of the geology, effect of different vegetation covers) should be regarded; (iii) reaction rates are often not well-constrained and should be investigated more closely by combining field studies with reactive transport modeling; and (iv) hydrochemical and geochemical studies should be jointly conducted where possible to link aquifer solute and solid-phase biogeochemistry.

12.3 Nutrients and Trace Metals in Subterranean Estuaries of Sandy Beach Sediments

The nutrient and trace metal distribution in the pore water of subterranean estuaries (STE) is strongly affected by salinity and redox gradients. Under reducing conditions, nitrogen may be lost through denitrification. Manganese and iron oxide reduction lead to the liberation of dissolved manganese and iron to the pore water and the subsequent release of phosphorus from the solid phase (Stal et al. 1996; Slomp and Malschaert 1997). Sulfate-reducing conditions result in the precipitation of iron as iron sulfide (Roy et al. 2011). The aeration of surface sediments, in turn, may lead to nitrification and the precipitation of manganese and iron oxides, the latter resulting in the removal of phosphate through adsorption (Slomp and Malschaert 1997; Spiteri et al. 2008). Increased pore water silica concentrations may result from the dissolution of biogenic opal, e.g., diatom frustules, or quartz dissolution (Anschutz et al. 2009; Ehlert et al. 2016).

Wave- and tide-dominated sandy beaches typically exhibit a salinity distribution being characterized by a surficial saline circulation cell in the intertidal zone. A freshwater discharge tube separates this upper saline plume from the saltwater wedge near the low water line (Robinson et al. 2007). The saline circulation cell may have a vertical extent of more than 15 meters (Robinson et al. 2007; Seidel et al. 2015; Beck et al. 2017). Low-salinity water discharging at the low water line has traveled a long way to its discharge point (Fig. 12.1) and may thus have a different redox state and nutrient and trace metal composition compared to young low-salinity groundwater found in the supratidal area of a beach site (Reckhardt et al. 2017). Seawater circulating through the beach face is usually also subject to changing redox conditions with increasing residence time in the sediment, i.e., along the flow path toward the low water line (O'Connor et al. 2015; Reckhardt et al. 2015).

On Spiekeroog Island (Germany), groundwater of the island's freshwater lens mixes with a small portion of seawater close to the dunes (Fig. 12.2). This pore water is characterized by high nitrate concentrations indicating oxic conditions. Around the mean high water level, oxic seawater infiltrates into the sediment. However, aerobic degradation of organic material is limited to the upper 1–2 m. Below, nitrate serves as terminal electron acceptor, and low concentrations of nitrite are found in the pore water. Toward the low water line, the oxic sediment layer stretches only some centimeters, whereas anoxic conditions and reduction of iron, manganese, and occasionally sulfate are found below (Fig. 12.2). This part close to the discharge point is characterized by high nutrient (dissolved inorganic carbon, ammonium, phosphate, and silica) and metal (manganese and iron) concentrations. Freshwater infiltration from below decreases the salinity near the low water line (Fig. 12.2). According to a modeling approach, this water originates from deep (>20 m) and anoxic parts of the island's freshwater lens (Beck et al. 2017; compare Fig. 12.1). This freshwater lens is characterized by low dissolved iron and manganese concentrations (Seibert et al. 2018).

In contrast, there are other STE where fresh groundwater is a significant iron and manganese source. These systems are typically characterized by metal oxide precipitation, where suboxic freshwater encounters oxic seawater (Charette et al. 2005; McAllister et al. 2015). A second source of manganese and iron may develop in the saline circulation cell. Such conditions were found in the STE of Waquoit Bay, Massachusetts, USA (Charette et al. 2005). Under sulfidic conditions, however, iron may be removed from the solution due to the precipitation of iron sulfides (McAllister et al. 2015).

At the Aquitanian coast (France), the saline circulation cell of the intertidal zone is also characterized by decreasing oxygen concentrations toward the discharge point. Redox conditions, however, do not reach a lower level than occasional denitrification, and the main nitrogen species being discharged remains to be nitrate, not ammonium (Charbonnier et al. 2013). Similarly, at Huntington Beach (California, USA), oxic freshwater meets oxic saline pore water, and nitrogen discharges primarily as nitrate. Low organic matter inputs and high advective pore water flow allow the groundwater pool of such systems to remain oxygenated (Santoro 2009).

Several studies suggest that pore water draining from sandy coastal sediments represents a source of nutrients and

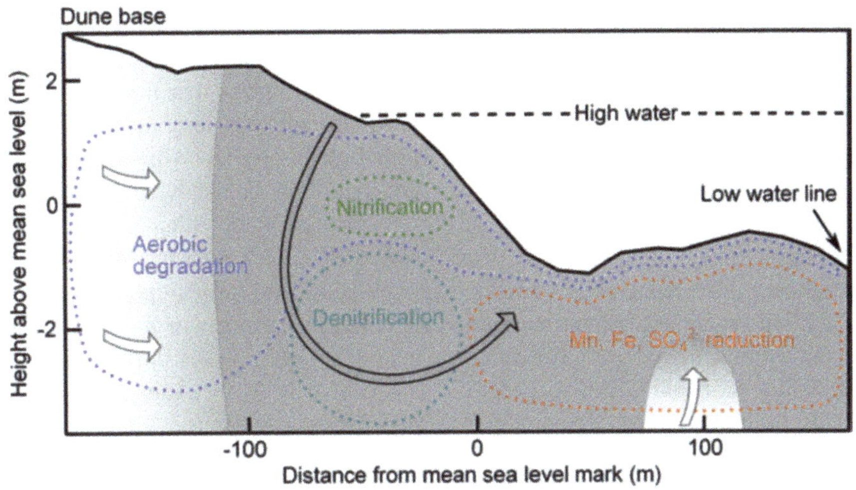

Fig. 12.2 Biogeochemical processes and redox zones at a beach site studied on Spiekeroog Island (based on the concept of Beck et al. 2017). The transect stretches from close to the dune base to the low water line. Aerobic degradation (blue) dominates close to the dune base, nitrification (green) and denitrification (cyan) were identified at the high water line and Mn/Fe oxide and sulfate reduction (red) take over toward the low water line. The shading indicates the gradient from fresh terrestrial groundwater (white) to saline seawater (gray). The arrows show main groundwater flow pathways, such as the upper saline plume (gray arrow) and terrestrial freshwater input from the island's freshwater lens and freshwater discharge tube (white arrows)

Table 12.1 Nutrient fluxes from subterranean estuaries into adjacent seawater

Nutrient	Flux (mmol d^{-1} $m_{shoreline}^{-1}$)	Reference
NO_3^-	296	Anschutz et al. (2009)
NH_4^+	117	Beck et al. (2017)
PO_4^{3-}	55	Beck et al. (2017)
	15.2	Anschutz et al. (2009)
$Si(OH)_4$	575	Beck et al. (2017)
Mn	48	Reckhardt et al. (2017)
Fe	185	Reckhardt et al. (2017)

trace metals to the surrounding surface seawater (Caetano et al. 1997; Kroeger and Charette 2008; Anschutz et al. 2009). This flux can be in the order of millimoles per day and meter of shoreline (Table 12.1). The examples listed above show that STE can be subject to different redox characteristics, which have a major influence on the pore water composition. Since organic matter degradation occurs in most STE, nutrient concentrations are higher in pore waters compared to adjacent seawater at many locations. Where aerobic degradation dominates, the beach may serve as source of dissolved inorganic carbon, nitrate, phosphate, and silica to the surrounding ocean. Under anaerobic conditions, manganese and iron may be added to the nutrient pool and nitrogen will be discharged as ammonium. STE may consequently play different roles in coastal carbon, nutrient, and trace metal cycling (Slomp and Van Cappellen 2004). Therefore, it is necessary to characterize redox conditions in many beach systems and to quantify solute fluxes into the adjacent ocean to finally be able to assess the global impact of SGD.

12.4 Dissolved Organic Matter in the Subterranean Estuary

Dissolved organic matter (DOM), operationally defined as any organic matter passing through a 0.2–0.7 μm filter, is the fuel that drives microbial respiration and production of the climate-relevant carbon dioxide in the subterranean estuary (STE). It is therefore a crucial link between the organic and inorganic carbon cycle at this land-ocean interface. Dissolved organic matter can be quantified as dissolved organic carbon (DOC), but DOM compounds may also contain hydrogen, oxygen, nitrogen, phosphorous, and sulfur. Furthermore, they can bind and transport trace metals in coordination complexes. Early quantitative investigations of bulk DOC in a STE showed conservative increases with increasing salinities, leading to the suggestion that DOC was not substantially altered on its passage through the coastal aquifer (Beck et al. 2007). However, Santos et al. (2008, 2009) found enrichment of DOC at mid-salinities compared to the low- and high-salinity end-members in the STE, analogous to surface estuaries (Osterholz et al. 2016). They attributed this pattern to microbial release of DOC from marine debris buried during tidal inundation, as well as autochthonous production, for example, by microphytobenthos (Santos et al. 2008, 2009; Chipman et al. 2010; Reckhardt et al. 2015). Despite reporting contrasting patterns in DOC distributions within the STE, studies on DOC indicated three overarching trends: (i) The majority of the DOC loads were derived from autochthonous and allochthonous marine sources, (ii) the marine-derived DOC drives microbial respiration successions (oxygen respiration, denitrification, and iron and manganese reduction) along the groundwater flow paths, and (iii) the tide- and wave-driven burial and hydrolysis of marine organic matter may turn terrestrial, fresh groundwater or brackish groundwater mixtures into transport vehicles for marine DOM (Santos et al. 2008, 2009; Reckhardt et al. 2015; Beck et al. 2017). As a consequence of the dominance of marine DOC in the STE, the resulting submarine groundwater discharge (SGD) delivers mainly recycled DOC back into the coastal water column. The input of this recycled DOC is particularly high during summer, when primary production in the water column and activity of the STE microbial reactor are highest (Santos et al. 2009; Kim et al. 2012).

DOC is merely a bulk proxy for DOM, containing no qualitative information about sources and reactivity of its components. Taking elemental combinations, size fractions, and steric properties into account, it is likely that DOM consists of hundreds of thousands of individual compounds at the pico- to nanomolar level (Zark et al. 2017). Fluorescence and absorbance analyses of chromophoric DOM (CDOM) provide swift and straightforward identification of organic compound groups with chemical characteristics indicative of their reactivity and origin (Coble 1996). These analyses can be combined with targeted measurements of molecular building blocks, for example, sugars and amino acids, as well as bulk determinations of $\delta^{13}C$ isotope signatures characteristic for terrestrial and marine primary producers. The most comprehensive method of DOM characterization to date is ultrahigh resolution Fourier transform ion cyclotron resonance mass spectrometry (FT-ICR-MS), which allows the assignment of unique molecular formulae to thousands of masses detected in a single DOM sample (Fig. 12.3).

So far, only two publications report FT-ICR-MS data of DOM composition in the STE (Seidel et al. 2015; Beck et al. 2017), whereas CDOM analyses are more widely applied (Kim et al. 2012, 2013; Suryaputra et al. 2015; Nelson et al. 2015; Couturier et al. 2016). Overall, these qualitative DOM studies confirmed supply of marine, labile DOM to the STE by tides and waves, indicated by enrichment patterns of labile organic matter, such as sugars, amino acids, and protein-like CDOM, at deposition sites of algal debris or in areas where seawater infiltration occurred (Fig. 12.3; Kim et al. 2012; Seidel et al. 2015). On the other hand, strong negative correlations of humic-like CDOM fluorescence and absorbance with salinity have been attributed to a terrestrial source of aromatic DOM compounds. As a result, humic-like CDOM fluorescence has been used as SGD tracer in areas with productive submarine springs (Fig. 12.3; Nelson et al. 2015; Kim and Kim 2017). Compared to aliphatic, protein-like CDOM, the aromatic, humic-like DOM compounds are less biodegradable and more photodegradable and absorb sunlight. Furthermore, humic-like DOM compounds form complexes with trace metals such as iron and copper, and iron oxide formation along redox gradients of the STE has been identified to sequester terrestrially derived organic matter (Abualhaija et al. 2015; Linkhorst et al. 2017; Sirois et al. 2018). This means that the quality of DOM in STE could vary substantially along the advective flow paths, with profound effects on DOM bioavailability, reactivity, and light attenuation of the receiving coastal water column.

Although overarching patterns of DOM processing in several STE have been derived from quantitative and qualitative studies, still surprisingly little is known about this important electron donor which drives the microbial reactor (see also Sect. 12.5) and is thus crucial in controlling retention and/or release of, for example, inorganic nutrients (nitrogen, phosphorous, and carbon dioxide) and redox-sensitive trace metals (manganese, iron). Export of organic matter from the STE into the coastal water column could be an important land-ocean pathway affecting the global carbon budget. However, to date even quantitative DOC data are only available for a handful of sites, although SGD is a ubiquitously occurring phenomenon along the global coastline (Zhang and Mandal 2012). While the first global estimates on SGD-driven nitrogen, phosphorous, and silicon fluxes have been published recently (Cho et al. 2018), not enough data exists to calculate potential contributions of SGD to land-ocean DOC inputs. Furthermore, qualitative data are still lacking for a larger diversity of STE. These data are imperative, however, to delineate recycled marine inputs from new terrestrial loads, which are likely to increase as a result of global climate change (higher temperatures and precipitation). Finally, ultrahigh resolution molecular data will be crucial to assess the information gathered by swift in situ methods such as CDOM tracing. For example, Suryaputra et al. (2015) found release of humic-like fluorescent DOM (FDOM) at mid-salinities in a STE and linked these signals to marine rather than terrestrial sources. This may prevent an unambiguous identification of DOM sources based on CDOM data alone. A fundamental understanding of DOM behavior throughout the STE is an essential prerequisite to develop mechanistic models for geochemical speciation, which currently neglect DOM complexity. Such adapted reactive transport models will ultimately be the key to predict the role of SGD as DOC source or sink in the light of global change.

12.5 Microbial Community Composition of the Subterranean Estuary

From a microbiological perspective, land-sea interfaces are of special interest, since end-members with different nutrient compositions and physicochemical parameters merge. These environmental settings are known to enhance microbial diversity by creating niches for specialists. Coastal seawater usually is rich in sulfate, oxygen, and organic carbon, but (temporarily) limited in nitrogen, while terrestrial ground-

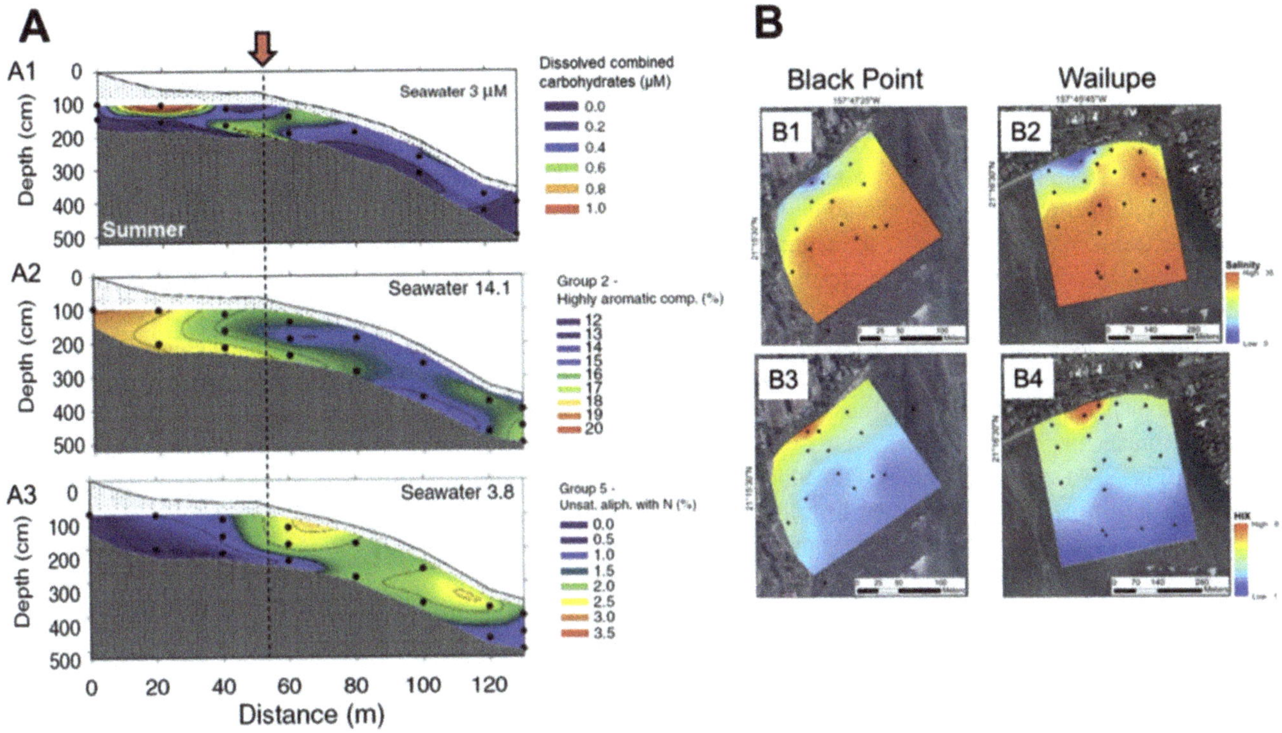

Fig. 12.3 (**a**) Distribution of carbohydrates (**a1**, concentration in µM) and DOM molecular groups indicative of terrestrial (aromatic compounds, **a2**, relative abundance in %) and marine (aliphatic peptide formula compounds, **a3**, relative abundance in %) origin in a STE on Spiekeroog Island during a summer campaign. The x axis shows distance in (m) from the dune base toward the low water line, and the y axis shows sampling depths in (cm). The red arrow indicates the mean high water line. White dotted areas represent de-saturated sediments, for which no pore water could be extracted, and grey areas indicate the water-saturated, unsampled deep sediment layer. (**b**) Co-variance of low salinity groundwater springs (**b1**, **b2**) and humic-like CDOM (represented by the "HIX" index, **b3**, **b4**) in two marine embayments, Black Point and Wailupe, on Hawaii. Low salinities are indicated by cool colors, and high relative abundances of terrestrially-derived humic substances (or HIX) are indicated by warm colors. Modified from (**a**) Seidel et al. (2015) and (**b**) Nelson et al. (2015) with permissions from Elsevier

water often can be a source of nitrogen, phosphorous, and iron. Garcés et al. (2011) have previously shown that nutrient input by terrestrial sources can influence microbial abundance and diversity and even induce phytoplankton blooms.

In coastal systems, the sedimentation of decaying phytoplankton blooms stimulates the activity of heterotrophic bacteria by introducing large quantities of organic carbon. The high microbial activity leads to a rapid depletion of electron acceptors like oxygen and nitrogen, usually in the form of nitrate. In Wadden Sea tidal-flat sediments, which are mostly sand flats, this typically happens within the first millimeters to centimeters (Billerbeck et al. 2006; Böttcher et al. 2000). Following a thin zone of manganese and iron reduction, those sediments then show wide zones of sulfate reduction and methanogenesis (Wilms et al. 2007). A similar pattern, following the availability of electron acceptors, has been reported for other land-sea transition zones, e.g., for mangrove systems (Kristensen et al. 2000; Holguin et al. 2001), yet these environments are different from sandy beaches in many of their physicochemical characteristics. Furthermore, submarine groundwater discharge (SGD), which can transport nutrient quantities comparable to those of rivers (Kwon et al. 2014), influences microbial communities and biogeochemical cycling in the subterranean estuary (STE).

The solute composition of pore water being transported into the ocean is altered along the flow path due to the reduction of electron acceptors and the oxidation of organic matter by microbial communities (Slomp and Van Cappellen 2004). On the one hand, the way of alteration depends on the community composition of the STE. On the other hand, the microbial communities and thus their metabolic potential are influenced by the physicochemical properties (e.g., salinity, temperature) within the STE (Santoro et al. 2008; Lee et al. 2016). Understanding which microbes are driving specific biogeochemical cycles in the STE and the respective conditions that are necessary for them to thrive in this habitat will help to predict the effects of SGD on different coastal environments. So far, only little is known about microbial communities in STE at the land-sea transition zone and the key processes they are driving. Most studies on these kinds of aquifers were performed from a chemical point of view (Santos et al. 2008) but are lacking a molecular in-depth analysis on the microbial communities.

Furthermore, not many studies have looked at the microbial diversity of STE, using modern sequencing technologies. A previous molecular study from the island of Spiekeroog by Beck et al. (2017) was based on denaturing gradient gel electrophoresis (DGGE). However, DGGE is an early molecular technique and does not gain the detailed amount of information like next-generation sequencing. Despite the technological limitations, the authors found a higher diversity of the bacterial communities in comparison to those of the archaea, which confirms findings of Lipp et al. (2008). They also showed that aerobic organic matter degradation dominates the supratidal part of the beach, while suboxic and anoxic processes like manganese and iron reduction become dominant in sediments located closer to the low water line (Figs. 12.1 and 12.2). Sulfate reduction seems to be insignificant in the first 1.5 m of the investigated sediments. A similar microbial response was assumed by a geochemical study on salt marshes of Sapelo Island, Georgia, USA (Snyder et al. 2004). Here, iron reduction was identified as the main pathway of anaerobic organic matter degradation, despite the availability of sulfate. This finding is surprising, since if the investigated sediments are not very oligotrophic, metal oxides are usually reduced quickly, followed by sulfate reduction in deeper sediment layers. Yet none of those two studies could show why metal oxide reduction is preferred over sulfate reduction and which bacteria are involved in this process. A study conducted by McAllister et al. (2015) at Cape shores, Lewes, Delaware, USA, describes an iron-rich aquifer with a distinct zone characterized by iron sulfides at the discharge site, which precipitate due to active sulfate reduction. Within that aquifer, the microbial community is structured along the availability of different iron and sulfur species, and not only sulfate reduction but also iron oxidation was observed in different parts of the aquifer.

However, microbial community compositions are not only shaped by the availability of electron acceptors but also by the physicochemical properties of the investigated systems (Santoro et al. 2008; Lee et al. 2016). Santoro et al. (2008) investigated a sandy STE at Huntington Beach, California, USA, with a focus on the abundances of ammonium-oxidizing proteobacteria and ammonium-oxidizing archaea. While the archaea showed constant abundances independent of the salinity, abundances of bacterial ammonium oxidizers were much lower under freshwater than under saltwater conditions. They also observed that the shift between those two groups migrated with the brackish mixing zone during the year. Another study that found a correlation between microbial community structure and hydrology was performed at Jeju Island, Korea, by Lee et al. (2016), who discovered that the microbial community composition can be altered by the SGD rate within the different stages of a tidal cycle.

In summary, it is not yet coherently understood which microbial groups thrive in STE, how the geochemical conditions are shaping the community composition, and what hydrogeological factors are introduced by the heterogeneity of land-sea interfaces. Microbial diversity and activity are not only influenced by the availability of different electron acceptors and donors but also by hydrological parameters influencing the stability of redox conditions and salinities, such as waves, SGD discharge rate, and seasonal patterns. In order to understand if and how microbial processes influence coastal oceans by changing the chemical composition within STE, further in-depth research is needed to determine key species responsible for organic matter degradation and nutrient cycling. For instance, while SGD with an anthropogenic influence contributes to coastal ocean eutrophication (Beusen et al. 2013), the effect of SGD in pristine areas is fully unclear. Investigating this would give an understanding of how the chemical load transported within a STE is altered before being discharged into the oceans. Another question to be answered is which factors are responsible for the occurrence of sulfate reduction in STE, as it proceeds to a significant degree at some locations, while it cannot be detected at others.

12.6 Radiotracers: A Useful Toolbox for Quantifying Rates and Fluxes

12.6.1 Estimating Pore Water Residence Times

A key factor governing organic matter degradation efficiency and nutrient transformation pathways within subterranean estuaries (STE) is the pore water residence time, which is defined as the timespan between seawater infiltration and discharge. Numerical simulations showed that seawater residence times can range from orders of a few days in the upper saline plume to about 1000 days in the saltwater wedge (Robinson et al. 2007). The combination of residence times and microbial reaction rates determines nutrient and metal transformations as well as organic matter degradation efficiency. After oxygen has been consumed, other electron acceptors are favored, lowering the redox potential and affecting nitrogen transformation: As an example, short residence times (up to 1 week) in combination with a large supply of oxygen may enrich pore waters in nitrate by nitrifying organic bound nitrogen (Anschutz et al. 2009). Contrary, at a certain setup of residence times and hydrogeological forcing, denitrification can remove nitrate efficiently by producing N_2 (Heiss et al. 2017). Finally, long residence times in the range of months can lead to an enrichment in ammonium, enhancing the overall export of dissolved inorganic nitrogen (Beck et al. 2017; Tamborski et al. 2017). The residence time may therefore have profound influence on the release of degrada-

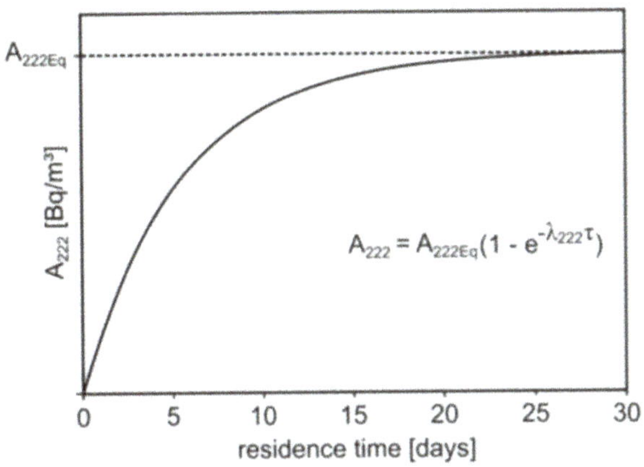

Fig. 12.4 Activity of dissolved ^{222}Rn (A_{222}) as function of A_{222Eq} and $\lambda_{222} = 0.1813$ day^{-1}

tion products like nutrients, metals, and inorganic carbon into the pore water and finally to the adjacent sea.

To estimate pore water residence times, the usage of dissolved ^{222}Rn (half-life of 3.8 days) has become a valuable method in both marine and terrestrial systems (Hoehn and Von Gunten 1989; Snow and Spalding 1997; Colbert et al. 2008; Goodridge and Melack 2014; Tamborski et al. 2017). In seawater, ^{222}Rn activities[1] are low due to its decay, degassing, and a high rate of dilution of pore water-derived ^{222}Rn. This results in the initial ^{222}Rn activity of infiltrating seawater to be almost zero. Within the sediment, dissolved ^{222}Rn is produced by the decay of its parent isotope ^{226}Ra, which is part of the uranium decay chain and occurs naturally in sediment minerals and in Fe-Mn oxide coatings on mineral surfaces (Swarzenski 2007). Because of the long half-life of the parent isotope ^{226}Ra (1602 years), its own decrease is negligible and the production rate of ^{222}Rn remains constant. After a certain time of increasing ^{222}Rn activity, it reaches an equilibrium activity A_{222Eq} (Fig. 12.4), and the production of ^{222}Rn is balanced by its own decay. Since the absolute decay rate of ^{222}Rn is proportional to its concentration, a certain level of ^{222}Rn has to grow in until its production rate is balanced. The calculation of residence times is based on the predictable ingrowth of ^{222}Rn until the equilibrium activity A_{222Eq} is reached (after 5–6 half-lives, or 20–25 days; Fig. 12.4). Because the half-life of ^{222}Rn (3.8 days) is much shorter than of ^{226}Ra (1602 years), it can be assumed that under equilibrium the ^{222}Rn activity equals its precursor's activity – in other terms, whenever a ^{226}Ra atom disintegrates, its daughter isotope ^{222}Rn is "immediately" decaying (Eq. 12.1):

$$A_{226} \approx A_{222\,Eq} \tag{12.1}$$

Based on the assumption of a constant production rate, the Bateman equation (Bateman 1910), describing activities in decay chains, can be transformed to describe the temporal evolution of the ^{222}Rn activity A_{222}. Thus, A_{222} is a function of its production rate, which is equal to A_{222Eq} and its own decay rate $\lambda_{222} = 0.1813$ day^{-1} (Eq. 12.2, Fig. 12.4):

$$A_{222} = A_{222\,Eq}\left(1 - e^{-\lambda_{222}\tau}\right) \tag{12.2}$$

If A_{222Eq} is known, the equation can easily be solved for residence time τ. Tamborski et al. (2017) additionally corrected for initially dissolved seawater ^{222}Rn and extended the equation to eliminate excess ^{222}Rn deriving from mixing with deep fresh groundwater, potentially delivering high loads of ^{222}Rn to the STE.

The application of this equation requires the boundary conditions of a closed system and the assumption that ^{226}Ra is equally distributed in the sediment. Because the ingrowth of ^{222}Rn approaches its equilibrium activity asymptotically (Fig. 12.4), the application is limited to a maximum residence time of four to five half-lives (15–20 days). While early studies investigated ^{222}Rn residence times in terrestrial aquatic systems (Hoehn and Von Gunten 1989; Snow and Spalding 1997), Colbert et al. (2008) used ^{222}Rn to deduce flow dynamics in a tidal beach. The authors assessed the potential of Rn as a geochronometer by comparing ^{222}Rn residence times with residence times derived from tidal wedge balance calculations of the same site. However, the authors considered Rn loss due to evasion to unsaturated sediments as a reason of age underestimations. In tidal systems with significant water table fluctuations, the characteristics of ^{222}Rn as dissolved gas need to be considered.

A detailed study of applying ^{222}Rn and Ra isotopes as indicators of seawater residence times in a tidal STE was published by Tamborski et al. (2017). The authors compared two sites and could deduce the influence of different hydrological forcing regimes on residence times of circulating seawater: A large seaward hydraulic gradient of fresh groundwater leads to short residence times of infiltrating seawater in the intertidal circulation cell, whereas longer seawater residence times were observed at a site, where freshwater supply was reduced and seawater percolation dominated the flow regime. Additionally, by investigating sediment geochemistry, the radionuclide behavior was linked to Fe and Mn cycles, as their oxyhydroxides may scavenge Ra. Due to the microbially mediated dissolution of its carrier phase, a redistribution of ^{226}Ra may change the production rate of ^{222}Rn (Tamborski et al. 2017). This is offending the necessary condition of a constant ^{222}Rn production rate, respectively, equilibrium activity A_{222Eq} (Eq. 12.2), creating uncertainties and restricting its applicability.

The studies listed above demonstrate that Rn disequilibrium approaches are a promising method in dynamic beach systems. Permeable sandy beaches are often fulfilling the requirements for ^{222}Rn applications, with respect to timescales and flow velocities. The knowledge about temporal hydrodynamics in beach systems is vital to understand the

[1]Radioactive compounds are typically expressed quantitatively by their activity (a combination of concentration and decay constant).

pathways and the efficiency of organic matter degradation and biogeochemical rates in groundwater.

12.6.2 Quantification of Submarine Groundwater Discharge

The importance of processes in the STE derives from the fluxes of ecologically relevant pore water constituents across the sediment water interface (Marsh Jr 1977; Moore 1996; Kim et al. 2005). Estimating these fluxes is challenging, because it requires the volumetric quantification of submarine groundwater discharge (SGD). Groundwater seepage is spatially and temporally highly variable and has been observed to occur patchy, diffuse, and variable on different timescales (seconds to seasons), depending on aquifer characteristics and driving forces (Burnett et al. 2006; Santos et al. 2012). Various methods have been developed and applied to quantify SGD: For example, a very simple method to quantify discharge includes the use of manual seepage meters – usually PVC or steel domes, which collect the discharging water in a plastic bag to measure its volume (Fig. 12.6). However, if SGD is assumed to have a patchy distribution, several seepage meters of this kind would be necessary to quantify SGD accurately (Burnett et al. 2006). Numerical modeling approaches are much more complex and can also be applied to calculate volumetric fluxes. In general, numerical models are calibrated, using field data to estimate, for instance, hydraulic conductivities, the unconfined storage coefficient, and to reproduce observed salinity distributions (e.g., Beck et al. 2017). The main challenge is to obtain these input variables representatively, because these may vary in space and time (Burnett et al. 2006). Numerical modeling requires a set of assumptions, boundary conditions, and simplifications. Hence, simple modeling approaches are not capable to capture aquifer heterogeneities sufficiently, leading to errors or limitations in SGD estimates.

Another option to estimate groundwater fluxes is to use natural radiotracers, which are highly enriched in groundwater relative to seawater and behave conservatively with respect to mixing and biological processes (Moore 1996). Unlike residence time calculations (cf. Sec. 12.6.1), which focus on the radioisotope behavior in the sediment subsurface prior to discharge, groundwater flux quantifications rely on the behavior of radioisotopes in the water column after discharge.

The application of Ra isotopes and ^{222}Rn has evolved to a well-established method to estimate groundwater input (Burnett and Dulaiova 2003; Kim et al. 2005; Burnett et al. 2006; Moore 2006; Peterson et al. 2008; Liu et al. 2018). The advantage of using a natural tracer approach is that it integrates overall groundwater input pathways and is not affected by small-scale spatial heterogeneities or temporal variations (Burnett et al. 2006). The disadvantage of these methods is that calculations are based on isotope mass balances that require the knowledge of all sinks and sources, which are often challenging to determine (Burnett et al. 2006).

The principle of SGD calculation is based on a steady-state mass balance approach: The tracer's addition to a box (e.g., shelf water volume) is balanced by its loss including decay, mixing loss, and atmospheric evasion for the gaseous Rn (Fig. 12.5) (Burnett and Dulaiova 2003). Therefore, all non-SGD sources must be quantified, such as diffusion from sediments and river flux in case of Ra applications – often estimated by determining background activities. Additionally, exchange at the box edges (mixing loss) and residence time of water within that box must be well constrained (Burnett and Dulaiova 2003). Another prerequisite is that the end-member activity of SGD is known, which is still the most struggling step, and contributes to large proportions of uncertainties of that approach (Burnett and Dulaiova 2003; Burnett et al. 2006; Peterson et al. 2008).

In a detailed study, applying Rn and Ra isotopes, Peterson et al. (2008) determined an offshore transport rate, to better constrain the offshore mixing loss factor. The authors calculated apparent water mass ages on an offshore transect using the activity ratio of short-lived ^{224}Ra (A_{Ra224}; λ_{224} = 0.1909 day^{-1}) to long-lived ^{223}Ra (A_{Ra223}; λ_{223} = 0.0606 day^{-1}):

$$AR = \frac{A_{Ra224}}{A_{Ra223}} \tag{12.3}$$

Based on the assumption that the same activity ratio is constantly discharging into the water column by SGD, the temporal evolution of this ratio can be predicted due to their distinct half-lives. Consequently, the timespan from being discharged ($AR_{initial}$) to reaching its sampling site ($AR_{observed}$) can be calculated (Moore 2000):

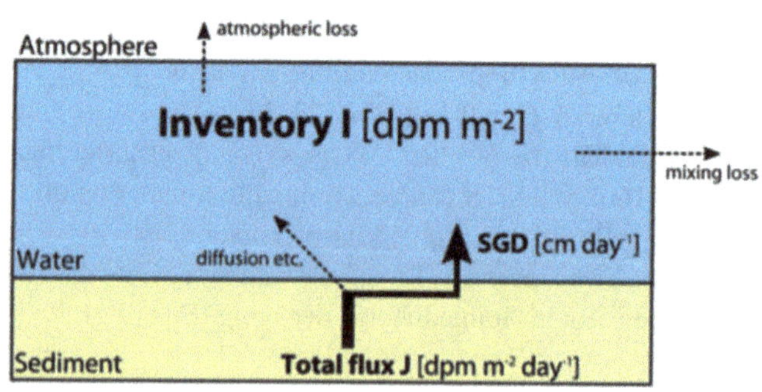

Fig. 12.5 Conceptual model of the radon (Rn) mass balance. Submarine groundwater discharge (SGD) and diffusion are supplying Rn, whereas mixing loss and atmospheric loss reduce the inventory of Rn in a defined volume. All non-SGD sources and sinks, which need to be well-constrained, are indicated by dashed lines. The figure was produced based on the concept of Burnett and Dulaiova (2003)

$$t = \ln\left(\frac{AR_{\text{initial}}}{AR_{\text{observed}}}\right)\frac{1}{\lambda_{224} - \lambda_{223}} \quad (12.4)$$

$AR_{initial}$ corresponds to the end-member concentration measured at seepage sites. Peterson et al. (2008) combined the transport rate, based on the water age calculation, and a 24 h time series sampling to calculate SGD rates in the Yellow River Delta, China. They found nitrate fluxes from SGD were two- to threefold higher compared to river fluxes to the adjacent sea. This often overlooked nutrient input mechanism can be a missing term in coastal nutrient budgets and, at worst, support the growth of harmful algal blooms (Hu et al. 2006).

Numerous studies have applied Ra isotopes and ^{222}Rn methods to calculate SGD contributions of pore water constituents to coastal waters – often conducted in bays which simplify the declaration of box boundaries (see listed in Burnett et al. 2006; Liu et al. 2018). The comparison of different SGD quantification methods has shown that each method addresses SGD in a specific way, and a combination of methods is the best way to constrain SGD in a comprehensive way (Burnett et al. 2006). Because Ra isotopes are preferentially desorbed from particles by water with high ionic strength (Li et al. 1977), estimations based on Ra isotopes are affected by the saline component of SGD. In contrast ^{222}Rn is not affected by pore water salinity and generally captures total SGD. The subtraction of both flux estimations could thus account for the fresh SGD component (Mulligan and Charette 2006; Peterson et al. 2008).

It is vital to understand the transformation and liberation of nutrients, carbon, and metals in the STE to determine the fate of pore water constituents at the land-sea interface. Residence time calculations can help to constrain these mechanisms. Additionally, gaining knowledge about the magnitude of SGD-derived inputs is essential to understand coastal (eco-)systems. Ra isotopes and ^{222}Rn are extremely useful tracers of SGD, and adapting their chemical behaviors and half-lives to the study site and the timescales of SGD flow paths will help to constrain hydrodynamic driving factors and their influence in the coastal realm.

12.7 Developing a New Type of Seepage Meter

For a better understanding of the role of submarine groundwater discharge (SGD) for solute fluxes and coastal environments, it is necessary to determine the flow rates and the physical parameters of discharge. The flow rate seems to vary during tidal cycles in direction and intensity, and, thereby, also the physical parameters, e.g., electrical conductivity or temperature, may change. To get a better understanding of temporal and spatial patterns of SGD in sandy subterranean estuaries (STE), high temporal resolution measurements during several tidal cycles are needed.

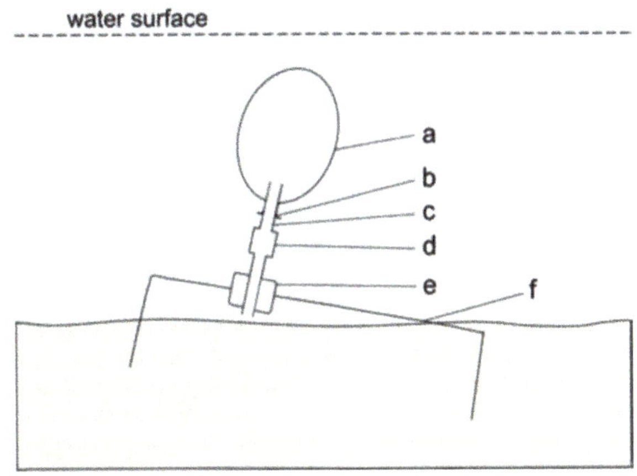

Fig. 12.6 Concept of the seepage meter based on Lee (1978), which is properly placed in the sediment. (**a**) 4 liter plastic bag; (**b**) rubber-band wrap; (**c**) polyethylene tube; (**d**) amber-latex tube; (**e**) one hole rubber stopper; (**f**) epoxycoated cylinder

A common device for measuring SGD is a seepage meter, built as a dome with an attached bag on top (Fig. 12.6). The dome collects outflowing groundwater and transfers it into the bag (Lee 1977, 1978). With this approach, it is possible to get an integrated signal for a defined area over a given time, yielding a flow rate. One disadvantage of this method is that the measurement time is limited due to the volume of the bag. Other seepage meter models are, for example, equipped with an ultrasonic sensor (Paulsen et al. 2001) or with an electromagnetic sensor (Rosenberry and Morin 2004; Waldrop and Swarzenski 2006) for measuring water flow of SGD. These models need a direct connection to a base station and an onshore power supply.

A study on SGD on Spiekeroog Island, Germany, shows that outflow rates are relatively low (between 2 and 4 l m^{-2} h^{-1}), and, due to this, it is difficult to accurately measure volumes (Grünenbaum et al. 2017). Furthermore, to obtain high-resolution measurements of the outflow without any influence of the mechanical measurement systems, it is necessary to measure the flow without direct contact. Therefore, a new generation of seepage meters is currently being developed (Fig. 12.7). This new type of seepage meter is equipped with sensors for several physical parameters. It operates autonomously with a battery pack and without a pumping system or mechanical applications, which would disturb the natural water flow. Two sensor systems are installed in the seepage meter: firstly, the RBRmaestro CTD (RBR Ltd., Canada) with sensors for conductivity, temperature, pressure, fluorescent dissolved organic matter (fDOM) (420 nm), turbidity, dissolved oxygen, and pH and, secondly, a FEH521 flow meter (ABB Ltd., Switzerland) for the flow rates in both directions (Figs. 12.7 and 12.8). The advantage

Fig. 12.7 Field application of the new type of seepage meter. The black box in the middle of the picture is the measurement chamber, containing the sensors inside. The control unit and the battery pack for the flow meter can be seen on the right side. Photography by K. Schwalfenberg

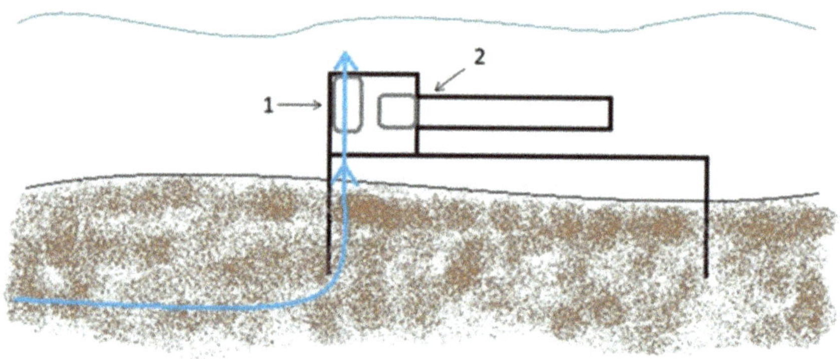

Fig. 12.8 Conceptual set-up of the new type of seepage meter. Blue arrow = flow direction of the groundwater outflow, 1 = EM flow meter, 2 = RBRmaestro

of the flow meter is that flow rates are measured in both directions with an electromagnetic field without disturbing the natural water flow.

First measurements show that the principle of the new type of seepage meter works well. After installation at the low water line at the Northern beach on Spiekeroog Island, the physicochemical parameters collected with the RBRmaestro sensor system indicate influence of terrestrial groundwater, i.e., fresh SGD. While the seepage meter is filled with pure seawater at the beginning directly after deploying at the low water line, the data demonstrates exchange of seawater by SGD after one tidal cycle. Our preliminary results show that conductivity and temperature are the most reliable parameters for a direct differentiation of the different water bodies (i.e., terrestrial groundwater, seawater, or a mixture of both). In combination with pressure data, we could infer whether groundwater exfiltrates or seawater infiltrates. The new type of seepage meter is, however, still in the process of development, with the flow meter sensor, control unit, and battery pack facing some issues related to the application underwater.

12.8 Outlook

The land-sea transition zone includes the subterranean estuary (STE), the zone where terrestrial fresh and marine saline groundwater as well as seawater mix. Sandy beaches are examples of land-sea transition zones. They are characterized by, among other things, steep physicochemical gradients manifested in salinity, inorganic chemistry, availability and characteristics of organic matter, microbial community composition, and flow regimes.

To better understand the environmental processes and complex system interactions at the land-sea transition zone, previous studies as well as our research activities at Spiekeroog Island have demonstrated that a broad scientific approach is needed. For instance, hydrogeologic studies are necessary to determine relevant hydrochemical processes, such as decalcification, cation exchange, and organic carbon remineralization, and to characterize the composition of freshwater that discharges into the STE. This will in part influence the biogeochemical processes occurring in pore waters and sediments of the STE, which ultimately govern the chemistry of submarine groundwater discharge (SGD).

The characterization of (dissolved) organic matter further is crucial to determine its origin and reactivity and to identify possible transformation processes. This is necessary to, for example, explain spatial trends of microbial respiration or to estimate the role of SGD for marine carbon budgets. As microbial communities are vital for most biogeochemical processes, a detailed investigation of microbial communities at different locations (e.g., marine, terrestrial, varying flow/discharge rates) and along biogeochemical gradients is required to link observed biogeochemical patterns to microbial activity. This will, for example, help to explain local signatures of the STE and improve our understanding regarding the microbiogeochemistry of natural sediments. Ultimately, measures of pore water residence times and SGD are needed to estimate and predict, for instance, rates of organic matter oxidation/transformations and constituent fluxes from the STE into the adjacent coastal ocean. As described above, natural radiotracers, such as Rn, can be successfully applied to determine pore water residence times and characterize SGD. Although recent studies on SGD benefit from advances in technology, high temporal and spatial variability can still cause large uncertainties, particularly when upscaling local fluxes to estimate the global imprint of SGD. The improvements of marine sensor systems are promising to more accurately (e.g., higher temporal resolution, increasing number of measurable parameters) quantify and characterize SGD.

While this review article only focuses on some aspects of the land-sea transition zone including hydrochemistry, microbiogeochemistry, tracer-based flux estimations, and marine sensor systems, it is clear that a multidisciplinary research approach is the key to disentangling the complexity and understanding the interrelations of the environmental processes in coastal areas. Besides the reviewed fields, multidisciplinary approaches may extend to further disciplines, such as biology (e.g., McLachlan 1983; Charbonnier et al. 2016), geophysics (e.g., Swarzenski et al. 2007; Stieglitz et al. 2008; Dimova et al. 2012), or social sciences (e.g., Defeo et al. 2009; Jones et al. 2007).

Acknowledgments The authors wish to thank B. Engelen, G. Massmann, and B. Schnetger for support during the writing process. H. Nicolai, G. Behrens, M. Friebe and H. Simon are thanked for their manifold support during field campaigns and lab work. Moreover, the authors are very thankful to the YOUMARES organization team, namely, V. Liebich and S. Jungblut. We would further like to thank M.-L. Paulsen and one anonymous reviewer for their very constructive comments, which helped us to improve the quality of the manuscript. The German Research Foundation (DFG, project number MA 3274/6-1, BR 775/33-1) and the Niedersächsisches Ministerium für Wissenschaft und Kultur (MWK) in the scope of project "BIME" (ZN3184) are thanked for project funding.

Appendix

This article is related to the YOUMARES 9 conference session no. 15: "Investigating the land-sea transition zone." The original Call for Abstracts and the abstracts of the presentations within this session can be found in the Appendix "Conference Sessions and Abstracts", Chapter "12 Investigating the Land-Sea Transition Zone", of this book.

References

Abualhaija MM, Whitby H, van den Berg CM (2015) Competition between copper and iron for humic ligands in estuarine waters. Mar Chem 172:46–56

Andersen MS, Nyvang V, Jakobsen R et al (2005) Geochemical processes and solute transport at the seawater/freshwater interface of a sandy aquifer. Geochim Cosmochim Acta 69(16):3979–3994

Anderson WP Jr (2002) Aquifer salinization from storm overwash. J Coast Res 18(3):413–420

Anschutz P, Smith T, Mouret A et al (2009) Tidal sands as biogeochemical reactors. Estuar Coast Shelf Sci 84:84–90

Appelo CAJ, Geirnaert W (1991) Processes accompanying the intrusion of salt water. Hydrogeology of salt water intrusion. A selection of SWIM papers. Int Contr Hydrogeol 11:291–303

Appelo CAJ, Postma D (2005) Geochemistry, groundwater and pollution, 2nd edn. CRC Press/Taylor & Francis Group, Amsterdam

Barbier EB (2015) Climate change impacts on rural poverty in low-elevation coastal zones. Estuar Coast Shelf Sci 165:A1–A13

Barbier EB, Hacker SD, Kennedy C et al (2011) The value of estuarine and coastal ecosystem services. Ecol Monogr 81(2):169–193

Basu AR, Jacobsen SB, Poreda RJ et al (2001) Large groundwater strontium flux to the oceans from the Bengal Basin and the marine strontium isotope record. Science 293(5534):1470–1473

Bateman H (1910) Solution of a system of differential equations occurring in the theory of radioactive transformations. Proc Camb Philos Soc 15:423–427

Beck AJ, Tsukamoto Y, Tovar-Sanchez A et al (2007) Importance of geochemical transformations in determining submarine groundwater discharge-derived trace metal and nutrient fluxes. Appl Geochem 22:477–490

Beck M, Reckhardt A, Amelsberg J et al (2017) The drivers of biogeochemistry in beach ecosystems: a cross-shore transect from the dunes to the low-water line. Mar Chem 190:35–50

Beekman HE, Appelo CAJ (1991) Ion chromatography of fresh-and salt-water displacement: laboratory experiments and multicomponent transport modelling. J Contam Hydrol 7(1–2):21–37

Beusen AHW, Slomp CP, Bouwman AF (2013) Global land-ocean linkage: direct inputs of nitrogen to coastal waters via submarine groundwater discharge. Environ Res Lett 8(3):034035

Billerbeck M, Werner U, Bosselmann K et al (2006) Nutrient release from an exposed intertidal sand flat. Mar Ecol Prog Ser 316:35–51. https://doi.org/10.3354/meps316035

Blackmon PD, Todd R (1959) Mineralogy of some foraminifera as related to their classification and ecology. J Paleontol 33(1):1–15

Böttcher ME (1999) The stable isotopic geochemistry of the sulfur and carbon cycles in a modern karst environment. Isot Environ Health Stud 35(1–2):39–61

Böttcher ME, Hespenheide B, Llobet-Brossa E et al (2000) The biogeochemistry, stable isotope geochemistry, and microbial community structure of a temperate intertidal mudflat: an integrated study. Cont Shelf Res 20:1749–1769. https://doi.org/10.1016/S0278-4343(00)00046-7

Bryan E, Meredith KT, Baker A et al (2017) Carbon dynamics in a late quaternary-age coastal limestone aquifer system undergoing saltwater intrusion. Sci Total Environ 607:771–785

Burnett WC, Dulaiova H (2003) Estimating the dynamics of groundwater input into the coastal zone via continuous radon-222 measurements. J Environ Radioact 69:1–2

Burnett WC, Bokuniewicz H, Huettel M et al (2003) Groundwater and pore water inputs to the coastal zone. Biogeochemistry 66:3–33

Burnett WC, Aggarwal PK, Aureli A et al (2006) Quantifying submarine groundwater discharge in the coastal zone via multiple methods. Sci Total Environ 367:498–543

Caetano M, Falcão M, Vale C et al (1997) Tidal flushing of ammonium, iron and manganese from inter-tidal sediment pore waters. Mar Chem 58:203–211

Chapelle FH, Knobel LL (1983) Aqueous geochemistry and the exchangeable cation composition of glauconite in the Aquia aquifer, Maryland. Groundwater 21(3):343–352

Chapelle FH, Knobel LL (1985) Stable carbon isotopes of HCO_3 in the Aquia Aquifer, Maryland: evidence for an isotopically heavy source of CO_2. Groundwater 23(5):592–599

Chapelle FH, Lovley DR (1992) Competitive exclusion of sulfate reduction by Fe(lll)-reducing bacteria: a mechanism for producing discrete zones of high-iron ground water. Groundwater 30(1):29–36

Charbonnier C, Anschutz P, Poirier D et al (2013) Aerobic respiration in a high-energy sandy beach. Mar Chem 155:10–21. https://doi.org/10.1016/j.marchem.2013.05.003

Charbonnier C, Lavesque N, Anschutz P et al (2016) Role of macrofauna on benthic oxygen consumption in sandy sediments of a high-energy tidal beach. Cont Shelf Res 120:96–105

Charette MA, Sholkovitz ER (2002) Oxidative precipitation of groundwater-derived ferrous iron in the subterranean estuary of a coastal bay. Geophys Res Lett 29(10):85–81

Charette MA, Sholkovitz ER, Hansel CM (2005) Trace element cycling in a subterranean estuary: Part 1. Geochemistry of the permeable sediments. Geochim Cosmochim Acta 69:2095–2109. https://doi.org/10.1016/j.gca.2004.10.024

Chipman L, Podgorski D, Green S et al (2010) Decomposition of plankton-derived dissolved organic matter in permeable coastal sediments. Limnol Oceanogr 55:857–871

Cho HM, Kim G, Kwon EY et al (2018) Radium tracing nutrient inputs through submarine groundwater discharge in the global ocean. Sci Rep 8:2439. https://doi.org/10.1038/s41598-018-20806-2

Coble PG (1996) Characterization of marine and terrestrial DOM in seawater using excitation-emission matrix spectroscopy. Mar Chem 51:325–346

Colbert SL, Berelson WM, Hammond DE (2008) Radon-222 budget in Catalina Harbor, California: 2. Flow dynamics and residence time in a tidal beach. Limnol Oceanogr 53:659–665

Couturier M, Nozais C, Chaillou G (2016) Microtidal subterranean estuaries as a source of fresh terrestrial dissolved organic matter to the coastal ocean. Mar Chem 186:46–57

Defeo O, McLachlan A, Schoeman DS et al (2009) Threats to sandy beach ecosystems: a review. Estuar Coast Shelf Sci 81(1):1–12

Deines P, Langmuir D, Harmon RS (1974) Stable carbon isotope ratios and the existence of a gas phase in the evolution of carbonate ground waters. Geochim Cosmochim Acta 38(7):1147–1164

Dimova NT, Swarzenski PW, Dulaiova H et al (2012) Utilizing multichannel electrical resistivity methods to examine the dynamics of the fresh water–seawater interface in two Hawaiian groundwater systems. J Geophys Res Ocean 117(C2):C02012

Drabbe J, Badon Ghijben W (1889) Nota in verband met de voorgenomen put-boring nabij Amsterdam (Notes on the results of the proposed well drilling in Amsterdam). Tijdschrift van het Koninklijk Instituut van Ingenieurs, Verhandelingen, Instituutjaar 1888/1889. pp 8–22

Ehlert C, Reckhardt A, Greskowiak J et al (2016) Transformation of silicon in a sandy beach ecosystem: Insights from stable silicon isotopes from fresh and saline groundwaters. Chem Geol 440:207–218. https://doi.org/10.1016/j.chemgeo.2016.07.015

Fetter CW (1972) Position of the saline water interface beneath oceanic islands. Water Resour Res 8(5):1307–1315

Flemming BW, Davis RA Jr (1994) Holocene evolution, morphodynamics and sedimentology of the Spiekeroog barrier island system (southern North Sea). Senck Marit 24(1):117–155

Garcés E, Basterretxea G, Sánchez AT (2011) Changes in microbial communities in response to submarine groundwater input. Mar Ecol Prog Ser 438:47–58. https://doi.org/10.3354/meps09311

Goodridge BM, Melack JM (2014) Temporal evolution and variability of dissolved inorganic nitrogen in beach pore water revealed using radon residence times. Environ Sci Technol 48:14211–14218

Grünenbaum N, Greskowiak J, Harms A et al (2017) Investigation of the recirculation on the beach of Spiekeroog from in- to exfiltration. Poster presented at the European Geophysical Union (EGU) General Assembly 2017, Vienna, 23–28 April 2017

Hays RL, Ullman WJ (2007) Direct determination of total and fresh groundwater discharge and nutrient loads from a sandy beach face at low tide (Cape Henlopen, Delaware). Limnol Oceanogr 52(1):240–247

Heiss JW, Post VEA, Laattoe T et al (2017) Physical controls on biogeochemical processes in intertidal zones of beach aquifers. Water Resour Res 53:9225–9244

Herzberg A (1901) Die Wasserversorgung einiger Nordseebäder. J Gasbeleuchtung Wasserversorgung 44:815–819

Hild A (1997) Geochemie der Sedimente und Schwebstoffe im Rückseitenwatt von Spiekeroog und ihre Beeinflussung durch biologische Aktivität. Dissertation, Forschungszentrum Terramare, Wilhelmshaven, University Oldenburg

Hoehn E, Von Gunten HR (1989) Radon in groundwater: a tool to assess infiltration from surface waters to aquifers. Water Resour Res 25:1795–1803

Holguin G, Vazquez P, Bashan Y (2001) The role of sediment microorganisms in the productivity, conservation, and rehabilitation of mangrove ecosystems: an overview. Biol Fertil Soils 33:265–278. https://doi.org/10.1007/s003740000319

Holt T, Seibert SL, Greskowiak J et al (2017) Impact of storm tides and inundation frequency on water table salinity and vegetation on a juvenile barrier island. J Hydrol 554:666–679

Holt T, Greskowiak J, Seibert SL et al (2019) Modeling the evolution of a freshwater lens under highly dynamic conditions on a currently developing barrier island. Accepted by Geofluids

Houben GJ, Koeniger P, Sültenfuß J (2014) Freshwater lenses as archive of climate, groundwater recharge, and hydrochemical evolution: insights from depth-specific water isotope analysis and age determination on the island of Langeoog, Germany. Water Resour Res 50(10):8227–8239

Hu C, Muller-Karger FE, Swarzenski PW (2006) Hurricanes, submarine groundwater discharge, and floridas red tides. Geophys Res Lett 33(11):L11601

Huettel M, Ziebis W, Forster S et al (1998) Advective transport affecting metal and nutrient distributions and interfacial fluxes in permeable sediments. Geochim Cosmochim Acta 62:613–631

Jakobsen R, Postma D (1999) Redox zoning, rates of sulfate reduction and interactions with Fe-reduction and methanogenesis in a shallow sandy aquifer, Rømø, Denmark. Geochim Cosmochim Acta 63(1):137–151

Jeandel C, Oelkers EH (2015) The influence of terrigenous particulate material dissolution on ocean chemistry and global element cycles. Chem Geol 395:50–66

Jones A, Gladstone W, Hacking N (2007) Australian sandy-beach ecosystems and climate change: ecology and management. Aust Zool 34(2):190–202

Kim J, Kim G (2017) Inputs of humic fluorescent dissolved organic matter via submarine groundwater discharge to coastal waters off a volcanic island (Jeju, Korea). Sci Rep 7:7921. https://doi.org/10.1038/s41598-017-08518-5

Kim G, Ryu JW, Yang HS et al (2005) Submarine groundwater discharge (SGD) into the Yellow Sea revealed by Ra-228 and Ra-226 isotopes: implications for global silicate fluxes. Earth Planet Sci Lett 237:156–166

Kim TH, Waska H, Kwon E et al (2012) Production, degradation, and flux of dissolved organic matter in the subterranean estuary of a large tidal flat. Mar Chem 142–144:1–10

Kim TH, Kwon E, Kim I, Lee SA, Kim G (2013) Dissolved organic matter in the subterranean estuary of a volcanic island, Jeju: Importance of dissolved organic nitrogen fluxes to the ocean. J Sea Res 78:18–24

Kristensen E, Andersen FØ, Holmboe N et al (2000) Carbon and nitrogen mineralization in sediments of the Bangrong mangrove area, Phuket, Thailand. Aquat Microb Ecol 22:199–213

Kroeger KD, Charette MA (2008) Nitrogen biogeochemistry of submarine groundwater discharge. Limnol Oceanogr 53:1025–1039. https://doi.org/10.4319/lo.2008.53.3.1025

Kwon EY, Kim G, Primeau F et al (2014) Global estimate of submarine groundwater discharge based on an observationally constrained radium isotope model. Geophys Res Lett 41:62–68. https://doi.org/10.1002/2014GL061574

Lee DR (1977) A device for measuring seepage flux in lakes and estuaries. Limnol Oceanogr 22:140–147

Lee DR (1978) A field exercise on groundwater flow using seepage meter measurement of the flow using seepage meters and mini-piezometers. J Geol Educ 27:6–20

Lee E, Shin D, Hyun SP et al (2016) Periodic change in coastal microbial community structure associated with submarine groundwater discharge and tidal fluctuation. Limnol Oceanogr 62:437–451. https://doi.org/10.1002/lno.10433

Li YH, Mathieu G, Biscaye P et al (1977) The flux of 226-Ra from estuarine and continental shelf sediments. Earth Planet Sci Lett 37:237–241

Linkhorst A, Dittmar T, Waska H (2017) Molecular fractionation of dissolved organic matter in a shallow subterranean estuary: the role of the iron curtain. Environ Sci Technol 51:1312–1320

Lipp JS, Morono Y, Inagaki F et al (2008) Significant contribution of Archaea to ex-tant biomass in marine subsurface sediments. Nature 454:991–994. https://doi.org/10.1038/nature07174

Liu J, Du J, Wu Y et al (2018) Nutrient input through submarine groundwater discharge in two major Chinese estuaries: the Pearl River Estuary and the Changjiang River Estuary. Estuar Coast Shelf Sci 203:17–28

Lovley DR, Phillips EJ (1987) Competitive mechanisms for inhibition of sulfate reduction and methane production in the zone of ferric iron reduction in sediments. Appl Environ Microbiol 53(11):2636–2641

Luijendijk A, Hagenaars G, Ranasinghe R et al (2018) The state of the world's beaches. Sci Rep 8(1):6641

Marsh Jr JA (1977) Terrestrial inputs of nitrogen and phosphorus on fringing reefs of Guam. In: Proceeding of the 2nd international coral reef symposium, vol 1, pp 332–336

McAllister SM, Barnett JM, Heiss JW et al (2015) Dynamic hydrologic and biogeochemical processes drive microbially enhanced iron and sulfur cycling within the intertidal mixing zone of a beach aquifer. Limnol Oceanogr 60:329–345. https://doi.org/10.1111/lno.10029

McGranahan G, Balk D, Anderson B (2007) The rising tide: assessing the risks of climate change and human settlements in low elevation coastal zones. Environ Urban 19(1):17–37

McLachlan A (1983) Sandy beach ecology—a review. In: McLachlan A, Erasmus T (eds) Sandy beaches as ecosystems. Springer, Dordrecht, pp 321–380

Milliman JD, Müller G, Förstner F (2012) Recent sedimentary carbonates: part 1 marine carbonates. Springer, Berlin

Moore WS (1996) Large groundwater inputs to coastal waters revealed by Ra-226 enrichments. Nature 380:612–614

Moore WS (1999) The subterranean estuary: a reaction zone of ground water and sea water. Mar Chem 65(1–2):111–125

Moore WS (2000) Ages of continental shelf waters determined from 223Ra and 224Ra. J Geophys Res Ocean 105:22117–22122

Moore WS (2006) Radium isotopes as tracers of submarine groundwater discharge in Sicily. Cont Shelf Res 26:852–861

Moore WS (2010) The effect of submarine groundwater discharge on the ocean. Annu Rev Mar Sci 2:59–88

Mulligan AE, Charette MA (2006) Intercomparison of submarine groundwater discharge estimates from a sandy unconfined aquifer. J Hydrol 327:411–425

Nelson CE, Donahue MJ, Dulaiova H et al (2015) Fluorescent dissolved organic matter as a multivariate biogeochemical tracer of submarine groundwater discharge in coral reef ecosystems. Mar Chem 177:232–243

Nicholls RJ, Cazenave A (2010) Sea-level rise and its impact on coastal zones. Science 328(5985):1517–1520

O'Connor AE, Luek JL, McIntosh H et al (2015) Geochemistry of redox-sensitive trace elements in a shallow subterranean estuary. Mar Chem 172:70–81. https://doi.org/10.1016/j.marchem.2015.03.001

Osterholz H, Kirchman DL, Niggemann J et al (2016) Environmental drivers of dissolved organic matter molecular composition in the Delaware Estuary. Front Earth Sci 4:95. https://doi.org/10.3389/feart.2016.00095

Paulsen RJ, Smith CF, O'Rourke D et al (2001) Development and evaluation of an ultrasonic ground water seepage meter. Groundwater 39(6):904–911

Peterson RN, Burnett WC, Taniguchi M et al (2008) Radon and radium isotope assessment of submarine groundwater discharge in the Yellow River delta. China J Geophys Res 113:C09021

Post VEA (2005) Fresh and saline groundwater interaction in coastal aquifers: is our technology ready for the problems ahead? Hydrogeol J 13(1):120–123

Post VEA, Houben GJ (2017) Density-driven vertical transport of saltwater through the freshwater lens on the island of Baltrum (Germany) following the 1962 storm flood. J Hydrol 551:689–702

Postma D, Jakobsen R (1996) Redox zonation: equilibrium constraints on the Fe(III)/SO_4-reduction interface. Geochim Cosmochim Acta 60(17):3169–3175

Reckhardt A, Beck M, Seidel M et al (2015) Carbon, nutrient and trace metal cycling in sandy sediments: a comparison of high-energy beaches and backbarrier tidal flats. Estuar Coast Shelf Sci 159:1–14. https://doi.org/10.1016/j.ecss.2015.03.025

Reckhardt A, Beck M, Greskowiak J et al (2017) Cycling of redox-sensitive elements in a sandy subterranean estuary of the southern North Sea. Mar Chem 188:6–17

Rickard D, Morse JW (2005) Acid volatile sulfide (AVS). Mar Chem 97(3):141–197

Riedel T, Lettmann K, Beck M et al (2010) Tidal variations in groundwater storage and associated discharge from an intertidal coastal aquifer. J Geophys Res 115:1–10

Robinson C, Li L, Barry D (2007) Effect of tidal forcing on a subterranean estuary. Adv Water Resour 30:851–865. https://doi.org/10.1016/j.advwatres.2006.07.006

Röper T, Kröger KF, Meyer H et al (2012) Groundwater ages, recharge conditions and hydrochemical evolution of a barrier island freshwater lens (Spiekeroog, Northern Germany). J Hydrol 454:173–186

Röper T, Greskowiak J, Freund H et al (2013) Freshwater lens formation below juvenile dunes on a barrier island (Spiekeroog, Northwest Germany). Estuar Coast Shelf Sci 121:40–50

Röper T, Greskowiak J, Massmann G (2015) Instabilities of submarine groundwater discharge under tidal forcing. Limnol Oceanogr 60(1):22–28

Rosenberry DO, Morin RH (2004) Use of an electromagnetic seepage meter to investigate temporal variability in Lake Seepage. Groundwater 42(1):68–77

Roy M, Martin JB, Smith CG et al (2011) Reactive-transport modeling of iron diagenesis and associated organic carbon remineralization in a Florida (USA) subterranean estuary. Earth Planet Sci Lett 304:191–201. https://doi.org/10.1016/j.epsl.2011.02.002

Santoro A (2009) Microbial nitrogen cycling at the saltwater-freshwater interface. Hydrogeol J 18:187–202. https://doi.org/10.1007/s10040-009-0526-z

Santoro AE, Francis CA, de Sieyes NR et al (2008) Shifts in the relative abundance of ammonia-oxidizing bacteria and archaea across physicochemical gradients in a subterranean estuary. Environ Microbiol 10(4):1068–1079. https://doi.org/10.1111/j.1462-2920.2007.01547.x

Santos IR, Burnett WC, Chanton J et al (2008) Nutrient biogeochemistry in a Gulf of Mexico subterranean estuary and groundwater-derived fluxes to the coastal ocean. Limnol Oceanogr 53:705–718. https://doi.org/10.4319/lo.2008.53.2.0705

Santos IR, Burnett WC, Dittmar T et al (2009) Tidal pumping drives nutrient and dissolved organic matter dynamics in a Gulf of Mexico subterranean estuary. Geochim Cosmochim Acta 73:1325–1339

Santos IR, Eyre BD, Huettel M (2012) The driving forces of porewater and groundwater flow in permeable coastal sediments: a review. Estuar Coastal Shelf Sci 98:1–15

Schoonen MA (2004) Mechanisms of sedimentary pyrite formation. Geol Soc Spec Pap 379:117–134

Seibert SL, Holt T, Reckhardt A et al (2018) Hydrochemical evolution of a freshwater lens below a barrier island (Spiekeroog, Germany): the role of carbonate mineral reactions, cation exchange and redox processes. Appl Geochem 92:196–208

Seibert SL, Böttcher ME, Schubert F et al (2019) Iron sulfide formation in young and rapidly-deposited permeable sands at the land-sea transition zone. Sci Total Environ 649:264–283

Seidel M, Beck M, Greskowiak J et al (2015) Benthic-pelagic coupling of nutrients and dissolved organic matter composition in an intertidal sandy beach. Mar Chem 176:150–163

Sirois M, Couturier M, Barber A et al (2018) Interactions between iron and organic carbon in a sandy beach subterranean estuary. Mar Chem 202:86–96

Slomp CP, Malschaert J (1997) Iron and manganese cycling in different sedimentary environments on the North Sea continental margin. Cont Shelf Res 17:1083–1117

Slomp CP, Van Cappellen P (2004) Nutrient inputs to the coastal ocean through submarine groundwater discharge: controls and potential impact. J Hydrol 295(1–4):64–86

Snow DD, Spalding RF (1997) Short-term aquifer residence times estimated from ^{222}Rn disequilibrium in artificially-recharged ground water. J Environ Radioact 37:307–325

Snyder M, Taillefert M, Ruppel C (2004) Redox zonation at the saline-influenced boundaries of a permeable surficial aquifer: effects of physical forcing on the biogeochemical cycling of iron and manganese. J Hydrol 296(1):164–178

Spiteri C, Slomp CP, Charette MA et al (2008) Flow and nutrient dynamics in a subterranean estuary (Waquoit Bay, MA, USA): field data and reactive transport modeling. Geochim Cosmochim Acta 72(14):3398–3412

Stal LJ, Behrens SB, Villbrandt M et al (1996) The biogeochemistry of two eutrophic marine lagoons and its effect on microphytobenthic communities. Hydrobiologia 329:185–198. https://doi.org/10.1007/BF00034557

Stieglitz T, Taniguchi M, Neylon S (2008) Spatial variability of submarine groundwater discharge, Ubatuba, Brazil. Estuar Coast Shelf Sci 76(3):493–500

Stuyfzand PJ (1993) Hydrochemistry and hydrology of the coastal dune area of the Western Netherlands. Dissertation, Vrije Universiteit of Amsterdam

Stuyfzand PJ (1998) Decalcification and acidification of coastal dune sands in the Netherlands. In: Arehart H (eds) Proceedings of the 9th international symposium on water-rock interaction, Taupo, New Zealand, pp 79–82

Stuyfzand PJ (1999) Patterns in groundwater chemistry resulting from groundwater flow. Hydrogeol J 7(1):15–27

Stuyfzand PJ (2017) Observations and analytical modeling of freshwater and rainwater lenses in coastal dune systems. J Coast Conserv 21(5):577–593

Suryaputra IG, Santos IR, Huettel M et al (2015) Non-conservative behavior of fluorescent dissolved organic matter (FDOM) within a subterranean estuary. Cont Shelf Res 110:183–190

Swarzenski PW (2007) U/Th series radionuclides as coastal groundwater tracers. Chem Rev 107:663–674

Swarzenski PW, Simonds FW, Paulson AJ et al (2007) Geochemical and geophysical examination of submarine groundwater discharge and associated nutrient loading estimates into Lynch Cove, Hood Canal, WA. Environ Sci Technol 41(20):7022–7029

Tait DR, Erler DV, Santos IR et al (2014) The influence of groundwater inputs and age on nutrient dynamics in a coral reef lagoon. Mar Chem 166:36–47

Tamborski JJ, Cochran JK, Bokuniewicz HJ (2017) Application of ^{224}Ra and ^{222}Rn for evaluating seawater residence times in a tidal subterranean estuary. Mar Chem 189:32–45

United Nations Environment Programme (UNEP) (2014) The UNEP environmental data explorer, as compiled from UNEP/DEWA/GRID-Geneva. UNEP, Geneva. http://geodata.grid.unep.ch

Vacher HL (1988) Dupuit-Ghyben-Herzberg analysis of strip-island lenses. Geol Soc Am Bull 100(4):580–591

Waldrop WR, Swarzenski PW (2006) A new tool for quantifying flux rates between surface water and ground water. In: Singh VP, Xu YJ (eds) Coastal hydrology and processes 24. Water Resources Publications, Highlands Ranch, pp 305–312

Werner AD, Simmons CT (2009) Impact of sea-level rise on sea water intrusion in coastal aquifers. Groundwater 47(2):197–204

Wilms R, Sass H, Köpke B et al (2007) Methane and sulfate profiles within the subsurface of a tidal flat are reflected by the distribution of sulfate-reducing bacteria and methanogenic archaea. FEMS Microbiol Ecol 59:611–621

Windom HL, Moore WS, Niencheski LFH et al (2006) Submarine groundwater discharge: a large, previously unrecognized source of dissolved iron to the South Atlantic Ocean. Mar Chem 102(3–4):252–266

Zark M, Christoffers J, Dittmar T (2017) Molecular properties of deep-sea dissolved organic matter are predictable by the central limit theorem: evidence from tandem FT-ICR-MS. Mar Chem 191:9–15

Zhang J, Mandal AK (2012) Linkages between submarine groundwater systems and the environment. Curr Opin Environ Sustain 4:219–226

13 Fisheries and Tourism: Social, Economic, and Ecological Trade-offs in Coral Reef Systems

Liam Lachs and Javier Oñate-Casado

Abstract

Coastal communities are exerting increasingly more pressure on coral reef ecosystem services in the Anthropocene. Balancing trade-offs between local economic demands, preservation of traditional values, and maintenance of both biodiversity and ecosystem resilience is a challenge for reef managers and resource users. Consistently, growing reef tourism sectors offer more lucrative livelihoods than subsistence and artisanal fisheries at the cost of traditional heritage loss and ecological damage. Using a systematic review of coral reef fishery reconstructions since the 1940s, we show that declining trends in fisheries catch and fish stocks dominate coral reef fisheries globally, due in part to overfishing of schooling and spawning-aggregating fish stocks vulnerable to exploitation. Using a separate systematic review of coral reef tourism studies since 2013, we identify socio-ecological impacts and economic opportunities associated to the industry. Fisheries and tourism have the potential to threaten the ecological stability of coral reefs, resulting in phase shifts toward less productive coral-depleted ecosystem states. We consider whether four common management strategies (unmanaged commons, ecosystem-based management, co-management, and adaptive co-management) fulfil ecological conservation and socioeconomic goals, such as living wage, job security, and maintenance of cultural traditions. Strategies to enforce resource exclusion and withhold traditional resource rights risk social unrest; thus, the coexistence of fisheries and tourism industries is essential. The purpose of this chapter is to assist managers and scientists in their responsibility to devise implementable strategies that protect local community livelihoods and the coral reefs on which they rely.

Keywords

Sustainable development · Adaptive co-management · Systematic review · Ecological impacts · Economic shift

13.1 Context

Coral reef ecosystems are considered one of the most productive and economically valuable ecosystems on Earth, providing habitat for a highly diverse species assemblage (Roberts et al. 2002). Various global and local stressors threaten coral reefs, from global warming-induced heat stress to tourism- and fisheries-induced ecological stresses. The result of overuse and overexploitation by either of these industries can be disastrous for the reef ecosystem (Hodgson and Dixon 1988; Hawkins and Roberts 1994; Cesar et al. 2003; Fenner 2012; Jackson et al. 2014; Gil et al. 2015). While both industries present economic opportunities necessary for coastal communities in the vicinity of coral reefs (Cesar et al. 2003), they often compete for the same operational spaces (Fabinyi 2008). This review draws on the history of tourism and fisheries industries from around the world to answer questions about how best to manage these growing industries in the future. We unravel the different ecological threats posed by fisheries and tourism and discuss the trade-offs managers make to minimize coral reef degradation. Considering the benefits and pitfalls of various management strategies, we

L. Lachs (✉)
Marine Biology, Ecology and Biodiversity, Vrije Universiteit Brussel, Brussel, Belgium

Institute of Oceanography and Environment, Universiti Malaysia Terengganu, Kuala Terengganu, Terengganu, Malaysia

Department of Biology, University of Florence, Sesto Fiorentino, Italy

J. Oñate-Casado (✉)
Department of Biology, University of Florence, Sesto Fiorentino, Italy

Sea Turtle Research Unit (SEATRU), Universiti Malaysia Terengganu, Kuala Terengganu, Terengganu, Malaysia

School of Biological Sciences, University of Queensland, St Lucia, QLD, Australia

S. Jungblut et al. (eds.), *YOUMARES 9 - The Oceans: Our Research, Our Future*,
https://doi.org/10.1007/978-3-030-20389-4_13

compare the social, ecological, and economic trade-offs that coral reef stakeholders must make to successfully tread the path of sustainable socioeconomic development. We also highlight various tools available for the benefit of local communities in coral reef systems.

Although we do not consider the effects of global change on coral reef social-ecological systems in this review, it is important to frame our discussion and management recommendations on the backdrop of a changing world. Coral bleaching occurs when excessively high water temperatures invoke decoupling of coral host tissue and symbiotic algal zooxanthellae (Bessell-Browne et al. 2014). With a reduced metabolism, bleached corals have higher probabilities of falling victim to starvation, disease, predation, or competition (Bellwood et al. 2006). Mass bleaching events occurred around the world in 1998, 2002, 2010, and 2016, whilst individual coral reefs are experiencing ever more frequent bleaching events (Heron et al. 2016). During the 2016 bleaching event in the Great Barrier Reef (GBR), less than 8.9% of reefs escaped without bleaching, compared to 42.4% in 2002 and 44.7% in 1998 (Hughes et al. 2017). Similarly, coral reefs in the Maldives bleached extensively in 2016, with live coral cover dropping below 6% in the southern Maldivian reefs (Perry and Morgan 2017). Mass coral bleaching has the potential to wipe out wide swathes of coral reefs, transitioning the ecosystem toward degraded states (Fig. 13.1) with detrimental impacts to global biodiversity and both coastal tourism and fisheries economies. Therefore, we must frame our arguments on the trade-offs between fisheries and tourism against a backdrop of unprecedented global change and the worst-case scenario.

13.2 Ecosystem Services

As the most biodiverse of marine habitats, coral reefs provide a wide range of ecosystem services, from fisheries and recreation/tourism to coastal protection and potential medical innovation, which in turn drive the social, ecological, and economic trade-offs discussed in this chapter. Coral reef fisheries provide a key source of income and livelihood to coastal communities, are a non-substitutable source of protein for many island populations (Laurans et al. 2013), and are key to culturally significant local traditions (McClanahan 1999; Bruggemann et al. 2012; Fenner 2012). Growing tourism industries, based on recreational activities such as snorkeling, diving, whale watching, and recreational fishing (Asafu-Adjaye and Tapsuwan 2008; Young et al. 2015; Chen et al. 2016a) require different skill sets than traditional livelihoods and offer alternative income to coastal communities (Hicks et al. 2013; Harvey and Naval 2016; Outra et al. 2016). The structure of carbonate reefs directly protects coastal areas, especially in tsunami- and storm-prone tropical regions of the Indian and Pacific Oceans (Ferrario et al. 2014), and indirectly protects these areas through supply of carbonate sand to beaches and mangrove ecosystems (Wells et al. 2006). Coral reef biodiversity, from coral and algae to cone shells and sponges, provide many novel compounds useful to medical science including painkillers and antiviral, antimicrobial, and anticancer drugs (Kelman et al. 2001; Knowlton et al. 2010).

Valuing coral reef ecosystem services in a monetary way can be a useful tool to aid public decision-making. While valuation methods provide wildly different results (Cesar et al. 2003; Brander et al. 2007; Craig 2008; Laurans et al. 2013), using standardized methods, Cesar et al. (2003) have provided insight on the relative importance of four major ecosystem services (biodiversity maintenance, coastal protection, tourism, and fisheries) which were estimated to be worth US$ 30 billion in net benefits in goods and services to world economies annually. The annual value of coastal protection from surging oceans (i.e. the cost of rebuilding if the protective function was lost) has been estimated at US$ 9 billion (Cesar et al. 2003). Reef biodiversity, through research, conservation, and medical value, was estimated at US$ 5.5 billion. Tourism was valued at US$ 9.6 billion, almost twice the estimated value of reef fisheries (US$ 5 billion) (Cesar et al. 2003), a finding also reflected by other valuation studies (Van Beukering et al. 2006; Craig 2008). For example, the US Commission on Ocean Policy (USCOP) indicates higher value of tourism over fisheries on non-coral reef industries, US$ 60 to 31 million, respectively (Craig 2008; Spalding et al. 2017). Given the high growth of the coral reef tourism sector (Outra et al. 2016; Harvey and Naval 2016) that we detail further on in this chapter, new opportunities offered by tourism are underpinned by social, economic, and ecological trade-offs for scientists, managers, and fishers alike (Hicks et al. 2013).

13.3 Impacts and Trends of Fisheries and Tourism

13.3.1 Impacts of Fisheries

Although coral reef fisheries are a major source of local income and are socially and economically integral to coastal communities (McClanahan 1999; Cesar et al. 2003; Bruggemann et al. 2012; Fenner 2012), overfishing and destructive reef fisheries can jeopardize fish resources (Cesar et al. 2003; Fox 2004) and the resilience of entire reef ecosystems (Mumby et al. 2006; Fenner 2012; Bozec et al. 2016). Coral and their larvae, the seed stock of future coral reefs, can be outcompeted by macroalgae for space (Smith et al. 1981; Hunter and Evans 1995; Mumby et al. 2007; Doropoulos et al. 2017). Hence, overfishing of key func-

Fig. 13.1 Effects of the 2016 mass coral bleaching event in the central Maldives shown by the transition from healthy pre-bleaching coral reefs in the beginning of March (**a**) to a bleached coral state in the end of March (**b**) at a reef crest in eastern Baa Atoll and finally to a post-bleaching macroalgal colonization at a propagation project on the reef flat of the nearby North Male Atoll (**c**). Photo credit: Stephen Bergacker

tional groups of reef organisms such as herbivorous fish can reduce grazing pressure on macroalgae, promoting phase shifts toward less productive coral-depleted ecosystem states (Mumby et al. 2016; Doropoulos et al. 2017). Overfishing of top predators can induce trophic cascades that also the coral reef ecosystem (Mumby et al. 2006). A study across the Northern Line Islands by Sandin et al. (2008) characterizes the systemic ecological effects of fishing on coral reefs. At Palmyra and Kingsman, uninhabited atolls where fishing pressure is low, top predators dominate the fish assemblage, the fish biomass pyramid is inverted, and coral coverage is very high. Conversely, at inhabited atolls Tabuaeran and Kiritimati where fishing pressure is high, there are far fewer large long-lived fish, a bottom-heavy food web, greater prevalence of coral disease, less coral recruitment, and generally more degraded reefs with higher algal overgrowth (Fig. 13.2) and lower coral coverage. Degraded overfished reefs are less productive for local fisheries causing conflicts for ever-limited resources (Bruggemann et al. 2012).

To understand long-term overfishing trends that underpin trade-offs affecting coral reef fishers, we conducted a systematic literature review in Web of Science® using the following study topic search string: ("coral reef" *or* "coral reefs") *and* ("fisheries" *or* "fishery" *or* "fishing") *and* ("historic" *or* "reconstruction" *or* "reconstruct"). Of the 250 results, 12 studies met our two relevance criteria, namely, a main focus on coral reef fisheries and a reconstruction period <25 years. A key reconstruction by Zeller et al. (2015) that did not show in the search results was also included for this review.

As many coral reef fisheries lack historic data on catch size, catch composition, fishing gear use or catch per unit effort (CPUE) (Sadovy de Mitcheson et al. 2008) alternative methods for estimating fisheries trends are useful. Traditional ecological knowledge of fishing communities can be used to understand prominent ecological changes (Lavides et al. 2010), but such assessments are limited to the period of living memory, approximately 50 years pre-publication (Golden et al. 2014). By combining anecdotal evidence from semi-

Fig. 13.2 General fore reef habitats with characteristic fish communities (top row: **a**, **c**, **e**) and representative 0.5 m^2 photos of the reef substrate (bottom row: **b**, **d**, **e**) at Kingsman (**a**, **b**), Tabuaeran (**c**, **d**), and Kiritimati (**e**, **f**), Northern Line Islands, showing a degradation gradient – from reefs with numerous top predators and high coral coverage to reefs with few large predators, only small herbivorous fish, and dominated by fleshy macroalgae in place of coral. (Adapted from Sandin et al. (2008) with permission from PLoS One)

structured interviews with available fisheries catch or human population data we can gain insight into temporal trends in fish biomass, catch size and composition, extinction date or CPUE (Hardt 2008; Claro et al. 2009; Lavides et al. 2010; Young et al. 2015; Samoilys et al. 2017). As shown in the schematic timeline (Table 13.1), the 1950s–1970s was a period characterized by high yields of large reef fish such as the herbivorous green bumphead parrotfish (*Bolbometopon muricatum*) (Lavides et al. 2016). By the 1980s–2000s, large schooling or spawning fish began to be replaced by small reef fish and invertebrates (Sadovy de Mitcheson et al. 2008).

13.3.1.1 Anecdotal Reconstructions

Due to observer bias, using semiquantitative anecdotal evidence for fisheries reconstructions is less reliable than using landing data. Golden et al. (2014) reported on changes to ecosystem dynamics and fish catch based on 22 semi-structured interviews and a spearfishing survey. Only 11% of the recorded fish community composition was shared by both survey methods, and only three out of 14 species declines were reported by more than one respondent. The other 78% of species declines were reported by no more than one out of 22 respondents (4.5%). Hence, these results may be heavily biased by individual experience or change in attitude, and thus should be interpreted with caution. A larger interview study by Lavides et al. (2010) (n = 232) reported a similar proportion of rare species declines (82%), also reported by less than 4.5% of the sample size (<11 reports). These studies exemplify the difficulty in detecting subtle ecological changes with nonquantitative or semiquantitative methodological techniques.

Fisheries reconstructions based on anecdotal evidence can be useful in identifying larger ecological perturbations and trends (Sadovy de Mitcheson et al. 2008; Lavides et al. 2016). Larger-scale dynamics are more likely to be detected by many people, increasing congruence between respondents. Lavides et al. (2016) identified declining trends in mean perceived CPUE for five species of reef fish, including the green bumphead parrotfish (*B. muricatum*) which declined 88% compared to 1950s' levels. As the largest of its kind, this widespread schooling fish was probably fished before the 1950s and is particularly vulnerable to heavy fishing with widespread declines in their once-common populations (Dulvy et al. 2004). Spawning aggregations for most reef fish occur in a short breeding season of up to 3 months making them highly vulnerable to fishing pressure. Through interview techniques Sadovy de Mitcheson et al. (2008) identified that most reef fish spawning aggregations in the Indo-Pacific and West Atlantic are in decline, with increasing aggregations only occurring where effective management strategies are in place.

13.3.1.2 Quantitative Reconstructions

Fisheries reconstructions using quantitative data mining of catch data provide more detailed information than those using anecdotal evidence; however, the spatial and temporal

Table 13.1 Reconstructed reef fishery trends over the last century from 12 relevant publications, referring to the study and study country, target organisms (a, b, or c), net fish stock change throughout the reconstruction period, habitat (*R* reef, *C* coastal, *P* pelagic, or inshore), fishery type (*S* subsistence, *A* artisanal, *I* industrial), and the methodology

Reference	Country	Change unit	Target organisms	1900s	'40s	'50s	'60s	'70s	'80s	'90s	'00s	'10s	Net change	Habitat	Fishery type	Methods
Cheung and Sadovy (2004)	Hong Kong	Biomass relative to 1950s level (%)	a) Small benthic fish & crustacean	–	–							–	0	R; C	A; I	148 semi-structured interviews; Catch data
			b) Large demersal & pelagic fish	–	–							–	-			
			c) Small pelagic fish; Cephalopods	–	–							–	+			
Claro et al. (2009)	Cuba	Catch (MT $year^{-1}$)	a) *Epinephelus striatu*	–	–								-	R	S; A; I	Catch data
			b) *Lutjanus synagris*	–	–								-			
			c) 4 other snapper species	–	–								0			
Hardt (2008)	Jamaica	Abundance (%)	a) All reef fish										-	R	S; A	Literature 600AD+
Lachica-Alino et al. (2009)	Philippines	Biomass (no units given)	a) Large, high-value fish	–	–	–							-	R; C	I	Catch data; ECOSIM models
			b) Small reef carnivore; Cephalopod	–	–	–							+			
Lavides et al. (2010)	Philippines	Time of zero catch	a) *Epinephelus anceolatus*; +8 species	–	–	–							-	R	S; A	232 semi-structured interviews
			b) *Thunnus albacares*; +6 species	–	–	–							-			
			c) 6 other species	–	–	–							-			
Lavides et al. (2016)	Philippines	Perceived CPUE (kg day^{-1})	a) *Bolbometopon muricatum*	–	–								-	R	S; A; I	2655 semi-structured Interviews
			b) *Epinephelus Anceolatus*; +2 species	–	–								-			
			c) *Alectis ciliaris*	–	–								-			
McClenachan and Kittinger (2013)	Hawaii; Florida	Net Catch (T km^{-2} reef)	a) All Hawaiian fish								–	–	+	R; C; P	S; A; I	Meta-analysis 1300AD+
			b) All Florida Keys fish								–	–	+			
Sadovy de Mitcheson et al. (2008)	Indo-Pacific; West Atlantic	Catch (kg $trip^{-1}$) & Spawn-aggregate status	a) Palau grouper aggregations	–	–							–	-	R	S; A	377 semi-structured interviews
			b) Indo-Pacific aggregations	–	–							–	-			
			c) West Atlantic aggregations	–	–							–	-			
Samoilys et al. (2017)	Kenya	CPUE (kg $trip^{-1}$)	a) All fish catch	–	–	–	–	–				–	-	R	A	Meta-analysis
Weijerman et al. (2016)	Guam	Catch (T $year^{-1}$); CPUE (kg hr^{-1})	a) All reef organisms	–	–	–	–	–					-	Inshore	S; A	Creel survey; Catch data
			b) All reef organisms	–	–	–	–	–					-			
Young et al. (2015)	Australia	Reef catch (%); Fish weight (kg)	a) Reef fish	–	–				–			–	+	R; C; P	R	Meta-analysis, magazine reports
			b) Reef fish	–	–				–			–	-			
Zeller et al. (2015)	Pacific Islands	Est. Catch (T $year^{-1}$)	a) All fish catch	–	–							–	+	R; C; P	S; A; I	FAO Catch data

Cell color indicates the sign of net fisheries trends in that period, either declining (dark red), stable (medium grey), or increasing (light blue). Note that different units are used in each study. Fisheries trends of Cheung and Sadovy (2004), Lachica-Alino et al. (2009), McClenachan and Kittinger (2013), and Zeller et al. (2015) use combined habitats analyses that mask underlying reef-specific fisheries trends

availability of catch data is predominantly limited to the most commonly fished areas in more recent times (Pauly 1995; Cheung and Sadovy 2004). Declines in reef fisheries since the 1950s are commonplace (Claro et al. 2009; Lachica-Alino et al. 2009; Weijerman et al. 2016); however more complex population dynamics between different groups of reef organisms obscure these net trends. In the Philippines, overfishing from trawl fisheries is shown to have reduced large high-value fish stocks. The concurrent effects to the food web structure of this marine system have resulted in increased biomass of small reef carnivores and cephalopods (Lachica-Alino et al. 2009). Similar results for Hong Kong were shown by Cheung and Sadovy (2004), where large fish species become replaced by small fish species and invertebrates. Bottom trawling fisheries will avoid reef areas as the nets, which are very expensive, can catch and tear on these hard substrates. In Guam, although there was a small increase in annual catch caused by successful spear fisheries in the late 1990s, the average catch from shore fisheries declined from 100 T $year^{-1}$ in 1985–1990 to 37 T $year^{-1}$ in 2007–2012. This was consistent with non-fisheries surveys which also show depleted shallow reef populations (Weijerman et al. 2016). Landings data allowed high-resolution assessments of six commercial reef fish in Cuba from 1955 to 2005. While four snapper species underwent no net change in catch biomass, Nassau grouper (*Epinephelus striatus*) and lane snapper (*Lutjanus synagris*) both experienced large declines in average catch over the 50-year study period: 1600 and 800 MT $year^{-1}$ to less than 100 and 450 MT $year^{-1}$, respectively. In the early 1960s there were sharp increases in catch biomass for all commercial species mainly driven by the development of bottom trawl and fish trap fisheries (Claro et al. 2009). Coral reef fish population trends vary depending on a balance between biological life cycles and fishing gear and effort. Generally, schooling species such as green bumphead parrotfish and lane snapper or spawning species such as Nassau grouper are much more vulnerable to overfishing than cryptic reef organisms more inaccessible to fishers such as moray eel (Muraenidae) or octopus (Octopoda).

While large-scale studies are useful, they can often lose fine-scale resolution. A recent study on national catch reconstructions in 25 Pacific island nations and territories showed increasing fishing trends throughout the Pacific from 1950 to 2010 (Zeller et al. 2015), while a reconstruction of Hawaiian and Florida Keys fisheries showed similar increasing trends (Table 13.1) (McClenachan and Kittinger 2013). The growth of huge pelagic fisheries over the last century masks the relatively smaller coral reef fishery declines reported in this synthesis. Historic reef fish declines reported by Hardt (2008) who focussed solely on reef fisheries was lost in the large-scale studies by Zeller et al. (2015) and McClenachan and Kittinger (2013). In summary, coral reef fish declines are not ubiquitous but are the dominant global trend. Appropriate fisheries reconstructions using quantitative data mining rather than anecdotal evidence are useful for improving global fisheries catch datasets and hence inform fishing communities and governments on long-term trends lost in official records (Zeller et al. 2015). Economic pressures associated with such declining reef fisheries can influence the trade-offs fishers make when considering alternate sources of income such as tourism.

13.3.2 Tourism Trends

Alongside fisheries declines and global population rise, the last half century has been characterized by the technical revolution with huge advances in transportation efficiency and cost, allowing economic shifts toward a globally multimillion-dollar tourism industry (Craig 2008; Spalding et al. 2017). The number of visitors to Asia has increased more than 60% in the last 15 years with growth expected to reach 75% in the next decade (Outra et al. 2016). Of all global regions, the Asia-Pacific is experiencing the fastest growth in international tourism, closely followed by the Americas (Harvey and Naval 2016). This growth trend has been mirrored by the scuba diving industry which was once the fastest-growing recreational activity in the world (Tabata 1992), characterized by huge increases in the number of certified scuba divers since the 1970s (Fig. 13.3).

To understand the recent opportunities and impacts of coral reef tourism relevant to trade-offs made by coral reef resource users, we conducted a systematic literature search in Web of Science® targeting all studies on coral reefs since 2013. The search string combined the following three categories with *and* operators: (1) coral reef synonyms ("coral reefs" *or* "coral reef"), (2) current topics in coral reef ecology ("ecotourism" *or* "tourism" *or* "social ecological system" *or* "ecosystem-based management" *or* "ecosystem management" *or* "connectivity" *or* "replantation" *or* "keystone species" *or* "flagship species" *or* "invasive species" *or* "global warming" *or* "ocean acidification" *or* "climate change" *or* "fisheries"), and (3) a comprehensive list of coastal tropical countries from the United Nations (2018) and overseas territories from nationsonline.org, separated by *or* operators.

Based on the title and abstract, the 1043 search results were categorized by relevance to coral reefs, relevance to tourism, study country, and theme of main impact. Therefore, the resultant dataset of 36 tourism-related studies is a randomly sampled, spatially explicit representative of current research on coral reef tourism. This database was characterized by four major impact topics, referred to throughout this chapter and shown in Fig. 13.4 alongside the proportion of studies focused on scuba-diving compared to other tourism related topics. Socioeconomic and environmental impacts of tourism are discussed in this section, while socio-ecological and social perceptions and preferences are discussed in the next section.

13.3.2.1 Economic Impacts

Reef tourism provides major employment to coastal communities (Murray 2007; Lopes et al. 2015). The success of this industry rests on its high economic value (Cesar et al. 2003; Craig 2008), contributed to by on-reef tourism activities including diving, snorkeling, and glass-bottom boating, as well as reef-adjacent tourism attractions such as seafood, scenery, and beaches (Spalding et al. 2017). An extensive meta-analysis of 166 reef valuation studies from the 1980s until 2007 revealed that the combination of diving, viewing, and snorkeling had the highest mean value (approx. US$ 300), followed by diving alone (approx. US$ 200), compared to snorkeling which was valued at less than 15% of mean

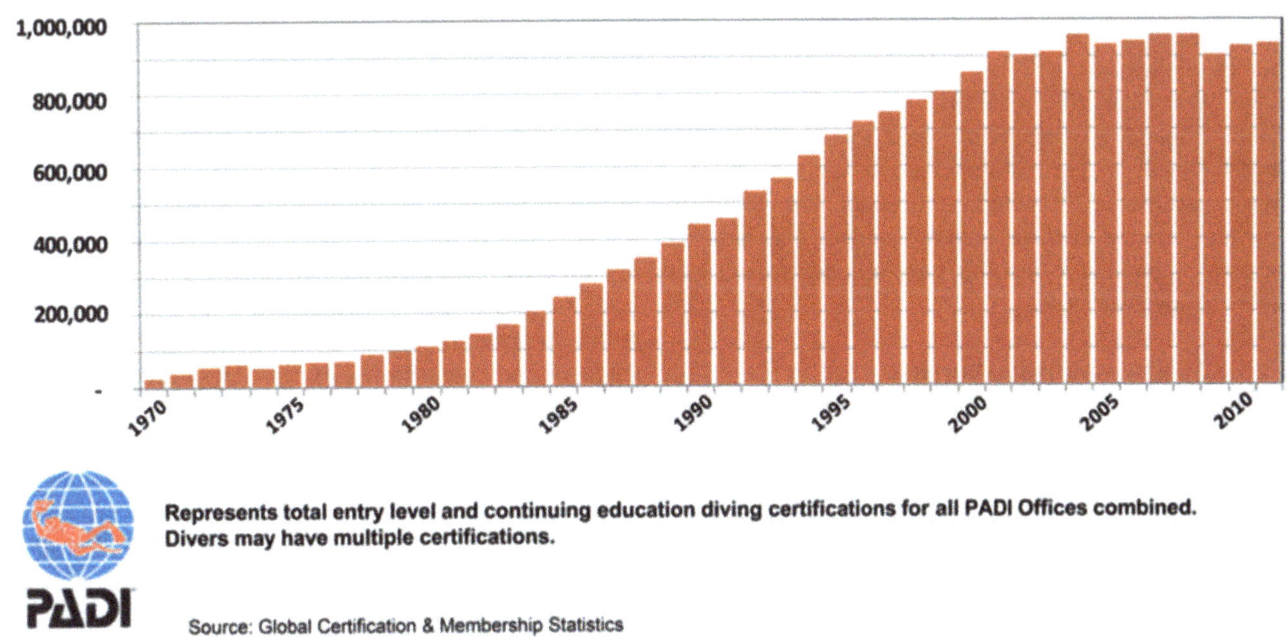

Fig. 13.3 Number of PADI diving certifications obtained worldwide from 1970 to 2011. (Adapted from PADI global certification and membership statistics (http://www.padi.co.kr/images/Statistics-Kor.pdf accessed 21/05/2018) with permission from PADI Worldwide)

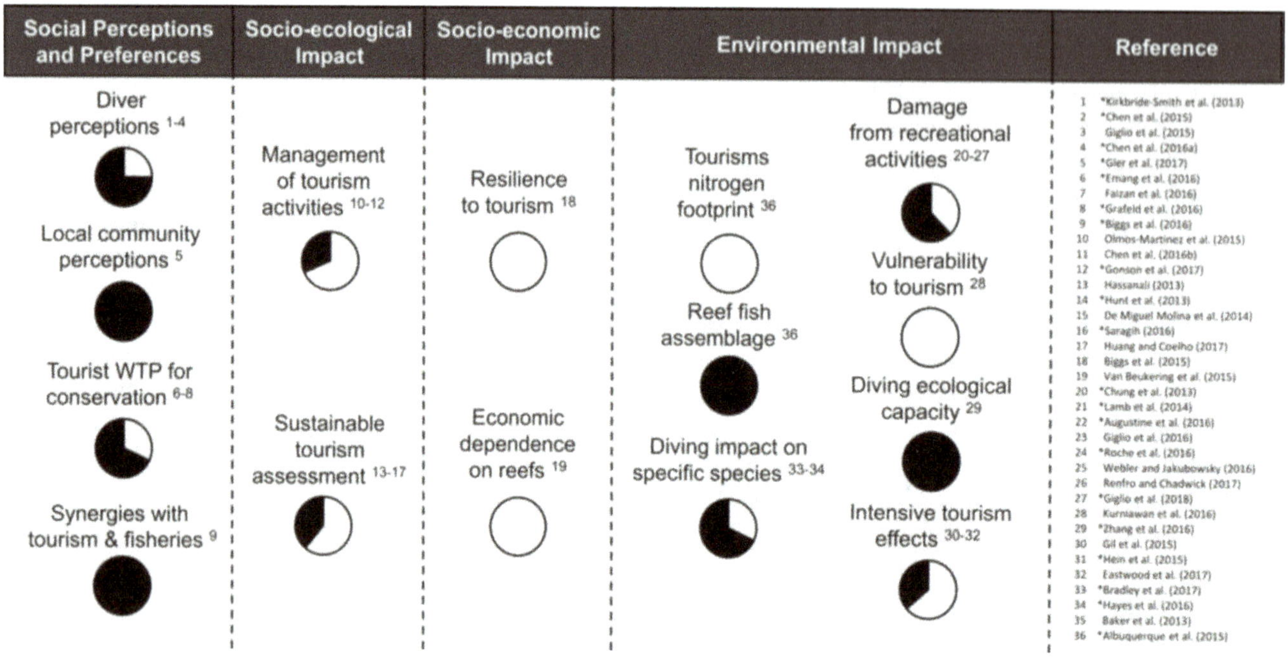

Fig. 13.4 From systematic review, a representative overview of 36 coral reef tourism studies since 2013, under four major impacts, and several sub-topics. Pie charts show the proportion of studies within each sub-topic relevant to diving (black), with diving-related studies marked (*) in the references of this figure

diving value (Fig. 13.5b, Brander et al. 2007). Diving and scenery are some of the most important activities for coral reef tourism (Hsui and Wang 2013). Brander et al. (2007) also showed that the economic value of coral reefs varies by global region (Fig. 13.5a). Coral reefs were valued highly across all global regions except the United States, with high median valuations for Australia and East Africa but lower median valuations for Southeast Asia and the Caribbean. As shown in our systematic review, the majority of coral reef tourism publications in the last 5 years have been conducted in the West Atlantic (Caribbean), Southeast Asia, and the Pacific (Fig. 13.5c), regions that have also undergone the largest growth in reef tourism over the past two decades (Harvey and Naval 2016; Outra et al. 2016). Therefore, as a growing industry in these regions, tourism may provide lucrative opportunities causing trade-offs for fishers and other employment sectors.

13.3.2.2 Environmental Impacts

Employee livelihoods are often heavily reliant on reef tourism and its ability to attract tourists to healthy coral reefs (Hunter et al. 2018). However, tourism-related threats such as enhanced sedimentation from changes in land use, loss of habitat due to land reclamation, expulsion of sewage and solid waste, and overuse by snorkelers and divers (Fig. 13.4) can contribute to reduced ecosystem resilience or phase shifts away from coral-dominated ecosystem states (Hawkins and Roberts 1994; Redding et al. 2013; Lamb et al. 2014; Renfro and Chadwick 2017), thereby jeopardizing tourism-based livelihoods (Smith et al. 1981).

Corals are controlled on a large scale by sedimentation. In areas further away from sources of runoff, with lower concentrations of sediment in overlying waters, reefs are generally more diverse, are better developed, and have higher framework accretion rates (Rogers 1990). Coral responses to moderate sedimentation include synchronous polyp pulsations, cleaning with tentacles or cilia, and concentration and excretion of sediment in mucus (Hubbard and Pocock 1972; Lirman and Manzello 2009), while complete covering by sediment leads to coral death within hours (Mayer 1918; Rogers 1990; Hunte and Wittenberg 1992). Phase shift theory suggests that the tipping point moving away from the coral-dominated state is not the same as the threshold on the return succession (Hughes et al. 2010). Therefore, fully degraded coral-dominated reefs can fail to recover even at much lower levels of sedimentation, due to repressed recruitment of sensitive juvenile corals (Hughes et al. 2010; Doropoulos et al. 2016). Enhanced sedimentation from tourism development has already caused substantial degradation of inshore reefs in the Egyptian Red Sea (Hawkins and Roberts 1994).

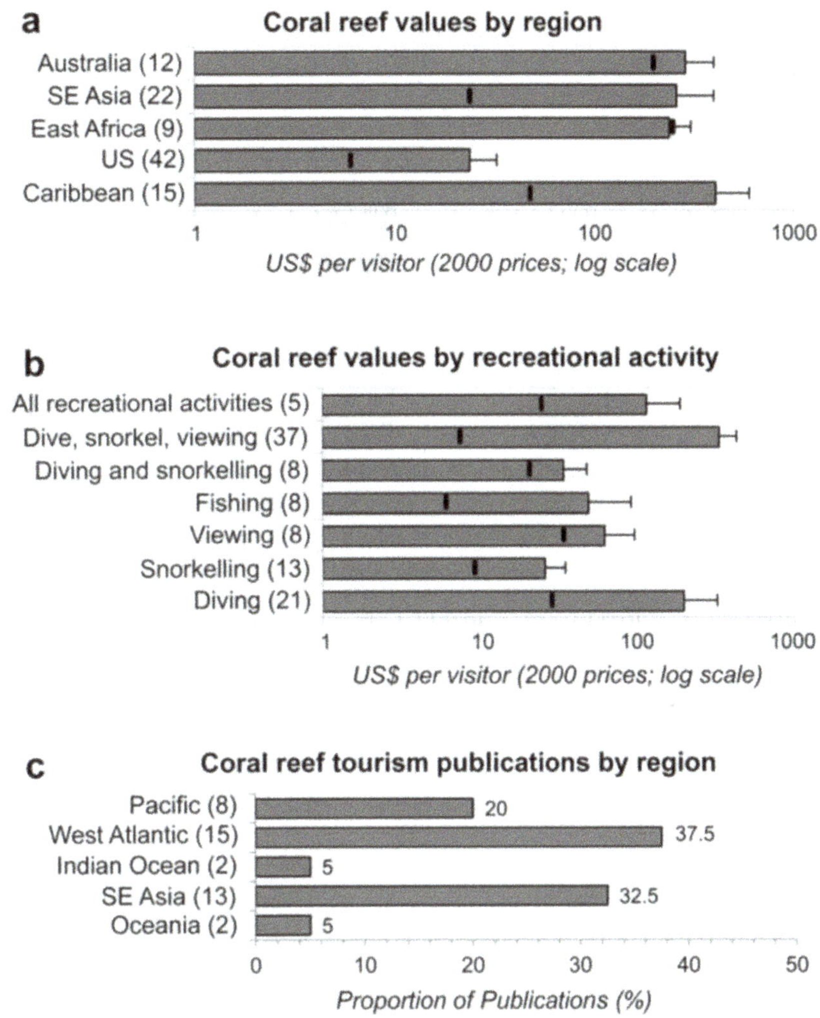

Fig. 13.5 Coral reef valuations from the 1980s until 2007 by (**a**) global region and (**b**) recreational activity, showing mean and median value (bar and dot) with standard error bars (Brander et al. 2007). For comparison, the proportion of reef tourism studies published since 2013, derived from our systematic review dataset (n = 36), are shown for each global region (**c**). Sample size of each region/activity is shown in brackets. Regional labels differ between by our systematic review and Brander et al. (2007): Australia within Oceania, East Africa within Indian Ocean, US split between Hawaii in Pacific and Florida Keys in West Atlantic, Caribbean within West Atlantic. (Adapted from Brander et al. (2007) with permission from Elsevier)

Discharge of untreated or partially treated effluent is a higher priority threat to coral reefs, with the potential to decrease coral coverage and promote overgrowth of other spatial benthic competitors such as macroalgae (Johannes 1975; Lapointe et al. 2005; Gil et al. 2015) or Zoantharia, soft-bodied benthic Cnidaria (Hunter and Evans 1995; Smith et al. 1981; Lapointe et al. 2010; Hernández-Delgado et al. 2008; Acosta et al. 2001; Lachs unpublished data). Field experiments and surveys show that nutrient enrichment and sewage pollution can jeopardize coral reef resilience by increasing the severity of diseases such as aspergillosis or yellow-band disease in common gorgonian sea fans (*Gorgonia ventalina*) and reef-building corals (*Montastraea* sp. and *Porites* sp.) (Bruno et al. 2003; Baker et al. 2007; Redding et al. 2013). Results of coral damage, disease advancement, and coral tissue loss (Fig. 13.6) are consistent from the Caribbean and Pacific Oceans. In Guam, the highest sewage signals were consistently measured at Tumon Bay which is the center of tourism, showing the specific risks of tourism-derived sewage on coral reef ecosystems (Redding et al. 2013). Given the global rise in population and tourism intensity, ecological impacts from sewage release should be closely monitored and considered by coral reef managers.

13.4 Sector Overlap and Trade-Offs

Managers and conservationists should consider the ecological trade-offs between tourism and fisheries industries. Overuse through heavy fishing, land-use change, or poor waste management can all lead to coral reef degradation, phase shifts, and even reef fishery collapse (Hawkins and Roberts 1994; Cesar et al. 2003; Mumby et al. 2006; Fenner 2012; Redding et al. 2013; Lamb et al. 2014; Bozec et al. 2016; Renfro and Chadwick 2017). While balancing the ecological trade-offs between coral reef fisheries and tourism, management strategies must also align with the social and economic interests of workers. Between industries, these interests are often in opposition, with regular disputes over spatial planning and zonation rights, varying education/skill set requirements and levels of salary/job security, and different world views and ecosystem service priorities (Brown

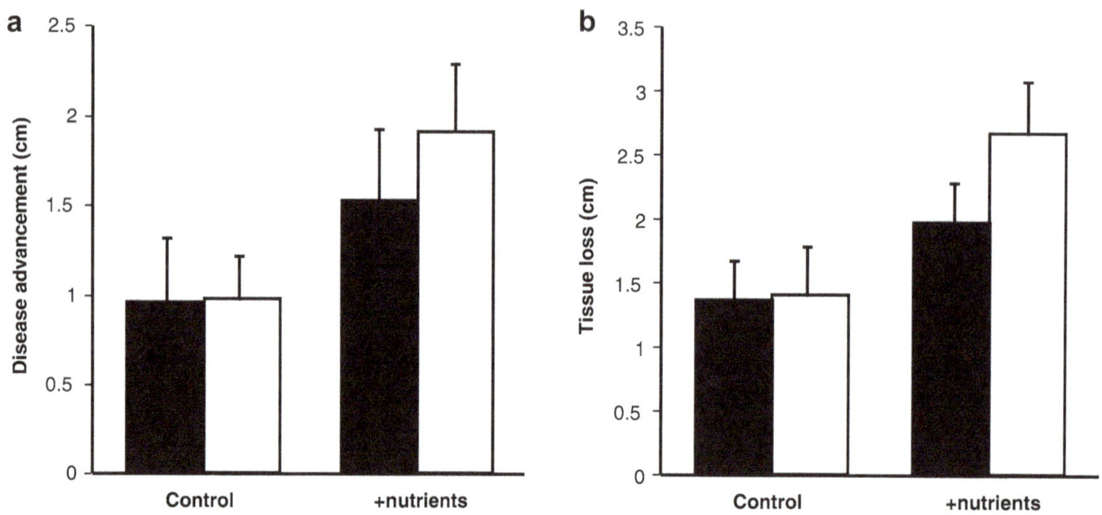

Fig. 13.6 Effect of experimental nutrient enrichment on (**a**) front advancement of yellow band disease and (**b**) coral tissue loss in the Caribbean reef building corals *Montastraea annularis* (white bars) and *Montastraea franksii* (black bars) during a 90-day in situ experiment (mean ± standard error). (Adapted from Bruno et al. (2003), with permission from John Wiley and Sons)

et al. 1997; Fabinyi 2008; Hicks et al. 2013; Nejati et al. 2014; Lopes et al. 2015; de Andrade and de Oliveira Soares 2017; Hunter et al. 2018).

13.4.1 Zoning Issues

Inevitably, fishers and dive/snorkel tourism operators both need to work at coral reef sites. However, they cannot work alongside each other for obvious reasons. There is a potential gap in our current understanding of the perceptions of dive operators and fishers on the coexistence of their activities (Barker and Roberts 2004). Several recent studies agree that the motivations and principles of fishers and dive operators are distant, partially due to different educational backgrounds and ecosystem service priorities (Satria et al. 2004; Fabinyi 2008; de Andrade and de Oliveira Soares 2017). Despite this, both stakeholders agree on the importance of establishing marine protected areas (MPAs) (Fabinyi 2008; de Andrade and de Oliveira Soares 2017). Conflicts among these two sectors have been reported from developing countries such as Kenya and the Philippines (Hodgson and Dixon 1988; Samoilys et al. 2017). Divers and fishers repeatedly compete for space and resources in locations where zoning rules are not well established (Fabinyi 2008; de Andrade and de Oliveira Soares 2017), resulting in both parties blaming each other for negative ecosystem impacts in these areas. One hand, many large resort operators have conservation-oriented perceptions (Hein et al. 2018), promoting the protection of coral reefs to maintain the high biodiversity that attracts tourists, allowing them to enjoy greater underwater experiences. On the other hand, fishers defend all ecosystem services that involve exploitation opportunities and support their livelihood (Fabinyi 2008; de Andrade and de Oliveira Soares 2017), especially those related to food security (Fisher et al. 2014). Accordingly, the role of MPAs in coral reef ecosystems may be less effective than they are designed to be. Fragile government regulations demonstrate that certain MPAs only exist on paper, enhancing zonation conflicts between tourism and traditional fishers (Satria et al. 2004).

13.4.2 Livelihoods

Despite the conflicts between tourism and fisheries industries, their coexistence is a persistent component of coral reef socioeconomic systems. As a seasonal industry, tourism cannot provide year-round employment, bringing with it a suite of social and economic challenges (Brown et al. 1997). Fisheries can provide an alternative livelihood in the tourist low season, causing a bidirectional flow of workers between both industries with seasonal cycles. However, the long term fisheries are relying on ever-dwindling fish stocks (Bruggemann et al. 2012; Zeller et al. 2015), influencing a residual flow of workers from fisheries to tourism where opportunities are more plentiful (Yacob et al. 2007). For instance, skippers can renovate and adapt their fishing boats to accommodate tourists or divers instead. Workers often transition from traditional livelihoods to tourism-based employment due to better wages and job security (Murray 2007; Lopes et al. 2015). Employee wages are consistently higher within the tourism industry than in fisheries (Nejati et al. 2014; Lopes et al. 2015; Hunter et al. 2018). For instance, in Malaysia, the employment of the local population on Redang Island is quite equally divided between tourism (50%) and fisheries (45%) (Nejati et al. 2014), but the

Table 13.2 Literature summary of access value for coral reef MPAs since 1987

Reference	Location	WTP US$	Purpose of valuation
Sloan (1987)	Heron Island, Great Barrier Reef, Australia	27 person^{-1} day^{-1}	Recreation
Dixon et al. (1995)	Bonaire National Marine Park, Caribbean	18 diver^{-1} year^{-1}	Maintain dive quality
Arin et al. (2002)	Anilo Marine Sanctuary, Philippines	3.70 diver^{-1} day^{-1}	Support marine sanctuaries
Yeo (2004)	Pulau Payar Marine Park, Malaysia	4.20 person^{-1} year^{-1}	Recreation
Asafu-Adjaye et al. (2008)	Mu Ko Similan arine Park, Thailand	27-63 diver^{-1} year^{-1}	Diving in MPAs
Yacob et al. (2009)	Redang Island Marine Parks, Malaysia	1.96-2.67 person^{-1} year^{-1}	Ecotourism resources
Thur (2010)	Bonaire National Marine Park, Caribbean	61-134 diver^{-1} year^{-1}	Diving in MPAs
Mamat et al. (2013)	Pulau Redang Marine Park, Malaysia	2.73-7.20 person^{-1} visit^{-1}	Environment protection
Faizan et al. (2016)	Cape Rachado, Malaysia	0.75* person^{-1} visit^{-1}	Coral reef management
Grafeld et al. (2016)	Guam	10 person^{-1} visit^{-1}	Coastal and watershed management

Willingness to pay (WTP) units vary between studies, and values above ten are rounded to the nearest unit. Currency conversions were calculated using average annual exchange rates from the year of publication (www.ecb.europa.eu/stats, accessed 04 October 2018). The US$ value by Faizan et al. (2016) (∗) was converted from MYR. Adapted from Asafu-Adjaye and Tapsuwan (2008)

difference in monthly income is heavily in favor of tourism (MYR 500–700 or US$ 106–149) over fisheries (MYR 350–450 or US$ 74–96) (Yacob et al. 2007) (Table 13.2 describes currency conversion methods). Tourism can provide higher wages up to double or triple that of fisheries in some regions (Lopes et al. 2015; Hunter et al. 2018).

13.4.3 Ecosystem Service Priorities

Fishers, tourism operators, scientists, and conservationists inherently value ecosystem services differently; however, there is an overlap in their priorities. Using a combination of interviews and network analysis in the Western Indian Ocean, Hicks et al. (2013) aimed to identify the key trade-offs in how fishers, managers, and scientists prioritize coral reef ecosystem services. While scientists and managers' ecosystem service priorities were more aligned, all three stakeholder groups agreed that fisheries, education and habitat are highly important services. However, The order of ecosystem priorities was different between stakeholder groups, whereby scientists agreed least with fishers leading to difficulties in balancing stakeholder value. Network analyses identified concerning trade-offs not immediately clear from the respondent's ecosystem service priorities – for fishers maximization of recreation and tourism was not possible without a loss in education and legacy of local cultural traditions. As the long-term shift from traditional livelihoods to tourism-based industry proceeds (Murray 2007; Yacob et al. 2007; Lopes et al. 2015) tourism is considered to threaten local culture by offering a tempting and profitable alternative to embracing local cultural heritage (Brown et al. 1997) resulting in a loss of culture, traditional knowledge and even language, especially in younger generations.

13.5 Management Strategies: Benefits and Pitfalls

13.5.1 The Unmanaged Commons

Long-standing fishing traditions, low tourism potential, and poor governance can cause mismanagement of reef resources and maximization of fishery intensity (Hardt 2008). The conceptual "tragedy of the unmanaged commons" is a problem described by Hardin (1968) where individual resource users aim to maximize their own benefit from an open access resource, resulting in the complete exhaustion of that resource. Open commons may benefit reef fishers temporarily, but long-term overfishing, depletion, or exhaustion of fish resources can lead to reduced ecological resilience, enhanced economic pressure, and concurrent social tension for subsistence income families that may be on the poverty line (Mumby et al. 2006; Walmsley et al. 2006; Fenner 2012; Teh and Sumaila 2013). Strategies to manage coral reef resources are necessary and vary widely. Top-down approaches by government, using ecosystem-based MPAs and fisheries embargos, are generally more suited to tourism-based coastal economies

(Oracion et al. 2005; Yacob et al. 2007; Munga et al. 2012). Comparatively, bottom-up initiatives using collaborative management frameworks empower small-scale reef fishers and tourism operators to self-regulate (Cinner et al. 2012; Weeks and Jupiter 2013; Hunter et al. 2018). However, large coral reef tourism businesses or resorts can monopolize decision-making with strong financial backing and hence threaten co-management initiatives (Levine and Richmond 2014). Under the ever-changing world of international mobility, economic shifts, and climate-driven mass coral bleaching, adaptive co-management strategies supported by governments may provide the most resilient basis for management of coral reef resources (Cinner et al. 2012; Plummer et al. 2013; Weeks and Jupiter 2013).

13.5.2 Ecosystem-Based Management

To ensure a sustained resilience of coral reefs, management decisions must account for trends in ecosystem functionality (Bozec et al. 2016). On both ends of the spectrum between fishing intensity and tourism intensity, there are increased risks of ecological collapse and phase shifts away from the coral-dominated stable state (Van Beukering and Cesar 2004; Bozec et al. 2016). The importance of an ecological framework in decision-making is exemplified in the case of Bacuit Bay, Palawan, Philippines, in the 1970s (Hodgson and Dixon 1988). At this time Palawan was one of the last unspoiled areas of the Philippines with very low population density and plentiful marine and terrestrial resources. Throughout the 1980s and onward there was extensive immigration to Palawan, and unused resources became the subject of exploitation, with a 20% decline in forest area in 7 years alongside declines in yellowfin and skipjack tuna from intense fishery activities. Environmental degradation of the previously pristine coral reef and other marine ecosystems was further confounded by heavy siltation from forestry logging combined with dynamite and poison fisheries. An economic model was developed to test the effects of two management solutions: (1) to ban logging entirely in the bay's watershed or (2) to allow logging to continue as planned. The results of the economic analysis predicted a "reduction in gross revenue of more than US$ 40 million over a 10-year period with continued logging of the Bacuit Bay watershed as compared with gross revenue given implementation of a logging ban" (Hodgson and Dixon 1988). This case study was resolved by the banning of logging in Palawan by the national government alongside the declaration of marine park status for the bay. The predictions about tourism growth were correct, however, overfishing has severely reduced populations of most high-value fish (Hodgson and Dixon 2000). This case highlights economic risks of coral reef degradation and the importance for policymakers and environmental managers to heed and incorporate scientific recommendations on ecological trends into ecosystem-based management policies.

Another ecosystem-based management approach is the use of MPAs. Theoretically, MPAs fulfil the requirements of conservation scientists, tourism managers, and artisanal fishers (Fabinyi 2008) by promoting conservation, management, and protection of natural resources and positively influencing fish diversity and abundance, including that of commercially valuable fish (Munga et al. 2012). However, marine park gazettements are often combined with legislation to ban coral reef fisheries or allow only minor fishing activities (Yacob et al. 2007; Lopes et al. 2015; Samoilys et al. 2017). Therefore, MPAs solve the tragedy of the common dilemma at the expense of resource users; not all stakeholders benefit equally from MPA management (Lopes et al. 2015; Samoilys et al. 2017). This is due to combinations of the following effects: competition between different resource users for the same resource, weak management regulations, ineffective governance, scarcity of funding, and nonproportionality of stakeholder representation in decision-making positions (McClanahan 1999; Tupper et al. 2015; Zimmerhackel et al. 2016). MPAs in the tropics are typically designed around coral reefs, where marine-based tourism plays an important and potentially disproportionately strong role in MPA management. Increasingly marine tourism causes conflict in local communities where traditional fishers who are not well-suited to tourism are excluded from their livelihoods. Foreign tourists pay high prices that produce positive responses in some local groups but negative responses in other social groups such as artisanal fishers who do not benefit from tourism (Satria et al. 2004; Hicks et al. 2013). A lack of participative management and communication between stakeholders fosters divided perceptions and a lack of management policy uptake. Hence the drawback of ecosystem-based management is the uneven distribution of benefits.

13.5.3 Co-management

Collaborative management, also coined co-management, describes a decision-making system that combines top-down institutional frameworks and advice with bottom-up decision-making and empowerment of all local stakeholder groups (Roberts and Hawkins 2000; Cinner et al. 2012; Weeks and Jupiter 2013). Moving away from the top-down approach to management, such as in MPAs where some resources users are excluded, co-management employs community-scale local knowledge to work toward common goals (Levine and Richmond 2014). Increasing local involvement in MPA and resource use decision-making allows more balanced management solutions that fulfil the goals of tourism, fisheries, and other stakeholder groups, ensuring benefit-sharing from reef resources (Roberts and Hawkins 2000; de Andrade and de Oliveira Soares 2017). When executed successfully with local institutions, co-management initiatives provide various social benefits and can promote more culturally relevant policies (Cinner et al. 2012; Levine and Richmond 2014). Governments that lack financial resources can pair with local partners to implement activities that would be otherwise unfeasible (Techera 2007). Various studies show that co-management can also influence the revitalization and sustainable use of marine resources maintaining livelihoods (Cinner et al. 2012; Weeks and Jupiter 2013).

Linking themes underpinning success include government and legislative support frameworks, government encouragement of local leadership, distinct community boundaries, unified village perceptions and representative leadership, the right to exclude outsiders from resource exploitation, and community-level enforcement of local laws (Levine and Richmond 2014). However, without these necessary components, co-management initiatives can fail and waste financial resources (Schultz et al. 2011; Cinner et al. 2012; Levine and Richmond 2014). This is shown by the Malagasy case described by Bruggemann et al. (2012). In Madagascar, coral reef resources are managed under legally recognized local-scale governing bodies known as gelose (gestion locale sécurisée) and by local groups without legal status. This system is defined by a lack of government involvement or support. Resource use regulations are built locally using customary concepts including *fady* – activities that are taboo in certain areas, and *dina* – local laws. While this was a previously successful co-management system, recently, reefs have become overfished due to increased human migration from inland areas to the coast, increasing the number of fishers breaking *fady* and *dina* rules (Bruggemann et al. 2012). Co-management initiatives require some top-down government organization and influence to support the adaptive capacity of local institutions (Plummer et al. 2013; Weeks and Jupiter 2013; Levine and Richmond 2014; Hunter et al. 2018).

13.5.4 Adaptive Co-management

While co-management initiatives have extensive societal benefits, extensive field surveys around the Indian Ocean and Indo-Pacific suggest that co-management initiatives do not significantly improve fish biomass or ecosystem resilience, "indeed, people may collectively organize to exploit resources rather than to sustain them" (Cinner et al. 2012). Adaptive co-management may present a more progressive sustainable approach to resource use (Cinner et al. 2012; Weeks and Jupiter 2013; Hunter et al. 2018) that is relevant to the Anthropocene and recent unprecedented bleaching of coral reefs across the world (Hughes et al. 2017). This decision-making system combines the government-local format of co-management with an additional evaluation and adaptation framework that includes environmental scientists in decision-making, using scientific advice to also promote long-term ecological sustainability (Weeks and Jupiter 2013).

13.6 Tools to Manage Trade-Offs

Due to complexity of multi-stakeholder decision-making and the wide range of factors affecting management success, *sustainable adaptive co-management* may seem an insurmountable challenge. However, various implementable management tools exist that can aid in balancing the trade-offs between fisheries, tourism, and other stakeholder groups and support coral reef socioeconomic systems (Stolk et al. 2007; Bozec et al. 2016; Faizan et al. 2016).

13.6.1 Ecological Fisheries Regulations

Combining ecosystem-based management and co-management empowers local fishers while also managing for ecological sustainability (Hunter et al. 2018). Using scientific knowledge of ecosystem functioning to give fisheries recommendations can balance the ecological trade-offs of fisheries without excluding resource users (Sary et al. 1997; Bozec et al. 2016). From fish-exclusion mesocosms at the inner Great Barrier Reef, we know that 70–90% reductions in herbivorous fish biomass can induce phase shifts away from the coral-dominated ecosystem state to a dense algal stable state with >90% maximum algal coverage (Hughes et al. 2007). A fully calibrated fishery model developed by Bozec et al. (2016) suggests that harvesting parrotfish at maximum sustainable yield (40% of exploitable biomass) can lead to long-term reductions (75%) in unfished biomass similar to those shown in Hughes' fish-exclusion experiments. Given these results, phase shifts to algal-dominated ecosystem states are a realistic outcome from overfishing of grazing fish in coral reef ecosystems. Bozec et al. (2016) combined functional ecology and resilience theory to provide implementable management solutions to avoid ecosystem-breakdown scenarios; a minimum catch length of >30 cm for parrotfish fisheries can provide a win-win scenario for fisheries and environmental interests in the short term. Fisheries yields are predicted to benefit due to a higher proportion of large size-class fish, while grazing pressure is maintained, leading to more resilient coral reefs. Such win-win scenarios have also been shown empirically. A fish trap exchange program which replaced small mesh-size traps with larger mesh-size traps in Discovery Bay, Jamaica led to a recovery of local reef fish populations alongside a increased catch of larger more valuable fish and increased CPUE (Sary et al. 1997). Therefore, small changes in fishing practice can lead to reductions in fishing pressure needed to allow recovery of reef fish populations and even increase catch. Such a strategy can be used to alleviate overfishing, without compromising local livelihoods and traditions.

13.6.2 Iconic Species

Shark, schooling fish, rays, and sea turtles are used by snorkel and dive operators throughout the world to promote tourism (Fisher et al. 2008; Vianna et al. 2012; Zimmerhackel et al. 2016). Diving tourism related to marine megafauna is a stable industry and has increased in popularity immensely around the world over the last decades (Higham and Lück 2008). While all divers have a strong preference to see charismatic megafauna, experienced divers have more interest in cryptic fauna (Giglio et al. 2015). Therefore, even coral reefs without megafauna have tourism potential, and adapting to diver preferences can increase consumer satisfaction and revenue (Giglio et al. 2015). Vianna et al. (2012) showed that shark-based tourism and shark-diving were worth US$ 18 million per year to the economy of Palau, 24 times that of total fisheries revenue. The chance to view sharks was the principal reason chosen by visitors to come to Palau. Thus, shark diving is the main economic activity, generating employment opportunities for boat drivers, hotels and restaurant workers, and civil engineers. Promoting iconic species tourism can help support biodiversity, improve tourism revenues, and provide local populations with alternate employment opportunities than fisheries (Vianna et al. 2012; Higham and Lück 2008).

13.6.3 Tourist Fees

Implementing marine park and beach access fees for leisure activities is another method to increase tourism revenue while subsidizing losses in fisheries revenue. We present a summary of the willingness to pay (WTP) of tourists visiting coral reefs over the last 30 years, adapted from Asafu-Adjaye and Tapsuwan (2008) (Table 13.2). From this summary, Faizan et al. (2016) found that local visitors and tourists had WTP for fees of MYR 3 (US$ 0.65) for improving coral reef conservation in Malacca, Malaysia, which equates to over US$ 150,000 per annum. In Guam, diver WTP for reef conservation could contribute over US$ 8 million to annual revenues (Grafeld et al. 2016). As overseas divers' WTP is more than that of local divers, increasing prices for foreign divers is a likely way to increase revenues. Consequently, more visitors would not be needed to compensate for the cost of maintaining MPAs (Asafu-Adjaye and Tapsuwan 2008). While most marine park rangers are not part of the fisheries sector, the additional revenues from tourist fees could be used to employ fishers to assist rangers in patrolling, an option that has already shown large public interest from local fishing communities (Elliott et al. 2001).

13.6.4 Artificial Reefs and Restoration

Artificial reefs, restored reefs, and recent efforts to *reskin* artificial or dead corals with live coral are ecologically relevant techniques to promote reef resilience, support fish populations, provide employment, enhance tourism opportunities, and promote public awareness on coral reef loss (Grossman et al. 1997; Lirman and Schopmeyer 2016; Hein et al. 2018). Therefore, such projects have an applied use as a management tool, to offer alternative tourism-based employment to fishers (Lirman and Schopmeyer 2016). Although it is debated, evidence suggests that large communities of fish can be sustained on artificial reefs (Stolk et al. 2007; Smith et al. 2016). Artificial reefs were developed in the United States, Canada, Japan, Australia, and Europe (Coutin 2001) and were utilized up to 100 years ago by coastal fishing communities to boost fish catch around these structures (McGurrin et al. 1989). Improved fisheries from such aggregations have been well documented (McGurrin et al. 1989); however, it is not known if attracting and concentrating fish are effects of increased biomass or just a redistribution of biomass (Polovina and Sakai 1989; Polovina 1990; Stolk et al. 2007; Ajemian et al. 2015; Scott et al. 2015; Smith et al. 2016). Therefore, there is an urgent need for scientific assessments on the true effect of artificial reefs on fish stocks. SCUBA diving is the main commercial activity in coral reef areas (Hsui and Wang 2013). Recently, the use of artificial reefs has shifted toward tourism-based activities like diving, snorkeling, recreational fishing, nature preservation, and science (Seaman and Jensen 2000; Jakšić et al. 2013). It is important to consider the attitudes, perceptions, and satisfaction levels of scuba divers in the design of artificial reefs to guarantee good dives with a high level of biodiversity and wildlife photographic opportunities (Kirkbride-Smith et al. 2013). In Barbados, novice divers have a greater preference for artificial reef dive sites than experienced divers (Polak and Shasnar 2012). Artificial reefs can be used to reduce the physical damage of novice diving at sensitive natural sites while maintaining economic benefits by attracting an increasing number of advanced divers with specific diving requirements to less degraded natural reefs (Dearden et al. 2006; Kirkbride-Smith et al. 2013). Again, this shows how artificial reefs are an ecologically sensitive and enriching method of building resilience in MPAs.

13.7 Recommendations for Management

Weighing up the various costs and benefits of different industrial practices in coral reef ecosystems is a continual challenge. As resource rights, political situations and natural environments change new conflicts arise between conservationists, scientists, fishers, tourism operators and local employees of other coastal industries. Proposed and implemented management strategies are rarely one-fits-all solutions. Management plans tend to push for consensus in identifying the most important ecosystem service and then manage for that service; however, this approach does not accommodate complex interactions between stakeholders' opinions or ecosystem service priorities.

We recommend holistic and effective resource use by developing adaptive co-management systems that combine top-down strategic frameworks with bottom-up decision-making. The tools and theories outlined in this review have been developed to promote the effectiveness of management actions, and some have good potential. Ecosystem-based fisheries modelling or long-term fisheries reconstructions can help direct fisheries regulations toward resilience, while the use of artificial reefs, tourist fees, and the promotion of iconic species can promote tourism and provide alternative livelihoods to fishers. Determining different stakeholder opinions and understanding trade-offs between different stakeholder priorities, as shown by Hicks et al. (2013), may lead to more integrated management decisions likely to represent the needs of local stakeholders proportionally. However, we point out that such co-management strategies should be framed by scientific ecological knowledge on the state and stability of coral reef ecosystems in the face of growing anthropogenic pressures. Hence, there is a need for extensive long-term ecological monitoring data. Comprehensive economic valuations of tourism and fisheries industries (e.g. those provided in the development of Palawan tourism in the Philippines. Hodgson and Dixon 1988, have the power to make real change and are a central component needed to convince governments to implement sustainable policies that promote the maintenance of healthy coral reef ecosystems, economies, and livelihoods.

Acknowledgments First, we would like to thank Sailee Sakhalkar, Farid Dahdouh-Guebas, Erin Lachs, and our reviewers for their critical readings of the manuscript and suggestions to layout, style, and structure. Their contributions were invaluable to the construction of this chapter. We would like to thank Seh Ling for her knowledgeable insights into the social dynamics of coastal coral reef communities in Peninsular Malaysia. We are thankful to Simon Jungblut, Farid Dahdouh-Guebas, and Zainudin Bachok for their general advice on the publication process. Last but not least, we would like to thank Stephen Bergacker for sharing the photographs of his humbling first-hand experience with coral bleaching in the Maldives.

Appendix

This article is related to the YOUMARES 9 conference session no. 16: "Tropical Marine Research Mosaic: combining small studies to reveal the bigger picture." The original Call for Abstracts and the abstracts of the presentations within this session can be found in the Appendix "Conference Sessions and Abstracts", Chapter "12 Tropical Marine Research Mosaic: combining small studies to reveal the bigger picture", of this book.

References

Acosta A, Sammarco PW, Duarte LF (2001) Asexual reproduction in a zoanthid by fragmentation: the role of exogenous factors. Bull Mar Sci 68:363–381

Ajemian MJ, Wetz JJ, Shipley-Lozano B et al (2015) An analysis of artificial reef fish community structure along the northwestern Gulf of Mexico Shelf: potential impacts of 'rigs-to-reefs' programs. PLoS One 10:e0126345. https://doi.org/10.1371/journal.pone.0126354

Albuquerque T, Loiola M, Nunes J et al (2015) In situ effects of human disturbances on coral reef-fish assemblage structure: temporary and persisting changes are reflected as a result of intensive tourism. Mar Freshw Res 66:23–32. https://doi.org/10.1071/MF13185

Arin T, Kramer RA (2002) Divers' willingness to pay to visit marine sanctuaries: an exploratory study. Ocean Coast Manage 45:171–183

Asafu-Adjaye J, Tapsuwan S (2008) A contingent valuation study of scuba diving benefits: case study in mu Ko Similan marine National Park, Thailand. Tourism Manage 29:1122–1130. https://doi.org/10.1016/j.tourman.2008.02.005

Augustine S, Dearden P, Rollins R (2015) Are changing diver characteristics important for coral reef conservation? Aquat Conserv Mar Freshw Ecosyst 26:660–673. https://doi.org/10.1002/aqc.2574

Baker DM, MacAvoy SE, Kim K (2007) Relationship between water quality, δ15N, and aspergillosis of Caribbean Sea fan corals. Mar Ecol Prog Ser 343:123–130. https://doi.org/10.3354/meps06937

Baker DM, Rodríguez-Martínez RE, Fogel ML (2013) Tourism's nitrogen footprint on a Mesoamerican coral reef. Coral Reefs 32:691–699. https://doi.org/10.1007/s00338-013-1040-2

Barker NHL, Roberts CM (2004) Scuba diver behavior and the management of diving impacts on coral reefs. Biol Conserv 120:481–489

Bellwood DR, Hoey AS, Ackerman JL et al (2006) Coral bleaching, reef fish community phase shifts and the resilience of coral reefs. Glob Chang Biol 12:1587–1594. https://doi.org/10.1111/j.1365-2486.2006.01204.x

Bessell-Browne P, Stat M, Thomson D et al (2014) *Coscinaraea marshae* corals that have survived prolonged bleaching exhibit signs of increased heterotrophic feeding. Coral Reefs 33:795–804. https://doi.org/10.1007/s00338-014-1156-z

Biggs D, Hicks CC, Cinner JE et al (2015) Marine tourism in the face of global changes: the resilience of enterprises to crises in Thailand and Australia. Ocean Coast Manage 105:65–74. https://doi.org/10.1016/j.ocecoaman.2014.12.019

Biggs D, Amar F, Valdebenito A et al (2016) Synergies between nature-based tourism and sustainable use of marine resources: insights from dive tourism in territorial user rights for fisheries in Chile. PLoS One 11:e0148862. https://doi.org/10.1371/journal.pone.0148862

Bozec YM, O'Farrell S, Bruggemann JH et al (2016) Tradeoffs between fisheries harvest and the resilience of coral reefs. Proc Natl Acad Sci U S A 113:4536–4541. https://doi.org/10.1073/pnas.1601529113

Bradley D, Papastamatiou YP, Caselle JE (2017) No persistent behavioural effects on scuba diving on reef sharks. Mar Ecol Prog Ser 567:173–184. https://doi.org/10.3354/meps12053

Brander LM, Van Beukering PJH, Cesar HSJ (2007) The recreational value of coral reefs: a meta-analysis. Ecol Econ 63:209–218. https://doi.org/10.1016/j.ecolecon.2006.11.002

Brown K, Turner RK, Hameed H et al (1997) Environmental carrying capacity and tourism development in the Maldives and

Nepal. Environ Conserv 24:316–325. https://doi.org/10.1017/S0376892997000428

Bruggemann JH, Rodier M, Guillaume MMM et al (2012) Wicked social-ecological problems forcing unprecedented change on the latitudinal margins of coral reefs: the case of Southwest Madagascar. Ecol Soc 17:47. https://doi.org/10.5751/ES-05300-170447

Bruno JF, Petes LE, Harvell CD et al (2003) Nutrient enrichment can increase the severity of coral diseases. Ecol Lett 6:1056–1061. https://doi.org/10.1046/j.1461-0248.2003.00544.x

Cesar HSJ, Burke L, Pet-Soede L (2003) The economics of worldwide coral reef degradation. Cesar Environmental Economics Consulting, Arnhem, and World Wildlife Fund, Zeist, Dordrecht

Chen TC, Ying TC, Ku KC (2015) Advertising coral reefs with underwater panoramas: an application study on presenting information to prospective divers. J Mar Sci Tech 23:127–132. https://doi.org/10.6119/JMST-014-0416-1

Chen TC, Ku KC, Chen CS (2016a) Collaborative adaptive management for bigfin squid applied to tourism-related activities in coastal waters of Northeast Taiwan. Ocean Coast Manage 119:208–216. https://doi.org/10.1016/j.ocecoaman.2015.10.010

Chen TC, Ho CT, Jan RQ (2016b) The differentiation of common species in a coral-reef fish assemblage for recreational scuba diving. Springer Plus 5:1758. https://doi.org/10.1186/s40064-016-3467-8

Cheung WWL, Sadovy Y (2004) Retrospective evaluation of data-limited fisheries: a case from Hong Kong. Rev Fish Biol Fisher 14:181–206. https://doi.org/10.1007/s11160-004-5422-y

Chung S, Au A, Qiu JW (2013) Understanding the underwater behaviour of scuba divers in Hong Kong. Environ Manag 51:824–837. https://doi.org/10.1007/s00267-013-0023-y

Cinner JE, McClanahan TR, MacNeil MA et al (2012) Comanagement of coral reef social-ecological systems. Proc Natl Acad Sci U S A 109:5219–5222. https://doi.org/10.1073/pnas.1121215109

Claro R, Sadovy Y, Mitcheson D et al (2009) Historical analysis of Cuban commercial fishing effort and the effects of management interventions on important reef fishes from 1960–2005. Fish Res 99:7–16. https://doi.org/10.1016/j.fishres.2009.04.004

Coutin PC (2001) Artificial reefs: applications in Victoria from a literature review. In: Marine and Freshwater Resources Institute, report No. 31, Queenscliff, Victoria

Craig RK (2008) Coral reefs, fishing, and tourism: tensions in U.S. ocean law and policy reform. Stanf Environ Law J 27:3–41

de Andrade AB, de Oliveira SM (2017) Offshore marine protected areas: divergent perceptions of divers and artisanal fishers. Mar Policy 76:107–113. https://doi.org/10.1016/j.marpol.2016.11.016

Dearden P, Bennet M, Rollins R (2006) Implications for coral reef conservation of diver specialization. Environ Conserv 33:353–363. https://doi.org/10.1017/S0376892906003419

de-Miguel-Molina B, de-Miguel-Molina M, Rumiche-Sosa ME (2014) Luxury sustainable tourism in small island developing states surrounded by coral reefs. Ocean Coast Manag 98:86–94. https://doi.org/10.1016/j.ocecoaman.2014.06.017

Dixon JA, Scura LF, van't Hof T (1995) Ecology and microeconomics as 'joint products': the Bonaire Marine Park in the Caribbean. In: Perrings CA, Mäler KG, Folke C et al (eds) Biodiversity conservation, Ecology, economy and environment, vol 4. Springer, Dordrecht, pp 127–145

Doropoulos C, Roff G, Bozec YM et al (2016) Characterizing the ecological trade-offs throughout the early ontogeny of coral recruitment. Ecol Monogr 86:20–44. https://doi.org/10.1890/15-0668.1

Doropoulos C, Roff G, Visser MS et al (2017) Sensitivity of coral recruitment to subtle shifts in early community succession. Ecology 98:304–314. https://doi.org/10.1002/ecy.1663

Dulvy NK, Polunin NV, Mill AC et al (2004) Size structural change in lightly exploited coral reef fish communities: evidence for weak indirect effects. Can J Fish Aquat Sci 61:466–475. https://doi.org/10.1139/f03-169

Eastwood EK, Clary DG, Melnick DJ (2017) Coral reef health and management on the verge of a tourism boom: a case study from Miches, Dominican Republic. Ocean Coast Manage 138:192–204. https://doi.org/10.1016/j.ocecoaman.2017.01.023

Elliott G, Mitchell B, Wiltshire B et al (2001) Community participation in marine protected area management Wakatobi National Park, Sulawesi, Indonesia. Coast Manag 29:295–316. https://doi.org/10.1080/089207501750475118

Emang D, Lundhede TH, Thorsen BJ (2016) Funding conservation through use and potentials for price discrimination among scuba divers at Sipadan, Malaysia. J Environ Manage 182:436–445. https://doi.org/10.1016/j.jenvman.2016.07.033

Fabinyi M (2008) Dive tourism, fishing and marine protected areas in the Calamianes Islands, Philippines. Mar Policy 32:898–904. https://doi.org/10.1016/j.marpol.2008.01.004

Faizan M, Sasekumar A, Chenayah S (2016) Estimation of local tourists willingness to pay. Reg Stud Mar Sci 7:142–149. https://doi.org/10.1016/j.rsma.2016.06.005

Fenner D (2012) Challenges for managing fisheries on diverse coral reefs. Diversity 4:105–160. https://doi.org/10.3390/d4010105

Ferrario F, Beck MW, Storlazzi CD et al (2014) The effectiveness of coral reefs for coastal hazard risk reduction and adaptation. Nat Commun 5:3794. https://doi.org/10.1038/ncomms4794

Fisher JB, Nawaz R, Fauzi R et al (2008) Balancing water, religion and tourism on Redang Island, Malaysia. Environ Res Lett 3:024005. https://doi.org/10.1088/1748-9326/3/2/024005

Fisher JA, Patenaude G, Giri J et al (2014) Understanding the relationship between ecosystems services and poverty alleviation: a conceptual framework. Ecosyst Serv 7:34–45. https://doi.org/10.1016/j.ecoser.2013.08.002

Fox HE (2004) Coral recruitment in blasted and unblasted sites in Indonesia: assessing rehabilitation potential. Mar Ecol Prog Ser 269:131–139. https://doi.org/10.3354/meps269131

Gier L, Christie P, Amolo R (2017) Community perceptions of scuba dive tourism development in Bien Unido, Bohol Island, Philippines. J Coast Conserv 21:153–166. https://doi.org/10.1007/s11852-016-0484-2

Giglio VJ, Luiz OJ, Schiavetti A (2015) Marine life preferences and perceptions among recreational divers in Brazilian coral reefs. Tour Manag 51:49–57. https://doi.org/10.1016/j.tourman.2015.04.006

Giglio VJ, Luiz OJ, Schiavetti A (2016) Recreational diver behaviour and contacts with benthic organisms in the Abrolhos National Marine Park, Brazil. J Environ Manag 57:637–648. https://doi.org/10.1007/s00267-015-0628-4

Giglio VJ, Luiz OJ, Barbosa M et al (2018) Behaviour of recreational spearfishers and its impacts on corals. Aquat Conserv Mar Freshw Ecosyst 28:167–174. https://doi.org/10.1002/aqc.2797

Gil MA, Renfro B, Figueroa-Zavala B et al (2015) Rapid tourism growth and declining coral reefs in Akumal, Mexico. Mar Biol 162:2225–2233. https://doi.org/10.1007/s00227-015-2748-z

Golden AS, Naisilsisili W, Ligairi I et al (2014) Combining natural history collections with fisher knowledge for community-based conservation in Fiji. PLoS One 9:e98036. https://doi.org/10.1371/journal.pone.0098036

Gonson C, Pelletier D, Alban F et al (2017) Influence of settings management and protection status on recreational issues and pressures in marine protected areas. J Environ Manag 200:170–185. https://doi.org/10.1016/j.jenvman.2017.05.051

Grafeld S, Oleson K, Barnes M et al (2016) Divers' willingness to pay for improved coral reef conditions in Guam: an untapped source of funding for management and conservation? Ecol Econ 128:202–213. https://doi.org/10.1016/j.ecolecon.2016.05.005

Grossman GD, Jones GP, Seaman WJ Jr (1997) Do artificial reefs increase regional fish production? A review of existing data. Fisheries 22:17–23. https://doi.org/10.1577/1548-8446(1997)022<0017:DARIRF>2.0.CO;2

Hardin G (1968) The tragedy of the commons. Science 162:1243–1248 https://doi.org/10.1126/science.162.3859.1243

Hardt MJ (2008) Lessons from the past: the collapse of Jamaican coral reefs. Fish Fish 10:143–158. https://doi.org/10.1111/j.1467-2979.2008.00308.x

Harvey C, Naval J (2016) Green fins – a proven approach for managing marine tourism industry growth. Visit Econ Bull. Available at: https://pata.org/store/wp-content/uploads/2016/11/PATA-VE-Bulletin_November-2016.pdf. Accessed 18 May 2018

Hassanali K (2013) Towards sustainable tourism: the need to integrate conservation and development using the Buccoo reef Marine Park, Tobago, West Indies. Nat Resour Forum 37:90–102. https://doi.org/10.1111/1477-8947.12004

Hawkins JP, Roberts CM (1994) The growth of coastal tourism in the Red Sea: present and future effects on coral reefs. Ambio 23:503–508. https://doi.org/10.1016/0006-3207(96)83261-7

Hayes CT, Baumbach DS, Juma D et al (2016) Impacts of recreational diving on hawksbill sea turtle (*Eretmochelys imbricata*) behaviour in a marine protected area. J Sustain Tour 25:79–95. https://doi.org/10.1080/09669582.2016.1174246

Hein MY, Lamb JB, Scott C et al (2015) Assessing baseline levels of coral health in a newly established marine protected area in a global scuba diving hotspot. Mar Environ Res 103:56–65. https://doi.org/10.1016/j.marenvres.2014.11.008

Hein MY, Couture F, Scott CM (2018) Ecotourism and coral reef restoration. In: Prideaux B, Pabel A (eds) Coral reefs: tourism, conservation and management. Earthscan Oceans, Oxford

Hernández-Delgado EA, Sandoz B, Bonkosky M et al (2008) Impacts of non-point source sewage pollution on Elkhorn coral, *Acropora palmata* (Lamarck), assemblages of the southwestern Puerto Rico shelf. In: Proceedings of the 11th international coral reef symposium, Fort Lauderdale, Florida, 7–11 July 2018

Heron SF, Maynard JA, Van Hooidonk R et al (2016) Warming trends and bleaching stress of the world's coral reefs 1985–2012. Sci Rep 6:38402. https://doi.org/10.1038/srep38402

Hicks CC, Graham NAJ, Cinner JE (2013) Synergies and tradeoffs in how managers, scientists, and fishers value coral reef ecosystem services. Glob Environ Chang 23:1444–1453. https://doi.org/10.1016/j.gloenvcha.2013.07.028

Higham JES, Lück M (2008) Marine wildlife and tourism management: in search of scientific approaches to sustainability. In: Higham JES, Lück M (eds) Marine wildlife and tourism management: insights from the natural and social sciences. CAB International, Wallingford, pp 1–16

Hodgson G, Dixon JA (1988) Logging versus fisheries and tourism in Palawan: an environmental and economic analysis. Occasional papers of the East-West Environment and Policy Institute (Paper no 7). East-West Center, Honolulu

Hodgson G, Dixon J (2000) El Nido revisited: ecotourism, logging and fisheries. In: Cesar HSJ (ed) Collected essays on the economics of coral reefs. Cordio, Kalmar, pp 55–68

Hsui CY, Wang CC (2013) Synergy between fractal dimensions and lacunarity index in design of artificial habitat for alternative Scuba diving site. Ecol Eng 53:6–14. https://doi.org/10.1016/j.ecoleng.2013.01.014

Huang Y, Coelho VR (2017) Sustainability performance assessment focusing on coral reef protection by the tourism industry in the coral triangle region. Tour Manag 59:510–527. https://doi.org/10.1016/j.tourman.2016.09.008

Hubbard JAEB, Pocock YP (1972) Sediment rejection by recent scleractinian corals: a key to palaeo-environmental reconstruction. Geol Rundsch 61:598–626. https://doi.org/10.1007/BF01896337

Hughes TP, Rodrigues MJ, Bellwood DR et al (2007) Phase shifts, herbivory, and the resilience of coral reefs to climate change. Curr Biol 17:360–365. https://doi.org/10.1016/j.cub.2006.12.049

Hughes TP, Graham NAJ, Jackson JBC et al (2010) Rising to the challenge of sustaining coral reef resilience. Trends Ecol Evol 25:633–642. https://doi.org/10.1016/j.tree.2010.07.011

Hughes TP, Kerry JT, Álvarez-Noriega M et al (2017) Global warming and recurrent mass bleaching of corals. Nature 543:373–377. https://doi.org/10.1038/nature21707

Hunt CV, Harvey JJ, Miller A et al (2013) The green fins approach for monitoring and promoting environmentally sustainable scuba diving operations in South East Asia. Ocean Coast Manag 78:35–44. https://doi.org/10.1016/j.ocecoaman.2013.03.004

Hunte W, Wittenberg M (1992) Effects of eutrophication and sedimentation on juvenile corals. Mar Biol 112:131–138. https://doi.org/10.1007/BF00349736

Hunter CL, Evans CW (1995) Coral reefs in Kaneohe Bay, Hawaii: two centuries of western influence and two decades of data. Bull Mar Sci 57:501–515

Hunter CE, Lauer M, Levine A et al (2018) Maneuvering towards adaptive co-management in a coral reef fishery. Mar Policy 98:77–84. https://doi.org/10.1016/j.marpol.2018.09.016

Jackson JBC, Donovan MK, Cramer KL et al (2014) Status and trends of Caribbean coral reefs 1970–2012. Global Coral Reef Monitoring Network, IUCN, Gland

Jakšić S, Stamenković I, Đorđević J (2013) Impacts of artificial reefs and diving tourism. Turizam 17:155–165

Johannes RE (1975) Pollution and degradation of coral reef communities. In: Ferguson Wood EJ, Johannes RE (eds) Tropical marine pollution. Oceanogr Ser 12:13–51

Kelman D, Kashman Y, Rosenberg E et al (2001) Antimicrobial activity of the reef sponge *Amphimedon viridis* from the Red Sea: evidence for selective toxicity. Aquat Microb Ecol 24:9–16. https://doi.org/10.3354/ame024009

Kirkbride-Smith AE, Wheeler PM, Johnson ML (2013) The relationship between diver experience levels and perceptions of attractiveness of artificial reefs – examination of a potential management tool. PLoS One 8:e68899. https://doi.org/10.1371/journal.pone.0068899

Knowlton N, Brainard RE, Fisher R et al (2010) Coral reef biodiversity. In: McIntyre AD (ed) Life in the world's oceans. Blackwell, pp 65–77. https://doi.org/10.1002/9781444325508.ch4

Kurniawan F, Adrianto L, Bengen DG et al (2016) Vulnerability assessment of small islands to tourism: the case of the Marine Tourism Park of the Gili Matra Islands, Indonesia. Glob Ecol Conserv 6:308–326. https://doi.org/10.1016/j.gecco.2016.04.001

Lachica-Alino L, David LT, Wolff M et al (2009) Distributional patterns, habitat overlap and trophic interactions of species caught by trawling in the Ragay Gulf, Philippines. Philipp Agric Sci 92:46–65

Lamb JB, True JD, Piromvaragorn S et al (2014) Scuba diving damage and intensity of tourist activities increases coral disease prevalence. Biol Conserv 178:88–96. https://doi.org/10.1016/j.biocon.2014.06.027

Lapointe BE, Barile PJ, Littler MM et al (2005) Macroalgal blooms on southeast Florida coral reefs II. Cross-shelf discrimination of nitrogen sources indicates widespread assimilation of sewage nitrogen. Harmful Algae 4:1106–1122. https://doi.org/10.1016/j.hal.2005.06.002

Lapointe BE, Langton R, Bedford BJ et al (2010) Land-based nutrient enrichment of the Buccoo Reef complex and fringing coral reefs of Tobago, West Indies. Mar Pollut Bull 60:334–343. https://doi.org/10.1016/j.marpolbul.2009.10.020

Laurans Y, Pascal N, Binet T et al (2013) Economic valuation of ecosystem services from coral reefs in the South Pacific: taking stock of recent experience. J Environ Manag 116:135–144. https://doi.org/10.1016/j.jenvman.2012.11.031

Lavides MN, Polunin NVC, Stead SM et al (2010) Finfish disappearances around Bohol, Philippines inferred from traditional. Environ Conserv 36:235–244. https://doi.org/10.1017/S0376892909990385

Lavides MN, Molina EPV, de la Rosa Jr. GE et al (2016) Patterns of coral-reef finfish species disappearances inferred from fishers' knowledge in global epicentre of marine shorefish diversity. PLoS One 11:e0155752. doi:https://doi.org/10.1371/journal.pone.0155752

Levine AS, Richmond LS (2014) Examining enabling conditions for community-based fisheries comanagement: comparing efforts in Hawai'i and American Samoa. Ecol Soc 19:24. https://doi.org/10.5751/ES-06191-190124

Lirman D, Manzello D (2009) Patterns of resistance and resilience of the stress-tolerant coral Siderastrea radians (Pallas) to sub-optimal salinity and sediment burial. J Exp Mar Biol Ecol 369:72–77. https://doi.org/10.1016/j.jembe.2008.10.024

Lirman D, Schopmeyer S (2016) Ecological solutions to reef degradation: optimizing coral reef restoration in the Caribbean and Western Atlantic. PeerJ 4:e2597. https://doi.org/10.7717/peerj.2597

Lopes PFM, Pacheco S, Clauzet M et al (2015) Fisheries, tourism, and marine protected areas: conflicting or synergistic interactions? Ecosyst Serv 16:333–340. https://doi.org/10.1016/j.ecoser.2014.12.003

Mamat MP, Yacob MR, Radam A et al (2013) Willingness to pay for protecting natural environments in Pulau Redang Marine Park, Malaysia. Afr J Bus Manage 1:2420–2426. https://doi.org/10.5897/AJBM10.752

Mayer AG (1918) Ecology of the Murray Island coral reef. Carnegie Inst Wash Pub 213:3–48

McClanahan TR (1999) Is there a future for coral reef parks in poor tropical countries? Coral Reefs 18:321–325. https://doi.org/10.1007/s003380050205

McClenachan L, Kittinger JN (2013) Multicentury trends and the sustainability of coral reef fisheries in Hawai'i and Florida. Fish Fish 14:239–255. https://doi.org/10.1111/j.1467-2979.2012.00465.x

McGurrin JM, Stone RB, Sousa RJ (1989) Profiling United States artificial reef development. Bull Mar Sci 44:1004–1013

Mumby PJ, Dahlgren CP, Harborne AR et al (2006) Fishing, trophic cascades, and the process of grazing on coral reefs. Science 311:98–101. https://doi.org/10.1126/science.1121129

Mumby PJ, Hastings A, Edwards HJ (2007) Thresholds and the resilience of Caribbean coral reefs. Nature 450:98–101. https://doi.org/10.1038/nature06252

Mumby PJ, Steneck RS, Adjeroud M et al (2016) High resilience masks underlying sensitivity to algal phase shifts of Pacific coral reefs. Oikos 125:644–655. https://doi.org/10.1111/oik.02673

Munga CN, Mohamed MOS, Amiyo N et al (2012) Status of coral reef fish communities within the Mombasa marine protected area, Kenya, more than a decade after establishment. West Indian Ocean J Mar Sci 10:169–184

Murray G (2007) Constructing paradise: the impacts of big tourism in the Mexican coastal zone. Coast Manag 35:339–355. https://doi.org/10.1080/08920750601169600

Nejati M, Mohamed B, Omar SI (2014) Locals' perceptions towards the impacts of tourism and the importance of local engagement: a comparative study of two islands in Malaysia. Tourism 62:135–146

Olmos-Martínez E, Arizpe-Covarubias OA, Ibañez Perez RM et al (2015) Ecosystem services with tourism potential of the Espíritu Santo Archipelago natural park in Baja California Sur, México. Teoría y Praxis at: www.redalyc.org/html/4561/456144904009/index.html. Accessed 16 May 2018

Oracion EG, Miller ML, Christie P (2005) Marine protected areas for whom? Fisheries, tourism, and solidarity in a Philippine community. Ocean Coast Manag 48:393–410. https://doi.org/10.1016/j.ocecoaman.2005.04.013

Outra MIH, Sari SK, Sukandar H et al (2016) Engaging dive tourism in sustainable financing and coral reef data collection for better management of Karimunjawa National Park, Indonesia. In: Thirteenth international coral reef symposium. Honolulu, Hawaii, 19–14 June 2016

Pauly D (1995) Anecdotes and the shifting baseline syndrome of fisheries. Trends Ecol Evol 10:430. https://doi.org/10.1016/S0169-5347(00)89171-5

Perry CT, Morgan KM (2017) Post-bleaching coral community change on southern Maldivian reefs: is there potential for rapid recovery? Coral Reefs 36:1189–1194. https://doi.org/10.1007/s00338-017-1610-9

Plummer R, Armitage DR, de Loë RC (2013) Adaptive comanagement and its relationship to environmental governance. Ecol Soc 18:21. https://doi.org/10.5751/ES-05383-180121

Polak O, Shasnar N (2012) Can a small artificial reef reduce diving pressure from a natural coral reef? Lessons learned from Eilat, Red Sea. Ocean Coast Manag 55:94–100

Polovina JJ (1990) Assessment of biological impacts of artificial reefs and FADS. In: Symposium on artificial reefs and fish aggregating devices as tools for the management and enhancement of marine fishery resources. Colombo, Sri Lanka, 14–17 May 1990

Polovina JJ, Sakai I (1989) Impacts of artificial reefs on fishery production in Shimamaki, Japan. Bull Mar Sci 44:997–1003

Redding JE, Myers-Miller RL, Baker DM et al (2013) Link between sewage-derived nitrogen pollution and coral disease severity in Guam. Mar Pollut Bull 73:57–63. https://doi.org/10.1016/j.marpolbul.2013.06.002

Renfro B, Chadwick NE (2017) Benthic community structure on coral reefs exposed to intensive recreational snorkeling. PLoS One 12:e0184175. https://doi.org/10.1371/journal.pone.0184175

Roberts CM, Hawkins JP (2000) Fully protected marine reserves: a guide. WWF Endangered Seas Campaign, Washington, DC. Available at: assets.panda.org/downloads/marinereservescolor.pdf. Accessed 24 April 2018

Roberts CM, Mcclea CJ, Veron JEN et al (2002) Marine biodiversity hotspots and conservation priorities for tropical reefs. Science 295:1280–1284. https://doi.org/10.1126/science.1067728

Roche RC, Harvey CV, Harvey JJ et al (2016) Recreational diving impacts on coral reefs and the adoption of environmentally responsible practices within the scuba diving industry. Environ Manag 58:107–116. https://doi.org/10.1007/s00267-016-0696-0

Rogers CS (1990) Responses of coral reefs and reef organisms to sedimentation. Mar Ecol Prog Ser 62:185–202. https://doi.org/10.3354/meps062185

Sadovy de Mitcheson Y, Cornish A, Domeier M et al (2008) A global baseline for spawning aggregations of reef fishes. Conserv Biol 22:1233–1244. https://doi.org/10.1111/j.1523-1739.2008.01020.x

Samoilys MA, Osuka K, Maina GW et al (2017) Artisanal fisheries on Kenya's coral reefs: decadal trends reveal management needs. Fish Res 186:177–191. https://doi.org/10.1016/j.fishres.2016.07.025

Sandin SA, Smith JE, DeMartini EE et al (2008) Baselines and degradation of coral reefs in the northern Line Islands. PLoS One 3:e1548. https://doi.org/10.1371/journal.pone.0001548

Saragih H (2016) Marketing artificial reef as recreational scuba diving resources: feasibility study for sustainable tourism. In: Proceedings of the 2016 global conference on business, management and Entrepreneurship, Bandung, Indonesia, 08 August 2016

Sary Z, Oxenford HA, Woodley JD (1997) Effects of an increase in trap mesh size on an overexploited coral reef fishery at Discovery Bay, Jamaica. Mar Ecol Prog Ser 154:107–120. https://doi.org/10.3354/meps154107

Satria A, Matsuda Y, Sano M (2004) Multilevel conflicts in community based coral reef management systems: case study in West-Lombok, Indonesia. In: IIFET 2004 Japan proceedings, Tokyo, Japan, 20–30 July 2004

Schultz L, Duit A, Folke C (2011) Participation, adaptive co-management, and management performance in the World Network

of Biosphere Reserves. World Dev 39:662–671. https://doi.org/10.1016/j.worlddev.2010.09.014
Scott ME, Smith JA, Lowry MB et al (2015) The influence of an off-shore artificial reef on the abundance of fish in the surrounding pelagic environment. Mar Freshw Res 66:429–437. https://doi.org/10.1071/MF14064
Seaman W, Jensen AC (2000) Purposes and practices of artificial reef evaluation. In: Seaman W (ed) Artificial reef evaluation with application to natural marine habitats. CRC Press, Boca Raton, pp 1–20
Sloan K (1987) Valuing Heron Island: preliminary results. In: 16th conference of economists. Surfers Paradise, Queensland, Australia, 23–27 August 1987
Smith SV, Kimmerer WJ, Laws EA et al (1981) Kaneohe Bay sewage diversion experiment: perspectives on ecosystem responses to nutritional perturbation. Pac Sci 35:279–395. https://hdl.handle.net/10125/616
Smith JA, Lowry MB, Champion C et al (2016) A designed artificial reef is among the most productive marine fish habitats: new metrics to address 'production versus attraction'. Mar Biol 163:188. https://doi.org/10.1007/s00227-016-2967-y
Spalding M, Burke L, Wood SA et al (2017) Mapping the global value and distribution of coral reef tourism. Mar Policy 82:104–113. https://doi.org/10.1016/j.marpol.2017.05.014
Stolk P, Markwell K, Jenkins JM (2007) Artificial reefs as recreational scuba diving resources: a critical review of research. J Sustain Tour 15:331–350. https://doi.org/10.2167/jost651.0
Tabata RS (1992) Scuba-diving holidays. In: Weiler B, Hall CM (eds) Special interest tourism. Belhaven Press, New York, pp 171–184
Techera E (2007) Customary law and community based conservation of marine areas in Fiji. In: 6th Global conference in environmental justice and global citizenship. Oxford, UK, 2–5 July 2007
Teh LCL, Sumaila UR (2013) Contribution of marine fisheries to worldwide employment. Fish Fish 14:77–88. https://doi.org/10.1111/j.1467-2979.2011.00450.x
Thur SM (2010) User fees as sustainable financing mechanisms for marine protected areas: an application to the Bonaire National Marine Park. Mar Policy 34:63–69. https://doi.org/10.1016/j.marpol.2009.04.008
Tupper M, Asif F, Garces LR et al (2015) Evaluating management effectiveness of marine protected areas at seven selected sites in the Philippines. Mar Policy 56:33–42
United Nations (2018) Country classifications. In: World economic situation and prospects, New York. Available at: https://www.un.org/development/desa/dpad/wp-content/uploads/sites/45/publication/WESP2018_Full_Web-1.pdf. Accessed 7 May 2018
Van Beukering PJH, Cesar HSJ (2004) Ecological economic modelling of coral reefs: evaluating tourist overuse at Hanauma Bay and algae blooms at the Kihei Coast. Pac Sci 58:243–260. https://doi.org/10.1353/psc.2004.0012
Van Beukering PJH, Haider W, Wolfs E et al (2006) The economic value of the coral reefs of Saipan, Commonwealth of the Northern Mariana Islands. CEEC Report, p 153
Van Beukering PJH, Sarkis S, van der Putten L et al (2015) Bermuda's balancing act: the economic dependence of cruise and air tourism on healthy coral reef. Ecosyst Serv 11:76–86. https://doi.org/10.1016/j.ecoser.2014.06.009
Vianna GMS, Meekan MG, Pannell DJ et al (2012) Socio-economic value and community benefits from shark-diving tourism in Palau: a sustainable use of reef shark populations. Biol Conserv 145:267–277. https://doi.org/10.1016/j.biocon.2011.11.022
Walmsley S, Purvis J, Ninnes C (2006) The role of small-scale fisheries management in the poverty reduction strategies in the Western Indian Ocean region. Ocean Coast Manag 49:812–833. https://doi.org/10.1016/j.ocecoaman.2006.08.006
Webler T, Jakubowsky K (2016) Mitigating damaging behaviors of snorkelers to coral reefs in Puerto Rico through a pre-trip media-based intervention. Biol Conserv 197:223–228. https://doi.org/10.1016/j.biocon.2016.03.012
Weeks R, Jupiter SD (2013) Adaptive Comanagement of a marine protected area network in Fiji. Conserv Biol 27:1234–1244. https://doi.org/10.1111/cobi.12153
Weijerman M, Williams I, Gutierrez J et al (2016) Trends in biomass of coral reef fishes, derived from shore-based creel surveys in Guam. Fish Bull 114:237–257. https://doi.org/10.7755/FB.114.2.9
Wells S, Ravilious C, Corcoran E (2006) In the front line: shoreline protection and other ecosystem services from mangroves and coral reefs. In: UNEP-WCMC, Cambridge, UK. Available at: https://www.icriforum.org/sites/default/files/in_front_line.pdf. Accessed 12 May 2018
Yacob MR, Shuib A, Mamat MF et al (2007) Local economic benefits of ecotourism development in Malaysia: the case of Redang Island Marine Park. Int J Econ Manage 1:365–386
Yacob MR, Radam A, Shuib A (2009) A contingent valuation study of marine parks ecotourism: the case of Pulau Payar and Pulau Redang in Malaysia. J Sustain Dev 2:95–105. https://doi.org/10.5539/jsd.v2n2p95
Yeo BH (2004) The recreational benefits of coral reefs: a case study of Pulau Payar Marine Park, Kedah, Malaysia. In: Economic valuation and policy priorities for sustainable management of coral reefs. WorldFish Center. Available at: http://www.worldfishcenter.org/Pubs/coral_reef/pdf/section2-7.pdf. Accessed 25 April 2018
Young MAL, Foale S, Bellwood DR (2015) Dynamic catch trends in the history of recreational spearfishing in Australia. Conserv Biol 29:784–794. https://doi.org/10.1111/cobi.12456
Zeller D, Harper S, Zylich K et al (2015) Synthesis of underreported small-scale fisheries catch in Pacific island waters. Coral Reefs 34:25–39. https://doi.org/10.1007/s00338-014-1219-1
Zhang LY, Chung S, Qiu J (2016) Ecological carrying capacity assessment of diving site: a case study of Mabul Island, Malaysia. J Environ Manag 183:253–259. https://doi.org/10.1016/j.jenvman.2016.08.075
Zimmerhackel JS, Pannell DJ, Meekan M (2016) Diving tourism and fisheries in marine protected areas: market values and new approaches to improve compliance in the Maldives Shark Sanctuary. Working paper 1610. School of Agricultural and Resource Economics, University of Western Australia, Crawley, Australia

14 Progress in Microbial Ecology in Ice-Covered Seas

Tobias R. Vonnahme, Ulrike Dietrich, and Brandon T. Hassett

Abstract

Sea ice seasonally covers 10% of the earth's oceans and shapes global ocean chemistry. The unique physical processes associated with sea ice growth and development shape the associated biological diversity and ecosystem function. Microbes make up the base of all marine food webs and the overwhelming majority of biomass in the sea ice ecosystem. Despite their biomass, microbial processes are not fully integrated into marine ecosystem models. Recent applications of novel molecular biology technologies to studies of marine ecology have elucidated numerous microbial-mediated processes interfaced by previously unknown organisms and processes. These discoveries are yielding more in-depth studies on the relevance of mixotrophy, the ecology of fungi, and the interplay between major microbial clades. In ecosystem studies, the basis of the food web is frequently neglected even though the accessibility of energy, recycling of nutrients, and parasitism are crucial factors shaping the environment for grazers and higher trophic levels. In this review, we focus on the species composition, abundance, and functions of microalgae, bacteria, archaea, fungi, and viruses in the sea ice-covered seas throughout the year. A strong emphasis will be put on advances in molecular methods that empower scientists to further investigate microorganisms in more detail. Since microbes make up the majority of all oceanic biomass, we believe that it is impossible to accurately forecast the biological fate of polar marine ecosystems without placing a proportional emphasis on microbes relative to their biomass.

Keywords

Mixotrophy · Autotrophy · Heterotrophy · Omics · Fungi

T. R. Vonnahme (✉) · U. Dietrich (✉) · B. T. Hassett
UiT- The Arctic University of Norway, Tromsø, Norway
e-mail: tobias.vonnahme@uit.no; Ulrike.dietrich@uit.no

S. Jungblut et al. (eds.), *YOUMARES 9 - The Oceans: Our Research, Our Future*,
https://doi.org/10.1007/978-3-030-20389-4_14

14.1 Introduction

Sea ice seasonally covers approximately 10% of the global ocean surface and is responsible for altering global ocean chemistry. The main ice-covered marine ecosystems are found in the Arctic Ocean, the Southern Ocean, and the Baltic Sea. Within the polar marine ecosystems, sea ice formation and subsequent coverage influence light transmittance that seasonally governs under-ice primary production and the associated heterotrophic biological community. Specifically, sea ice can support 50% of total primary productivity in permanently ice-covered ecosystems (Gosselin et al. 1997; Fernández-Méndez et al. 2015) and constitutes a habitat and feeding ground for various organisms. It provides microhabitats for microalgae, chemoautotrophic and heterotrophic bacteria, archaea, viruses, fungi, and multicellular organisms (Bluhm et al. 2018) that inhabit the hypersaline brine channels (Hunt et al. 2016). The sea ice habitat is characterized by strong gradients in temperature, salinity, nutrients, and light. The small-scale spatial distribution of sea ice-associated (sympagic) biota is determined to a large extent by these physical properties (Krembs et al. 2011). Organisms within the brine channels are exposed to extreme temperatures from 0 °C in summer to below −15 °C in winter with associated brine salinities ranging from 0 to over 200 (Gradinger 2001). Most of the biomass within brine channels is localized near the warmer ice-water interface, where temperatures are about −1.8 °C, the brine-volume fraction is greatest, and a continuous exchange of nutrients from the water below takes place.

The two major ice types found in polar environments provide different habitat characteristics (e.g., thickness, ice bulk salinities, age, and albedo), relevant for the associated biological processes (Weeks and Ackley 1986). Multiyear ice (MYI) persists at least one melting season, whereas first-year ice (FYI) follows a seasonal pattern of ice formation and melt. On average, the surface salinity of FYI is typically around 10–12, whereas MYI typically has surface salinities that approach 0 (Weeks and Ackley 1986). FYI is often structurally less complex, characterized by greater light penetration through the ice that is prone to an earlier onset of seasonal melt (Moline et al. 2008). Contrasting studies between FYI and MYI indicate that FYI hosts a higher number of organisms, but a less rich microbial community. Changes within the ice biological system might have cascading effects on the ice-associated ecosystem (Secretariat of Arctic Council 2017).

The open water in sea ice-covered seas is a special system in itself. At the marginal ice zone or in open leads, a system with high levels of light and nutrients may support ice edge phytoplankton blooms dominated by different species compared to sea ice (Assmy et al. 2017). With climate change, these areas are expected to increase, changing the microbial community structure in sea-ice covered seas (Oziel et al. 2017). The consequences for higher trophic levels and carbon export are a topic of recent studies. In the Arctic and Antarctic, a large part of the ocean is ice-free during the polar night. The absence of light challenges the pelagic microbial food web due to a lack of photosynthetic primary production. Nevertheless, microbes have been found to be active throughout the polar night and different biogeochemical cycles may be dominant (Zhang et al. 2003; Berge et al. 2015; Nguyen et al. 2015).

The polar marine environment is in a state of rapid transition with tremendous changes in the abiotic environment. In the Arctic Ocean, air and surface-layer temperatures are increasing faster than the global average (Serreze and Francis 2006; Holding et al. 2015) and is driving the replacement of MYI with thinner FYI (Maslanik et al. 2011; Perovich et al. 2014; Barber et al. 2015). This replacement has contributed to an earlier onset of seasonal ice melt, an increased duration of ice melt (Stroeve et al. 2014), and persistent open water conditions in the seasonal ice zone (Lange et al. 2016). A strong reduction in overall Arctic sea ice extent occurred over the last two decades, with the lowest summer minimum ice extent in 2012 (3.61×10^6 km^2), which had been 18% below the previous low of 2007 (Beitler 2012). In contrast to the Arctic, the Antarctic is characterized by a large extent of seasonally forming ice that grows from 4×10^6 km^2 in summer to approximately 19×10^6 km^2 in late winter (Cavalieri et al. 1999). Specifically during autumn, the surface of the ocean surrounding the Antarctic continent begins to freeze, forming sea ice of about 0.4 m thickness (up to 1 m; Worby et al. 2001). Overall, the ice extent in the Antarctic has been much less impacted compared to the Arctic. In the Antarctic Peninsula and Bellinghausen Sea region, the ice-free summer season is extended by three months, whereas in the western Ross Sea region, the ice-free season is shortened by two months (Lange et al. 2016). Due to strong wind events, large quantities of heat are extracted from the surface ocean, facilitating rapid formation of frazil ice (Eicken 2003).

The Baltic Sea is one of the world's largest brackish water basins with a surface area of 422,000 km^2 and a mean depth of only 55 m. Surface salinities vary from 9 in the southern part to below 1 in the innermost parts (Voipio 1981). Annually, sea ice covers about 40% of the Baltic Sea (Kaartokallio et al. 2007). Even though the seasonal ice of the Baltic Sea has many similarities with the seasonal ice in the polar areas, fresher water results in sea ice with lower bulk salinities and smaller brine channels, despite the comparably high temperatures (Meiners et al. 2002). Low brine volumes reduce the rate of seawater exchange across the ice-water interface that affects rates of nutrient replenishment, convective heat transport, and desalination processes (Lytle and Ackley 1996). Due to milder climate in the Baltic Sea region, snow and freeze-melt cycles occur throughout winter, leading to a greater contribution (up to 35%) of sea ice mass in the form of metamorphic snow (Granskog et al. 2006). The high dissolved organic matter (DOM) content in Baltic Sea water and ice leads to different chemical characteristics and causes increased absorption of solar radiation at shorter wavelengths than are utilized for photosynthesis (Granskog et al. 2006). During early winter most of the Baltic Sea is ice free and below the Arctic Circle, where daylight is available for photosynthesis throughout the year in contrast to the polar night in polar regions. As a result, the microbial community differs considerably from polar sea ice environments.

14.2 Advances in Microbial Ecology

Advances in marine microbial ecology are driven by methodological advances in understanding both, the environment, as well as the biological taxa that inhabit the environment. Microbial methodology and associated observations advanced marginally from the late 1800s during Nansen's First Fram Expedition to the 1960s and 1970s, where light microscopy and cultivation-based studies of microbes shaped science's understanding of microbial ecology (e.g., Hobbie et al. 1977; reviewed by Baross and Morita 1978). With advancing resolution in microscopy, it was possible to get a better understanding of microbial diversity and abundances that has now ushered in the -omics era. These microbial methodologies have evolved in parallel with in situ technologies. Early approaches measured primary production in slices of an ice core incubated in surrounding ice (Mock and Gradinger 1999). Since these early studies, technological advancements have allowed for in situ measurements of primary production; oxygen microsensors have been used successfully in artificial sea ice experiments to measure in situ ecosystem production (Mock et al. 2002). However, the standard method is to still work on melted sea ice, which may be an underestimation of primary production (Søgaard et al.

2010). Stable- and radioisotope incubations allowed estimates of microbial activities and associated organic matter utilization. The future of methods for studying biogeochemistry in sea ice may be in situ technologies (reviewed by Miller et al. 2015). Methods for water sampling and biogeochemical studies have been similar to traditional work in other pelagic systems.

The application of molecular fingerprinting methods (e.g., denaturing gradient gel electrophoresis, restriction fragment length polymorphism), clone library sequencing, and in recent years, metagenomics have generated a detailed understanding of microbial phylogeny, taxonomy, and more recently, function and ecology in the seasonal ice zone. Amplicon-based sequencing of the taxonomically informative small ribosomal subunit became a standard genetic barcode, which allowed the identification of microbial taxa down to the level of ecotypes. Advancing sequencing technologies are generating more sequence reads at a lower cost, affording high spatial and temporal resolution of microbial (primarily bacterial) communities and subsequent investigation of their connectivity, seasonal successions, and biogeography (e.g., Brown and Bowman 2001; Brinkmeyer et al. 2003; Collins et al. 2010; Hatam et al. 2016; Yergeau et al. 2017; Rapp et al. 2018). Novel sequencing tools, such as the Nanopore MinION have the potential to be used for in-field sequencing, and have been used in remote polar regions (e.g., Johnson et al. 2017), but not yet in sea ice. Novel sequencing technologies are evolving in parallel with bioinformatic tools that can identify small, yet significant community differences. When used together, these novel sequencing technologies and bioinformatic tools are yielding novel ecological insights that are, in turn, shifting sciences' understanding of microbial community complexity. Consequently, there is an emerging trend away from the traditional 97% to 98% similarity cutoff that defines microbial taxa toward network- and nucleotide entropy-based clustering methods (e.g., Rapp et al. 2018).

Recent studies are focusing more on full genome, metagenomic shotgun sequencing approaches. Approaches, which simultaneously generate taxonomic and functional gene information. So far, only several studies have applied metagenomic shotgun sequencing to sea ice samples (e.g., Bowman et al. 2014; Yergeau et al. 2017), but the potential to elucidate complex polar microbial ecology questions is generally unrealized. Metagenomic sequencing efforts have demonstrated bacterial-mediated chemical cycling in frost flowers (Bowman et al. 2014) and the importance of select photoreceptors in Antarctic and Arctic sea ice (Koh et al. 2010; Vader et al. 2018). With increasing throughput of sequence generation and decreasing costs, deep sequencing (i.e., sequencing the same locus multiple times) of the environment should allow comparative studies of full metagenomes to describe the metabolic potential (including uncultured strains) and strain level microbial diversity (e.g., Delmont et al. 2017) in sea ice ecosystems.

Other -omics studies that target byproducts of protein synthesis and secondary metabolism are rare in sea ice. While metagenomics can demonstrate the genetic potential of microbes, RNA-based studies, such as metatranscriptomics, can show whether the genes are expressed. For example, Koh et al. (2010) and Vader et al. (2018) showed that the genes for proteorhodopsin are actively transcribed in Antarctic and Arctic sea ice, indicating an active phototrophic bacterial community. Interdisciplinary research with biochemists identified the functions of translated proteins and ascribed a functional purpose for gene products used in survival and metabolism in sea ice (reviewed by Feller and Gerday 2003; Feng et al. 2014). Metaproteomics is not only possible for cultured bacteria, but can be used for understanding the biochemical functions of the in situ community (Junge et al. 2019). Ultimately, combined -omics studies are important for a thorough understanding of microbial ecology and biogeochemistry (Junge et al. 2019). In cultures, the potential to combine proteomics and genomics has already been shown to help understanding key genes for a life in subzero temperatures (Feng et al. 2014). To date, metaproteomics and metabolomics studies of the whole community have yet to be applied to studies of sea ice.

Ribosomal gene sequencing data have been used to develop fluorescently labeled nucleotide probes that target taxonomically informative genetic loci, namely, fluorescence in situ hybridization (FISH) (Pernthaler et al. 2002). The application of FISH has informed analyses of spatial interactions and abundances of specific taxa, without the known biases associated with DNA sequencing (De Corte et al. 2013), nonspecific fluorescent stains (e.g., 4′,6-diamidino-2-phenylindole), or cultivation (e.g., Brinkmeyer et al. 2003; Baer et al. 2015). Combined with isotope probing methods, catalyzed reporter deposition-FISH (CARD-FISH) has been used to identify microbial taxa responsible for the uptake of specific organic compounds (e.g., Alonso-Sáez et al. 2008; Nikrad et al. 2012). Consequently, CARD-FISH is a robust method that should be used to supplement DNA sequencing analysis. Only a few of the metabolic capacities mentioned in this chapter have been measured and a common limitation is still the separation of biogeochemical rate measurements and investigations of the genetic potential of communities, or organisms. Studies coupling the function, activity, and diversity of bacteria are lacking in sea ice systems, but their potential has been shown in other marine systems. RNA stable isotope probing is one recent method, which could be used to overcome these limitations. For example, Fortunato and Huber (2016) coupled stable isotope probing with metatranscriptomics to identify taxa and pathways involved in chemolithotrophic processes at hydrothermal vents. Methods for visualization of radioisotope (Microautoradiography,

(Nierychlo et al. 2016) or stable isotope (Nanoscale secondary ion mass spectrometry, Gao et al. 2016) enrichments in single cells coupled to CARD-FISH could be another method to quantify biogeochemical fluxes of certain taxonomic groups.

One of the major applications of novel ecological data is the incorporation into ecosystem models. Despite the increasing computational power, the representation of microbial interactions in ecosystem models is still rudimentary. For example, bacterial activities are often hidden in functions for organic matter remineralization and respiration (e.g., Tedesco et al. 2010; Wassmann et al. 2010; Vancoppenolle and Tedesco 2017). A recent ecosystem model in the Baltic Sea started realizing for the first time the importance of bacteria beyond nutrient remineralization. Specifically, aerobic and anaerobic bacterial taxa were separately considered, both as crucial for remineralization processes and for generating anaerobic conditions linked to algal production (Tedesco et al. 2017). Linking metabolic pathway models, bacterial functions (such as denitrification and nitrogen fixation), and viral lysis may further improve the accuracy of models with increasing data availability and computational power. In most ecosystem models and discussions, the role of sea ice bacteria and archaea is seen in the heterotrophic aerobic remineralization of DOM (e.g., Tedesco et al. 2010; Wassmann et al. 2010; Vancoppenolle and Tedesco 2017).

14.3 Sea Ice-Associated Microorganisms

The base of polar food webs is comprised of microbial organisms allied to multiple clades of life. Polar organisms are well adapted to the seasonality of light, nutrient/food availability, and cold temperatures. Additional challenges arise for ice-associated biota, with extreme cold temperatures, highly variable salinities and only temporary existence of their habitat (Meier et al. 2014). The balance between producers and consumers seasonally shifts with light availability which drives taxa-specific abundances. Diatoms and other microalgae (haptophytes, prasinophytes, dinoflagellates) are some of the most common eukaryotic producers that support a diverse heterotrophic community of prokaryotes, fungi, and fungal-like organisms, ciliates, and larger multicellular organisms. These organisms are all presumably susceptible to viral infection, which can rapidly shunt organic material into the available dissolved organic material pool. Together, these organisms cycle carbon and exchange genes that maintain ecosystem function and support the feeding needs of higher trophic levels.

14.3.1 Microalgae

The microalgae community in polar sea ice is dominated by diatoms that comprise the most biomass and greatest species richness, including up to 170 species predominated by *Nitzschia* sp., *Thalassiosira* sp., *Fragilariopsis* sp., and *Navicula* sp. (Arrigo 2010). Pennate diatoms dominate the spring ice algal bloom in Arctic FYI, as well as in Antarctic sea ice due to the nutrient-rich Southern Ocean (Arrigo et al. 2014). Sea ice associated phytoplankton blooms are often dominated by aggregates of *Phaeocystis* sp., capable of producing large biomasses and drawing down large amounts of nutrients (Assmy et al. 2017). Other algae groups in the pico- and nanoplankton-size fraction contribute substantially to the pelagic and sympagic winter community. *Micromonas* sp., *Cyanobacteria*, and *Ostreococcus* sp. have been found to be abundant phytoplankton species in the polar night (Joli et al. 2017; Amargant Arumí 2018) but only constitute a small fraction of the biomass during spring and summer (Riedel et al. 2008; Niemi et al. 2011; Vader et al. 2018) and therefore have not been studied in more detail until recently. Still, reliable identification and quantification of pico- and nanosized eukaryotes are lacking (Piwosz et al. 2013) or are purely based on sequencing (Vader et al. 2018).

As a consequence of the lower water salinity and corresponding small-sized brine ice channels, the Baltic Sea ice is dominated by smaller protists (Kaartokallio et al. 2007). In early spring, centric diatoms dominate under-ice biomass. These centric diatoms are supplemented by large contributions of *Melosira arctica* and the cyanobacterium *Aphanizomenon* sp. that can predominate abundances in the brine channels (Majaneva et al. 2017). In contrast to Arctic and Antarctic sea ice communities, dinoflagellates and green algae contribute to a large fraction of the biomass in Baltic Sea ice and open water (Kaartokallio et al. 2007; Piiparinen et al. 2010). Furthermore, the surface-layer algal biomass can significantly contribute to the overall sea ice algal biomass (Meiners et al. 2002; Piiparinen et al. 2010). So far, the knowledge of species composition and distribution is limited, and there is only little known on the overwintering of cyanobacteria, which are typical for the Baltic Sea (Laamanen 1996).

14.3.2 Bacteria

The most common orders found in sea ice are *Alteromonadales* (*Gammaproteobacteria*) and *Flavobacteriales* (*Bacteroidetes*) with the most common genera *Pseudoalteromonas*, *Colwellia*, *Shewanella*, *Flavobacterium*, and *Polaribacter* (Bowman et al. 2012, 2014; Boetius et al. 2015; Yergeau et al. 2017). Rarer phyla are the *Alphaproteobacteria*, *Betaproteobacteria*,

Actinobacteria, and *Firmicutes* (Bowman et al. 2012, 2014; Boetius et al. 2015; Yergeau et al. 2017). Archaea are mainly found in autumn and winter; they consist primarily of the genus *Nitrosopumilus*, known for its nitrification capability (Brinkmeyer et al. 2003; Collins et al. 2010). However, most studies are biased toward sampling in summer and spring, but a few studies in winter indicate differences in communities (Collins et al. 2010). On the operational taxonomic unit (OTU) level (97% cutoff), there seem to be no endemic species for sea ice in certain ice zones so far (reviewed by Deming and Collins 2017), but further studies focusing on strain variability may find differences. It has been shown that the bacterial OTUs are more variable in seasonal sea ice and more related to temperate communities compared to MYI (Hatam et al. 2016). Several bacteria found are known to be psychrophilic (e.g., Feng et al. 2014), and a large fraction could be cultured (up to 60%, Junge et al. 2002). The bacterial and archaeal communities in the water column of sea ice systems are significantly different with *Nitrosopumilus* sp., *Pelagibacter* sp., *Flavobacteriales* sp., and *Oceanosprillaceae* sp. as dominating taxa (e.g., Bowman et al. 2012, 2014; Yergeau et al. 2017), indicating a strong selection of potentially endemic sea ice bacteria.

14.3.3 Fungi

Fungi are eukaryotic, spore-bearing, heterotrophic organisms that secrete extracellular enzymes used for interfacing symbiosis and facilitating osmotrophy. Within this ecological definition, fungi are a polyphyletic functional group that include the *Labyrinthulomycota*, *Mesomycetozoea*, *Oomycota*, select *Amoebozoa*, the True Fungi, and several additional clades. True Fungi are distinct from ecological fungi (fungal-like organisms) by possessing cell walls made of chitin and forming a molecular monophyletic clade among the opisthokonts. The True Fungi include many prominent mycelial-producing members, such as the *Ascomycota*, *Basidiomycota*, and *Mucoromycota*, as well as reduced zoosporic varieties, such as the *Blastocladiomycota*, *Chytridiomycota*, and *Neocallimastigomycota*. In this review, fungi are explored within their ecological definition, unless otherwise noted. The often inconspicuous morphology of fungi has challenged the easy identification and subsequent integration of mycological data into ecosystem ecology. As a result, the relevance of fungi remains unrealized in ecosystem modeling efforts globally. Historical culturing-based studies have resulted in the description of hundreds of marine fungal species (Johnson and Sparrow 1961; Kohlmeyer and Kohlmeyer 2013), whose global distribution remains largely unexplored. The more-recent application of molecular methods to studies of marine ecosystem ecology helps to circumvent challenges associated with visual classification and have identified an abundant and dynamic fungal community in subseafloor sediment (Orsi et al. 2013), in association with pelagic marine snow (Bochdansky et al. 2017), in coastal marine habitats (Ueda et al. 2015; Picard 2017) as parasites of phytoplankton (Hanic et al. 2009; Lepelletier et al. 2014; Hassett and Gradinger 2016; Jephcott et al. 2016; Scholz et al. 2017a, b) and metazoans (Polglase 1980; Mclean and Porter 1982; Bower 1987; Shields 1990; Rahimian 1998). Relative to lower latitudes, knowledge of Arctic marine fungi is considerably less developed, in part due to the logistical constraints and inaccessibility of sampling sites. The state of ecological knowledge on Arctic marine True Fungi has largely centered on establishing presence-absence data, supplemented with baselines of diversity and richness, currently estimated at several hundred species (Rämä et al. 2017), with a low success rate of culturing (Bubnova and Nikitin 2017). DNA sequence-based analysis identified overlapping True Fungi taxa from the Bering Sea region and Svalbard, demonstrating a broad distribution of fungal taxa across the Arctic Ocean (Hassett et al. 2017) that are selectively predominated by the *Chytridiomycota* (Terrado et al. 2011; Hassett and Gradinger 2016) and comprise a novel, uncharacterized branch of life (Comeau et al. 2016; Hassett et al. 2017). The diversity and distribution of Arctic marine fungal-like organisms is currently unknown and unreported in assessments of unicellular eukaryotic biodiversity (Poulin et al. 2011).

14.3.4 Viruses

Historically, the study of viral diversity was limited by the co-cultivation of the virus and its host (Borriss et al. 2003; Wells and Deming 2006b). Sea ice viruses are cold-adapted (Luhtanen et al. 2018) and may be less host-specific than in more temperate regions (Wells and Deming 2006b). Cultivated viruses only include *Siphoviridae* and *Myoviridae* as sea ice-specific taxa (Borriss et al. 2003; Wells and Deming 2006a; Sencilo et al. 2015). With increasing -omic efforts, more viruses could be found indicating a higher diversity than previously thought, and taxa such as *Podoviridae*, *Nodaviridae* (RNA), *Iridoviridae* (DNA), and *Caudovirales* have been detected (Allen et al. 2017).

14.4 General Ecology of Sea Ice-Associated Microbes

14.4.1 Autotrophy

14.4.1.1 Photoautotrophy

The activity of the microbial food web follows the fixation of carbon and its subsequent turnover into the DOM pool

(Arrigo and Thomas 2004). While sea ice algal annual primary production rates are generally low compared to the phytoplankton fraction, they are often the main source of fixed carbon for higher trophic levels in ice-covered seas. During winter, ice algae are of special importance, when other sources of food are lacking (Lizotte 2003). Chlorophyll *a* (Chl *a*) concentrations in sea ice vary by region, ice type and season and covers a range somewhat typical for oceanic values up to the highest concentrations found in aquatic environments (Arrigo 2010). In the Arctic, the balance between annual phytoplankton to ice algal primary production differs regionally. In the northern Barents Sea, ice algae account for about 20% of total primary production (Hegseth 1998), whereas in more heavily ice-covered areas like the Central Arctic Ocean, ice algae can contribute more than 50% to the total primary production (Gosselin et al. 1997). Chl *a* concentrations vary between 22 mg m^{-2} in Allen Bay, Nunavut, during spring (Campbell et al. 2014) and 0.3–8 mg m^{-2} in the Central Arctic in summer (Fernández-Méndez et al. 2015). In ice-covered waters of the Antarctic, sea ice algae account for up to 25% of total annual primary production (Arrigo and Thomas 2004). Chl *a* concentrations vary from 1 to 50 mg m^{-2} in the Weddel Sea region (Ackley et al. 1979), whereas sea ice attached to the coast of Antarctica (fast ice) accumulates biomass of up to 2120 mg m^{-2} (Arrigo and Sullivan 1992). On average, under sufficient light intensities for photosynthesis, Chl *a* concentrations exceed 200 mg m^{-2} during spring and summer (Palmisano and Sullivan 1983; Trenerry et al. 2002). In contrast, productivity by sea ice algae in the Baltic Sea is much lower, contributing about 10% to the primary production during the ice-covered season (Haecky et al. 1999). Average Chl *a* values range between 0.2 and 5.5 mg m^{-2} (Haecky et al. 1999; Kaartokallio 2001, 2004).

Based on sequencing results and detection of photopigments, sea ice and sea ice-associated bacteria have also been speculated to be capable of photoautotrophic carbon fixation (Petri and Imhoff 2001; Koh et al. 2011, 2012; Boetius et al. 2015). Despite their high abundance in other cold environments, such as glaciers (Vonnahme et al. 2016), phototrophic cyanobacteria and anoxygenic phototrophs (e.g., purple sulfur bacteria, Chloroflexi) are not as abundant as eukaryotic sea ice algae, but are frequently detected in the Arctic (Petri and Imhoff 2001; Boetius et al. 2015; Yergeau et al. 2017) and Antarctic (Koh et al. 2011, 2012). Their pigments (phycobiliproteins and bacteriochlorophyll), as well as their genes (e.g., 16S rRNA genes), have been found (Cottrell and Kirchman 2009; Koh et al. 2012; Boetius et al. 2015). However, a proof of their phototrophic activity in sea ice is, yet, lacking. Cyanobacteria are more abundant in the snow layer of sea ice suggesting aeolian origin in the Antarctic (Koh et al. 2012). It appears that cyanobacteria are more abundant in fresher systems, such as melt ponds and the Baltic Sea (Petri and Imhoff 2001; Rintala et al. 2014) and that they become more abundant in winter (Cottrell and Kirchman 2009). In the water column, anoxygenic phototrophs may contribute to up to around 15% of the bacterial communities in the Arctic, which is 1000 times more than cyanobacteria such as *Synechococcus* sp. and may indicate a high importance of this pathway in addition to photosynthesis by sea ice algae (Cottrell and Kirchman 2009). In contrast to cyanobacteria, anoxygenic phototrophs appear to be more abundant in summer (Cottrell and Kirchman 2009). Proteorhodopsin, a pigment for using light energy to create a proton motive force, which can be used for energy production, or nutrient transport, is commonly found in seasonal sea ice in the Arctic and Antarctic and can be seen as another way of phototrophic carbon fixation by *Alphaproteobacteria*, *Gammaproteobacteria*, and *Flavobacterium* (Koh et al. 2010, 2012; Yergeau et al. 2017). A combination of these alternative photosynthetic pathways using different wavelengths efficiently may be an important component of the primary production in the seasonal ice zones, but their phototrophic activity has not been quantified in sea ice, yet.

14.4.1.2 Chemoautotrophy

In the water column, chemoautotrophy (inorganic carbon uptake, using chemical energy), appears to be an important autotrophic carbon acquisition process. During nitrification, ammonium is used as energy source, which delivers energy during reduction to nitrite and nitrate for inorganic carbon fixation. The ammonium originates commonly from primary production by sea ice algae and phytoplankton. The nitrate produced via nitrification can be used as recycled inorganic nitrogen for primary production. Archaea constitute a large fraction of potential nitrifiers in the water column, with significantly higher abundances than in sea ice (Yergeau et al. 2017). Nitrifying bacteria appear to be important in coastal areas of the Arctic seasonal ice zone, potentially due to a high supply of ammonium from the bottom water (Damashek et al. 2017). Nitrifying taxa have also been found in deeper stations, but nitrifying archaea are more abundant (e.g., Yergeau et al. 2017). Other autotrophic pathways have not been described yet and are rather unlikely due to the lack of sources for reduced ions in sea ice.

Aerobic bacterial production may become high enough to leave anoxic pockets in the sea ice, where anaerobic processes become energetically favorable. Denitrification and anaerobic ammonia oxidation (Anammox) rates comparable to sediments have been measured in sea ice, reducing the overall nitrogen availability for primary production (Rysgaard and Glud 2004; Rysgaard et al. 2008). New production of nitrogen is possible via upwelling from nutrient-rich bottom waters, bacterial recycling of organic matter, or N_2 fixation. So far, the *nifH* gene for nitrogen fixation has been detected in Arctic seasonal sea ice connected to a diverse group of cyanobacteria (Diez et al. 2012). The potential for nitrogen fixation has also been found in the central Arctic Ocean, but the genes are mainly related to heterotrophic bacteria indicating different communities for N_2 fixation in the Arctic (Fernández-Méndez et al. 2016). However, the importance of nitrogen fixation in sea ice remains unclear until nitrogen fixation has been measured directly or the gene expression has been assessed via omics approaches. Nitrogen fixation measurements from open water in the seasonal ice zone suggest that 27.1% of the nitrogen lost via denitrification can be resupplied via nitrogen fixation (Sipler et al. 2017).

14.4.1.3 Others

Other biogeochemical cycles have been discovered in metagenomic and metatranscriptomic datasets (reviewed by Bowman 2015), indicating the potential for mercury cycling and dimethyl sulfide (DMS) production (Bowman et al. 2014), hydrocarbon degradation (Gerdes et al. 2005), and vitamin B12 synthesis (Taylor and Sullivan 2008) in and under sea ice, all processes which can effect primary production in sea ice-covered seas.

14.4.2 Mixotrophy

Mixotrophy in algae is the combination of a heterotrophic (phagotrophic and/or osmotrophic) and phototrophic nutritional mode within a single cell (Sanders 1991). The awareness of the importance of mixotrophic behavior in aquatic systems has increased tremendously (Hansen 2011) and has been reported to be widespread among flagellate algal groups such as dinoflagellates, prymnesiophytes, and cryptophytes in the marine system (Ballen-Segura et al. 2017). Many bloom-forming algal species have been recognized to be mixotrophs causing an increased interest in this field. The potential benefits of particle ingestion include the acquisition of organic carbon, energy, major nutrients, vitamins, and trace metals (Caron et al. 1993). Mixotrophy is particularly beneficial when there is a limitation in inorganic nutrients (Unrein et al. 2014) or light availability (Hansen 2011). Under oligotrophic conditions, flagellated algae can account for up to 80% of total bacterial grazing (Unrein et al. 2007; Sanders and Gast 2012). Predation by flagellated algae is among the primary mortality factors of prokaryotes in planktonic communities, constituting an important selective pressure (Ballen-Segura et al. 2017).

The Arctic nanoplankton species *Micromonas pusilla* is abundant in polar waters throughout the year and was identified as being independent of the availability of light based on a mixotrophic life style (Unrein et al. 2007; Sanders and Gast 2012). Under thick ice cover in the Canadian Arctic, *Micromonas* sp. contributed up to 93% of autotrophic cell abundance (Sherr et al. 1997). *Micromonas pusilla* ingests higher rates of fluorescently labeled bacteria at oligotrophic conditions, but only if exposed to light. This suggests that the ingestion supplemented nitrogen and/or phosphorus supply to allow for balanced growth when photosynthesis rate is high (McKie-Krisberg and Sanders 2014). In the dark, the tested strain would take up less fluorescently labeled bacteria, which points to an osmotrophic uptake of carbon and energy.

Although the potential importance of mixotrophy within the sea ice community has been recognized, comparably few studies have focused on this environment. Facultative heterotrophy and energy storage have been suggested to be the main processes enabling winter survival in sea ice (Syvertsen 1991; Zhang et al. 2003). Piwosz et al. (2013) found bacterial cells in the food vacuoles of picoeukaryotes from various trophic groups in FYI of the Arctic. Phagotrophic ingestion was investigated in mixotrophic nanoflagellates (MNF) of Antarctic sea ice during spring where they comprised 5–10% of the autotrophic nanoflagellates (Moorthi et al. 2009). Mixotrophy has been proposed to be an important mode of winter survival in sea ice algae. However, Horner and Alexander (1972) only found low uptake rates of organic substances by sea ice diatoms. Osmotrophy is widespread among pennate and centric diatoms from Antarctic and Arctic marine environments. The uptake rate of organic material is dependent on solar radiation and shows great interspecific variability (Ruiz-González et al. 2012). Algae are able to take up a variety of organic substrates such as pyruvate, acetate, lactate, ethanol, saturated fatty acids, glycerol, urea, and amino acids (Parker et al. 1961; Lewin and Hellebust 1976; Amblard 1991; Bronk et al. 2007). Ruiz-Gonzáles et al. (2012) concluded that osmotrophy together with phagotrophy suggest that algae may play a more diverse role in aquatic biogeochemical cycles than only supplying heterotrophs with photosynthetically fixed organic matter.

14.4.3 Cryptic Carbon Cycling and Underrepresented Microbes

Both ecological and phylogenetic fungi are prominent parasites of cyanobacteria, macro-algae, and animals that also biogeochemically cycle nutrients and degrade recalcitrant molecules. Fungal-like organisms in the *Oomycota* have been reported as parasites on algae in the Canadian Arctic (Küpper et al. 2016) and have been observed in the Norwegian Sea and Svalbard region (Fig. 14.1). The detection of fungi on Arctic marine bird feathers (Singh et al. 2016) and in association with driftwood (Rämä et al. 2014) suggests additional ecological niches occupied by the fungi that currently remain unexplored. Arctic members of the *Labyrinthulomycota* can exceed 10^5 cells L^{-1} (Naganuma

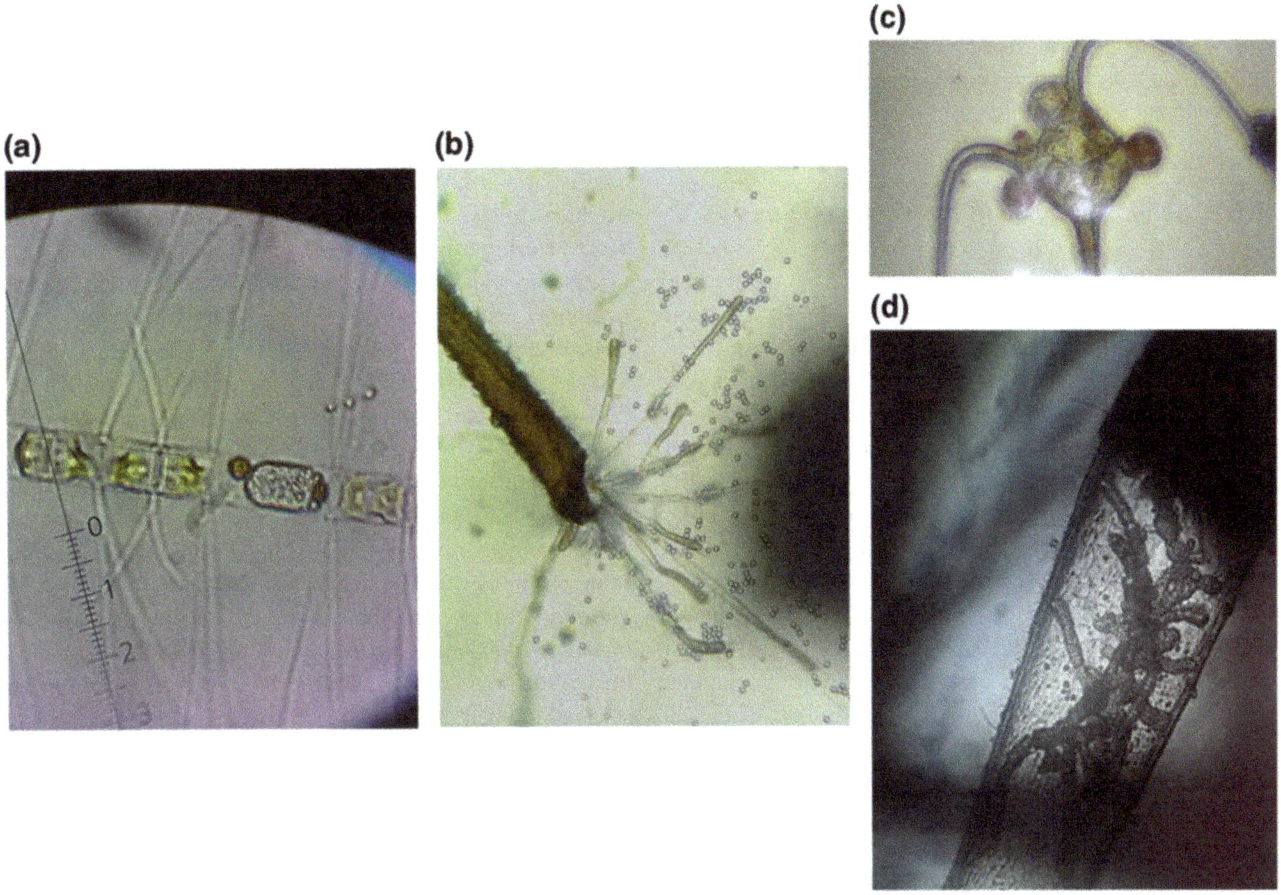

Fig. 14.1 Unidentified fungi and fungal-like pathogens. (**a**) A probable oomycete pathogen of *Chaetoceros* sp. captured in southern Norway (Drøbak). (**b**) A probable oomycete pathogen (*Saprolegnia* sp.) fruiting at the tip of benthic red macro-algae (image from samples collected in Drøbak by the authors). (**c**) Unknown pathogen parasitizing the dinoflagellate *Tripos* sp. in a Tromsø fjord (photo provided by Richard Ingebrigtsen, UiT). (**d**) Unknown pathogen with extensive branching inside a benthic red algae. (Image from Drøbak)

et al. 2006) and are capable of degrading pine pollen (Hassett and Gradinger 2018), suggesting that these organisms might be seasonally important in degradative processes. Members of the *Mesomycetozoea* (namely, the genus *Ichthyophonus*) primarily exist as parasites of Arctic fish (Klimpel et al. 2006) but whose abundance, distribution, and relevance to other ecosystem processes remain unknown.

Ecologically, the Arctic marine *Chytridiomycota* are seasonally abundant parasites that can infect approximately 1% of all diatoms in the near-shore sea ice environment (Hassett and Gradinger 2016) and 25% of a single diatom species (Hassett et al. 2017). Specifically, the *Chytridiomycota* parasitize light-stressed diatoms within sea ice brine channels (Horner and Schrader 1982; Hassett and Gradinger 2016). True Fungi have been reported in sea ice as far north as the North Pole (Bachy et al. 2011) and across the Western Arctic (Hassett et al. 2017). Beyond these observations, there is little information detailing the sea ice ecology of Fungi and fungal-like organisms. Seawater advection into sea ice and entrainment processes of particulates (Eicken et al. 2005; Gradinger et al. 2009) suggest that many fungi and fungal-like organisms observed in sediments and seawater are likely present in sea ice. However, little to no known specific empirical evidence exists to suggest this phenomenon.

Viruses play an important role in nutrient recycling via the viral shunt. Especially in winter and autumn, viral lysis is known to be the most important mortality factor for bacteria (Krembs et al. 2002b). Besides, viruses may have important roles for horizontal gene transfer. In an extensive virome study by Allen et al. (2017), a diverse virus community was found, including viruses with photosystem genes. Data mining of published metagenomes and metatranscriptomes revealed similar patterns of evolutionary important genes (e.g., photosystem) in viruses. The sea ice metagenome from Cottrell and Kirchman (2012), for example, showed cyanophages, synechococcus phages, and prochlorococcus phages with photosystem genes in their genome. The high bacterial density in sea ice, the close spatial connection, and low host specificity may allow a rather quick evolution via viral gene transfer compared to other systems. This may help to develop hypotheses on the evolution of life and life on other planets, such as Europa. With further studies of transposons in metagenome assembled genomes, these hypotheses can be developed and tested further.

14.5 Seasonal Cycle in Ice-Covered Seas

Sea ice-covered systems are characterized by highly seasonal variabilities of temperature and light. In sea ice, brine channels spatially constrain biota and increase interactions between consumers (bacteria and fungi) and primary producers. Therefore, the sea ice habitat is more comparable to biofilms and processes taking place within biofilms (Krembs et al. 2000). Ice algae and bacteria release extracellular polymeric substances (EPS) that protect cells from high salinities and low temperatures, foster the adhesion to surfaces, and alter the microstructure and corresponding desalination processes of sea ice (Krembs et al. 2011). Thus, sympagic microalgae and bacteria can cause physical changes to their immediate environment, improving sea ice habitability (Krembs et al. 2002a, 2011). Concentration and chemical composition of EPS significantly differed in response to gradients in temperature and salinity in Antarctic sea ice and simulated sea ice formation experiments with cultures of the bi-polar diatom *Fragilariopsis cylindrus* (Aslam et al. 2018). Under combined conditions of low temperature and high salinity, the relative contribution of EPS to total carbohydrates and their monosaccharide composition changed significantly, both in the field and lab. Increased concentrations of uronic acids and mannose at low temperatures increase the stiffness of EPS gels (Aslam et al. 2012), needed to produce protective cell coatings as observed in natural sea ice brines.

The initial colonization stage during sea ice formation is followed by a low-productive, heterotrophic winter stage, dominated by pelagic organisms (Fig. 14.2) (Grossmann and Gleitz 1993; Joli et al. 2017; Amargant Arumí 2018). Biomass accumulation follows the seasonal increase in solar radiation during winter-spring transition. The sea ice algal bloom is terminated either by nutrient depletion or sea ice melt in late spring/early summer. At the ice edge phytoplankton blooms of *Phaeocystis* sp. may contribute to high primary production (Assmy et al. 2017). The post-bloom stage is dominated again by heterotrophic processes in the sea ice and water column (Haecky et al. 1999; Kaartokallio 2004).

14.5.1 Autumn

In autumn, sea ice starts forming as frazil ice, which can form at the surface or in deeper water layers (Petrich and Eicken 2010). During ice formation, larger particles, such as eukaryotes or sediment particles, can be transported to the water surface and incorporated into the ice. Ice formation begins in autumn when there are still substantial microbial populations left over in surface waters from the preceding spring bloom (Arrigo and Thomas 2004). As the frazil crystals rise to the surface, particles such as microalgae, heterotrophic protists, and bacteria are scavenged (Garrison et al. 1989). During wave movement, the frazil ice may further act as a sieve, concentrating biomass from the water column (Reimnitz et al. 1993; Weissenberger and Grossmann 1998). This process is thought to select for sticky bacteria, diatoms and organisms attached to larger organisms or particles (Grossmann and Dieckmann 1994; Weissenberger and Grossmann 1998). Another concentration process is the brine exclusion during the ice formation, concentrating nutrients, salts, and organisms in small and densely packed brine channels (reviewed by Deming and Collins 2017). The brine channels are very rich in nutrients and organic matter but may limit bacterial and algal activities at low temperatures and high salinities, which may fluctuate greatly. Thus, the organisms that are concentrated into the ice are further selected by their capability to survive or grow in the extreme physical conditions (Grossmann and Dieckmann 1994). Survival may be enhanced by EPS production or cryoprotectants and changes in biochemical structures to withstand osmotic shock and freezing (Krembs et al. 2011). Studies in the autumn are rare, but the overall bacterial production is low compared to the rest of the year and exceeds that by microalgae (Grossmann and Dieckmann 1994). Grazing seems to be a minor impact, while viruses and phages are enriched 10–100 times (e.g., Gowing et al. 2002; Collins and Deming 2011). Knowledge on bacterial and algal diversity and functions are rare, but the community structures seem to differ from the other seasons and are more similar to the seawater below the newly forming sea ice (Collins et al. 2010). Archaea and oligotrophic bacteria (e.g., OM182) become more dominant, while *Alteromonadales* become less abundant (Collins et al. 2010). Throughout the winter, large centric diatoms are gradually reduced, shifting to a dominance of small pennate species such as *Fragilariopsis* sp. (Lizotte 2003). A water column bloom may start due to the nutrient upwelling after autumn storms which are increasingly abundant with the effects of climate change and retreating sea ice (Ardyna et al. 2014).

14.5.2 Winter

In winter, the temperature in sea ice drops and the brine volume may decrease giving little space for eukaryotes and high concentrations of DOM for bacteria (reviewed by Deming and Collins 2017). Absence of light, extreme salinities, and cold temperatures challenge the microbial survival and

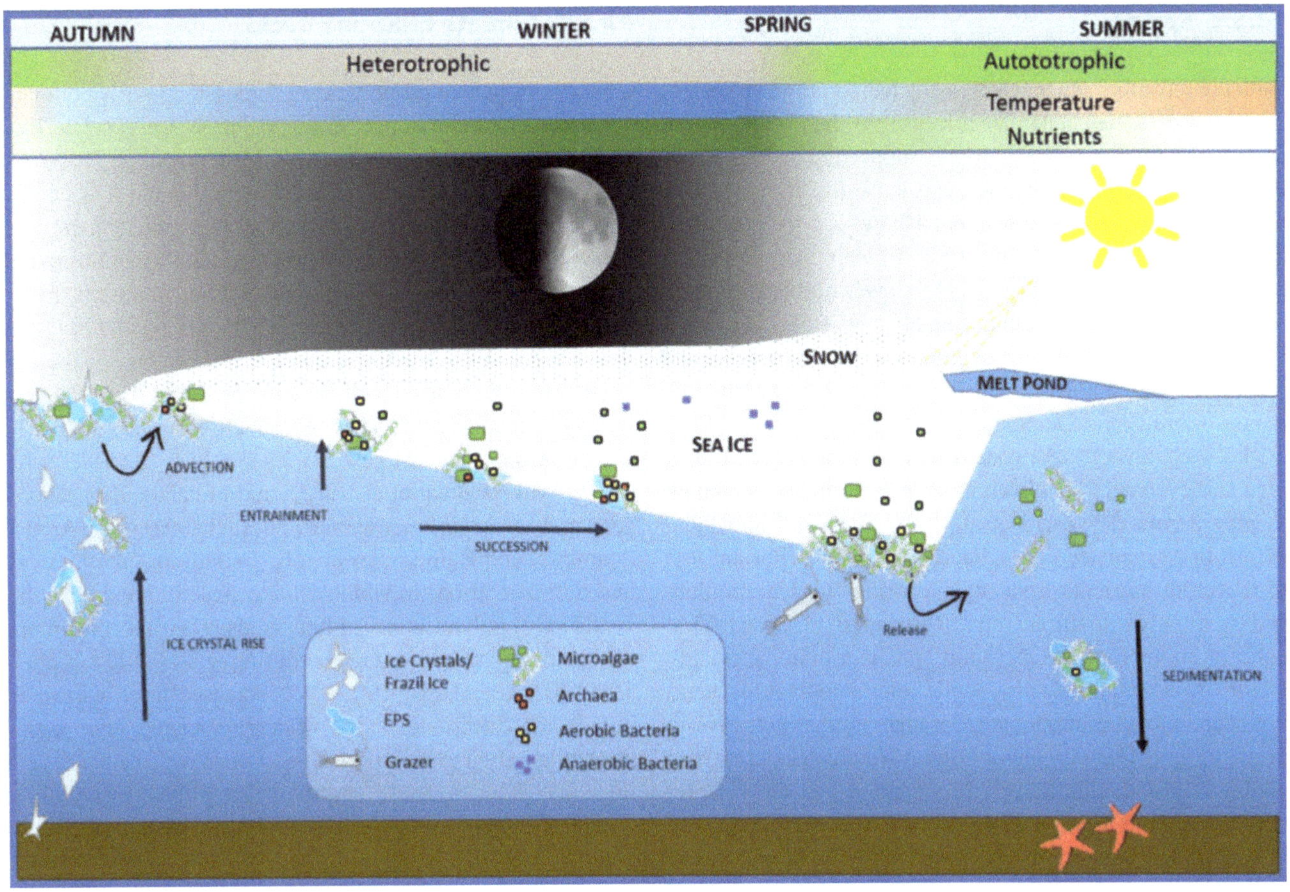

Fig. 14.2 Sea ice and microbial dynamics in the seasonal ice zone. Figure produced based on references in the text

activities. Grazers are mostly absent in winter and bacterial mortality is mainly caused by viral lysis (Krembs et al. 2002a). Overall, bacterial production is low, but certain bacteria are active allowing viruses to produce (Wells and Deming 2006a, b). Up to 86% of winter sea ice bacteria have been found to be active in FISH counts and 4% showed respiration during 5-Cyano-2,3-ditolyl tetrazolium chloride staining (Junge et al. 2004). Bacterial activity is crucial for surviving the harsh conditions, ion transporters have to be active and the membranes intact to compensate for large osmotic differences between the brine water and cell interior transportation of ions and compatible solutes (Collins and Deming 2013). EPS has been found in winter sea ice, which can be important as cryoprotectant and which is mostly produced by bacteria due to the low abundances of eukaryotic algae (Krembs et al. 2002b, 2011). As in autumn, heterotrophic bacterial production is higher than primary production, making the system net heterotroph (Kottmeier and Sullivan 1987). Little is known about the community structure, but the community becomes more similar to typical heterotrophic sea ice communities. Archaea appear more abundant in winter and autumn sea ice, compared to spring and summer indicating their potential importance for autotrophic nitrification (Brinkmeyer et al. 2003; Collins et al. 2010). The microalgal winter diversity in the Beaufort Sea is similar to spring with pennate diatoms dominating, and *Nitzschia frigida* being the most abundant species (Niemi et al. 2011). The dominance of pennate diatoms during the spring algal bloom indicates a competitive advantage, due to their high production rates of EPS and potential for heterotrophy as survival strategies (Niemi et al. 2011). In the water column, picophytoplankton becomes dominant with *Micromonas* sp., *Ostreococcus* sp., and cyanobacteria as main phototrophic organisms (Vader et al. 2015; Joli et al. 2017; Amargant Arumí 2018). In the absence of light, mixotrophic pathways may be important for their survival and explain their dominance (Sanders and Gast 2012; Vader et al. 2015). Nitrifier abundances based on bacterial and archaea amoA genes in the Arctic water column may become 30–115 times more abundant in winter indicating that nitrification may be an important autotrophic process (Christman et al. 2011).

14.5.3 Spring

With the returning sun and increasing temperatures in spring, microbial activities generally increase. At the bottom of the ice, sea ice algae may form blooms supporting bacterial production. The under-ice production is generally higher than in autumn and winter but only about 10% of sea ice algal production (Deming and Collins 2017). In the upper ice layers, bacterial activity is still increasing to levels higher than algal primary production (Gradinger and Zhang 1997). The production can be fueled by degradation of cryoprotectants (e.g., EPS) which are not needed anymore, or by old organic matter (Collins and Deming 2013; Arrigo et al. 2014; Firth et al. 2016). Overall, the system may become net heterotrophic and even anoxic, allowing denitrification and annamox to play a role (Rysgaard et al. 2008; Rysgaard and Glud 2004). In oxygenated zones, bacteria play an important role in recycling nutrients from organic matter into ammonium and eventually into nitrate (e.g., Rysgaard et al. 2008). Viral concentrations increase, but viral lysis is a minor mortality factor of bacterial mortality, while microzooplankton grazing becomes more important (Maranger et al. 1994; Piwosz et al. 2013). The community structure changes toward a community with significantly less archaea and more representatives of the *Alteromonadales* clade (Collins et al. 2010), indicating a system in favor of faster-growing heterotrophic bacteria. Primary production in the water column is limited by light absorbed through the sea ice.

14.5.4 Summer

In summer, the sea ice starts melting from the bottom releasing concentrated organic matter and algal biomass. Large algae aggregate together with their bacterial community may sink to the bottom seeding sediment communities and feeding deep-sea animals (Rapp et al. 2018). In the deep central basins, this process may be a permanent loss, while seeding of new sea ice with the sediment communities is possible down to depth of about 30 m. With a retreating seasonal ice zone toward the central Arctic Ocean, this may become problematic. The communities are dominated by *Flavobacteriia* and *Gammaproteobacteria*, but *Flavobacteriia* are only a minor fraction in sediment associated communities and autumn communities (Collins et al. 2010; Rapp et al. 2018). Archaea are mostly absent (Collins et al. 2010). The water column becomes more heterotrophic with increased bacterial production with decreasing phytoplankton biomass (Kirchman et al. 2009) and a community shift toward heterotrophs and mixotrophs (e.g., Stoecker et al. 2014).

14.5.5 Climate Change Effects

Changes in sea ice properties are forecasted to drive climate feedback loops and impact biological processes, such as element cycling (Holding et al. 2015; Tremblay et al. 2015). As sea ice melting starts earlier in the year, habitat loss occurs before solar radiation is sufficient for algal productivity and growth in the high Arctic (Arrigo et al. 2008) and near the Antarctic Peninsula. Sea ice algae are an important high-quality energy source for grazers in early spring and can seed the pelagic phytoplankton bloom. A shift or loss of the sea ice bloom is forecasted to change the productivity of the system and negatively affect the reproduction success of highly synchronized grazers like *Calanus glacialis* (Søreide et al. 2010), krill (Hoshiai et al. 1987), and of calanoid copepods such as *Acartia bifilosa* (Werner and Auel 2004) in the Arctic, Antarctic, and Baltic, respectively. In addition, the nutrient and light conditions in which sea ice algae thrive induce the synthesis of polyunsaturated fatty acids, a crucial constituent of the diet of grazers, especially during winter (Nichols et al. 1999; Arrigo and Thomas 2004). Increased light intensities reduce the nutritional value of ice-associated (sympagic) microalgae (Leu et al. 2010). Due to warming, the water column may become more stratified (Tremblay et al. 2015), reducing the vertical supply of nutrients to the euphotic zone. On the other hand, a longer ice-free period might counteract stratification through re-mixing of nutrients by wind and storms, but only minor effects on vertical mixing could be observed (Toole et al. 2010; Lincoln et al. 2016). Extended ice-free periods and thinner sea ice in the Arctic increases light availability to pelagic and sympagic microalgae, enhancing annual primary production (Wassmann et al. 2011). The community structure may shift to more pelagic algae species, such as *Phaeocystis* sp. (Assmy et al. 2017). However, the overall consequences of sea ice retreat on primary production remain under debate (Arrigo et al. 2014; Tremblay et al. 2015).

14.6 Conclusions

Knowledge on the distribution and abundance of sea ice protists is crucial to understand and foresee possible changes in microbial communities that potentially have great impacts on food webs of the ice-covered oceans. Even though -omics provide thorough insights into species composition and processes, one should not neglect the importance of assessing their respective abundances. New methods have the potential to shed more light onto the functions of sea ice microorganisms. However, one should not overlook the limitations and restrictions when it comes to sea ice sampling. Melting of sea ice material exposes organisms to changes in physical and chemical conditions, to which microorganisms might react instantaneously. Good care needs to be taken to manipulate the community as little as possible. In the future, the development of good practices and standards is crucial to ensure intercomparable observations of ice biota and properties. A strong seasonality in the seasonal ice zone determines the community structures and functions of the sea ice ecosystem and the associated water column. With climate change, some of the annual processes change (e.g., timing of ice formation), while others will stay the same (length of the polar night). This will eventually change the communities and functions in sea ice, as we know it now. We believe that a focus on microbes, and a proportional representation of their activity (relative to their biomass contributions) in ecosystem models will help to forecast the biological fate of polar marine ecosystems under forecasted climate change scenarios.

Appendix

This article is related to the YOUMARES 9 conference session no. 17: "Bridging disciplines in the seasonal ice zone (SIZ)." The original Call for Abstracts and the abstracts of the presentations within this session can be found in the Appendix "Conference Sessions and Abstracts", Chapter "13 Bridging disciplines in the seasonal ice zone (SIZ)", of this book.

References

Ackley SF, Buck KR, Taguchi S (1979) Standing crop of algae in the sea ice of the Weddell Sea Region. Deep-Sea Res I 26:269–281. https://doi.org/10.1016/0198-0149(79)90024-4

Allen LZ, McCrow JP, Ininbergs K et al (2017) The Baltic Sea virome: diversity and transcriptional activity of DNA and RNA viruses. Msystems 2:e00125–e00116. https://doi.org/10.1128/mSystems.00125-16

Alonso-Sáez L, Sanchez O, Gasol JM et al (2008) Winter-to-summer changes in the composition and single-cell activity of near-surface Arctic prokaryotes. Environ Microbiol 10:2444–2454. https://doi.org/10.1111/j.1462-2920.2008.01674.x

Amargant Arumí M (2018) Arctic marine microbial ecology during the Svalbard Polar Night. Master thesis, UiT Norges arktiske universitet

Amblard C (1991) Carbon heterotrophic activity of microalgae and cyanobacteria – ecological significance. Ann Biol-Paris 30:6–107

Ardyna M, Babin M, Gosselin M et al (2014) Recent Arctic Ocean sea ice loss triggers novel fall phytoplankton blooms. Geophys Res Lett 41:6207–6212. https://doi.org/10.1002/2014gl061047

Arrigo KR (2010) Marine microalgae in Antarctic sea ice. Integr Comp Biol 50:E5–E5

Arrigo KR, Sullivan CW (1992) The influence of salinity and temperature covariation on the photophysiological characteristics of Antarctic sea ice microalgae. J Phycol 28:746–756. https://doi.org/10.1111/j.0022-3646.1992.00746.x

Arrigo KR, Thomas DN (2004) Large scale importance of sea ice biology in the Southern Ocean. Antarct Sci 16:471–486. https://doi.org/10.1017/S0954102004002263

Arrigo KR, van Dijken G, Pabi S (2008) Impact of a shrinking Arctic ice cover on marine primary production. Geophys Res Lett 35:L19603. https://doi.org/10.1029/2008GL035028

Arrigo KR, Perovich DK, Pickart RS et al (2014) Phytoplankton blooms beneath the sea ice in the Chukchi Sea. Deep-Sea Res II 105:1–16. https://doi.org/10.1016/j.dsr2.2014.03.018

Aslam SN, Strauss J, Thomas DN et al (2018) Identifying metabolic pathways for production of extracellular polymeric substances by the diatom *Fragilariopsis cylindrus* inhabiting sea ice. ISME J 12:1237–1251. https://doi.org/10.1038/s41396-017-0039-z

Aslam SN, Cresswell-Maynard T, Thomas DN et al (2012) Production and characterization of the intra- and extracellular carbohydrates and polymeric substances (EPS) of three sea-ice diatom species, and evidence for a cryoprotective role for EPS. J Phycol 48:1494–1509. https://doi.org/10.1111/jpy.12004

Assmy P, Fernández-Méndez M, Duarte P et al (2017) Leads in Arctic pack ice enable early phytoplankton blooms below snow-covered sea ice. Sci Rep 7:40850. https://doi.org/10.1038/srep40850

Bachy C, Lopez-Garcia P, Vereshchaka A et al (2011) Diversity and vertical distribution of microbial eukaryotes in the snow, sea ice and seawater near the North Pole at the end of the polar night. Front Microbiol 2:106. https://doi.org/10.3389/fmicb.2011.00106

Baer SE, Connelly TL, Bronk DA (2015) Nitrogen uptake dynamics in landfast sea ice of the Chukchi Sea. Polar Biol 38:781–797. https://doi.org/10.1007/s00300-014-1639-y

Ballen-Segura M, Felip M, Catalan J (2017) Some mixotrophic flagellate species selectively graze on archaea. Appl Environ Microbiol 83:e02317–e02316. https://doi.org/10.1128/AEM.02317-16

Barber DG, Hop H, Mundy CJ et al (2015) Selected physical, biological and biogeochemical implications of a rapidly changing Arctic Marginal Ice Zone. Prog Oceanogr 139:122–150. https://doi.org/10.1016/j.pocean.2015.09.003

Baross J, Morita R (1978) Microbial life at low temperatures: ecological aspects. In: Kushner DJ (ed) Microbial life in extreme environments. Academic Press, London, pp 9–71

Beitler J (2012) Arctic sea ice extent settles at record seasonal minimum. National Snow and Ice Data Center. http://nsidc.org/arcticseaicenews/2012/09/arctic-sea-ice-extent-settles-at-record-seasonal-minimum/. Accessed 1 Feb 2016

Berge J, Daase M, Renaud P et al (2015) Unexpected levels of biological activity during the polar night offer new perspectives on a warming Arctic. Curr Biol 25:2555–2561. https://doi.org/10.1016/j.cub.2015.08.024

Bluhm BA, Hop H, Vihtakari M et al (2018) Sea ice meiofauna distribution on local to pan-Arctic scales. Ecol Evol 8:2350–2364. https://doi.org/10.1002/ece3.3797

Bochdansky AB, Clouse MA, Herndl GJ (2017) Eukaryotic microbes, principally fungi and labyrinthulomycetes, dominate biomass on bathypelagic marine snow. ISME J 11:362–373. https://doi.org/10.1038/ismej.2016.113

Boetius A, Anesio AM, Deming JW et al (2015) Microbial ecology of the cryosphere: sea ice and glacial habitats. Nat Rev Microbiol 13:677–690. https://doi.org/10.1038/nrmicro3522

Borriss M, Helmke E, Hanschke R et al (2003) Isolation and characterization of marine psychrophilic phage-host systems from Arctic sea ice. Extremophiles 7:377–384. https://doi.org/10.1007/s00792-003-0334-7

Bower SM (1987) *Labyrinthuloides haliotidis* n.sp (Protozoa, Labyrinthomorpha), a pathogenic parasite of small juvenile abalone in a British-Columbia mariculture facility. Can J Zool 65:1996–2007. https://doi.org/10.1139/z87-304

Bowman JS (2015) The relationship between sea ice bacterial community structure and biogeochemistry: a synthesis of current knowledge and known unknowns. Elem Sci Anth 3:000072. https://doi.org/10.12952/journal.elementa.000072

Bowman JS, Rasmussen S, Blom N et al (2012) Microbial community structure of Arctic multiyear sea ice and surface seawater by 454 sequencing of the 16S RNA gene. ISME J 6:11–20

Bowman JS, Berthiaume CT, Armbrust EV et al (2014) The genetic potential for key biogeochemical processes in Arctic frost flowers and young sea ice revealed by metagenomic analysis. FEMS Microbiol Ecol 89:376–387. https://doi.org/10.1111/1574-6941.12331

Brinkmeyer R, Knittel K, Jurgens J et al (2003) Diversity and structure of bacterial communities in Arctic versus Antarctic pack ice. Appl Environ Microbiol 69:6610–6619. https://doi.org/10.1128/Aem.69.11.6610-6619.2003

Bronk DA, See JH, Bradley P et al (2007) DON as a source of bioavailable nitrogen for phytoplankton. Biogeosciences 4:283–296. https://doi.org/10.5194/bg-4-283-2007

Brown MV, Bowman JP (2001) A molecular phylogenetic survey of sea-ice microbial communities (SIMCO). FEMS Microbiol Ecol 35:267–275. https://doi.org/10.1016/S0168-6496(01)00100-3

Bubnova EN, Nikitin DA (2017) Fungi in bottom sediments of the Barents and Kara Seas. Russ J Mar Biol 43:400–406. https://doi.org/10.1134/S1063074017050029

Campbell K, Mundy CJ, Barber DG et al (2014) Remote estimates of ice algae biomass and their response to environmental conditions during spring melt. Arctic 67:375–387. https://doi.org/10.14430/arctic4409

Caron DA, Sanders RW, Lim EL et al (1993) Light-dependent phagotrophy in the freshwater mixotrophic chrysophyte *Dinobryon cylindricum*. Microb Ecol 25:93–111. https://doi.org/10.1007/BF00182132

Cavalieri DJ, Parkinson CL, Gloersen P et al (1999) Deriving long-term time series of sea ice cover from satellite passive-microwave multisensor data sets. J Geophys Res Oceans 104:15803–15814. https://doi.org/10.1029/1999jc900081

Christman GD, Cottrell MT, Popp BN et al (2011) Abundance, diversity, and activity of ammonia-oxidizing prokaryotes in the Coastal Arctic Ocean in summer and winter. Appl Environ Microbiol 77:2026–2034. https://doi.org/10.1128/Aem.01907-10

Collins RE, Deming JW (2011) Abundant dissolved genetic material in Arctic sea ice Part II: viral dynamics during autumn freeze-up. Polar Biol 34:1831–1841. https://doi.org/10.1007/s00300-011-1008-z

Collins RE, Deming JW (2013) An inter-order horizontal gene transfer event enables the catabolism of compatible solutes by *Colwellia psychrerythraea* 34H. Extremophiles 17:601–610. https://doi.org/10.1007/s00792-013-0543-7

Collins RE, Rocap G, Deming JW (2010) Persistence of bacterial and archaeal communities in sea ice through an Arctic winter. Environ Microbiol 12:1828–1841. https://doi.org/10.1111/j.1462-2920.2010.02179.x

Comeau AM, Vincent WF, Bernier L et al (2016) Novel chytrid lineages dominate fungal sequences in diverse marine and freshwater habitats. Sci Rep 6:30120. https://doi.org/10.1038/srep30120

Cottrell MT, Kirchman DL (2009) Photoheterotrophic microbes in the Arctic Ocean in summer and winter. Appl Environ Microbiol 75:4958–4966. https://doi.org/10.1128/Aem.00117-09

Cottrell MT, Kirchman DL (2012) Virus genes in Arctic marine bacteria identified by metagenomic analysis. Aquat Microb Ecol 66:107–116. https://doi.org/10.3354/ame01569

Damashek J, Pettie KP, Brown ZW et al (2017) Regional patterns in ammonia-oxidizing communities throughout Chukchi Sea waters from the Bering Strait to the Beaufort Sea. Aquat Microb Ecol 79:273–286. https://doi.org/10.3354/ame01834

De Corte D, Sintes E, Yokokawa T et al (2013) Comparison between MICRO-CARD-FISH and 16S rRNA gene clone libraries to assess the active versus total bacterial community in the coastal Arctic. Environ Microbiol Rep 5:272–281. https://doi.org/10.1111/1758-2229.12013

Delmont TO, Kiefl E, Kilinc O et al (2017) The global biogeography of amino acid variants within a single SAR11 population is governed by natural selection. BioRxiv 170639. https://doi.org/10.1101/170639

Deming JW, Collins RE (2017) Sea ice as a habitat for bacteria, archaea and viruses. In: Thomas ND (ed) Sea ice. Wiley, New York, pp 326–351

Diez B, Bergman B, Pedros-Alio C et al (2012) High cyanobacterial nifH gene diversity in Arctic seawater and sea ice brine. Environ Microbiol Rep 4:360–366. https://doi.org/10.1111/j.1758-2229.2012.00343.x

Eicken H (2003) From the microscopic, to the macroscopic, to the regional scale: growth, microstructure and properties of sea ice. In: Thomas DN, Dieckmann GS (eds) Sea ice: an introduction to its physics, chemistry, biology and geology. Blackwell Sci, Oxford, pp 22–81

Eicken H, Gradinger R, Gaylord A et al (2005) Sediment transport by sea ice in the Chukchi and Beaufort Seas: increasing importance due to changing ice conditions? Deep-Sea Res II 52:3281–3302. https://doi.org/10.1016/j.dsr2.2005.10.006

Feller G, Gerday C (2003) Psychrophilic enzymes: hot topics in cold adaptation. Nat Rev Microbiol 1:200–208. https://doi.org/10.1038/nrmicro773

Feng S, Powell SM, Wilson R et al (2014) Extensive gene acquisition in the extremely psychrophilic bacterial species *Psychroflexus torquis* and the link to sea-ice ecosystem specialism genome. Biol Evol 6:133–148. https://doi.org/10.1093/gbe/evt209

Fernández-Méndez M, Katlein C, Rabe B et al (2015) Photosynthetic production in the central Arctic Ocean during the record sea-ice minimum in 2012. Biogeosciences 12:3525–3549. https://doi.org/10.5194/bg-12-3525-2015

Fernández-Méndez M, Turk-Kubo KA, Buttigieg PL et al (2016) Diazotroph diversity in the sea ice, melt ponds, and surface waters of the Eurasian Basin of the Central Arctic Ocean. Front Microbiol 7:1884. https://doi.org/10.3389/fmicb.2016.01884

Firth E, Carpenter SD, Sorensen HL et al (2016) Bacterial use of choline to tolerate salinity shifts in sea-ice brines. Elem Sci Anth 4:000120. https://doi.org/10.12952/journal.elementa.000120

Fortunato CS, Huber JA (2016) Coupled RNA-SIP and metatranscriptomics of active chemolithoautotrophic communities at a deep-sea hydrothermal vent. ISME J 10:1925–1938. https://doi.org/10.1038/ismej.2015.258

Gao DW, Huang XL, Tao Y (2016) A critical review of NanoSIMS in analysis of microbial metabolic activities at single-cell level. Crit Rev Biotechnol 36:884–890. https://doi.org/10.3109/07388551.2015.1057550

Garrison DL, Close AR, Reimnitz E (1989) Algae concentrated by frazil ice – evidence from laboratory experiments and field measurements. Antarct Sci 1:313–316. https://doi.org/10.1017/S0954102089000477

Gerdes B, Brinkmeyer R, Dieckmann G et al (2005) Influence of crude oil on changes of bacterial communities in Arctic sea-ice. FEMS Microbiol Ecol 53:129–139. https://doi.org/10.1016/j.femsec.2004.11.010

Gosselin M, Levasseur M, Wheeler PA et al (1997) New measurements of phytoplankton and ice algal production in the Arctic Ocean. Deep Sea Res II 44:1623–1644. https://doi.org/10.1016/S0967-0645(97)00054-4

Gowing MM, Riggs BE, Garrison DL et al (2002) Large viruses in Ross Sea late autumn pack ice habitats. Mar Ecol Prog Ser 241:1–11. https://doi.org/10.3354/meps241001

Gradinger RR (2001) Adaptation of Arctic and Antarctic ice metazoa to their habitat. Zool-Anal Complex Sy 104:339–345. https://doi.org/10.1078/0944-2006-00039

Gradinger RR, Zhang Q (1997) Vertical distribution of bacteria in Arctic sea ice from the Barents and Laptev Seas. Polar Biol 17:448–454. https://doi.org/10.1007/s003000050139

Gradinger RR, Kaufman MR, Bluhm BA (2009) Pivotal role of sea ice sediments in the seasonal development of near-shore Arctic fast ice biota. Mar Ecol Prog Ser 394:49–63. https://doi.org/10.3354/meps08320

Granskog M, Kaartokallio H, Kuosa H et al (2006) Sea ice in the Baltic Sea – a review. Estuar Coast Shelf Sci 70:145–160. https://doi.org/10.1016/j.ecss.2006.06.001

Grossmann S, Gleitz M (1993) Microbial responses to experimental sea-ice formation – implications for the establishment of Antarctic sea-ice communities. J Exp Mar Biol Ecol 173:273–289. https://doi.org/10.1016/0022-0981(93)90058-V

Grossmann S, Dieckmann GS (1994) Bacterial standing stock, activity, and carbon production during formation and growth of sea-ice in the Weddell Sea, Antarctica. Appl Environ Microbiol 60:2746–2753

Haecky P, Jonson S, Andersson A (1999) Influence of sea ice on the composition of the spring phytoplankton bloom in the northern Baltic Sea. Polar Biol 20:1–8. https://doi.org/10.1007/s003000050270

Hanic LA, Sekimoto S, Bates SS (2009) Oomycete and chytrid infections of the marine diatom *Pseudo-nitzschia pungens* (Bacillariophyceae) from Prince Edward Island, Canada. Botany 87:1096–1105. https://doi.org/10.1139/B09-070

Hansen PJ (2011) The role of photosynthesis and food uptake for the growth of marine mixotrophic dinoflagellates. J Eukaryot Microbiol 58:203–214. https://doi.org/10.1111/j.1550-7408.2011.00537.x

Hassett BT, Gradinger RR (2016) Chytrids dominate arctic marine fungal communities. Environ Microbiol 18:2001–2009. https://doi.org/10.1111/1462-2920.13216

Hassett BT, Gradinger RR (2018) New species of saprobic Labyrinthulea (=Labyrinthulomycota) and the erection of a *gen. nov* to resolve molecular polyphyly within the Aplanochytrids. J Eukaryot Microbiol 65:475–483. https://doi.org/10.1111/jeu.12494

Hassett BT, Ducluzeau ALL, Collins RE et al (2017) Spatial distribution of aquatic marine fungi across the western Arctic and sub-arctic. Environ Microbiol 19:475–484. https://doi.org/10.1111/1462-2920.13371

Hatam I, Lange B, Beckers J et al (2016) Bacterial communities from Arctic seasonal sea ice are more compositionally variable than those from multi-year sea ice. ISME J 10:2543–2552. https://doi.org/10.1038/ismej.2016.4

Hegseth EN (1998) Primary production of the northern Barents Sea. Polar Res 17:113–123. https://doi.org/10.1111/j.1751-8369.1998.tb00266.x

Hobbie JE, Daley RJ, Jasper S (1977) Use of Nuclepore filters for counting bacteria by fluorescence microscopy. Appl Environ Microbiol 33:1225–1228

Holding JM, Duarte CM, Sanz-Martin M et al (2015) Temperature dependence of CO_2-enhanced primary production in the European Arctic Ocean. Nat Clim Chang 5:1079–1082. https://doi.org/10.1038/Nclimate2768

Horner R, Alexander V (1972) Algal populations in Arctic sea ice – investigation of heterotrophy. Limnol Oceanogr 17:454–458. https://doi.org/10.4319/lo.1972.17.3.0454

Horner R, Schrader GC (1982) Relative contributions of ice algae, phytoplankton, and benthic microalgae to primary production in nearshore regions of the Beaufort Sea. Arctic 35:485–503

Hoshiai T, Tanimura A, Watanabe K (edited by Hoshiai T) (1987) Ice algae as food of an Antarctic ice-associated copepod, *Paralabidocera antarctica* (IC Thompson). In: Proceedings of the NIPR Symposium on Polar Biology, Citeseer, pp 105–111

Hunt GL, Drinkwater KF, Arrigo K et al (2016) Advection in polar and sub-polar environments: impacts on high latitude marine ecosystems. Prog Oceanogr 149:40–81. https://doi.org/10.1016/j.pocean.2016.10.004

Jephcott TG, Alves-De-Souza C, Gleason FH et al (2016) Ecological impacts of parasitic chytrids, syndiniales and perkinsids on populations of marine photosynthetic dinoflagellates. Fungal Ecol 19:47–58. https://doi.org/10.1016/j.funeco.2015.03.007

Johnson TW, Sparrow FK (1961) Fungi in oceans and estuaries. Hafner Publishing, New York

Johnson SS, Zaikova E, Goerlitz DS et al (2017) Real-time DNA sequencing in the Antarctic dry valleys using the Oxford Nanopore sequencer. J Biomol Tech 28:2. https://doi.org/10.7171/jbt.17-2801-009

Joli N, Monier A, Logares R et al (2017) Seasonal patterns in Arctic prasinophytes and inferred ecology of *Bathycoccus* unveiled in an Arctic winter metagenome. ISME J 11:1372–1385. https://doi.org/10.1038/ismej.2017.7

Junge K, Imhoff F, Staley T et al (2002) Phylogenetic diversity of numerically important arctic sea-ice bacteria cultured at subzero temperature. Microbial Ecol 43:315–328. https://doi.org/10.1007/s00248-001-1026-4

Junge K, Eicken H, Deming JW (2004) Bacterial activity at −2 to −20 degrees C in Arctic wintertime sea ice. Appl Environ Microbiol 70:550–557. https://doi.org/10.1128/Aem.70.1.550-557.2004

Junge K, Cameron K, Nunn B (2019) Chapter 12 – diversity of psychrophilic bacteria in sea and glacier ice environments—insights through genomics, metagenomics, and proteomics approaches. In: Das S, Dash HR (eds) Microbial diversity in the genomic era. Academic Press, Cambridge, pp 197–216. https://doi.org/10.1016/B978-0-12-814849-5.00012-5

Kaartokallio H (2001) Evidence for active microbial nitrogen transformations in sea ice (Gulf of Bothnia, Baltic Sea) in midwinter. Polar Biol 24:21–28. https://doi.org/10.1007/s003000000169

Kaartokallio H (2004) Food web components, and physical and chemical properties of Baltic Sea ice. Mar Ecol Prog Ser 273:49–63. https://doi.org/10.3354/meps273049

Kaartokallio H, Kuosa H, Thomas DN et al (2007) Biomass, composition and activity of organism assemblages along a salinity gradient in sea ice subjected to river discharge in the Baltic Sea. Polar Biol 30:183–197. https://doi.org/10.1007/s00300-006-0172-z

Kirchman DL, Hill V, Cottrell MT et al (2009) Standing stocks, production, and respiration of phytoplankton and heterotrophic bacteria in the western Arctic Ocean. Deep-Sea Res II 56:1237–1248. https://doi.org/10.1016/j.dsr2.2008.10.018

Klimpel S, Palm HW, Busch MW et al (2006) Fish parasites in the Arctic deep-sea: poor diversity in pelagic fish species vs. heavy parasite load in a demersal fish. Deep-Sea Res I 53:1167–1181. https://doi.org/10.1016/j.dsr.2006.05.009

Koh EY, Atamna-Ismaeel N, Martin A et al (2010) Proteorhodopsin-bearing bacteria in Antarctic sea ice. Appl Environ Microbiol 76:5918–5925. https://doi.org/10.1128/Aem.00562-10

Koh EY, Phua W, Ryan KG (2011) Aerobic anoxygenic phototrophic bacteria in Antarctic sea ice and seawater. Env Microbiol Rep 3:710–716. https://doi.org/10.1111/j.1758-2229.2011.00286.x

Koh EY, Martin AR, McMinn A et al (2012) Recent advances and future perspectives in microbial phototrophy in Antarctic sea ice. Biology 1:542–556. https://doi.org/10.3390/biology1030542

Kohlmeyer J, Kohlmeyer E (2013) Marine mycology: the higher fungi. Elsevier, Amsterdam

Kottmeier ST, Sullivan CW (1987) Late winter primary production and bacterial production in sea ice and seawater west of the Antarctic Peninsula. Mar Ecol Prog Ser 36:287–298. https://doi.org/10.3354/meps036287

Krembs C, Gradinger R, Spindler M (2000) Implications of brine channel geometry and surface area for the interaction of sympagic organisms in Arctic sea ice. J Exp Mar Biol Ecol 243:55–80. https://doi.org/10.1016/S0022-0981(99)00111-2

Krembs C, Eicken H, Junge K et al (2002a) High concentrations of exopolymeric substances in Arctic winter sea ice: implications for the polar ocean carbon cycle and cryoprotection of diatoms. Deep-Sea Res I 49:2163–2181. https://doi.org/10.1016/S0967-0637(02)00122-X

Krembs C, Tuschling K, von Juterzenka K (2002b) The topography of the ice-water interface – its influence on the colonization of sea ice by algae. Polar Biol 25:106–117. https://doi.org/10.1007/s003000100318

Krembs C, Eicken H, Deming JW (2011) Exopolymer alteration of physical properties of sea ice and implications for ice habitability and biogeochemistry in a warmer Arctic. Proc Natl Acad Sci USA 108:3653–3658. https://doi.org/10.1073/pnas.1100701108

Küpper FC, Peters AF, Shewring DM et al (2016) Arctic marine phytobenthos of northern Baffin Island. J Phycol 52:532–549. https://doi.org/10.1111/jpy.12417

Laamanen M (1996) Cyanoprokaryotes in the Baltic Sea ice and winter plankton. Algological Studies/Archiv für Hydrobiologie, Supplement Volumes 83:423–433

Lange BA, Katlein C, Nicolaus M et al (2016) Sea ice algae chlorophyll a concentrations derived from under-ice spectral radiation profiling platforms. J Geophys Res Oceans 121:8511–8534. https://doi.org/10.1002/2016jc011991

Lepelletier F, Karpov SA, Alacid E et al (2014) *Dinomyces arenysensis* gen. et sp nov (Rhizophydiales, Dinomycetaceae fam. nov.), a chytrid infecting marine dinoflagellates. Protist 165:230–244. https://doi.org/10.1016/j.protis.2014.02.004

Leu E, Wiktor J, Søreide JE et al (2010) Increased irradiance reduces food quality of sea ice algae. Mar Ecol Prog Ser 411:49–60. https://doi.org/10.3354/meps08647

Lewin J, Hellebust JA (1976) Heterotrophic nutrition of marine pennate diatom *Nitzschia angularis var affinis*. Mar Biol 36:313–320. https://doi.org/10.1007/Bf00389192

Lincoln BJ, Rippeth TP, Simpson JH (2016) Surface mixed layer deepening through wind shear alignment in a seasonally stratified shallow sea. J Geophys Res Oceans 121:6021–6034. https://doi.org/10.1002/2015jc011382

Lizotte MP (2003) The microbiology of sea ice. In: Thomas DN, Dieckmann GS (eds) Sea ice: an introduction to its physics, chemistry, biology and geology. Blackwell Sci, Oxford, pp 184–210

Luhtanen AM, Eronen-Rasimus E, Oksanen HM et al (2018) The first known virus isolates from Antarctic sea ice have complex infection patterns. Fems Microbiol Ecol 94:fiy028. https://doi.org/10.1093/femsec/fiy028

Lytle VI, Ackley SF (1996) Heat flux through sea ice in the western Weddell Sea: convective and conductive transfer processes. J Geophys Res Oceans 101:8853–8868. https://doi.org/10.1029/95jc03675

Majaneva M, Blomster J, Müller S et al (2017) Sea-ice eukaryotes of the Gulf of Finland, Baltic Sea, and evidence for herbivory on weakly shade-adapted ice algae. Eur J Protistol 57:1–15. https://doi.org/10.1016/j.ejop.2016.10.005

Maranger R, Bird DF, Juniper SK (1994) Viral and bacterial dynamics in Arctic sea-ice during the spring algal bloom near Resolute, Nwt, Canada. Mar Ecol Prog Ser 111:121–127. https://doi.org/10.3354/meps111121

Maslanik J, Stroeve J, Fowler C et al (2011) Distribution and trends in Arctic sea ice age through spring 2011. Geophys Res Lett 38:L13502. https://doi.org/10.1029/2011gl047735

Mclean N, Porter D (1982) The yellow-spot disease of *Tritonia diomedea* Bergh, 1894 (Mollusca, Gastropoda, Nudibranchia) – encapsulation of the thraustochytriaceous parasite by host amebocytes. J Parasitol 68:243–252. https://doi.org/10.2307/3281182

McKie-Krisberg ZM, Sanders RW (2014) Phagotrophy by the picoeukaryotic green alga Micromonas: implications for Arctic Oceans. ISME J 8:1953–1961. https://doi.org/10.1038/ismej.2014.16

Meier WN, Hovelsrud GK, van Oort BEH et al (2014) Arctic sea ice in transformation: a review of recent observed changes and impacts on biology and human activity. Rev Geophys 52:185–217. https://doi.org/10.1002/2013rg000431

Meiners K, Fehling J, Granskog MA et al (2002) Abundance, biomass and composition of biota in Baltic sea ice and underlying water (March 2000). Polar Biol 25:761–770. https://doi.org/10.1007/s00300-002-0403-x

Miller LA, Fripiat F, Else BGT et al (2015) Methods for biogeochemical studies of sea ice: the state of the art, caveats, and recommendations. Elem Sci Anth 3:000038. https://doi.org/10.12952/journal.elementa.000038

Mock T, Gradinger RR (1999) Determination of Arctic ice algal production with a new in situ incubation technique. Mar Ecol Prog Ser 177:15–26. https://doi.org/10.3354/meps177015

Mock T, Dieckmann GS, Haas C et al (2002) Micro-optodes in sea ice: a new approach to investigate oxygen dynamics during sea ice formation. Aquat Microb Ecol 29:297–306. https://doi.org/10.3354/ame029297

Moline MA, Karnovsky NJ, Brown Z et al (2008) High latitude changes in ice dynamics and their impact on polar marine ecosystems. Ann N Y Acad Sci 1134:267–319. https://doi.org/10.1196/annals.1439.010

Moorthi S, Caron DA, Gast RJ et al (2009) Mixotrophy: a widespread and important ecological strategy for planktonic and sea-ice nanoflagellates in the Ross Sea, Antarctica. Aquat Microb Ecol 54:269–277. https://doi.org/10.3354/ame01276

Naganuma T, Kimura H, Karimoto R et al (2006) Abundance of planktonic thraustochytrids and bacteria and the concentration of particulate ATP in the Greenland and Norwegian Seas. Polar Biosci 20:37–45

Nguyen D, Maranger R, Balague V et al (2015) Winter diversity and expression of proteorhodopsin genes in a polar ocean. ISME J 9:1835–1845. https://doi.org/10.1038/ismej.2015.1

Nichols D, Bowman J, Sanderson K et al (1999) Developments with Antarctic microorganisms: culture collections, bioactivity screening, taxonomy, PUFA production and cold-adapted enzymes. Curr Opin Biotechnol 10:240–246. https://doi.org/10.1016/S0958-1669(99)80042-1

Niemi A, Michel C, Hille K et al (2011) Protist assemblages in winter sea ice: setting the stage for the spring ice algal bloom. Polar Biol 34:1803–1817. https://doi.org/10.1007/s00300-011-1059-1

Nierychlo M, Nielsen JL, Nielsen PH (2016) Studies of the ecophysiology of single cells in microbial communities by (quantitative) microautoradiography and fluorescence in situ hybridization (MAR-FISH). In: McGenity TJ, Timmis KN, Nogales B (eds) Hydrocarbon and lipid microbiology protocols: ultrastructure and imaging. Springer, Berlin, pp 115–130. https://doi.org/10.1007/8623_2015_66

Nikrad MP, Cottrell MT, Kirchman DL (2012) Abundance and single-cell activity of heterotrophic bacterial groups in the Western Arctic Ocean in summer and winter. Appl Environ Microbiol 78:2402–2409. https://doi.org/10.1128/Aem.07130-11

Orsi W, Biddle JF, Edgcomb V (2013) Deep sequencing of subseafloor eukaryotic rRNA reveals active fungi across marine subsurface provinces. PLoS One 8:e56335. https://doi.org/10.1371/journal.pone.0056335

Oziel L, Neukermans G, Ardyna M et al (2017) Role for Atlantic inflows and sea ice loss on shifting phytoplankton blooms in the Barents Sea. J Geophys Res Oceans 122:5121–5139. https://doi.org/10.1002/2016jc012582

Palmisano AC, Sullivan CW (1983) Sea ice microbial communities (SIMCO) – distribution, abundance, and primary production of ice microalgae in McMurdo Sound, Antarctica in 1980. Polar Biol 2:171–177. https://doi.org/10.1007/Bf00448967

Parker BC, Bold HC, Deason TR (1961) Facultative heterotrophy in some chlorococcacean algae. Science 133:761–763. https://doi.org/10.1126/science.133.3455.761

Pernthaler A, Pernthaler J, Amann R (2002) Fluorescence in situ hybridization and catalyzed reporter deposition for the identification of marine bacteria. Appl Environ Microbiol 68:3094–3101. https://doi.org/10.1128/Aem.68.6.3094-3101.2002

Perovich D, Richter-Menge J, Polashenski C et al (2014) Sea ice mass balance observations from the North Pole Environmental Observatory. Geophys Res Lett 41:2019–2025. https://doi.org/10.1002/2014gl059356

Petri R, Imhoff JF (2001) Genetic analysis of sea-ice bacterial communities of the Western Baltic Sea using an improved double gradient method. Polar Biol 24:252–257. https://doi.org/10.1007/s003000000205

Petrich C, Eicken H (2010) Growth, structure and properties of sea ice. In: Thomas DN, Dieckmann SD (eds) Sea ice, 2nd edn. Wiley, Oxford, pp 23–77

Picard KT (2017) Coastal marine habitats harbor novel early-diverging fungal diversity. Fungal Ecol 25:1–13. https://doi.org/10.1016/j.funeco.2016.10.006

Piiparinen J, Kuosa H, Rintala JM (2010) Winter-time ecology in the Bothnian Bay, Baltic Sea: nutrients and algae in fast ice. Polar Biol 33:1445–1461. https://doi.org/10.1007/s00300-010-0771-6

Piwosz K, Wiktor JM, Niemi A et al (2013) Mesoscale distribution and functional diversity of picoeukaryotes in the first-year sea ice of the Canadian Arctic. ISME J 7:1461–1471. https://doi.org/10.1038/ismej.2013.39

Polglase JL (1980) A preliminary report on the thraustochytrid(s) and labyrinthulid(s) associated with a pathological condition in the Lesser octopus Eledone cirrhosa. Bot Mar 23:699–706. https://doi.org/10.1515/botm-1980-1106

Poulin M, Daugbjerg N, Gradinger RR et al (2011) The pan-Arctic biodiversity of marine pelagic and sea-ice unicellular eukaryotes: a first-attempt assessment. Mar Biodivers 41:13–28. https://doi.org/10.1007/s12526-010-0058-8

Rahimian H (1998) Pathology and morphology of *Ichthyophonus hoferi* in naturally infected fishes off the Swedish west coast. Dis Aquat Org 34:109–123. https://doi.org/10.3354/dao034109

Rämä T, Norden J, Davey ML et al (2014) Fungi ahoy! Diversity on marine wooden substrata in the high North. Fungal Ecol 8:46–58. https://doi.org/10.1016/j.funeco.2013.12.002

Rämä T, Hassett BT, Bubnova E (2017) Arctic marine fungi: from filaments and flagella to operational taxonomic units and beyond. Bot Mar 60:433–452. https://doi.org/10.1515/bot-2016-0104

Rapp JZ, Fernández-Méndez M, Bienhold C et al (2018) Effects of ice-algal aggregate export on the connectivity of bacterial communities in the central Arctic Ocean. Front Microbiol 9:1035. https://doi.org/10.3389/fmicb.2018.01035

Reimnitz E, Clayton JR, Kempema EW et al (1993) Interaction of rising frazil with suspended particles – tank experiments with applications to nature. Cold Reg Sci Technol 21:117–135. https://doi.org/10.1016/0165-232x(93)90002-P

Riedel A, Michel C, Gosselin M et al (2008) Winter-spring dynamics in sea-ice carbon cycling in the coastal Arctic Ocean. J Mar Syst 74:918–932. https://doi.org/10.1016/j.jmarsys.2008.01.003

Rintala JM, Piiparinen J, Blomster J et al (2014) Fast direct melting of brackish sea-ice samples results in biologically more accurate results than slow buffered melting. Polar Biol 37:1811–1822. https://doi.org/10.1007/s00300-014-1563-1

Ruiz-González C, Gali M, Sintes E et al (2012) Sunlight effects on the osmotrophic uptake of DMSP-Sulfur and leucine by polar phytoplankton. PLoS One 7:45545. https://doi.org/10.1371/journal.pone.0045545

Rysgaard S, Glud RN (2004) Anaerobic N-2 production in Arctic sea ice. Limnol Oceanogr 49:86–94. https://doi.org/10.4319/lo.2004.49.1.0086

Rysgaard S, Glud RN, Sejr MK et al (2008) Denitrification activity and oxygen dynamics in Arctic sea ice. Polar Biol 31:527–537. https://doi.org/10.1007/s00300-007-0384-x

Sanders RW (1991) Mixotrophic protists in marine and freshwater ecosystems. J Protozool 38:76–81. https://doi.org/10.1111/j.1550-7408.1991.tb04805.x

Sanders RW, Gast RJ (2012) Bacterivory by phototrophic picoplankton and nanoplankton in Arctic waters. FEMS Microbiol Ecol 82:242–253. https://doi.org/10.1111/j.1574-6941.2011.01253.x

Scholz B, Küpper FC, Vyverman W et al (2017a) Chytridiomycosis of marine diatoms-The role of stress physiology and resistance in parasite-host recognition and accumulation of defense molecules. Mar Drugs 15:26. https://doi.org/10.3390/md15020026

Scholz B, Vyverman W, Küpper FC et al (2017b) Effects of environmental parameters on chytrid infection prevalence of four marine diatoms: a laboratory case study. Bot Mar 60:419–431. https://doi.org/10.1515/bot-2016-0105

Secretariat of Arctic Council (2017) State of the arctic marine biodiversity report. Conservation of Arctic Flora and Fauna, Fairbanks

Sencilo A, Luhtanen AM, Saarijarvi M et al (2015) Cold-active bacteriophages from the Baltic Sea ice have diverse genomes and virus-host interactions. Environ Microbiol 17:3628–3641. https://doi.org/10.1111/1462-2920.12611

Serreze MC, Francis JA (2006) The Arctic amplification debate. Clim Chang 76:241–264. https://doi.org/10.1007/s10584-005-9017-y

Sherr EB, Sherr BF, Fessenden L (1997) Heterotrophic protists in the Central Arctic Ocean. Deep-Sea Res II 44:1665–1673. https://doi.org/10.1016/S0967-0645(97)00050-7

Shields JD (1990) *Rhizophydium littoreum* on the eggs of *Cancer anthonyi* – parasite or saprobe. Biol Bull 179:201–206. https://doi.org/10.2307/1541770

Singh SM, Tsuji M, Gawas-Sakhalker P et al (2016) Bird feather fungi from Svalbard. Arctic Polar Biol 39:523–532. https://doi.org/10.1007/s00300-015-1804-y

Sipler RE, Gong D, Baer SE et al (2017) Preliminary estimates of the contribution of Arctic nitrogen fixation to the global nitrogen budget. Limnol Oceanogr Lett 2:159–166. https://doi.org/10.1002/lol2.10046

Søgaard DH, Kristensen M, Rysgaard S et al (2010) Autotrophic and heterotrophic activity in Arctic first-year sea ice: seasonal study from Malene Bight, SW Greenland. Mar Ecol Prog Ser 419:31–45. https://doi.org/10.3354/meps08845

Søreide JE, Leu E, Berge J et al (2010) Timing of blooms, algal food quality and *Calanus glacialis* reproduction and growth in a changing Arctic. Glob Chang Biol 16:3154–3163. https://doi.org/10.1111/j.1365-2486.2010.02175.x

Stoecker DK, Weigel AC, Stockwell DA et al (2014) Microzooplankton: abundance, biomass and contribution to chlorophyll in the Eastern Bering Sea in summer. Deep-Sea Res II 109:134–144. https://doi.org/10.1016/j.dsr2.2013.09.007

Stroeve JC, Markus T, Boisvert L et al (2014) Changes in Arctic melt season and implications for sea ice loss. Geophys Res Lett 41:1216–1225. https://doi.org/10.1002/2013gl058951

Syvertsen EE (1991) Ice algae in the Barents Sea – types of assemblages, origin, fate and role in the ice-edge phytoplankton bloom. Polar Res 10:277–287. https://doi.org/10.1111/j.1751-8369.1991.tb00653.x

Taylor GT, Sullivan CW (2008) Vitamin B-12 and cobalt cycling among diatoms and bacteria in Antarctic sea ice microbial communities. Limnol Oceanogr 53:1862–1877. https://doi.org/10.4319/lo.2008.53.5.1862

Tedesco L, Vichi M, Haapala J et al (2010) A dynamic biologically active layer for numerical studies of the sea ice ecosystem. Ocean Model 35:89–104. https://doi.org/10.1016/j.ocemod.2010.06.008

Tedesco L, Miettunen E, An BW et al (2017) Long-term mesoscale variability of modelled sea-ice primary production in the northern Baltic Sea. Elem Sci Anth 5:29. https://doi.org/10.1525/elementa.223

Terrado R, Medrinal E, Dasilva C et al (2011) Protist community composition during spring in an Arctic flaw lead polynya. Polar Biol 34:1901–1914. https://doi.org/10.1007/s00300-011-1039-5

Toole JM, Timmermans ML, Perovich DK et al (2010) Influences of the ocean surface mixed layer and thermohaline stratification on Arctic sea ice in the central Canada Basin. J Geophys Res Oceans 115. https://doi.org/10.1029/2009jc005660

Tremblay JE, Anderson LG, Matrai P et al (2015) Global and regional drivers of nutrient supply, primary production and CO_2 drawdown in the changing Arctic Ocean. Prog Oceanogr 139:171–196. https://doi.org/10.1016/j.pocean.2015.08.009

Trenerry LJ, McMinn A, Ryan KG (2002) In situ oxygen microelectrode measurements of bottom-ice algal production in McMurdo Sound, Antarctica. Polar Biol 25:72–80. https://doi.org/10.1007/s003000100314

Ueda M, Nomura Y, Doi K et al (2015) Seasonal dynamics of culturable thraustochytrids (Labyrinthulomycetes, Stramenopiles) in estuarine and coastal waters. Aquat Microb Ecol 74:187–204. https://doi.org/10.3354/ame01736

Unrein F, Massana R, Alonso-Sáez L et al (2007) Significant year-round effect of small mixotrophic flagellates on bacterioplankton in an oligotrophic coastal system. Limnol Oceanogr 52:456–469. https://doi.org/10.4319/lo.2007.52.1.0456

Unrein F, Gasol JM, Not F et al (2014) Mixotrophic haptophytes are key bacterial grazers in oligotrophic coastal waters. ISME J 8:164–176. https://doi.org/10.1038/ismej.2013.132

Vader A, Marquardt M, Meshram AR et al (2015) Key Arctic phototrophs are widespread in the polar night. Polar Biol 38:13–21. https://doi.org/10.1007/s00300-014-1570-2

Vader A, Laughinghouse HD, Griffiths C et al (2018) Proton-pumping rhodopsins are abundantly expressed by microbial eukaryotes in a high-Arctic fjord. Environ Microbiol 20:890–902. https://doi.org/10.1111/1462-2920.14035

Vancoppenolle M, Tedesco L (2017) Numerical models of sea ice biogeochemistry. In: Thomas DN (ed) Sea ice. Wiley, New York, pp 492–515

Voipio A (1981) The Baltic Sea. Elsevier, Amsterdam

Vonnahme TR, Devetter M, Zarsky JD et al (2016) Controls on microalgal community structures in cryoconite holes upon high-Arctic glaciers, Svalbard. Biogeosciences 13:659–674. https://doi.org/10.5194/bg-13-659-2016

Wassmann P, Slagstad D, Ellingsen I (2010) Primary production and climatic variability in the European sector of the Arctic Ocean prior to 2007: preliminary results. Polar Biol 33:1641–1650. https://doi.org/10.1007/s00300-010-0839-3

Wassmann P, Duarte CM, Agusti S et al (2011) Footprints of climate change in the Arctic marine ecosystem. Glob Chang Biol 17:1235–1249. https://doi.org/10.1111/j.1365-2486.2010.02311.x

Weeks WF, Ackley SF (1986) The growth, structure, and properties of sea ice. In: Untersteiner N (ed) The geophysics of sea ice. Springer US, Boston, pp 9–164. https://doi.org/10.1007/978-1-4899-5352-0_2

Weissenberger J, Grossmann S (1998) Experimental formation of sea ice: importance of water circulation and wave action for incorporation of phytoplankton and bacteria. Polar Biol 20:178–188. https://doi.org/10.1007/s003000050294

Wells LE, Deming JW (2006a) Characterization of a cold-active bacteriophage on two psychrophilic marine hosts. Aquat Microb Ecol 45:15–29. https://doi.org/10.3354/ame045015

Wells LE, Deming JW (2006b) Modelled and measured dynamics of viruses in Arctic winter sea-ice brines. Environ Microbiol 8:1115–1121. https://doi.org/10.1111/j.1462-2920.2005.00984.x

Werner I, Auel H (2004) Environmental conditions and overwintering strategies of planktonic metazoans in and below coastal fast ice in the Gulf of Finland (Baltic Sea). Sarsia 89:102–116. https://doi.org/10.1080/00364820410003504

Worby AP, Bush GM, Allison I (2001) Seasonal development of the sea-ice thickness distribution in East Antarctica: measurements from upward-looking sonar. Ann Glaciol 33:177–180. https://doi.org/10.3189/172756401781818167

Yergeau E, Michel C, Tremblay J et al (2017) Metagenomic survey of the taxonomic and functional microbial communities of seawater and sea ice from the Canadian Arctic. Sci Rep 7:42242. https://doi.org/10.1038/srep42242

Zhang Q, Gradinger R, Zhou QS (2003) Competition within the marine microalgae over the polar dark period in the Greenland Sea of high Arctic. Acta Oceanol Sin 22:233–242

15 Complex Interactions Between Aquatic Organisms and Their Chemical Environment Elucidated from Different Perspectives

Mara E. Heinrichs, Corinna Mori, and Leon Dlugosch

Abstract

Ecosystems form a complex network of interactions regarding energy and material transfers between the living and nonliving environment. Phytoplankton supports all life in the ocean as it converts inorganic compounds into organic constituents. This autotrophically produced biomass presents the foundation of the marine food web. A central part of this food web is the concept of the microbial loop. It describes the prokaryotic degradation and remineralization of organic and inorganic matter and its recycling within the pelagic food web or its return to the nonliving environment. In this review, we describe the composition and functioning of the different compartments of the involved organisms (phytoplankton, prokaryotes, and viruses) and their chemical environment (dissolved organic and inorganic matter) in the ocean, particularly emphasizing their interactions. The aim of this chapter is, therefore, to demonstrate the various ways in which these compartments are connected and how they shape each other. We further emphasize the importance of interdisciplinary research approaches to increase the understanding of the complex interactions within marine ecosystems.

Keywords

Biogeochemical cycling · Microbial loop · Ecosystem functioning · Interdisciplinary research approach · Primary production

15.1 Introduction

> Few of us may ever live on the sea or under it, but all of us are making increasing use of it either as a source of food and other materials, or as a dump. As our demands upon the ocean increase, so does our need to understand the ocean as an ecosystem. Basic to the understanding of an ecosystem is knowledge of its food web, through which energy and materials flow. (Pomeroy 1974)

An ecosystem is defined as a structural and functional unit of the biosphere, where organisms and the abiotic substances interact with each other to produce and exchange material and energy between the living and nonliving environment (Fig. 15.1). The functional properties of an ecosystem emerge from these diverse interactions. The traditional subdivision comprises four basic constituents: (1) *abiotic substances* – including organic and inorganic matter, (2) *producers* – consisting of autotrophic organisms, capable to fix light energy and inorganic nutrients to build up complex organic substances, (3) *consumers* – consisting of heterotrophic organisms of higher trophic levels, which are able to consume dissolved (DOM) and particulate organic matter (POM), and (4) *decomposers* – consisting of heterotrophic organisms (e.g., bacteria, archaea) which are able to remineralize organically bound abiotic substances and make them available for the autotrophs (Odum 1959).

A central part of the organic and inorganic matter cycling is the concept of the "microbial loop" (Azam et al. 1983). It describes the trophic pathway in the marine microbial food web where mainly plankton-derived organic matter is recycled by prokaryotes or integrated into the microbial biomass

M. E. Heinrichs (✉)
Paleomicrobiology Group, Institute for Chemistry and Biology of the Marine Environment, Carl von Ossietzky University of Oldenburg, Oldenburg, Germany
e-mail: mara.elena.heinrichs@uni-oldenburg.de

C. Mori
Microbiogeochemistry Group, Institute for Chemistry and Biology of the Marine Environment, Carl von Ossietzky University of Oldenburg, Oldenburg, Germany

L. Dlugosch
Biology of Geological Processes Group, Institute for Chemistry and Biology of the Marine Environment, Carl von Ossietzky University of Oldenburg, Oldenburg, Germany

S. Jungblut et al. (eds.), *YOUMARES 9 - The Oceans: Our Research, Our Future*,
https://doi.org/10.1007/978-3-030-20389-4_15

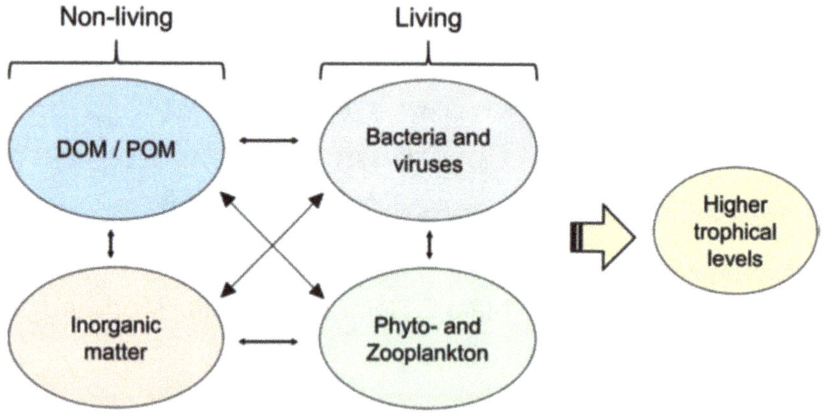

Fig. 15.1 Connections between the different compartments of the living (bacteria/viruses and phyto–/zooplankton) and the nonliving (DOM/POM and inorganic matter) environment

and subsequently either transferred to larger organisms or subjected to continuous recycling (Pomeroy 1974; Azam and Godson 1977; Azam and Malfatti 2007). It is considered to be responsible for the energy and carbon transfer from the DOM pool to higher trophic levels. In this review, the ecosystem will be considered as a complex network of microorganism communities and their environment, interacting as a functional group.

A major part of the nonliving environment is represented by inorganic constituents, which may serve as essential macro- and micronutrients for the living environment (i.e., bacteria and phytoplankton). Nutrient cycling exerts a major control over the overall patterns of primary production in aquatic systems, as it forms the basis for key biological processes, such as the formation of organic matter (Chester and Jickells 2012). In coastal areas, nutrients are mainly derived from processes such as river runoff/erosion, submarine groundwater discharge, glacial weathering, upwelling of nutrient-rich deep water, and atmospheric deposition (Moore 2010; Chester and Jickells 2012). In offshore regions, however, internal microbial recycling is the main control of the nutrient budget in the euphotic zone (Chester and Jickells 2012).

Another major compartment of the nonliving environment is represented by DOM. The global DOM pool contains approximately 662 ± 32 Gt C, a size comparable to the amount of carbon fixed in all living biomass (Hansell et al. 2009; Dittmar and Stubbins 2014). It constitutes one of the largest exchangeable reservoirs of organic carbon in the ocean as it is actively involved in biological processes (Hedges 1992; Hansell and Carlson 2001; Hansell et al. 2009). Due to its enormous size and its importance as substrate, it has a huge ecological significance, playing a central role in marine biogeochemical cycles (Azam et al. 1983; Dittmar and Paeng 2009). DOM is mainly produced by photosynthetic primary production, and it forms the basis of microbial life in the ocean as it supports the metabolic energy demands of carbon and nutrient by heterotrophic marine organisms.

An important fraction of the living environment is represented by phytoplankton. As they are dependent on light and inorganic nutrient supply to perform oxygenic photosynthesis, this pool of primary production is limited to the euphotic zone, which includes the part of the water column where sufficient visible light for photosynthesis penetrates the water body. As primary producers, phytoplankton lives at the interface between the abiotic and biotic realm in the ocean and have a key role in linking both compartments (e.g., Falkowski et al. 1998; Buchan et al. 2014). They represent the basis of the marine food web, supporting most of the heterotrophic production in the ocean as the photosynthetically fixed carbon flows through all other trophic levels (Azam 1998). Moreover, they influence the oceanic chemistry by impacting key biogeochemical cycles, for example, through their involvement in export of carbon and nutrients to the deep sea and by the formation of deep sea sediment through organismic skeletal remains of calcite and opal A (Billet et al. 1983; Nelson et al. 1995; Guidi et al. 2016). They have a global significance for climate regulation due to the fixation of carbon dioxide and generation of oxygen during photosynthesis (Siegenthaler and Sarmiento 1993). Furthermore, some algae species (e.g., *Phaeocystis* spp.) release the precursor of the climate active gas dimethylsulfide, significantly influencing the sulfur cycle (Charlson et al. 1987; Liss et al. 1994; Archer et al. 2011; Moran et al. 2012).

Bacteria and archaea represent the second key component of the living environment. Their global estimated number even exceeds the amount of stars estimated for our universe (Whitman et al. 1998; Pomeroy et al. 2007). For a long time, members of the domain Archaea were thought to be exclusively restricted to inhabit extreme environments. As culture-independent studies in more recent years found that archaea can comprise a significant fraction of the bacterioplankton throughout the ocean (Fuhrman et al. 1992; DeLong et al. 1993; Karner et al. 2001), they can considerably contribute to the global element cycles. In this review, we therefore use the term "prokaryotes" inclusive for both, bacteria and archaea. Prokaryotes live in all parts of the ocean, using nutrients and energy from diverse sources. Autotrophic prokaryotes use energy in the form of light (photoautotroph) and/or oxidize inorganic nutrients (chemolithotroph) for their energy demands and build up organic matter, which serves as the basis for food webs in diverse ecosystems (Azam et al. 1983; Pomeroy et al. 2007). Heterotrophic prokaryotes, on the other hand, are able to enzymatically decompose and remineralize biomass and make it available again for autotrophs (Azam et al. 1983). Not only bacteria and archaea but also viruses influence the living as well as the nonliving environment. During the last two decades, scientists discovered and described the process of the so called viral shunt (Wilhelm and Suttle 1999), which describes a short cut in the marine food web where viral infections of host cells lead to the release of cellular DOM and POM, which then can be reused by the prokaryotic community. Thus, the viral shunt redirects carbon and energy fluxes from higher trophic levels toward the microbial domain promoting prokaryotic respiration and production on a community level (Fuhrman 1999; Wilhelm and Suttle 1999). As prokaryotic microorganisms represent a large fraction of the marine biomass and are characterized by relatively high active metabolic rates, they are capable of dominating the flux of organic and inorganic matter and create sustained cycles of production, decomposition, and remineralization in the ocean (Pomeroy et al. 2007).

From different scientific points of view, each of the above-described compartments has the potential to be considered to be fundamental for marine life. However, ecosystems are highly interactive systems in which these described living and nonliving compartments exert a controlling influence on the other. They are inseparably linked to each other and none of these compartments is able to stand alone—thus, it is necessary to study them in combined research approaches with scientists from various disciplines to unravel the complex interactions between these compartments. The aim of this review is to emphasize the importance of interdisciplinary research approaches by pointing out fundamental connections between these fields, which are traditionally separated into different subdisciplines. In the following, we will highlight the interconnections between the different compartments of the pelagic ecosystem and emphasize their connections from different scientific point of views.

15.2 Nonliving Environment

15.2.1 Inorganic Components

15.2.1.1 Macronutrients

Traditionally, nitrogen, silicon, and phosphorus have been regarded as the main nutrient elements in the ocean (Chapman 1986). There are major differences between the three macronutrient cycles in the ocean. Especially, nitrogen cycling forms a complicated web of mainly microbially mediated processes (Chapman 1986; Arrigo 2005). Nitrate and phosphate are essential for a wide range of organisms as they are involved in nutritional processes and are directly incorporated into the soft tissue of autotrophic microorganisms (e.g., via photosynthesis) (Chapman 1986; Arrigo 2005). In contrast, silicon is growth-limiting to a distinct fraction of phyto- and zooplankton (diatoms, silicoflagellates, and radiolarians) and, unlike nitrogen and phosphorous, is involved in the building of hard skeleton parts (Chapman 1986). The assimilation of these macronutrients is mainly dependent on their abundance in the environment and the amount of "cellular machinery" available for their transport into the cell (Arrigo 2005). However, micronutrients (mainly trace metals) are essential to the functioning of this machinery as they are required for the assimilation and transport of macronutrients (Arrigo 2005; Morel and Price 2003).

Nitrogen Nitrogen is considered as the principal limiting nutrient in marine waters (Elser and Hassett 1994; Arrigo 2005). Thus, the human alteration of the global nitrogen cycle has considerable consequences including eutrophication, hypoxia, and harmful algae blooms (Dobson and Frid 2008). Due to its various inorganic species and oxidation states present in the ocean, the nitrogen cycle is often considered as the most complex of the macronutrient cycles. Furthermore, it builds up part of the organic tissue of organism in the form of proteins, amino acids, sugars and enzymes (Aluwihare and Meador 2008; Gruber 2008). By far the largest and least reactive pool of nitrogen is represented by nitrogen gas, which is considered to maintain a constant equilibrium with the atmosphere (Carritt 1954). Other important components of the marine nitrogen pool are (1) fixed inorganic salts, nitrate, nitrite, and ammonium; (2) organic nitrogen species, amino acids, urea, and degradation products of organic matter; and (3) particulate nitrogen (Arrigo 2005). Its cycling and redox chemistry is mainly mediated by phytoplankton and bacterial as well as probably archaeal mediated metabolic pathways and redox reactions (Arrigo 2005) (Fig. 15.2).

Most oceanic phytoplankton species (except for some cyanobacteria taxa) are not capable of direct nitrogen fixation but rely on prefixed nitrogen in the form of dissolved species such as nitrate, nitrite, and ammonium. Thus, the input of nitrogen by in situ fixation and allochthonous sources are important drivers of marine productivity. Most of the nitrogen required for photosynthesis is supplied by internal nutrient cycling in the ocean (Chester and Jickells 2012). The utilization of fixed nitrogen by phytoplankton is limited to the euphotic zone. Part of the organically bound nitrogen is already remineralized by prokaryotes and released back to

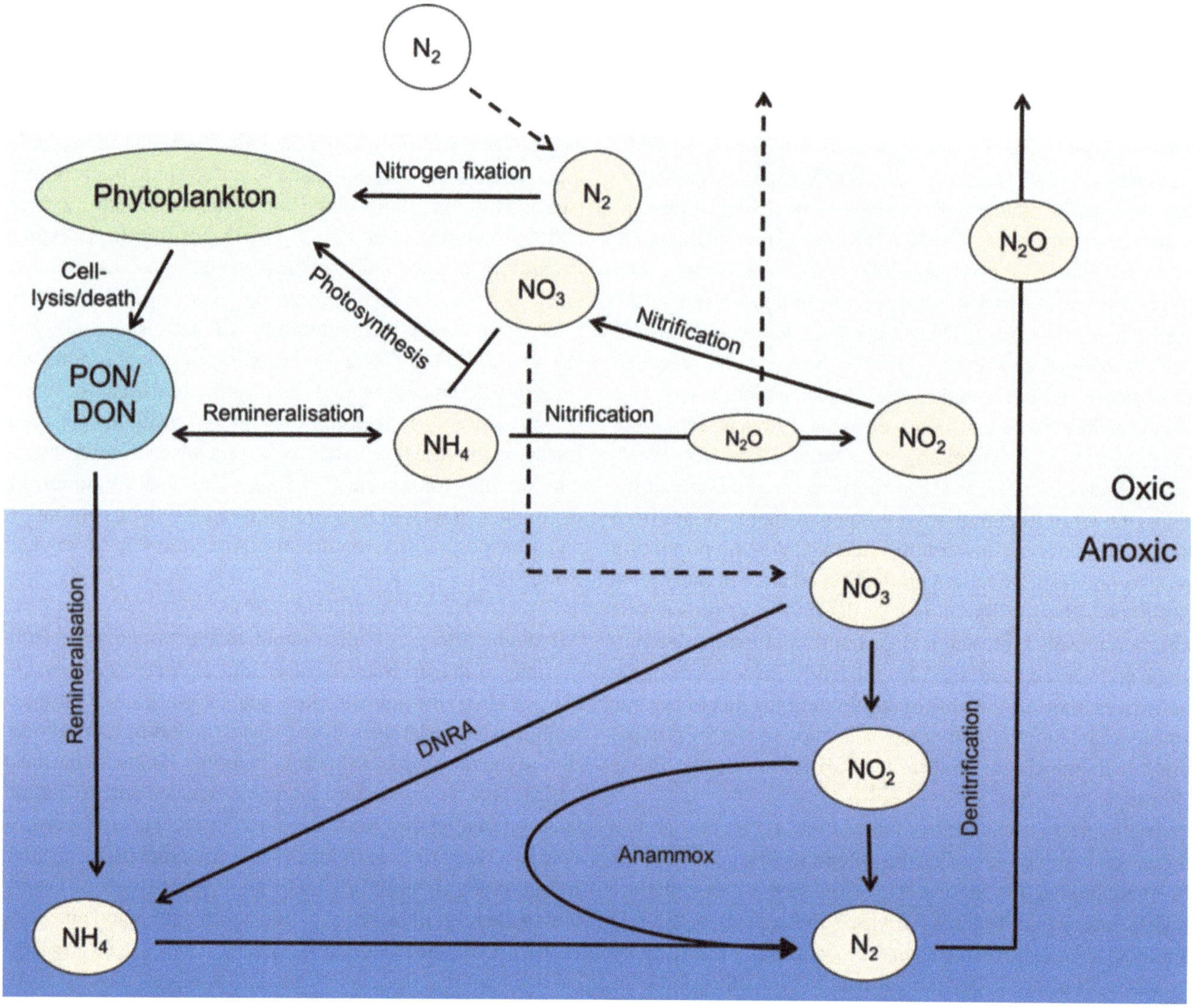

Fig. 15.2 Nitrogen cycle in the upper and lower water column under oxic and anoxic conditions. Chemically and biologically mediated reaction pathways (remineralization, nitrification, dissimilatory nitrate reductase to ammonium (DNRA), anammox, denitrification, nitrogen fixation, and photosynthesis) between the different nitrogen pools (nitrogen gas (N_2), nitrate (NO_3), nitrite (NO_2), ammonium (NH_4), and nitrous oxide (N_2O)) and their link to the living and nonliving organic matter (dissolved (DON) and particulate organic nitrogen (PON)) are represented by the solid arrows. Dashed arrows implicate transport by physical forcing. Figure produced based on Arrigo (2005)

solution within the euphotic zone. The other fraction, particulate organic matter-bound nitrogen, is exported to the deep ocean, where it is subject to continuous prokaryotic degradation. Nitrification, which describes the oxidation of ammonium to nitrite and then nitrate, is mediated by nitrifying bacteria and archaea (see Sect. 15.3.2). The reverse process (denitrification) is carried out by denitrifying microorganisms in anoxic or oxygen-limited environments (e.g., the deep sea, upwelling regions, and highly productive coastal regions). It is considered as the dominant mechanism responsible for the removal of fixed nitrogen from the biosphere and carried out by denitrifying prokaryotes (Arrigo 2005; Chester and Jickells 2012). An intermediate product of these redox reactions is nitrous oxide, which is a greenhouse gas 298 times stronger than carbon dioxide and produced by anoxic prokaryotes during a process called anaerobic oxidation of ammonium (anammox) (Solomon et al. 2007). The final inorganic end product of this prokaryotic mediated redox cascade is nitrate, the thermodynamically stable form of nitrogen in the oxygenated water column (Arrigo 2005).

Phosphorus Unlike nitrogen, phosphorus is not considered as a major limiting nutrient to primary production on a global scale (Paytan and McLaughlin 2007). However, previous studies showed that even if an ecosystem as a whole is limited by nitrogen, certain species within can still be phosphate limited (e.g., Nicholson et al. 2006). Furthermore, in times of anthropogenic pressure and alteration of nutrient cycling, especially in the coastal oceans, detailed studies on the marine phosphorus cycle and its role in the marine food web receive more attention (Nicholson et al. 2006; Paytan and McLaughlin 2007). Phosphorus is considered as an important functional as well as structural component, because it is involved in DNA and RNA formation, energy transmission of ATP molecules, and part of cellular components like proteins, lipids, and cell membranes (Paytan and McLaughlin 2007). Thus, its availability not only influences the primary production rates directly in the ocean, but also the species distribution and ecosystem structures (Paytan and McLaughlin 2007). The marine phosphorus pool contains inorganic, organic, and particulate phosphorus. The major fraction (~87%) is represented by the dissolved reactive orthophosphate, whereby the relative abundance of the different species is pH dependent (Paytan and McLaughlin 2007). Marine phytoplankton and autotrophic prokaryotes are able to take up orthophosphate for their metabolic needs (Paytan and McLaughlin 2007). Heterotrophic prokaryotes are mainly responsible for the reverse process; the hydrolysis of dissolved organic phosphorus back to dissolved inorganic phosphorus. While the uptake of orthophosphate is limited to the euphotic zone, the recycling of dissolved organic phosphorus takes place throughout the water column. Nevertheless, the majority of dissolved organic phosphorus is hydrolyzed in the surface layers where it is rapidly taken up by autotrophs and just a small fraction is transferred to the deep ocean (Paytan and McLaughlin 2007). A large fraction of the organic phosphorus pool is considered nonbioavailable as it cannot be taken up into the cell prior to its conversion/hydrolysis to orthophosphate (Cotner and Wetzel 1992). However, in response to phosphate limitation, some phytoplankton species are able to produce specific enzymes, which catalyze the separation of phosphate from organic matter (Cotner and Wetzel 1992). Furthermore, newer studies found a discrepancy in high molecular weight dissolved organic phosphorus concentrations between the mixed and pelagic layer and thus concluded that this pool must be bioavailable on time scales of months to years (Dyhrman et al. 2007).

Another important process, which influences the oceanic reservoir of bioavailable phosphorus, is its burial into marine sediments. Postdepositional processes transform labile phosphorus phases (e.g., adsorbed-, iron-, or organic-bound phosphorus) via amorphous calcium-phosphate phases to diagenetically stable authigenic carbonate fluorapatite (Ruttenberg and Berner 1993; Benitez-Nelson 2000). This process was found in most marine environments and leads to increased burial of reactive phosphorus.

Silicon In the water column, dissolved silicon is mostly present in the form of orthosilicic acid, which is often referred to as silica. Its supply and availability is only crucial for certain families of phyto- and zooplankton such as diatoms and radiolarians, which require silica for their opal shell formation (Ragueneau et al. 2000). Thus, silica availability is an important factor in regulating the species composition, but not necessarily the overall productivity in the euphotic zone (Ragueneau et al. 2000). However, especially in coastal waters, silicon limitation induced by eutrophication or riverine input has important consequences for the functioning of the ecosystem. Several studies documented a shift from siliceous phytoplankton to less desirable planktonic organisms (e.g., harmful algae blooms) due to a disrupted silicon supply and budget (e.g., Conley et al. 1993; Heisler et al. 2008). The fate of the biogenic opal produced in the euphotic layer is controlled by physical, chemical, and biological factors, which mediate the competition between the export and recycling in the surface waters (Ragueneau et al. 2000). Since the ocean usually is undersaturated with respect to the biologically formed opal A (siliceous ooze), the biogenic silica from the skeletal parts of the organisms is released back into the water column by simple physical dissolution without any prokaryotic involvement. However, as the coating of opal by biological material (e.g., residual cell membranes) is known to inhibit its dissolution, even this chemical process is mediated by prokaryotes (Ragueneau et al. 2000). The fraction of the biogenic opal which is not remineralized finds its final fate in the diagenesis of opal A via opal-CT (porcelanite) to chalcedony or cryptocrystalline quartz (chert) (Kastner et al. 1977; Williams et al. 1985).

15.2.1.2 Micronutrients

Even though most ecological studies focus on macronutrient cycling, there are several elements (e.g., manganese, iron, molybdenum, cobalt, nickel, copper, zinc, and cadmium), which are known to be biologically active. They can be (co-) limiting as they catalyze macronutrient uptake and transformation processes (e.g., species transformation within the nitrogen cycle) as cofactors or part of cofactors in enzymes (e.g., nitrogenase) (Fig. 15.3). As part of proteins, they are able to influence the productivity and shape planktonic and prokaryotic communities (Sunda 1989, 1994; Morel and Price 2003; Arrigo 2005; Twining and Baines 2013). As a result of the planktonic uptake, most of the essential trace metals (except for manganese, molybdenum, and nickel) are depleted in the surface ocean and show the typical nutrient-type element distribution along the water-column (Bruland and Lohan 2003). These concentration profiles are the result of a steady downward flux of mainly planktonic biomass, which is balanced by an advective/diffusive upward flux of prokaryotic recycled elements (Bruland et al. 2013). Most of the small plankton (<2 μm; mainly picoplankton) is already decomposed by heterotrophic prokaryotes in the surface waters and, along with formerly organically bound nutrients

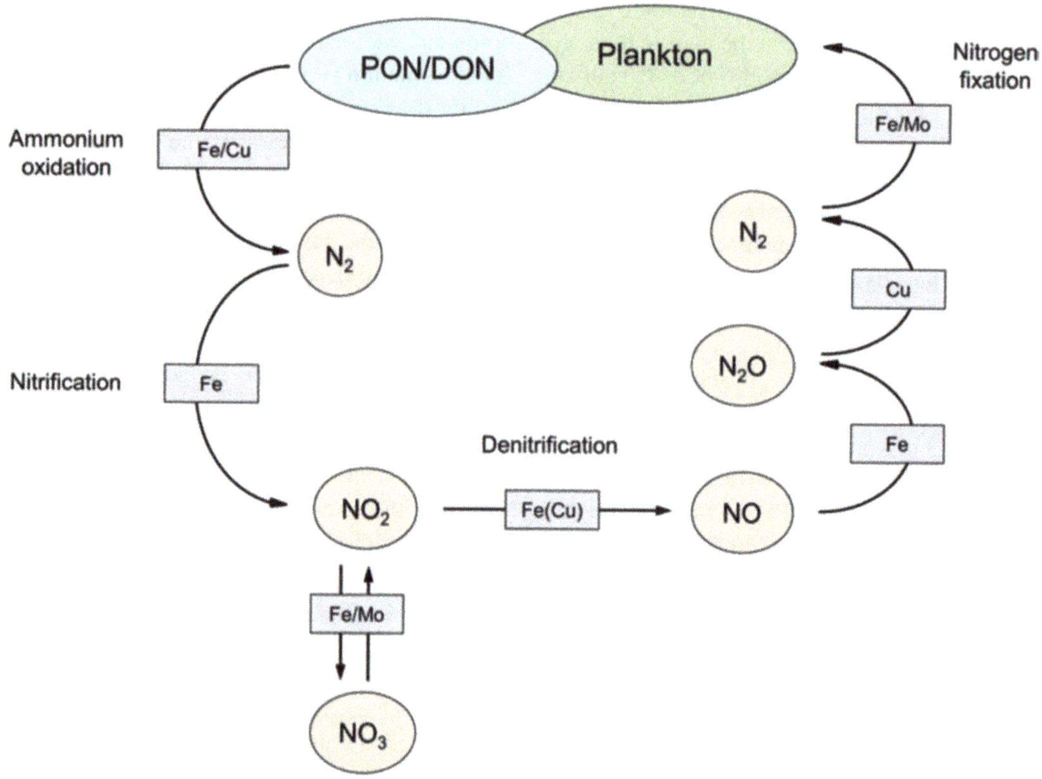

Fig. 15.3 Nitrogen cycle, illustrating the cycling of particulate (PON) and dissolved organic nitrogen (DON), nitrite (NO_2), nitrate (NO_3), nitrous oxide (N_2O), and nitric oxide (NO) and the involved metal cofactors in each enzymatically catalyzed step. Produced based on Morel and Price (2003)

and trace metals, released back into solution (Morel and Price 2003).

The uptake of trace metals by prokaryotic microorganisms is dependent on the metal's chemical speciation. As not all essential trace metals (iron, cobalt, copper, zinc, and cadmium) are present in a bioavailable form, microorganisms developed mechanisms to make them available or rather adjust their environment according to their specific needs (Morel and Price 2003; Sunda 2012; Boyd et al. 2017). Prokaryotic microorganisms are able to release chelators, which are complexing agents (e.g., siderophores and cobalophores) that break bonds of, for example, organic matter and transfer trace metals to accessible forms (Maldonado and Price 2001; Saito et al. 2002). Culture studies showed that some cyanobacteria are able to release, for example, copper and cadmium, complexing agents to detoxify their environment (Moffett and Brand 1996). Besides being the object of strong organic/biogenic ligand binding, trace metals are present in several oxidation states and are thus involved in dynamic redox cycling. To access trace metals, which would be otherwise nonbioavailable, some plankton species are actively involved in these redox transformation processes. A prominent example for microbiological-mediated redox transformation is the oxidation of manganese (II) by a number of prokaryotes via extracellular copper oxidase (Francis and Tebo 2001; Tebo et al. 2004). Furthermore, there have been comprehensive studies on the microbial use of manganese oxides as an alternative electron acceptor during the early diagenesis in sediments (Burdige 1993). Another prominent example is the enzymatically catalyzed extracellular reduction of iron(III) from complexes such as siderophores by the diatom *Thallassiosira oceanica* (Maldonado and Price 2001). These examples show that the interaction between trace metals and planktonic, bacterial, and archaeal microorganisms are reciprocal. As trace metals are part of essential microbial processes (e.g., macronutrient uptake and photosynthesis), they are able to influence the productivity as well as community composition. In turn, the microorganisms have a profound effect on the chemistry and cycling of these metals in the ocean.

15.2.2 Dissolved Organic Matter

15.2.2.1 Composition and Biogeochemical Cycling of DOM

Operationally defined as the fraction of organic matter that passes through filters with nominal pore sizes of 0.2–0.7 μm (Ogawa and Tanoue 2003), DOM is a highly complex mixture and is mainly composed of reduced carbon, oxygen, and hydrogen bound to heteroatoms such as nitrogen, phosphorus, and sulfur (Hansell 2013; Repeta 2015). The presence of these elements substantiates the role of DOM as nutrient and energy source for marine organisms and its involvement in many biological processes such as photosynthesis and respiration. By binding a wide range of bioactive trace metals along with various other trace elements (e.g., manganese, iron, copper) onto its colloidal fraction, DOM further plays a role in the biogeochemical cycling and the availability of trace metals (Aiken et al. 2011; Sunda 2012). These metal-colloid complexes can influence the biological production of the ocean by affecting the growth and species composition of marine phytoplankton (Sunda 1989, 1994).

Most of marine DOM is formed in the euphotic zone of the ocean by photosynthetic primary production at an annual rate of approximately 50 Gt C (Behrenfeld and Falkowski 1997; Carlson 2002). A major part of the fixed carbon is respired, while the remaining fraction is either incorporated by organisms of higher trophic levels, and thus transformed to POM, or enters the DOM pool via a variety of processes (Fig. 15.4). A common mechanism of algae to produce DOM is the extracellular release (e.g., Wetz and Wheeler 2007; Thornton 2014) by which about 20% of the photosynthetically fixed carbon can be excreted from actively growing algae cells (Mague et al. 1980; Maranon et al. 2005). A number of other processes are furthermore responsible for the production of DOM including viral and microbial-mediated lysis, natural decay of cells (Wilhelm and Suttle 1999; Martin 2002), herbivore grazing (e.g., by sloppy feeding; Nagata and Kirchman 1992), or the transformation of POM (Azam and Malfatti 2007; Smith et al. 1992).

The production and consumption of DOM is tightly coupled in the surface of the oceans. As the basis of the microbial life in the ocean, DOM supports the demands of

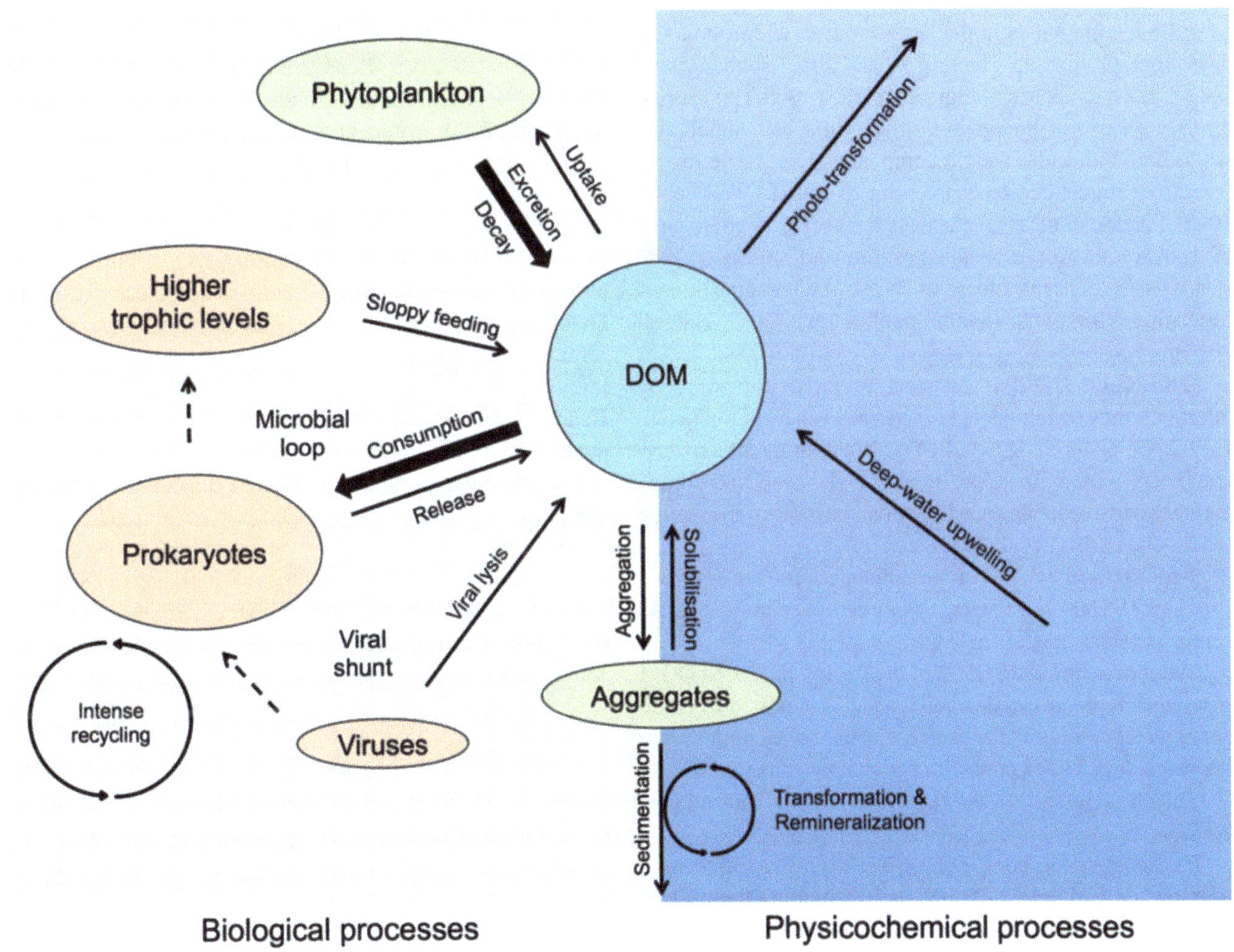

Fig. 15.4 Schematic representation of the fate of DOM in the ocean. Arrows indicate the various production (arrowhead pointing toward DOM pool) and removal processes of DOM (arrowhead pointing away), while the dashed arrows represent dominant biological processes involved in the transfer of DOM. Due to these processes, the fraction of labile DOM decreases rapidly with depth, whereas the refractory character of the DOM pool considerably increases during its export to the deep ocean. Produced based on Carlson (2002). DOM, dissolved organic matter.

heterotrophic prokaryotic microorganisms for metabolic energy, carbon, and nutrients. Prokaryotes play a central role in the cycling of DOM within the microbial loop (Azam et al. 1983; Pomeroy et al. 2007; Logue et al. 2016). Additionally, also marine eukaryotes can directly utilize certain compounds of the DOM pool, though they make up a relatively small proportion of DOM consumption relative to heterotrophic prokaryotes (Michelou et al. 2007; First and Hollibaugh 2009; Flynn et al. 2013).

15.2.2.2 Bioavailability of DOM

DOM represents a continuum regarding biological availability from very labile to highly inaccessible material. Based on its residence time and thus reactivity, DOM can be classified into five categories:

Labile DOM includes highly bioreactive compounds such as amino acids, short-chain organic acids, vitamins, and easily hydrolyzable biopolymers such as proteins or homopolysaccharides. Prokaryotic microorganisms rapidly turn over these compounds in the surface ocean within hours or days after production (Teeling et al. 2012; Buchan et al. 2014). Low molecular weight material (<600 Da) can be directly taken up through the prokaryotic cell membrane, while high molecular weight compounds have to be hydrolyzed by extracellular enzymes prior to uptake (Weiss et al. 1991; Carlson et al. 2007; Arnosti 2011). The high demand of prokaryotes for organic substrate and nutrients keep steady-state concentrations of labile DOM constituents exceedingly low with a global pool of <0.2 Gt C (Hansell 2013).

Semi-labile DOM is resistant to rapid utilization by microbes, thus residing longer in surface waters compared to very labile material. It is turned over within months to several years within the upper mesopelagic zone and forms, together with the labile DOM fraction, the basis of the marine food web (Carlson et al. 2004).

Semi-refractory DOM accumulates in mesopelagic zones (500–1000 m), where it is turned over in decades to centuries (Hansell et al. 2012).

About one third (~16 Gt C) of the freshly produced DOM is resistant to microbial degradation and is exported from the upper water column to the deep sea, where it accumulates in the **refractory DOM** pool with a global inventory of 630 Gt C (Falkowski et al. 1998; Jiao et al. 2010; Hansell and Carlson 2013).

The most stable pool of DOM is the **ultra-refractory** fraction amounting to >12 Gt C and accounting for ~2% of the concentration of bulk dissolved organic carbon (DOC) in the deep sea (Dittmar and Paeng 2009; Hansell 2013).

Average radiocarbon ages between 4000 and 6000 years and a mean residence time of several thousands of years, until its reintroduction into the surface ocean via ocean circulation, already indicate the limited reactivity and slow cycling of deep sea DOM (Williams and Druffel 1987; Bauer et al. 1992). Mean concentrations of DOC in surface waters vary between 60 and 90 μ μmol L^{-1}, while the concentrations of deep sea DOC are uniformly low ranging from 35 to 45 μmol L^{-1} (Ogawa and Tanoue 2003). The stoichiometric ratio of C:N:P in DOM is depleted in nitrogen and phosphorus in comparison to the average marine phytoplankton composition of C:N:P = 106:16:1 (Redfield et al. 1963) with an elemental relationship of C:N:P = 300:22:1 in surface waters and even more depleted with C:N:P = 444:25:1 in deep water DOM (Benner 2002). This indicates the preferential consumption of nitrogen- and phosphorus-rich compounds within the DOM pool and a more degraded status of DOM in deeper water relative to the diagenetically younger surface DOM.

Some biotic and abiotic mechanisms are attributed to be potential processes for the generation of biologically inaccessible DOM, such as the production by prokaryotic microorganisms (Ogawa et al. 2001; Jiao et al. 2010; Lechtenfeld et al. 2015; Osterholz et al. 2015), photochemical reactions (Mopper et al. 1991; Benner and Biddanda 1998; Rossel et al. 2013), or heat-induced condensation processes (Dittmar and Paeng 2009; Rossel et al. 2015). Abiotic processes like sorption onto particles (Druffel et al. 1996), thermal decomposition (Lang et al. 2006; Hawkes et al. 2015), and incorporation into marine aerosols (Kieber et al. 2016) are proposed as removal mechanisms of refractory DOM. Furthermore, photochemical processes can remove biologically resistant DOM (Stubbins et al. 2010) but was found to also transform refractory to bioavailable compounds leading to further microbial utilization (Kieber et al. 1989; Mopper et al. 1991; Benner and Biddanda 1998; Gonsior et al. 2014; Medeiros et al. 2015a).

Despite the importance of DOM as substrate for prokaryotes, it is still an enigma why DOM persists in the ocean over these long time spans without being degraded by prokaryotes. Current hypotheses about the recalcitrance of DOM are summarized by Dittmar (2015). Either environmental factors such as lack of essential metabolites or electron acceptors could hinder the microbial decomposition. Furthermore, specific molecular structures may be difficult to assimilate or metabolize by prokaryotic microorganisms. Due to its huge compositional and structural diversity, individual DOM compounds could be present at too low concentrations (Koch et al. 2005; Hertkorn et al. 2013), limiting an efficient assimilation of energy and processing by prokaryotes (Kovarova-Kovar and Egli 1998; LaRowe et al. 2012; Stocker 2012; Arrieta et al. 2015).

15.2.2.3 Biological Imprint on the DOM Composition

All production, removal, and transformation processes leave an imprint on the composition of DOM and influence both, the ecological and biogeochemical significance of the resulting DOM compounds. Phytoplankton communities synthesize and excrete hundreds of different organic compounds in the surrounding water (Becker et al. 2014; Bittar et al. 2015; Longnecker et al. 2015; Medeiros et al. 2015b). The chemical composition and concentration of phytoplankton-derived DOM is highly variable and depends on the growth stage of the cells (Myklestad 1974; Carlson et al. 1998) as well as the producing taxa or nutrient availability (Fuhrman et al. 2008). Therefore, DOM represents a highly diverse substrate for the prokaryotic populations. The huge molecular diversity of the DOM pool is further intensified by subsequent prokaryotic production and transformation processes (Grossart et al. 2006; Rink et al. 2007; Lechtenfeld et al. 2015), making DOM one of the most complex molecular mixtures on Earth with possible hundreds of thousands of different molecules (Dittmar and Stubbins 2014; Zark et al. 2017). The chemical diversity of DOM is reflected by the phylogenetic diversity as well as the metabolic versatility and potential, which is encoded in marine microbial communities (Becker et al. 2014). Both compartments are deeply interconnected and shape the composition of each other (Alonso-Sáez et al. 2012; Kujawinski et al. 2016).

15.3 Living Environment

15.3.1 Phytoplankton

> The importance of phytoplankton is beyond question. (Reynolds 1984)

15.3.1.1 Diversity and Ecological Function(ing) of Phytoplankton

In general, plankton is defined as free-floating, unicellular organisms or colonies of organisms that are suspended in the water column. The term phytoplankton underlies no systematic concept, but is rather functionally defined by the ability to synthesize complex organic biomass using inorganic matter and solar energy via oxygenic photosynthesis (Buchan et al. 2014).

Phytoplankton is an extremely diverse polyphyletic group, including both prokaryotic and eukaryotic species (Sournia et al. 1991; Falkowski and Raven 2007). Diatoms, dinoflagellates, and coccolithophorides are among the most prevalent eukaryotic phytoplankton taxa, contributing significantly to the bulk diversity of phytoplankton communities (Simon et al. 2009). Bloom-forming diatoms are globally distributed and often predominant in temperate waters of higher latitudes, accounting for up to 40% of the total primary production in the ocean (Nelson et al. 1995; Brzezinski et al. 1997; Armbrust 2009). All species are covered by bivalved frustules made of biogenic opal A (see Sect. 15.2.1.1) that increases the density of the cells. Thus, diatom abundance is dependent on mixing conditions that keep them in the euphotic zone of the water column, and the availability of silicate for the construction of their shells (Armbrust 2009; Amin et al. 2012). Unicellular dinoflagellates have the ability of active swimming owing to their two flagella and mixotrophic lifestyle, which contributes to their success in the ocean (Smayda 1997; Smayda and Reynolds 2003). Coccolithophores produce external shells composed of calcium carbonate plates and have therefore an influence on the ocean's alkalinity and carbon budget (Archer et al. 2000). The most abundant species in the contemporary ocean is *Emiliania huxleyi*, which is known to form enormous blooms, thus representing an important source of biologically produced calcite (Westbroek et al. 1989; Brown and Yoder 1994). An impressive example is the formation of the white chalk cliffs of the island of Rügen in the Baltic Sea by fossil coccoliths (Hjuler and Fabricius 2009).

Prokaryotic picoplankton is almost exclusively represented by the unicellular Cyanobacteria genera *Synechococcus* and *Prochlorococcus* (Partensky et al. 1999). They are abundant and widespread in many ocean regions, contributing considerably to marine primary production due to their ability of efficient nutrient and light acquisition and their high adaptability to different environmental conditions (Moore et al. 1995; Flombaum et al. 2013; Biller et al. 2015; Callieri 2017).

Although phytoplankton biomass accounts for only 1% of the biomass of all photosynthetic organisms on earth, they are responsible for roughly half of the global net primary production with 45–50 Gt C per year (Longhurst et al. 1995; Antoine et al. 1996; Field et al. 1998). About half of the fixed carbon is directly processed by heterotrophic prokaryotes (Cole et al. 1988; Ducklow et al. 1993). The remaining carbon either enters the classic marine food web or is transported from surface waters to the deep ocean as sinking particles via the biological pump (Fig. 15.5) (Eppley and Peterson 1979; Passow 2002; Amin et al. 2012). During export, the aggregates are colonized by prokaryotes and subject to transformation, i.e., microbial remineralization, adsorption or desorption processes leading to nutrient regeneration (Smith et al. 1992).

15.3.1.2 Spatial and Seasonal Effects on Primary Production

Marine net primary production varies over a wide range of time scales including tidal, daily, and seasonal cycles as well as decadal oscillations (Behrenfeld et al. 2006; Cloern and Jassby 2010; Chavez et al. 2011). Abundance and composition of phytoplankton, and thus the annual cycles of primary production, are influenced by various physical, chemical, and biological factors. Nutrient limitation (bottom-up regulation), grazing, and parasitism by prokaryotes and viruses

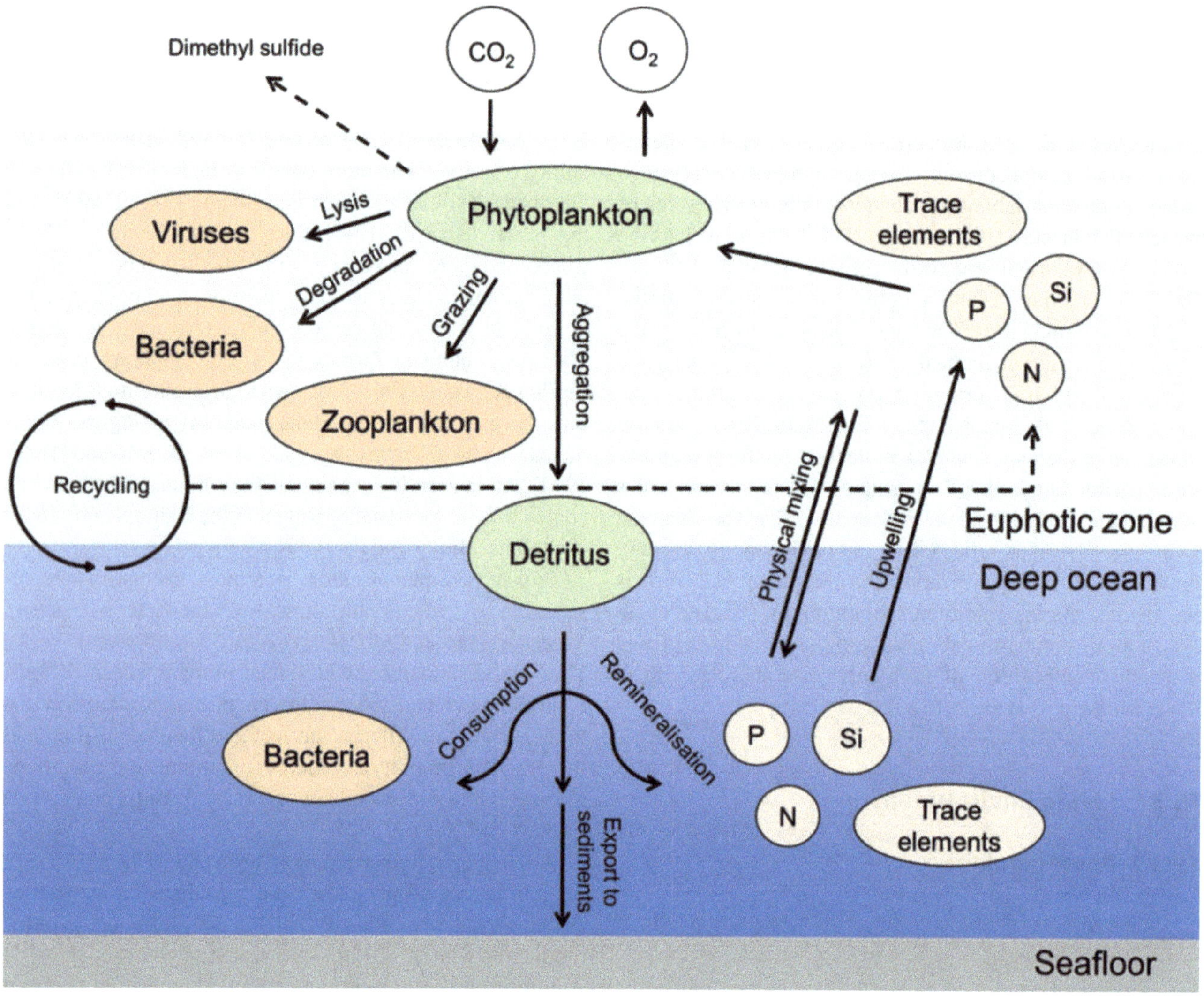

Fig. 15.5 Overview of the role of phytoplankton on various compartments of the marine environment including the atmospheric gas composition, inorganic nutrients, and trace element fluxes as well as the transfer and cycling of organic matter via biological processes. The photosynthetically fixed carbon is rapidly recycled and reused in the surface ocean, while a certain fraction of this biomass is exported as sinking particles to the deep ocean, where it is subject to ongoing transformation processes, e.g., remineralization. Figure based on references in the text

(top-down forces), competition, and seasonal and spatial variations of environmental drivers, such as sunlight or turbulence of the water body, are examples of these factors (Egge and Aksnes 1992; Bratbak et al. 1996; Cloern 1996; Metaxas and Scheibling 1996; Cloern and Dufford 2005; Loeder et al. 2011; Landa et al. 2016). In polar regions with their very short ice-free and sunlight period, brief pulses of phytoplankton abundance occur during summer, when light becomes sufficient for a net increase in primary production (Longhurst et al. 1995; Søreide et al. 2010; Kahru et al. 2011). The intense solar radiation in tropic regions produces a permanent thermocline or nutricline, which separates the warmer surface water layer from the deeper water body, preventing vertical mixing of the surface with the subeuphotic nutrient-rich water masses (Fiedler and Talley 2006; Pennington et al. 2006). Hence, tropical regions are characterized by low rates of primary production and intense nutrient recycling of the permanently nutrient-limited phytoplankton (Pennington et al. 2006). Picoplankton such as *Prochlorococcus* spp. dominate the phytoplankton biomass in these oligotrophic zones (Landry and Kirchman 2002; Reynolds 2006) due to their relatively high surface to volume ratios, which is an advantage to more efficiently exploit resources at low nutrient concentrations (Chisholm 1992; Flombaum et al. 2013; Biller et al. 2015).

At higher latitudes, typically two annual bloom events are observed, namely, in autumn and during the transition from winter to spring (Pennington et al. 2006). In winter, nutrients become replenished in the upper water, due to intense mixing of the water body and erosion of the thermocline by wind and cooling (e.g., Sommer and Lengfellner 2008). The combination of increasing light intensity, higher sea surface temperatures, and high nutrient concentrations, but low grazing pressure in early spring allow the phytoplankton to develop high cell densities (Lessard et al. 2005; Smetacek and Cloern 2008; Behrenfeld 2010; Taylor and Ferrari 2011). Phytoplankton blooms last from several weeks to months and are accompanied by a succession of different phytoplankton species as well as grazers, prokaryotes, and viruses (Riemann et al. 2000; Yager et al. 2001; Chang et al. 2003; Loeder et al. 2011; Teeling et al. 2012). Diatoms account for most of the net primary production during spring blooms in the cold and nutrient-rich waters of higher latitudes, as they have both high nutrient uptake and growth rates (Chisholm 1992; Litchman et al. 2006). Because of their large size, the density of their skeleton, and the tendency to form aggregates, they significantly contribute to the export of organic matter to the deep sea and loss of nutrients bound in their biomass (Smetacek 1985; Riebesell 1991; Smetacek 1999; Armstrong et al. 2002; Amin et al. 2012). During summer, intense nutrient recycling occurs and the autotrophic community is mainly dominated by dinoflagellates, which are better adapted to the present conditions. Falling temperatures and increasing storm events in autumn destabilize the thermal stratification of the water column resulting in the upwelling of nutrients to the surface. The still sufficient light intensities in combination with the nutrient input allow a second blooming event of mainly diatoms and dinoflagellates, before conditions become unfavorable for photosynthesis in winter.

During a bloom event, the release of photosynthetically produced organic compounds by autolysis of dead, decaying, and healthy cells (extracellular release, see Sect. 15.2.2) to their surroundings fills up the DOM and POM pool and provides a wide range of substrate, which is readily assimilated by prokaryotes (Fogg et al. 1965; Myklestad 2000; Arnosti 2011; Kujawinski 2011; Teeling et al. 2012; Buchan et al. 2014). The chemical composition and concentration of the phytoplankton-derived material depend and considerably shape the composition and fate of the organic carbon pool (see Sect. 15.2.2; Bjørrisen 1988; Biddanda and Benner 1997). The interactions between prokaryotes and plankton are quite complex and have evolved over evolutionary time scales, ranging from commensalism to parasitism (e.g., Amin et al. 2012; Lima-Mendez et al. 2015; Seymour et al. 2017). Not only prokaryotic microorganisms depend on phytoplanktonic primary production such as DOM as substrate, but they also support the growth of autotrophs by the provision of nutrients via recycling (Azam et al. 1983) or fixation of nitrogen (Foster et al. 2011), the synthesis of vitamins (Kazamia et al. 2012), or the detoxification of metabolic byproducts (see Sect. 15.2.1.2; Hünken et al. 2008). However, prokaryotes can also compete with phytoplankton for limited essential nutrients, which can manifest in the direct lysis of phytoplankton cells or the production of algicidal and antimicrobial molecules, respectively (Amaro et al. 2005; Seyedsayamdost et al. 2011; Cude et al. 2012; Bertrand et al. 2015). The form of interaction also depends on the environmental conditions, e.g., the nutrient availability (Danger et al. 2007).

15.3.1.3 Nutritional Requirements and Elemental Stoichiometry

Major nutrients (i.e., nitrogen, phosphorus, and silicate for silicified diatoms and silicoflagellates) and trace elements (e.g., iron, cobalt, zinc, nickel, selenium) are required for phytoplankton growth as they are integral parts of phytoplankton biomass and are involved in various internal processes (see Sect. 15.2.1; Morel 1986; Lessard et al. 2005; Schoemann et al. 2005; Twining and Baines 2013). Both compartments, plankton and nutrients, influence each other (e.g., Sunda 2012). Bio-limiting nutrients (nitrogen, phosphorus, and iron) can regulate community structures and restrict phytoplankton biomass and productivity by influencing both the initiation and termination of bloom events (Boyd et al. 2000; Sunda 2012). Plankton, on the other hand, can influence the concentration, redox chemistry, and the biogeochemical cycling of major and trace metal nutrients (Morel 2008; Sunda 2012). As well as prokaryotes, phytoplankton is able to actively shape its trace element environment toward its needs by the release of metal binding chelators for detoxification or trace metal uptake (see Sect. 15.2.1.1; Moffett and Brand 1996; Hutchins et al. 1999; Sunda 2012).

The canonical Redfield ratio (see Sect. 15.2.2; Fleming 1940; Redfield 1958) represents the average elemental composition of planktonic organisms, which can also be extended to trace element ratios (Quigg et al. 2003; Ho et al. 2004). Ratios of trace elements in the deep sea make an exception due to the adsorption onto particles (Morel 2008). The ratios of major nutrients in the ocean are similar to the Redfield ratio. Redfield (1958) hypothesized that this stoichiometric uniformity of plankton and major nutrients is controlled by phytoplankton. This is due to the fact that the ability of phytoplankton to fix carbon and nitrogen leaves an imprint on the oceanic stoichiometry during the remineralization of phytoplankton-derived organic matter. However, the stoichiometric ratio of phytoplankton varies across regions, seasons, and between different species, which leads to deviations from the Redfield ratio (Karl et al. 2001; Michaels et al. 2001). Inter- and intraspecific variations in the overall stoichiometry depend on the growth conditions, precisely the difference between the optimal ratio and the environmental conditions, to which the phytoplankton is actually exposed to (Hillebrand and Sommer 1999). Phytoplankton populations can, for example, produce and excrete carbon-rich

DOM, which is depleted in nitrogen and phosphorus in relation to the original stoichiometry under nutrient limitation (Carlson et al. 1998; Hopkinson and Vallino 2005; Conan et al. 2007). Thus, the interaction between the given environmental conditions and cellular growth impact the stoichiometric ratio, which influences the bioreactivity and therefore the fate of the phytoplankton-derived organic matter during subsequent prokaryotic transformation processes (Azam and Malfatti 2007; Moreno and Martiny 2018).

15.3.2 Prokaryotes and Viruses

15.3.2.1 Prokaryotes

An estimated number of 10^{29} bacterial and archaeal cells reside ubiquitously in all habitats of the world's oceans (Whitman et al. 1998) and drive many of the underlying processes of the geochemical macro- and micronutrient cycles, by the remineralization or channeling of organic matter via the microbial loop (Azam et al. 1983). The chemical structure and abundance of organic matter, which is used as substrate and energy source by marine heterotrophic bacterioplankton, shape the prokaryotic community and its function in the respective ecosystem, from the nutrient-depleted pelagic ocean to nutrient-rich coastal regions and estuaries.

Copiotrophic bacteria (such as the classes Gammaproteobacteria, Flavobacteriia, and some Alphaproteobacteria) harbor a great genomic potential to react fast and efficient to short-term environmental changes and are well adapted to nutrient-rich environments (Lauro et al. 2009). This lifestyle is commonly found in lineages associated with phytoplankton bloom degradation, where different taxa are grown in substrate-controlled successions (see Sect. 15.3.1; Teeling et al. 2012, 2016). In contrast, oligotrophic bacteria (such as the SAR11- or SAR86-clusters) exhibit slower growth but are found in high relative abundances in the open ocean where nutrients are a limiting factor. Genomic streamlining and enzymatic adaption to low substrate concentration (Giovannoni et al. 2014) and therefore a more efficient use of resources give them a competitive advantage in low-nutrient environments (see Sect. 15.3.1). Since the establishment of the microbial loop concept by Azam et al. (1983), our understanding of the underlying prokaryotic processes has grown tremendously. Newly developed approaches to assess bacterial activity (Simon and Azam 1989; Pernthaler et al. 2002a; Alonso and Pernthaler 2005), community structure (Pernthaler et al. 2002b; Caporaso et al. 2012) and especially genetic and functional diversity (Tringe and Rubin 2005; Bashiardes et al. 2016) helped our understanding of how microbes biochemically shape the environment for millions of years (e.g., Canfield 2005). Yet metagenomics studies regularly reveal large amounts of unknown protein families (Sunagawa et al. 2015), suggesting a tremendous unexplored functional potential of microbes in the global oceans.

15.3.2.2 Microbially Mediated Nutrient Cycling

Many of the processes within the geochemical cycles have developed exclusively in bacteria or archaea and are performed by multiple species occupying a wide variety of ecological niches (DeLong et al. 1993; Sogin et al. 2006; Teeling et al. 2012). Biologically mediated nutrient cycling is carried out by enzymes catalyzing constrained redox reactions. These reactions are thermodynamically successive and involve identical or highly similar pathways that can be used as electron donor (oxidation) or electron acceptor (reduction) by different species (Falkowski et al. 2008). Dissimilatory processes are used by microorganisms to conserve energy, through oxidizing molecules leading to a net energy yield, which is released as heat or can be stored in ATP for later usage. In contrast, small molecules can also be reduced during assimilatory processes and used to build up complex molecules or biomass by using up chemical energy. For example, heterotrophic bacteria use the citric acid cycle to gain energy in form of ATP from the stepwise oxidation of acetate to carbon dioxide and water (Hederstedt 1993). The reverse reaction takes place under anoxic conditions, where phototrophic green sulfur bacteria grow by assimilating carbon dioxide in the same pathway under a net energy loss (Evans et al. 1966). Although substantial amounts of carbon are directly respired, up to two thirds are used for biomass production (Rivkin and Legendre 2001) and are thus available for further degradation within the microbial loop or for consumption by higher trophic levels. Only small amounts of carbon are channeled to the oceanic long-term carbon storage by either sinking, via the microbial carbon pump or by transformation to more resistant DOM (e.g., Volk and Hoffert 1985; Osterholz et al. 2015). In most cases, complete remineralization is temporally and spatially separated and is carried out by an array of diverse bacterial species. These different reaction steps are commonplace within biologically mediated geochemical cycling and can occur over large spatial distances (sea surface to deep sea) or in micro-gradients such as the phycosphere (Amin et al. 2012) or sinking particles (Alldredge and Cohen 1987). One of the many prokaryotic-mediated processes is the cycling of nitrogen in the oceans. Compounds containing nitrogen are essential for the synthesis of amino- and nucleic acids, yet the most abundant form of nitrogen, nitrogen gas, is highly inert and not bioavailable (see Sect. 15.2.1.1). Approximately, half of the global nitrogen fixation is carried out in the euphotic zone, mostly by cyanobacteria (e.g., *Trichodesmium* sp.), which transform nitrogen gas to ammonium (Fowler et al. 2013; Karl et al. 2002). Due to the extremely stable triple bond, the activation of nitrogen gas is a very energy-demanding reaction catalyzed by nitrogenases (Karl et al. 2002). This process is carried out in specialized cyanobacterial cells, called heterocysts. In these cells, the oxygen concentration is minimized by specific adaptations: thicker cell walls, the degradation of the photosynthesis apparatus, the upregulation of glycolytic processes, and specialized oxygen scavenging enzymes to prevent oxygen-induced enzyme damage to the nitrogenases (Gallon 1981). Ammonia itself is an essential part in the prokaryotic and eukaryotic biosynthesis of amino acids and is thus incorporated into biomass. Chemolithotrophic nitrifying bacteria (e.g., *Nitrobacter* sp., *Nitrosomonas* sp.) oxidize ammonium to nitrite and nitrate (Gruber 2008). Also archaea (e.g., the Crenarcheon *Nitrosopumilus* sp.) are found to catalyze these reactions (Könneke et al. 2005). In suboxic and anoxic environments (oxygen minimum zones or sediments), nitrate is reduced via a small number of intermediate products to nitrogen gas by denitrifying bacteria, thereby closing the marine nitrogen cycle (see Figs. 15.2 and 15.3). Additionally, some bacteria belonging to the *Planctomycetes* phylum are capable of direct oxidation of ammonium coupled to nitrite reduction, producing nitrogen gas via a hydrazine intermediate. Although the exact reaction mechanics are still under investigation, it is estimated that 30–50% of the nitrogen gas release by the oceans is produced by anammox bacteria (Devol 2003).

15.3.2.3 Viruses

Many publications about marine viruses emphasize that they are the most abundant biological entities in the marine environment (e.g., Wommack and Colwell 2000; Suttle 2007; Weynberg 2018). Viruses play a crucial role in controlling the composition of prokaryotic and phytoplankton populations, and as such shape and alter the function of marine food webs and global biogeochemical cycles (Weinbauer 2004; Short 2012; Weitz et al. 2015). By being ubiquitously distributed and the most abundant predatory agents in the sea, viruses ultimately control the diversity and biomass of their host populations (Thingstad 2000; Suttle 2007; Winter et al. 2010). The majority of marine viruses or phages is assumed to infect prokaryotic organisms (Fuhrman 1999; Suttle 2007; Breitbart 2011), removing approximately 5–40% of the microbial standing stock on a daily basis (Suttle 2007; Middelboe 2008). These lytic phages can keep abundances of dominant prokaryotic species in balance, favoring less competitive species, thus maintaining the biodiversity of prokaryotic communities (Thingstad and Lignell 1997). Viruses also infect eukaryotic phytoplankton (e.g., Wilhelm and Suttle 1999; Brussaard 2004; Wilson et al. 2005), influencing the occurrence or the termination of bloom events (Bratbak et al. 1993).

The interaction between viruses and their hosts is, however, far more sophisticated than parasitism and predation (Martiny et al. 2014). Viruses can confer new metabolic and morphological traits to their hosts through various mechanisms (Weinbauer and Rassoulzadegan 2004; Paul 2008; Rohwer and Thurber 2009; Hurwitz and U'Ren 2016), which may increase the fitness and, therefore, the survival of their hosts. This ultimately can impact the life history and evolution of the affected organisms (e.g., Sullivan et al. 2005).

Viruses also influence carbon, nutrient, and trace metal cycles. The production of viral progenies, through lytic infection, provokes lysis of the host cells, which results in the release of the cellular material into the environment. Within the viral shunt, most of the previously cell-bound DOM and POM can be readily recycled by noninfected prokaryotic communities or exported to the deep sea (Wilhelm and Suttle 1999; Middelboe and Lyck 2002; Weinbauer 2004). Approximately one quarter of the annual primary production in the ocean flows through the viral shunt and generates annually between 3 and 20 Gt C (Wilhelm and Suttle 1999). By viral lysis, bio-limiting inorganic nutrients (e.g., phosphorus, ammonium) and other micronutrients, such as the organically complexed iron, are also regenerated and made available for the biologically utilization (Gobler et al. 1997; Poorvin et al. 2004; Shelford and Suttle 2018).

15.4 Summary and Conclusion

Both, biota and their chemical environment co-evolved over geological time periods, which led to sophisticated nets of mutual influence and dependence as outlined in this chapter. Autotrophic organisms require energy and inorganic electron donors to fix carbon dioxide for building up biomass, which in turn feeds the pool of DOM and POM. Without the work of heterotrophic organisms, however, the nonliving pools required by autotrophs would soon be exhausted. Thus, one compartment alone cannot be considered as the basis of marine life and cannot alone sufficiently explain the intertwined interactions within the ocean.

> To understand the scope of ecology, the subjects must be considered in relation to other branches of biology and to 'ologies' in general. In the present age of specialization in human endeavors, the inevitable connections between different fields are often obscured by the large masses of knowledge within the fields. (Odum 1959)

As stated by Odum (1959), scientists run the risk of losing themselves in the details of their respective scientific discipline(s). An ecosystem can also be considered as a complex organism, such as the human body, where different organs are required to ensure its survival. Thus, also researchers should not just focus on their specific point of interest, but work together in multidisciplinary research approaches, such as doctors in hospitals, to unravel the complex network of an ecosystem.

Authors Contribution MH and CM designed the concept of this review and contributed equally to the manuscript. MH wrote the virus, DOM and the phytoplankton section and parts of the introduction, while CM wrote the introduction, the inorganic matter section and conclusion. LD wrote the prokaryotes section.

Appendix

This article is related to the YOUMARES 9 conference session no. 18: "Crossing traditional scientific borders to unravel the complex interactions between organisms and their nonliving environment." The original Call for Abstracts and the abstracts of the presentations within this session can be found in the Appendix "Conference Sessions and Abstracts", Chapter "14 Crossing traditional scientific borders to unravel the complex interactions between organisms and their nonliving environment", of this book.

References

Aiken GR, Hsu-Kim H, Ryan JN (2011) Influence of dissolved organic matter on the environmental fate of metals, nanoparticles, and colloids. Environ Sci Technol 45:3196–3201

Alldredge AL, Cohen Y (1987) Can microscale chemical patches persist in the sea? Microelectrode study of marine snow, fecal pellets. Science 235:689–691

Alonso C, Pernthaler J (2005) Incorporation of glucose under anoxic conditions by bacterioplankton from coastal North Sea surface waters. Appl Environ Microbiol 71:1709–1716

Alonso-Sáez L, Sánchez O, Gasol JM (2012) Bacterial uptake of low molecular weight organic in the subtropical Atlantic: are major phylogenetic groups functionally different? Limnol Oceanogr 57:798–808

Aluwihare LI, Meador T (2008) Chapter 3 – Chemical composition of marine dissolved organic nitrogen. In: Capone DG, Bronk DA, Mulholland MR et al (eds) Nitrogen in the marine environment, 2nd edn. Academic Press, San Diego, pp 95–140

Amaro AM, Fuentes MS, Ogalde SR et al (2005) Identification and characterization of potentially algal-lytic marine bacteria strongly associated with the toxic dinoflagellate *Alexandrium catenelly*. J Eukaryot Microbiol 52:191–200

Amin SA, Parker MS, Armbrust EV (2012) Interactions between diatoms and bacteria. Microbiol Mol Biol Rev 76:667–684

Antoine D, Andre JM, Morel A (1996) Oceanic primary production. 2. Estimation at global scale from satellite (coastal zone color scanner) chlorophyll. Glob Biogeochem Cycles 10:57–69

Archer DE, Gidon E, Arne W et al (2000) Atmospheric pCO_2 sensitivity to the biological pump in the ocean. Glob Biogeochem Cycles 14:1219–1230

Archer SD, Tarran GA, Stephens JA et al (2011) Combining cell sorting with gas chromatography to determine phytoplankton group-specific intracellular dimethylsulphoniopropionate. Aquat Microb Ecol 62:109–121

Armbrust EV (2009) The life of diatoms in the world's oceans. Nature 459:185–192

Armstrong RA, Lee C, Hedges JI et al (2002) A new, mechanistic model for organic carbon fluxes in the ocean based on the quantitative association of POC with ballast minerals. Deep-Sea Res II 49:219–236

Arnosti C (2011) Microbial extracellular enzymes and the marine carbon cycle. Annu Rev Mar Sci 3:401–425

Arrieta JM, Mayol E, Hansman RL et al (2015) Dilution limits dissolved organic carbon utilization in the deep ocean. Science 348:331–333

Arrigo KR (2005) Marine microorganisms and global nutrient cycles. Nature 437:349

Azam F (1998) Microbial control of oceanic carbon flux: the plot thickens. Science 280:694–696

Azam F, Godson R (1977) Size distribution and activity of marine microheterotrophs. Limnol Oceanogr 22:492–501

Azam F, Malfatti F (2007) Microbial structuring of marine ecosystems. Nat Rev Microbiol 5:782–791

Azam F, Fenchel T, Field JG et al (1983) The ecological role of water-column microbes in the sea. Mar Ecol Prog Ser 10:257–263

Bashiardes S, Zilberman-Schapira G, Elinav E (2016) Use of metatranscriptomics in microbiome research. Bioinf Biol Insights 10:19–25

Bauer JE, Williams PM, Druffel ER (1992) ^{14}C activity of dissolved organic carbon fractions in the north-central Pacific and Sargasso Sea. Nature 357:667–670

Becker JW, Berube PM, Follett CL et al (2014) Closely related phytoplankton species produce similar suites of dissolved organic matter. Front Microbiol 5:111

Behrenfeld MJ (2010) Abandoning sverdrup's critical depth hypothesis on phytoplankton blooms. Ecology 91:977–989

Behrenfeld MJ, Falkowski PG (1997) Photosynthetic rates derived from satellite-based chlorophyll concentration. Limnol Oceanogr 42:1–20

Behrenfeld MJ, O'Malley RT, Siegel DA et al (2006) Climate-driven trends in contemporary ocean productivity. Nature 444:752–755

Benitez-Nelson CR (2000) The biogeochemical cycling of phosphorus in marine systems. Earth-Sci Rev 51:109–135

Benner R (2002) Chapter 3 – Chemical composition and reactivity. In: Hansell DA, Carlson CA (eds) Biogeochemistry of marine dissolved organic matter. Academic Press, San Diego, pp 59–90

Benner R, Biddanda B (1998) Photochemical transformations of surface and deep marine dissolved organic matter: effects on bacterial growth. Limnol Oceanogr 43:1373–1378

Bertrand EM, McCrow JP, Moustafa A et al (2015) Phytoplankton-bacterial interactions mediate micronutrient colimitation at the coastal Antarctic sea ice edge. Proc Natl Acad Sci U S A 112:9938–9943

Biddanda B, Benner R (1997) Carbon, nitrogen, and carbohydrate fluxes during the production of particulate and dissolved organic matter by marine phytoplankton. Limnol Oceanogr 42:506–518

Biller SJ, Berube PM, Lindell D et al (2015) *Prochlorococcus*: the structure and function of collective diversity. Nat Rev Microbiol 13:13–27

Billett DSM, Lampitt RS, Rice AL et al (1983) Seasonal sedimentation of phytoplankton to the deep-sea benthos. Nature 302:520–522

Bittar TB, Vieira AA, Stubbins A et al (2015) Competition between photochemical and biological degradation of dissolved organic matter from the cyanobacteria *Microcystis aeruginosa*. Limnol Oceanogr 60:1172–1194

Bjørrisen PK (1988) Phytoplankton exudation of organic-matter – why do healthy cells do it. Limnol Oceanogr 33:151–154

Boyd PW, Watson AJ, Law CS et al (2000) A mesoscale phytoplankton bloom in the polar Southern Ocean stimulated by iron fertilization. Nature 407:695–702

Boyd PW, Ellwood MJ, Tagliabue A et al (2017) Biotic and abiotic retention, recycling and remineralization of metals in the ocean. Nat Geosci 10:167–173

Bratbak G, Egge J, Heldal M (1993) Viral mortality of the marine alga *Emiliania huxleyi* (Haptophyceae) and termination of algal blooms. J Eukaryot Microbiol 93:39–48

Bratbak G, Wilson W, Heldal M (1996) Viral control of *Emiliania huxleyi* blooms? J Mar Syst 9:75–81

Breitbart M (2011) Marine viruses: truth or dare. Annu Rev Mar Sci 4:425–448

Brown CW, Yoder JA (1994) Coccolithophorid blooms in the global ocean. J Geophys Res-Ocean 99:7467–7482

Bruland KW, Lohan MC (2003) Controls of trace metals in seawater. In: Elderfield H (ed) Treatise on geochemistry, vol 6. Elsevier, Amsterdam, pp 23–47

Bruland KW, Middag R, Lohan MC (2013) Controls of trace metals in seawater. In: Mittle MJ, Elderfield H (eds) Treatise on geochemistry, 2nd edn. Elsevier, Philadelphia, pp 19–51

Brussaard CPD (2004) Viral control of phytoplankton populations – a review. J Eukaryot Microbiol 51:125–138

Brzezinski MA, Phillips DR, Chavez FP et al (1997) Silica production in the Monterey, California, upwelling system. Limnol Oceanogr 42:1694–1705

Buchan A, LeCleir GR, Gulvik CA et al (2014) Master recyclers: features and functions of bacteria associated with phytoplankton blooms. Nat Rev Microbiol 12:686–698

Burdige DJ (1993) The biogeochemistry of manganese and iron reduction in marine sediments. Earth-Sci Rev 35:249–284

Callieri C (2017) *Synechococcus* plasticity under environmental changes. FEMS Microbiol Lett 364:fnx229

Canfield DE (2005) The early history of atmospheric oxygen: homage to Robert M. Garrels. Annu Rev Earth Planet Sci 33:1–36

Caporaso JG, Lauber CL, Walters WA et al (2012) Ultra-high-throughput microbia community analysis on the Illumina HiSeq and MiSeq platforms. ISME J 6:1621–1624

Carlson CA (2002) Production and removal processes. In: Hansell DA, Carlson CA (eds) Biogeochemistry of marine dissolved organic matter. Academic Press, San Diego, pp 91–151

Carlson CA, Ducklow HW, Hansell DA et al (1998) Organic carbon partitioning during spring phytoplankton blooms in the Ross Sea polynya and the Sargasso Sea. Limnol Oceanogr 43:375–386

Carlson CA, Giovannoni SJ, Hansell DA et al (2004) Interactions among dissolved organic carbon, microbial processes, and community structure in the mesopelagic zone of the northwestern Sargasso Sea. Limnol Oceanogr 49:1073–1083

Carlson CA, Del Giorgio PA, Herndl GJ (2007) Microbes and the dissipation of energy and respiration: from cells to ecosystems. Oceanography 20:89–100

Carritt DE (1954) Atmospheric pressure changes and gas solubility. Deep-Sea Res 2:59–62

Chang FH, Zeldis J, Gall M et al (2003) Seasonal and spatial variation of phytoplankton assemblages, biomass and cell size from spring to summer across the north-eastern New Zealand continental shelf. J Plankton Res 25:737–758

Chapman P (1986) Nutrient cycling in marine ecosystems. J Limnol Soc S Afr 12:22–42

Charlson RJ, Lovelock JE, Andreae MO et al (1987) Oceanic phytoplankton, atmospheric sulfur, cloud albedo and climate. Nature 326:655–661

Chavez FP, Messié M, Pennington JT (2011) Marine primary production in relation to climate variability and change. Annu Rev Mar Sci 3:227–260

Chester R, Jickells T (2012) Marine geochemistry. Wiley, Chichester

Chisholm SW (1992) Phytoplankton size. In: Falkowski PG, Woodhead AD, Vivirito K (eds) Primary productivity and biogeochemical cycles in the sea. Springer US, Boston, pp 213–237

Cloern JE (1996) Phytoplankton bloom dynamics in coastal ecosystems: a review with some general lessons from sustained investigation of San Francisco Bay, California. Rev Geophys 34:127–168

Cloern JE, Dufford R (2005) Phytoplankton community ecology: principles applied in San Francisco Bay. Mar Ecol Prog Ser 285:11–28

Cloern JE, Jassby AD (2010) Patterns and scales of phytoplankton variability in estuarine-coastal ecosystems. Estuar Coast 33:230–241

Cole JJ, Findlay S, Pace ML (1988) Bacterial production in fresh and saltwater ecosystems – a cross-system overview. Mar Ecol Prog Ser 43:1–10

Conan P, Sondergaard M, Kragh T et al (2007) Partitioning of organic production in marine plankton communities: the effects of inorganic nutrient ratios and community composition on new dissolved organic matter. Limnol Oceanogr 52:753–765

Conley DJ, Schelske CL, Stoermer EF (1993) Modification of the biogeochemical cycle of silica with eutrophication. Mar Ecol Prog Ser 101:179–192

Cotner JB, Wetzel RG (1992) Uptake of dissolved inorganic and organic phosphorus compounds by phytoplankton and bacterioplankton. Limnol Oceanogr 37:232–243

Cude WN, Mooney J, Tavanaei AA et al (2012) Production of the antimicrobial secondary metabolite indigoidine contributes to competitive surface colonization by the marine roseobacter *Phaeobacter* sp. strain Y4I. Appl Environ Microbiol 78:4771–4780

Danger M, Leflaive J, Oumarou C et al (2007) Control of phytoplankton-bacteria interactions by stoichiometric constraints. Oikos 116:1079–1086

DeLong EF, Franks DG, Alldredge AL (1993) Phylogenetic diversity of aggregate-attached vs. free-living marine bacterial assemblages. Limnol Oceanogr 38:924–934

Devol AH (2003) Nitrogen cycle: solution to a marine mystery. Nature 422:575

Dittmar T (2015) Chapter 7 – Reasons behind the long-term stability of dissolved organic matter. In: Hansell DA, Carlson CA (eds) Biogeochemistry of marine dissolved organic matter. Academic Press, San Diego, pp 369–388

Dittmar T, Paeng J (2009) A heat-induced molecular signature in marine dissolved organic matter. Nat Geosci 2:175–179

Dittmar T, Stubbins A (2014) Dissolved organic matter in aquatic systems. In: Birrer B, Falkowski P, Freemann K (eds) Treatise of geochemistry, 2nd edn. Elsevier, Oxford, pp 125–156

Dobson M, Frid C (2008) Ecology of aquatic systems. Oxford University Press, Oxford

Druffel ERM, Bauer JE, Williams PM et al (1996) Seasonal variability of particulate organic radiocarbon in the northeast Pacific ocean. J Geophys Res-Oceans 101:20543–20552

Ducklow HW, Kirchman DL, Quinby HL et al (1993) Stocks and dynamics of bacterioplankton carbon during the spring bloom in the eastern North Atlantic Ocean. Deep-Sea Res Pt II 40:245–263

Dyhrman ST, Ammerman JW, van Mooy BAS (2007) Microbes and the marine phosphorus cycle. Oceanography 20:110–116

Egge JK, Aksnes DL (1992) Silicate as regulating nutrient in phytoplankton competition. Mar Ecol Prog Ser 83:281–289

Elser JJ, Hassett RP (1994) A stoichiometric analysis of the zooplankton–phytoplankton interaction in marine and freshwater ecosystems. Nature 370:211–213

Eppley RW, Peterson BJ (1979) Particulate organic-matter flux and planktonic new production in the deep ocean. Nature 282:677–680

Evans M, Buchanan BB, Arnon DI (1966) A new ferredoxin-dependent carbon reduction cycle in a photosynthetic bacterium. Proc Natl Acad Sci USA 55:928–934

Falkowski PG, Raven R (2007) Aquatic photosynthesis. Princeton University Press, Princeton

Falkowski PG, Barber RT, Smetacek V (1998) Biogeochemical controls and feedbacks on ocean primary production. Science 281:200–206

Falkowski PG, Fenchel T, Delong EF (2008) The microbial engines that drive Earth's biogeochemical cycles. Science 320:1034–1039

Fiedler PC, Talley LD (2006) Hydrography of the eastern tropical Pacific: a review. Prog Oceanogr 69:143–180

Field CB, Behrenfeld MJ, Randerson JT et al (1998) Primary production of the biosphere: integrating terrestrial and oceanic components. Science 281:237–240

First MR, Hollibaugh JT (2009) The model high molecular weight DOC compound, dextran, is ingested by the benthic ciliate *Uronema marinum* but does not supplement ciliate growth. Aquat Microb Ecol 57:79–87

Fleming RH (1940) The composition of plankton and units for reporting population and production. Proc Pacif Sci Congr 3:535–540

Flombaum P, Gallegos JL, Gordillo RA et al (2013) Present and future global distributions of the marine Cyanobacteria *Prochlorococcus* and *Synechococcus*. Proc Natl Acad Sci USA 110:9824–9829

Flynn KJ, Stocker DK, Mitra A et al (2013) Misuse of the phytoplankton-zooplankton dichotomy: the need to assign organisms as mixotrophs within plankton functional types. J Plankton Res 35:3–11

Fogg GE, Nalewajko C, Watt WD (1965) Extracellular products of phytoplankton photosynthesis. Proc R Soc B Biol Sci 162:517–534

Foster RA, Kuypers MMM, Vagner T et al (2011) Nitrogen fixation and transfer in open ocean diatom-cyanobacterial symbiosis. ISME J 5:1484–1493

Fowler D, Coyle M, Skiba U et al (2013) The global nitrogen cycle in the twenty-first century. Philos Trans R Soc B 368:20130164

Francis CA, Tebo BM (2001) *cumA* multicopper oxidase genes from diverse Mn(II)-oxidizing and non-Mn(II)-oxidizing *Pseudomonas* strains. Appl Environ Microbiol 67:4272–4278

Fuhrman JA (1999) Marine viruses and their biogeochemical and ecological effects. Nature 399:541–548

Fuhrman JA, McCallum K, Davis AA (1992) Novel major archaebacterial group from marine plankton. Nature 356:148–149

Fuhrman JA, Steele JA, Hewson I et al (2008) A latitudinal diversity gradient in planktonic marine bacteria. Proc Natl Acad Sci USA 105:7774–7778

Gallon JR (1981) The oxygen sensitivity of nitrogenase: a problem for biochemists and micro-organisms. Trends Biochem Sci 6:19–23

Giovannoni SJ, Thrash JC, Temperton B (2014) Implications of streamlining theory for microbial ecology. ISME J 8:1553–1565

Gobler CJ, Hutchins DA, Fisher NS et al (1997) Release and bioavailability of C, N, P Se, and Fe following viral lysis of a marine chrysophyte. Limnol Oceanogr 42:1492–1504

Gonsior M, Hertkorn N, Conte MH et al (2014) Photochemical production of polyols arising from significant photo-transformation of dissolved organic matter in the oligotrophic surface ocean. Mar Chem 163:10–18

Grossart HP, Czub G, Simon M (2006) Algae-bacteria interactions and their effects on aggregation and organic matter flux in the sea. Environ Microbiol 8:1074–1084

Gruber N (2008) Chapter 1 – The marine nitrogen cycle: overview and challenges. In: Capone DG, Bronk DA, Mulholland MR et al (eds) Nitrogen in the marine environment, 2nd edn. Academic Press, San Diego, pp 1–50

Guidi L, Chaffron S, Bittner L et al (2016) Plankton networks driving carbon export in the oligotrophic ocean. Nature 532:465–470

Hansell DA (2013) Recalcitrant dissolved organic carbon fractions. Annu Rev Mar Sci 5:421–445

Hansell DA, Carlson CA (2001) Marine dissolved organic matter and the carbon cycle. Oceanography 14:41–49

Hansell DA, Carlson CA (2013) Localized refractory dissolved organic carbon sinks in the deep ocean. Global Biogeochem Cycles 27:705–710

Hansell DA, Carlson CA, Repeta DJ et al (2009) Dissolved organic matter in the ocean a controversy stimulates new insights. Oceanography 22:202–211

Hansell DA, Carlson CA, Schlitzer R (2012) Net removal of major marine dissolved organic carbon fractions in the subsurface ocean. Global Biogeochem Cycles 26:GB1016

Hawkes JA, Rossel PE, Stubbins A et al (2015) Deep-ocean dissolved organic matter in hydrothermal vents. Nat Geosci 8:856–860

Hederstedt L (1993) The Krebs citric acid cycle. In: Sonenshein A, Hoch J, Losick R (eds) *Bacillus subtilis* and other gram-positive bacteria. ASM Press, Washington, DC, pp 181–197

Hedges JI (1992) Global biogeochemical cycles: progress and problems. Mar Chem 39:67–93

Heisler J, Glibert PM, Burkholder JM et al (2008) Eutrophication and harmful algal blooms: a scientific consensus. Harmful Algae 8:3–13

Hertkorn N, Harir M, Koch BP et al (2013) High-field NMR spectroscopy and FT-ICR mass spectrometry: powerful discovery tools for the molecular level characterization of marine dissolved organic matter. Biogeosciences 10:1583–1624

Hillebrand H, Sommer U (1999) The nutrient stoichiometry of benthic microalgal growth: redfield proportions are optimal. Limnol Oceanogr 44:440–446

Hjuler ML, Fabricius IL (2009) Engineering properties of chalk related to diagenetic variations of Upper Cretaceous onshore and offshore chalk in the North Sea area. J Pet Sci Eng 68:151–170

Ho TY, Quigg A, Finkel Z et al (2004) The elemental composition of some marine phytoplankton. J Phycol 39:1145–1159

Hopkinson CS Jr, Vallino JJ (2005) Efficient export of carbon to the deep ocean through dissolved organic matter. Nature 433:142–145

Hünken M, Harder J, Kirst GO (2008) Epiphytic bacteria on the Antarctic ice diatom *Amphiprora kufferathii* Manguin cleave hydrogen peroxide produced during algal photosynthesis. Plant Biol 10:519–526

Hurwitz BL, U'Ren JM (2016) Viral metabolic reprogramming in marine ecosystems. Curr Opin Microbiol 31:161–168

Hutchins DA, Witter AE, Butler A et al (1999) Competition among marine phytoplankton for different chelated iron species. Nature 400:858–861

Jiao N, Herndl GJ, Hansell DA et al (2010) Microbial production of recalcitrant dissolved organic matter: long-term carbon storage in the global ocean. Nat Rev Microbiol 2010:593–599

Kahru M, Brotas V, Manzano-Sarabia M et al (2011) Are phytoplankton blooms occurring earlier in the Arctic? Glob Chang Biol 17:1733–1739

Karl DM, Bjoerkman K, Dore EJ et al (2001) Ecological nitrogen-to-phosphorus stoichiometry at Station ALOHA. Deep-Sea Res II 48:1529–1566

Karl DM, Michaels A, Bergman B et al (2002) Dinitrogen fixation in the world's oceans. In: Boyer EW, Howarth RW (eds) The nitrogen cycle at regional to global scales. Springer, Dordrecht, pp 47–98

Karner MB, DeLong EF, Karl DM (2001) Archaeal dominance in the mesopelagic zone of the Pacific Ocean. Nature 409:507–510

Kastner M, Keene JB, Gieskes JM (1977) Diagenesis of siliceous oozes. Chemical controls on the rate of opal-A to opal-CT transformation – an experimental study. Geochim Cosmochim Acta 41:1041–1059

Kazamia E, Czesnick H, Nguyen TT et al (2012) Mutualistic interactions between vitamin B12-dependent algae and heterotrophic bacteria exhibit regulation. Environ Microbiol 14:1466–1476

Kieber DJ, McDaniel J, Mopper K (1989) Photochemical source of biological substrates in sea-water – implications for carbon cycling. Nature 341:637–639

Kieber DJ, Keene WC, Frossard AA et al (2016) Coupled ocean-atmosphere loss of marine refractory dissolved organic carbon. Geophys Res Lett 43:2765–2772

Koch BP, Witt MR, Engbrodt R et al (2005) Molecular formulae of marine and terrigenous dissolved organic matter detected by electrospray ionization Fourier transform ion cyclotron resonance mass spectrometry. Geochim Cosmochim Acta 69:3399–3308

Könneke M, Bernhard AE, de la Torre JR et al (2005) Isolation of an autotrophic ammonia-oxidizing marine archaeon. Nature 437:543–546

Kovarova-Kovar K, Egli T (1998) Growth kinetics of suspended microbial cells: from single-substrate-controlled growth to mixed-substrate kinetics. Microbiol Mol Biol Rev 62:646–666

Kujawinksi EB, Longnecker K, Barott KL et al (2016) Microbial community structure affects marine dissolved organic matter composition. Front Mar Sci 3:45

Kujawinski EB (2011) The impact of microbial metabolism on marine dissolved organic matter. Annu Rev Mar Sci 3:567–599

Landa M, Blain S, Christaki U et al (2016) Shifts in bacterial community composition associated with increased carbon cycling in a mosaic of phytoplankton blooms. ISME J 10:39–50

Landry MR, Kirchman DL (2002) Microbial community structure and variability in the tropical Pacific. Deep-Sea Res II 49:2669–2693

Lang SQ, Butterfield DA, Lilley MD et al (2006) Dissolved organic carbon in ridge-axis and ridge-flank hydrothermal systems. Geochim Cosmochim Acta 70:3830–3842

LaRowe DE, Dale AW, Amend JP et al (2012) Thermodynamic limitations on microbially catalyzed reaction rates. Geochim Cosmochim Acta 90:96–109

Lauro FM, McDougald D, Thomas T et al (2009) The genomic basis of trophic strategy in marine bacteria. Proc Natl Acad Sci USA 106:15527–15533

Lechtenfeld OJ, Hertkorn N, Shen Y et al (2015) Marine sequestration of carbon in bacterial metabolites. Nat Commun 6:6711

Lessard E, Merico A, Tyrrell T (2005) Nitrate: phosphate ratios and *Emiliania huxleyi* blooms. Limnol Oceanogr 50:1020–1024

Lima-Mendez G, Faust K, Henry N et al (2015) Top-down determinants of communitiy structure in the global plankton interactome. Science 348:6237

Liss P, Malin G, Turner SM et al (1994) Dimethyl sulphide and *Phaeocystis*: a review. J Mar Sci 5:41–53

Litchman E, Klausmeier CA, Miller JR et al (2006) Multi-nutrient, multi-group model of present and future oceanic phytoplankton communities. Biogeosciences 3:585–606

Loeder M, Meunier C, Wiltshire K et al (2011) The role of ciliates, heterotrophic dinoflagellates and copepods in structuring spring plankton communities at Helgoland Roads, North Sea. Mar Biol 158:1551–1580

Logue JB, Stedmon CA, Kellerman AM et al (2016) Experimental insights into the importance of aquatic bacterial community composition to the degradation of dissolved organic matter. ISME J 10:533–545

Longhurst A, Sathyendranath S, Platt T et al (1995) An estimate of global primary production in the ocean from satellite radiometer data. J Plankton Res 17:1245–1271

Longnecker K, Kido Soule MC, Kujawinski EB (2015) Dissolved organic matter produced by *Thalassiosira pseudonana*. Mar Chem 168:114–123

Mague TH, Friberg E, Hughes DJ et al (1980) Extracellular release of carbon by marine phytoplankton – a physiological approach. Limnol Oceanogr 25:262–279

Maldonado MT, Price NM (2001) Reduction and transport of organically bound iron by *Thalassiosira oceanica* (Bacillariophyceae). J Phycol 37:298–310

Maranon E, Cermeno P, Perez V (2005) Continuity in the photosynthetic production of dissolved organic carbon from eutrophic to oligotrophic waters. Mar Ecol Prog Ser 299:7–17

Martin MO (2002) Predatory prokaryotes: an emerging research opportunity. J Mol Microb Biotechnol 4:467–477

Martiny JB, Riemann L, Marston MF et al (2014) Antagonistic coevolution of marine planktonic viruses and their hosts. Annu Rev Mar Sci 6:393–414

Medeiros PM, Seidel M, Powers LC et al (2015a) Dissolved organic matter composition and photochemical transformations in the northern North Pacific Ocean. Geophys Res Lett 42:863–870

Medeiros PM, Seidel M, Ward ND et al (2015b) Fate of the Amazon River dissolved organic matter in the tropical Atlantic Ocean. Global Biogeochem Cycles 29:677–690

Metaxas A, Scheibling R (1996) Top-down and bottom-up regulation of phytoplankton assemblages in tidepools. Mar Ecol Prog Ser 145:161–177

Michaels A, Karl DM, Capone D (2001) Element stoichiometry, new production and nitrogen fixation. Oceanography 14:68–77

Michelou VK, Cottrell MT, Kirchman DL (2007) Light-stimulated bacterial production and amino acid assimilation by cyanobacteria and other microbes in the North Atlantic Ocean. Appl Environ Microbiol 73:5539–5546

Middelboe M (2008) Microbial disease in the sea: effects of viruses on carbon and nutrient cycling Infections Disease Ecology. Princeton University Press, Princeton, pp 242–260

Middelboe M, Lyck PG (2002) Regeneration of dissolved organic matter by viral lysis in marine microbial communities. Aquat Microb Ecol 27:187–194

Moffett JW, Brand LE (1996) Production of strong, extracellular Cu chelators by marine cyanobacteria in response to Cu stress. Limnol Oceanogr 41:388–395

Moore WS (2010) The effect of submarine groundwater discharge on the ocean. Annu Rev Mar Sci 2:59–88

Moore LR, Goericke R, Chisholm SW (1995) Comparative physiology of *Synechococcus* and *Prochlorococcus*: influence of light and temperature on growth, pigments, fluorescence and absorptive properties. Mar Ecol Prog Ser 116:259–275

Mopper K, Zhou XL, Kieber RJ et al (1991) Photochemical degradation of dissolved organic carbon and its impact on the oceanic carbon cycle. Nature 353:60–62

Moran MA, Reisch C, Kiene R et al (2012) Genomic insights into bacterial DMSP transformations. Annu Rev Mar Sci 4:523–542

Morel FMM (1986) Trace metals – phytoplankton interactions: an overview. Elsev Oceanogr Serie 43:177–189

Morel FMM (2008) The co-evolution of phytoplankton and trace element cycles in the oceans. Geobiology 6:318–324

Morel FMM, Price N (2003) The biogeochemical cycles of trace metals in the oceans. Science 300:944–947

Moreno AR, Martiny AC (2018) Ecological stoichiometry of ocean plankton. Annu Rev Mar Sci 10:43–69

Myklestad SM (1974) Production of carbohydrates by marine planktonic diatoms. 1. Comparison of 9 different species in culture. J Exp Mar Biol Ecol 15:261–274

Myklestad SM (2000) Dissolved organic carbon from phytoplankton. In: Wangersky PJ (ed) Marine chemistry. Springer, Berlin, pp 111–148

Nagata T, Kirchman D (1992) Release of dissolved organic matter by heterotrophic protozoa: Implications for microbial food webs. Arch Hydrobiol 35:99–109

Nelson DM, Tréguer P, Brzezinski A et al (1995) Production and dissolution of biogenic silica in the ocean: revised global estimates, comparison with regional data and relationship to biogenic sedimentation. Global Biogeochem Cycles 9:359–732

Nicholson D, Dyhrman S, Chavez F et al (2006) Alkaline phosphatase activity in the phytoplankton communities of Monterey Bay and San Francisco Bay. Limnol Oceanogr 51:874–883

Odum EP (1959) Fundamentals of ecology, vol 2. WB Saunders Company, Philadelphia

Ogawa H, Tanoue E (2003) Dissolved organic matter in oceanic waters. J Oceanogr 59:129–147

Ogawa H, Amagai Y, Koike I et al (2001) Production of refractory dissolved organic matter by bacteria. Science 292:917–920

Osterholz H, Niggemann J, Giebel H-A et al (2015) Inefficient microbial production of refractory dissolved organic matter in the ocean. Nat Commun 6:7422

Partensky F, Blanchot J, Vaulot D (1999) Differential distribution and ecology of *Prochlorococcus* and *Synechococcus* in oceanic waters: a review. In: Loic C, Larkum AWD (eds) Marine cyanobacteria, Bulletin-Insitut Oceanographique de Monaco, Paris, pp 457–475

Passow U (2002) Production of transparent exopolymer particles (TEP) by phyto-and bacterioplankton. Mar Ecol Prog Ser 236:1–12

Paul JH (2008) Prophages in marine bacteria: dangerous molecular time bombs or the key to survival in the seas? ISME J 2:579–589

Paytan A, McLaughlin K (2007) The oceanic phosphorus cycle. Chem Rev 107:563–576

Pennington JT, Mahoney KL, Kuwahara VS et al (2006) Primary production in the eastern tropical Pacific: a review. Prog Oceanogr 69:285–317

Pernthaler A, Pernthaler J, Amann R (2002a) Fluorescence in situ hybridization and catalyzed reporter deposition for the identification of marine bacteria. Appl Environ Microbiol 68:3094–3101

Pernthaler A, Pernthaler J, Schattenhofer M et al (2002b) Identification of DNA-synthesizing bacterial cells in coastal North Sea plankton. Appl Environ Microbiol 68:5728–5736

Pomeroy LR (1974) The ocean's food web, a changing paradigm. Bioscience 24:499–504

Pomeroy LR, Williams PJ, Azam F et al (2007) The microbial loop. Oceanography 20:28–33

Poorvin L, Rinta-Kanto JM, Hutchins DA et al (2004) Viral release of iron and its bioavailability to marine plankton. Limnol Oceanogr 49:1734–1741

Quigg A, Finkel Z, Irwin A et al (2003) The evolutionary inheritance of elemental stoichiometry in marine phytoplankton. Nature 425:291–294

Ragueneau O, Tréguer P, Leynaert A et al (2000) A review of the Si cycle in the modern ocean: recent progress and missing gaps in the application of biogenic opal as a paleoproductivity proxy. Glob Planet Chang 26:317–365

Redfield AC (1958) The biological control of chemical factors in the environment. Am Sci 46:205–221

Redfield AC, Ketchum BH, Richards FA (1963) The Influence of organisms on the composition of sea-water. In: Hill MN (ed) The composition of seawater: comparative and descriptive oceanography. The sea: ideas and observations on progress in the study of the sea, vol 2. Wiley Interscience, New York, p 26–77

Repeta DJ (2015) Chapter 2 – Chemical characterization and cycling of dissolved organic matter. In: Hansell DA, Carlson CA (eds) Biogeochemistry of marine dissolved organic matter, 2nd edn. Academic Press, Boston, pp 21–63

Reynolds CS (1984) The ecology of freshwater phytoplankton. Cambridge University Press, Cambridge

Reynolds CS (2006) Ecology of phytoplankton. Cambridge University Press, Cambridge

Riebesell U (1991) Particle aggregation during a diatom bloom. II. Biological aspects. Mar Ecol Prog Ser 69:281–291

Riemann L, Steward G, Azam F (2000) Dynamics of bacterial community composition and activity during a mesocosm diatom bloom. Appl Environ Microbiol 66:578–587

Rink B, Seeberger S, Martens T et al (2007) Effects of phytoplankton bloom in a coastal ecosystem on the composition of bacterial communities. Aquat Microb Ecol 48:47–60

Rivkin RB, Legendre L (2001) Biogenic carbon cycling in the upper ocean: effects of microbial respiration. Science 291:2398–2400

Rohwer F, Thurber RV (2009) Viruses manipulate the marine environment. Nature 459:207–212

Rossel PE, Vähätalo AV, Witt M et al (2013) Molecular composition of dissolved organic matter from a wetland plant (*Juncus effusus*) after photochemical and microbial decomposition (1.25 years): common features with deep sea dissolved organic matter. Org Geochem 60:62–71

Rossel PE, Stubbins A, Hach P et al (2015) Bioavailability and molecular composition of dissolved organic matter from a diffuse hydrothermal system. Mar Chem 177:257–266

Ruttenberg KC, Berner RA (1993) Authigenic apatite formation and burial in sediments from non-upwelling, continental margin environments. Geochim Cosmochim Acta 57:991–1007

Saito MA, Moffett JW, Chisholm SW et al (2002) Cobalt limitation and uptake in *Prochlorococcus*. Limnol Oceanogr 47:1629–1636

Schoemann V, Becquevort S, Stefels J et al (2005) *Phaeocystis* blooms in the global ocean and their controlling mechanisms: a review. J Sea Res 53:43–66

Seyedsayamdost MR, Case RJ, Kolter R et al (2011) The Jekyll-and-Hyde chemistry of *Phaeobacter gallaeciensis*. Nat Chem 3:331–335

Seymour JR, Amin SA, Raina JB et al (2017) Zooming in on the phycosphere: the ecological interface for phytoplankton-bacteria relationships. Nat Microbiol 30:17065

Shelford EJ, Suttle CA (2018) Virus-mediated transfer of nitrogen from heterotrophic bacteria to phytoplankton. Biogeosciences 15:809–819

Short SM (2012) The ecology of viruses that infect eukaryotic algae. Environ Microbiol 14:2253–2271

Siegenthaler U, Sarmiento J (1993) Atmospheric carbon dioxide and the ocean. Nature 365:119–125

Simon M, Azam F (1989) Protein content and protein synthesis rates of planktonic marine bacteria. Mar Ecol Prog Ser 51:201–213

Simon N, Cras AL, Foulon E et al (2009) Diversity and evolution of marine phytoplankton. C R Biol 332:159–170

Smayda TJ (1997) Harmful algal blooms: their ecophysiology and general relevance to phytoplankton blooms in the sea. Limnol Oceanogr 42:1137–1153

Smayda TJ, Reynolds CS (2003) Strategies of marine dinoflagellate survival and some rules of assembly. J Sea Res 49:95–106

Smetacek V (1985) Role of sinking in diatom life-history cycles: ecological, evolutionary and geological significance. Mar Biol 84:239–251

Smetacek V (1999) Diatoms and the ocean carbon cycle. Protist 150:25–32

Smetacek V, Cloern J (2008) Oceans – on phytoplankton trends. Science 319:1346–1348

Smith DC, Simon M, Alldredge AL et al (1992) Intense hydrolytic enzyme-activity on marine aggregates and implications for rapid particle dissolution. Nature 359:139–142

Sogin ML, Morrison HG, Huber JA et al (2006) Microbial diversity in the deep sea and the underexplored "rare biosphere". Proc Natl Acad Sci USA 103:12115–12120

Solomon S, Qin D, Manning M et al (2007) Climate change 2007-the physical science basis: working group I contribution to the fourth assessment report of the IPCC. Cambridge University Press, New York

Sommer U, Lengfellner K (2008) Climate change and the timing, magnitude, and composition of the phytoplankton spring blooms. Glob Change Biol 14:1199–1208

Søreide JE, Leu E, Berge J et al (2010) Timing of blooms, algal food quality and Calanus glacialis reproduction and growth in a changing Arctic. Glob Change Biol 16:3154–3163

Sournia A, Chrdtiennot-Dinet MJ, Ricard M (1991) Marine phytoplankton: how many species in the world ocean? J Plankton Res 13:1093–1099

Stocker R (2012) Marine microbes see a sea of gradients. Science 338:628–633

Stubbins A, Spencer RGM, Chen H et al (2010) Illuminated darkness: molecular signatures of Congo River dissolved organic matter and its photochemical alteration as revealed by ultrahigh precision mass spectrometry. Limnol Oceanogr 55:1467–1477

Sullivan MB, Coleman ML, Weigele P et al (2005) Three *Prochlorococcus* cyanophage genomes: signature features and ecological interpretations. PLoS Biol 3:e144

Sunagawa S, Coelho LP, Chaffron S et al (2015) Structure and function of the global ocean microbiome. Science 348:261359

Sunda WG (1989) Trace metal interactions with marine phytoplankton. Biol Oceanogr 6:411–442

Sunda WG (1994) Trace metal/phytoplankton interactions in the sea. In: Bidoglio G, Stumm W (eds) Chemistry of aquatic systems: local and global perspectives. Springer, Dordrecht, pp 213–247

Sunda WG (2012) Feedback interactions between trace metal nutrients and phytoplankton in the ocean. Front Microbiol 3:204

Suttle CA (2007) Marine viruses—major players in the global ecosystem. Nat Rev Microbiol 5:801–812

Taylor JR, Ferrari R (2011) Shutdown of turbulent convection as a new criterion for the onset of spring phytoplankton blooms. Limnol Oceanogr 56:2293–2307

Tebo BM, Bargar JR, Clement BG et al (2004) Biogenic manganese oxides: properties and mechanisms of formation. Annu Rev Earth Planet Sci 32:287–328

Teeling H, Fuchs BM, Becher D et al (2012) Substrate-controlled succession of marine bacterioplankton populations induced by a phytoplankton bloom. Science 336:608–611

Teeling H, Fuchs BM, Bennke CM et al (2016) Recurring patterns in bacterioplankton dynamics during coastal spring algae blooms. elife 5:e11888

Thingstad TF (2000) Elements of a theory for the mechanisms controlling abundance, diversity, and biogeochemical role of lytic bacterial viruses in aquatic systems. Limnol Oceanogr 45:1320–1328

Thingstad TF, Lignell R (1997) Theoretical models for the control of bacterial growth rate, abundance, diversity and carbon demand. Aquat Microb Ecol 13:19–27

Thornton DCO (2014) Dissolved organic matter (DOM) release by phytoplankton in the contemporary and future ocean. Eur J Phycol 49:20–46

Tringe SG, Rubin EM (2005) Metagenomics: DNA sequencing of environmental samples. Nat Rev Genet 6:805–814

Twining BS, Baines SB (2013) The trace metal composition of marine phytoplankton. Annu Rev Mar Sci 5:191–215

Volk T, Hoffert MI (1985) Ocean carbon pumps: Analysis of relative strengths and efficiencies in ocean driven atmospheric CO_2 changes. Geophys Monogr Ser 32:99–110

Weinbauer MG (2004) Ecology of prokaryotic viruses. FEMS Microbiol Rev 28:127–181

Weinbauer MG, Rassoulzadegan F (2004) Are viruses driving microbial diversification and diversity? Environ Microbiol 6:1–11

Weiss MS, Abele U, Weckesser J et al (1991) Molecular architecture and electrostatic properties of a bacterial porin. Science 254:1627–1630

Weitz JS, Stock CA, Wilhelm SW et al (2015) A multitrophic model to quantify the effects of marine viruses on microbial food webs and ecosystem processes. ISME J 9:1352–1364

Westbroek P, Young JR, Linschooten K (1989) Coccolith production (biomineralization) in the marine alga *Emiliania huxleyi*. J Protozool 36:368–373

Wetz MS, Wheeler PA (2007) Release of dissolved organic matter by coastal diatoms. Limnol Oceanogr 52:798–807

Weynberg KD (2018) Viruses in marine ecosystems: From open waters to coral reefs. Adv Virus Res 101:1–38

Whitman WB, Coleman DC, Wiebe WJ (1998) Prokaryotes: the unseen majority. Proc Natl Acad Sci USA 95:6578–6583

Wilhelm SW, Suttle CA (1999) Viruses and nutrient cycles in the sea: viruses play critical roles in the structure and function of aquatic food webs. Bioscience 49:781–788

Williams PM, Druffel ERM (1987) Radiocarbon in dissolven organic-matter in the central North Pacific Ocean. Nature 330:246–248

Williams LA, Parks GA, Crerar DA (1985) Silica diagenesis. Solubility controls. J Sediment Res 55:301–311

Wilson WH, Schroeder DC, Allen MJ et al (2005) Complete genome sequence and lytic phase transcription profile of a Coccolithovirus. Science 309:1090–1092

Winter C, Bouvier T, Weinbauer MG et al (2010) Trade-offs between competition and defense specialists among unicellular planktonic organisms: the 'killing the winnter' hypothesis revisited. Microbiol Mol Biol Rev 74:42–57

Wommack KE, Colwell RR (2000) Virioplankton: viruses in aquatic ecosystems. Microbiol Mol Biol Rev 64:69–114

Yager P, Connelly T, Mortazavi B et al (2001) Dynamic bacterial and viral response to an algal bloom at subzero temperatures. Limnol Oceanogr 46:90–801

Zark M, Christoffers J, Dittmar T (2017) Molecular properties of deep-sea dissolved organic matter are predictable by the central limit theorem: evidence from tandem FT-ICR-MS. Mar Chem 191:9–15

Appendixes

Appendix 1: List of Conference Participants

Listed are participants who agreed to be listed.

Last name	First name	Presentation type/function
Aepfler	Rebecca Felicitas	Oral Presentation
Agyekum	Michael K.	Oral Presentation
Ahrens	Janis	Listener
Alex	Antje S.	Oral Presentation
Allan	Nicola	Oral Presentation
Al-Saadi	Athraa	Oral Presentation
Amargant-Arumí	Martí	Oral Presentation
Arts	Milou G.I.	Oral Presentation
Baali	Ayoub	Oral Presentation
Bank	Rose M.	Poster Presentation
Barba-Herrera	Sonia	Oral Presentation
Barrett	Chris J.	Session Host, Oral Presentation
Barz	Fanny	Session Host
Basconi	Laura	Session Host
Beck	Kristina K.	Oral Presentation
Benkens	Andreas	Listener
Bertoni	Matteo	Poster Presentation
Bharadwaj Tatipamula	Vinay	Oral Presentation
Bick	Berenike	Listener
Blomenkamp	Carolina	Listener
Boelmann	Jan	Listener
Bonzi	Lucrezia C.	Oral Presentation
Bourceau	Patric	Oral Presentation
Brünjes	Jonas	Oral Presentation
Brüwer	Jan	Organization Team
Bunse	Carina	Invited Oral Presentation
Burgoa	Javier	Listener
Burgoa Cardas	Javier	Oral Presentation
Buschbaum	Christian	Invited Oral Presentation
Cadier	Charles	Organization Team, Session Host
Cerbule	Kristine	Poster Presentation
Cheronet	Alexandrine	Workshop
Choquet	Marvin	Invited Oral Presentation
Clay	Megan	Oral Presentation
Curdt	Franziska	Poster Presentation

(continued)

Last name	First name	Presentation type/function
de Goeij	Jasper M.	Invited Oral Presentation
Degenhardt	Julius M.	Session Host
Déniel	Maureen	Oral Presentation
Dibke	Christopher	Poster Presentation
Diener	Stefan	Listener
Dietrich	Ulrike	Session Host
Düsedau	Luisa	Listener
Earp	Hannah S.	Organization Team, Poster Presentation
Eichsteller	Angelina	Listener
Elías Ilosvay	Xochitl E.	Oral Presentation
El-Khaled	Yusuf C.	Oral Presentation
Emrich	Maria	Listener
Engel	Julian	Oral Presentation
Esser	Ferdinand	Organization Team
Fahning	Jana	Industry Representative
Faisal	Muhammad R.	Organization Team, Poster Presentation
Ferse	Sebastian C.A.	Invited Oral Presentation
Fischer	Marten	Session Host
Folkers	Mainah	Session Host
Fontán Alende	Elena	Oral Presentation
Gajendra	Niroshan	Oral Presentation
Gamaza	MariAngeles	Oral Presentation
Gaviria Lugo	Nestor	Poster Presentation
Gerlach	Nadine	Oral Presentation
Glöckner	Frank-Oliver	Oral Presentation
Götz	Lisa	Industry Representative
Granger	Audrey J.	Oral Presentation
Greife	Anna	Listener
Grist	Hannah	Invited Oral Presentation
Guerrero Limon	Gustavo	Session Host
Gustavs	Lydia	Workshop
Gutsfeld	Sebastian	Oral Presentation
Hagan	James G.	Session Host
Halbach	Maurits	Session Host
Hammerl	Constanze	Poster Presentation
Hara	Jenevieve P.	Poster Presentation
Hartmann	Daniel	Workshop
Heel	Lena	Organization Team
Heidenreich	Marie	Workshop
Heidrich	Kristina	Oral Presentation
Heinrichs	Mara	Session Host

(continued)

S. Jungblut et al. (eds.), *YOUMARES 9 - The Oceans: Our Research, Our Future*,
https://doi.org/10.1007/978-3-030-20389-4

Last name	First name	Presentation type/function
Helten	Oliver	Poster Presentation
Hennings	Laura	Organization Team
Hewitt	Olivia H.	Oral Presentation
Heyen	Simone	Oral Presentation
Hildebrandt	Lars	Oral Presentation
Hintz	Nils Hendrik	Oral Presentation
Hohensee	Dorothee	Organization Team
Jabinski	Stanislav	Poster Presentation
Jones	Jessica B.	Poster Presentation
Josphat	Nguu	Poster Presentation
Jung	Julia	Oral Presentation
Jung	E. Maria U.	Oral Presentation
Jungblut	Simon	Organization Team
Jupe	Lucy	Oral Presentation
Kaiser	Patricia	Organization Team
Kamyab	Elham	Organization Team, Session Host
Kamyab	Elham	Session Host
Kappas	Lea	Listener
Karambelkar	Amruta	Oral Presentation
Karcher	Denis B.	Oral Presentation
Käse	Laura	Oral Presentation
Kingston	Dale	Oral Presentation
Kirchner	Julia	Listener
Kittu	Leila	Oral Presentation
Klose	Christina	Listener
Knoke	Melina	Oral Presentation
Koester	Anna	Oral Presentation
Konijnenberg	Rebecca	Invited Oral Presentation
Koopmann	Inga	Listener
Koptelova	Katerina	Poster Presentation
Korez	Špela	Session Host
Kornder	Niklas	Oral Presentation
Kraak	Sarah B.M.	Invited Oral Presentation
Kramer	Annemarie	Oral Presentation
Kriegl	Michael	Oral Presentation
Kunze	Charlotte	Organization Team
Lachs	Liam	Session Host
Laeseke	Philipp	Session Host
Leonida	Jerry Ian	Oral Presentation
Letschert	Jonas	Organization Team, Session Host
Li	Huiru	Oral Presentation
Liconti	Arianna	Organization Team, Oral Presentation
Liebich	Viola	Organization Team, Session Host
Lishchenko	Fedor	Invited Oral Presentation
Logemann	Anna	Listener
Luypaert	Thomas	Session Host
Ma	Cenling	Poster Presentation
Madeira	Carolina	Oral and Poster Presentation
Mańko	Maciej	Session Host
Marques	Márcia	Oral and Poster Presentation
Martinez-Hernandez	Benjamin	Oral Presentation
Mascarenhas	Velosia	Organization Team
McCarthy	Morgan L.	Session Host, Oral Presentation
McGavin	Katherine	Oral Presentation
Merlo	Guenda	Oral Presentation
Metcalf	Olivia	Poster Presentation
Meyer	Lara	Poster Presentation
Meyerjürgens	Jens	Oral Presentation
Mitschke	Nico	Listener
Moeller	Mareen	Oral Presentation
Mohar	Thomas	Industry Representative
Mönnich	Julian	Oral Presentation
Morales-de-Anda	Diana Elizabeth	Oral Presentation
Morales-Guadarrama	Adrían Andrés	Poster Presentation
Morganti	Teresa Maria	Oral Presentation
Mori	Corinna	Session Host, Oral Presentation
Moses	Sonya R.	Poster Presentation
Moye	Fabian	Oral Presentation
Mühlena	Lukas	Oral Presentation
Müller	Carolin	Oral Presentation
Nauen	Cornelia	Workshop
Nietzer	Samuel	Oral Presentation
Nugroho	Avianto	Oral Presentation
O'Mahony	Éadin	Oral Presentation
Oehler	Till	Invited Oral Presentation
Onate	Javier	Session Host
Orfanoudaki	Maria	Oral Presentation
Ort	Mara	Session Host
Paffrath	Ronja	Oral Presentation
Palecek	Dragana	Listener
Parrondo Lombardía	Marina	Oral Presentation
Pauli	Nora-Charlotte	Organization Team, Oral Presentation
Paulischkis	Eva	Poster Presentation
Petersen	Lars-Erik	Session Host
Pineda Metz	Santiago E.A.	Session Host
Pisternick	Timo	Oral Presentation
Poti	Meenakshi	Session Host, Oral Presentation
Prinz	Natalie	Session Host
Quitzau	Marita	Poster Presentation
Rayon-Viña	Fernando	Oral Presentation
Rebelein	Anja	Listener
Reckhardt	Anja	Listener
Rehse	Saskia	Invited Oral Presentation
Riesbeck	Sarah	Oral Presentation
Rodrigues Mateus	David José	Poster Presentation
Rogenhagen	Johannes	Industry Representative
Rohlfs	Nina	Listener
Rölfer	Lena	Organization Team
Rombouts	Titus	Session Host
Roscher	Lisa	Listener
Ross	Lukas	Organization Team
Ruigendijk	Ester	Vice President of University Oldenburg
Rupp	Ann-Sophie	Poster Presentation
Sadd	Daniel	Listener

(continued)

Last name	First name	Presentation type/function
Salinas Akhmadeeva	Irene Antonina	Oral Presentation
Schadewell	Yvonne	Organization Team, Workshop
Schätzle	Philipp-Konrad	Listener
Scheel	Maria	Oral Presentation
Schellenberg	Lisa	Listener
Schick	Veronika	Listener
Schiller	Jessica	Session Host
Schmitz	Jana	Listener
Schnier	Jannik	Listener
Schöneich-Argent	Rosanna	Session Host
Schröder	Tim	Workshop
Schumacher	Linda	Oral Presentation
Schwalfenbach	Kai	Poster Presentation
Schwanck	Tanja	Oral Presentation
Schwermer	Heike	Session Host
Scribano	Giovanni	Listener
Seibert	Stephan L.	Session Host
Senff	Paula	Organization Team, Oral Presentation
Sievers	Antje Frederike	Listener
Singh	Pradeep A.	Session Host
Sinnecker	Tristan	Industry Representative
Slaby	Beate M.	Oral Presentation
Smith Sanchez	Nicolas Juan	Oral Presentation
Sondej	Greta	Oral Presentation
Song	Min	Oral Presentation
Speidel	Linn	Listener
Stehouwer	Peter Paul	Listener
Steinberg	Ronny	Listener
Stévenne	Chloé	Oral Presentation
Stevens	Kevin	Oral Presentation
Suárez	Marcos	Oral Presentation
Suaria	Giuseppe	Oral Presentation
Turov	Polina	Invited Oral Presentation
Twesten	Merle	Oral Presentation
Txurruka Alberdi	Estibalitz	Poster Presentation
van der Sprong	Joëlle	Poster Presentation
Varma	Devika	Listener
Vogel	Christopher	Oral Presentation
Vonnahme	Tobias R.	Session Host
Wagner	Gretchen	Oral Presentation
Walczyńska	Katarzyna	Session Host
Walker	Zoe	Oral Presentation
Ware	Jessica	Oral Presentation
Welch	Amelia	Oral Presentation
Wibowo	Joko Tri	Organization Team, Oral Presentation
Willenbrink	Nils Tobias	Organization Team, Workshop
Winkels	Konrad	Oral Presentation
Wölfelschneider	Mirco	Organization Team
Wu	Yu-Chen	Oral Presentation
Zhu	Qingzeng	Poster Presentation
Zielinski	Oliver	Director of ICBM, University Oldenburg

Appendix 2: Conference Sessions and Abstracts

In the following appendix, the Calls for Abstracts and the abstracts of the oral and poster presentations of each session of YOUMARES 9 are listed. The appendix chapters here are ordered according to the corresponding proceedings chapters of the book, not according to the session numbers during the conference. While the appendix Chapters 1–14 have a corresponding proceedings chapter, Chapters 15–18 do not.

1 Could Citizen Scientists and Voluntourists Be the Future for Marine Research and Conservation?

Hannah S. Earp[1] and Arianna Liconti[1]

[1]School of Ocean Sciences, Bangor University, Askew Street, Menai Bridge, LL59 5AB

Hannah S. Earp: h.earp@bangor.ac.uk

Arianna Liconti: osu6a2@bangor.ac.uk

1.1 Call for Abstracts

Estimates show that over 10 million per year are involved in citizen science and more recently "voluntourism" projects, which extend from the ocean floor to the Milky Way and cover almost everything in-between. Public participation in scientific research has the potential to broaden the scope of research, enhance the ability to collect data, and foster increased public awareness of research importance. This session aims to explore marine research conducted by volunteers around the world, highlight successes and challenges across different participatory research projects, and discuss the role of public participation in driving future marine research, conservation, and management.

1.2 Abstracts of Oral Presentations

1.2.1 Marine Citizen Science: Challenges and Opportunities

Hannah Grist[1,2*]

[1]Scottish Association for Marine Science, European Marine Science Park, Oban, Argyll, PA37 1QA, Scotland

[2]Full Capturing our Coast project author list available at: www.capturingourcoast.co.uk

*invited speaker, corresponding author: hannah.grist@sams.ac.uk

Keywords: Citizen science, Scientific literacy, Volunteer

Citizen science is the current hot method for research, but how can it be applied in marine environments? Can it really provide answers to our growing list of intractable marine science questions? And what do the volunteers get out of it? In this talk, I will introduce the methods of citizen science, and discuss how they have and can be applied in the marine environment. I'll present some lessons learned, and when citizen science might be the best (or worst!) approach. Finally, I'll talk about the results of a research project that interviewed citizen scientists, practitioners, and academics about their experiences across five different countries, sharing ideas on scientific literacy as well as activism, philosophy of science, and the opportunities of working together in the future.

1.2.2 Addressing Distribution Questions of Habitat-Forming Corals in Italian Coastal Waters Using Volunteer Collected Data

Arianna Liconti[1*], Stuart Jenkins[1], Massimo Ponti[2]

[1]School of Ocean Sciences, Bangor University, Menai Bridge, LL59 5AB, Wales

[2]School of Sciences, Università di Bologna, Via Zamboni 33, 40126, Bologna, Italy

*corresponding author: arianna.liconti@outlook.it

Keywords: Marine citizen science, Octocorals, Community-based monitoring, Environmental policy, Ecological surveying

Habitat-forming coral assemblages are fragile, and highly ecologically and economically important elements of Mediterranean subtidal marine ecosystems. However, these valuable coral species are threatened by human exploitation, including climate change, ocean acidification, and a number of anthropogenic activities (i.e., anchoring, fishing, and recreational scuba diving). In Italian coastal waters, little is known about the patterns of distribution and abundance of corals and gorgonian species, or the effectiveness of Marine Protected Areas (MPAs) in protecting these fragile and valuable organisms. This study analyzed data collected by Reef Check Italia's (RCI) "eco-divers" in Italian coastal waters over an 11-year period (2006–2016), to address abundance and distribution questions regarding five habitat-forming coral species (*Corallium rubrum*, *Eunicella cavolini*, *Eunicella verrucosa*, *Eunicella singularis*, and *Paramuricea clavata*). The geographical distribution and abundance over time, bathymetric distribution, and the abundance inside and outside of MPAs of the considered coral species were investigated using ArcMap GIS and univariate statistical analyses. Findings from this study represent the first quantitative evidence of habitat-forming coral recovery in Italian coastal waters, and of Italian MPAs effectiveness in protecting coral assemblages. The evidence obtained also highlights the need to establish conservation strategies to further protect the overexploited red coral *C. rubrum*, to limit the decreasing population of the white gorgonian *E. singularis*, and to protect vulnerable sites such as Imperia-Diano Marina. In addition, this study demonstrates the value and effectiveness of the RCI Underwater Coastal Environmental Monitoring (U-CEM) protocol in addressing questions of macrobenthic species distribution and abundance across large spatial and temporal scales within Italian coastal waters.

1.2.3 Marine Research and Conservation in the Seychelles: The Role of Citizen Science in Data Collection, Enhanced Resource Management, and Local Capacity Building

Katherine McGavin[1,2*], David Smith[3], Josephine Head[1], Toos van Nordwijk[1]

[1]Earthwatch Institute, 256 Banbury Road, Oxford, United Kingdom

[2]Faculty of Technology, Design and Environment, Oxford Brookes University, Headington Campus, Headington, Oxford, United Kingdom

[3]Coral Reef Research Unit, University of Essex, Wivenhoe Park, Colchester, Essex, United Kingdom

*corresponding author: kmcgavin@earthwatch.org.uk

Keywords: Coral communities, Reef resilience, Socioeconomic impacts, Citizen science

Understanding how present day and future coral reef systems respond to environmental change is key to their effective management. The "Coral Communities in the Seychelles" program – a collaboration between the Earthwatch Institute, Mitsubishi Corporation, University of Essex, and the Seychelles National Park Authority – has been running citizen science expeditions in Curieuse Marine National Park, and the islands of Praslin, Mahe, and La Digue, for 11 years. Research program goals are to provide managers and policy makers with key information to mitigate the combined threats to local coral reefs, specifically the resilience of coral reef ecosystems to increasing environmental changes threatening their survival, and the socioeconomic impacts of reef degradation on local communities. To achieve this, the project has quantified coral resilience across species and environments; characterized resilient holobionts; harvested resilient corals via selective breeding; and quantified fishing-related pressures and the nature of casual fishing associated with local communities. The long-term involvement of over 130 citizen scientists over 160 days has contributed substantially to this scientific research. Key findings include: diversity within the microbiome is associated with local environment and coral host species; some coral genotypes may be naturally adapted to withstand environmental pressures; establishing nurseries of translocated resilient corals should be a future priority to increase resilient species biomass. A unique aspect of this program is the involvement of local (Western Indian Ocean-based) early career researchers alongside international volunteers. This approach has multiple advantages, including increasing long-term project impact through building local capacity for marine research conservation, improving the programs environmental sustainability through more localized recruitment, and increasing data collection by involving skilled volunteers. Here we present the successes and challenges associated with the "Coral Communities in the Seychelles" program and highlight the importance of citizen science in driving marine research, increasing impact, and influencing local conservation management policy.

1.2.4 Joint Efforts to Save a Species: "Expedición Vaquita 2017" a Crowdfunding Experience

Benjamin Martinez-Hernandez[1*]

[1]Explorando la Vida, 03730, Ciudad de México, México

∗corresponding author: benmar@explorandolavida.com

Keywords: Awareness, *Phocoena sinus*, Extinction, Expedition, Explorandolavida

Described for the first time in 1958, the vaquita (*Phocoena sinus*) has remained virtually anonymous and despite national and international efforts, populations have shown a constant decline from 600 individuals in 1997, to today's estimates of just a few tens, meaning it is currently on the verge of extinction. In 2017 "Explorando la Vida" joined in efforts to protect the vaquita through the creation of "Expedición Vaquita 2017" which was financed through crowdfunding and various fundraising activities. The project's purpose was to improve the conservation of this species by fostering increased public and policy maker's awareness, supporting the running of population studies and cleaning the vaquita habitat from ghost and illegal fishing nets. Throughout the project, we made allies and worked with policymakers, museums, schools, universities, NGO's, scientist, science disseminators, and others passionate people, who helped in making the vaquita known and promoted the crowdfunding through didactics and ludic activities. The support of the media and a group of celebrities allowed us to receive donations around the globe. From August 17–22, 2017, we cruised the vaquita refuge "Roca Consag" (31° 07' N, 114° 29' W) aboard the "Museo de la Ballena y Ciencias del Mar's" (La Paz, Baja California) scientific vessel "Narval." Main activities were tracking to locate ghost nets, revision, data recovery, and reinstallation of hydrophones. Survey efforts involved power binoculars (big eyes), video and photographic documentation and the testing of a new tool; a drone with a thermal camera. A highlight of expedition was the rescue of shipwrecked fishermen 3 days after their engine failed. The project faced several challenges but sets a precedent showing how people interested in protecting nature can support science through a crowdfunding platform in Mexico.

1.3 Abstracts of Poster Presentations

1.3.1 Do You See What I See? Quantifying Interobserver Variability in an Intertidal Disturbance Experiment

Hannah S. Earp[1*], Siobhan Vye[1], Victoria West[1], Hannah Grist[2], Peter Lamont[2], Michael Burrows[2], Jacqueline Pocklington[3], Stephanie Dickens[3], Jade Chenery[3], Jane Delany[3], Nicola Dobson[4], Jane Pottas[4], Sue Hull[4], Ruth Dunn[4], Katrin Bohn[5], Abbi Scott[5], Zoe Morrall[5], Sarah Long[5], Gordon Watson[5], Stuart Jenkins[1]

[1]School of Ocean Sciences, Bangor University, Menai Bridge, LL59 5AB, Wales

[2]Scottish Association for Marine Science, Scottish Marine Institute, Oban, PA37 1QA, Scotland

[3]School of Marine Science & Technology, Newcastle University, North Shields, NE30 4PZ, England

[4]School of Environmental Sciences, University of Hull, Hull, HU6 7RX, England

[5]Institute of Marine Sciences, University of Portsmouth, Portsmouth, PO4 9LY, England

∗corresponding author: hannahsearp@hotmail.com

Keywords: Marine citizen science, Algal community, Recovery, Data verification, Quality assurance

Citizen scientists play an important role in generating extensive datasets on marine environments and their associated flora and fauna. However, errors resulting from misidentification and over-/underestimation of abundances may reduce the accuracy of these datasets, and consequently, perceptions regarding data validity are one of the greatest challenges facing the use of citizen-collected data. This study used a novel method to test for differences in percentage cover estimates between observer units, including between citizen and professional scientists, as well as within both of these groups. We assessed the variation in observer units as part of a manipulative intertidal experiment conducted within the UK citizen science project, Capturing our Coast. This experiment aimed to identify and quantify variation in the rate and nature of algal community recovery from physical disturbance. Citizen and professional scientists estimated and photographed the recovery of key algal groups, and interobserver variability was assessed by comparing reported algal coverage values to those generated through photo analysis using Coral Point Count with Excel Extensions (CPCe). This made it possible to establish whether reported changes were attributed to differences in ecological communities as opposed to differences in observer units. This study highlights the importance of considering interobserver variation in datasets collected by both citizen and professional scientists in order to promote confidence in the accuracy of experimental findings.

2 Toward a Sustainable Management of Marine Resources: Integrating Social and Natural Sciences

Fanny Barz[1] and Heike Schwermer[2]
[1]Thünen-Institute of Baltic Sea fisheries, Rostock, Germany
[2]Institute for Hydrobiology and Fisheries Science, University of Hamburg, Germany
Fanny Barz: fanny.barz@thuenen.de
Heike Schwermer: heike.schwermer@uni-hamburg.de

2.1 Call for Abstracts

Regarding political frameworks, that is, MSFD (Marine Strategy Framework Directive), the holistic approach on managing marine resources becomes more and more important. Facing the problem of overfishing and mismanagement to different degrees, there is a strong need for research and implementation of the consideration of social criteria.

We invite you to present:

- Studies related to management of marine resources worldwide, highlighting the implementation of social aspects
- Studies focusing on marine research using an interdisciplinary approach
- Examples of research approaches using socio-based methodology in marine biology
- Examples of how marine research questions can benefit from social science

2.2 Abstracts of Oral Presentations

2.2.1 Management Plan Evaluation Ignoring Fisher Behavior: The Case of the EU 2008–2015 Cod Plans

Sarah B. M. Kraak[1*]
[1]Thünen Institut für Ostseefischerei, Alter Hafen Süd 2, 18069 Rostock, Germany
*invited speaker, corresponding author: sarah.kraak@thuenen.de

Keywords: Management strategy evaluation, Simulations, Fisher behavior, Mixed fishery, EU cod stocks

In 2008 management plans for four European cod stocks were adopted. These cod are caught in mixed fisheries, often as bycatch. In mixed fisheries, problems commonly occur when quotas of some stocks have been fully caught while portions of quotas remain to be caught for other stocks. Under the landings quotas prevalent in the EU until 2015, fishers were allowed to continue fishing until all quotas were caught, resulting in over-quota catches of the stocks whose quotas were exhausted first and discarding of those catches because landing of them was prohibited. In 2008/2009 ICES had carried out impact assessments of the plans. However, they only reported simulation scenarios predicting the consequences for yield and SSB assuming that the fishing mortality (F) specified in the plan would be fully achieved, without regarding the over-quota catches. Known issues of stakeholder support, compliance, fleet behavior, and discards were not addressed. The impact assessments did not comment on how likely the plan was to achieve its objectives in the mixed-fisheries context; they assumed that the intended F would be achieved, without discussing how. In fact the F implied by the plan could only be achieved by changes in fisher behavior. While the plan contained provisions aiming to incentivize changes in fisher behavior, these had not been considered in the impact assessments. Because these provisions were novel, it was difficult to predict how well they would achieve the intended changes in fisher behavior. Nevertheless, social scientists could have explored the likeliness of these incentives to change fisher behavior. In 2011 it was apparent that the cod plans had failed to achieve their objectives, as cod catches continued to be in excess of quotas. It would have been better if the impact assessments had included knowledge of fisher responses to regulations and incentives for behavioral change.

2.2.2 Social Response of the Gulf of Cádiz Fisheries to EU Regulations: The Discard Ban Policy

MariAngeles Gamaza[1*], Karim Erzini[2], Ignacio Sobrino[1]
[1]Instituto Español de Oceanografia, Puerto pesquero, muelle de levante s/n, 11006, Cádiz, Spain
[2]Centre of Marine Sciences, University of the Algarve, Campus de Gambelas, 8005-139 Faro, Portugal
*corresponding author: mari.gamaza@gmail.com

Keywords: Discards, Trawling, Stakeholders, Holistic, Co-management

The European Union Regulation 1380/2013 revising the Common Fisheries Policy (CFP) aims at addressing the bycatch and discarding problem in commercial fisheries by implementing the "Landing Obligation" of the regulated species by 2019. At present, the Gulf of Cádiz (GoC) landing ports are not ready to compile with this regulation. Within this framework, the present study focused on the challenges of finding relevant and efficient management solutions, considering all relevant agents within this fishery. The approach taken involved participatory meetings, questionnaires, and interviews with different stakeholders, including government administrators, national inspectors, fishers, and other representatives of the trawling fleet from the main ports in the GoC. To achieve the goal of improving the management of this specific fishery and to bring it in line with the requirements of the CFP, we conclude that economic incentives and the participation of the fishing industry at all stages are of vital importance. Therefore, opportunities and entrepreneurial ideas for an efficient use of the discards generated by this fleet are promoted with the aim of contributing to the establishment of alternative management strategies for this fishery.

2.2.3 Sustainable Artisanal Fisheries Under Risk: Is the Minimum Legal Size Regulation Working for the Chilean Abalone *Concholepas concholepas*?

Michael Kriegl[1,2,3*], Maria Dulce Subida[1], Miriam Fernández[1]

[1]Estación Costera de Investigaciones Marinas, Departamento de Ecologia, Facultad de Ciencias Biológicas, Pontificia Universidad Católica de Chile, Santiago, Chile

[2]Ghent University, Campus Sterre S8, Krijgslaan 281, B-9000 Gent, Belgium

[3]Sorbonne Université, UPMC Paris VI, Paris, France

*corresponding author: mkriegl@outlook.com

Keywords: Illegal fishing, Small-scale benthic fisheries, Co-management, Territorial use rights, Tragedy of the commons

The majority of marine fisheries, in particular small-scale fisheries, are overexploited or collapsed. A common management tool to mitigate the impact of (over-)exploitation for fished stocks and ensure future catches is the definition of a minimum legal size of capture (MLS). Its effectiveness to promote a sustainable exploitation of marine resources, however, depends on compliance by resource users. The extent of illegal fishing resulting from violations to the MLS-regulation was assessed in the artisanal fishery for Chile's economically most important shellfish, the Chilean abalone *Concholepas concholepas*. This benthic resource is harvested in areas operating under two distinct management regimes: (a) areas with exclusive extraction-rights for designated fishermen ("TURFs") and (b) areas without entry-restrictions ("open-access"), where the *C. concholepas* fishery is officially banned. At 10 fishing coves located in central Chile, the shell length of individuals harvested in both management regimes was assessed to estimate the percentage of individuals smaller than the MLS in the catch. Fishermen extracted significantly larger-sized individuals in TURFs (median = 11.0 cm, IQR = 1.2) compared to open-access areas (median = 10.1 cm, IQR = 1.7) ($p < 0.001$). While 14% (0–66.2%) of the individuals harvested from TURFs were smaller than the MLS, almost half of the individuals (47%; 16.8–80.8%) extracted in open-access areas were undersized. Considerable variability in MLS-compliance was observed among sites. Furthermore, a significantly larger share of individuals smaller than the size of sexual maturity was extracted from open-access areas (mean ± SD = 27 ± 23%) compared to TURFs (mean ± SD = 6 ± 11%) ($p < 0.001$). These results suggest that the total *C. concholepas* landings in Chile contain a substantial fraction (approximately one-third) of undersized individuals. In the light of these findings, the need to develop better-informed management strategies toward "optimal" and sustainable exploitation patterns is discussed.

2.2.4 Alternative or Additional? Modelling Perceptions of Tourism and Fishing

Julian Engel[1*], Judith Almonacid[2], Annette Breckwoldt[3], Filipina Sotto[4], Marie Fujitani[1]

[1]Leibniz Zentrum für Marine Tropenforschung, Fahrenheitstraße 6, 28359 Bremen, Deutschland

[2]People and the Sea, JV's Pass - Barrio Street, Malapascua Island, 6013 Cebu, Philippines

[3]Alfred-Wegener-Institut, Am Handelshafen 12, 27570 Bremerhaven, Deutschland

[4]University of San Carlos Marine Research Station, Maribago, Lapu-Lapu City, Provinz Cebu, Philippines

*corresponding author: Julian.engel@stop-finning.com

Keywords: Social-ecological systems, Stakeholder analysis, Alternative livelihoods, Mental models, Fuzzy logic cognitive maps

The Philippines, residing within the Coral Triangle, harbors extensive and biodiverse small island ecosystems. However, destructive fishing practices, e.g., shark and dynamite fishing, have decreased ecosystem functions and services. On Malapascua Island, shark tourism generates higher revenue than extractive uses of the ecosystem. These economic benefits lead to the potential for alternative livelihoods. In this study fuzzy logic cognitive mapping (FCM) was used to analyze the perceptions of individuals and groups across four stakeholder groups (tourists, businesses, and fisherfolk from Malapascua and Cebu) either with direct or intrinsic use value toward marine resources. A total of 53 (63 interviewees) interviews and 99 Maps (188 interviewees) were constructed. Our findings suggest tourism to be perceived as a solution to destructive fishing practices. FCM analysis indicated tourism is related to environmental awareness, marine conservation, economic benefits, and alternative livelihoods in the community. However, tourism is also perceived to negatively impact the island, for example, the natural water reservoir is overexploited and contaminated. An increase in pollution, due to consumption and population, threatens the natural resource system and the island community. Temporary (e.g., fishing) or permanent migrations to Malapascua create additional pressure. A potential management strategy emerged from the analysis of the perceived benefits and pressures. Especially regulations toward illegal fishing and resource limitations could potentially be enhanced with the allocation of tourism operations to the larger neighboring island Cebu. This shift may not be a solution to overuse and resource depletion, but creates opportunities through user diversification and ideally compensates the community in the shape of new livelihoods. The analysis suggests that implementation of this management strategy can only be successful, if stakeholder participation, community-based management regimens and law enforcement are incorporated. If thoughtfully managed, decentralization of tourism for pressure and benefit distributions could have a potential for comparable tourism operations in the Philippines and around the world.

2.2.5 Mangrove Firewood Usage and Indoor Air Pollution from Traditional Cooking Fires in Gazi Bay, Kenya

Julia Jung[1*], Mark Huxham[1], Agnes Mukami[2]

[1]Edinburgh Napier University, 9 Sighthill Ct, Edinburgh EH11 4BN, UK

[2]KMFRI – Kenya Marine and Fisheries Research Institute, KMFRI Gazi substation, Gazi Village, Kenya

∗corresponding author: 40178829@live.napier.ac.uk

Keywords: Improved cook stoves, Women, Labor

Mangrove forests are increasingly being recognized for the important ecosystem services they provide including carbon fixation, shoreline protection, and fisheries habitats. In addition, they provide typical forest goods such as timber and firewood; harvesting these can cause forest degradation and loss. In Kenya, a large proportion of the rural population cook using firewood on traditional, inefficient three-stone fires (TSF). Although harvesting mangrove wood is illegal, the high poverty rate, and lack of alternative fuels and law enforcement mean it is likely to remain widespread, with consequent pressure on the forests. The use of TSF has been associated with high levels of indoor air pollution (IAP), causing adverse health impacts. This project aimed to determine a baseline of wood usage and health burden caused by IAP at a mangrove-dependent community in Gazi Bay, southern Kenya. Basic information about fuel usage and perceived health problems related to IAP was collected using a questionnaire. Wood usage patterns were recorded for 28 days to establish the average daily wood consumption and main species used. Passive diffusion tubes were used to assess CO concentrations over 24 hours. Particulate pollution for the size fraction PM2.5 was measured during cooking using a DustTrak aerosol monitor. Mean daily per capita wood consumption was 1.2 kg although this varied significantly depending on household size, with larger households using less per capital wood. The mangrove *Rhizophora mucronata* made up 10% of wood used and people spent, on average, 22 hours per month collecting wood. The mean 24-hour CO concentration was 5.9 ppm. The average level of PM 2.5 during cooking was 10 mg/m^3, respectively. Chronic exposure at those levels is expected to cause significant health impacts of the kinds indicated by symptoms reported from the questionnaires. The introduction of improved cook stoves is recommended here, and could decrease wood consumption and IAP.

2.2.6 Bringing My Thesis Back to Life: A Stakeholder Workshop Informed by Scientific Findings on Pond Aquaculture

Paula Senff[1*], Ocha Buhari[2]

[1]Leibniz Centre for Tropical Marine Research, Bremen, Germany

[2]University of Mataram, Lombok, Indonesia

∗corresponding author: paula.senff@leibniz-zmt.de

Keywords: Mangrove aquaculture, Development, Knowledge transfer, Science-society gap

The plight of many ecological studies is that their results never make the transition from peer-reviewed literature to application. In 2016, we researched the use of an aquaculture pond area in a coastal community in Indonesia. The results pinpoint key challenges such as a lack of organization of the upkeep of infrastructure, the degradation of the mangrove ecosystem, and the threat of possible mercury contamination in ponds from artisanal gold mining. These identified problems are locally specific and most relevant to stakeholders who will never access the published findings. In July 2018, we returned to the study community to hold a stakeholder workshop presenting management recommendations drawn from our field research. Workshop participants included members of the local community, government extension workers, as well as students and government researchers. We discussed the current obstacles to improving aquaculture yield, methods to monitor water quality and possibilities to increase production and improve sustainability by increasing mangrove cover in the ponds. In addition, a future study to test mercury levels in the pond environment and in the food output was introduced. The response of stakeholders was positive, but community members also voiced contempt over the lack of government action and a general mismatch between government programs and local needs became apparent. This case provides an example of bridging the science-society gap and gives insights into challenges and lessons learned. We believe that it is important to return relevant results to local communities both for their benefit and to acknowledge their cooperation and hospitality.

2.2.7 The Fight Against Beach Litter: A Matter of Perception and Awareness

Fernando Rayon-Viña[1*], Laura Miralles[1], Marta Gómez-Agengo[1], Eduardo Dopico[1], Eva García-Vazquez[1]

[1]University of Oviedo, 33071 Oviedo, Spain

∗corresponding author: frv777@gmail.com

Keywords: Marine debris, Plastic waste, ALDFG, Social awareness, Litter perception

Marine litter is one of the biggest threats to marine ecosystems. It is defined as all solid materials of anthropogenic origin that are discarded or reach the sea, commonly due to beachgoers and local residents' ocean/coastal dumping. Understanding people's perception and awareness is decisive toward management, since the public awareness enhancement drives to a reduction on littering. In this study, we developed and validated a questionnaire to measure for the first time people's public consciousness and willingness to take actions against marine litter in Asturias (Bay of Biscay, northwest Spain). Macro litter was classified and quantified from nine beaches in the region using standardized transect sampling methodology. The beachgoers perception regarding the waste dispersed on the sand was also quantified. This data was used to compare the state of the beaches in the region, to characterize these spaces based on the potential origins of the waste thrown away, and to analyze the surveyors' perception and their role in the litter deposition and withdrawal. Significant differences were found among beaches for litter amount and types, the closest to the ports being the most thrashed. Plastic represented approximately the 64% of the litter materials, followed by fishing gear (ALDFG) (12%). Litter perception and awareness of beachgoers were significantly correlated with in litter amount in the studied beaches. Litter perception was positively correlated with people's higher frequency of visits to the beach. Significant gender differences were found, men taking more actions against litter than women regardless how much litter they perceived. These results can be employed for designing campaigns of beach garbage reduction and awareness raising. Finally, our study results can be also used as a point of departure for regional administrations for better action and management campaigns on coastal ecosystems.

2.2.8 Conservation and Consumption of Sea Turtle Eggs in Redang Island, Peninsular Malaysia

Meenakshi Poti[1,2]*, Seh Ling Long[3], Khaulah Zakaria[2], Jean Hugé [1,4,5,6,7], Mohd Uzair Rusli [2,8], Jarina Mohd Jani[2], Farid Dahdouh-Guebas[1,4]

[1]Systems Ecology and Resource Management Laboratory, Université Libre de Bruxelles (ULB), 1050, Brussels, Belgium

[2]School of Marine and Environmental Sciences (PPSMS), Universiti Malaysia Terengganu (UMT), 21300, Kuala Terengganu, Malaysia

[3]Institute of Oceanography and Environment (INOS), UMT, 21300, Kuala Terengganu, Malaysia

[4]Laboratory of Plant Biology and Nature Management, Vrije Universiteit Brussel (VUB), 1050, Brussels, Belgium

[5]Centre for Environmental Science, Hasselt University, 3500, Hasselt, Belgium

[6]Centre for Sustainable Development, Ghent University, 9000, Ghent, Belgium

[7]KLIMOS ACROPOLIS Research Platform on Climate Change & Development Cooperation

[8]Sea Turtle Research Unit (SEATRU), UMT, 21300, Kuala Terengganu, Malaysia

*corresponding author: meenakshipoti@gmail.com

Keywords: Human-sea turtle interactions, Local communities, Natural resources, Traditional food, Awareness raising

Engaging with social science has become increasingly relevant in the realm of sea turtle conservation, especially due to the anthropogenic nature of threats. It helps provide insights on the complexity of human-sea turtle interactions. In Redang Island, a primary nesting site for green sea turtles (*Chelonia mydas*), turtle eggs have been a traditional food source and medicine for decades. In the past, the local community directly depended on turtle eggs for their livelihood through a licensed egg collection system. The overexploitation of turtle eggs caused a serious decline in sea turtles. As a result of this, the main nesting beaches in Redang were gazetted as turtle sanctuaries in 2004, prohibiting egg collection from these sites. Our study investigates the current prevalence and drivers of egg consumption. Furthermore, we focus on understanding the local community perceptions toward sea turtles and the protection status of the island. Semi-structured interviews and questionnaires were conducted in 73 local households, with one adult household representative. Interviews allowed the exploration of respondent perceptions and individual consumption behavior. Meanwhile, questionnaires helped in gathering demographic information and egg consumption practices of all household members. We found that the consumption of turtle eggs has decreased due to their low availability in the market and increased price. Through a mixed-effect logistic regression model, we found that age was the main driver of egg consumption: older age groups were found to be more likely to consume turtle eggs. Awareness programs resulted in a significant decrease in egg consumption in younger age groups. The perception toward the protection status of beaches was positive, primarily due to the rapid growth of tourism as an alternative livelihood. Through this social science approach, we hope to influence conservation managers to increase local awareness-raising efforts and to integrate local perspectives in current policies related to turtle conservation.

2.3 Abstracts of Poster Presentations

2.3.1 Looking Behind the Curtains: Using Expert Interviews to Evaluate the Problem of Overfishing From Different Perspectives

Constanze Hammerl[1]*, Caroline Blomenkamp[1], Heike Schwermer[1]

[1]University of Hamburg, Institute of Marine Ecology and Fisheries Science

*corresponding author: constanze.hammerl@studium.uni-hamburg.de

Keywords: Fisheries management, Fuzzy cognitive mapping, Interdisciplinarity, Stakeholder engagement, Western Baltic cod

Sustainable management of marine resources benefits from the participation of a multitude of stakeholders (SH) representing science, politics, environmental conservation, tourism as well as commercial and recreational fisheries. Typically, however, conflicts between these SHs result in mutual distrust and unaccepted management decisions. Solving such conflicts requires knowledge of each stakeholder regarding their perception on detected conflict fields. We conducted 16 structured expert interviews with national and international stakeholders based on 30 questions evaluating the present conflict on the management of Western Baltic cod (*Gadus morhua*). Using fuzzy cognitive mapping, we studied how stakeholders' perceptions on management, ecology, economy, and communication vary. Focusing on management and communication as the conflict hotspots, we detected major differences between fisheries and environmental conservation SHs concerning stock status. Whereas representatives from eNGOs call for a fisheries moratorium, interview partners from the fisheries sector advocated for an increase in total allowable catches (TACs). In contrast to a conflicting appreciation of stock status, all SHs described the communication between stakeholder groups as disturbed or even missing. The findings of our study represent a starting point for further research on decision-making processes between stakeholder groups within the social-ecological system (SES) of Western Baltic Sea fisheries.

3 Law and Policy Dimensions of Ocean Governance

Pradeep A. Singh[1] and Mara Ort[2]

[1]University of Bremen, Faculty of Law, Bremen, Germany

[2]University of Bremen, Sustainability Research Center (Artec), Bremen, Germany

Pradeep A. Singh: pradeep@uni-bremen.de

Mara Ort: ort@uni-bremen.de

3.1 Call for Abstracts

Extending across boundaries, the governance of the seas is an intricate and highly contested matter. National, regional, and global solutions in the form of regulatory frameworks are essential in ensuring sustainable utilization of marine resources. However, decision-making, planning, and governance are often influenced by, e.g., geostrategic interests or the wish to ensure access to resources.

To explore and analyze marine regulations, governance, politics, and institutions, we invite submissions on a broad range of issues, from fields of marine governance and planning, blue growth and the sustainable development agenda, science/policy/decision-making interaction, as well as other related areas. We welcome (critical) contributions from law, social sciences, and humanities.

3.2 Abstracts of Oral Presentations

3.2.1 Coastal Governance in the Era of Rising Sea Levels: The Role of Law as an Instrument of Coastal Adaptation

Linda Schumacher[1,2,3*]

[1]University of Bremen, 28359, Bremen, Germany

[2]Centre for Marine Environmental Science (MARUM), 28359, Bremen, Germany

[3]University of Waikato, 3216, Hamilton, New Zealand

*corresponding author: schumali@uni-bremen.de

Keywords: Adaptation, Sea level rise, Sustainability

At the coast, ocean and land intersect and impact each other. For instance, land use may contaminate the ocean and the sea may impact land through eroding the coastline. In fact, most pollution of the marine environment is caused by land-based sources. At the same time, human population and the global economy are largely concentrated in coastal areas. Hence, the role of coastal governance is to reconcile the impacts and interests arising from this intersection. In the light of climate change-induced sea level rise, coastal communities face increasing coastal erosion and coastal inundation. Adapting to rising sea levels is, therefore, one of the most important challenges for coastal communities worldwide. In my presentation, I will analyze law as a key instrument of transformation, in particular for coastal adaptation to sea level rise. Using examples from Germany and New Zealand, I will demonstrate how legal regulations can, for instance, set requirements for sustainability in spatial planning decisions or demand that buildings in flood-prone areas are flood-resistant or relocatable. This is critical because spatial and land use planning can promote risk avoidance or reserve areas for future coastal defenses. From this assessment, it will become clear that legal obligations are necessary to ensure that coastal hazard risks are evaluated, taken into account and counteracted with appropriate strategies. Furthermore, my talk will examine the conflict between property rights and adaptation strategies. In this context, I will address how the law needs to reconcile and provide for a fair trade-off between the different interests of, e.g., coastal protection and the environment. In conclusion, it will become evident that legal intervention is necessary to provide a framework that enables coastal adaptation to sea level rise, incorporating a long-term strategy while also allowing to take a flexible, robust approach that promotes resilient coastal communities.

3.2.2 Of Cooperation and Conflict: UNCLOS and Ocean Governance in the Indian Ocean and the South China Sea

Amruta Karambelkar[1]

[1]Centre for Indo-Pacific Studies, School of International Studies, Jawaharlal Nehru University, 110067, New Delhi, India

*corresponding author: amruta.karambelkar@live.com

Keywords: India, China, Geopolitics, Maritime security, Navy

The idea of maritime security is finding place in the security calculus of several nations, whose earlier strategic orientation was primarily continental. This shift is a result of globalization, liberalizing economies and ever expansion in international trade. But UNCLOS is a significant causal factor for it provided nations with EEZ which necessitated its protection. This has given rise to complications particularly in the South China Sea where a Code of Conduct is yet to see light of the day. The Indian Ocean, an emerging geopolitical hotspot, has IORA & IONS for management of regional affairs. But it has had a limited success. This paper would present the situations in both the maritime theatres, each dominated by India and China, and attempt to explain why we see a certain kind of tension in each of these- what worked and what has not and the role of UNCLOS therein. In conclusion, it is apparent that UNCLOS is not entirely adequate to address realpolitik matters and that regional and subregional mechanisms seem more preferable.

3.2.3 Cross-Border MPA in the Northwest Iberian Coast

Márcia Marques[1*], Cristina Cervera-Núñez[2], Adriano Quintela[1], Lisa Sousa[1], Ana Silva[1], Fátima L. Alves[1], María Gómez-Ballesteros[2], Carla Murciano[3], Ana Lloret[3]

[1]Centre for Environmental and Marine Studies (CESAM) & Department of Environment and Planning, University of Aveiro, Portugal

[2]Spanish Institute of Oceanography (IEO), Madrid, Spain

[3]Center of Studies and Experimentation of Public Works (CEDEX), Madrid, Spain

*corresponding author: marcia.marques@ua.pt

Keywords: Transboundary, Co-management, Governance, Conservation, EBM

Cross-border marine conservation initiatives constitute a real challenge for many European countries taking into account their governmental arrangements and national maritime spatial planning processes. However, they have something in common: the MSP Directive and the Marine Strategy Framework Directive encourage Member States to follow the ecosystem-based approach when implementing them. This paper conceptualizes the creation of a cross-border MPA in the Northwest sector of Iberian Peninsula, and comprises the Portuguese and Spanish EEZs. The study area includes the Spanish Marine Protected Area of Galicia Bank and the Vigo and Vasco da Gama seamounts located in the western limit of the geologic continental platform and on the northern limit of the Portuguese jurisdictional area. The area comprised between them and the coast is also considered in the analysis in order to take into account all the activities that might represent a risk for conservation. The exercise advance the background work that aim to pin point the methodology to create and manage a cross-border MPA between Portugal and Spain. The three main objectives of this case study are identification of competent authorities and regulatory frameworks and governance; the definition of steps to propose a cross-border MPA; and the development of a proposal of share management regarding governance, monitoring, and measures for emergency response.

3.2.4 Can MPA Networks Be Used as a Tool to Improve a Sea-Basin Conservation Strategy?

Márcia Marques[1*], Fátima Lopes Alves[1]

[1]Centre for Environmental and Marine Studies (CESAM) & Department of Environment and Planning, University of Aveiro, Portugal

*corresponding author: marcia.marques@ua.pt

Keywords: Atlantic Ocean, MPA networks, Conservation

Ocean is governed through several instruments (e.g., UNCLOS, CBD, RSC, RFMO, etc.) addressing various problems at different scales. The interaction and coordination among and between different ocean actors across scales is not always very clear. Moreover, the international guidelines for marine sustainable development largely depend on partnerships in various dimensions: vertical (global-regional-national), horizontal (across sectors), and multi-stakeholders partnerships (including private sectors and society). The Atlantic Ocean as a shared resource, linking European, African, and American continents, presents an arena to develop, apply, and improve a sea-basin governance framework and an opportunity to build capacity and find synergies through actors involved in the Atlantic Ocean governance and management, namely, through MPA networks (OSPAR, RAMPAO, CaMPAM), providing an opportunity for a regionally coordinated effort. This study presents an integrated analysis of the Atlantic governance dynamic, an evaluation of the governance mechanisms that most contribute to the performance of the Atlantic MPA networks and a screen of the scope and scale of possible trade-offs and synergies between and among MPA networks. The results reveals that the Atlantic MPA networks work as regional hubs of marine conservation actions and their performance largely depends on human networks that follows the ecologic ones. Regional mechanisms, as OSPAR, also determine MPA network performance offering an opportunity to moving forward to an Atlantic coordination.

3.2.5 Conflicting Goals and Best Strategies for Reaching Good Environmental Status in the Baltic Sea

Kristina Heidrich[1*], Christian Möllmann[1], Saskia Otto[1]

[1]Institute for Hydrobiology and Fisheries Science, Hamburg University, Große Elbstraße 133, 22767 Hamburg, Germany

*corresponding author: kristina.heidrich@studium.uni-hamburg.de

Keywords: Ecosystem-based management, Good environmental status, Indicator-based approach, Decision-support tool, Bayesian Belief Networks

Recently there has been growing interest in the need for a sustainable utilization of marine resources and sound ecosystem-based management, which is reflected in Europe by the establishment of various legal frameworks such as the Helsinki Convention for the Baltic Sea, the OSPAR convention in and around the Northeast Atlantic, or the EU Marine Strategy Framework Directive. All of these frameworks enforce an integrated indicator-based approach to managing human activities based on the best available scientific knowledge about the ecosystem. However, the highly stochastic, interlinked and complex dynamics in marine ecosystems make it particularly challenging for marine managers to evaluate best strategies for maintaining ecosystem integrity and achieving Good Environmental Status (GES) across various system components. The latest computational developments show a great deal of potential as a tool for decision-support and management strategy evaluations but have not been widely used on marine indicators. Here, we present a framework that allows the evaluation of potential trade-offs between objectives and introduces best strategies for achieving GES based on a suite of robust indicators. This framework will simplify the science/policy/management interactions. As a first step, indicator candidates are screened for adequate performance, thresholds are developed and the current status is assessed. Subsequently, management strategies are evaluated using so-called Bayesian Belief Networks (BNN), which allow the combination of pre-existing knowledge and new data in a mathematically transparent way. We demonstrate the framework in the context of the Baltic Sea and show the implications of different management strategies in tackling the key problems of eutrophication and overfishing as well as potential cascading effects through the food web. This approach will provide managers a simple decision-support tool in their mission to achieve healthy and sustainable European marine waters.

3.3 Abstracts of Poster Presentations

3.3.1 Maritime Spatial Planning in Spain: Pilot Projects and Transboundary Cooperation

Cristina Cervera-Núñez[1*], María Gómez-Ballesteros[1] and MSP Working Group[#]

[#]MSP Working Group: Pablo Abaunza[1], Olvido Tello[1], Sonsoles González-Gil[1], Adriano Quintela[2], Márcia Marques[2], Lisa Sousa[2], Fatima Alves[2], Carla Murciano[3], Ana Lloret[3], Antonin Gimard[4], Ana De Magalhães[4], Neil Alloncle[4], Cecile Nys[5], Sybill Henry[5], Denis Bailly[5]

[1]Spanish Institute of Oceanography (IEO), Madrid, Spain

[2]Department of Environment and Planning & CESAM - Centre for Environmental and Marine Studies, University of Aveiro, University Campus of Santiago, 3810-193 Aveiro, Portugal

[3]Center of Studies and Experimentation of Public Works (CEDEX), Madrid, Spain

[4]Agence Française pour la Biodiversité, Brest, France

[5]UMR AMURE, Center for Law and Economics of the Sea - European Institute for Marine Studies, University of Western Brittany - Rue Dumont D'Urville, Plouzané, France

*corresponding author: cristina.cervera@externos.ieo.es

Keywords: MSP, Stakeholders, SIMNORAT, SIMWESTMED, TPEA

Marine/Maritime Spatial Planning (MSP) is becoming an emerging marine management paradigm in many places around the globe. In Europe, the Directive 2014/89/EU of the European Parliament and of the Council obliges coastal Member States to develop maritime spatial plans at the latest by March 31, 2021. This Directive was transposed to the Spanish legal system through Royal Decree 363/2017 and linked to Act 41/2010 for protection of the marine environment. After the transposition of the Directive, a MSP Working Group was created under the Interministerial Commission of Marine Strategies as a forum for work, exchange, and advice. It does not have a decision-making role, and its ultimate objective is to support the technical and negotiation work of the MSP and to present a proposal for MSP plans to the Interministerial Commission. On the other hand, the transnational nature of MSP is undeniable, therefore, the Directive urge Member States to collaborate in order to develop coherent plans along the sea regions. So far, in Spain, the Spanish Institute of Oceanography and the Centre of Studies and Experimentation of Public Works have been working in co-funded European projects as "Transboundary Planning in the European Atlantic (TPEA)," "Supporting the Implementation of Maritime Spatial Planning in the Northern Atlantic Region (SIMNORAT)," and "Supporting the Implementation of Maritime Spatial Planning in the Western Mediterranean

Region (SIMWESTMED)." These projects seek to promote collaboration among institutions of neighboring countries exchanging tools and developing methods and guidelines in order to support the implementation of the Directive in the Member States, conducted through in-desk analysis, case study development, and stakeholder engagement actions, among others.

4 Species on the Brink: Navigating Conservation in the Anthropocene

Morgan L. McCarthy[1,2,3,4], Thomas Luypaert[1,2,3,5], Meenakshi Poti[1,2,3,5] and James G. Hagan[1,2,3,5]

[1]Vrije Universiteit Brussel (VUB), Faculty of Sciences and Bioengineering Sciences, Department of Biology, Pleinlaan 2, 1050 Brussels, Belgium

[2]Université libre de Bruxelles (ULB), Faculty of Sciences, Department of Biology of Organisms, Av. F.D. Roosevelt 50, 1050 Brussels, Belgium

[3]Università degli Studi di Firenze (UniFi), Faculty of Maths, Physics and Natural Sciences, Department of Biology, Via Madonna del Piano 6, 50019 Sesto Fiorentino, Italy

[4]The University of Queensland, School of Biological Sciences, St. Lucia, 4072 Queensland, Australia

[5]University of Malaysia Terengganu, School of Marine and Environmental Sciences, Terengganu, 21300 Kuala Terengganu, Malaysia

Morgan L. McCarthy: m.l.mccarthy@uq.net.au

Thomas Luypaert: luypaert.thomas@vub.be

Meenakshi Poti: meenakshipoti@gmail.com

James G. Hagan: james_hagan@outlook.com

4.1 Call for Abstracts

In what scientists are calling "The 6th mass extinction," the future of our marine organisms is increasingly uncertain. Although there are more known extinctions in terrestrial ecosystems, researchers warn that the number of marine extinctions could rise rapidly as the oceans are industrialized for food, fossil fuels, minerals, energy, and transportation. Since its inception in the late 1970s, conservation biology has been integral in addressing threats to biodiversity and the implementation of policies to conserve unique species and ecosystems. We call for abstracts highlighting conservation research for marine species and the oceans they inhabit.

4.2 Abstracts of Oral Presentations

4.2.1 Predictive Habitat Modeling of the Elusive South American Burmeister's Porpoise (*Phocoena spinipinnis*) in Northern Patagonia, Chile

Rebecca Konijnenberg[1*], Sophie Smout[1], Sonja Heinrich[1]

[1]Scottish Oceans Institute, School of Biology, University of St Andrews, KY16 8LB, St Andrews, Scotland

*invited speaker, corresponding author: konijnenberg.uwcim@gmail.com

Keywords: Spatial distribution, Statistical regression analysis, Salmon aquaculture, Marine conservation, Cryptic species

The spatial distributions of many marine species, especially mobile ones such as small cetaceans, are still poorly known and difficult to study. Coastal species, while generally easier to access for study, are also facing an unprecedented increase in pressures from multiple anthropogenic stressors. The endemic South American Burmeister's porpoise (*Phocoena spinipinnis*) is subject to high levels of fisheries-related mortality off the coast of Peru and northern Chile while also facing expanding fish farming activities in southern Chile. In order to support effective conservation and management efforts, this study provides the first in-depth analysis of Burmeister's porpoise habitat use in the context of anthropogenic coastal activities in southern Chile. We used a three-stage approach to determine potential conservation hot spots for cryptic and rarely seen Burmeister's porpoises. First, we modelled relatively fine-scale habitat use patterns of Burmeister's porpoises off the Chiloé archipelago. For this, we used statistical regression models on the only existing systematically collected sighting dataset available for this species (spanning 13 years) and relevant geostatic environmental variables. We then used the model that best explains porpoise distribution to predict porpoise occurrence across the entire northern Patagonian fjords and archipelagos, restricting predictions to sampled environmental space. Finally we mapped anthropogenic activities from various sources onto the derived predictive habitat map for the porpoises and identified multiple areas of potential conservation importance. Burmeister's porpoises preferred coastal waters deeper than 40 m both in channels and along open coasts, which lead to substantial overlap with existing and proposed fish farm concessions, particularly in the poorly studied and vast Patagonian fjord systems. We show that predictive habitat modelling can be a powerful tool to identify areas of importance for further research and management action when survey data are spatially limited or species are cryptic and extremely expensive to study at large scales.

4.2.2 The Road to Sustainable Sea Turtle Tourism Is Paved by Science-Based Management Interventions: A Case Study of Apo Island Protected Landscape and Seascape in the Philippines

J. M. Micklem[1,2*], S. A. Ong[2], M. J. Lamoste[2], G. Araujo[2], N. Koedam[1], A. Ponzo[2]

[1]Vrije Universiteit Brussel, 1050 Ixelles, Belgium

[2]Large Marine Vertebrates Research Institute Philippines, 6308 Bohol, Philippines

*corresponding author: jes.micklem@gmail.com

Keywords: Sea turtle tourism, Compliance, Green turtles, *Chelonia mydas*, Apo Island

Green turtles (*Chelonia mydas* L.) experienced a historical population decline in the Philippines and conservation efforts have been taken to mitigate threats and protect key areas. With a growing tourism industry, green turtles have become the target species for many wildlife tourism endeavors. While the economic benefits for local stakeholders are substantial, it has dominated pro-conservation arguments, eclipsing the possible sublethal impacts of high levels of human disturbance. Apo Island Protected Landscape and Seascape, a marine protected area in the Philippines, is an important foraging ground for over 90 resident green turtles. The Protected Areas Management Board and the Apo Island Snorkel and Equipment Rental Guide Association have worked together to regulate snorkeling by establishing best practice behaviors for tourist-turtle interactions. To assess the effectiveness of current management strategies in minimizing potential impacts, 143 in-water observations of tourist behavior were recorded, and 509 visitor surveys were collected. Some of the surveyed snorkelers admitted to having touched a turtle (7%) and to approaching to within 1 m or less from a turtle (49%), both of which are regulation breaches and were higher than in-water observations indicated. In 2017, visitor numbers to Apo Island reached 72,600, of which at least half participate in snorkeling, these data suggest a staggering 2,541 touches that year. Factors influencing the non-compliance behaviors of tourists include low levels of awareness, inconsistent briefings, and low enforcement by guides. These issues reflect the need for improvements in the management strategies, possibly through a compulsory interpretation program for tourists, improved guide training and further research into the effects of human disturbance on green turtles.

4.2.3 Ecological Niche Modelling of Humpback Whale Mother-Calf Pairs on a Northern British Columbian Feeding Ground

Éadin O'Mahony[1,2*]

[1]Centre for Biodiversity and Environment Research, Department of Genetics, Evolution and Environment, University College London, London, WC1E 6BT, UK

[2]North Coast Cetacean Society, 446 Hayimiisaxaa Way, Hartley Bay, BC, V0V 1A0, Canada

*corresponding author: eadinkerkhoff@gmail.com

Keywords: Conservation, Habitat use, MaxEnt, *Megaptera novaeangliae*

It is widely accepted that humans have major environmental impacts on marine ecosystems. The marine realm is particularly threatened by pollution, overexploitation, disease, and habitat destruction, while climate warming and ocean acidification also have underlying impacts. Large migratory apex predators are crucial components of marine ecosystems worldwide, but due to their migratory nature, these species are often difficult to both study and protect. However, novel techniques such as ecological niche modelling (ENM) can reveal the habitat use of such species and thus positively influence conservation management. Here, the habitat preferences and life-history trends of humpback whale, *Megaptera novaeangliae,* mothers and calves were investigated on a northern British Columbian feeding ground, Canada. This comprised 233 mother and calf sightings recorded by the North Coast Cetacean Society between 2006 and 2016 in Gitga'at First Nation territory, in addition to a suite of environmental variables. The results indicate that mothers are returning to the area with new calves most commonly every 2–3 years and the mother-calf population is increasing (rho = 0.62, P = 0.044). Using MaxEnt ENM, I found that mother-calf pairs prefer nearshore habitats: MaxEnt models weighted distance from shore as the highest contributor to model predictions (at 44.8% contribution, AUC = 0.709). Such trends have been documented on tropical breeding grounds. However, this study shows that these preferences also apply to mothers and calves post-migration. The results highlight the importance of these waters for humpback mothers and calves. In light of Canada's commitment to protect 10% of marine systems by 2020 and a proposed liquefied natural gas tanker route threatening the region, the study area is recommended for increased protection.

4.2.4 Population Connectivity and Gene Flow Among *Hypanus americanus* in the Central and Western Bahamas

Tanja Schwanck[1,2*], Kathrin Lampert[3], Maximilian Schweinsberg[1], Tristan Guttridge[5], Ralph Tollrian[1], Owen O'Shea[2,4]

[1]Department of Animal Ecology, Evolution and Biodiversity, University of Bochum, Bochum, Germany

[2]Shark Research and Conservation Program, the Cape Eleuthera Institute, Rock Sound, Eleuthera, Bahamas

[3]Institute for Zoology, University of Cologne, Cologne, Germany

[4]CORE Sciences, Gregory Town, Eleuthera, Bahamas

[5]Bimini Biological Field Station Foundation, South Bimini, Bahamas

*corresponding author: tanja.schwanck@rub.de

Keywords: Coastal batoid, Microsatellites, Population genetics, Conservation, Dispersal

The southern stingray (*Hypanus americanus*, Hildebrand & Schroeder, 1928) is a commonly occurring coastal batoid particularly known from the Caribbean region. As an epibenthic predator, *H. americanus* influences the biophysical dynamics of its habitat, resulting in an engineering role within the benthoscape. Additionally, the southern stingray is highly prized as an ecotourism commodity throughout its range. Despite their value and broad distribution, limited data is available pertaining to life history and population structures. In order to determine potential vulnerabilities of *H. americanus* populations, an analysis of their dispersal and genetic variability is required. Being k-selected, batoids are especially vulnerable to decline and loss of genetic diversity through exploitation or isolation. In this study, we sampled 200 individual stingrays encompassing 9 sites around Cape Eleuthera over a 2.5-year period. Of the sampled individuals, more than a third were recaptured. No seasonal patterns were found and the home range appears to be very restricted, which supports the notion of high site residency. Furthermore, as resident populations of stingrays could suffer from a lack of population connectivity and be predestined for genetic isolation and local extirpation, this study also investigated the genetic connectivity of four sample sites in the central and western Bahamas. The analysis of 242 individuals with five microsatellites loci revealed not only high degrees in genotypic variability, but also low population differentiation. An additional haplotype analysis from the polymorphic mitochondrial d-loop region confirms these findings and suggests gene flow by both males and females.

4.2.5 Assessing Restocking Actions of the Sea Urchin *Paracentrotus lividus* in the Central Area of the Bay of Biscay (Northern Spain) Using Microsatellites Loci

M. Parrondo[1], R. Torres-Fernández-Soria[1], S. de la Uz[2], L. Miralles[1], C. Rodríguez[2], J.F. Carrasco[2], E. Garcia-Vazquez[1], L. García-Flórez[2], Y.J. Borrell[1*]

[1]University of Oviedo, 33006, Oviedo, Spain

[2]Centro de Experimentación Pesquera, 33212, Gijón, Spain

*corresponding author: borrellyaisel@uniovi.es

Keywords: Overexploitation, Multiplex PCR, Artisanal fisheries, Population structure

The sea urchin *Paracentrotus lividus* is an echinoderm of the family Parechinidae. This species is distributed throughout the Mediterranean Sea and eastern Atlantic Ocean on rocky bottoms. The artisanal sea urchin fishery is a fundamental part of the culture traditions in Asturias (Northern Spain) and, therefore, of the economy of the region. In recent years, there has been a significant decline in the abundance of populations on the Asturias coasts. This seems to be due to both the overexploitation of the resource and the disappearance of kelp forests. Consecutive campaigns have been conducted in the last 3 years to reverse this shortage of sea urchins throughout supplementing the sea urchin aggregates with captive juveniles in the Asturian coast. In this work, eight variable microsatellite markers aggregated in two new multiplex PCRs have been used to evaluate these programs in terms of its effects/consequences on restocked populations. Methods have included parentage assignments using breeders, juveniles, and samples from restocked populations and genetic variation comparative analyses including wild and restocked populations from Asturias, Galicia (Atlantic Ocean), and Catalonia (Mediterranean Sea). Microsatellites loci showed high polymorphic information content (PIC = 0.89) and allowed to obtain 100% of correct allocation of kinship between juveniles and their parents with strict and very high levels of confidence. This study also enabled the evaluation of the relative success of supplementing campaigns and its genetic effects on restocked populations.

4.2.6 Tooth Wear in Free-Ranging Killer Whales (*Orcinus orca*): A Function of Ecotype and Age

Audrey J. Granger[1*], Patrick J. White[1]

[1]Edinburgh Napier University, School of Applied Sciences, Sighthill Campus, EH11 4BN, UK

*corresponding author: granger.audrey.j@gmail.com

Keywords: Cetacean, Dentition, Interdigitation, Teeth, Wear

Sympatric forms of killer whales (*Orcinus orca*) occur globally, varying in size, morphology, behavior, genetics, and diet. Their teeth, as with most mammalian dentition, are worn down by food properties, mastication, behavior, and disease factors. So far, investigations into the occurrence and trends of tooth wear in killer whales have been limited, and none have been compared across all ecotypes. The objective of this study was to investigate the extent and causes of variation in tooth wear between killer whale ecotypes. Tooth wear scores from individual specimens from different ecotypes were compared in relation to age stage, sex, and body length. Teeth from 36 killer whales from seven global ecotypes were scored. Frequencies of tooth wear were high, with 70% of all teeth analyzed moderately to severely worn. Variation in tooth wear was related to ecotype and, to a lesser degree, age. Tooth wear in offshore killer whales was significantly greater than Crozet, North Atlantic, transient, and resident ecotypes. The repeatable scoring system we developed for assessing tooth wear in killer whales will allow for the continued monitoring of wear in stranded animals, and can be adapted to other cetaceans. This study provides the first steps in understanding differences in tooth wear between ecologically distinct killer whales and the impact of differing diets on teeth and feeding behaviors.

4.2.7 Developing New Molecular Tools for Assessing Connectivity and Recruitment Patterns in the Stalked Barnacle (*Pollicipes pollicipes*) (Gmelin, 1790)

M. Parrondo[1], P. Morán[2], J. Chiss[3], M. Ballenghien[3], A.S. LePort[3], C. Lejeusne[3], D. Jollivet[3], E. García-Vázquez[1], L. García-Flórez[4], J.L. Acuña[1], Y.L. Borrell[1*]

[1]University of Oviedo, 33006, Oviedo, Spain

[2]University of Vigo, 36310, Vigo, Spain

[3]Station Biologique de Roscoff, Sorbonne University, 29680, Roscoff, France

[4]Centro de Experimentación Pesquera, 33212, Gijón, Spain

*corresponding author: borrellyaisel@uniovi.es

Keywords: Microsatellites, Gooseneck barnacle, Multiplex PCR, Artisanal fisheries, Sustainability

The stalked barnacle (*Pollicipes pollicipes*) is a commercially exploited pedunculate cirripede that inhabits rocky shores from the Atlantic coasts of France to Senegal. It is an important target for local fisheries in Spain and Portugal giving rise to exorbitant high prices in Christmas markets. Scientists and stakeholders share a common interest on the long-term sustainability of this resource since it is part of the European traditions and culture. In order to establish a sustainable fishery management for this stalked barnacle species it is important to assess the genetic connectivity of its subpopulations to estimate the real patterns of present-days migration and recruitment at both local and regional scales using fine-tune genetic markers. During this work, a set of already available microsatellites loci was tested but without any reliable results. For this reason, we have developed a new GT enriched library of microsatellites and sequencing it using Ion Torrent method. A total of 10,781 raw sequences were obtained and analyzed and 123 primers pairs were designed for regions containing di- and tetra- microsatellites. Twelve polymorphic loci and three multiplex PCRs were newly designed for *P. pollicipes*. Allelic variation was screened for 140 individuals from three different regions: Atlantic Galicia (Northwestern Spain), West Coast of Asturias (North Spain) and South Brittany (North of the Bay of Biscay, France). The new microsatellites loci showed a mean number of alleles ranging from 4 to 56 and mean observed heterozygosities ranging from 0.310 to 0.867. The development of this new set of microsatellite markers for the stalked barnacle will help to provide new understanding on its population dynamics and adequate criteria to establish sustainable fishery management plans.

4.2.8 Reading the Bones: How Archived Skulls Can Inform Temporal Genetic Variation in Central Queensland's Dugong Population

Morgan L. McCarthy[1,2,3*], Marc Kochzius[2], Jennifer M. Seddon[3], Janet M. Lanyon[1]

[1]The University of Queensland, School of Biological Sciences, St. Lucia, Queensland 4072, Australia

[2]Marine Biology, Ecology and Biodiversity, Vrije Universiteit Brussel (VUB), Pleinlaan 2, 1050 Brussel, Belgium

[3]The University of Queensland, School of Veterinary Science, Gatton, Queensland 4343, Australia

*corresponding author: m.l.mccarthy@uqconnect.edu.au

Keywords: Queensland shark-net program, Genetic bottleneck, Dugong, *Dugong dugong*, Mitochondrial DNA

The dugong (*Dugong dugong*) is a long-lived marine mammal distributed throughout the Indo-Pacific region. A dependence on seagrass puts the dugong at risk because natural catastrophes disturb and destroy their feeding grounds. While some individuals have been recorded undertaking regional migrations, most dugongs normally live in relatively restricted home ranges, creating a series of genetically distinct populations along the eastern Queensland coast. In Townsville, a major urban center and historically significant feeding ground for dugongs, a shark-netting program was established in 1964, resulting in mass mortality of dugongs. This program, followed by serial cyclones in the 1970s, pressures of modern-day coastal urbanization, and El Niño-Southern Oscillation (ENSO)-associated flooding and storms, placed these dugongs under considerable stress. This study documented changes in population structure of dugongs in the Townsville region of central Queensland over time, by comparing genetic variation in mitochondrial DNA from historical museum samples from the time of inception of the shark-netting program to genetic variation in contemporary populations, i.e., 30–40 years later. While there was a notable decrease in haplotype richness, haplotype and nucleotide diversity measures remained constant between the two populations, potentially an artifact of insufficient time lapse, overlapping generations, and highly differentiated ancient haplogroups. The findings of this study provide the first record of successful mitochondrial DNA extraction from dugong cheek teeth, tusks, and periotic bone as well as novel insight into how population disturbance (through storms, netting, and human development) appears to impact genetic structure and long-term viability of dugong populations.

5 The Challenge of Marine Restoration Programs: Habitats-Based Scientific Research as a Key to Their Success

Laura Basconi[1], Charles Cadier[2] and Gustavo Guerrero Limon[2]
[1]University of Salento-Lecce, Apulia, Italy
[2]MER Consortium, UPV, Bilbao, Spain
Laura Basconi: l.basconi92@gmail.com
Charles Cadier: charlescadier@hotmail.fr
Gustavo Guerrero Limon: guerrerolimong@gmail.com

5.1 Call for Abstracts

Active restoration is the new trend to mitigate effects of climate change. Over the last decades, marine habitats of primary importance have been estimated to suffer globally high decrease of their covers since subjected to multiple stressors. However, few restoration programs have been successfully developed in the marine environment, with disparities among habitats. Hence, there is a general need of scientific research to understand habitats' specific complexity, services they provide, and restoration outcomes. In few words, "Are marine habitats' state-of-art knowledge robust enough to successfully restore them?" We welcome any research that could contribute to solve this issue.

5.2 Abstracts of Oral Presentations

5.2.1 Relationship Between the Size of Naturally Recruited Coral Colonies and Fish Diversity, as a Guide for Reef Restoration

Irene Antonina Salinas Akhmadeeva[1,2*], Héctor Reyes Bonilla[2]
[1]Escuela Nacional de Estudios Superiores, UNAM, Unidad Morelia. Antigua Carretera a Pátzcuaro No. 8701, Col. Ex Hacienda de San José de la Huerta, C.P., 58190 Morelia, Michoacán, México
[2]Laboratorio de Sistemas Arrecifales, Universidad Autónoma de Baja California Sur, Carretera al sur km. 5.5, Colonia El Mezquitito, C.P., 23080 La Paz, Baja California Sur, México
*corresponding author: ireneasalinas@gmail.com

Keywords: *Pocillopora*, Gulf of California, Biodiversity indexes, Ecological function, Coral growth

Coral reefs suffer multiple environmental and anthropic threat, and restoration is a useful tool to increase its resilience. The success of these efforts is usually measured on the basis of coral survival and growth rates, but that approach may be improved by taking in consideration specific evaluations of the ecological role that transplanted colonies play in the reef. Corals are ecological engineers and offer shelter to numerous species, but also provide food in form of mucus; these functions are in direct relation to the size of the colony, thus, it is relevant to determine the minimum size a fragment needs to fulfill its role in a reef. The aim of the study was to detect the minimum size that colonies of naturally settled *Pocillopora* spp. require to house a full set of associated reef fish. The study was carried out at Espiritu Santo and Cabo Pulmo National Parks, in the Gulf of California, Mexico. We measured the largest diameter, smallest diameter, and height of each colony (in cm) to calculate its volume, and simultaneously, visual censuses of associated fish were conducted. Fish richness and abundance were also calculated, and these response variables were used to perform nonlinear regression models to determine the minimum volume that a colony should have to reach a saturation of richness and number of individuals. The minimum volume that a *Pocillopora* colony requires to have full functionality for the fish fauna is 100,000 cm^3, which, according to published estimations of growth rate, represents a minimum age of approximately 6 years, depending on environmental conditions. Nonlinear models calculations showed that colonies of *Pocillopora* can be transplanted at a size of at least 50,000 cm^3, when the fish assemblages have reached half of its ecological function, rated as 50% of the maximum species richness and abundance.

5.2.2 The Application of Artificial Floating Islands in Saline Environments

Jessica Ware[1*], Ruth Callaway[1], Kam Tang[1]
[1]Swansea University, Biosciences, Singleton Campus, Swansea, SA2 8PP
*corresponding author: 808169@swansea.ac.uk

Keywords: Coastal habitat creation, Wildlife, Wetlands

Approximately 30% of UK electricity generation must be produced via renewable energy, to meet ambitious 2020 carbon emissions targets. Due to such pressures for decarbonization, the marine energy sector is gaining momentum, as they seek to develop, construct, and operate tidal and wave energy projects across the UK. Such hardening of the coastal landscape is anticipated to cause considerable changes in coastal ecosystem structure and community assemblages. However, the scale of the impact is still unknown and further research is required on ecosystem scale enhancement measures such as Artificial Floating Islands (AFIs). AFIs have primarily been used in freshwater habitats such as reservoirs, ponds, and river systems for water quality improvement and habitat creation for breeding birds. To assess the potential application of AFIs in marine environments, this comparative study focuses on both the floral species suitable for island installation and the fauna associated with the islands including birds, fish, and macroinvertebrate populations. Sea rush (*Juncus maritimus*), common cordgrass (*Spartina anglica*), sea purslane (*Halimione portulacoides*) and sea aster (*Tripolium pannonicum*) are halophytes that have been utilized in a controlled laboratory experiment. The experiment assessed their stem, leaf, and root growth when hydroponically grown in two salinities of water: 15 PSU and 30 PSU. Fish activity has been monitored using remote underwater video technology and sonar, birds via vantage point surveying techniques and water sampling has provided information on the mobile macroinvertebrate communities. The faunal activity in association with the AFI has been compared with hard structures such as pontoons and "unshaded" areas within the local habitat. By addressing gaps in current research on artificial habitat creation, this study aims to support future ecosystem enhancement programs that seek to mitigate the loss of coastal habitats.

5.3 Abstracts of Poster Presentations

5.3.1 Mangrove Restoration Through Participatory Forestry, Mombasa County, Kenya

Nguu Josphat[1,2*]

[1]Kenya Marine and Fisheries Research Institute P.O. Box 080100-81651 Mombasa

[2]Vrije Universiteit Brussels, Schoofslaan 12-214 Brussels, Belgium

*corresponding author: nguugachoki@gmail.com

Keywords: Silviculture skills, Training, Techniques, Community

Worldwide mangrove forests are among the most degraded ecosystems. In Western Indian Ocean, over-harvesting of wood resources is a main threat for mangroves. This loss leads to negative effects on ecosystem services including nursery for fish, shoreline stabilization, and resource sustainability. There are concerted actions to restore denuded areas, but the success rate of these projects is still dallied due to lack of proper silviculture skills. The locals who have the best chance to restore and monitor the forest are sidelined in training on the appropriate methods of restoration. In efforts to close this knowledge gap, a project to train community was carried out in Mombasa County in Kenya with main activities involving species identification and zonation, nursery establishment, transplanting, and monitoring techniques. The community members were taken through a series of initial reconnaissance assessments in degraded areas to study and understand the steps that are needed before actual restoration. An initial valuation on communities' familiarity with restoration techniques revealed fragmented knowledge of the practice. At the end of a 5-day workshop, a total of 30 community members, 15 men and 15 women, were trained on ecological mangrove restoration techniques. The future aim is to continue training more locals with additional financial support.

5.3.2 Production Techniques of Stony Corals and Perspectives for Reef Restoration at the North of Quintana Roo, Mexico

A. Claudia Padilla-Souza[1*], A. A. Morales-Guadarrama[1], E. Ramírez-Mata[1], D. J. González-Vázquez[1], A. D. Santana-Cisneros[1], A. Romero-Nava[1]

[1]National Fisheries and Aquaculture Institute (INAPESCA), Matamoros 7, 77580, Puerto Morelos, México

*corresponding author: klaus.padilla@gmail.com

Keywords: Aquaculture, Coral gene bank, Micro-fragmentation, Reef-building corals, Ecological restoration

Coral reefs in Quintana Roo are part of the second largest barrier reef of the world, the Mesoamerican Reef System. These ecosystems provide several important environmental goods and services. Nevertheless, coastal development, marine traffic, global warming, and climate change, among others, threaten them. Since 2009, the National Fisheries and Aquaculture Institute of Mexico (INAPESCA) has been developing biotechnology for ecological restoration of damaged zones due to ship groundings and/or hurricanes. This includes the production of coral reef builders at indoor hatcheries as well as marine nurseries for outplanting them later. The production techniques include Assisted Reproductive Technology to obtain sexual recruits of the Elkhorn coral (*Acropora palmata*) with unique genotypes and also asexually propagated fragments and microfragments of six more species. The coral production process is modular: the corals are cultivated at indoor facilities, then moved to semi-controlled tanks and finally taken into marine nurseries to finish growth. Through outplanting these corals, the coral coverage has been reestablished, and old niches have been recovered quickly (within 4 years); also the usage of sexual recruits would enhance the genetic diversity once they reach sexual maturity (approximately 6 years). The new technique of micro-fragmentation has shown an exponential growth of the tissue (up to 250%) allowing the production of tissue in shorter periods of time (6 months). The generated tissue can be used for reskinning skeletons or artificial structures of great volumes to obtain reproductive individuals in less time (2 years), thus generating substrate rugosity and environmental heterogeneity and enabling new niches. Nowadays, the research facilities hold 25 k microfragments of 22 different genotypes and 7 species. These genotypes will be artificially selected taking into account specific traits (i.e., higher thermal tolerance) to enhance the success of the restoration program and the preservation of the coral reef environment.

6 Submerged in Plastic: Impacts of Plastic Pollution on Marine Biota

Natalie Prinz[1,2] and Špela Korez[3]

[1]University of Bremen, Faculty of Biology and Chemistry, Bremen, Germany

[2]Leibniz Centre for Tropical Marine Research, Bremen, Germany

[3]Alfred Wegener Institute, Helmholtz Centre for Polar and Marine Research, Germany

Natalie Prinz: nprinz@uni-bremen.de

Špela Korez: spela.korez@awi.de

6.1 Call for Abstracts

Nowadays, plastic waste is found everywhere in the marine realm. Regardless of their size, plastics are therefore affecting a range of marine organisms. Through degradation mechanisms and/or biofouling, plastics become bioavailable, may be ingested, enter the marine food webs, or become vectors for organisms to travel large distances. Fortunately, these issues are increasingly gaining attention in the scientific world, allowing us to understand impacts on affected organisms. Are you trying to eliminate the knowledge gaps or uncertainties of plastic pollution interacting with marine biota? We invite you to share your innovative ideas, improved methodologies, and novel results with fellow young researchers.

6.2 Abstracts of Oral Presentations

6.2.1 Does Microplastic Shape Have an Effect on the Rejection Ability in the Hard Coral *Pocillopora damicornis*?

Anna Feuring[1,2]*, Valeska Diemel[1,2], Sonia Bejarano[2]

[1]University of Bremen, Bibliothekstraße 1, 28359 Bremen, Germany

[2]Leibniz Center for Tropical Marine Research, Fahrenheitstraße 6, 28359 Bremen, Germany

*corresponding author: afeuring@uni-bremen.de

Keywords: Coral-microplastics interaction, Particle shape, Short-term exposure, *Pocillopora damicornis*, Rejection mechanisms

Increasing litter deposits in the oceans result in large accumulations of plastics along shorelines. Wave exposure and high UV radiation degrade particles into smaller pieces called microplastics (< 5 mm). As a variety of different polymer types make their way into the oceans, microplastic particles differ greatly in shape and size. Once reaching negative buoyancy, particles sink and sediment onto benthic organisms. Hence, the ability of particle rejection plays a key role for sessile species such as corals. This study investigates the effect of particle shape on the rejection ability and health status in the hard coral *Pocillopora damicornis*. Therefore, fluorescently labeled PET particles (0.2 mm) of different shapes are applied onto fragments of *P. damicornis*. Using a macro underwater camera, coral-particle interactions are recorded for the time needed to reject all particles (rejection capacity). Behavioral patterns will be studied by analyzing video material and the concentration of coral mucus. Moreover, chemical analysis of coral fragments will add information about potential short-term cellular responses. The talk will present and discuss results obtained to this point of time, as this study is conducted within the scope of an ongoing master thesis.

6.2.2 Insights on Potential Effects of Microplastics and Associated Pollutants on Freshwater Organisms

Saskia Rehse[1,2]*, Werner Kloas[1,3], Christiane Zarfl[2]

[1]Institute of Freshwater Ecology and Inland Fisheries, 12587, Berlin, Germany

[2]Center for Applied Geosciences, Eberhard Karls Universität Tübingen, 72074, Tübingen, Germany

[3]Department of Endocrinology, Institute of Biology, Humboldt-Universität Berlin, 10115 Berlin, Germany

*invited speaker, corresponding author: rehse@igb-berlin.de

Keywords: Zooplankton, Amphibians, Physical effects, Chemical effects

Freshwater systems are widely polluted with plastics and act as emission source of plastics to the oceans. During the last decade, research started focusing on the presence and potential threats of small plastics, i.e., microplastics (< 5 mm), on marine and freshwater organisms. Some studies consider microplastics as potential vector for chemical pollutants adsorbed to and leached from microplastics, and others hypothesize that microplastics are negligible compared to further uptake pathways, e.g., via water. The focus of this study was to disentangle these potential "origins" of observed effects and to analyze systematically how microplastics potentially affect freshwater organisms. Both, effects of the pristine microplastic material itself (physical effects) and the influence of microplastics on the effects of chemical pollutants (chemical effects) were included. Water fleas, *Daphnia magna*, and tadpoles, *Xenopus laevis*, were selected as model species. Exposure of polyethylene beads (PE, 1 μm) led to immobilization of daphnids at very high concentrations (10^1–10^2 mg L^{-1}), while bigger PE beads (100 μm) and polyamide fragments (PA, 20–50 μm) did not induce immobilization. When bisphenol A (BPA) was presented to the daphnids in combination with the PA fragments, immobilization caused by BPA was lower, compared to the exposure with BPA alone. The negative impact of BPA was buffered by the presence of the PA fragments, rather than being enhanced. In tadpoles, the PA fragments alone did not affect general development and biomarker activities. In combination with the oral contraceptive 17α-ethinylestradiol (EE2), the PA fragments slightly increased the specific effects of EE2, that is, for the most sensitive biomarker (vitellogenin mRNA synthesis). The results of this study indicate that microplastics have limited physical and chemical effects on daphnids and tadpoles, but depending on the organisms, the outcome of microplastics on biological effects is difficult to predict and depends on the biology of the exposed animals.

6.2.3 Monitoring Micro- and Nanoplastic Interaction with Microalgae Using Spectroscopic Tools

Maureen Déniel[1*], Aurore Caruso[2], Nicolas Errien[1], Fabienne Lagarde[1]

[1]Institut des Molécules et Matériaux du Mans, UMR 6283, Le Mans Université, Avenue Olivier Messiaen, Le Mans, France

[2]Laboratoire Mer, Molécules, Santé, EA 2160, Avenue Olivier Messiaen, Le Mans, France

∗corresponding author: maureen.deniel@univ-lemans.fr

Keywords: Infrared spectroscopy, Microalgae, Plastic particles, Confocal microscopy

Plastic particles are widespread in the aquatic environment, becoming a nonliving integral environment. Their possible interaction with aquatic organisms is complex and not completely understood. Microalgae, primary producers, play an important role at the base of the aquatic trophic chain. Different pathways of interaction between microalgae and particles can be envisaged depending on particle's characteristics (ion release, adsorption, absorption). The interaction can cause different levels of stress effects on microalgae, such as photosynthesis inhibition, cell wall degradation, and cell death. In this research, the interaction between microplastic and nanoparticles with microalgae were studied and characterized, using traditional methods coupled to infrared spectroscopy as a sensitive tool. Among other advantages, infrared spectroscopy can rapidly provide the biochemical composition of large amounts of microalgae in their living medium. Particle nature, size, and surface chemistry are three potentials factors influencing the patterns of interaction. Several experiments were performed in order to monitor the influence of each factor on microalgae *Chlamydomonas reinhardtii,* a freshwater model organism. The particles studied were gold and polystyrene nanoparticles and polyethylene microparticles with different surface coatings. All infrared spectra were compared using principal component analysis and data clustering by different stress conditions was observed. We could also discriminate microalgae composition variation between the two types of particles. It was notably shown that nanoplastics have much more impact on microalgae than gold nanoparticles. In parallel, confocal microscopy imaging was performed to visualize and confirm interaction of nanoplastic with microalgae. In conclusion, infrared spectroscopy provides a very useful tool to improve knowledge on microalgae and their interaction with the nonliving aquatic environment.

6.2.4 Fate and Effects of Microplastics in the Shrimp *Palaemon varians*

Sarah Riesbeck[1*], Lars Gutow[2], Reinhard Saborowski[2]

[1]Technische Universität Darmstadt, Department of Biology, Schnittspahnstraße 10, 64287 Darmstadt, Germany

[2]Alfred Wegener Institute, Helmholtz Centre for Polar and Marine Research, Am Handelshafen 12, 27570 Bremerhaven, Germany

∗corresponding author: sarah.riesbeck@gmail.com

Keywords: Crustaceans, Histology, Oxidative stress, NADPH oxidase, Enzyme activity

The invention of plastics in the twentieth century yielded durable and cheap materials, which meanwhile are integral components in almost all areas of human life. Production of plastics increased substantially in the last decades. To the same extent, environmental pollution by plastic litter rose. Only recently, we start to realize environmental consequences. Marine litter can have adverse effects on animals of all sizes. Degradation of plastic items generates a continuously increasing number of smaller sized particles, dispersed throughout the oceans. Microplastics, finally ranging in the μm-size classes, may have detrimental effects on marine invertebrates. These effects can be attributed to the cellular level and may lead to an imbalance of the cells' redox state upon microplastic ingestion. This emerging of oxidative stress has adverse effects on cell membranes, proteins, or DNA. The ingestion of microplastics by marine invertebrates and associated effects were examined in the Atlantic ditch shrimp, *Palaemon varians*. This species inhabits coastal regions, estuaries, and brackish water systems, which are strongly affected by anthropogenic pollution. Fluorescent polystyrene microbeads of different sizes were offered as food and served as a tracer within the digestive organs. Uptake into the digestive tract of *P. varians* was analyzed by fluorescence microscopy and histological cryostat sections. The formation of reactive oxygen species (ROS) and the activity of the membrane-bound NADPH oxidase, a superoxide (O_2^-) catalyzing enzyme, served as oxidative stress markers. Activity levels after microplastic incubation were analyzed with the chemiluminescent superoxide-specific reagent lucigenin. The expression of NADPH oxidase in *P. varians* was verified by PCR amplification of an isoform transcript. The outcome of this work may help identify cellular reactions after exposure to microplastics and indicate toxicological impacts on cells, organs, and whole organisms.

6.2.5 Growing Up in a Plastic Ocean: The Impact of Microplastic Uptake in Juvenile Sea Bream

Carolin Müller[1]*, Karim Erzini[2], Werner Ekau[1]
[1]Leibniz Centre for Tropical Marine Research (ZMT), Fahrenheitstraße 6, 28359 Bremen, Germany
[2]Centro de Ciências do Mar (CCMAR), Universidade do Algarve, Campus de Gambelas, 8005-139 Faro, Portugal
*corresponding author: carolin.mueller@leibniz-zmt.de
Keywords: Marine litter, Ichthyoplankton, Early life stages of fish, Ingestion, Fish condition

Coastal ecosystems are known to face severe exposure to microplastic particles as a result of riverine input in combination with the continuously increasing urbanization of on- and offshore regions. Conflicting with their acknowledged ecological, economic, and social importance, these ecosystems nowadays represent the gateway of microplastic pollution to the global oceans. Accounting for the high spatial diversity of coastal habitats, it is hypothesized that early life stages (ELS) of fish, using near shore habitats as nursery grounds and showing a high site fidelity, encounter a gradient of habitat quality and pollution within the coastal environment. In relation to the ingestion and bioaccumulation of marine pollution, microplastic has recently become the focus of scientific and public attention: on the one hand, due to their size ranges allowing the interaction with plankton at the base of the food web and on the other hand, due to their cumulative application resulting in increasing quantities in marine habitats. Although the impacts of ingestion likely differ depending on the size relation between the organism affected and the particles encountered, the exposure of ELS to microplastic and its potential effects on ELS condition have not been analyzed in situ to a sufficient extent. Therefore, the aim of this research project is to assess the spatial and temporal variability of microplastic pollution along with different habitat quality parameters in East Atlantic coastal ecosystems, to compare this with the uptake of microplastic by ELS of sea bream (Sparidae) and to evaluate the physiological effects on survival and growth. In a holistic scientific approach, the field studies are complemented by feeding experiments and biochemical and microbiological analyses. Initial results indicate a high exposure of ELS to microplastic particles in vital recruitment areas; accordingly, the results of this study have the potential to contribute to future fisheries management and environmental protection initiatives.

6.2.6 The Development, and Utilization, of a Novel Microplastic Extraction Technique to Identify Microplastics in the Liver and Blood of European Sea Bass (*Dicentrarchus labrax*)

Gretchen Wagner[1]*, Sinem Zeytin[1], Matthew Slater[1]
[1]Alfred Wegener Institute, Helmholtz Centre for Polar and Marine Research, Am Handelshafen 12, 27570 Bremerhaven, Germany
*corresponding author: gretchen.wagner@awi.de
Keywords: Microplastic, Translocation, KOH, Liver, Circulatory system

With an increased demand for plastics, we see a concurrent increase in plastic waste accumulating in marine systems. This plastic waste breaks down into microplastics which then spread throughout the water column and interact with all marine trophic levels. Alongside our increasing knowledge regarding microplastic accumulation and distribution is the concern that microplastics are impacting marine biota. Despite an abundance of research examining these impacts, there is a knowledge gap that has yet to be filled, that is, examining the potential of microplastic translocation into the liver and blood of marine organisms. To identify these translocated microplastics, a methodology to break down the biogenic material of the marine organism is essential. As of yet, there is no agreed-upon standard methodology, but with the aim of developing a cost-effective, time-saving, and straightforward methodology to break down the biogenic material of fish and extract microplastics for identification, we developed a novel technique utilizing 10% KOH. Using this technique, we focused on the fillet, blood, and liver of European sea bass, *Dicentrarchus labrax,* and achieved digestion efficiencies of 99.72%, 98.24%, and 99.68%, respectively. To test the effectiveness of our method, we exposed juvenile European sea bass to 5 μm fluorescent microplastics incorporated into their diet for 14 weeks and then examined the blood and liver of 20 individual fish to identify translocation. Using our method, we were able to show that 15% of sea bass contained microplastics in their liver and that 70% contained microplastics in their blood. These results build on the limited material published, regarding the translocation of microplastics into the liver of fish. More importantly, this research provides a novel method to digest various forms of biogenic material to extract and identify microplastics and is the first example showing that microplastics can translocate into the blood of a marine organism.

6.3 Abstracts of Poster Presentations

6.3.1 Ingestion of Microplastic Fibers by the Shrimp *Palaemon varians*

Eva Paulischkis[1,2*], Lars Gutow[2], Reinhard Saborowski[2]

[1]Free University of Berlin, Königin-Luise-Straße 1-3, 14195 Berlin, Germany

[2]Alfred-Wegener-Institute, Helmholtz Centre for Polar and Marine Research, Am Handelshafen 12, 27570 Bremerhaven, Germany

∗corresponding author: eva.paulischkis@fu-berlin.de

Keywords: Microplastic pollution, Feeding experiment, Ingestion, Food availability, Regurgitation

Fibers often constitute the major fraction of microplastic debris in coastal marine systems. However, the effects of microplastic fibers upon ingestion by marine biota are widely disregarded in ecotoxicological studies. Therefore, we investigated whether microplastic fibers were ingested by the shrimp *Palaemon varians*, where they remained in the organism, and whether simultaneous availability of natural food influenced fibrinous microplastic uptake. Animals were exposed for 3 hours to a suspension of 5 mg L^{-1} of fluorescent acrylic fibers (300 ± 200 μm) and different concentrations of food (0, 2.5, 5, and 10 mg $animal^{-1}$). After the feeding experiment, the digestive tracts were dissected, and the amount of fibers was counted under a fluorescence binocular. *P. varians* readily ingested microfibers. The number of ingested fibers was lower in animals without food than in animals which received intermediate food concentrations. Moreover, microplastic fibers ingested by *P. varians* remained in their stomachs and were not present in the guts. The day after the feeding experiment, fibers were found agglutinated in gel-like structures within the water, but not in the feces of the animals. The ingestion of microplastic fibers by *P. varians* does not seem to be an active process to assuage hunger because the animals were not eating fibers when the food was absent. They were rather ingesting them by chance, together with their regular food. Further research is required to investigate whether the fibers were egested by regurgitation and, if so, whether this mechanism also appears in other marine organisms, representing an efficient way to eliminate fibrinous microplastic debris from the digestive tract.

7 Biodiversity of Benthic Holobionts: Chemical Ecology and Natural Products Chemistry in the Spotlight

Elham Kamyab[1] and Lars-Erik Petersen[1]

[1]ICBM, AG Umweltbiochemie, Schleusenstr. 1, 26382 Wilhelmshaven, Germany

Elham Kamyab: elham.kamyab@uni-oldenburg.de

Lars-Erik Petersen: lars-erik.petersen1@uni-oldenburg.de

7.1 Call for Abstracts

Oceans cover most of our planet's surface, and certain marine ecosystems, including coral reefs and deep seafloors, have been reported to have a higher biodiversity than tropical rain forests. Marine systems play important roles in maintaining benthic communities and affect humanity by being rich and versatile sources for new drug discoveries. Many benthic communities are characterized by environmentally harsh conditions in matters of pressure, temperature, nutrient availability, and salinity. Hence, reproduction and survival often depend on the formation of bioactive secondary metabolites. We encourage contributions emphasizing community functioning, chemical interactions between marine (micro)organisms and sustainable exploitation of natural products as sources for new drugs.

7.2 Abstracts of Oral Presentations

7.2.1 Molecular Enzymatic Recycling of Sulfated Algal Polysaccharides by Marine Microbes

Nadine Gerlach[1,2*], Andreas Sichert[1,2], Tatjana von Rosen[3], Craig S. Robb[1,2], Jan-Hendrik Hehemann[1,2]

[1]Center for Marine Environmental Sciences, University of Bremen, Bremen, Germany

[2]Max Planck Institute for Marine Microbiology, Bremen, Germany

[3]present address: Department of Biology, ETH Zurich, Switzerland

∗corresponding author: ngerlach@marum.de

Keywords: Fucoidan, Glycoside hydrolase, Glycobiology, X-ray crystallography

An important step in the (marine) carbon cycle is the remineralization of algal polysaccharides by heterotrophic bacteria. Marine algae consist of up to 70% of polysaccharides, which physically support the thallus of macroalgae, have storage properties, or are part of extracellular released organic compounds. Polysaccharide-degrading bacteria specialize on glycans resulting in a pronounced niche partitioning with respect to algal polysaccharide degradation during and after algal blooms. Studying the degradation mechanisms of abundant key degraders will contribute to our knowledge on the ecophysiology of marine bacteria, and thus to our understanding of the biogeochemical cycles in the oceans. Here, we study the utilization of the anionic polysaccharide fucoidan, the main cell wall component of brown seaweed. In the past, fucoidan has been considered as recalcitrant for carbon in the marine environment due to its biochemical and structural features. To date, only a few studies report bacterial enzymatic activity on fucose-containing sulfated polysaccharides (FCSP), or fucoidans. In this study, we invest the molecular mechanisms of carbohydrate-active enzymes (CAZymes) and associated enzymes from a Verrucomicrobia. Currently, we are working on the cloning and expression of putative endoacting glycoside hydrolases (GHs) using recombinant plasmids with an N-terminal hexa histidine-tag and expression in *E. coli* strains. So far, we solved the structure of an exoacting α-L-fucosidase GH29 and its associated sulfatase by protein crystallization and X-ray crystallography. GH29 is thermal stable until 37 °C and was most active on *4*-nitrophenyl-α-L-fucopyranoside in Na-citrate pH 5.8. We predicted the putative catalytic residues by multiple-sequence alignment as classic nucleophile and acid/base pair. Side-directed mutagenesis of both residues resulted into an activity loss by 70–80%. The aim of this study is to gain insights into the extraordinary carbohydrate-degrading capacity of this bacterial isolate as well as its role in the marine carbon cycle.

7.2.2 Characterizing the Exometabolomic Fingerprint of Coral Reefs

Milou G.I. Arts[1,2*], Craig Nelson[3], Linda Wegley Kelly[4], Daniel Petras[5], Irina Koester[6], Zachary Quinlan[4], Pieter C. Dorrestein[5], Andreas F. Haas[1,2]

[1]NIOZ Royal Netherlands Institute for Sea Research, 1797 SH, Den Hoorn Texel, The Netherlands

[2]Utrecht University, 3584 CB Utrecht, The Netherlands

[3]University of Hawai'i at Mānoa, School of Ocean and Earth Science and Technology, Honolulu, HI, United States

[4]San Diego State University, Department of Biology, San Diego, CA, United States

[5]University of California San Diego, Collaborative Mass Spectrometry Innovation Center, La Jolla, CA, United States

[6]Scripps Institution of Oceanography, La Jolla, CA, United States

∗corresponding author: milou.g.i.arts@gmail.com

Keywords: Dissolved organic matter, LC-MS/MS, Molecular networking, Environmental metabolomics, Marine microbial communities

One of the most complex and abundant exometabolomes on earth is marine dissolved organic matter (DOM). Benthic primary producers significantly add to the coastal DOM pool by releasing photosynthates and other metabolites. The quantity and quality of the available DOM has significant effects on the composition and metabolism of the surrounding microbial community in the water column. Changes in the microbial community have in turn implications for the health of coral holobionts and thus affect the benthic community composition, including the sudden change to algae overgrowing coral reefs (coral to algae phase shift). The underlying mechanisms and especially the molecular makeup of the DOM that governs these processes remain largely unknown. Solid phase extraction of DOM and high-throughput, high-resolution liquid chromatography tandem mass spectrometry (LC-MS/MS) allow for the characterization (molecular fingerprint) of DOM on a large scale. Molecular networking (Global Natural Product Social Molecular Networking, GNPS) makes it further possible to more precisely identify DOM components or molecular groups through comparison against a rapidly increasing library of mass spectra and connections of molecules with similar structure. Here we describe the metabolomic signal of five different primary producers (two corals *Pocillopora verrucosa* and *Porites lobate,* CCA, turf algae, and the macro algae *Dictyota* spp.) from the coral reef system of Moorea, French Polynesia. Molecular network analysis of all 9101 identified features created 5283 distinct clusters of molecular groups. Results show that the exometabolomic signal distinctively clusters by functional groups. Difference in potential energy (ΔG) stored in the produced exometabolomes might be an explanation for the previously identified metabolic shift in the microbial community during reef decline. The presented data shows a comprehensive overview on coral reef primary producer exudates. This provides a first stepping-stone in unraveling the molecular underpinnings of microbially mediated, algal-induced coral reef phase shifts.

7.2.3 Antimicrobial Activity of Bacteria Isolated from the Sea Cucumber *Stichopus vastus*

Joko Tri Wibowo[1,2*], Matthias Y. Kellermann[1], Dennis Versluis[1], Masteria Yunovilsa Putra[2], Tutik Murniasih[2], Kathrin I. Mohr[3], Joachim Wink[3], Peter Schupp[1]

[1]Institute for Chemistry and Biology of the Marine Environment (ICBM) Terramare, Schleusenstraße 1, 26382 Wilhelmshaven, Germany

[2]Research Center for Oceanography LIPI, Jl. Pasir Putih Raya 1, Pademangan, Jakarta Utara, 14430, Indonesia

[3]Helmholtz Centre for Infection Research, Inhoffenstraße 7, 38124 Braunschweig, Germany

∗corresponding author: joko.tri.wibowo@uni-oldenburg.de; joko018@lipi.go.id

Keywords: Sea cucumber-associated bacteria, *Stichopus vastus*, Antimicrobial, Bioactive compounds, *Streptomyces*

Marine bacteria produce unique and bioactive compounds. Associated with marine invertebrates, bacteria are often identified as the active producer of bioactive compounds. Sea cucumbers are one of the marine invertebrates that are known not only for its nutritious values but also for its potent source of bioactive substances. However, there are only a few studies that investigated the presence and function of sea cucumber-associated bacteria. The aim of this study is to investigate the bioactive compounds produced not by the host itself but by the sea cucumber-associated bacteria. Therefore, the sea cucumber *Stichopus vastus*, known to contain bioactive compounds, was collected in Lampung (Indonesia). The associated bacteria were isolated from both, the inner and outer body part of the sea cucumber using either Marine Agar, M1 or M2 media. Selected axenic bacteria were identified by 16s rRNA, grown to high density followed by organic solvent extraction. The activity of potential bioactive compounds was tested against a set of test microorganisms and the structure of known and potentially novel compounds elucidated using high-resolution mass spectrometry (HRMS). Here we identified 49 different bacteria (via 16s rRNA) that consist of *Actinobacteria* (40.8%), *Firmicutes* (40.8%), and *Proteobacteria* (18.4%). Among the detected bacteria, 4.1% had a sequence similarity of less than 97%, indicating the discovery of potential novel bacterial strains. An antimicrobial assay of the bacterial extracts showed that 45.8% of isolated bacteria were active against the Gram-positive bacteria *Bacillus subtilis*. Furthermore, the isolation of bioactive compounds from larger fermentation experiments followed by compound analysis via HRMS and nuclear magnetic resonance (NMR) spectroscopy indicated that one bioactive compound, isolated from *Streptomyces* sp., JK 21, was related to valinomycin. Valinomycin is a depsipeptide that has antimicrobial bioactivities and has been previously found in *Streptomyces*.

7.2.4 Antarvedisides A–B from Manglicolous Lichen, *Dirinaria consimilis* (Stirton) D.D. Awasthi and Their Pharmacological Profile

Vinay Bharadwaj Tatipamula[1]*, Girija Sastry Vedula[1]

[1]AU College of Pharmaceutical Sciences, Andhra University, 530003, Visakhapatnam, India

*corresponding author: vinaybharadwajt@gmail.com

Keywords: Depsides, Antioxidants, Anti-inflammatory, Cytotoxicity, Acute toxicity studies

Aquatic environment is teeming with several life forms which produce a remarkable variety of structurally exciting natural products. Mangals are one such form that host a number of animals and microorganisms. As mangroves survive under stressful surroundings such as intense environments, high concentration of salt and moisture, and low and high tidal water, the epiphytes (lichens – mutualistic existence of algae and fungi) habituated on these species may vary in their chemical constituents. Lichens betide to mangroves are noted as manglicolous lichens. In addition, there are very few studies reported from manglicolous lichens. Keeping in mind of the aforementioned aspects, we have established the chemical and biological profile of an under-investigated species, *Dirinaria consimilis* using standard models and procedures. The chemical examination of the acetone extract yields two new depsides, antarvedisides A-B (1 and 2) along with five known compounds atranorin (3), divaricatic acid (4), usnic acid (5), scrobiculin (6), ethyl everninate (7) and 2'-O-methyldivaricatic acid (8). The structures of all the metabolites were elucidated by using 1D and 2D NMR, IR, HRMS, ESI-MS, GC-MS spectral data. The acute toxicity studies of acetone extract of *D. consimilis* depicted no signs of toxicity up to 2000 mg/Kg body weight. Compounds 1 and 2 showed potent inhibitory profile against DPPH, ABTS, and superoxide free radicals, with better IC_{50} values than that of the standard (ascorbic acid). The compound 2 and 3 depicted better inhibition of protein denaturation with IC50 values of 0.81 and 0.39 mg/mL, respectively, while standard (indomethacin) with 0.11 mg/mL. The outcomes of MTT assay revealed that the compounds 1, 2, 4, 6, and 8 have prominent degree of specificity toward MCF-7, DLD-1, HeLa, FADU, and A549 cancer cell lines and very less degree of specificity toward normal human epithelial cell lines.

7.2.5 From Discovery to Production: *Microascus brevicaulis* Strain LF580 as a Case Study

Annemarie Kramer[1]*

[1]Flensburg University of Applied Sciences, Kanzleistraße 91-93, 24943 Flensburg, Germany

*corresponding author: annemarie.kramer@hs-flensburg.de

Keywords: Marine biotechnology, Marine natural products, Marine fungi

The discovery of new natural products is essential for the development of new drugs. In order to translate the immense bio- and chemodiversity of marine habitats into products, marine biotechnology plays an important role at several steps in the discovery-to-development pipeline. Furthermore, the implementation of biotechnological production processes based on state-of-the-art technologies enables the sustainable use of marine resources. The fungus *Microascus brevicaulis* strain LF580 was isolated from the marine sponge *Tethya aurantium*. Its capability to produce the two scopularides A and B was already discovered in 2008. Both cyclodepsipeptides show distinct cytotoxic activity against tumor cell lines. Several methods were adapted and utilized in order to bridge gaps in the drug discovery pipeline of marine natural products. The focus was set on, e.g., comparative metabolome analysis, transfer of cultivation into stirred tank reactor, development and validation of screening processes of mutant libraries, and qualitative and quantitative proteome analysis. On the one hand, the integrated approach provided fundamental knowledge on filamentous fungi and their biology and, on the other hand, enlarged the toolbox suitable for non-model fungal producer strains. Moreover, the impact of physiology on the biotechnological process design was demonstrated. The work on the fungal strain LF580, which was conducted in the frame of the project MARINE FUNGI (May 2011–April 2014, FB7/265926, coordinated by GEOMAR Helmholtz Centre for Ocean Research Kiel, Germany), is discussed as a case study to highlight steps toward the implementation of non-model organisms of marine origin as producer strains.

7.3 Abstracts of Poster Presentations

7.3.1 Characterization of Actinomycetes with Antibiotic-Producing Potential from the Rhizosphere of Mangrove Plant *Camptostemon philippinensis*

Jenevieve P. Hara[1*], Jhonamie Mabuhay-Omar[1]

[1]Western Philippines University-Puerto Princesa Campus, Puerto Princesa City 5300, Palawan, Philippines

*corresponding author: jenevieve.hara@imbrsea.eu

Keywords: Mangroves, Microorganisms, Isolation, Enumeration, Antibacterial test

Actinomycetes are among the microorganisms of special interest since they are prolific producers of microbial bioactive secondary metabolites for potential agricultural, pharmaceutical, and industrial applications. This study aimed to enumerate, isolate, and characterize antibiotic-producing culturable actinomycetes associated with the rhizosphere of mangrove plant, *Camptostemon philippinensis*, in Turtle Bay, Puerto Princesa City. The dilution plate count technique was used for enumeration and abundance, while streak plate method was used to determine the antibiotic-producing potential. Characterization was done following standard protocols for cultural, physiological, and biochemical assays. Three sample replicates yielded a total number of 135 x 10^6 colony forming units (CFUs). Out of these CFUs, 14 distinct colonies differing in cultural morphology (coded A1 to A14) were chosen and isolated for antibacterial tests. Nine of 14 isolates showed zones of inhibition in any of *Escherichia coli* and *Staphylococcus aureus*. The analysis of variance and post hoc tests showed that A12 had the highest antibacterial potential against both *E. coli* and *S. aureus*. A12 was proven to be catalase positive but negative for fermentation of carbohydrates and hydrolysis of starch. *Camptostemon philippinensis* do have abundant associated actinomycetes with some antibiotic-producing potential.

7.3.2 The Potent Antimicrobial Activity of Brominated Compounds Extracted from the Marine Sponge *Lamellodysidea* sp. (Seribu Islands, Indonesia)

Muhammad R. Faisal[1*], Matthias Y. Kellermann[1], Masteria Y. Putra[2], Tutik Murniasih[2], Kathrin I. Mohr[3], Joachim Wink[3], Peter J. Schupp[1]

[1]Carl-von-Ossietzky University Oldenburg, Institute for Chemistry and Biology of the Marine Environment (ICBM), Schleusenstr. 1, 26382 Wilhelmshaven, Germany

[2]Research Center for Oceanography (RCO), LIPI, Pasir Putih Raya Street No.1, Jakarta, 14430, Indonesia

[3]Helmholtz Centre for Infection Research, Inhoffenstraße 7, 38124 Braunschweig

*corresponding author: muhammad.reza.faisal@uni-oldenburg.de

The marine sponge *Lamellodysidea* sp. has been reported to contain bioactive compounds that function as antimicrobial, antitumor, and anti-HIV, as well as enzyme inhibition and increased cytotoxicity against different cancer cell lines. This sponge and/or their associated symbionts may hold additional information on novel bioactivities that are in high demand for human health and society. Here we study the metabolome of *Lamellodysidea* sp., via high-performance liquid chromatography-mass spectrometry (HPLC-MS), with a strong focus on the abundance, diversity, and bioactivity of the diverse brominated compounds. We collected the sponge in Seribu Islands (Indonesia) and extracted sponge, together with their associated bacteria, three times with MeOH:EtOAc (1:1). A clinical microbial assay was conducted on the crude extract at the Helmholtz Centre for Infection Research (HZI) in Braunschweig using the qualitative minimum inhibition concentration (MIC) method. The result showed that among the eight tested pathogenic isolates, four showed tremendous activity, including two pathogenic bacteria (*Bacillus subtilis* and *Staphylococcus aureus*) and two pathogenic fungi (*Rhodotorula glutinis* and *Mucor hiemalis*). To further determine the active components of the crude extract, the complex organic mixture was further fractionated using liquid-liquid, solid phase extraction and preparative HPLC separation protocols. For tentative peak identification, HPLC-MS was applied on the semi-pure and purely isolated target peaks. Up until now, we identify six different brominated compounds and fully isolated the two most abundant ones (2.4 mg and 8.2 mg). The structure of the latter two was determined via nuclear magnetic resonance (NMR). This study aims to uncover novel structural information and their bioactivities of the highly abundant brominated compounds found within *Lamellodysidea* sp.

8 Sponges (*Porifera*): Fantastic Filter Feeders

Mainah Folkers[1] and Titus Rombouts[1]

[1]Institute for Biodiversity and Ecosystem Dynamics, University of Amsterdam, Science Park 904, 1098 XH Amsterdam, The Netherlands

Mainah Folkers: mainahfolkers@gmail.com

Titus Rombouts: titusrombouts@gmail.com

8.1 Call for Abstracts

Over 8000 marine and freshwater sponge species are distributed over polar, temperate, and tropical ecosystems as well as intertidal, photic, and abyssal zones. Recent research shows that sponges are important contributors to ecosystem functioning in these environments. Nevertheless, sponges are often underrepresented in research, conservation, and monitoring programs. We would like young marine sponge researchers worldwide to share their results with us. Sharing of knowledge on different sponge disciplines will lead to new insights and potential new collaborations on future sponge research.

8.2 Abstracts for Oral Presentations

8.2.1 Do Sponges Bring Life – or Destruction – to Shallow and Deep Reef Ecosystems?

Jasper M. de Goeij[1*]

[1]Department of Freshwater and Marine Ecology, Institute for Biodiversity and Ecosystem Dynamics, University of Amsterdam, PO Box 94248, 1090 GE Amsterdam, Netherlands

*invited speaker, corresponding author: j.m.degoeij@uva.nl

Keywords: Food web, Coral reef biogeochemistry, Stable isotope probing, Ecosystem engineers, Eukaryote-prokaryote interactions

Coral reefs are iconic examples of biological hotspots, highly appreciated because of their ecosystem services. Yet, they are threatened by human impact and climate change, highlighting the need to develop tools and strategies to curtail changes in these ecosystems. Remarkably, ever since Darwin's first descriptions of coral reefs, it has been a mystery how one of Earth's most productive and diverse ecosystems thrives in oligotrophic seas, as an oasis in a marine desert. The common view on how highly productive systems cope with oligotrophic conditions has changed completely with the discovery of the sponge loop. Sponges are now increasingly recognized as key ecosystem engineers, efficiently retaining and transferring energy and nutrients on the shallow reef and in the deep sea. As a result, current reef food web models, lacking sponge-driven resource cycling, are incomplete and need to be redeveloped. However, mechanisms that determine the capacity of sponge "engines," how they are fueled, and drive communities are unknown. In this perspective I will discuss how sponges integrate within the novel reef food web framework. Sponges will be evaluated on functional traits in the processing of food (e.g., morphology, associated microbes, pumping capacity), and to what extent these different traits are a driving force in structuring shallow- to deep-sea reef ecosystems. Finally, as climate change causes the onset of alterations in the community structure and food web of reef ecosystems, there is evidence accumulating that certain biological pathways are triggered, such as the sponge loop and the microbial loop that may shift reef ecosystems faster than their original stressors (e.g., warming oceans and ocean acidification). Unfortunately, these biological pathways receive much less attention at present, which seriously hampers our ability to predict future changes within reef ecosystems.

8.2.2 Assessing Coral Reef Biomass Reveals Importance of Marine Sponges

N. Kornder[1*], S. J. Martinez[1], F. J. Pocino[1], M. J. L. Zalm[1], B. Mueller[1], M. J. A. Vermeij[1,2], J. M. de Goeij[1]

[1]Department of Freshwater and Marine Ecology, Institute for Biodiversity and Ecosystem Dynamics, University of Amsterdam, PO Box 94248, 1090 GE Amsterdam, the Netherlands

[2]Carmabi Foundation, P.O. Box 2090, Willemstad, Curacao

*corresponding author: n.a.kornder@uva.nl

Keywords: Coral reef communities, Marine sponges, Photogrammetry

Monitoring the biomass and distribution of organisms in coral reef communities is an essential component of understanding these complex, three-dimensional (3D) ecosystems. Traditionally, reef community composition assessments are done on the basis of two-dimensional (2D) projections. This offers fast and reliable data for surface cover estimates of dominant benthic species, including most Caribbean corals, but may underestimate the contribution of cryptic or erect organisms, such as sponges. Also, the preferred ecological currency to study energy flows on coral reef ecosystems is biomass which cannot always be derived from projected abundance estimates. This is especially true for cryptic communities whose biomass can exceed that of exposed organisms generally included in 2D reef surveys (Hutchings 1974, Proc 2nd Int Coral Reef Symp, Brisbane, Australia). In this study, we estimated the biomass of all major benthic reef biota in exposed *and* cryptic habitats to provide an integrated estimate of the total biomass on a Caribbean reef. Using a combination of in situ measurements, laboratory analyses, and structure-from-motion (SfM) photogrammetry, we show that biomass estimates from 3D methods differ enormously from those generated with 2D methods, especially severely underestimating the dominance of sponges in terms of biomass on reefs. We also found a positive correlation between the biomass of cryptic sponges and reef-building corals, which was not visible in 2D estimates. Considering the key contribution of sponges to the retention of carbon on Caribbean coral reefs (de Goeij et al. 2013, Science 342:108-110), we show that our 3D assessment of benthic biomass is essential to study ecological and biogeochemical processes on coral reefs.

8.2.3 Energy Sources of the Bacteriosponge *Geodia* Forming Giant Landscapes at Langseth Ridge, Arctic, 87°N

T. Morganti[1*], A. De Kluijver[2], A. Purser[3], S. Beate[4], J. Middelburg[2], A. Boetius[1,3]

[1]Max Planck Institute for Marine Microbiology, Celsiusstr. 1, 28359 Bremen, Germany

[2]University of Utrecht, Vening Meineszgebouw A, Princetonlaan 8a, 3584 Utrecht, The Netherlands

[3]Alfred Wegener Institute Helmholtz Center for Polar and Marine Research, Am Handelshafen 12, 27570 Bremerhaven, Germany

[4]GEOMAR Helmholtz Centre for Ocean Research Kiel, Düsternbrooker Weg 20, 24105 Kiel, Germany

*corresponding author: tmorgant@mpi-bremen.de

Keywords: Arctic deep sea, Stable isotopes, Sponge grounds, Food web, Seamounts

During the *Polarstern* PS101 expedition to the Central Arctic, Langseth Ridge 85–87°N, dense populations of the bacteriosponge *Geodia* sp. were discovered, covering a chain of three ice-covered seamounts (Karasik, Central, and Northern mounts) at 500–1000 m water depth. This community, dominated by *Geodia parva* and *G. hentscheli*, represents the densest accumulation of sponges known from Northern seas. This sponge landscape represents a biodiversity hotspot in the Arctic Ocean, raising the question as to its food supplies and the role of its microbial symbionts. To investigate the carbon sources and trophic position of the sponges, we measured the stable carbon and nitrogen isotope signatures ($\delta^{13}C$ and $\delta^{15}N$ respectively) of sponge tissues, sediments, associated macrofauna, and particulate organic matter (POM) in water. $\delta^{13}C$ and $\delta^{15}N$ signatures were similar among sponge species and locations. The sponge isotope signatures indicate that planktonic POM is not a main food source for these species, but the sponges also did not match typical chemolithoautotrophic profiles of symbiotic life from venting seamounts. They were observed to sit on a layer of dead tubeworms and mussel shells, and display a $\delta^{13}C$ similar to the tubes, but slightly lower in $\delta^{15}N$. Overall, the low $\delta^{15}N$ signature of the bacteriosponges compared with glass deep-sea sponge species, lacking of microbial symbionts, indicates that they might obtain nutrient differently. We currently test the hypotheses that the giant sponge accumulations are nourished by sedimenting sea-ice algae, or live from the remnants of past vent communities with the help of their microbial symbionts.

8.2.4 Creatures of the Cold and Deep: The Sponge Microbiota of Langseth Ridge

Beate M. Slaby[1*], Kathrin Busch[1], Autun Purser[2], Teresa Morganti[3], Hans Tore Rapp[4], Antje Boetius[2], Ute Hentschel[1,5]

[1]GEOMAR Helmholtz Centre for Ocean Research Kiel, Düsternbrooker Weg 20, 24105 Kiel, Germany

[2]Alfred Wegener Institute, Am Handelshafen 12, 27570 Bremerhaven, Germany

[3]Max Planck Institute for Marine Microbiology, Celsiusstr. 1, 28359 Bremen, Germany

[4]University of Bergen, Department of Biological Sciences, Thormøhlensgt. 53 A/B, PB 7803, 5020 Bergen, Norway

[5]Christian-Albrechts University of Kiel, Christian-Albrechts-Platz 4, 24118 Kiel, Germany

*corresponding author: bslaby@geomar.de

Keywords: Metagenomics, Metatranscriptomics, Amplicon sequencing, Ecosystem, Host-associated microbiome

On RV *Polarstern* cruise PS101 to the Central Arctic Langseth Ridge, the demosponge *Geodia parva* was found to dominate the summits of three adjacent seamounts: Karasik seamount and the newly discovered Central mount and Northern mount, as well as the saddle between Karasik and Central mount. In their megafaunal community composition, the different sites appeared to be similar, but differences in density and relative abundances were apparent during Ocean Floor Observation System (OFOS) transects. Thus, our main research questions are: Are there differences also in the sponge microbiota? How are the microbes involved in the sponges' nutrition? Which roles do the different microbes play in the system? By 16S rRNA gene amplicon sequencing, differences in sponge microbial community composition were addressed and the sponge microbiota were compared to those of associated seawater and sediment. While the demosponges contained very similar microbial communities with no significant differences between habitats or host species, the sponge microbiota were clearly different from the surrounding seawater and sediment references. A high-resolution metagenomic and metatranscriptomic study of the dominant *Geodia parva* was conducted to assess the functional role of the sponge microbiome and to discover possible differences between the habitats. Taking into account the large sponge biomass at Langseth Ridge, the microbial involvement in nutrient turnover could affect the sponge host and consequently, also have cascading effects on the sponge grounds ecosystem. SponGES has received funding from the European Union's Horizon 2020 research and innovation program under grant agreement No 679849. This output reflects only the authors' view and the European Union cannot be held responsible for any use that may be made of the information contained therein. The funders had no role in study design, data collection and analysis, decision to publish, or preparation of the manuscript.

8.2.5 Response of a Deep-Sea Sponge Holobiont, *Geodia barretti*, to Acute Crude Oil Exposure: A Mesocosm Experiment

Chloé Stévenne[1,2*], Sonnich Meier[1], Bryan Wilson[3], Raymond John Bannister[1]

[1]Institute of Marine Research, Postbox 1870, Bergen, Norway

[2]University of Liège, 27 Bld du Rectorat, Liège, Belgium

[3]University of Bergen, Thormøhlensgate 53B, Bergen, Norway

*corresponding author: chloe.stevenne@student.uliege.be

Keywords: Sponges, Oil spills, Contamination, Microbiome, Deep-sea

Exploration and extraction of petroleum reserves in deep-sea marine ecosystems are ongoing activities worldwide and are continuously expanding with the identification of new reservoirs. While a wealth of knowledge is known on the effects of oil spills on fish and other organisms in shallow water environments, the effects of subsurface oil spills (blowout events like the Deepwater Horizon) on deep-sea ecosystems have been less well studied. With exploration activities in deep-sea ecosystems predicted to increase, there are concerns regarding the accidental release of hydrocarbons into deep-sea ecosystems and the subsequent cascading effects on associated fauna. Sponges are abundant and ecologically valuable, known to be highly efficient filter feeders, contributing significantly to benthic pelagic coupling and providing habitat for a suite of organisms. However, sponges, which are sessile in nature, accumulate contaminants present in their ambient environments, making them vulnerable to oil spills. Surprisingly, the impacts of oil on deep-sea sponges remain unexplored, despite sponges being particularly dominant around oil and gas exploration locations in the Northern Atlantic. Here we present key findings from a mesocosm study, where we exposed the locally abundant deep-sea sponge *Geodia barretti* to three ecologically relevant oil concentrations for a duration of 8 days, followed by a recovery period of 30 days. A holistic approach to elucidate the effects of oil on *G. barretti* focused on measuring changes in physiology (respiration and filtration efficiencies), cellular stress (lysosomal membrane stability), and the structure of the sponge-associated microbiome using high-throughput sequencing of 16s rRNA gene amplicons. This study enhances our understanding of the vulnerability of deep-sea sponges to hydrocarbon exposure, providing useful data for managing risks associated with oil and gas exploration in the Northern Atlantic.

8.2.6 Reef Specific Microbiomes in a Caribbean Sponge

Megan Clay[1,2], Thomas Swierts[1], Roy Belderok[1,2], Daniel Cleary[3], Nicole de Voogd[1]

[1]Naturalis Biodiversity Centre, Leiden, Netherlands

[2]University of Amsterdam, Amsterdam, Netherlands

[3]University of Aveiro, Aveiro, Portugal

*corresponding author: megan.clay@student.auc.nl

Keywords: Microbial ecology, *Porifera*, Biogeography, Amplicon sequencing

The Caribbean giant barrel sponge, *Xestospongia muta*, is a high microbial abundance sponge with a variable microbiome. Intercontinental comparison of the microbiome of its Indo-Pacific sister species, *X. testudinaria* has revealed that unlike many sponges, its microbiome varies in accordance to geography, not genotype. Here, 16s analysis of endosymbiotic communities demonstrates that small-scale local geography (>1 km) and environmental conditions correlate with distinct clustering of microbiomes in Curaçaoan giant barrel sponges. The main environmental factors which appear to relate to microbiome signature are the mean altered surface of the watershed, mean delta N15, the density of coral recruits, macroalgal cover of the benthos, turf algae height, and carnivorous fish biomass. These factors all relate to human impact through pollution and overfishing, two problems the waters of Curaçao suffer from. The microbiome of the sponges is potentially being altered by anthropogenic forcing. Anthropogenic activities which lead to soil erosion may also affect light levels on reefs. A pilot study which manipulated light levels sought to test whether a shading altered sponge feeding and sponge microbiome, using InEx and 16s rRNA analysis. No correlation between shading, feeding, and microbial community was found, although genomic analyses pointed to a temporal shift in microbiome. This temporal shift must be further studied as it may be important in light of the current sampling techniques used for demographic analyses.

8.2.7 Larvae Under the Light: Analyzing the Microbial Community of the Coral Reef Sponge, *Amphimedon queenslandica*, Larvae in Response to Varying Environmental Light Regimes

Olivia H. Hewitt[1,2*], Sandie M. Degnan[1]

[1]The University of Queensland, Brisbane, Queensland 4072, Australia

[2]Université Libre de Bruxelles, Bruxelles, 1050, Belgium

*corresponding author: o.hewitt@uq.net.au

Keywords: Microbiome, Bacteria, *Porifera*, Maturation, Stability

Marine sponges are one of the most basal of the extant metazoan lineages. They have neither muscles nor a nervous system and live as sessile filter feeders on the ocean floor. Yet, sponges are a significant and dominant element of marine ecosystems. In nutrient-poor tropical reef systems, they help solve "Darwin's paradox," and they function as reef builders and are prominent nutrient recyclers. They achieve these ecosystem engineering roles with the help of a rich community of symbiotic microbes, collectively known as their microbiome. In adult sponges, microbe functions appear broadly to include defense, nutrition, and nutrient assimilation. However, the role of microbes during earlier life cycle stages remains poorly understood. We begin to address the role of the microbiome during the larval life phase of the coral reef sponge, *Amphimedon queenslandica*, by assessing the stability of its microbial community under experimentally varying light regimes that are known to affect larval settlement. *A. queenslandica* is a low microbial abundance sponge, dominated by three proteobacterial symbiont species that are vertically inherited from the maternal adult. Light regulates circadian rhythms in *A. queenslandica*, and a light-to-dark transition provides a critical environmental cue for larval maturation and settlement, but it is unknown if, and how, this environmental light affects the bacterial symbionts in the larvae. Based on 16s rRNA sequences, we quantified the activity and abundance of the dominant bacterial symbionts in larvae of different ages that had been exposed to varied environmental light regimes. We investigated changes in spatial distribution of the bacterial symbionts using fluorescent in situ hybridization (FISH) imagery. Our analyses reveal the nature of light-induced effects on the larval microbiome for *A. queenslandica*, thus furthering our insight into environmental regulation of symbiont function during this critical life stage.

8.2.8 The Response of the Mediterranean Sponge *Aplysina aerophoba* to Grazing by the Sea Slug *Tylodina perverse*

Yu-Chen Wu[1,2*], María García-Altares[3], Marta Ribes[4], Ute Hentschel[1,2], Lucia Pita Galán[1]

[1]GEOMAR Helmholtz Centre for Ocean Research, RD3 Marine Microbiology, Düsternbrooker Weg 20, D-24105 Kiel, Germany

[2]Christian-Albrechts University of Kiel, Düsternbrooker Weg 20, D-24105 Kiel, Germany

[3]Leibniz Institute for Natural Product Research and Infection Biology – Hans Knöll Institute (HKI), Biomolecular Chemistry Department, Adolf-Reichwein-Straße 23, D-07745 Jena, Germany

[4]Institut de Ciències del Mar, Departament de Biologia Marina i Oceanografia, Pg. Marítim de la Barceloneta, 37-49, 08003 Barcelona, Catalonia, Spain

∗corresponding author: yuwu@geomar.de

Keywords: Sponge, Chemical defense, Secondary metabolites, Spherulous cells, MALDI imaging

Sponges (phylum *Porifera*) are sessile invertebrates that produce chemical defenses (i.e., secondary metabolites) to defend against predators, competition, and fouling. Nevertheless, certain sea slugs (Phylum Mollusca: Class Opistobranchia) have specialized in grazing on specific sponge species, which leaves the sponge tissue exposed. To date, no study has examined the response of sponges upon grazing. Here, we investigated the Mediterranean sponge *Aplysina aerophoba* which produces brominated alkaloids with antibacterial and deterrence properties against generalist predators. These brominated compounds are accumulated in a particular cell type, so-called spherulous cells. However, *A. aerophoba* is the main food source for a specialist grazer – the sea slug *Tylodina perversa* – which tolerates brominated compounds and also exploits them for its own defense. By ways of microscopical technologies and matrix-assisted laser desorption/ionization (MALDI) imaging, we aim to investigate the cellular mechanisms and the potential role of secondary metabolites in the response of the sponge to grazing. We hypothesized that injury can induce a recruitment of spherulous cells with secondary metabolites. Three treatments were applied: control, grazing, and mechanical damage. Samples were collected 3 hours, 1 day, 3 days, and 6 days after treatment. Our results showed that spherulous cells are recruited to the injured site in a time-dependent manner. MALDI imaging showed that aeroplysinin-1, a bioconversion product from brominated precursors, localizes usually at the sponge surface. Upon injury, this compound accumulates at the damaged surface. Moreover, complementary sponge transcriptome sequencing was performed to identify the molecular response against grazing versus injury. As spherulous cells are common in many members of the class Demospongiae, the recruitment of defensive cells may also occur in other sponges and contribute to the chemical defense of these filter feeders. This study contributes to understanding the evolutionary mechanisms that provide selective advantages for the sponge to survive under grazing pressure.

8.3 Abstracts of Poster Presentations

8.3.1 Unraveling the DOM Uptake and Assimilation in Sponges: A Guide for Cell Separation and Stable Isotope Identification for Sponge Holobionts in Coral Reef Ecosystems

Joëlle van der Sprong[1]

[1]University of Amsterdam, Postbox 19268, Amsterdam, The Netherlands

∗corresponding author: joelle.vandersprong@student.uva.nl

Keywords: Metabolic interactions, DOM-shunt, Sponge Loop

Sponges efficiently shunt dissolved organic matter (DOM), which is one of the most important carbon sources on coral reefs, but largely unavailable to most heterotrophic reef organisms to higher trophic levels in the form of detritus. This process is better known as the sponge loop. However, the mechanisms that drive this process are not entirely understood. It is assumed that sponge-associated microbes mainly contribute to these complex metabolic processes. Nevertheless, previous studies have shown that sponges with relatively low numbers of microorganisms are just as effective in the processing of DOM as sponges with high numbers of microorganisms. The aim of this project is to understand whether the sponge cells themselves are responsible for DOM uptake and assimilation or that their microbial associates make an important contribution. For this study a modified cell separation protocol will be developed for efficiently separating the sponge and sponge-associated microbial cells in the sponge holobiont. This technique will then be applied combined with tracer incubation experiments using ^{13}C- and ^{15}N-labeled DOM to identify where the DOM is being processed. Therefore, this study will be an important step in unraveling the complex metabolic interactions involved with nutritional processing in sponges.

8.3.2 Impact of Shading on the Chemistry of a Cyanosponge

Franziska Curdt[1]*, Matthias Y. Kellermann[1], Sven Rohde[1], Peter J. Schupp[1]

[1]Carl von Ossietzky University Oldenburg, Institute for Chemistry and Biology of the Marine Environment (ICBM), Schleusenstr. 1, 26382 Wilhelmshaven, Germany

∗corresponding author: franziskacurdt@posteo.de

Keywords: Cyanosponge, Symbiosis, PAM, *Synechococcus spongiarum*

Despite the large prevalence of sponge-associated cyanobacteria, we still lack important information on the nature and metabolic processes involved in the sponge microorganism relationship. Different light conditions should affect the symbiosis of photosynthetic cyanobacteria and the host sponge. This project aimed to study the impact of light availability on the amount of cyanobacteria in the sponge tissue, their spatial distribution, as well as the photosynthetic activity and pigment concentration. For this purpose, clones of a cyanosponge of the family Thorectidae were analyzed before and after exposition to four different light levels for nine weeks. During the experiment, the sponge growth rates were determined photographically, the photosynthetic activity was measured using a pulse amplitude modulation (PAM) fluorometer and pigment concentrations were measured after extraction with light absorbance spectrometry. The number of bacteria was evaluated using epifluorescence microscopy imaging of cryosections. Concentrations of different pigments were done by fluorescence absorption emission. Distribution of chlorophyll a was confirmed by liquid chromatography-mass spectrometry (LC/MS) and fluorescence absorption/emission spectroscopy. It was found that a rod-shaped cyanobacteria, very likely *Synechococcus spongiarum*, is present with a density of 8,000–18,000 bacteria per mm^2 in the entire sponge tissue. Our analysis showed that the abundance of cyanobacteria decreased at dark conditions. At the same time, photoactivity (yield) and chlorophyll a concentrations declined as did phyocyanin and phycoerythrin concentrations. LC/MS analyses revealed that additionally, the chemistry and visual appearance of the sponge changed with manipulated light conditions, which can give further information on the physiological and ecological role of cyanobacteria in sponges.

9 Higher Temperatures and Higher Speed: Marine Bioinvasions in a Changing World

Philipp Laeseke[1], Jessica Schiller[1] and Jonas Letschert[2]

[1]Marine Botany, University of Bremen, Leobener Straße NW2, 28359 Bremen, Germany

[2]Leibniz Centre for Tropical Marine Research (ZMT), Bremen, Germany

Philipp Laeseke: philipp.laeseke@uni-bremen.de

Jessica Schiller: j.schiller@uni-bremen.de

Jonas Letschert: jonas_letschert@web.de

9.1 Call for Abstracts

Global warming leads to both loss of habitat and newly accessible space for species. Likewise, ship traffic, aquaculturing efforts, and the amount of floating nondegradable debris are increasing. Thereby, new pathways across the oceans are becoming available, leading to migrations and possible displacement of native species. Analyzing biotic interactions and effects within the new distributional range pose a challenging task. However, species distribution modelling provides a powerful tool to assess potential habitat under past, present, and future climate conditions. We invite you to present and discuss your work on marine bioinvasions and/or species distribution modelling under climate change.

9.2 Abstracts of Oral Presentations

9.2.1 Alien Species Change our Coasts Forever

Christian Buschbaum[1*]

[1]Alfred Wegener Institute Helmholtz Centre for Polar and Marine Research, Wadden Sea Station Sylt, Hafenstrasse 43, 25992 List/Sylt, Germany

*invited speaker, corresponding author: christian.buschbaum@awi.de

Keywords: Non-native bioengineers, Invasive parasites, Species interactions, Global warming

Worldwide, coastal ecosystems are invaded by an increasing number of alien species, which cause effects on native ecosystems including existing species interactions. In the European Wadden Sea, especially introduced non-native ecosystem engineers such as Pacific oysters *Crassostrea gigas* show the most obvious effects on native communities. The invasion of Pacific oysters has turned resident blue mussel beds *Mytilus edulis* into mixed reefs of mussels and oysters. Initially, it was feared that the alien oysters might outcompete resident mussels but the current situation is much more complex. It changed from competitive displacement to accommodation of mussels underneath a canopy of oysters where they are well protected from predation and detrimental barnacle overgrowth. Additionally, oyster reefs attract further non-native species using this newly developed habitat: The Japanese seaweed *Sargassum muticum* settles on oyster shells and its thallus is colonized by a highly diverse species community. Pacific shores crabs *Hemigrapsus takanoi* hide between the oysters and compete with native shore crabs. Non-native parasites infect oysters and mussels and manipulate predation on their bivalve hosts. Thus, the successful establishment of alien oysters on native mussel beds caused cascading processes in the Wadden Sea. The consequences are irreversible effects and a faster changing ecosystem than ever before.

9.2.2 Easy Riders: Nonindigenous Peracarids Travelling Across Mediterranean Marinas

Guenda Merlo[1,2*], Giovanni Scribano[1], Aylin Ulman[1,3], Agnese Marchini[1]

[1]Department of Earth and Environmental Sciences, University of Pavia, Italy

[2]ICBM Terramare, Carl von Ossietzky University, Oldenburg, Germany

[3]Sorbonne Université, CNRS, Laboratoire d'Ecogeochimie des Environnements Benthiques, France

*corresponding author: guenda.merlo01@universitadipavia.it

Keywords: Non-native species, Peracarida, Mediterranean Sea, Recreational boating, Biofouling

Recreational boating has proven to be a vector of introduction for nonindigenous species (NIS), by carrying organisms as part of the biofouling on the boat hull. However, this subject has been poorly studied in the Mediterranean Sea thus far. In our work, we focused on peracarid crustaceans as a model for studying dispersal and range expansion of NIS. In fact, peracarids commonly occur in biofouling assemblages and have weak autonomous dispersal capability; thus their presence in non-native regions can be confidently attributed to human transport. Twenty-four marinas were selected from seven countries across the Mediterranean Sea considering their relevance as touristic hotspots and maritime connectivity. Fouling assemblage samples were collected in each marina from submerged structures as well as from the hulls of moored vessels. Peracarids were identified to species level and then classified by status as "native," "cryptogenic," or "nonindigenous." Nine NIS were recorded, four of which were new location records, highlighting range expansion, and one NIS constituted a new Mediterranean record. About 1/3 of sampled boats were found hosting at least one nonindigenous peracarid species. Community similarities were found in marinas very distant from one another, suggesting that marinas are artificially connected by boating travel patterns. Additionally, the most common species in marinas and on boats were either cryptogenic, or NIS suggesting that the colonization process of NIS is favored by continuous boat-mediated introductions and a lack of well-structured native communities inside the marinas. International strategies to avoid marine NIS dispersal via recreational boating are currently lacking. Further studies and increased awareness on this topic are required to better understand the strength of the recreational boating vector in order to design efficient strategies to control NIS introductions and spreading to help preserve Mediterranean marine biodiversity.

9.2.3 Environmental DNA for Biodiversity Assessment and Environmental Health of Coastal Lagoons

Marcos Suárez[1*], Serge Planes[2,3], Eva García-Vázquez[1], Alba Ardura[1]

[1]Department of Functional Biology, University of Oviedo, C/ Julian Claveria s/n, 33006 Oviedo, Spain

[2]USR 3278-CRIOBE-CNRS-EPHE, Laboratoire d'excellence "CORAIL," University of Perpignan-CBETM, 58 Rue Paul Alduy, 66860 Perpignan CEDEX, France

[3]Laboratoire d'Excellence "Corail," Centre de Recherche Insulaire et Observatoire de l'Environment (CRIOBE), BP 1013, 98 729 Papetoai, Moorea, French Polynesia

∗corresponding author: marcos.sume@gmail.com

Keywords: Continental waters, Mediterranean Sea, Metabarcoding, Nonindigenous species (NIS), Environmental protection

Coastal lagoons within the Mediterranean Sea have a rich biodiversity, large human populations nearby, and support important economic activities such as aquaculture, fisheries, and leisure activities. However, their fragile status together with pollution and accelerated eutrophication from human activities (e.g., agriculture, aquaculture, and waste management) is threatening these ecosystems. Moreover, climate change and increased maritime transport open the doors for the introduction of nonindigenous species (NIS) that can worsen the state of such important and rich ecosystems, most of them under environmental protection programs. Reliable assessment of biodiversity is crucial for conservation and management strategies of coastal lagoons. Traditional methods used for biodiversity surveys and detection of NIS can prove costly and unsuccessful for the detection and identification of species with early development stages, cryptic, elusive, and new coming species with low population densities. The recent development of metabarcoding techniques offers new opportunities for more reliable biodiversity surveillance and early detection of NIS. The objective of this study was to assess the environmental health of 10 Mediterranean coastal lagoons within the Gulf of Lion, analyzing water samples as a source of environmental DNA and employing molecular tools. The results served to evaluate the current management protocols efficiency and recommend new ones.

9.2.4 Ecophysiology of Recently Introduced Shrimp Species in the Tropical Atlantic Enlightens Present and Future Invasion Potential of Tropical Regions Under Ocean Warming

Carolina Madeira[1,2*], Vanessa Mendonça[1], Miguel C. Leal[1,3], Augusto A.V. Flores[4], Henrique N. Cabral[1], Mário S. Diniz[2], Catarina Vinagre[1]

[1]MARE – Marine and Environmental Sciences Centre, Faculdade de CIências da Universidade de Lisboa, Campo Grande, 1749-016 Lisboa, Portugal

[2]UCIBIO – REQUIMTE, Departamento de Química, Faculdade de Ciências e Tecnologia da Universidade Nova de Lisboa, 2829-516 Caparica, Portugal

[3]Department of Fish Ecology & Evolution, Centre for Ecology, Evolution and Biogeochemistry, Swiss Federal Institute of Aquatic Science and Technology (Eawag) Seestrasse 79, 6047 Kastanienbaum, Switzerland

[4]CEBIMar – Centro de Biologia Marinha, Universidade de São Paulo, Rod. Manoel Hipólito do Rego, Km 131.5, São Sebastião, SP, Brazil

∗corresponding author: scmadeira@fc.ul.pt, carolbmar@gmail.com

Keywords: Tropical shrimp, Invasive species, Warming oceans, Rocky reef, Thermal biology

Climate change, particularly ocean warming, is thought to benefit the spread of invasive species due to their increased tolerance to temperature fluctuations when compared to native species. However, the physiological tolerance of invasive species as a potential mechanism driving invasion success has been overlooked. Here, we experimentally evaluated the physiological responses of a recent invader in the tropical Atlantic Ocean, the shrimp *Lysmata lipkei*, under a warming ocean scenario. Adult shrimps were collected from rocky shores in southeastern Brazil and subjected to experimental trials under a control and a +3°C scenario temperature. Molecular biomarkers (in gills and muscle), upper thermal limits, acclimation response ratio, thermal safety margins, mortality, body condition, and energy reserves were all measured throughout 1 month of experimental trials. Results suggest that higher temperatures elicit physiological adjustments at the molecular level, which underpin the high thermal tolerances observed. Additionally, results show that this invasive shrimp has significant acclimation capacity, without negative performance consequences under an ocean warming scenario. Thermal safety margins were low for intertidal habitat but considerably high for subtidal habitat. We conclude that this shrimp possesses the ability to continue its invasion in subtropical waters of the Atlantic Ocean (mainly in subtidal habitats) both under present conditions and future climate warming scenarios.

9.2.5 Local Adaptation to a Latitudinal Temperature Gradient in Two Species of Littorinids with Contrasting Dispersal Capabilities

Lucy Jupe[1*], Antony Knights[1]

[1]University of Plymouth, PL4 8AA, Plymouth, United Kingdom

∗corresponding author: lucylouisejupe@gmail.com

Keywords: Thermal tolerance, Climate change, Species distributions, Gene flow, Metabolism

The redistribution of life on earth is emerging as one of the most profound impacts of anthropogenic climate change on global biodiversity, with temperature thought to be the major driver. Most projections of species redistributions are based upon the climate variability hypothesis, which proposes that species' latitudinal ranges reflect their thermal tolerance limits, leading to predictions of uniform poleward shifts in species ranges with rising temperatures. However, evidence of variation in thermal niche breadths between populations of the same species suggests that the reality will be far less straightforward. Adaptation of thermal tolerance to local temperature regimes between populations of marine species has not been adequately described due to the engrained perception that marine species comprise of demographically open populations interconnected by high levels of gene flow, and thus possess low propensities for producing locally adapted populations. In order to assess the extent to which gene flow (via dispersal capacity) influences adaptive differentiation in thermal traits, the acute thermal sensitivity of metabolism was compared between populations for two congeneric species of marine gastropod (*Littorina littorea* and *L. saxatilis*) with vastly different dispersal capacities across a latitudinal temperature gradient. The scale of adaptive differentiation in population thermal optima and thermal breath conformed largely with expectations based upon supposed levels of interpopulation gene flow: *L. saxatilis* exhibited latitudinal patterns in both traits, while *L. littorea* exhibited little evidence of differentiation in either. However, maximal metabolic performance showed a strong negative correlation with latitude in both species, supporting the emerging consensus that high dispersal capacities may not constrain local adaptation in all instances. Establishing patterns of local adaptation within marine species will allow us to move beyond climate envelope model predictions, which may underestimate extinction risk in species where individual populations have much narrower thermal tolerance windows than the species as a whole.

9.3 Abstracts of Poster Presentations

9.3.1 Red Sea Marinas as Potential Source of Marine Alien Species

Mohamed El-Metwally[1], Matteo Bertoni[2*], Jasmine Ferrario[2], Cesare Bogi[3], Joachim Langeneck[4], Agnese Marchini[2], Anna Occhipinti-Ambrogi[2]

[1]National Institute of Oceanography and Fisheries (NIOF), Hurghada, Egypt

[2]Department of Earth and Environmental Sciences, University of Pavia, Italy

[3]Gruppo Malacologico Livornese, Livorno, Italy

[4]Department of Biology, University of Pisa, Italy

∗corresponding author: matteo.bertoni01@universitadipavia.it

Keywords: Fouling species, Shipping, Artificial substrates, Suez Canal, Indian Ocean

The arrival of Indo-Pacific species in the Mediterranean Sea is a process not yet fully understood. Species may have been introduced by two distinct vectors: active movement through the Suez Canal and hitchhiking on ships that pass through the Canal. When a new Indo-Pacific species appears in the Mediterranean, its vector of introduction is often only a guess. In this work we analyze fouling communities in Red Sea marinas and other artificial moorings, in order to unveil the possible source and pathway of alien species that already entered in the Mediterranean Sea, and identify potential new invaders. Within a CICOPS fellowship (University of Pavia), a collaboration project between the Department of Earth and Environmental Sciences (Italy) and NIOF (Egypt) was carried out. A survey on the fouling species colonizing artificial substrates in six Red Sea localities from Qusier to Hurghada (NW Red Sea, Egypt) was conducted in summer 2017 by scraping docks, floating pontoons, and buoys. In many cases bryozoans, crustaceans, mollusks, and polychaetes that are considered alien species in the Mediterranean (about 15 species) were found, thus supporting the hypothesis that they might have been introduced by boats travelling from there. The fouling communities observed are also composed by some Indo-Pacific species that have not yet been reported in the Red Sea. These results raise a new question: is the Red Sea part of their natural biogeographic range, or have they been brought there artificially by boats? For a better knowledge of fouling communities of the Red Sea and the ongoing alteration of the biogeography of marine biota caused by human activities, scientific cooperation among countries should be promoted.

9.3.2 Risks Analysis of Biological Invasions in the Port of Gijón (Asturias, Spain), Based on a Geographical Information System (GIS)

D. Menéndez Teleña[1*], J. García Maza[1], J.M. Cuetos Megido[1], H.J. Montes Coto[1], M. Bartolome Saez[1]

[1]University of Oviedo, 33203, Gijón, Spain

∗corresponding author: UO244452@uniovi.es

Keywords: Bio pollution, Ballast water discharge, Hull fouling, Ship traffic, Nonindigenous species

Gijón is a Port City located in the Principality of Asturias in the north of Spain. Our port is specialized in the discharge of solid bulks (iron ore, coal, and others), liquid bulks (petroleum derivatives and gases), different loads (bulk cement, steel products, containers), and others. Ships that arrive to Gijón come from many places of the world and thus carry along the risk of invasive species introductions into our waters, by means of ballast water discharges and biofouling on the hull. Our work is to analyze these risks based on the statistics of entries and exits from the port of Gijón, starting in January 1, 2004 to October 10, 2017. Main operations are the unloading of solid bulks, which does not require vessels to perform discharges of ballast water. This is why we suggest hull fouling as the major threat for introduced species; however, ballast water discharging occurred as the consequence of loading operations. In our analysis, we used dynamic Excel graphs and data from the "Global Invasive Species Database" to create a risk map based on a GIS to locate possible invasive species.

10 Connecting the Bentho-pelagic Dots

Santiago E.A. Pineda Metz[1,2]

[1]Alfred-Wegener-Institut Helmholtz-Zentrum für Polar- und Meeresforschung, Columbusstrasse, 27568 Bremerhaven, Germany

[2]University of Bremen, Fachbereich 02 Biologie/Chemie, Leobener Straße NW2, 28359 Bremen, Germany

Santiago E.A. Pineda Metz: santiago.pineda.metz@awi.de

10.1 Call for Abstracts

When thinking of biotic and abiotic parameters as points, we start noticing lines (i.e., processes) connecting them. These processes connect the benthic and pelagic realms; this is known as bentho-pelagic coupling and includes exchange of matter between both realms, e.g., flux of sinking organic matter, and vice versa, e.g., release of larvae into the water column. Recognizing these processes and understanding their functioning will be essential to assess the impact of climate change.

10.2 Abstracts of Oral Presentations

10.2.1 Pelagic-Benthic Coupling Over Winter in a Northern Norwegian Fjord

Zoe Walker[1,2*]

[1]UiT The Arctic University of Norway, Tromsø, 9008, Tromsø, Norway

[2]Univresity of Akureyri, 600, Akureyri, Iceland

*corresponding author: zwyukon@gmail.com

Keywords: Seasonality, Zooplankton, Autotrophic biomass, Total particulate matter flux, Aquaculture management

Global aquaculture is projected to double by 2050 to meet the demand of a growing human population. Norway has stated its interest in expanding its aquaculture sector to supply this growing international and domestic demand. The environmental impact of aquaculture by-products is determined by their concentration and distribution, which are affected by seasonal signals in pelagic-benthic coupling. Spring and summer studies of pelagic-benthic coupling in Norwegian fjords are significantly more abundant than those focusing on late autumn and winter. This study compared meteorology, hydrography, concentration of suspended and sinking biomass, and total particulate matter flux from October 2017 to February 2018 in Kaldfjorden, Norway (69.746°N, 18.683°E) to explore the physical and biological drivers of pelagic-benthic coupling. Stratification of the water column in Kaldfjorden weakened between October and December before disappearing completely in January and February, identifying winter as a time of high mixing. Changes to the physical environment coincided with a steep decline in suspended chlorophyll *a* concentration (October, 0.09–3.15 mg m^{-3}; December–February, 0.03–0.12 mg m^{-3}) and zooplankton abundance (November, 4502.83 ind. m^{-3}; January–February, <101.72 ind. m^{-3}). Sinking material was sampled using short-term sediment traps (24 h). The downward biomass flux decreased throughout winter and particulate matter became more degraded, most likely due to zooplankton grazing. Sediment trap samples also showed evidence of resuspension following episodic winds throughout winter. The observed decrease in stratification and biological activity in this study is considered characteristic of a Northern Norwegian fjord, and supports the importance of including seasonally appropriate environmental baselines in the management of open circuit aquaculture to mitigate environmental impacts.

10.2.2 Simultaneous Measurements of Flow Fields and Oxygen Concentrations in Microscale Applications

Patric Bourceau[1*], Soeren Ahmerkamp[1], Marcel Kuypers[1]

[1]Max Planck Institut für Marine Mikrobiologie, Celsiusstr. 1, 28359 Bremen, Germany

*corresponding author: pbourcea@mpi-bremen.de

Keywords: Particle image velocimetry, LED-induced fluorescence, Aggregates, Micro niches

In many biologically active systems, the consumption of solutes is limited by transport processes. Typical examples are marine aggregates – millimeter-sized coagulated remains of plankton blooms – which are responsible for the export of organic carbon from the sunlit ocean to the seafloor. During the descent the aggregates are colonized by highly active microbial communities which remineralize a large fraction of the organic carbon before it eventually gets sequestered. The fate of carbon is, therefore, regulated by a tight interaction of microbial reactions and transport processes. To gain new insights into these processes, methods are needed that cover spatial resolutions in the micrometer-range at temporal resolutions far below seconds. However, most applied methods typically only draw a bulk picture of the system by averaging in time and space. To overcome this limitation and to measure oxic remineralization and transport processes at the same time, we used a novel lifetime-based LED-induced fluorescence (τLIF) technique and combined it with particle image velocimetry (PIV). Nanoparticles coated with an oxygen-sensitive dye were illuminated by a LED sheet and recorded with a high-speed camera. As indicator for dissolved oxygen, the fluorescence lifetime (in the range of 50 μs–100 μs) was estimated for each nanoparticle – providing hundreds of measuring points for each field of view. To measure transport, the fluorescent particles were tracked between subsequent images, and the flow field was calculated using particle image velocimetry – literally connecting dots. In a first application and as validation of the method, we simultaneously determined the flow field and oxygen concentration around a settling aggregate.

11 Investigating the Land-Sea Transition Zone

Stephan L. Seibert[1] and Julius M. Degenhardt[2]

[1]Hydrogeology and Landscape Hydrology Group, Institute for Biology and Environmental Sciences, Carl von Ossietzky University of Oldenburg, Ammerländer Heerstraße 114-118, D-26129 Oldenburg, Germany

[2]Paleomicrobiology Group, Institute for Chemistry and Biology of the Marine Environment (ICBM), Carl von Ossietzky University of Oldenburg, Carl von Ossietzky-Str. 9-11, D-26129 Oldenburg, Germany

Stephan L. Seibert: stephan.seibert@uni-oldenburg.de

Julius M. Degenhardt: julius.degenhardt@uni-oldenburg.de

11.1 Call for Abstracts

The land-sea transition zone is the location where terrestrial and marine environments merge and interact. It recently received much attention due to the importance for coastal ecosystems, nutrient fluxes to the oceans and impacts on coastal aquifers in the context of climate change. While the complex interaction of both environments offers many great research opportunities, a broad scientific approach is often required. We therefore highly encourage students and researchers from all different kinds of fields (e.g., biogeochemistry, microbiology, hydrogeology, marine sensor systems, soil sciences, coastal biology, etc.) to present their research related to the land-sea transition zone in this session.

11.2 Abstracts of Oral Presentations

11.2.1 Societal Use of Submarine Groundwater Discharge: Worldwide Examples of an Underresearched Water Resource

Till Oehler[1*]

[1]Leibniz Center for Tropical Marine Research (ZMT), Bremen

*invited speaker, corresponding author: Till.Oehler@leibniz-zmt.de

Keywords: Submarine springs, Submarine groundwater discharge, Water resource

Fresh submarine groundwater discharge (fresh SGD) has been researched in the recent years in the context of local- and global-scale land-sea matter fluxes. However, it may also be used as a water resource, which has been nearly entirely neglected by scientific studies. In my talk I will present examples from various places around the world where fresh SGD is used as water resource. It is used as drinking water but also for hygiene (bathing, laundry) and agriculture (irrigation, cattle feeding), while its effects are used, e.g., by fishermen and dive schools, or they are relevant for ship navigation. Furthermore, fresh SGD locations often play a substantial role in the spiritual life of the local communities. Examples are that in Lombok (Indonesia) fresh SGD is used for drinking, on Kode Island (Fiji) for bathing, and in Kiveri (Greece) for irrigation. Vrulja Bay (Croatia) is the type locality for submarine springs, called "vruljas" among karst geologists, and attracts divers and tourists. Alexander von Humboldt noted that Manatees gathered around a submarine spring in Cuba and were hunted by fishermen. Fresh SGD, freshwater in the ocean, is likely to inspire spiritual thoughts, and it is part of Hawaiian legends, gives insight about the History of Tarento (Italy) and is used to bless 2 million visitors of an intertidal spring each year on Bali. However, fresh SGD is sensitive to sea level rise, and coastal groundwater pumping. Submarine springs off Bahrein, which were used by divers to deliver freshwater to the city as drinking water until the 1950s, have dried out because of coastal pumping. It is important to recognize the relevance of fresh SGD to local communities before taking management decisions that could endanger the existence of this hidden water resource.

11.2.2 Rare Earth Element Behavior in a Sandy Subterranean Estuary of the Southern North Sea

Ronja Paffrath[1*], Katharina Pahnke[1], Bernhard Schnetger[1], Hans-Jürgen Brumsack[1]

[1]Institute for Chemistry and Biology of the Marine Environment (ICBM), University of Oldenburg, Carl von Ossietzky-Str. 9-11, 26129 Oldenburg, Germany

*corresponding author: ronja.paffrath@uni-oldenburg.de

Keywords: Rare earth elements, Barrier island, North Sea, Subterranean estuary, Gd anomaly

Rare earth element (REE) concentrations in the ocean track input and transport pathways of trace elements and provide information on marine geochemical processes. To use REEs as tracers in the ocean, their sources and sinks need to be identified and quantified. Recently, subterranean estuaries (STE) have been suggested to be the "missing source" of REEs to the ocean and to account for about the same REE input as rivers. To investigate the behavior of REEs in a STE of a rapidly changing, tidally influenced sandy beach system and to assess its potential contribution to coastal waters, we measured dissolved REE concentrations in pore waters along a beach transect, in coastal seawater, and fresh groundwater from the Barrier Island Spiekeroog, German North Sea. The pore waters show variable REE concentrations (Nd between 7 and 469 pmol kg^{-1}), in general increasing toward the shoreline. Lowest concentrations landwards are due to conservative mixing of seawater with rainwater, exhibiting a linear relationship with salinity. Highest concentrations are found close to the low water line in accordance with sub-/anoxic conditions and elevated Fe, Mn, and DOC concentrations. Therefore, release of REE from particles or Fe-Mn phases is likely to occur, resulting in much higher REE concentrations compared to seawater (33 pmol kg^{-1} for Nd). Net submarine groundwater discharge (SGD) of both, fresh groundwater and recirculated seawater, significantly adds all REEs to the ocean, light REEs in larger quantities than heavy REEs. The total amount of REEs added to the North Sea by SGD at this location, however, is more than one order of magnitude smaller than the input from nearby major rivers. An anthropogenic Gd signal, higher than previously reported, is present throughout the pore waters and seawater. This indicates the increasing accumulation of anthropogenic Gd in the marine environment and the short residence time of discharged waters.

11.2.3 Biogeochemical Processes Influencing the Molecular Composition and Spatial Distribution of Dissolved Organic Matter in the Amazon River Plume

Melina Knoke[1*], Michael Seidel[1], Thorsten Dittmar[1]

[1]ICBM, University of Oldenburg, Oldenburg, Germany

*corresponding author: melina.knoke@uni-oldenburg.de

Keywords: Amazon River plume, Mangroves, DOM, River-to-ocean continuum, High discharge

The Amazon River provides 20% of the global freshwater discharge and delivers large amounts of nutrients and riverine dissolved organic matter (DOM) to the ocean. DOM remains one of the most elusive parts of biogeochemical cycling, because of its great diversity. In estuarine systems it can be derived from multiple sources such as terrestrial material or algal organic matter. The time scales of DOM production and removal in estuaries can vary on broad time scales which further complicates the identification of its sources and sinks. This study aims to explore the main removal processes for riverine DOM and trace metals in estuarine systems, i.e., photo- and sorption-based flocculation as well as biodegradation. We sampled the Amazon River to ocean continuum covering the full salinity gradient from the river mouth to the tropical North Atlantic Ocean during high discharge conditions in 2018. The samples were taken along a south-to-north transect from Belém (Brazil) to French Guiana, covering also the coastal mangrove areas south of the river mouth to serve as a brackish/saline reference ecosystem, which is dominated by intertidal water exchange. River water was incubated under light and dark conditions to test for the effects of photo- and bio-flocculation. Flocculation due to estuarine mixing was simulated by mixing filtered as well as unfiltered river and seawater covering different salinities. We hypothesize that the removal via estuarine mixing, photochemical flocculation, and microbial transformations will leave characteristic DOM molecular imprints and we aim to identify the spatial distribution of these DOM molecular signatures in the river plume along the salinity gradient. The samples are currently being analyzed regarding their DOM molecular composition using non-targeted Fourier transform ion cyclotron resonance mass spectrometry (FT-ICR-MS). Our study will provide important molecular-level insights into the distribution and biogeochemical cycling of DOM along the salinity gradient of tropical river plumes.

11.2.4 The Effect of Non-native Kelp Species on Detrital Processing in the UK Strandline

Dale Kingston[1*], Andy Foggo[1]

[1]University of Plymouth, PL4, Plymouth, England

*corresponding author: dale.kingston@students.plymouth.ac.uk

Keywords: Range expansion, Laminariales, Nutrient input, Adjacent ecosystems, Trophic dynamics

Subtidal kelp is a conspicuous feature of the UK's temperate coastline. Kelps are sensitive to environmental conditions, and with current warming trends, the contraction of geographic ranges of cold-temperate and expansion of warm-temperate species have been predicted. Energy and material export from subtidal kelp forest ecosystems is substantial in sustaining both intertidal and supra-littoral organisms in low productivity environments. *Laminaria ochroleuca* de la pylaie (Laminariales) is a warm-temperate kelp of Lusitanian origin, and has recently proliferated along the south coast of England. The presence of *Laminaria ochroleuca* is hypothesized to increase the overall processing of detritus in the strandline, because of its high palatability compared to its native congeners, *Laminaria hyperborea* and *Laminaria digitata*. A mesocosm approach was developed in which a detrital consumer was offered, single, paired, and three species combinations of kelp containing mixtures of native and non-native species. Results show that interaction between species was significant; however detrital processing in three-species treatments compared to monocultures was greatly reduced. The observed trends indicated that the presence *L. hyperborea* was the most preferred kelp species, which also coincided with its seasonal timing of growth. The kelps used have been shown to display complementary differences in the timing of maximum biomass accumulation suggesting temporal energy subsidies within the water column and adjacent productivity-poor areas. *Laminaria ochroleuca* may contribute to UK nutrient cycling by growing and degrading at different times of the year to native species contributing to the stability of nutrient input into the water column and on beaches around the UK. This prediction requires further testing but may provide a basis for the future of UK kelp beds under the expansion of a non-native species.

11.3 Abstracts of Poster Presentations

11.3.1 The Relative Importance of the Presence of Conspecifics, Predation, and Wave Action on the Recruitment of the Stalked Barnacle *Pollicipes pollicipes*

D. Mateus[1*], D. Jacinto[1], J.N. Fernandes[1], T. Cruz[1,2]

[1]MARE – Marine and Environmental Sciences Centre, Laboratório de Ciências do Mar, Universidade de Évora, Apartado 190, 7520-903 Sines, Portugal

[2]Departamento de Biologia, Escola de Ciências e Tecnologia, Universidade de Évora, Portugal

*corresponding author: david.mateus5@gmail.com

Keywords: Intertidal rocky shores, Cirripedia, Post-settlement processes, Sines, SW of Portugal

Pollicipes pollicipes (*Pp*) is restricted to very exposed rocky shores where its recruitment is very intense on conspecifics. On less exposed shores, recruitment has been observed on conspecifics protected from predation. Past studies showed that recruitment of *Pp* on artificial substratum ("barticle") occurs on very exposed sites. In this study, the relative importance of the presence of conspecifics and wave action on recruitment of *Pp* was studied by doing an intertidal field experiment in which "barticles" were deployed at two distances (5 and 20 cm) from natural/transplanted groups of adult conspecifics in different locations: very exposed sites in a headland – the Cape of Sines (CS) (*Pp* abundant); less exposed sites of CS (*Pp* rare); and in the lee of CS (Port of Sines, PS) at a more exposed and at a less exposed site (*Pp* absent/rare in both). The predation effect on recruitment on "barticles" was studied in cage/no cage treatments. Recruitment of *Pp* was estimated as the number of cyprids and juveniles attached to "barticles" or to conspecifics protected from predators. Calcein marking of conspecifics at transplant time (tt) was used to distinguish between "new" (after tt) from "old" recruitment (before tt). In very exposed sites, recruitment on conspecifics is high and the presence of conspecifics and predation do not limit recruitment of *Pp*, as recruitment on "barticles" occurred independently of the deployment distance to conspecifics and no predation effect was detected. In less exposed sites (of CS and in PS in general), the presence of conspecifics limits recruitment of *Pp*, as recruitment was only observed on conspecifics and not on "barticles." Considering recruitment on conspecifics, wave action had an effect, as "new" and "old" recruitment were lower on less exposed sites, namely in PS, which might indicate a lower settlement and a higher mortality in less exposed sites.

11.3.2 Simultaneous Sulfate Reduction and Nitrate Consumption in Sandy Sediments

Olivia Metcalf[1*], Hannah Marchant[1]

[1]Max Plank Institute for Marine Microbiology, Celsiusstraße 1, 28359 Bremen, Germany

*corresponding author: ometcalf@mpi-bremen.de

Keywords: Sand, Sulfur, Nitrogen

Recent research has revealed that sandy sediments are extremely dynamic environments housing microbial communities that seem to behave very differently from both the water column above and comparable muddy sediments. In muddy sediments electron acceptors are depleted in order of their oxidizing power, so that electron acceptors which allow microbes to generate the most energy per molecule of substrate are depleted first. This pattern is less strictly observed in sandy sediments exposed to environmental variability, where denitrification was observed to proceed in the presence of oxygen. Understanding the rate and order of acceptor usage is important for developing climate modules to evaluate and predict the resilience of costal aquatic ecosystems to anthropogenic nitrogen input and eutrophication. Here, sandy sediments from the Jannsand tidal flat on the German coast of the North Sea were exposed to 9-day regimes of either variable or constant percolation with fresh, filtered anoxic seawater either amended or unamended with nitrate. Sand was subsequently subsampled and amended with fresh anoxic water, 15N nitrate, and radiolabeled 35S sulfate to investigate the co-occurrence of nitrate consumption and sulfate reduction. Subsamples were killed at intervals over 12–24 hours. Nitrate, nitrite, and $15N_2$ concentrations were subsequently determined to calculate nitrate reduction and N_2 production rates. Acid-chromium-labile reduced sulfur was distilled and the radioactive fraction measured to calculate sulfate reduction. Sulfate reduction was found to proceed simultaneously with nitrate consumption in sand exposed to all three regimes. However, denitrification proceeding completely to N_2 was observed only in the sand previously supplied with constant nitrate, where sulfate reduction proceeded at lower rates. In the other cores, dissimilatory nitrate reduction to ammonium appeared to occur. The simultaneous respiration of sulfate and nitrate is surprising given that denitrification has a substantially higher energy yield than sulfate reduction and may be common in sandy sediments.

12 Tropical Marine Research Mosaic: Combining Small Studies to Reveal the Bigger Picture

Javier Onate-Casado[1,2,3] and Liam Lachs[1,4,5]

[1]Department of Biology, University of Florence, Sesto Fiorentino, Italy

[2]Sea Turtle Research Unit (SEATRU), Universiti Malaysia Terengganu, Kuala Terengganu, Terengganu, Malaysia

[3]School of Biological Sciences, University of Queensland, Australia

[4]Marine Biology, Ecology and Biodiversity, Vrije Universiteit Brussel, Pleinlaan 2, Brussel B-1050, Belgium

[5]Institute of Oceanography and Environment, Universiti Malaysia Terengganu, Kuala Terengganu, Terengganu, Malaysia

Javier Onate: javiatocha@gmail.com

Liam Lachs: liamlachs@gmail.com

12.1 Call for Abstracts

Tropical marine ecosystems are some of the most biodiverse and complex ecosystems on the planet; however these ecosystems are under threat from unmanaged fisheries, untreated wastewater outflows, tourism exploitation, and increasingly extreme weather events due to climate change. Our understanding of the effects of such processes is especially limited in tropical regions due to general data deficiencies. Long-term monitoring studies are rare and ecosystem mapping is non-extensive in comparison to temperate regions. By combining small-scale studies on flora, fauna, or abiotic factors, coming from multiple disciplines including genetics, chemistry, classical ecology, behavioral ecology, and ecosystem management, we can consider a research mosaic which may provide insights into the overall state of tropical ecosystems.

12.2 Abstracts of Oral Presentations

12.2.1 The Future Is Now: Anthropogenic Impacts Are Reshaping Coral Reef Ecosystems

Sebastian C.A. Ferse[1*]

[1]Department of Marine Ecology, FB2 Biology/Chemistry, University of Bremen, 28359 Bremen, Germany

∗invited speaker, corresponding author: sebastian.ferse@leibniz-zmt.de

Keywords: Anthropocene, Ecosystem services, Functional ecology, Human impacts, Novel ecosystems, Reef restoration, Resilience, Social-ecological systems

At the beginning of the twenty-first century, coral reefs have become under widespread pressure from anthropogenic stressors. The recognition that humans have become a primary driver of natural processes and ecosystems has led to the characterization of the present geological epoch as the Anthropocene. The majority of reefs nowadays are within reach of human populations, and very few, if any, remain free from human impacts. This poses particular challenges for coral reef ecology. Environmental stressors are affecting reefs in non-random ways, leading to changes and potential homogenization in community composition and functioning. In many cases, there is no historical precedent for the environmental conditions and composition of coral reef ecosystems. This presentation will review the current state of coral reefs under the impact of multiple anthropogenic stressors, tracing the effects of human activities on reefs, and discuss challenges and trends of coral reef research in the Anthropocene, where reef systems are becoming increasingly homogenized as a result of non-random environmental filtering.

12.2.2 Response of Benthic Reef Communities to Manipulated In situ Eutrophication in the Central Red Sea

Denis B. Karcher[1*], Florian Roth[1,2], Arjen Tilstra[1], Yusuf El-Khaled[1], Susana Carvalho[2], Burton Jones[2], Christian Wild[1]

[1]Marine Ecology Department, Faculty of Biology and Chemistry, University of Bremen, Bremen, Germany

[2]Red Sea Research Center, King Abdullah University of Science and Technology (KAUST), Thuwal, Saudi Arabia

∗corresponding author: dkarcher@uni-bremen.de

Keywords: Coral reefs, Nutrients, Stable isotopes, Phase shifts, Picture analysis

Local eutrophication may heavily affect coral reef communities in the oligotrophic Red Sea, but related knowledge is scarce. As such, we investigated how a simulated 2-month eutrophication event influenced coral reef communities in an in situ experimental approach using a slow release fertilizer. Community development and elemental (C and N) composition of four major functional groups (hard corals, soft corals, turf algae, sediments) were analyzed pre- and post-nutrient manipulation using digital photographs along with elemental and stable isotope analyses. With this approach, we investigated which functional groups were best in utilizing introduced nutrients and how this is related to community development. Water samples confirmed successful in situ eutrophication with fourfold increased dissolved inorganic nitrogen (DIN) relative to background. Preliminary results (based on the photographs) show that the reef community structure was affected by eutrophication. Soft coral abundance declined in 75% of the communities where they previously had a high coverage, while turf algae increased with eutrophication in 71% of the observed communities, including all previously turf algae-dominated communities. We also occasionally observed a decreased pigmentation of hard corals and increased relative benthic coverage of macro algae (*Halimeda* sp.) under eutrophication. According to these preliminary results, turf algae and macro algae are the winners of eutrophication, while hard- and soft corals are the losers. Elemental and isotopic analyses are in progress now, and related results will be presented during YOUMARES. This analysis will provide mechanistic explanations for the observed community changes. The present study is instrumental in understanding the ecological and biogeochemical consequences of eutrophication on Red Sea coral reef communities.

12.2.3 Plasticity of Nitrogen Fixation and Denitrification of Key Reef Organisms in Response to Short-Term Eutrophication

Yusuf C. El-Khaled[1*], Florian Roth[2], Denis B. Karcher[1], Nils Rädecker[2], Claudia Pogoreutz[2], Arjen Tilstra[1], Burton Jones[2], Christian R. Voolstra[2], Christian Wild[1]

[1]Marine Ecology Department, Faculty of Biology and Chemistry, University of Bremen, 28359 Bremen, Germany

[2]Red Sea Research Center, King Abdullah University of Science and Technology (KAUST), 23995 Thuwal, Saudi Arabia

*corresponding author: yek2012@uni-bremen.de

Keywords: Nitrogen cycling, Coral reefs, Local factors of global change

An efficient uptake and recycling of nutrients is necessary to maintain high primary production in coral reefs. Particularly, the cycling of nitrogen (N) is of fundamental importance, since coral reefs are usually N-limited. However, a comprehensive understanding of the involved processes is missing to date. We thus investigated major benthic functional groups (hard corals, soft corals, biotic rock, turf algae, reef sands) of a central Red Sea coral reef. We quantified their N_2 fixation and denitrification rates under ambient and short-term (24 h) eutrophic (5 μM nitrate-enriched seawater) conditions using a combined acetylene reduction assay. Findings revealed that all functional groups were performing N_2 fixation and denitrification under both ambient control and eutrophic conditions. Under ambient conditions, N_2 fixation of turf algae was sixfold higher than that of biotic rock and tenfold higher than for all other functional groups. Denitrification was highest for biotic rock and 30% higher than for turf algae, twofold higher than for reef sands, and sixfold higher than for hard and soft corals. Under eutrophication, no significant changes relative to controls were observed for N_2 fixation, while denitrification rates were twofold higher for sands and increased by 15% for turf algae and biotic rock compared to controls. Denitrification rates for hard and soft corals under nutrient-enriched conditions were comparable to rates at ambient conditions. These findings suggest a high and rapid acclimatization and metabolic plasticity of N_2-fixing and denitrifying microbes associated with reef sands, turf algae, and biotic rock. In contrast, responses of N cycling microbes associated with corals, which are highly adapted to nutrient-poor conditions, may only become apparent at greater time scales (e.g., days to weeks). These results may have important implications for N cycling and resilience of impacted coral reefs

12.2.4 Increased Abundance and Biodiversity of Reef Fish in a Lagoonal System Affected by Submarine Groundwater Discharge

Timo Pisternick[1,2,3*], Yashvin Neehaul[3], Danishta Dumur-Neelayya[3], Julian Döring[2], Werner Ekau[2], Nils Moosdorf[2]

[1]University of Bremen, Faculty 2 Biology/Chemistry, Bibliothekstraße 1, 28359 Bremen, Germany

[2]Leibniz Centre for Tropical Marine Research (ZMT), Fahrenheitstraße 6, 28359 Bremen, Germany

[3]Mauritius Oceanography Institute, Avenue des Anchois, Morcellement de Chazal, Albion, Mauritius

*corresponding author: timo.pisternick@leibniz-zmt.de

Keywords: Submarine groundwater discharge, Nutrients, Fish abundance, Coral reef

Submarine inflow of freshwater from land into the ocean (fresh submarine groundwater discharge, FSGD) is increasingly recognized as an important source of local nutrient and pollutant influx to coastal ecosystems. Still, very little is known about the effects of FSGD on ecosystem functioning. While there is evidence of a positive link between FSGD-derived nutrients and phytoplankton abundances, further research is needed to assess the effects of FSGD on the productivity of higher trophic levels such as ichthyofaunal communities. To follow up on this relationship, we sampled two sites in a tropical coastal lagoon (Trou-aux-Biches, Mauritius) during early summer months (October–December 2017). The hydrology in the lagoon's southern part is highly influenced by six distinct freshwater springs, whereas the lagoon's northern part served as a control site. At each of the two sites, abiotic parameters, nutrient concentrations, as well as the dry weight of total suspended solids (TSS), composition of benthic cover and fish abundances were recorded. Water nutrient (PO_4^{3-}, SiO_4^{4+}, NO_3^-, NO_2^-, NH_3) levels were significantly higher, whereas salinity and pH values were significantly lower in the SGD-affected southern part when compared to the northern, strictly marine part of the lagoon. TSS and coral cover were significantly higher at the SGD site while macroalgae cover was significantly higher at the control site. In addition, the SGD site exhibited significantly higher fish species richness and diversity. Also, abundances of commercially important fish species were significantly higher in the SGD-affected part of the lagoon. Our results provide first evidence of a SGD-driven positive relationship between high nutrient loadings, elevated primary production, and enhanced secondary consumer abundances in coral reef lagoons. These findings will have implications for coastal management advice and secondary consumer productivity appraisal on tropical islands.

12.2.5 Functional Redundancy and Rarity of Coral Reef Fish Among Three National Parks in Central Mexican Pacific

Diana Morales-de-Anda[1], Amílcar Leví Cupul-Magaña[1*], Alma Paola Rodríguez-Troncoso[1], Fabián Rodríguez-Zaragoza[2], Consuelo Aguilar-Betancourt[3], Gaspar González-Sansón[3]

[1]Departamento de Ciencias Biológicas CUCosta Universidad de Guadalajara, 48280, Puerto Vallarta, México

[2]Departamento de Ecología CUCBA Universidad de Guadalajara, 45110, Zapopan, México

[3]Departamento de Estudios para el Desarrollo Sustentable de la Zona Costera CUCSur Universidad de Guadalajara, 48980, Cihuatlán, México

*corresponding author: amilcar.cupul@gmail.com

Keywords: Fish assemblages, Functional traits, Functional vulnerability, Functional entities

Reef fish from the Central Mexican Pacific (CMP) represent an interesting scenario to understand how sites with lower diversity respond to multiple stressors (high dynamism, marked hydroclimatic periodicity, overall low coral cover, and anthropogenic pressure) and are still able to maintain ecosystem processes and functions. Therefore, we evaluated the functional component of reef fish communities and their spatial variation across three national parks in CMP (Islas Marietas, Isla Isabel, and Isla Cleofas). Our aim was to identify island's functional redundancy and vulnerability, evaluate rarity of species and functional entities (FEs), and analyze changes in fish functional community composition. We used data from underwater visual census; fish species and interval length records were used to create FEs matrix with species data and fish traits (length, aggregation, diet, mobility, position in water column, and activity period). To evaluate functional component, we performed functional indices (redundancy, FR; vulnerability, FV; and rarity) and evaluated differences in indices and FEs composition among islands. We found 80 species distributed in 54 FEs; FR was low for all islands (less than 1.5 species/FE) being significantly lower for Islas Marietas, the recently popularized island closest to shore. On the other hand, FV was markedly high for all islands (76% of FEs represented by one species). Across islands, rarity was common with 75% of rare FEs and 85% of rare species in FEs; meanwhile 7–16% of species and FEs had only one individual in all islands. Overall, our study shows that despite the low redundancy present across the three islands, we also found a large number of functions among reef fish; however, many functions remained vulnerable and rare. Each island had specific FEs and marked dissimilarities in FEs composition; these characteristics make each island unique and highlight the importance of special protection for the three islands, particularly those with constant anthropogenic pressure.

12.2.6 Population and Reproductive Biology Aspects of the Vinegar Crab *Episesarma mederi* H. Milne Edwards, 1853 (Decapoda, Sesarmidae) from a Tropical Mangrove Area in Capiz, Philippines

Jerry Ian Leonida[1*], Juliana Baylon[2], Shaira Ballon[2], Elilyn Farrah Belle Barredo[2], Ma. Shirley Golez[2]

[1]Institute of Marine Fisheries and Oceanology, College of Fisheries and Ocean Sciences, University of the Philippines Visayas, Miagao, Iloilo 5023, Philippines

[2]Division of Biological Sciences College of Arts and Sciences, University of the Philippines Visayas, Miagao, Iloilo 5023, Philippines

*corresponding author: jlleonida@up.edu.ph

Keywords: Mangrove crab, Sesarmid, Ovarian development, Histology, Oocyte maturation, Spawning period

Vinegar crabs of the *Episesarma* genus are among the dominant crab groups in estuarine and mangrove areas in the tropics. These burrow-dwelling crabs play vital roles in the nutrient cycling and substrate biochemistry of their inhabited ecosystems. *Episesarma mederi* is a traditional and a growing fishery resource in the Philippines that may have aquaculture potential. However, a constraint in managing this crab species is the lacking knowledge about its reproductive biology. This holds true for the other members of the genus. This study was the first to analyze aspects of the population structure and gonadal maturation stages of *E. mederi* in Capiz, Philippines (11° 26' 34" N, 122° 55' 23" E). Samples were randomly collected by hand during low tide periods every first week of the month from February 2016 to January 2017. In the laboratory, morphometric analysis and sexual determination were performed. Staging of gonadal development was established by complementing morphological features (i.e., color, volume) with histological analysis (i.e., cell type, size). A total of 448 crabs (184 males, 264 females, 105 ovigerous females) were collected. Males were fewer but larger and heavier than females. Five stages of ovarian development were identified: immature, developing, maturing, mature (two substages – early mature and late mature), and spent. Among ovigerous females, the smallest size has 29 mm carapace width (CW), and the size class range of 30–34 mm CW had the highest prevalence of spawning samples. Monthly ovarian development indicates a continuous breeding cycle. However, two periods of higher reproductive activity were identified in females, which coincided with the onset and culmination of the rainy season in the Philippines. This suggests that *E. mederi* exhibits a seasonal-continuous reproduction strategy. Accordingly, these reproductive features must be accounted in management strategies to prevent the overexploitation of the wild stocks of this species.

12.2.7 The Overlooked Contribution to Organic Matter Sequestration in Mangrove Forests: *Acrostichum aureum*, an Understory Fern

Michael K. Agyekum[1*], Martin Zimmer[1], José M. Riascos[2]

[1]Leibniz Centre for Marine Tropical Research (ZMT), Bremen

[2]Estuaries & Mangroves Research group, Universidad del Valle, Cali, Colombia

∗corresponding author: michaelk@uni-bremen.de

Keywords: Mangrove fern, Organic matter content, Organic carbon, Macrofauna, Sediment characteristics, Detritivore

Understory plants have been long assumed an odd exception in mangrove forest; thus we ignore the role that some common floristic elements may play. Here we studied the cosmopolitan fern *Acrostichum aureum* in a neotropical mangrove. Experimental manipulations of the fern litter fall (litter exclusion, the potential effect of the net, control and fern areas) were set up to test the hypothesis that fern-derived organic matter influences the species composition of the fauna and sediment characteristics in the mangrove forest of Bahía Málaga, Colombia. Sediment samples were collected in October 2016 and March 2017 from the experimental plots and analyzed for pH, total dissolved solids, carbon-nitrogen ratio, organic matter content, and organic carbon. Macrofauna species were sampled after 5 months of experimental manipulations of the sediment. The results indicated no significant differences for the macrofauna species composition in either the stations or treatments. In contrast, there was a significant difference in the sediment characteristics among treatments. The results suggest the magnitude of change in organic matter and organic carbon is small yet quite consistent in this study to support the hypothesis that the fern-derived organic matter is important in influencing the sediment characteristics. Therefore, more time is required for the integration of organic matter in the sediment of the mangrove forest.

12.2.8 Lipidomic Characterization of Microbial Communities of the Extremely Acidic Shallow Water Hydrothermal System of Kueishantao (Taiwan)

R.F. Aepfler[1*], M. Elvert[1], K.U. Hinrichs[1], Y. Lin[2], S.I. Bühring[1]

[1]University of Bremen, MARUM, Postbox 330 440, Bremen, Germany

[2]National Sun Yat-Sen University, Kaohsiung, Taiwan

∗corresponding author: rebecca.aepfler@uni-bremen.de

Keywords: Extreme environment, Polar lipid-derived fatty acids, Intact polar lipids, Campylobacteria, Reverse tricarboxylic acid cycle

The Kueishantao shallow water hydrothermal vent system located off Taiwan's NE coast (121°57' E, 24°50' N) is characterized by world-record low pH conditions (pH < 1) and fluid temperatures of up to 116°C. High CO_2 discharge provides a natural laboratory to study the effect of ocean acidification on microbial communities. In 2015, sediment cores and hydrothermal fluids were collected from a hot (116°C, pH = 2.88) and a warm temperature vent (58°C, pH = 4.51), in order to study the lipidome of the microbial community. The polar lipid-derived fatty acid (PLFA) pattern of both vent sediments showed dominating $C_{16:1\omega7c}$, $C_{18:1\omega9}$, and $C_{18:1\omega7c}$, revealing enriched $\delta^{13}C$-values of -12‰, -19‰, and -12‰ in the warm sediments and of -10‰, -14‰, and -15‰ in the hot sediments, respectively. PLFAs in the warm vent fluid likewise showed $C_{16:1\omega7c}$ and $C_{18:1\omega7c}$ being enriched in 13C with $\delta^{13}C$ values of -13‰ and -8‰, respectively. These positive $\delta^{13}C$ values are best explained by the activity of campylobacteria, which have been identified metagenomically at the study site. These bacteria possess the reverse tricarboxylic acid cycle for carbon fixation, a metabolic pathway that discriminates less against ^{13}C than the Calvin-Benson-Bassham cycle. Complementing analyses of intact polar lipids (IPLs) revealed the presence of sphingolipids and hydroxylated ornithine lipids in the warm vent sediments, with the latter being assumed to increase the stress tolerance toward high temperatures and acidic pH values. That way, bacteria are able to adjust their membrane properties by modifying already existing membrane lipids without the need to synthesize new lipids. Sphingolipids on the other hand, might reduce the membrane permeability through strong intramolecular hydrogen bonding, counterbalancing configurations that lead to a stress-induced fluidization of the membrane.

12.2.9 Patterns and Drivers of Coral Reef Resilience at Aldabra Atoll, Seychelles

Anna Koester[1*], April J. Burt[2], Nancy Bunbury[3], Amanda K. Ford[4], Valentina Migani[5], Cheryl Sanchez[3], Frauke Fleischer-Dogley[3], Christian Wild[1]

[1]Marine Ecology Department, Faculty of Biology & Chemistry, University of Bremen, Germany

[2]The Queens College, High Street, Oxford, OX1 4AW, United Kingdom

[3]Seychelles Islands Foundation, Postbox 853, Mont Fleuri, Victoria, Mahé, Seychelles

[4]Leibniz Center for Tropical Marine Research (ZMT), Fahrenheitstrasse 6, 28359 Bremen, Germany

[5]Institute for Ecology, Faculty of Biology & Chemistry, University of Bremen, Germany

∗corresponding author: anna.koester@uni-bremen.de

Keywords: Coral bleaching, Climate change, Recovery, Marine monitoring, Remote

The recent global coral bleaching event in 2016 caused mass mortality of corals worldwide and affected even the best protected and most remote reefs. This challenges the common notion that reefs far removed from local human impacts are more resilient to the effects of climate change. To understand reef resilience under the exclusive influence of global impacts, we study the susceptibility to, and recovery since, the 2016 global bleaching event at Aldabra Atoll, a UNESCO World Heritage site in the Western Indian Ocean. We combine the annual monitoring of benthic reef assemblages with additional observations of coral larvae settlement, sedimentary oxygen consumption, and turf algae height to assess post-bleaching trajectories (i.e., stability/recovery/degradation) and how these differ on an atoll-wide scale. Aldabra lost 50% of its hard corals during the 2016 bleaching event and experienced reductions of taxonomic and morphological diversity within the benthic assemblages. Initial results of recent field work (December 2017–March 2018) reveal differing post-bleaching trajectories, with the two most easterly reefs indicating further reductions in hard coral cover and increasing cover of calcifying macroalgae (*Halimeda* spp.). At the remaining ten sites, hard coral cover remained stable or increased, while turf algae cover decreased to pre-bleaching levels. Additional observations indicate atoll-wide similarities in turf algae height and sedimentary oxygen consumption, but marked differences in the density of settled coral larvae, which was substantially higher within Aldabra's lagoon. These results suggest that prevailing environmental conditions at the individual reefs around the atoll are important drivers of reef resilience. Further analysis is now needed to examine the atoll-wide shifts in benthic and coral community composition in response to various biological and environmental variables. The results of this study will contribute to our understanding of the natural drivers of coral reef resilience, thereby aiding the identification of priority areas for conservation.

12.2.10 Recovery Capacity and Physiology of Kimberley Corals After Unprecedented Bleaching

E. Maria U. Jung[1,2*], Malcolm T. McCulloch[1,3], Verena Schoepf[1,3]

[1]ARC Centre of Excellence for Coral Reef Studies, UWA Oceans Institute and School of Earth Sciences, The University of Western Australia, Perth, WA, Australia

[2]Department of Biology and Chemistry, University of Bremen, Bremen, Germany

[3]The Western Australian Marine Science Institution, Perth, WA, Australia

*corresponding author: emu.jung@uni-bremen.de

Keywords: Heat tolerance, Energy reserves, Symbiont dynamics, Extreme reef environments

Ocean warming is one of the major threats to coral reefs today and leads to global mass bleaching events of increasing severity and frequency. In 2016, a marine heatwave caused unprecedented bleaching in the extreme macrotidal Kimberley region in northwest Australia. We report both recovery and the extent of coral mortality six months after the peak bleaching as well as the physiological mechanisms underlying both heat tolerance and recovery capacity of these corals. Coral cover and health from both heat-tolerant intertidal (IT) and heat-sensitive subtidal (ST) environments were investigated via photo-quadrat analyses. Samples of *Acropora aspera* were collected from both environments during and after the bleaching event and assessed for symbiont density, chlorophyll *a*, and energy reserves (lipid, protein, and carbohydrate). Despite being exposed to similar heat stress during peak bleaching (~4.6 degree heating weeks), bleaching was more severe in the subtidal than intertidal. Furthermore, ST corals had a much lower recovery capacity with 71% mortality, whereas 91% of IT corals were visibly healthy after 6 months of recovery. Analyses of symbiont dynamics confirmed visually observed differences in bleaching susceptibility and severity. Only bleached IT corals catabolized energy reserves during bleaching, demonstrating that maintaining energy reserves during bleaching cannot guarantee survival. This suggests that other factors influenced the low recovery capacity of ST corals. However, the fast recovery of IT corals gives hope for reef habitats that suffered from extensive mortality during the 2016 bleaching event in the Kimberley. These findings demonstrate that corals from extreme temperature environments can provide important insights into the mechanisms underlying coral heat tolerance.

12.2.11 Neuroactive Compounds Induce Larval Settlement in the Scleractinian Coral *Leptastrea purpurea*

Mareen Moeller[1*], Samuel Nietzer[1], Peter J. Schupp[1]

[1]Carl-von-Ossietzky Universität Oldenburg, ICBM, Environmental Biochemistry, Schleusenstrasse 1, 26382 Wilhelmshaven, Germany

*corresponding author: mareen.moeller@uni-oldenburg.de

Keywords: Recruitment, Settlement, Metamorphosis, Brooder, Cnidarian

Settlement of pelagic coral larvae is commonly induced by chemical cues that originate from biofilms. These natural settlement cues initiate signal pathways leading to attachment and metamorphosis of the coral larva. In order to investigate the settlement process and natural inducers, it is necessary to gain a better understanding of these signal pathways. At present, the pathways and neurotransmitters involved in this signal transduction are still widely unknown.

In this study, we exposed larvae of the brooding coral *Leptastrea purpurea* to five neuroactive compounds known to be present in cnidarians, and K+ Ions. All compounds were applied at different dilutions and settlement behavior of the larvae was documented over 48 h. Dopamine, glutamic acid, and epinephrine significantly induced settlement in the coral larvae. Exposure to dopamine resulted in 58% in the strongest metamorphosis response. Serotonin, L-DOPA, and K+ ions did not have an influence on settlement behavior in our experiments. Exposing larvae to settlement-inducing neurotransmitters and thus bypassing the initial induction could be utilized in coral aquaculture. The active neurotransmitters could also be used to study the settlement process in greater detail.

12.2.12 Coral Larvae Every Day: *Leptastrea purpurea*, a Brooding Species That Could Accelerate Coral Research

Samuel Nietzer[1*], Mareen Moeller[1], Makoto Kitamura[2], Peter J. Schupp[1]

[1]Carl-von-Ossietzky Universität Oldenburg, ICBM Terramare, Schleusenstrasse 1,

26382 Wilhelmshaven, Germany

[2]Okinawa Environment Science Centre, Urasoe, Okinawa 901-2111, Japan

*corresponding author: samuel.nietzer@uni-oldenburg.de

Keywords: Coral reproduction, Brooding, Recruitment, Settlement cues, Faviid

Sexually produced larvae are used in various fields of coral research. Because the vast majority of scleractinians reproduce only on one or few occasions per year through simultaneous release of gametes, and ex situ spawning induction is still very hard to achieve, high efforts are required to obtain planula larvae. Brooding corals have been used to harvest planulae, but their larvae oftentimes differ in various traits, e.g., settlement behavior, from most spawning corals. Other cnidarians, such as *Aiptasia* spp., have been substituting scleractinians in many aspects of coral research. However, organisms such as *Aiptasia* differ strongly from scleractinians limiting the transferability of obtained results. This study examines the potential of *Leptastrea purpurea* as a reliable source of larvae for coral research. Larval output throughout the year as well as settlement behavior of planulae was investigated. Our results show that *L. purpurea* releases larvae on a daily basis, thus allowing permanent access to planula larvae. Larval settlement is induced by the same and similar cues as in many spawning species which increases the transferability of conclusions. We discuss the aptitude of *L. purpurea* for research on scleractinian physiology, ecology, and larval settlement and conclude that *L. purpurea* is a well-suited organism to accelerate progress in many fields of coral research.

12.2.13 Development of a Coral Reef Resilience Index (CRRI): A Spatial Planning Tool for Managers and Decision-Makers Applicable in Caribbean Coral Reef Ecosystems

Sonia Barba-Herrera[1,2,3,4*], Edwin Hernández-Delgado[2,3,4]

[1]Department of Biological Sciences, School of Sciences, University of Málaga, Av. de Cervantes, 2, 29016 Málaga, España

[2]Department of Environmental Sciences, University of Puerto Rico, Av. Dr. José N. Gándara, San Juan, 00931, Puerto Rico

[3]CATEC: Centre of Applied Tropical Ecology and Conservation, Av. Dr. José N. Gándara, San Juan, 00931, Puerto Rico

[4]SAM: Sociedad Ambiente Marino, Marine Environment Society, San Juan, 00931, Puerto Rico

*corresponding author: sonia5bh@hotmail.com

Keywords: Ecosystem health, Coral reefs, Marine management, Marine biodiversity, Tropical ecosystems

Coral reefs worldwide are in severe decline due to a combination of local human stressors, and large-scale climatic stressors, such as sea surface warming. However, there is often a general lack of information regarding the status and resilience of coral reef ecosystems. Timely information is critical to implement sustainable management measures. The main objective of this project was to develop a Coral Reef Resilience Index (CRRI) and provide a GIS-coupled decision-making tool for reef managers applicable for Caribbean coral reef ecosystems. The CRRI is based in a 5-point scale parameterized from quantitative documentation regarding benthic assemblages. Separate sub-indices such as the Coral Index, the Threatened Species Index, and the Algal Index provide specific information regarding targeted benthic components. This case study was based on assessments conducted in 2014 on 11 reef sites located across 3 geographic zones and 3 depth zones along the southwestern shelf of the island of Puerto Rico, Caribbean Sea. Results showed a significant spatial and bathymetric gradient in the distribution of CRRI values that indicated higher degradation of inshore coral reefs, in comparison to mid-shelf and offshore sites. Mean global CRRI within inshore reefs ranged from 2.78 to 2.87, ranking them as fair. Mean CRRI within mid-shelf localities ranged from 2.97 to 3.17, ranking them between fair and good. CRRI ranged from 3.07 to 3.16 within outer shelf localities, ranking them as good. The coral and algal indices showed declining trends toward inshore reefs, and the threatened corals index showed a general cross-shelf poor state, with the exception of two localities. CRRI has provided an important, solid management and decision-making tool for Caribbean coral reefs. A future re-evaluation is recommended to be conducted to determine the effects of the 2017 Hurricane María and compare CRRI in a multi-temporal scale.

12.2.14 Factors Driving Benthic Community Change on the Mexican Caribbean Over the Last 12 Years (2005–2016)

Xochitl E. Elías Ilosvay[1*], Ameris I. Contreras-Silva[1], Lorenzo Alvarez-Filip[2], Christian Wild[1]

[1]Marine Ecology Group, Faculty of Biology and Chemistry, University of Bremen, Leobener Straße UFT, 28359, Bremen, Germany

[2]Biodiversity and Reef Conservation Laboratory, Unidad Académica de Sistemas Arrecifales, Instituto de Ciencias del Mar y Limnología, Universidad Nacional Autónoma de México, Puerto Morelos, México

*corresponding author: xoel@uni-bremen.de

Keywords: Coral reef, Drivers of change, Meta-analysis

The Mexican Caribbean experiences community changes in coral reefs, but knowledge on potential global and/or local drivers of these changes is scarce. This study thus describes through random effects meta-analysis, the relative effects of sea surface temperature (SST), chlorophyll water concentration, coastal human population development, reef distance to shore, water depth, and geographical latitude on the hard coral and algae cover in 50 reef sites along the coast of Quintana Roo, Mexico. Findings revealed that against our expectations, there was a significant increase of both hard coral cover (by ca. 5 %) and algae cover (by ca. 14 %, i.e., almost three times the increase of corals) over the last 12 years. These results on the one hand reflect partial coral recovery after the 2005 Caribbean mass coral bleaching event but on the other hand rapid invasion of algae in local reefs. Surprisingly, none of the selected factors correlated positively or negatively with changes in coral cover. However, latitude and human population density exhibited significant effects on reef algae cover increase. Against our expectation, there was more algae cover increase in the southern part of the Yucatán Peninsula coast where population density was lower, but recent major coastal and tourism development took place. This study is important for monitoring and management of coral reefs in the Caribbean, because by using simple methods and existing data it gives indications for the drivers of change in reef communities.

12.3 Abstracts of Poster Presentations

12.3.1 Linking Shapes and Functions: A Morphological Comparison of Tropical Coral Reef Fish

Marita Quitzau[1*], Romain Frelat[1], Christian Möllmann[1], Sonia Bejarano[2]

[1]Institute for Marine Ecosystem and Fisheries Science, University of Hamburg, Große Elbstraße 133, Hamburg, Germany

[2]Leibniz-Centre for Tropical Marine Ecology, Fahrenheitstraße 6, Bremen, Germany

*corresponding author: quitzaumarita@gmail.com

Keywords: Morphology, Herbivorous fish, Coral reef, Elliptical Fourier transform, Functional diversity

Coral reef fish diversity is at risk due to fishing pressure, local pollution, and habitat loss driven by coastal development and climate-related phenomena. Reducing diversity may result in the loss of ecosystem functions that can push coral reef ecosystems into impoverished configurations that are difficult to reverse. While fish counts at the species level are informative of taxonomic community structure, linking species counts to species functions has the potential to reveal whether vital ecosystem functions are eroded, reorganized, or replaced along environmental gradients or following perturbations. Quantifying fish functional traits is however not a trivial task for the megadiverse coral reefs. Fish morphology is an integrative trait that combines functional and evolutionary information. Indeed, shapes of herbivorous fish reflect their behavior and interactions with the environment, and thus can reveal their functional role in ecosystem processes. However, the description of shapes is not straightforward and different methods are usually applied. Traditional morphometrics rely on the measurement of length, areas, and ratios of selected aspects of shape. In contrast, modern morphometric methods (e.g., Elliptical Fourier Analysis) consider the entire shape, yet appear to be more challenging to interpret. In our study, we compared traditional and modern morphometric approaches and identify their complementarity on a dataset of 100 herbivorous coral reef fish. The two methods were able to identify the major components of morphological diversity: the elongation and the development of caudal and pelvic fins. The main advantage of modern morphometrics is its objectivity and its ability to reconstruct the original shape from its descriptors. Traditional morphometrics rely on a long history of studies connecting the shapes to the functions of organisms. Therefore, the two methods are highly complementary and offer a good understanding of the functional diversity in coral reef fish assemblages.

13 Bridging Disciplines in the Seasonal Ice Zone (SIZ)

Tobias R. Vonnahme[1] and Ulrike Dietrich[1]

[1]UiT Norges Arktiske Universitet, Department of Arctic Marine Systems Ecology

Postboks 6050 Langnes, N-9037 Tromsø, Norway

Tobias R. Vonnahme: tobias.vonnahme@uit.no

Ulrike Dietrich: ulrike.dietrich@uit.no

13.1 Call for Abstracts

The SIZ is the most affected environment under climate change. Temperatures are significantly rising, the ice edge retreats beyond the continental shelves, terrestrial inputs of organic matter change drastically, and changes in stratification alter nutrient concentrations. These physical factors will have strong impacts on the ecosystem. Shifting community structures on multiple trophic levels from whales to viruses have been observed. Species move northwards, which may cause hazards or novel ecosystem services. For understanding the effects on the ecosystem level, bridging interdisciplinary research, ranging from meteorology, and biogeochemistry to the physiology of key organisms is crucial.

13.2 Abstracts of Oral Presentations

13.2.1 Arctic Marine Microbial Ecology in the Svalbard Polar Night

Martí Amargant-Arumí[1*], Rolf Gradinger[1], Aud Larsen[2], Lena Seuthe[1]

[1]UiT - The Arctic University of Norway, Postbox 6050, Tromsø, Norway

[2]Uni Research Environment, Bergen, Norway

*corresponding author: martiamargant@gmail.com

Keywords: Lower food web ecology, Polar night, Picoplankton, Nanoplankton, Viruses

This study investigated the presence and activity of the components of the microbial food web (specifically viruses, heterotrophic bacteria and nanoflagellates, and autotrophic cyanobacteria and pico-nanoflagellates) in the waters around the Svalbard archipelago (Norway) during the polar night period. The study focused on two major questions – are there differences in the community composition in different water masses? And, are there significant changes occurring during the polar night period? Two cruises in January and November 2017 with a total of 11 stations offered the opportunity to test these hypotheses. Flow cytometry was used to determine cell abundances in the uppermost 100 m of the water column, and eight serial dilution experiments were conducted to estimate their growth and grazing rates. All studied organism groups occurred in all samples in low abundances in both January and November. Comparison to the hydrographic regime revealed strong linkages between community structure and hydrography with higher abundances in Atlantic Water samples. Heterotrophic nanoflagellates and autotrophic pico-nanoplankton were markedly less present in January, whereas bacteria and viruses displayed steady concentrations in both months. This supported the hypothesis of succession in the microbial network throughout the polar night, and the possible role of mixotrophy and resting stages are discussed. No significant growth or grazing was detected in the experiments, which could be caused, e.g., by low substrate availability and resting strategies. This study demonstrated that all members of the microbial food web organisms persist throughout the polar night in the major water masses around Svalbard. Future studies using alternative approaches are suggested to further study these processes during times of low activity.

13.2.2 Implications of a Dominance Shift from Krill to Salps to the Biological Carbon Pump in the Southern Ocean

Nora-Charlotte Pauli[1,2*], Morten H. Iversen[2,3], Bettina Meyer[1,2]

[1]Institute for Chemistry and Biology of the Marine Environment, Carl von Ossietzky University Oldenburg, Carl von Ossietzky-Straße 9-11, 26111 Oldenburg, Germany

[2]Alfred-Wegener Institute Helmholtz Centre for Polar and Marine Research, Am Handelshafen 12, 27570 Bremerhaven, Germany

[3]MARUM, University Bremen, Leobener Str. 8, 28359 Bremen, Germany

*corresponding author: nora-charlotte.pauli@uni-oldenburg.de

Keywords: Carbon flux, Fecal pellets, POC, Molecular fingerprinting

In the Southern Ocean, krill and salps are among the most important grazers and able to consume large amounts of primary production. In recent years, salp abundances have been increasing, while krill populations declined, seemingly leading to a dominance shift from krill to salps. This dominance shift will cause cascading effects in the Southern Ocean food web and also affect the carbon cycle. Krill and salps both produce fast sinking fecal pellets rich in organic matter; however, detailed studies on pellet production rates, carbon turnover, and export are so far lacking. As part of the POSER (Population Shift and Ecosystem Response – Krill vs. Salps) project, we are investigating the contribution of krill and salps to the biological carbon pump in terms of grazing, pellet production and attenuation, and export of carbon to the deep sea. Sampling was conducted around the Western Antarctic Peninsula where krill and salps are co-occurring and which is one of the fastest warming regions on our planet. In situ collections of krill and salp pellets were conducted using marine snow catcher and drift traps. Drifting sediment traps were deployed at three depths (100, 200, 300 m) to account for attenuation of pellets through the water column. These studies were completed by camera systems, CTD, and acoustic data to survey krill and salp aggregations. In addition, size-specific sinking velocities and respiration of the pellet-associated microbial community of krill and salp pellets were measured using a vertical flow chamber. Molecular analyses of gut content and fecal pellets will provide valuable data on grazing activity of krill and salps. On the basis of these data, we are aiming to quantify the role of krill and salps in the biological carbon pump and to predict the impact of a dominance shift from krill to salps in a future ocean.

13.3 Abstracts of Poster Presentations

13.3.1 Importance of Mixotrophy in the Polar Night

Rose M. Bank[1*], Tobias R. Vonnahme[2], Rolf Gradinger[2], Ronnie Glud[1], Ulrike Dietrich[2], Dolma Michellod[3]

[1]SDU, University of Southern Denmark, Odense, Denmark

[2]UiT, The Arctic University of Norway, Tromsø, Norway

[3]MPI, Max Planck Institute for Marine Microbiology, Bremen, Germany

∗corresponding author: roban14@student.sdu.dk

Keywords: Polar night, Mixotrophy, Algae, Sea ice, stable isotopes

Sea ice is mostly located in high latitudes, characterized by the absence of light during the polar night. During this time photosynthesis cannot occur. Thus, microalgae cells need to find a way to survive. Multiple survival strategies have been discussed; resting stages, use of storage compounds, and heterotrophic carbon uptake (mixotrophy). Our hypothesis is that mixotrophy is an important mechanism for survival and biogeochemical cycling in sea ice. I will test the hypothesis using two field studies: (1) during a seasonal study in Ramfjord in northern Norway, Tromsø, during August 2018–February 2019, the potential for mixotrophy and nitrification will be investigated using stable isotope uptake experiments. Besides the ecosystem will be characterized by cell counts of bacteria and algae, nutrients, chlorophyll, POC, and CTD data. (2) Another stable isotope probing experiment will be done with melted sea ice from the Central Arctic Ocean, in November 2018, on a cruise onboard Kronprins Haakon. The experiments combined will show the importance of heterotrophic carbon uptake in the polar night on a quantitative basis.

14 Crossing Traditional Scientific Borders to Unravel the Complex Interactions Between Organisms and Their Nonliving Environment

Corinna Mori[1] and Mara Heinrichs[1]

[1]Paleomicrobiology Group, Institute for Chemistry and Biology of the Marine Environment (ICBM), Carl von Ossietzky University of Oldenburg, Carl von Ossietzky-Str. 9-11, D-26129 Oldenburg, Germany

Corinna Mori: corinna.mori@uni-oldenburg.de

Mara Heinrichs: mara.elena.heinrichs@uni-oldenburg.de

14.1 Call for Abstracts

Organic and inorganic constituents play a crucial role in mediating the interactions between organisms and the associated flow of matter and energy in aquatic environments. The enormous molecular complexity of the nonliving environment results in diverse interactions with organisms that has not been considered in traditional ecological or geochemical studies yet. Integrative approaches that combine these classical research areas will greatly improve our understanding of the complex networks between the nonliving and living environment in the ocean. We invite young scientists from all fields to present innovative methodologies and multidisciplinary research approaches to bridge the gap between traditional scientific disciplines.

14.2 Abstracts of Oral Presentations

14.2.1 Does *Pseudo-nitzschia subcurvata* Produce Domoic Acid Under Iron-Limited Conditions?

Jana Geuer[1*], Scarlett Trimborn[1], Tina Brenneis[1], Bernd Krock[1], Florian Koch[1], Jan Tebben[1], Boris Koch[1]

[1]Alfred Wegener Institute, Helmholtz Centre for Polar and Marine Research, Postbox 120161, Bremerhaven, Germany

∗corresponding author: jana.geuer@awi.de

Keywords: Incubation, Organic ligands, Dissolved organic matter, Trace elements, Bioavailability

Domoic acid is a potent neurotoxin primarily produced by the marine diatom *Pseudo-nitzschia* spp. and can have harmful effects on the environment by accumulating in the food web. Its actual ecological function, though, is still unknown. Several studies support the hypothesis that domoic acid should act as a ligand and bind to iron to make it more bioavailable to its producing organisms. It was previously shown that domoic acid can form complexes with iron and is produced to a larger extent in iron-rich environments, thus likely helping organisms in iron-limited areas. Domoic acid can be found in its dissolved form in ocean water and should bind and solubilize extracellular iron. Our aim was to test if *Pseudo-nitzschia subcurvata* produces and releases domoic acid under iron-limited or iron-rich conditions and if the presence of domoic acid in the medium itself would facilitate to trigger further domoic acid production. *Pseudo-nitzschia subcurvata* cultures were incubated in four different conditions: in iron-limited seawater, iron-limited seawater that contained dissolved domoic acid, and in seawater containing iron in sufficient concentrations with and without dissolved domoic acid. Cellular growth rates were faster when enough iron was present. No release of domoic acid into the medium was observed. It is thus likely that the species *Pseudo-nitzschia subcurvata* does not produce domoic acid to increase iron bioavailability.

14.2.2 Changes in Nutrient and Trace Metal Cycling Induced by a Phytoplankton Spring Bloom: A High-Resolution Mesocosm Study

Corinna Mori[1]*, Nils Hintz[1], Leon Dlugosch[1], Bernhard Schnetger[1], Katharina Pahnke[1], Hans-Jürgen Brumsack[1]

[1]Institute for Chemistry and Biology of the Marine Environment, Carl von Ossietzky University of Oldenburg, Carl von Ossietzky Straße 9-11, Postbox 2503, 26129 Oldenburg, Germany

∗corresponding author: corinna.mori@uni-oldenburg.de

Keywords: ICP-MS, Bio-cycling, Micronutrients, Nonconservative behavior, Multidisciplinary research approach

Many trace metals are actively involved in bio-cycling processes, as they serve as essential micronutrients. Some control the uptake of macronutrients as part of cofactors in enzymes and thus the course of phytoplankton and associated bacteria blooms, whereas others are considered as non-bioactive or conservative. Even though not actively involved in bio-cycling processes, trace metals can still be passively influenced by biota, e.g., by biologically induced reduction/oxidation processes and/or adsorption to organic matter. Furthermore, recent studies showed that even trace metals considered as conservative (e.g., Mo, Tl) deviate, positively and negatively, from conservative behavior during certain time intervals, which are associated with the degradation of phytoplankton material. While significant research has been conducted on the role of trace metal cycling in open ocean settings, the cycling of trace metals during plankton blooms in near shore areas is still poorly understood. The main motivation of this study was to generate a high-resolution nutrient and trace metal dataset and link it to the different stages of a phytoplankton and associated bacteria bloom. In a highly interdisciplinary research approach, we conducted an indoor mesocosm experiment (fully controlled 600 L stainless steel tanks), where we studied a spring bloom over 6 weeks, in four replicates and three plankton-free controls. We inoculated artificial seawater with a natural phytoplankton and bacteria community from the southern North Sea and grew it under natural light and temperature conditions. We show in our high temporal resolution study that the cycling between particulate and dissolved phases of bioactive as well as non-bioactive major- and trace elements are linked to the different stages of the blooms and to shifts in phytoplankton and bacteria abundances and communities. This study emphasizes the importance of interdisciplinary research approaches to unravel the complex network of interactions between the living and nonliving environments in aquatic systems.

14.2.3 Nitrogen Fixation in the Coastal Peruvian Upwelling Zone Following a Simulated Upwelling Event

Leila Kittu[1]*, Allanah Paul[1], Ulf Riebesell[1]

[1]GEOMAR Helmholtz Centre for Ocean Research Kiel, Düsternbrooker Weg 20, 24105 Kiel, Germany

∗corresponding author: lkittu@geomar.de

Keywords: Diazotrophy, N:P ratio, OMZ, Simulated upwelling

The oxygen minimum zone (OMZ) in the eastern tropical South Pacific (ETSP) promotes the microbial loss of nitrogen from the water column accompanied by release of sediment bound phosphate. Consequently, subsurface water masses with a low N:P ratio are upwelled to the surface layer off Peru. Geochemical tracer studies and biogeochemical models suggest the excess of phosphate (P∗) in the upwelled waters provides a niche for marine nitrogen fixation in surface waters above the OMZ. Two mechanisms have been suggested as to how nitrogen fixation in the surface waters could be stimulated by the upwelled waters: (1) Redfield nutrient assimilation by phytoplankton which leaves behind P∗ that could be consumed by diazotrophs (nitrogen fixers) and (2) non-Redfield nutrient uptake by non-diazotrophs where surplus P from biomass is released as dissolved organic phosphorus (DOP). DOP may provide an additional source of phosphorus that could enhance nitrogen fixation. These two mechanisms could also act simultaneously to stimulate nitrogen fixation. There has been mixed evidence on nitrogen fixation in addition to scarcity of data on nitrogen fixation rates in the surface waters of the Peruvian upwelling zone. To better understand the impact of upwelled waters with OMZ influenced N:P ratios and corresponding feedbacks on nitrogen fixation, we measured nitrogen fixation rates during an offshore mesocosm study conducted in the coastal upwelling zone off Peru from February to April 2017. An upwelling event featuring two water masses of different OMZ influenced N:P ratios was simulated to the surface of the mesocosms. Stable isotope incubations of the surface (oxic) and bottom (anoxic) mesocosm waters were used to quantify nitrogen fixation rates. We present the temporal development of nitrogen fixation rates from this study and relate this to inorganic nutrient concentrations, stoichiometry, and other relevant biogeochemical parameters.

14.2.4 Integrating Marine Time Series to Unravel Marine Microbial Food Web Dynamics

Carina Bunse[1,2*], Jarone Pinhassi[1]

[1]Linnaeus Centre for Ecology and Evolution in Microbial model Systems (EEMiS), Department of Biology and Environmental Science. Linnaeus University, Sweden

[2]current address: Helmholtz Institute for Functional Marine Biodiversity at the University of Oldenburg (HIFMB), Oldenburg, Germany

*invited speaker, corresponding author: carina.bunse@hifmb.de

Keywords: Bacterioplankton, Seasons, Environmental drivers

Marine microbial communities are major drivers in organic matter production, turnover, and inorganic nutrient cycles. Among spatial and temporal scales, the community structures are influenced by biotic and abiotic environmental drivers. To unravel community interactions, functionalities and the environmental drivers influencing microbial communities across time and space, a multitude of research fields and expertise is needed. In recent years, especially the microbial oceanography and marine biogeochemistry research fields made great progress in the attempt to find interactions between marine microbial organisms. Time series and functional profiling offer a great opportunity to study such microbial interactions, functional properties of the communities and nutrient dynamics. At the Linnaeus Microbial Observatory (LMO) in the Baltic Sea proper, the microbial community composition and its functionality were studied at high frequency (up to twice weekly) over 4 consecutive years. A further multitude of parameters were measured to integrate the biotic and abiotic ecosystem with the microbial community dynamics. Phytoplankton blooms were observed during spring and summer, whereas heterotrophic bacterioplankton was most active in assimilating different substrates during summer. Bacterioplankton ecotypes and their abundance patterns could further be associated with environmental niches, such as specific phytoplankton communities, nutrients, or temperatures. This dataset offers the possibility to integrate data and expertise from different scientific disciplines, to retrieve an overview of the microbial food web and its drivers at different scales.

14.2.5 Targeted Analysis of Organic Iron Complexes (i.e., Siderophores) by Chromatography Coupled with Elementary Analysis (HPLC-ICP-OES)

Fabian Moye[1*], Tim Leefmann[2], Ingrid Stimac[2], Walter Geibert[2], Tilmann Harder[1,2], Jan Tebben[2]

[1]University of Bremen, P.O. Box 330440, 28334 Bremen, Germany

[2]Alfred-Wegener-Institut, Helmholtz Centre for Polar and Marine Research, Am Handelshafen 12, 27570 Bremerhaven, Germany

*corresponding author: fmoye@uni-bremen.de

Keywords: Iron ligands, ICP-OES, Siderophores, Trace metal quantification

Iron is an essential micronutrient required by most organisms for important metabolic processes, e.g., the respiratory chain or nitrogen fixation. Iron has very low solubility in oxygenic ocean surface waters. More than 99% of dissolved iron is complexed by chemically diverse organic ligands ranging from macromolecular complexes such as humic substances to low molecular weight metabolites such as bacterial siderophores. Siderophores are secondary metabolites mainly produced by bacteria to sequester Fe from the environment. The majority of marine siderophores have been characterized from culturable bacterial isolates, resulting in a limited and potentially biased spectrum of bacterial siderophores. Presumably, the natural diversity of siderophores is much higher than currently known. Recent developments in analytical methods allow the direct analysis of siderophores concentrated from seawater by liquid chromatography coupled with electrospray ionization mass spectrometry (ESI-MS), to characterize their molecular mass, or coupled with inductively coupled plasma mass spectrometry (ICP-MS), to directly quantify iron bound to the siderophore. I will present the development of an instrumental method for the quantification of iron bound to siderophores by HPLC coupled with inductively coupled plasma-optical emission spectrometry (ICP-OES). The detection by ICP-OES is based on the elements' individual emission spectra as measured by an optical sensor. This detector is more robust than ICP-MS which may allow a high throughput of concentrated organic samples when the sensitivity is sufficient. The limit of detection of the new method is 7.2 pmol iron bound by a siderophore (LoD 4.7 ng) injected on column. I will discuss the potential of this method for the quantification of organically complexed iron in concentrated samples from various sources such as marine dissolved organic matter or bacterial cultures.

14.2.6 Development of a Solid Phase Extraction Method for the Analysis of Organic Acids in the Exometabolomes of Bacteria

Simone Heyen[1*], Barbara Scholz-Böttcher[1], Ralf Rabus[1], Heinz Wilkes[1]

[1]Institute for Chemistry and Biology of the Marine Environment, Carl von Ossietzky University Oldenburg, Postbox 2503, Oldenburg, Germany

*corresponding author: simone.heyen@uni-oldenburg.de

Keywords: Gas chromatography-mass spectrometry, Sulfate-reducing bacteria, Denitrifying bacteria, Carboxylic acids, Biodegradation

Anaerobic bacteria like sulfate-reducing or denitrifying bacteria play a central role in carbon cycling in diverse marine habitats. They have the ability for complete oxidation combined with a broad nutritional versatility. Therefore, they are expected to have a significant influence on the compositional shaping of organic matter in these environments. However, systematic studies investigating the exometabolome of these bacteria and their contribution to the dissolved organic matter pool are lacking. Typical approaches to analyze dissolved organic matter include solid phase extraction on a styrene-divinylbenzene-based polymer followed by FT-ICR-MS measurement. Due to the limitations of this distinct stationary phase as well as of the instrumental method, it is almost impossible to detect small organic acids using this approach. However, these compounds play a key role in processing of organic matter. Their quantification can therefore provide valuable insights into biogeochemically relevant metabolic pathways. In this study we developed a method to analyze small mono-, di-, and tricarboxylic acids based on a solid phase extraction on an anionic exchange cartridge for the enrichment of organic acids followed by GC-MS identification and quantification. The method was then applied to characterize exometabolomes of sulfate-reducing as well as denitrifying bacteria grown with different substrates and harvested at the exponential as well as the stationary growth phase. We used targeted and non-targeted analysis to prove if types of bacterial strains, types of growth substrates and the growth phase will imprint on the amount and composition of the exometabolome and therefore on the contribution to dissolved organic matter.

14.2.7 Microbial Degradation and Transformation of Petroleum from a Natural Seep Asphalt

Jonas Brünjes[1*], Michael Seidel[2], Min Song[1], Florence Schubotz[1]

[1]MARUM – Center for Marine Environmental Sciences, Bremen, Germany

[2]ICBM, University of Oldenburg, Oldenburg, Germany

*corresponding author: jbruenje@uni-bremen.de

Keywords: Oil, Biodegradation, Microcosm, DOM, Stable isotope probing

Marine environments are heavily affected by oil spills and anthropogenic pollution. However, a remarkable input of oil in the marine realm also occurs through natural seepage, often establishing unique oil-adapted ecosystems. Here, oil-derived organic compounds are biodegraded by microorganisms, but they are not always completely oxidized to CO_2. Instead, metabolic intermediates such as water-soluble hydrocarbons can remain in seawater as dissolved organic matter (DOM). Its molecular composition and potential hazardous impact on the environment are mostly unknown. Our hypothesis is that the microbial degradation of asphalt releases oil-derived DOM to the water column. We simulated microbial hydrocarbon degradation under laboratory conditions in DOM-free, artificial seawater using natural asphalt (heavy oil) samples derived from the Chapopote asphalt volcano, a natural deep-sea hydrocarbon seep in the Southern Gulf of Mexico. The molecular composition and changes in DOM were monitored with two complementary analytical methods: ultrahigh-resolution mass spectrometry (FT-ICR-MS) with electrospray ionization to determine the molecular composition of the DOM and excitation-emission matrix (EEM) spectroscopy to characterize and quantify fluorescent organic compounds. By using these analytical approaches, we will gain an overview on changes in the molecular speciation of the DOM which will provide insights into the natural cycling of oil-derived organic compounds and their (potential) long-lasting environmental effects. In our microcosm approach, we will furthermore monitor microbially mediated transformation of two ^{13}C isotopically labeled representative aliphatic (hexadecane) and aromatic (naphthalene) hydrocarbons. For this we will track the complete remineralization of these substrates to either $^{13}CO_2$ or the transformation to other ^{13}C-labeled organic intermediates such as volatile fatty acids by using liquid chromatography coupled to isotope ratio mass spectrometry (LC-IRMS). Our study presents a comprehensive geochemical approach to study natural petroleum pollution and hydrocarbon cycling in the environment.

14.2.8 A Comprehensive Survey on the Production and Consumption of Short-Chain Hydrocarbons in Guaymas Basin Hydrothermal Sediments

Min Song[1*], Florence Schubotz[1], Matthias Y. Kellermann[1,2], Andreas Teske[3], Kai-Uwe Hinrichs[1]

[1]MARUM-Center for Marine Environmental Sciences and Department of Geosciences, University of Bremen, Leobenerstr. 13, D-28359 Bremen, Germany

[2]present address: Institute for Chemistry and Biology of the Marine Environment, University of Oldenburg, Schleusenstr. 1, D-26382 Wilhelmshaven, Germany

[3]Department of Marine Sciences, University of North Carolina at Chapel Hill, Chapel Hill, NC 27599-3300, USA

*corresponding author: msong@marum.de

Keywords: Thermogenesis, Methanogenesis, Alkane oxidation

Short-chain hydrocarbons such as methane (C_1), ethane (C_2), propane (C_3), butane (C_4) and pentane (C_5) support microbial life on a global scale and thus play a substantial role in microbial carbon cycling. An ideal study area to explore the production and consumption of short-chain hydrocarbons is the organic-rich and hydrothermally heated sediment of Guaymas Basin (Gulf of California, Mexico), given the wide spectrum of hydrocarbons and steep geochemical and thermal gradients characteristic for these sediments. While to date the sources and sinks of methane are well documented, our understanding of the cycling of C_2–C_5 alkanes is less clear. Here, through measurements of the concentrations and $\delta^{13}C$ isotopic compositions of C_1–C_5 alkanes, porewater geochemistry, and temperature gradients in the sediment cores, we present their distribution and investigate the pathways of production and consumption from 12 different sites at the Guaymas Basin. We observe a strong temperature-dependent distribution of C_1–C_5 alkanes and suggest that at sites with low temperatures (<30°C at 40 cmbsf), methane is produced through methanogenesis and consumed by anaerobic methanotrophy (AOM), while at sites with medium or elevated temperatures (>70°C at 40 cmbsf), methane carries a mixed biogenic and thermogenic signature, and C_2–C_5 alkanes are mostly thermogenically formed. Alkanes are microbially oxidized in the surficial sediments. We propose that the distribution of short-chain hydrocarbons as well as the microbial communities that catalyze the oxidation of short-chain hydrocarbons is strongly controlled by in situ temperature and ultimately by hydrothermal activities. This study gives an overview on the natural distribution, formation, and oxidation of C_1–C_5 hydrocarbons through a widespread region in the Guaymas Basin and contributes to our understanding of the carbon cycling at hydrothermal systems.

14.2.9 Unraveling Amino Acid Turnover in a Shallow Water Hydrothermal Environment Applying Position-Specific Stable Isotope Probing of Fatty Acids

Christopher Vogel[1*], Rebecca Aepfler[1], Solveig I. Bühring[1], Marcus Elvert[1]

[1]MARUM – Center for Marine Environmental Sciences, University of Bremen, PO Box 330440, 28334 Bremen, Germany

*corresponding author: christopher.vogel@uni-bremen.de

Keywords: Extreme environment, Milos (Eastern Mediterranean), Thermophilic heterotrophic processes, Fatty acids, Labeling study

The shallow water hydrothermal system of Milos is located on the Hellenic Volcanic Arc, which is the most seismically active part of Europe. This system is characterized by steep temperature gradients of up to 8.8°C cm^{-1} and oxygen and hydrogen sulfide profiles that indicate the activity of sulfur oxidizers. However, the microbial community reveals abundant populations of heterotrophic *Bacteroidetes* and *Chloroflexi*, but little is known about their in situ metabolic activity and carbon substrate turnover capabilities. Here, we performed sediment incubation experiments of up to 3 months following a temperature transect (30, 60, and 85°C). We used two position-specific labeled amino acids (AAs) of the pyruvate family (i.e., alanine-3-^{13}C and leucine-3-^{13}C), thus targeting diagnostic fatty acid synthesis routes of heterotrophs. Labeling of the C_3-position of AAs has the advantage of attenuation of autotrophic activity caused by decarboxylation and degradation reactions of AAs labeled at the C_1- or C_2-positions and significantly increases recoveries of ^{13}C-label in targeted fatty acids. We determined the uptake into fatty acids and monitored total organic carbon, dissolved inorganic carbon, and critical volatile fatty acids excreted during AA metabolism. A fast onset of metabolic activity recorded via the specific uptake ($\Delta\delta^{13}C$) into fatty acids of up to 220‰ after 1.5 h, combined with characteristically different fatty acid labeling patterns, were observed for both AA experiments at 35°C. In contrast, alanine turnover at 60°C significantly increased only after 1 month ($\Delta\delta^{13}C \approx$ 12000‰), while metabolic activity during leucine addition was generally lowered by a factor of 40. At 85°C, no significant uptake of AAs into fatty acids could be recorded, likely due to high temperature limitation of bacterial activity. Based on our results, we suggest a dedicated substrate and temperature-dependent metabolism of the heterotrophic bacterial community.

14.2.10 The Stable Carbon Isotopic Composition of Novel Butanetriol-Based Tetraether Membrane Lipids Suggests Their Production by Distinct Sedimentary Microbial Communities

L. Mühlena[1*], S. Coffinet[1], T.B. Meador[1], K.W. Becker[1], J. Schröder[1], V.B. Heuer[1], J.S. Lipp[1], K.-U. Hinrichs[1]

[1]MARUM – Center for Marine Environmental Sciences, University of Bremen, Germany

*corresponding author: luk_mhl@uni-bremen.de

Keywords: Archaea, Biogeochemistry, Carbon cycle, Deep biosphere, Methanogenesis

The chemical structure of archaeal membrane lipids is highly distinct from those of bacteria and eukaryotes. It consists of isoprenoid chains, called biphytanes, connected via ether bonds to a glycerol backbone. Glycerol is generally considered to be the universal backbone involved in the biosynthesis of membrane lipids among the three domains of life. However, recently the pool of archaeal membrane lipids was expanded by the identification of unusual butanetriol dialkyl glycerol tetraethers (BDGTs) in anoxic marine sediments. BDGTs represent a novel class of membrane lipids, as they differ from common archaeal tetraether lipids by substituting one of the glycerol backbones with a butanetriol moiety. To further investigate the sources of BDGTs and metabolic pathways associated with their production in marine sediments, they were isolated from a set of samples covering diverse geochemical settings and converted into biphytanes, in order to determine their compound-specific stable carbon isotopic composition. $\delta^{13}C$ values of BDGT-derived biphytanes observed in sapropels from the eastern Mediterranean Sea are in good agreement with an archaeal source community sustained by sedimentary organic matter. By contrast, organic-rich samples from the Rhone Delta and Black Sea are characterized by a considerable depletion in ^{13}C, indicating either autotrophic carbon fixation or the utilization of methylated substrates, both typically associated with large isotopic fractionation factors. Considering that BDGTs have only been discovered in the lipid inventory of the archaeon *Methanomassiliicoccus luminyensis*, the only isolated representative of the seventh order of methanogens, the results suggest a link of these peculiar lipids with methanogens. Yet the observed variety in the stable carbon isotopic composition of BDGTs observed in this study is most likely explained by the involvement of at least two distinct metabolic pathways, leaving open questions regarding the importance of these pathways and the role of their source organisms in the sedimentary carbon cycle.

14.2.11 High-Resolution Imaging of Molecular Biomarker Distributions in Microbial Mats

Niroshan Gajendra[1*], Lars Wörmer[1], Florence Schubotz[1], Andreas Greve[1,2], Kai-Uwe Hinrichs[1]

[1]Organic Geochemistry Group, MARUM Center for Marine Environmental Sciences and Department of Geosciences, University of Bremen, 28359 Bremen, Germany

[2]Microsensor Group, Max Planck Institute for Marine Microbiology, 28359 Bremen, Germany

*corresponding author: niroshan@uni-bremen.de

Keywords: MALDI FT-ICR MS, Octopus Spring, Intact polar lipids, Quinones, Pigments

Microbial mats are small scaled systems composed of layers of microbes that can be intercalated by sediment and chemical precipitates. They occur in a vast diversity in the marine and terrestrial realm. In addition, they persist in extreme environments such as hot spring systems, polar lakes, and desert crusts. Hence, they are studied in order to understand microbial survival mechanisms in harsh environmental conditions. Moreover, microbial mats are a good reference to investigate early life conditions, for example, during the Archean. This eon is of particular interest regarding the onset of oxygenic photosynthesis and the first major increase in atmospheric oxygen. The fine scaled layers of a microbial mat are built up by a great diversity of different microorganisms. We aim at tracking changes in this community and its adaptation to their environment by micrometer scale molecular biomarker analysis. Therefore, the biomarker distribution of a microbial mat from the Octopus Spring (Yellowstone National Park, USA) was analyzed with a laser-based mass spectrometry imaging (MSI) approach. MSI enables to investigate the fine structure of the microbial mat by ionizing the target compounds within micrometer-sized laser spots. Spatial distributions of intact polar lipids, pigments, and quinones are analyzed to determine the chemotaxonomic affiliations of the Octopus Spring mat and, for example, to differentiate between aerobic and anaerobic lifestyles. Furthermore, microbial adaptation to environmental stresses like pH, high temperature, and UV light can be assessed. The high-resolution images of the chemotaxonomy of the Octopus Spring microbial mat improve our understanding of the geochemical regimes of microbial mats and the interaction among microbial communities and with the environment.

14.3 Abstracts of Poster Presentations

14.3.1 Comparing Microbial Lipid Biosignatures from Serpentinite-Hosted Systems

Lara Meyer[1*], Paraskevi Mara[2], Virginia Edgcomb[2], Ömer K. Coskun[3], William Orsi[3], Frieder Klein[2], Florence Schubotz[1]

[1]MARUM Center for Marine Environmental Sciences, University of Bremen, 28359, Bremen, Germany

[2]Woods Hole Oceanographic Institution, 02543, Massachusetts, USA

[3]Ludwig-Maximilians-University, 80539, Munich, Germany

*corresponding author: lar_mey@uni-bremen.de

Keywords: Serpentinization, Biomarkers, Diether lipids, Hydrothermal vents, Early life

Low-temperature hydrothermal systems, such as the Lost City hydrothermal vents, host primitive but thriving microbial communities that may help to map the beginning of early life on primordial Earth and on other celestial bodies. These systems are driven by exothermic serpentinization reactions in the seawater-exposed ultramafic oceanic mantle rock. Hereby, released molecular dihydrogen together with mantle carbon serves as energy supply for indigenous chemolithotrophic microorganisms. To date only a few of these systems have been examined which limits our understanding of the extent of microbial life in mantle and crustal rock, its composition and metabolic capabilities. The aim of this study is to detect and compare lipid biomarker signatures of two potentially active hydrothermal environments (St. Paul's Archipelago, Mid-Atlantic Oceanic Ridge, Expedition AL 170602 and Atlantis Bank, Southwest Indian Ridge, IODP Exp. 360) with a formerly active site within the passive margin of Iberia (ODP Leg 149). This study reports on the detection of a diverse suite of unique apolar bacterial diether lipids (DEGs), as well as intact polar and apolar archaeal tetraether and diether lipids analyzed by targeted multiple reaction monitoring using ultrahigh-pressure liquid chromatography coupled to mass spectrometry (UHPLC-MS). At Atlantis Bank, the detection of varying amounts of intact polar lipids indicates the presence of viable microorganisms, which is confirmed by DNA and RNA investigations of crustal rock. Samples of St. Paul's Rocks are highly overprinted by signals of water column organisms, but the presence of apolar archaeal and bacterial diether lipids hints to a formerly active serpentinite-associated microbial community. The lipid biomarker distribution at Iberia Margin of a formerly active fluid mixing zone is dominated by fossilized signals. This study presents a first attempt to characterize potential lipid biomarkers that may be used to detect present and past serpentinite-hosted microbial communities.

14.3.2 Impact of Temperature and Sedimentary Setting on Microbial Lipid Distribution

N. Gaviria-Lugo[1*], B. Viehweger[1], M. Elvert[1], K.-U. Hinrichs[1]

[1]University of Bremen, Department of Geosciences & MARUM – Center for Marine Environmental Sciences, Leobener Str. 8, 28359 Bremen, Germany

*corresponding author: ngaviria@uni-bremen.de

Keywords: Intact polar lipids, Marine sediments, Temperature, Guaymas Basin

Intact polar lipids (IPLs) have high diversity and chemotaxonomical potential. These characteristics make them reliable biomarkers to study microbial communities in marine environments. However, in marine sediments several factors shape microbial communities, with the sedimentary environment and temperature being two important agents. Of these, it has been demonstrated that temperature influences the IPL composition in cultured microorganisms, but the distribution of IPLs in sediments is complex, and the main controls are not fully understood. The main goal of this project is to constrain the impact of the sedimentary environment and temperature on the distribution of IPLs in Guaymas Basin, which is influenced by subsurface hydrothermal processes. Two sets of samples, one set with samples under the same temperature gradient from three depositional environments (i.e., shelf, slope, and basin) and another with samples from the central basin at different temperatures and gradients, have been collected during SO241 expedition. IPLs were extracted using a modified Bligh & Dyer protocol and analyzed on a high-performance liquid chromatography (HPLC) system connected to a quadrupole time-of-flight tandem mass spectrometer (qTOF-MS) equipped with an ESI ion source. Preliminary results suggest that bacterial phospholipid abundance decreases toward the basin and archaeal glycolipids predominate at the sites with higher temperature. Further results of the IPLs analyses will bring a previously unknown picture of the geomicrobiology from Guaymas Basin, and furthermore, will increase the knowledge about the environmental controls on IPL distribution and consequently microbial life in marine sediments.

14.3.3 Abiotic Substrate Formation from Organic Matter in the Deep Biosphere

O. Helten[1*], F. Schubotz[1], V. B. Heuer[1], C. T. Hansen[2], R. Stein[3], B. Viehweger[1], Y. Morono[4], W. Bach[1], K.-U. Hinrichs[1]

[1]MARUM-Center for Marine Environmental Sciences & Department of Geosciences, University of Bremen, 28359 Bremen, Germany

[2]ICBM, Carl von Ossietzky University of Oldenburg, 26133 Oldenburg, Germany

[3]Alfred Wegener Institute for Polar and Marine Research, 27570 Bremerhaven, Germany

[4]Kochi Institute for Core Sample Research, Yokosuka, Japan

*corresponding author: helteno@uni-bremen.de

Keywords: Biogeochemistry, Bio-Geo coupling, Limits of life, Nankai Trough

More than a decade ago, the Ocean Drilling Program (ODP) expedition Leg 190 showed that microbial life in the Nankai Trough, Japan, extends down to several hundreds of meters deep into the subseafloor. Yet, the processes that support or limit life in such a hostile environment are still widely unknown. Therefore, International Ocean Discovery Program (IODP) Expedition 370 set out to explore the deep subsurface temperature limit of life and its controls in the high heat flow area of the Nankai Trough. Organic matter contents in sediments at the newly established IODP Site C0023 are extremely low. High temperatures cause methane and other C_{2+} hydrocarbon gases to carry a thermogenic signal. The presence of larger quantities of acetate and other volatile fatty acids in sediments situated at temperatures >75°C is thought to be the result of thermal degradation of organic matter at Site C0023. Those substrates may be used to sustain the indigenous microbial community. Twelve sediment samples from depths of 250 mbsf to 1070 mbsf, corresponding to temperatures between 36°C and 110°C, were chosen to investigate abiotic-biotic interactions in the deep biosphere of the Nankai Trough. The main aims of this study are (i) to perform a detailed characterization of the quality and bioavailability of organic matter at Site C0023, and (ii) to identify potential substrates that may be released at elevated temperatures over time. To do so, we will inquire three different subfractions of organic matter (OM), (1) water-extractable OM, (2) solvent-extractable OM, and (3) kerogen, using gas chromatography coupled to mass spectrometry and various pyrolytic and spectroscopic techniques. Our results will be put into context of cellular concentrations and pore water geochemistry in order to gain a better understanding of the controls that organic matter has on the sustenance of deep hot subsurface microbial life.

14.3.4 Distribution of Unsaturated Archaeols in Suspended Particulate Matter from the Surface Water of the Northwestern Pacific Ocean

Cenling Ma[1,2*], Sarah Coffinet[1], Julius S. Lipp[1], Chuanlun Zhang[3], Kai-Uwe Hinrichs[2]

[1]State Key Laboratory of Marine Geology, Tongji University, Shanghai 20092, China

[2]Organic Geochemistry Group, MARUM Center for Marine Environmental Sciences & Department of Geosciences, University of Bremen, 28359 Bremen, Germany

[3]Department of Marine Science and Engineering, Southern University of Science and Technology, Shenzhen 518055, China

*corresponding author: 0106macenling@tongji.edu.cn

Keywords: Biomarker, Marine Group I Thaumarchaeota, Marine Group II Euryarchaeota

Unsaturated acyclic archaeols (unsARs) are archaeal membrane lipids that are commonly reported in a few cultivated extremophiles and methane seep sediments. Additionally, unsAR with four double bonds ($unsAR_{0:4}$) are observed in marine suspended particulate matter (SPM), where it potentially represents a biomarker for photoheterotrophic Marine Group II Euryarchaeota (MG II). Furthermore, only traces (<0.1%) of $unsAR_{0:1}$ and $unsAR_{0:2}$ have been found in the marine group I (MG I) representative *Nitrosopumilus maritimus*. In this study, SPM samples from surface water samples (from 2 to 5 m) of the northwestern Pacific Ocean were collected in different cruises and during different seasons (April and October). We aim to constrain the distribution of unsaturated archaeols in MG II-enriched samples. Sequencing results show that archaeal communities changed substantially between samples collected in different cruises. The relative abundance of MG I in samples collected in April ranged from 0.05% to 5.2% of total archaea, and MG II from 94.1% to 99.7%. By contrast, MG I in samples collected in October accounted for 54–98.8%, while MG II only accounted for 0.5–42%. Correspondingly, the absolute abundance of unsARs changed from 5.2E+03-6.1E+05 Area Units/L (au/L) (average 1.9E+05 au/L) to 1.3E+06-6.1E+06 au/L (average 2.8E+06 au/L). The relative abundance of unsARs in total core lipids changed from 0.2–37.3% (average 18%) in April to 11.8–28.2% (average 20%) in October. The composition of unsAR also changed with the increase of MG I by increasing in the relative abundance of $unsAR_{0:2}$ and $unsAR_{0:4}$ and decreasing in $unsAR_{0:6}$. Our results suggest that MG I and MG II may both produce unsARs. Moreover, the abundance and composition of unsARs may reflect the archaeal community composition.

14.3.5 L-Leucine Metabolism of Consortia of Anaerobic Methanotrophic Archaea and Sulfate-Reducing Bacteria

Qingzeng Zhu[1*], Gunter Wegener[2], Kai-Uwe Hinrichs[1], Marcus Elvert[1]

[1]MARUM - Center for Marine Environmental Sciences, University Bremen, 28359 Bremen, Germany

[2]Max-Planck Institute for Marine Microbiology, 28359 Bremen, Germany

*corresponding author: qzzhu@marum.de

Keywords: Anaerobic oxidation of methane, Isoprenoid lipid, Fatty acid, Stable isotopic probing

L-leucine is at present positioned as one of the most abundant amino acids in proteins. The basic isoprene unit in the leucine molecule is a conceivable source for the synthesis of archaeal lipids, which are characterized by diether- or tetraether-based isoprenoid lipids. Likewise, leucine as a member of the pyruvate family could be directly channeled into anabolic routes of bacteria thereby being used as chain-initiator (i.e., isovaleryl-CoA) potentially leading to the formation of odd iso-branched fatty acids. In order to test the capability of archaea and bacteria to utilize leucine as a precursor molecule for lipid biosynthesis, we incubated sediment-free enrichments performing anaerobic oxidation of methane (AOM) from the Guaymas Basin with position specifically L-3-^{13}C-leucine as substrate. After incubation, we measured the different microbial lipid pools derived from the consortia of anaerobic methanotrophic archaea (ANME-1) and sulfate-reducing bacteria (SRB, HotSeep-1) for changes in their carbon isotopic compositions. Our results show that SRB much more efficiently metabolize the leucine than its consortium partner ANME-1. Most of the ^{13}C from leucine was directly incorporated into fatty acids, especially branched-chain iso- and anteiso-$C_{15:0}$ and $C_{17:0}$ and unexpectedly monounsaturated $C_{18:1w9}$ and $C_{18:1w7}$, while ANME-1 is not directly assimilating leucine into tetraether lipids but rather utilizes carbon derived from the dissolved inorganic carbon (DIC) pool. This pool is over time affected by full degradation of labeled leucine by the SRB. Additional analysis of direct or indirect leucine incorporation into the protein pool of the consortia is underway and will help to obtain a deeper understanding of the fate of leucine in microbial networks. Such combined lipid and protein studies will provide a comprehensive picture of the metabolism of amino acids by microbial consortia, in culture experiments or natural environmental systems.

14.3.6 Tracking of Microbial Activity by Ultrahigh-Resolution MS Isotope Pattern Matching (IPM)

Stanislav Jabinski[1*], Julius Sebastian Lipp[1], Niroshan Gajendra[1], Martin Könneke[1,2], Kai-Uwe Hinrichs[1]

[1]Organic Geochemistry Group, MARUM Center for Marine Environmental Sciences & Department of

Geosciences, University of Bremen, 28359 Bremen, Germany

[2]Marine Archaea Group, MARUM Center for Marine Environmental Sciences, University of Bremen, 28359 Bremen, Germany

*corresponding author: jabinski@uni-bremen.de

Keywords: Dual-stable isotope probing (dual-SIP), Isotopologues, *M. barkeri*

Remineralization of carbon substrates in subsurface sediments is playing an important role in the global carbon cycle. However, it is still a challenge to link specific microorganisms to distinct biogeochemical processes, to determine microbial rates of organic matter cycling, and to identify central metabolites. Dual isotope labeling experiments using deuterated water and ^{13}C-labeled substrates have been used to determine uptake rates into lipids which are a direct measure of microbial activity. Furthermore, they provide information on total lipid production and carbon assimilation and can be used to infer the dominant carbon metabolism, i.e., distinguish heterotrophy vs. autotrophy. These time-consuming experiments have provided informative data about bacteria, but comparable information regarding archaea is still limited. The major goal of this MSc thesis project is thus to develop and validate a new protocol to drastically speed up this process and gather data for several model species of bacteria and archaea. The new method will use ultrahigh-resolution exact mass spectrometry (time-of-flight (TOF) and Fourier transform ion cyclotron resonance (FTICR)) and compare isotope patterns of isotopologues (isotope pattern matching (IPM)) of measured and computed theoretical mass spectra to calculate label strength. The method will first be compared to common GC-irMS analysis by measuring microbial membrane lipids such as core glycerol dibiphytanyl glycerol tetraethers (C-GDGT) and core archaeol (C-AR) in single SIP experiments on role model cultures. Afterward it will be further developed to target dual-SIP experiments and include intact polar lipids and other lipids such as hydroxy archaeol (OH-AR) which are not directly measurable by GC-irMS. The results of this project will greatly extend our knowledge of assimilation of different substrates by microorganisms and provide the foundation for widespread application of dual-SIP to environmental samples.

14.3.7 The Influence of a *Phaeocystis globosa* Bloom and Mn/Fe(hydr)oxide Formation on Mo Cycling in the Water Column

Ann-Sophie Rupp[1*], Corinna Mori[1], Rolf Weinert[1], Bernhard Schnetger[1], Hans-Jürgen Brumsack[1]

[1]Institute for Chemistry and Biology of the Marine Environment (ICBM), Carl von Ossietzky University of Oldenburg, POB 2503, 26111 Oldenburg, Germany

*corresponding author: ann-sophie.rupp@uni-oldenburg.de

Keywords: Trace metals, ICP-MS, Bio-cycling, Culture experiment, Nonconservative behavior

Molybdenum (Mo) is a redox-sensitive element and the most abundant trace metal (110 nM) in the ocean. Even though it is actively involved in biological cycles (e.g., nitrogen cycle) it is generally considered as conservative. However, in recent studies from the coastal region of the southern North Sea, a deviation of Mo from conservative behavior was observed. The main mechanisms suggested to be responsible for the nonconservative behavior are (1) the scavenging of Mo by organic material, derived from *Phaeocystis globosa* blooms, and (2) the precipitation of Mo along with inorganic manganese (Mn) and iron (Fe) (hydr) oxides. The aim of the experiment was to estimate whether scavenging along with *Phaeocystis globosa*-derived organic matter or the formation of Mn-Fe(hydr)oxides is responsible for the nonconservative behavior. We set up two non-axenic *Phaeocystis* cultures and grew them under controlled light, temperature, and nutrient conditions, as well as one additional plankton-free control. We generated a high-resolution multiparameter dataset and compared the changes in Mn, Fe, and Mo concentration to those observed in phytoplankton biomass and composition (single cell vs. colony formation). The results indicate that Mn and Fe cycling was highly coupled to the development of the *Phaeocystis* bloom, while Mo concentrations seemed nearly unaffected by organic as well as inorganic processes/cycles. As the conducted culture experiment just covered the lag, growth, stationary, and onset of the death phase of the bloom, we can however not exclude that degrading biomass has an influence on the Mo cycle. For this purpose we will conduct a follow-up project to study the influence of degrading *Phaeocystis globosa* material on Mo cycling over a period of 4 months.

15 Plastics in the Environment: Analyzing Sources, Pathways, Occurrence, and Means of Tackling This Form of Pollution

Rosanna Schöneich-Argent[1], Marten Fischer[2] and Maurits Halbach[2]

[1]ICBM, AG Geoökologie, Carl von Ossietzky University Oldenburg, Germany

[2]ICBM, AG Organische Geochemie, Carl von Ossietzky University Oldenburg, Germany

Rosanna Schöneich-Argent: rosanna.schoeneich-argent@uni-oldenburg.de

Marten Fischer: marten.fischer@uni-oldenburg.de

Maurits Halbach: maurits.halbach@uni-oldenburg.de

This session of the YOUMARES 9 conference does not have a corresponding proceedings article.

15.1 Call for Abstracts

Macro- and microplastics have become ubiquitous in the marine, freshwater, and terrestrial environment. Despite a recent increase in scientific research efforts on the reasons behind, the extent and the impacts of this form of environmental pollution, a lack of standardized methods and many knowledge gaps remain. This session invites young scientists to present their work that aims to quantify macro- and microplastics, analyze their sources, pathways and occurrence, and assess long-term trends of abundance and composition. Also of interest are studies that deal with littering behavior, that analyze ongoing means to reduce plastic waste, and that evaluate awareness campaigns.

15.2 Abstracts of Oral Presentations

15.2.1 Microplastics Transport Simulations in the German Bight

Florentina Münzner[1,2*], Claudia Lorenz[2], Mirco Scharfe[2], Sebastian Primpke[2], Gunnar Gerdts[2], Ulrich Callies[3]

[1]University of Bremen

[2]Alfred-Wegener-Institute, Helmholtz Centre for Polar and Marine Research, Biologische Anstalt Helgoland

[3]Helmholtz-Zentrum Geesthacht

*corresponding author: fl.muenzner@gmail.com

Keywords: Lagrangian particle tracking, Forward and backward trajectories, Environmental pollution, Coastal waters

Marine pollution with microplastics has already become a considerable problem in marine environments. Current research shows that the North Sea is no exception. However, there still is a large knowledge gap about the occurrence and sources of microlitter. Long-term monitoring programs and surveys are often not feasible, which is why particle transport modelling is a promising and economic way to gather relevant information. This study uses Lagrangian transport simulations in order to determine potential source regions of microplastic particles found in southern North Sea surface waters during a cruise with the RV *Heincke* in summer 2014. The two-way approach shows particle distributions in a backward and a forward in time setting. Backward in time simulations identified residual currents in the North Sea to be anticyclonic during and about 4 weeks before microplastics sampling was carried out. Forward in time simulations identified rivers discharging into the southern North Sea as a potential source for microplastics in coastal regions such as the Wadden Sea, harbors, and bays. Sampling stations of this cruise were rather unaffected by riverine inflow. Riverine particle distribution in the German Bight is influenced by tidal forces during weak northeasterly winds in summer and by residual currents during strong westerly winds in winter. It is assumed that sampled microplastics potentially originate from the English Channel whereas estuaries and deltas of rivers were not favored as potential source regions in this particular scenario. The strong variability of the North Sea, however, makes it difficult to determine a general answer to potential source regions.

15.2.2 Floating Marine Debris Movement in the North Sea: A Complex Interaction Between Currents, Tides, and Local Wind Effects

Jens Meyerjürgens[1*], Thomas Badewien[1], Oliver Zielinski[1], Jörg-Olaf Wolff[1]

[1]Institute for Chemistry and Biology of the Marine Environment (ICBM), Carl von Ossietzky University of Oldenburg, Oldenburg, Germany

*corresponding author: jens.meyerjuergens@uni-oldenburg.de

Keywords: Surface current field, Low-cost drifter design, Floating marine debris, North Sea

The investigation of observed transport patterns of floating marine debris (FMD) is crucial for reliable simulations using high-resolution ocean models. A series of studies have focused on global, large-scale transport modelling of FMD by using Lagrangian observations from satellite-tracked drifting buoys (drifters). The coverage of those globally deployed drifters is very poor for the North Sea region. In an attempt to fill this knowledge gap in the North Sea region, a new design of a low-cost drifter is presented, which is primarily intended for the investigation of submesoscale dynamics of surface velocity fields in strongly tidally influenced shallow water areas. The drifter motion represents the current flow of the upper 0.5 meter of the surface layer and provides surface current information in tidal inlets and in the surf zone with water depths less than 1 meter. A dataset of 21 drifter deployments in the North Sea, which were conducted in March 2017, October 2017, and February 2018, will be presented and discussed. The movements of the drifters are strongly affected by tides and wind-induced surface currents. The principal semidiurnal lunar tide (M_2) represents the predominant tidal constituent and shows a significant proportion of the zonal and meridional velocity components. Residual currents were calculated by applying a low-pass filter with a cutoff period of 25 hours to the drifter derived velocity time series data. The residual currents show a significant correlation with local wind field data. The meridional component exhibits a stronger dependence on the local wind field (1.3% of the wind speed) than the zonal component (1% of the wind speed). Ultimately, the pathways, the beaching location, and the travel duration of the drifters are strongly influenced by local wind effects in the North Sea.

15.2.3 Macro- and Microplastics Around Antarctica: Preliminary Results from the Antarctic Circumnavigation Expedition (ACE)

Giuseppe Suaria[1,2*], Jasmine Lee[3], Vonica Perold[4], Stefano Aliani[1], Peter G. Ryan[4]

[1]ISMAR-CNR, Institute of Marine Sciences, La Spezia, Italy

[2]IEO-COB, Instituto Español de Oceanografía, Palma de Mallorca, Spain

[3]Centre for Biodiversity Conservation Science, University of Queensland, Australia.

[4]FitzPatrick Institute of African Ornithology, University of Cape Town, South Africa

*corresponding author: giuseppe.suaria@sp.ismar.cnr.it

Keywords: Southern Ocean, Plastic pollution, Microplastics, Microfibers

The Antarctic Circumnavigation Expedition (ACE) sampled micro-, meso-, and macroplastic litter around Antarctica from December 2016 to March 2017, with the aim of providing the first comprehensive synoptic survey of the levels of plastic pollution in Antarctic waters. Preliminary results show that the Southern Ocean can be regarded as the ocean least polluted by plastics globally. Only small numbers of microfibers were found in 173 beach sediment samples collected from 12 Antarctic and sub-Antarctic islands visited during the expedition, and only 7 plastic particles (identified by FTIR analysis) were retrieved from 33 plankton samples collected around the continent using a 200 µm neuston net. In addition, only 22 macro-litter items (>2 cm) were visually observed floating south of the Subtropical Front in almost 15,000 km of transect counts. Nevertheless, anthropogenic litter was found in two seabed Agassiz trawls and macroplastic items were recovered from most beach landings, though quantity varied with location. Synthetic microfibers were detected in virtually all bulk and underway water samples collected around Antarctica. Surprisingly, there was no marked gradient in these fibers as we approached continental source areas. Confirmation of the chemical identity of these fibers is still pending, but if they prove to be plastic, they suggest that all the world's surface waters apparently carry low concentrations of microfiber pollutants, at a density of ~0.1–1 fibers per liter.

15.2.4 Plastics in Peru: Quantity, Composition, and Public Perceptions of Beach-Stranded Marine Anthropogenic Debris

Nicola Allan[1,2*], Joanna Alfaro-Shigueto[1,2], Laura Braunholtz[1,3], Jeffrey Mangel[1,2], Brendan Godley[1]

[1]University of Exeter, Penryn Campus, UK

[2]Pro Delphinus, Miraflores, Lima, Peru

[3]University of Newcastle, Newcastle, UK

*corresponding author: njsa201@exeter.ac.uk

Keywords: Plastic, Beach-sampling methodologies, Social surveys, SE Pacific

To generate a full spatiotemporal understanding of plastic dispersal and accumulation patterns, it is necessary to combine long-term abundance trend and socioeconomic data regarding population-level littering and recycling habits. Results from the second year of a multi-year study investigating beach-stranded marine anthropogenic debris in Peru are presented here. Surveys were focused in two northern regions, Piura and Lambayeque, and site selection was based on relation to the Natural Park of Lobos de Afuera Island, 93 km off the coast of Lambayeque. Sampling was undertaken on 18 sandy beaches for microplastics (1 >x< 5 mm), and 16 beaches for macro litter (>5 mm). Data were also collected from a citizen science-led clean-up of the island and anchorage. Sampling and litter sourcing methodologies were critically assessed to establish most effective practice for the project in the long term. A total of 7,405 macro litter items were enumerated, of which 90% was plastic. 517 microplastic particles were encountered in 81% of sediment samples (n = 69), of which foam was the most abundant category (44.9%). Highest densities of macro litter (2.38 m^{-2}) and microplastic (1.75 m^{-3}) were both in Piura. Stakeholder interviews were conducted at each survey site, targeting three groups: regional government, fishers, and members of the public. Questions were formulated to better understand public perception of and concern toward waste management. Forty-four interviews were conducted, highlighting a need for improved information dissemination regarding the effects and fate of discarded litter (56.8% of respondents were unable to define the negative impacts of plastic pollution for marine organisms). Results are discussed in the context of the SE Pacific and data recorded in 2017, during which time northern Peru experienced El Niño-induced catastrophic flooding. The combination of quantitative data and qualitative surveys allow for better, localized understanding of plastic pollution and source pathways, leading to tailored solutions and recommendations.

15.2.5 Microplastics in Coastal North Sea Sediments: Analyzed Using Fourier Transform Infrared Spectroscopy

Lars Hildebrandt[1,2*], Claudia Lorenz[1], Sebastian Primpke[1], Gunnar Gerdts[1]

[1]Alfred Wegener Institute Helmholtz Centre for Polar and Marine Research, Biologische Anstalt Helgoland, Kurpromenade 201, 27498 Helgoland, Germany

[2]present address: Helmholtz Centre for Materials and Coastal Research, Max-Planck-Str. 1, 21502 Geesthacht, Germany

*corresponding author: lars.hildebrandt@hzg.de

Keywords: Microplastics, North Sea, Marine sediments, FTIR microspectroscopy

Anthropogenic litter, especially highly persistent plastic litter, has become a global problem. It is present in almost all marine habitats and freshwater ecosystems. Microplastics (≤5 mm) are more challenging to handle than larger plastic debris and pose a threat to a wide spectrum of organisms. Once microplastics reach the seafloor, degradation can come to a nearly complete halt which means that marine sediments serve as an ultimate repository for microplastics. In the present study, microplastics isolated from 14 sediment samples from locations close to the Frisian Island, from the English Channel, and from offshore locations were quantified, measured, and assigned to polymer clusters by state-of-the-art methods. In contrast to studies that solely use visual identification, this study employed μ-FTIR imaging for detection. Density separation with the MicroPlastic Sediment Separator was used in combination with a $ZnCl_2$ solution ($\rho = 1.7$ g mL^{-1}) to separate microplastics from the sediment. Particles ≥500 μm were visually sorted and manually analyzed using ATR-FTIR spectroscopy, whereas particles ≤500 μm were enzymatically and chemically purified using recently developed microplastic-reactors. Afterwards, the samples were transferred onto inorganic membrane filters and automatically analyzed using μ-FTIR imaging. The concentrations of microplastics at the different stations ranged between 34 and 1457 particles per kg sediment (dry weight). All particles had a size ≤300 μm. The fraction ≥500 μm contained no marine microplastics. Most particles (69%) were between 11 μm and 25 μm in size, which indicates a high risk of ingestion, e.g., by filter-feeding marine organisms. The study provides a substantial contribution to the assessment of the level of microplastic contamination of the North Sea which the Marine Strategy Framework Directive targets. To date, data on the microplastic burden of North Sea sediments are scarce as only three studies exist, with inter-study comparability being hampered by the lack of a standard operation procedure.

15.2.6 Abundance of Microplastics in Estonian Part of the Baltic Sea

Polina Turov[1*], Kati Lind[1], Inga Lips[1]

[1]Tallinn University of Technology, Department of Marine Systems, Akadeemia tee 15a, 12618 Tallinn, Estonia

*invited speaker, corresponding author: polina.turov@ttu.ee

Keywords: Microplastic, Baltic Sea, Sediments, Sea surface, WWTP

Our study presents the abundance, distribution, and characteristics (type, shape, color, and size) of microplastic (MP) in the sea surface layer and bottom sediments in Estonian waters of the Baltic Sea. The selection of sampling areas in the coastal sea was based on the locations of the potential MP pollution sources – rivers and wastewater treatment plant (WWTP) outflows. Additionally, open sea areas were sampled to assess the broader distribution of MP in sub-basins around Estonia. Water samples from the surface layer were collected in eight regions using a 333 μm mesh manta trawl. In total, 58 samples were collected in 2016–2017. Microplastic pollution levels in sediments were investigated in 2017 in the Gulf of Finland. Samples (upper 5 cm) from six locations were collected using a Van Veen Grab and Gemax corer. The results showed the presence of MP in each water sample, the average amounts of MP remaining below 1 particle per m^3 both in 2016 (0.98 pa m^{-3}) and 2017 (0.58 pa m^{-3}). MP abundance in sediments varied between 137 and 1,452 particles per kg of dry weight sediment, being dependent on the structure and composition of the sediment. MP is made up of approximately 30% of the total microlitter particles in both sea surface layer and sediment samples. MP pollution in the sea surface layer and sediments was highest in the central part of the Gulf of Finland. The presence of fibers (both plastic and non-plastic) prevailed over particles; the prevalent color of fibers was blue and black. Samples were also collected from treated sewage water in front of the outfall tunnels at Sillamäe and Tallinn WWTP using manta trawl cod end (mesh size 333 μm). The results of this study indicate that the WWTPs can be considered an important source of the MP in the Estonian marine waters.

15.2.7 Isolation and Production Efficiency of Novel Bioplastic-Synthesizing Bacteria from Redang and Bidong Islands, Malaysia

Athraa Al-Saadi[1,2*], Kesaven Bhubalan[1], Alessio Mengoni[2], Eveline Peeters[3]

[1]School of Marine and Environmental Sciences, University Malaysia Terengganu, Terengganu, Malaysia

[2]Department of Biology, Universita Degli Studi Firenze, Florence, Italy

[3]Department of Bioengineering Sciences, Faculty of Sciences and Bioengineering Sciences, Vrije Universiteit Brussel, Belgium

*corresponding author: athraabasem09@gmail.com

Keywords: Biotechnology, Pollution, Plastic, Sustainability, Gas chromatography, Carbon sources

The ability of some microorganisms to synthesize plastic-like materials has received much attention from academia and the industry over the past 30 years, due to the environmental problems caused by industrial plastic pollution and solid waste disposal. Biodegradable plastic is a worldwide alternative to synthetic plastic, but its usage is limited because production costs are over 3 times that of industrial plastic. Polyhydroxyalkanoate (PHA) is a type of biodegradable plastic produced by some bacteria strains under specific environmental conditions of nutrients and carbon. We aim to isolate and identify novel bacteria strains responsible for producing this polymer (PHA) from marine sediments and coastal soils on the previously unstudied islands of Bidong and Redang in Terengganu, Malaysia. The main cost and limitation of biodegradable plastic production is the expensive carbon source growth media. Therefore, after isolation, we tested the PHA production ability of microorganisms grown on six different renewable carbon sources – glucose, fructose, sucrose, sweet water, glycerol, and glycerin pitch – to find the cheapest and most suitable carbon source for producing PHA. Glycerin pitch would be the sustainable option as it is an industrial waste product. Gas chromatography was used for polymer characterization, while PCR purification and sequencing will be used to identify bacterial strains responsible for PHA production. Five potential PHA-producing bacterial strains cultured on all carbon sources except glycerol were detected from both islands. Preliminary analysis of gas chromatography identified the easily degraded SCL-PHA as well as the more durable MCL-PHA. These results can contribute to the ongoing search for PHA-producing bacterial strains that may produce alternatives to conventional plastic and could enhance bioplastic industry by reducing production costs and thus reducing synthetic plastic usage and its negative impact on the environment.

15.3 Abstracts of Poster Presentations

15.3.1 Microplastic Prevalence in the River Weser

Maurits Halbach[1*], Sonya Ranita Moses[2*], Lisa Roscher[3*], Barbara Scholz-Böttcher[1], Martin G.J. Löder[2], Christian Laforsch[2], Sarmite Kernchen[2], Gunnar Gerdts[3]

[1]Institute for Chemistry and Biology of the Marine Environment (ICBM), Carl von Ossietzky University of Oldenburg, P.O. Box 2503, D-26111 Oldenburg, Germany

[2]University of Bayreuth, D-95440 Bayreuth, Germany

[3]Alfred Wegener Institute, Helmholtz Centre for Polar and Marine Research, D-27483 Helgoland, Germany

∗corresponding authors: Maurits.Halbach@uni-oldenburg.de, Sonya.Moses@uni-bayreuth.de, Lisa.Roscher@awi.de

Keywords: Microplastics, Environmental samples, River Weser, Pyrolysis-GC/MS, FTIR spectroscopy

As global plastic production continues to increase steadily, so does the amount of synthetic material that enters aquatic environments. In recent years, more and more studies have focused on the occurrence of microplastics (MPs), i.e., synthetic organic polymers with a size <5 mm, in the environment. These omnipresent and hardly degradable pollutants are easily accumulated in the environment and potentially ingested by a wide range of organisms which holds a high toxic potential. While most studies focus on MP in marine environments, only very little data on MP pollution of river systems are available, particularly those focusing on estuarine regions as transition zones. However, riverine and estuarine systems may represent the main transport routes for MP, connecting terrestrial and marine systems. This work aims to assess MP pollution throughout the entire German River Weser, which connects large urban and agricultural areas with the North Sea. Besides the major river compartments (sediment/water), the role of point sources, like wastewater treatment plants, and diffuse sources (aeolian/drainage) is investigated. Riverine and atmospheric samples were collected in April and May 2018, while the outlet of two wastewater treatment plants will be sampled monthly over 1 year from July 2018 to July 2019. Samples will then be treated following a novel purification protocol in order to remove organic and inorganic residues. The isolated MP fraction will be analyzed with state-of-the-art methods (FTIR, RAMAN, PY-GC/MS) which provide polymer-specific information about the particle count, size distribution, and mass. Knowledge gained in this study will contribute to a better understanding of MP sources and pollution in rivers and will particularly aid the understanding of transport mechanisms within interconnected aquatic systems, in addition to acting as a basis for future conservation and monitoring measures.

15.3.2 Microplastic Abundance and Composition in Southern North Sea Surface Waters

Christopher Dibke[1*], Marten Fischer[1], Barbara M. Scholz-Böttcher[1]

[1]Institute for Chemistry and Biology of the Marine Environment (ICBM), Carl von Ossietzky University of Oldenburg, P.O. Box 2503, D-26111 Oldenburg, Germany

∗corresponding author: christopher.dibke@uni-oldenburg.de

Keywords: Microplastic, Source attribution, Py-GC-MS/thermochemolysis, Method standardization

Plastic is a ubiquitous pollutant in the marine environment. Its bioavailability increases with decreasing particle size. Thus, information on the occurrence of plastic particles <5 mm, defined as microplastic (MP) in marine waters helps to estimate its environmental relevance. To gather such information for defined regions, an area-wide approach is necessary. Studies on the MP content of surface waters of the southern North Sea mainly focus on methodical aspects by using random samples. This study aims to comprehensively identify the abundance and composition of MP in surface waters of the southern North Sea as well as its potential sources and temporal variation. Therefore, water volumes of 1.3–335.7 L were filtered in estuaries and along transects (max. 58.4 km) from 2.5 m depth during the cruises HE473 and HE498 (October 2016 and 2017), creating >20<125 μm and >125 μm particle size fractions. The spatial distribution of the sampling sites will allow for the discussion of potential sources and pathways when detecting distinct MP signals, as samples represent areas with (a) riverine input, (b) industry, (c) high shipping traffic, (d) high tourist activity, or

(e) cross-section transects. The focus will be on the analysis of the most commonly found polymer types (e.g., PE, PP, PS, PET, PVC, PMMA, PC, PA6, PUR) by applying a short preconcentration procedure of MP and a chemically weak sample purification step (H_2O_2, 30%), before using Py-GC-MS/thermochemolysis. py-GC-MS is a well-established method due to its nonselective, qualitative, and trace-level polymer-specific quantitative character. This method is especially suitable for small MP and does not require preliminary optical detection or mechanical separation. Blank sample analyses will help identify and further avoid airborne MP contamination. Recovery experiments will allow for the evaluation of the applied procedure in the context of method standardization for MP analyses.

16 Trends in Plankton Ecology

Katarzyna Walczyńska[1] and Maciej Mańko[1]

[1]Department of Marine Plankton Research, Institute of Oceanography, University of Gdańsk, Al. Marszałka Piłsudskiego 46, 81-378 Gdynia, Poland

Katarzyna Walczyńska: katarzyna.walczynska@phdstud.ug.edu.pl

Maciej Mańko: mmanko@ug.edu.pl

This session of the YOUMARES 9 conference does not have a corresponding proceedings article.

16.1 Call for Abstracts

Oceans cover most of our planet's surface, and certain marine ecosystems, including coral reefs and deep seafloors, have been reported to have a higher biodiversity than tropical rain forests. Marine systems play important roles in maintaining benthic communities and affect humanity by being rich and versatile sources for new drug discoveries. Many benthic communities are characterized by environmentally harsh conditions in matters of pressure, temperature, nutrient availability, and salinity. Hence, reproduction and survival often depend on the formation of bioactive secondary metabolites. We encourage contributions emphasizing community functioning, chemical interactions between marine (micro-) organisms, and sustainable exploitation of natural products as sources for new drugs.

16.2 Abstracts of Oral Presentations

16.2.1 A Case Study to Describe the Putative Associated Bacterial Core Community on a Marine Diatom

Julian Mönnich[1*], Rebecca J. Case[2], Sylke Wohlrab[3], Jan Tebben[3], Tilmann Harder[1,3]

[1]University of Bremen, Leobener Strasse 6, 28359, Bremen, Germany

[2]University of Alberta, T6G 2E9, Edmonton, AB, Canada

[3]Alfred Wegener Institute, Am Handelshafen 12, 27570, Bremerhaven, Germany

*corresponding author: jmoennich@uni-bremen.de

Keywords: Microalga-bacteria interaction, Diatoms, Vitamin B_{12}, Microbial community analysis

Marine microalgae and bacteria have co-occurred over billions of years, resulting in distinct interactions between these two groups of organisms. For example, over half of microalgal species representing marine and freshwater habitats require the cofactor cobalamin (vitamin B_{12}) for growth, which is synthesized de novo by only certain bacteria and archaea. Elucidating the interactions between marine microalgae and bacteria is of prime importance to better understand oceanic nutrient and carbon flux and, hence, marine primary production. Prior studies revealed distinct bacterial phylotypes associated with individual genera of microalgae; however it is unclear if the community of algae-associated bacteria differentially supports algal growth and performance, e.g., via the provision of essential micronutrients such as vitamins. Different bacterial communities were dissociated from several diatom species and inoculated to the same acceptor cultures of the axenic diatom *Thalassiosira rotula*. Bacterial communities in both, donor and acceptor cultures were characterized with Illumina-MiSeq targeting the 16S rDNA V4 region. In parallel, growth and performance of the experimental acceptor cultures were analyzed with imaging flow cytometry and pulsed amplitude modulation fluorometry. This study aimed to test (a) if the bacterial community composition in the acceptor cultures differed from those in the donor cultures, and (b) if the acceptor cultures shared a certain core community of bacterial taxa? The outcomes of this study will be discussed in the context of the pending research questions if marine microalgae harbor or support functionally important bacteria in their phycosphere to support algal growth and performance.

16.2.2 Monitoring Phytoplankton Community and Functional Composition in a High-Resolution Large-scale Mesocosm Experiment

Nils Hendrik Hintz[1*], Carina Bunse[2], Leon Dlugosch[1], Maren Striebel[1]

[1]Institute for Chemistry and Biology of the Marine Environment ICBM, Carl von Ossietzky University of Oldenburg, Carl von Ossietzky Straße 9-11, 26129 Oldenburg, Germany

[2]Helmholtz Institute for Functional Marine Biodiversity at the University of Oldenburg (HIFMB), Ammerländer Heerstraße 231, 26129 Oldenburg, Germany

*corresponding author: nils.hendrik.hintz@uni-oldenburg.de

Keywords: North Sea, Phytoplankton bloom, Microbial loop, Succession, Pigments

Phytoplankton blooms in the North Sea are long known and acknowledged as fundamental drivers in marine ecology. However, water currents, weather, and navigation issues constrict the high frequency monitoring. Therefore, a large-scale indoor mesocosm experiment for in situ monitoring under highly controlled conditions and high sampling frequency was performed over 6 weeks. Eight 600 L indoor mesocosm replicates (Planktotrons) were used to initiate a natural North Sea phytoplankton bloom and to study the subsequent plankton succession. Artificial seawater was used to set up the experiment, while inoculum was retrieved by the German Bight (RV *Heincke*) and added to the mesocosms. Regular mixing of the water column, natural diurnal light cycles as well as spring temperature conditions were applied. This study focuses on the phytoplankton succession *via* daily observation of community composition as well as functional composition (pigment and elemental stoichiometry). A rapid phytoplankton bloom was observed within the first days in all mesocosms, consisting mainly of diatom species. A second bloom was observed later on, dominated by *Phaeocystis* and revealed a much higher variation in succession progress between the replicates. Further work will focus on synthesizing of all experimental results in order to understand the relationships and interactions between the biotic and abiotic parameters. This will bridge the gap between microbiological activity, nutrient fluxes, and cycling as well as the formation and composition of dissolved organic matter.

16.2.3 New Insights into the Phytoplankton Community at Helgoland Roads

Laura Käse[1*], Katja Metfies[2], Karen H. Wiltshire[3], Maarten Boersma[1], Bernhard M. Fuchs[4], Alexandra Kraberg[1]

[1]Biologische Anstalt Helgoland, Alfred-Wegener-Institute Helmholtz Centre for Polar and Marine Research, 27498 Helgoland, Germany

[2]Alfred Wegener Institute, Helmholtz Centre for Polar and Marine Research, 27570 Bremerhaven, Germany

[3]Wattenmeerstation Sylt, Alfred-Wegener-Institute, Helmholtz Centre for Polar and Marine Research, 25992 List/Sylt, Germany

[4]Max Planck Institute for Marine Microbiology, 28359 Bremen, Germany

∗corresponding author: laura.kaese@awi.de

Keywords: Next-generation sequencing, Nanoplankton, North Sea, LTER, Food web

Plankton time series are an important tool in the study of long-term changes in marine biodiversity. Most conventional time series like the Helgoland Roads time series still use traditional microscopy techniques. However, information on co-existence networks as indicators for potential food web interactions is scarce. This is particularly true for its smaller components in the nano- and pico-range, which are barely investigated due to the limited size resolution in light microscopy. Here, we present the next-generation sequencing as a good option to obtain data on the occurrence of these smaller organisms and to enhance the resolution at genus level over the whole size spectrum. For that, fifty samples were taken during the phytoplankton spring bloom between March and May 2016. Quality filtering of sequencing data resulted in 580 OTUs for phytoplankton genera of which various planktons belong to nano- and picoplankton not routinely investigated at Helgoland Roads, for example, *Micromonas* sp. *Chrysochromulina* sp., *Emiliania* sp., *Leucocryptos* sp., Choanoflagellida and Cryptophyta. Non-metric multidimensional scaling indicated rapid shifts from 1 week to the next on genus level. Several abiotic factors (e.g., temperature) could be found as main drivers influencing the phytoplankton community. The implementation of this new technique is therefore well suited to unveil new information on ecosystem services and increases observational power, which could potentially result in the establishment of early warning systems.

16.2.4 Suitability of Trait-based Zooplankton Indicators to Evaluate the Food Web Status for Management

Merle Twesten[1*], Christian Möllmann[2], Saskia A. Otto[2]

[1]University of Hamburg, 20146 Hamburg, Germany

[2]Institute of Marine Ecosystem and Fishery Sciences, 22767 Hamburg, Germany

∗corresponding author: merle.twesten@studium.uni-hamburg.de

Keywords: Central Baltic Sea, MSFD, ZooScan, GAM

Under the European legislative Marine Strategy Framework Directive (MSFD), Member States are requested to achieve a so-called Good Environmental Status (GES) in the European waters by 2020. GES is assessed through 11 qualitative descriptors that each describes a sub-component of the ecosystem such as food webs (descriptor D4). Food web dynamics plays a key role in the ecosystem as it determines the energy transfer across different trophic levels, ensures long-term abundance, and food source provision. To achieve healthy food webs, it is crucial to define indicators that are able to forecast developments or trends of marine habitats or species when subjected to certain pressures such as overfishing, eutrophication, or global warming. While currently proposed food web indicators often are based on abundances or biomass of a single species or a taxa groups, recent studies suggest trait-based indicators as better representing the status of ecosystem functioning. Here, we studied the suitability of the size-based indicators mean zooplankton size (MS) in the Baltic Sea measured automatically with digital imaging. Central Baltic Sea summer cruise zooplankton samples collected with a BabyBongo net of 150 mµ mesh size in the years 2007–2018 were scanned with the

ZooScan device. Samples were analyzed spatially and temporally. We validated the performance of this indicator and identified key pressures following an indicator-testing framework based on generalized additive models. We contrasted the indicator performance of MS derived from the ZooScan versus from manual counts multiplied with species-specific wet weights, and provide recommendations of sampling strategies for management.

16.2.5 Biogeography and Population Structure of the Key Marine Zooplankton *Calanus finmarchicus* Revealed by Molecular Tools

Marvin Choquet[1,2]*, Irina Smolina[1], Janne Søreide[2], Galice Hoarau[1]

[1]Nord University, 8049, Bodø, Norway

[2]The University Centre in Svalbard, Longyearbyen, Norway

*corresponding author: marvin.choquet@nord.no

Keywords: Population genomics, Next-Generation Sequencing, Molecular markers

Copepods of the genus *Calanus* play a key role in marine food webs as consumers of primary producers and as prey for many commercially important marine species. Within the genus, *Calanus finmarchicus* and *C. glacialis* are considered indicator species for Atlantic and Arctic waters, respectively, and changes in their distributions are frequently used as a tool to track climate change effects in the marine ecosystems of the northern hemisphere. However, discrimination between these two species is challenging due to their morphological similarity and most of the knowledge on these species has been built based almost exclusively on morphological identification leading to misidentification. Here, we used molecular markers as tools for species identification in order to redraw the distribution ranges of the different species within the *Calanus* genus in the North Atlantic and Arctic Oceans. This revealed wider and more overlapping distributions for each species than it was described before when using only morphology to separate species. With this knowledge, we selected a set of *C. finmarchicus* individuals from nine locations that span the species distributional range, and we applied a technique that we developed based on targeted-resequencing to investigate the genomic variability and gene flow among different locations.

16.2.6 A Race of Gene Regulation and Climate Change: Arctic and Atlantic Copepods in the Fram Strait

Maria Scheel[1,2]*, Gareth Pearson[1]

[1]Ghent University, Marine Biology Research group, 9000 Gent, Belgium

[2]University of Algarve, CCMAR, Faro, 8005-139, Portugal

*corresponding author: Maria.Scheel@UGent.be

Keywords: Arctic, Transcriptomics, *Calanus*, Thermal stress, Copepods

The Arctic Ocean is allegedly the most vulnerable ecosystem under the globally changing climate and is exposed to changing abiotic regimes due to increasing temperatures. The northward shift of aquatic isotherms is followed by polewards distribution range expansions, leading to replacement of polar by boreal-Atlantic species. The most interesting aspect for studies on the temperature sensitivity of zooplankton in Fram Strait is the fact that both boreal-Atlantic and polar plankton communities are dominated by congeneric species, which are characterized by different morphology, trophic relations and lipid content, and hence shifts bear potential of trophic cascades. We hypothesized that Atlantic congeners possess higher gene regulation plasticity than polar species under thermal stress of realistic experimental culturing temperatures up to 8 °C. Comparative transcriptomic analysis was executed on the boreal *Calanus finmarchicus* and Arctic *C. glacialis* and *C. hyperboreus* to investigate species-specific gene regulation responses. Based on next-generation sequencing technology with a sequencing depth of > 330 million base pairs per species, differential gene expression was found absent for *C. glacialis* with increasing temperature, though an increase of the heat shock protein 70 kDa was calculated. In contrast, the other congeners repressed metabolism pathways of lipids and fatty acids, amino acids, carbohydrates and typical cellular stress response pathways, indicating that 8 °C did not yet induce vital stress. But as the molecular chaperone *DnaJ* is downregulated in *C. finmarchicus* and upregulated in the polar species, varying temperature sensitivities for protein degradation might be a key for thermal stress responses. Furthermore, epigenetic gene silencing in the invading Atlantic *C. finmarchicus* indicated thermal stress resilience and adaptive gene regulation, not seen for genetic information processing in the other two species. Eventually, *C. hyperboreus* exhibited both adaptive heat-shock protein regulation and a complex array of other biological pathways, and hence might resist ambient increasing temperature better than its polar sibling *C. glacialis*.

16.2.7 Testing Hatching Success of Sea Breams and Sardines Under Different Temperature Regimes

N. Smith Sánchez[1,2]*, M. Sswat[1], C. Clemmesen[1], S. Garrido[3], P. M. Pousão Ferrerira[4], L. Ribeiro[4], U. Riebesell[1]

[1]GEOMAR Helmholtz Centre for Ocean Research Kiel, Düsternbrooker Weg 20, 24105 Kiel, Germany

[2]Christian Albrechts Universität zu Kiel, 24118 Kiel, Germany

[3]Portuguese Institute for the Sea and the Atmosphere, Av. Brasilia s/n, 1449-006 Lisboa, Portugal

[4]Portuguese Institute for the Sea and the Atmosphere, Av. 5 de Outubro s/n, 8700-305 Olhão, Portugal

*corresponding author: nicolassmithsanchez@hotmail.com

Keywords: Hatching success, Temperature, European pilchard, Sparidae, Artificial upwelling

The concept behind artificial upwelling is to simulate processes occurring in the natural upwelling areas. In the oligotrophic waters, upwelling of the nutrient-rich deep waters may trigger a shift from a more complex, multilevel pelagic food web to a simpler one. A reduction in the number of trophic levels and links should translate into increased energy transfer to top consumers, and thus more biomass of economically important species. To test whether the artificial upwelling concept applies, mesocosm experiments will be carried out in Gando Bay, Gran Canaria (Canary Islands) from October to December 2018. During the experiment, the effect of artificial upwelling rates will be tested in a linear regression setup on a whole-community scale, ranging from bacteria and phytoplankton all the way up to top consumers, in this case fish larvae. In our study, fish larvae will be provided by IPMA-EPPO (Olhão, Portugal), where the temperature regime is colder (around 18°C) than that in the Canaries in fall (around 23°C). Because temperature affects metabolic requirements and rates, it can increase egg and larval mortality. In the laboratory, we tested the effect of three different temperatures (control, 18°C; moderate, 22°C; and warm, 24.5°C) on the egg hatching success of the four candidate species: European pilchard *Sardina pilchardus*, gilthead sea bream *Sparus aurata*, white sea bream *Diplodus sargus*, and zebra sea bream *Diplodus cervinus*. Hatching success of *S. aurata*, *D. sargus*, and *D. cervinus* ranged between 70 and 100%, while that of *S. pilchardus* was comparatively low, between 20% and 40%, even at the control temperature. The low hatching rates of *S. pilchardus* may be related to low egg quality of the batch. This leads to the conclusion that sea breams are more robust and thus more suitable for introduction to the upcoming mesocosm study.

17 Advances in Cephalopod Research

Chris J. Barrett[1]

[1]Cefas, Pakefield Road, Lowestoft, Suffolk, NR33 0HT, UK

Chris J. Barrett: christopher.barrett@cefas.co.uk

This session of the YOUMARES 9 conference does not have a corresponding proceedings article.

17.1 Call for Abstracts

Dear young cephalopod scientists, I would like to invite you to present your science at the "Advances in Cephalopod Research" session at YOUMARES9. The session will explore, but is not limited to, recent progress in the studies of cephalopod physiology, ecology, population structures, evolution, behaviors, and taxonomy and promises a friendly audience to disseminate your findings to. Whether you are hoping to use your evidence to contribute to the Cephalopod International Advisory Council (CIAC), inform policy, or simply seek feedback on your postgraduate studies, this session will be the ideal platform to do so and network with like-minded scientists.

17.2 Abstracts of Oral Presentations

17.2.1 Hard Structure-Based Age Estimation of Coleoid Mollusks

F. Lishchenko[1]*

[1]FSBSI "VNIRO," Verkhnaya Krasnoselskaya str. 17, 107140, Moscow, Russia

*invited speaker, corresponding author: Fedor-LN@ya.ru

Keywords: Cephalopoda, Age estimation, Hard structures, Statoliths, Beaks, Vestigial shells, Eye lenses

Determination of total life span and estimation of growth and maturation rate are the components of the problem which is known as aging studies. First attempts to study cephalopod's age were made in the end of the nineteenth century by A.E. Verrill, who estimated the age of squid *Loligo pealei* studying length frequency distribution. Since that times the field undergone notable development, especially significant in the second part of the twentieth century – beginning of the twenty-first century, when aging studies became an inherent part of the rational stock exploitation. Currently, there are three basic approaches to age estimation – analytical methods, studies on animals in captivity and age estimation using hard structures. Being predominantly soft-bodied animals, cephalopods bear some hard, chitinous, or calcareous structures. These include beaks, shells, eye lenses, statoliths, hooks, and sucker rings. However, possibilities of application of these structures are not equal; moreover some of them are not suitable for age estimation at all (such as hooks and sucker rings). In particular, statoliths are the main structure used for age estimation of squids, but they can't be used for aging of octopuses. Shells allow not only to estimate the age but to determine individual growth as well; however they are not presented in some cephalopod taxas or don't bear periodic increments. Beaks represent a reliable and precise tool for age estimation presented in all cephalopod species, but their readability varies significantly among species. Finally, eye lenses are the most controversial structure; on the one hand, all cephalopods have well-developed eyes containing lenses, and the structure is easy to process; however, to date, there is no study undoubtedly confirming the periodicity of increment deposition. To sum up, most of cephalopod hard structures have specific benefits and limitations for aging. This review aims the discussion on the recent advances and challenges in the field of application of cephalopod hard structures for age estimation.

17.2.2 Age Estimation of *Loligo vulgaris* Using Beaks

Anastasiia Lishchenko[1*], Catalina Perales-Raya[2], Fedor Lishchenko[1]

[1]Russian Federal Research Institute of Fisheries and Oceanography (VNIRO), Russia, Moscow, Verkhnaya Krasnoselskaya, 17

[2]Instituto Español de Oceanografía, Centro Oceanográfico de Canarias, Spain, Santa Cruz de Tenerife, Dársena Pesquera, 8

∗corresponding author: gvajta@ya.ru

Keywords: Age, *Loligo vulgaris*, Beaks, Statoliths, Cephalopoda

Age estimation is one of the important directions in cephalopod life history studies. This study represents the first application of age estimation method using beaks in *Loligo vulgaris* (Lamarck, 1798). This species is one of the most economically valuable myopsid squid in Europe, which inhabits the northeastern Atlantic and the Mediterranean Sea. Samples (statoliths and beaks) were obtained from squids, bought at a market in Barcelona (Spain), and prepared using standard methods (Arkhipkin and Shcherbich 2011, J Mar Biol Assoc UK 92:1389-1398; Perales-Raya et al. 2014, J Shellfish Res 33:1-13). The comparison of the number of increments between statoliths and beaks was used as a validation method.

17.2.3 Beaks of *Berryteuthis magister* (Berry, 1913): A Promising Tool for Age Estimation

F. Lishchenko[1*], A. Lishchenko[1], A. Bartolomé[2], C. Perales-Raya[2]

[1]FSBSI "VNIRO," Verkhnaya Krasnoselskaya str. 17, 107140, Moscow, Russia

[2]Instituto Español de Oceanografía, Centro Oceanográfico de Canarias. Vía Espaldón Dársena Pesquera PCL 8, 38180, Sta. Cruz de Tenerife, Spain

∗corresponding author: Fedor-LN@ya.ru

Keywords: Far East seas, *Berryteuthis magister*, Methods of age estimation, Beaks, Statoliths

The schoolmaster gonate squid, *Berryteuthis magister* (Berry, 1913) is among the most important fishery resources in the Russian Far East seas. Precise and accurate age estimation is an essential basis of its assessment and management. Nowadays, the age of the squid is routinely estimated using statolith increment reading. In this study, alternative methods of squid age estimation are tested. Sets of statoliths and beaks were collected from 40 randomly selected squids, caught in the area of the Northern Kuril Islands in spring-summer of 2016. The right-hand statoliths from each pair were examined following standard methods. Beaks (both lower and upper) were processed, following methods applied initially to *Octopus vulgaris*. The age of each specimen was estimated by counting the periodical increments on the section of each hard structure (statolith and beak). Unfortunately, due to the elongated shape of upper mandible, increments on its section were too narrowly spaced and, in most cases, impossible to read. Thus, the lower mandibles were selected as a main tool for age estimation. The age, estimated using the statolith sections, ranged from 119 to 148 days in males (167 and 223 mm of dorsal mantle length (DML), respectively), and from 132 to 177 days in females (DML 189 and 232 mm, respectively). The age of these individuals, estimated from their beaks, were 175 and 291 days in males and 177 and 216 days in females. In total, the number of increments on the beak sections exceeded the same value on statolith sections by approximately 35%. We suppose that small size and complex morphological structure of cuttlefish statoliths complicate the preparation and analysis of the sections, which can lead to mistakes and age underestimations. On the other hand, the lower beaks are larger in general and bear wider-spaced increments than upper beaks.

17.2.4 Reconstructing the Life History of Argonauts (Octopoda: *Argonauta*): The Potential of Geochemical and Microscopic Methods

Kevin Stevens[1*], Yasuhiro Iba[2], Akihiko Suzuki[3], Jörg Mutterlose[1]

[1]Ruhr-Universität Bochum, 44801 Bochum, Germany

[2]Hokkaido University, Hokkaido 060-0810, Sapporo, Japan

[3]Hokkaido University of Education, Hokkaido 002-8502, Sapporo, Japan

∗corresponding author: kevin.stevens@rub.de

Keywords: Cephalopod, Stable isotopes, Ecology, Habitat, Biomineralization

Argonauts are enigmatic cephalopods that inhabit the epipelagic zone of tropical and subtropical seas. Female argonauts produce a thin-walled calcitic shell, which they use to deposit their eggs and to acquire neutral buoyancy. Argonaut life history, especially their migration paths, as well as the biomineralization of their unique shells are largely unknown. Shells of the greater argonaut (*Argonauta argo*), derived from a mass-stranding of argonauts in Hokkaido, Japan (Suzuki and Enya 2013, J Jap Drif Soc 11:1-6), were used for geochemical analyses (carbon and oxygen stable isotopes, Mg/Ca, Sr/Ca, Ba/Ca). Thin sections of the shells were applied to reconstruct their shell growth patterns with microscopic methods. Argonaut shells grow via the accretion of successive chevron-shaped elements at the margin. Growth patterns of the shells are only visible via cathodoluminescence microscopy. The geochemistry of the ontogenetically sampled shells suggests a potential northward drift/migration path of the studied argonauts along the west coast of

Japan, following the Tsushima Current. However, the exact reconstruction of argonaut migration via geochemical data is still hampered by "vital effects," which are deviations of the isotopic and elemental values from equilibrium with seawater (Stevens et al. 2015, Mar Biol 162:2203-2215). Use of geochemical analyses of argonaut shells to reconstruct their migrations must be complemented by further studies of their biomineralization process. Due to their potentially high sensitivity to ocean acidification, further information on their life history and ecology could aid in protection of these unique cephalopods, whose shells might also record anthropogenic changes of the epipelagic realm.

17.2.5 An Overview of Global Squid and Cuttlefish Responses to Climate Change

Chris J. Barrett[1]*

[1]Cefas, Pakefield Road, Lowestoft, Suffolk, NR33 0HT, UK

*corresponding author: christopher.barrett@cefas.co.uk

Squid and cuttlefish are fast-growing, short-lived, generalist cephalopods, which play an important role in marine ecosystems as both prey and predators. As their reproductive rates and plasticity are generally high, these cephalopods are quick to respond to climate change, though long-term impacts of ecosystem changes are uncertain. There is evidence of recent changes in the spatial extents of many cephalopods, although migratory abilities and survivability depends on their tolerances to factors such as temperature and pH. Their lifestyles influence their adaptability too, i.e., whether a specimen grows as planktonic or nektonic and whether the specimen invests more energy in growth or reproduction. Predicting cephalopod responses to climate change is therefore complex, and there is a need for standardization in recording of landings (i.e., to species level), which, together with tagging, age and growth analyses, and data gaps being filled from citizen science, can be fed into scenario models to reduce uncertainty in ecosystem predictions to help better inform management.

17.3 Abstracts of Poster Presentations

17.3.1 Male Morphotypes and Their Possible Roles in Population Structure in Three Loliginid Squid Species

Jessica B. Jones[1,2,3]*, Graham J. Pierce[3,4,5], Warwick H. H. Sauer[6], Frithjof C. Kuepper[3], Zhanna N. Shcherbich[1], Alexander I. Arkhipkin[1,3]

[1]Falkland Islands Fisheries Department, Stanley, Falkland Islands

[2]South Atlantic Environmental Research Institute (SAERI), Stanley, Falkland Islands

[3]School of Biological Sciences, University of Aberdeen, Aberdeen, UK

[4]Instituto de Investigacións Mariñas (CSIC), Vigo, Spain

[5]Universidade de Aveiro, Aveiro, Portugal

[6]Department of Ichthyology, Rhodes University, Grahamstown, South Africa

*corresponding author: jjones@fisheries.gov.fk

Keywords: Geometric morphometrics, *Doryteuthis gahi*, Squid, Morphotypes, Reproduction

Myopsid squid populations are characterized by the presence of two size-dependent behavioral types of mature males on spawning grounds; large dominant "bull" males forming mating pairs with females and small "sneaker" males attempting extra-pair copulations (EPCs). In some loliginids, males much larger and rarer than what could usually be considered a "bull" male have been caught on a sporadic basis. We aimed to determine whether "sneakers," "bulls," or the "super bulls" were morphologically distinguishable by investigating intrapopulation body shape variation with ontogeny, sex, and size using traditional and geometric morphometric analysis. *Doryteuthis gahi* in Falkland Islands waters showed a single mode in size-at-maturity for males. Smoothing curves from binomial generalized additive models (GAMs) suggested two sizes-at-maturity in females. Length-weight regression analysis of both sexes at different ontogenetic stages suggested no significant change in the slope for females, but a significant ontogenetic elongation in the mantle relative to body weight for males. There was a simultaneous increase in relative fin size with increasing body size and throughout ontogeny. The large "super bull" males had a substantially more elongated body, heavier fin, and relatively larger fin area compared to the rest of the population, a body shape associated with enhanced swimming performance. It is possible that these "super bull" males are extending their migratory routes, providing geographic connectivity by migrating between spawning areas. The phenomenon of "super bull" males was also investigated in *Loligo forbesii* in Scottish waters and *Loligo reynaudii* in South African waters, to explore whether they were typical within other loliginid squid populations.

17.3.2 Using Statolith Elemental Signatures to Confirm Ontogenetic Migrations of the Squid *Doryteuthis gahi* Around the Falkland Islands (Southwest Atlantic)

Jessica B. Jones[1,2,3]*, Alexander I. Arkhipkin[1,2], Andy L. Marriott[4], Graham J. Pierce[5,6]

[1]University of Aberdeen, King's College, Aberdeen, AB24 3FX, UK

[2]Falkland Islands Fisheries Department, Stanley, FIQQ 1ZZ, Falkland Islands

[3]South Atlantic Environmental Research Institute, Stanley, FIQQ 1ZZ, Falkland Islands
[4]Centre for Environmental Geochemistry, British Geological Survey, Nottingham, NG12 5GG, UK
[5]CESAM and Departamento de Biologia, Universidade de Aveiro, Av. Padre Fernão de Oliveira, 3810-193 Aveiro, Portugal
[6]Instituto de Investigacións Mariñas (CSIC), Eduardo Cabello 6, 36208 Vigo, Spain
*corresponding author: JJones@fisheries.gov.fk
Keywords: Trace elements, Laser ablation ICP-MS

The Patagonian long-finned squid *Doryteuthis gahi* is a commercially important species found within Falkland Island waters. The population consists of two temporally distinct spawning cohorts, inferred to have markedly different patterns of migration and timings of ontogenetic events. Ontogenetic migrations of each cohort were confirmed by analysis of the chemical composition of statoliths collected from both spawning cohorts in two consecutive years. Trace element concentrations (Sr/Ca and Ba/Ca) were quantified using laser ablation inductively coupled plasma mass spectrometry (LA ICP-MS), to determine temporal and cohort-specific variation. Statoliths and individual ablation spots were aged to produce high-resolution elemental chronologies, thought to reflect the chemical properties of ambient water at the time of element incorporation. Generalized additive mixed models (GAMM) indicated that cohort and life history stage had a significant effect on Sr/Ca and Ba/Ca ratios. Sr/Ca and Ba/Ca ratios were both negatively correlated with near-bottom water temperature, with Ba/Ca also potentially correlated to depth. Elemental chronologies for each cohort had a consistent pattern over two consecutive years. Statolith elemental chronologies have useful applications as natural tags, discriminating between spawning cohorts.

17.3.3 Age Structure of *Berryteuthis magister* (Berry, 1913) Commercial Aggregation in the Northern Kuril Islands Area in Autumn of 2015

E. Koptelova[1]*, F. Lishchenko[2], A. Lishchenko[2]
[1]Russian State Agrarian University - Moscow Timiryazev Agricultural Academy, Timiryazevskaya st., 49, 127550, Moscow, Russia
[2]FSBSI "VNIRO," Verkhnaya Krasnoselskaya str. 17, 107140, Moscow, Russia
*corresponding author: katerina.koptelova@ya.ru
Keywords: *Berryteuthis magister*, Fishery, Biological status, Age estimation, Statoliths

Berryteuthis magister is the most commercially exploited squid species inhabiting the Russian Far East seas. Annual catch of this species exceeds 100 thousand metric tons, more than half of which is caught in Northern Kuril Islands area. Such a significant rate of exploitation of resource determines the necessity of careful stock status monitoring. In this study we wish to describe biological status and age structure of *B. magister* commercial aggregation which was exploited in the Northern Kuril Islands area during the autumn fishing season of 2015. Data was collected by scientific observers of VNIRO on board of commercial vessel targeted *B. magister* aggregations in September 2015, following standard framework for data collection. Set of data collected included dorsal mantle length (DML), fin length (FL), fin width (FW) and arm length (AL) measurement, determination of sex and maturity, statoliths were collected for age estimation. Later statoliths were processed following recommendations of A. Arkhipkin: they were mounted on the glass slide, grinded and polished from two sides and read under transmitted light at 400X magnification using Micromed 2 compound microscope. Commercial aggregation of *B. magister* was composed mainly from immature and maturing individuals. DML ranged from 153 to 351 mm in females and from 124 to 267 mm in males. Minimum FL in females was 87 and maximum – 173 mm; in males this characteristic ranged from 68 to 139 mm. FW ranged from 114 to 228 mm in females and from 85 to 176 mm in males. Al ranged from 91 to 188 mm and from 88 to 194 mm in females and males, respectively. Age of studied females ranged from 153 to 273 days, age of studied males from 111 to 276 days. Our findings show that even at highest in modern history fishing pressure stock of *B. magister* remained sustainable.

18 Open Session

Viola Liebich[1]
[1]Envio Maritime, Berlin, Germany
Viola Liebich: enviomaritime@gmail.com

This session of the YOUMARES 9 conference does not have a corresponding proceedings article.

18.1 Call for Abstracts

Marine sciences are a vast and diverse field of research, and barely any conference is able to represent all topics with a separate session. The Open Session will summarize contributions of young marine scientists from all research fields which do not seem to fit into one of the other sessions.

18.2 Abstracts of Oral Presentations

18.2.1 Feeding Ecology of *Sardinella aurita* and *Sardinella maderensis* in the South of Atlantic Moroccan Coast

Ayoub Baali[1]*, Ikram Elqoraychy[1], Mouna Elquendouci[1], Ahmed Yahyaoui[1] and Khadija Amenzoui[2]

[1]Laboratory of Biodiversity, Ecology and Genom, Mohammed V University, Rabat, Morocco

[2]Pêche, Institut National de Recherche Halieutique, Morocco

∗corresponding author: ayoubbaali22@gmail.com

Keywords: *Sardinella* sp., Feeding, Copepods, Crustaceans, Morocco

The feeding preferences of fish species are important in classic ecological theory, mainly in identifying feeding competition, structure and stability of food webs, omnivory, and assessing predator-prey functional responses. Additionally, the key role of feeding studies for fisheries biology and ecology and, more importantly, for fisheries management was uncovered only the last decade with the use of trophic level in predicting the effects of fishing on the balance of marine food webs. The feeding of *Sardinella* sp. (*Sardinella aurita* and *Sardinella maderensis*) was investigated in the south of Atlantic Moroccan coast, during February 2015–January 2017. Samples were immediately fixed with 70% Ethanol solution. The total length (TL, cm) was measured and prey items in the stomach contents were identified to large taxonomic groups. For *Sardinella aurita*, the diet is composed of crustaceans as preferential prey and fish (larvae, scales, and eggs) as secondary prey, and the rest of the prey constitute an occasional diet of this species. The main groups of prey consumed are: the copepods belonging to the Euterpinidae and Calanidae families. For *Sardinella maderensis*, the diet is also composed of crustaceans as preferential prey, fish as secondary prey, and mollusks and detritus presenting as occasional diet. The main groups of prey consumed are: fish (fish eggs) and copepods belonging to the Euterpinidae and Calanidae families. The vacuity index was low throughout the study period and for the two species, which shows the availability of food in the study area. According to the evolution of the relative importance index depending on the size, the *Sardinella* population can be divided into two large size groups with different dietary habits.

18.2.2 From Desert to Sea: Arabian Killifish Acclimation to High Salinity

Lucrezia C. Bonzi[1,2*], Alison A. Monroe[1,2], Jodie L. Rummer[3], Michael L. Berumen[2], Celia Schunter[1], Timothy Ravasi[1]

[1]KAUST Environmental Epigenetic Program (KEEP), Division of Biological and Environmental Sciences and Engineering, King Abdullah University of Science and Technology, Thuwal, Kingdom of Saudi Arabia

[2]Red Sea Research Center (RSRC), Division of Biological and Environmental Sciences and Engineering, King Abdullah University of Science and Technology, Thuwal, Kingdom of Saudi Arabia

[3]ARC Centre of Excellence for Coral Reef Studies, James Cook University, Townsville, Queensland 4811, Australia

∗corresponding author: lucrezia.bonzi@kaust.edu.sa

Keywords: Salinity stress, Transcriptome, Arabian killifish, Red Sea

The Arabian killifish, *Aphanius dispar*, is a euryhaline fish able to tolerate a wide range of salinities. This cyprinodontid inhabits inland and coastal waters of the Arabian Peninsula, with populations living in desert ponds (0.7–2 ppt), and in coastal lagoons of the highly saline Red Sea (41–44 ppt). In a recent genetic population study, we found that Red Sea populations receive migrants from desert ponds, which are flushed out to the sea during flash floods. In order to survive to these rapid and abrupt events, the fish need to cope, acclimate and adapt to more than 40 ppt increase in salinity. With this in mind, we investigated the molecular pathways underlying the phenotypic plasticity of *A. dispar* when acclimating to increased salinities. Fish were collected from a desert pond and a Red Sea lagoon and habituated in aquaria at 1.5 and 42 ppt, respectively. Desert pond fish were then transferred from 1.5 ppt to 42 ppt, and gills were sampled pre-transfer (0 h), and at 6 h, 24 h, 3 days, 7 days and 3 weeks post-transfer. Gill samples were morphologically analyzed using scanning electron microscopy and their transcriptomes assembled. No mortality was recorded throughout the 3-week exposure, and no differences in gill filament morphology were detected across sampling intervals nor between desert pond and Red Sea populations. The changes in gene expression over the acclimation timeline resulted extremely rapid, with individuals from desert ponds showing similar expression to saltwater population after only 6 hours of high salinity exposure. These findings provide insights into the mechanisms of phenotypic plasticity that allow the Arabian killifish to establish long-term adapted populations in waters with different salinities and to thrive in some of the most extreme environments on earth.

18.2.3 Developmental and Transgenerational Acclimation in a Brackish-Water Amphipod

Amelia Welch[1*], Manuela Truebano[1]

[1]University of Plymouth, PL4 8AA, Plymouth, United Kingdom

∗corresponding author: amelia.welch@students.plymouth.ac.uk

Keywords: Phenotypic plasticity, Metabolism, Climate change, Global warming, Invertebrate

Anthropogenic carbon emissions are causing sea surface temperatures to rise at an increasing rate. For marine populations to survive, they must be able to adapt or acclimate to these elevated temperatures. Both the parental and embryonic environment are known to be important in determining species ability to thermally acclimate; however the relative importance of exposure at these critical life history stages is not as well characterized. Here, individuals of the amphipod, *Gammarus chevreuxi*, were either exposed to control (15°C) or elevated temperatures (20°C) for two generations (F0-F1). Later-life metabolic performance was compared in individuals whose parents had been exposed to elevated temperatures 2 weeks prior to and during reproduction with individuals that were exposed as embryos or as juveniles. Rate of oxygen uptake (μg O_2 mg^{-1} hr^{-1}) at rest (RMR) was measured by closed chamber respirometry, and used as a proxy for metabolic performance. Both embryonic and parental exposure treatments similarly approximated the RMR of individuals reared under control conditions suggesting that it is the embryonic environment that is critical in determining thermal acclimation, irrespective of parental exposure. These findings highlight the importance of distinguishing stage of exposure in transgenerational studies and have significant implications for predicting species responses to ocean warming.

18.2.4 Physiological and Transcriptomic Responses of *Saccharina latissima* from the Arctic to Temperature and Salinity Stress

Huiru Li[1,2*], Cátia Monteiro[2,6], Sandra Heinrich[3,4], Inka Bartsch[4], Klaus Valentin[4], Lars Harms[4], Gernot Glöckner[5], Erwan Corre[6], Kai Bischof[2]

[1]Fisheries College, Ocean University of China, Qingdao, China

[2]Marine Botany, Faculty Biology/Chemistry, University of Bremen, Bremen, Germany;

[3]University of Hamburg, Hamburg, Germany

[4]Alfred-Wegener-Institute, Helmholtz Centre for Marine and Polar Research, Bremerhaven, Germany

[5]Institute for Biochemistry I, Medical Faculty, University of Cologne, Germany

[6]FR2424—Sorbonne Universités CNRS UPMC, Station Biologique de Roscoff, Roscoff, France

*corresponding author: huiru@uni-bremen.de

Keywords: *Saccharina latissima*, Global warming, Transcriptomics, ROS, Physiological response

The Arctic region is currently facing substantial environmental changes. Melting of glaciers as a consequence of increasing temperature subsequently creates stressful environmental conditions, such as reduced salinity in coastal habitats of kelp beds. To date, there is no study about the molecular responses of kelps to temperature and salinity stress. We investigated the transcriptomic performance of the sugar kelp *Saccharina latissima* from Kongsfjorden (Svalbard, Norway) over a 24-hour exposure at two salinities (20 and 30 psu) after a 7-day pre-acclimation at three temperatures (0, 8, and 15°C). Meanwhile, a more detailed physiological measurement was taken with a prolonged salinity acclimation (11 days). The results demonstrated that the physiological variables at 15°C were significantly higher than at 0°C during the course of acclimation, while salinity only showed the effect at the last day of acclimation. The highest number of differentially expressed genes (DEGs; DESeq2 with log2Ratio≥2), compared to the control at 8°C and normal salinity, was found in the specimens at 8°C-20 psu treatment (1,374), followed by samples at 0°C-20 psu treatment (1,193). The lowest number of DEGs appeared in the individuals at 0°C-30 psu treatment (274). These results indicate that *Saccharina latissima* in the Arctic region will be benefited by increased temperature, but suffering on the decreased salinity in some fjords because of ice melting.

18.2.5 Seasonal Growth and Skeletal Composition of the Cold Water Coral *Desmophyllum dianthus* Along an In Situ Aragonite Saturation Gradient

Kristina K. Beck[1,2], Gertraud M. Schmidt[1,3], Gernot Nehrke[1], Grit Steinhoefel[1], Jürgen Laudien[1], Kathrin Vossen[1,2], Aurelia Reichardt[1,2], Lea Happel[1,4], E. Maria U. Jung[1,2], Vreni Häussermann[5], Claudio Richter[1,2]

[1]Alfred Wegener Institute, Helmholtz Center for Polar and Marine Research, Bremerhaven, Germany

[2]University of Bremen, Bremen, Germany

[3]Max Planck Institute for Marine Microbiology, Bremen, Germany

[4]University of Oldenburg, Oldenburg, Germany

[5]Fundación San Ignacio del Huinay, Huinay, Chile

*corresponding author: Kristina.Beck@awi.de

Keywords: Ocean acidification, Calcification, Skeletal linear extension, Boron isotopic composition, Internal pH upregulation

Cold-water corals (CWC) have long been considered particularly sensitive to ocean acidification (OA). However, a number of laboratory studies indicate that exposure to acidic waters does not affect CWC growth, but in situ OA studies on CWC are scarce. In the naturally acidified Comau Fjord (Chile), high densities of the cosmopolitan CWC *Desmophyllum dianthus* are found at or below aragonite saturation ($\Omega_{ar} \leq 1$), but it is not known if the corals' ability to upregulate the pH in the calcifying fluid (pH_{cf}) and calcify shows seasonal fluctuations due to changes in Ω_{ar} and/or food supply. In the present study, corals were sampled along both horizontal and vertical pH gradients in Comau Fjord (equivalent to $0.81 < \Omega_{ar} < 1.45$ and $0.65 < \Omega_{ar} < 1.45$, respectively). We compared *D. dianthus*' calcification rates (alkalinity anomaly technique), skeletal carbonate accretion (buoyant weight technique (BWT)), linear extension rates (fluorescent microscopy), and pH_{cf} (skeletal $\delta^{11}B$; LA-ICP-MS) with the physicochemical conditions in the water column (T, Ω_{ar}) in austral summer 2016/2017 and winter 2017. Growth rates (BWT) were higher in summer than in winter, with highest values, irritably, at $\Omega_{ar} < 1$. Cross-transplant experiments showed that *D. dianthus* is able to acclimatize to $\Omega_{ar} < 1$. A strong biological pH_{cf} upregulation of 1.13 pH units was found at low pH_T with $\delta^{11}B$ of 25.6‰ compared to ΔpH of 0.77 ($\delta^{11}B$ = 23.4 ‰) at higher pH_T. The present study shows that Ω_{ar} alone is a poor predictor of *D. dianthus*' pH upregulation and growth. They suggest a complex combination of biological and physical factors that need to be considered to constrain the future of CWC in an era of OA.

18.2.6 Co-use of Fisheries and Offshore Wind Farms: The Distribution of the Brown Crab (*Cancer pagurus*) in the Vicinity of Wind Farms in the German EEZ of the North Sea in Relation to Sustainable Fisheries Resources

Avianto Nugroho[1*]

[1]Christian-Albrechts-Universität zu Kiel, 24118 Kiel, Germany

*corresponding author: stu206039@mail.uni-kiel.de

Keywords: Marine spatial planning, Co-use, Wind farms, Brown crab (*Cancer pagurus*), GIS

The expansion of offshore renewable energy and traditional uses for fisheries and shipping are the sectors having the widest spatial spread in the southern North Sea. Offshore and coastal planning for wind energy development is now taking a high priority in conjunction to the realization of Renewable Energy Act where 35% of the electricity supply must be generated from renewable energy generation by 2020. In addition, the designation of marine conservation areas as well as some reserved areas for safety zones will force a reallocation of the spatial distribution of fishing effort. Accordingly, in order to use space efficiently, multiple uses of space should be sought where uses can be combined as spatial planning measure – co-use. Due to the restriction on fisheries activities within the area of wind farms, fishermen have to substitute from active into passive gears as well as change the main target species. Brown crab (*Cancer pagurus*) is now becoming the commercial target species especially in the area of wind farms. With the help of GIS, the habitat modelling which took into account several parameters such as depth, temperature, and salinity was conducted to map the distribution of brown crab and to reveal its preferred living environment. The average landings of brown crab in Helgoland have increased steadily since the last 10 years. The hard bottom structure of wind farm foundations is assumed to be the driving factor of the increase of brown crab, since it has a correlation with the habitat preference and also can serve as artificial reefs for species aggregation. Moreover, the use of passive gears will considerably reduce the degradation of environment and the number of by catch, due to its static type and high selectivity of target species.

18.2.7 Foraminifera Rocket Experiment: Microgravitational Influence on Cell Physiology and Motility

Greta Sondej[1*], Katrin Hättig[1], Christoph Kulmann[1], Scarlett Gac Caceres[1], Jan Blumenkamp[1], Nils Kunst[1], Johanna Hartmann[1,2], Niklas Kipry[1], Kay Menken-Siemers[1], Daniel Rippberger[1]

[1]University of Bremen, Bibliothekstraße 1, 28359 Bremen, Germany

[2]University of the Arts Bremen, Am Speicher XI 8, 28217 Bremen, Germany

*corresponding author: greta.sondej@gmail.com

Keywords: Foraminifera, Microgravity, Cell physiology and motility, Life support system, REXUS/BEXUS program

The ability to respond appropriately to external signals is vital to all organisms. As stimuli for orientation, primitive organisms use, e.g., light or temperature. However, there is a decisive disadvantage with almost all imaginable environmental stimuli: They are variable and can fluctuate greatly. The only reliable environmental parameter is gravity, which is constantly present in almost the same size and same direction everywhere on earth. The FORAminifera Rocket EXperiment (FORAREX) will be realized in the course of the REXUS/BEXUS program – an opportunity for university students to carry out scientific and technological experiments on board of sounding rockets or stratosphere balloons, respectively. We focus on the behavior of the Foraminifera *Amphistegina lobifera* under microgravity condition examining changes in cell motility and migration. The exceptional physical stress during rocket launch is also the purpose of investigation. In addition, our constructed life-support system for cultivating foraminifera will be tested for general mission approval. The goal of the experiment is to optimize the setup for a long-term experiment on board of the International Space Station (ISS), where we would like to investigate foraminifera test formation under extended microgravity influence. Investigations will be conducted using a miniaturized container with a flow cell examination chamber and a closed-circuit life support system with integrated LED-based illumination for photosynthesis. Foraminifera will be observed using a camera. Temperature, oxygen, and pH will be recorded using high-performance miniaturized sensors. Further, we will perform experiments using clinostats (simulated microgravity), vibration, and centrifugation (hypergravity). We will present preliminary results from our project and will give a general outline of our experiment within the framework of the REXUS 25/26 mission, which is scheduled to launch in March 2019.

18.2.8 Development of an Underwater Biofouling Sensor

Antje S. Alex[1], Jan Boelmann[1*]

[1]Hochschule Bremerhaven, An der Karlstadt 8, 27568 Bremerhaven, Germany

*corresponding author: jboelmann@hs-bremerhaven.de

Keywords: Biofouling, Underwater camera, Marine sensors, Image processing

Marine sensors and systems are already affected by biofouling after a short operation time in water, causing malfunction and the reduction of lifetime. Countermeasures such as continuous cleaning and antifouling coatings, respectively, cause not only an increase in maintenance efforts and costs but often take unspecific, partly environmentally hazardous effects. To reduce costs and efforts, it is essential to understand and, thus, monitor the formation of biofilms on underwater surfaces. Furthermore, the efficiency of countermeasures could be controlled in the same way. For this purpose a new in situ observation camera system was developed. The overgrowing of an exchangeable transparent sample plate attached to the camera's underwater housing is observed via time series photography. The remotely controlled camera allows for viewing and photographing the growth state via the camera's live view function. The camera system is connected to an onshore control system providing power and internet access for real-time data exchange. This allows for the remote control of the camera from anywhere at any time. For the analysis of the photographs an image processing and interpretation algorithm was developed. In this presentation, the system design as well as challenges and solutions during the process of construction and realization will be presented. Furthermore the results of the first long-term test deployment in a marine environment at the Sven Lovén Centre at Kristineberg, Sweden, are shown.

18.2.9 Validation and Quantification of Cardiac Parameters Obtained with CINE Magnetic Resonance Imaging in Crustaceans

Sebastian Gutsfeld[1,2*], Bastian Maus[2], Christian Bock[2]

[1]Technical University of Darmstadt, Department of Biology, Schnittspahnstraße 10, Darmstadt 64287, Germany

[2]Alfred-Wegener-Institute for Polar and Marine Research, Section of Integrative Ecophysiology, Am Handelshafen 12, 27570 Bremerhaven, Germany

*corresponding author: sgutsfeld@gmail.com

Keywords: Thermal tolerance, *Cancer pagurus*, Cardiovascular system, Decapoda, Energetic costs

Projections for the ongoing warming and CO_2 accumulation in the oceans predict synergistic effects on marine ectothermic animals. Physiological limits in an animal's thermal tolerance are set by their capacity to sufficiently supply oxygen to the tissues (oxygen- and capacity-limited thermal tolerance (OCLTT)). At upper pejus temperatures, physiological capacities of heart rate and ventilation are reached and cannot meet rising oxygen demands. This may cause potential shifts in geographical distribution or increased risks of local extinctions. With the cardiac system as a crucial component for blood and oxygen transport, this study investigated multiple performance parameters of the heart of the North Sea edible crab, *Cancer pagurus*, in response to ocean warming and acidification. Therefore, simultaneous noninvasive online measurements of heart rate and oxygen consumption were conducted through photo-plethysmography and flow-through respirometry, respectively. To gain insight into mechanistic capacities under mid-century climate conditions, animals were subjected to an acute temperature increase from 12 to 20°C over 5 days under control (400 µatm) and high CO_2 levels (1350 µatm). Photo-plethysmograph signal calculations also provided stroke volume and cardiac output proxies. All observed parameters displayed coordinated, cyclic variations matching previously reported fluctuations in hemolymph PO_2. Temperature-dependent elevations of mean levels were caused by increases in maximum values, at stable minimum levels. The effects of CO_2 on thermal capacities of cardiac performance parameters remain obscured by high interindividual variation. To verify the use of stroke volume proxies derived from plethysmograph signals, single heart strokes were noninvasively visualized using CINE magnetic resonance imaging. Combining measurements of high temporal resolution and temporal allocation with experimental temperature and CO_2 regimes more alike to natural conditions, we confirmed the highly flexible capacities of the heart of *C. pagurus* under climate change conditions. In addition, CINE magnetic resonance imaging offers unique possibilities for noninvasive investigations of climate change effects on animals.

18.2.10 Monosaccharide Diversity in Headgroups of Microbial Membrane Glycolipids

Konrad Winkels[1*], Sarah Coffinet[1], Martin Könneke[1], Julius S. Lipp[1], Ana de Santiago Torio[2], Lukas Mühlena[1], Roger E. Summons[2], Tanja Bosak[2], Florence Schubotz[1], Kai-Uwe Hinrichs[1]

[1]MARUM Center for Marine Environmental Sciences, University of Bremen, D-28359 Bremen, Germany

[2]Department of Earth, Atmospheric and Planetary Sciences, Massachusetts Institute of Technology, Cambridge, MA, USA

*corresponding author: konrad.winkels@gmail.com

Keywords: Glycolipids, Intact polar lipids, HPLC-MS, Microbial mats, Microbial cultures

Glycolipids consist of sugar-based polar headgroups and apolar lipid moieties. The wide structural diversity of glycolipids allows coarse chemotaxonomic characterization of microbial communities in the environment. While the diversity of lipid moieties has been extensively investigated, the composition of glycosidic headgroups both in cultures and in the environment is still largely unknown. The available methods to analyze lipid-derived sugars require intensive and time-consuming lab work. We now present an optimized high-performance liquid chromatography (HPLC) method equipped with a mass spectrometer (MS), which allows quick analysis of headgroups with a sensitive detection. 22 sugars representing a wide diversity of sugar types present in nature were used to optimize the chromatographic conditions. The low concentration of glycolipids in natural samples requires a low limit of detection (LOD). The use of triple quadrupole MS enabled the implementation of a multiple reaction monitoring (MRM) method, which ultimately led to LODs between 0.5 and 2.5 pmol. The new HPLC-MS method was applied to a selection of microbial cultures and natural samples. The monosaccharide composition of thermophilic and mesophilic archaea and bacteria was examined and compared to microbial mats from the Black Sea and Yellowstone National Park. The monosaccharide composition of the different microbial cultures is relatively similar: glucose is most abundant, followed by mannose, galactose, allose, and a diverse suite of pentoses detected in lesser proportions. Archaeal lipids were isolated via semi-preparative HPLC from the total lipid extract of microbial mats. Unlike the archaeal cultures, galactose and not glucose was the most abundant monosaccharide headgroup in the environmental archaeal glycolipids. These preliminary results show a high sugar diversity in headgroups of microbial glycolipids and a discrepancy between cultures and environmental samples. Further investigations are needed to understand the sugar headgroup biosynthetic strategies of environmental microbes.

18.2.11 Analysis of Mycosporine-Like Amino Acids in Red Algae

Maria Orfanoudaki[1]*, Anja Hartmann[1], Ulf Karsten[2], Markus Ganzera[1]

[1]Institute of Pharmacy, Pharmacognosy, University of Innsbruck, Innrain 80-82, Innsbruck 6020, Austria

[2]Institute of Biological Sciences, Applied Ecology & Phycology, University of Rostock,

Albert-Einstein-Str. 3, Rostock 18059, Germany

*corresponding author: Maria.Orfanoudaki@uibk.ac.at

Keywords: Isolation, Collagenase, LC-MS, Photo-protection

Mycosporine-like amino acids (MAAs) are small, water-soluble compounds, found in algae, cyanobacteria, lichens, fungi, and marine animals. They are known as some of the strongest UV-absorbing molecules and Rhodophyta produce a variety of these compounds as a defense system against the harmful effects of UV radiation. In this study a variety of red algae were investigated, and several compounds were extracted and isolated from three different species (*Gracilaria chilensis*, *Pyropia plicata,* and *Champia novae-zelandiae*). The extracts were fractionated using different chromatographic techniques for purification such as fast centrifugal partition chromatography (FCPC), column chromatography (CC), size exclusion chromatography (SEC), and preparative high-performance liquid chromatography (HPLC). The pure compounds were identified using nuclear magnetic resonance (NMR) and mass spectrometry, finally resulting in the isolation of 14 MAAs, 2 terpenoids, and 6 betaines. Five MAAs are reported for the first time in algae, and two are new natural products. In addition, a new liquid chromatography–mass spectrometry (LC-MS) method was developed for the separation of eleven MAAs and subsequently 23 red algae were assayed for their MAA content. All isolated compounds were also investigated for their collagenase inhibitory activity, and some of them showed a good dose-dependent inhibition of collagenase in vitro (IC50 lower than 100 μM). For deeper investigations, more metabolites, including possibly novel molecules will be isolated and tested for their biological properties focusing on photoprotective and UV-absorbing effects.

18.3 Abstracts of Poster Presentations

18.3.1 Salmon Lice (*Lepeophtheirus salmonis*) Control Method Effectiveness in Atlantic Salmon (*Salmo salar*) Aquaculture

Kristine Cerbule[1]*

[1]UiT the Arctic University of Norway, Tromsø, Norway

*corresponding author: kristine.cerbule@gmail.com

Keywords: Atlantic salmon, Salmon lice, Chemical treatment, Warm water treatment, Cleaner fish

Salmon lice (*Lepeophtheirus salmonis*) is an increasing limiting factor of Atlantic salmon (*Salmo salar*) aquaculture development in the northern hemisphere. Different types of treatments have been tested and used to control lice on farmed Atlantic salmon with varying results. The aim of this systematic review is to examine effectiveness expressed as the reduction of the number of lice and associated negative effects to fish health and welfare (Atlantic salmon and cleaner fish, if used) for three types of methods – chemical treatment, cleaner fish use, and warm water treatment. A systematic literature review was used to gather and analyze data related to each type of method reported in peer-reviewed documents. After applying inclusion criteria, 62 of 782 documents of two scientific databases combined were further analyzed. Most of the documents described chemical treatment which showed decreasing effectiveness combined with increasing concentrations due to the significant development of resistance. Documents describing the use of cleaner fish showed effectiveness toward salmon lice in all studies, with little or no negative associated effects, and did not show a decreased effectiveness over time. The lack of data related to warm water treatment did not allow to assess the effectiveness of this method. Due to the development of resistance in lice selected by chemical treatments, those methods cannot be considered sustainable practices in aquaculture. Cleaner fish use is preferred if fish health and welfare criteria are met. A lack of data related to warm water treatment was noted, which is a research gap.

18.3.2 3Qs for Quality: Development of New Devices and Techniques for Seafood Quality Assessment

Edgar Teixeira[1,2], João Paulo Noronha[1], Vera Barbosa[3,4], Patrícia Anacleto[3,4,5], Ana Maulvault[3,4], Maria Leonor Nunes[4], António Marques[3,4], Rui Rosa[5], Carolina Madeira[2,5*], Valentina Vassilenko[6], Mário Sousa Diniz[2]

[1]LAQV-REQUIMTE, Departamento de Química, Faculdade de Ciências e Tecnologia/Universidade Nova de Lisboa, 2829-516 Caparica

[2]UCIBIO-REQUIMTE, Department of Chemistry, Faculdade de Ciências e Tecnologia, Universidade NOVA de Lisboa. 2829-516 Caparica, Portugal

[3]Division of Aquaculture and Seafood Upgrading, Portuguese Institute for the Sea and Atmosphere, I.P. (IPMA). Rua Alfredo Magalhães Ramalho, 6, 1495-006 Lisboa, Portugal

[4]Interdisciplinary Centre of Marine and Environmental Research (CIIMAR), Universidade do Porto. Terminal de Cruzeiros do Porto de Leixões, Avenida General Norton de Matos S/N, 4450-208 Matosinhos, Portugal

[5]MARE – Marine and Environmental Sciences Centre, Laboratório Marítimo da Guia, Faculdade de Ciências da Universidade de Lisboa. Av. Nossa Senhora do Cabo, 939, 2750-374 Cascais, Portugal

[6]Laboratório de Instrumentação, Engenharia Biomédica e Física da Radiação (LIBPhys)-UNL (LIBPhys), Dept. of Physics, Faculdade de Ciências e Tecnologia, Universidade NOVA de Lisboa. 2829-516 Caparica, Portugal

*corresponding author: scmadeira@fc.ul.pt, carolbmar@gmail.com

Keywords: Seafood quality, Marine biotechnology and fisheries, VOCs, Biogenic amines, HPLC, GC-MS

The "3Qs for Quality" work uses a multiple approach to develop different types of methodologies for the assessment of seafood quality. The freshness of selected fish species (*Trachurus trachurus* and *Sarda sarda*) were analyzed for volatile organic compounds (VOCs) (without the need for any pre-treatment of the samples) by the ion mass spectrometry (IMS) technique (FlavourSpec), using a multi-capillary pre-separation (MCC), which allows to analyze wet samples and provide extreme sensitivity for detection of nitrogenous compounds (e.g., biogenic amines). Additionally, biogenic amines (BA) present in seafood samples were also analyzed by the conventional techniques (HPLC, GC-MS, and ELISA) for method optimization and validation of the results. In addition, samples were analyzed by performing the traditional sensory methods and the Quality Index Method (QIM) for the assessment of seafood quality. The results obtained from the different techniques provided valuable information about the quality and freshness of fish and were integrated for a better detection of fish degradation. Finally, taking in account that food degradation and (bio)contamination is a worldwide problem responsible for numerous poisoning and disease events and that health authorities asked for new methodologies and techniques, we believe that our study can promote the use of advanced techniques for seafood quality assessment.

18.3.3 Interannual Variations in the Relationships Between Life Weight and Content in Biochemical Compounds of Muscle of the Anchovy *Engraulis encrasicolus* in the Bay of Biscay as Likely Indicator of Changes in the Reproductive Effort

E. Txurruka[1*], U. Cotano[2], F. Villate[1], J.M. Txurruka[3]

[1]Dept. of Plant Biology and Ecology, University of the Basque Country, Spain

[2]AZTI. Herrera Kaia, Portualdea z/g, 20110 Pasaia, Gipuzkoa

[3]Dept. of Genetics, Physical Anthropology and Animal Physiology University of the Basque Country, Spain

*corresponding author: estibalitz.txurruka@ehu.eus

Keywords: Molecular analysis, Atlantic, Fishery, Temporal analysis, Sexual

Condition factors based in life weights (LW) have a limited utility when estimating the well-being status of a fish population. Quantification of major biochemical constituents is much more biologically informative, but it requires time and resources. In 2005, after several consecutive failures in annual recruitment, anchovy biomass in the Bay of Biscay dropped to dangerous levels so that fishery was closed, up to be reopened in 2010. During the last years of this closure, two surveys in spring of May 2008 and May 2009 were carried out in order to estimate biological parameters of adults. In contrast, anchovies sampled in May 2016 were sampled, with the fishery already open. 965 anchovies were fished and dissected; their LW were registered, and total biochemical compounds of muscle were quantified. ANOVA analyses showed that there were significant differences ($p < 0.05$) between years in the LW and dry weights (DW) of muscles, these being smallest in anchovies of May 2016 and biggest in anchovies of May 2008. Moreover, there were significant differences ($p < 0.05$) between years in organic matter (OM) content and biochemical compounds of the muscles, the anchovies of May 2016 having the highest values for OM (91%) and those of May 2008 the lowest ones (88%). In the same way, while the highest values for proteins (91% of the OM) were for anchovies of May 2016, in lipids the highest values were for those of May 2008 (11% of the OM) and in carbohydrates for those of May 2009 (2% of the OM). These results indicate that May 2008 anchovies spent more lipid resources to reproduction than the other ones, investing more energy in reproduction than in the other years.

www.ing[illegible]ent.com/pod-product-compliance
Lightning[illegible] LLC
LaVergn[illegible]
LVHW0[illegible]0926
84848[illegible]008B/1946
9[illegible]013275043